MW01447270

Second Edition

Advanced Textbook on

Gene Transfer, Gene Therapy and Genetic Pharmacology

Principles, Delivery and Pharmacological and
Biomedical Applications of Nucleotide-Based Therapies

Other Related Titles from World Scientific

Cellular Therapy of Cancer: Development of Gene Therapy Based Approaches
edited by Robert E Hawkins
ISBN: 978-981-4295-13-0

Engineering Stem Cells for Tissue Regeneration
edited by Ngan F Huang, Nicolas L'Heureux and Song Li
ISBN: 978-981-3147-74-4

A Guide to Human Gene Therapy
edited by Roland W Herzog and Sergei Zolotukhin
ISBN: 978-981-4280-90-7
ISBN: 978-981-3203-60-0 (pbk)

The World Scientific Encyclopedia of Nanomedicine and Bioengineering I — Nanotechnology for Translational Medicine: Tissue Engineering, Biological Sensing, Medical Imaging, and Therapeutics
(A 4-Volume Set)
editor-in-chief: Donglu Shi
edited by Yu Cheng, Jia Huang, Yarong Liu, Pin Wang and Bingbo Zhang
ISBN: 978-981-4667-65-4 (Set)

The World Scientific Encyclopedia of Nanomedicine and Bioengineering II — Bioimplants, Regenerative Medicine, and Nano-Cancer Diagnosis and Phototherapy
(A 3-Volume Set)
editor-in-chief: Donglu Shi
edited by Maoquan Chu, Wei Xia and Jiang Chang
ISBN: 978-981-4667-58-6 (Set)

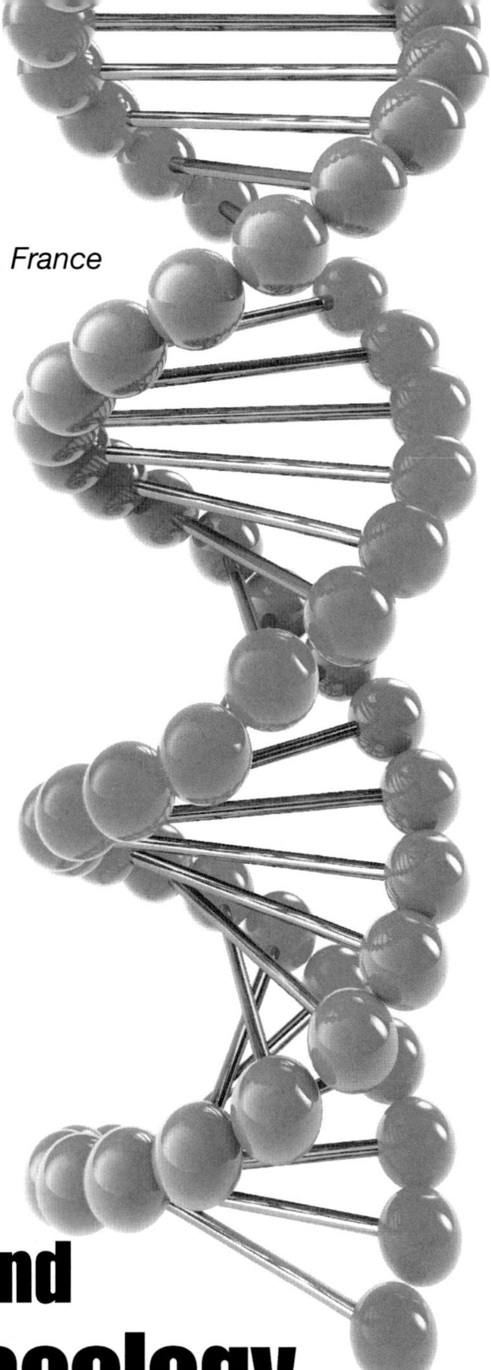

editor

Daniel Scherman

National Scientific Research Center (CNRS), France

Second Edition

Advanced Textbook on **Gene Transfer, Gene Therapy and Genetic Pharmacology**

Principles, Delivery and Pharmacological and Biomedical Applications of Nucleotide-Based Therapies

World Scientific

NEW JERSEY • LONDON • SINGAPORE • BEIJING • SHANGHAI • HONG KONG • TAIPEI • CHENNAI • TOKYO

Published by

World Scientific Publishing Europe Ltd.
57 Shelton Street, Covent Garden, London WC2H 9HE
Head office: 5 Toh Tuck Link, Singapore 596224
USA office: 27 Warren Street, Suite 401-402, Hackensack, NJ 07601

Library of Congress Cataloging-in-Publication Data
Names: Scherman, Daniel, editor.
Title: Advanced textbook on gene transfer, gene therapy and genetic pharmacology : principles, delivery and pharmacological and biomedical applications of nucleotide-based therapies / edited by Daniel Scherman (National Scientific Research Center (CNRS), France).
Description: 2nd edition. | New Jersey : World Scientific, [2019] | Includes bibliographical references.
Identifiers: LCCN 2019011949| ISBN 9781786346872 (hc : alk. paper) | ISBN 9781786347053 (pbk : alk. paper)
Subjects: LCSH: Genetic transformation--Technique. | Gene therapy. | Pharmacogenetics.
Classification: LCC RB155.8 .A38 2019 | DDC 615.8/95--dc23
LC record available at https://lccn.loc.gov/2019011949

British Library Cataloguing-in-Publication Data
A catalogue record for this book is available from the British Library.

For any available supplementary material, please visit
https://www.worldscientific.com/worldscibooks/10.1142/Q0205#t=suppl

Desk Editors: Herbert Moses/Jennifer Brough/Shi Ying Koe

Typeset by Stallion Press
Email: enquiries@stallionpress.com

Printed in Singapore

DEDICATION

This textbook is dedicated to my co-authors for their brilliant scientific contributions.

I would also like to pay a special tribute to Professors Claude Hélène, Jean-Bernard Le Pecq, and Bernard Roques, members of the French Academy of Sciences, who introduced me to the fields of gene therapy and of pharmacology.

My special thanks to my family, especially my wife Sylviane and my children, for supporting me through this achievement.

PREFACE TO THE FIRST EDITION

In medicinal chemistry, the ruling principle is: "*Corpora non agunt nisi fi xate,* meaning that active substances are only effective if they have a specific affinity for their target" Paul Ehrlich, 1913, in *Chemotherapy*.

The effects of medicinal active agents are based on the universal concept of "target recognition". Their history is characterized by a limited number of revolutionary advances. Except for the traditional medicines derived from the serendipity-driven use of natural plant decoctions, such conceptual revolutions or "quantum leaps" in drug discovery have always resulted from scientific advances.

The first revolution resulted from progresses in both analytical and synthetic chemistry, together with the Paul Ehrlich "magic bullet" concept. This has been at the basis of the discovery of chemical drugs directed to a molecularly defined "receptor" target. Indeed, before the 1913 Paul Ehrlich "lock and key" chemotherapy theory, the concept, today widely accepted, that molecular receptors in living organisms can be selectively affected by chemical compounds was hardly conceivable.

Biochemistry breakthroughs and the advances in both protein characterization and recombinant DNA technology paved the way to the use of proteins as therapeutic agents. Today, recombinant cytokines and hormones, such as erythropoietin (EPO), granulocyte–macrophage colony-stimulating factor (GMCSF), interferon, and monoclonal antibodies represent a considerable achievement and leading field of drug discovery.

As an overwhelming rule for both chemical and protein drugs, the molecular target in the patient is a protein, with rare exceptions including cytotoxic anticancer agents such as cisplatin which bind to DNA independently of the genetic sequence.

Representing the most recent revolutionary leap forward, both *genetic pharmacology* and *gene therapy* are based on the use of the genetic code. Genetic

Pharmacology represents the critical ultimate step of the Paul Ehrlich "lock and key" concept, in which the drug target is an intracellular genetic sequence within either a DNA or a RNA molecule which is recognized by Watson–Crick or Hogsteen base pairing. By contrast, in gene therapy, a gene is administered to the patient's cells, leading to the transcription by RNA polymerases of a RNA, which can be by itself a therapeutic agent, or most often represents an mRNA translated into a therapeutic protein by the ribosome machinery. In gene therapy, the administered gene can thus be considered as a "prodrug", with the amplification advantage resulting from the continuous intracellular production of the therapeutic RNA and eventually protein. This textbook describes the concepts and applications of genetic pharmacology and gene therapy. Extensive biological, pharmaceutical, and medicinal aspects are presented in a very simple and pedagogical way by world-recognized experts.

This rapidly developing field is finely introduced by Professor Thierry Vanden Driessche, former President of the European Society of Gene and Cell Therapy. The basic definitions and principles are presented in Part I, while the various vectors and gene-delivery techniques are introduced in Part II. Examples of Therapeutic Applications are described in Part III, and a more technological Part IV is dedicated to vector production. In addition to chapters written in a pedagogical "textbook style" advising the reader on a few key reviews for further reading, several chapters have been written as reviews including an extensive bibliography. Therefore, both newcomers and experienced readers might find in the present textbook the most helpful introductory information or more specialized answers to their specific needs.

I am honored and grateful that such an assembly of talents has accepted the Invitation to contribute to this textbook, which I hope will prove most useful to biological and medicinal students and scientists planning to use gene delivery for basic science or translational medical application, and to medical doctors wishing to be introduced to the exciting new frontline medical fields of genetic pharmacology and gene therapy.

PREFACE TO THE SECOND EDITION

Since the finalization of the first edition of this handbook, several sensational revolutionary advances have occurred in the field. These concern mainly genome editing (Chapter 8) and cancer immunotherapy through the generation of CAR T-cells (Chapter 21). Major future advances are expected from the two technologies, and the present Handbook's second edition could not miss the opportunity to present them.

In addition, two chapters have been added, extending the Part III: Therapeutic Applications. Chapter 22 concerns the gene therapy of severe combined immunodeficiencies, a most important cornerstone of the field, since it has represented the first historical clinical success of gene therapy. I have the greatest honor to have this chapter written by the French Team which has been at the origin of this historical success. Chapter 28 deals with the approaches to one of the most common monogenic disease: Cystic Fibrosis.

As for the first edition, I have no words to express my gratitude to the world leader scientists who have contributed to both editions of this handbook.

ABOUT THE EDITOR

Prof. Daniel Scherman is currently serving as Exceptional Class Director at the Centre National de la Recherche Scientifique (CNRS), National Scientific Research Center, France; Former Director of the Chemical and Biological Technologies for Health Laboratory and of the Centre de Recherche Pharmaceutique de Paris; European commission H2020 expert; President of the Non-Viral Gene Therapy Committee of the European Society of Cell and Gene Therapy (ESGCT); and a member of the Scientific Committee of the American Society for Gene and Cell Therapy (ASGCT). He was the President of the CNRS French National Research Scientific Committee (Section 28) on Pharmacology, Bioengineering, Bioimaging and Biotechnology from 2013 to 2016. Dr. Daniel Scherman obtained his PhD in Biochemistry in 1979 and Doctorat d'Etat és-Sciences in 1984 from University Paris VII. He has worked on several published books and prestigious journals, and holds 40 independent patent applications in the fields of gene therapy, drug delivery, drug discovery and bio-imaging, and is the former Editor of the series *Advances in Behavioral Biology*. He is the Editor for the first edition of *Advanced Textbook on Gene Transfer, Gene Therapy and Genetic Pharmacology*. He has obtained numerous scientific awards such as the Valori Grand Prize of the French National Academy of Science in 2017 and Academic Palms National Medal in 2004. His research interests are in the areas of gene and cell therapy, biotechnology, biomaterials, non-viral delivery of DNA and SiRNA, plasmid optimization, chemical DNA vectors, physical DNA delivery, drug delivery, cancer therapy, lysosomal storage disorders, and bioimaging using photonic, SPECT and IRM techniques. The author can be reached at daniel.scherman@parisdescartes.fr.

CONTENTS

PART I

BASIC DEFINITIONS AND PRINCIPLES

1

INTRODUCTION

Thierry VandenDriessche[a]

Gene therapy offers unprecedented opportunities to treat or cure disease and alleviate human suffering. It has been more than ten years now since the first clinical successes in gene therapy were first reported for the treatment of severe combined immunodeficiency (SCID-X1). The recurrent opportunistic and life-threatening infections in patients suffering from these types of hereditary diseases are like a sword of Damocles that constantly reminds them of their tragic fate. The majority of the infants that were afflicted by this devastating disease and that were subsequently treated by gene therapy can now essentially lead normal lives. These remarkable and historic achievements formally prove that gene therapy, moving forward, can be a realistic and successful medical intervention. Most importantly, these successes offer new therapeutic options for patients who are currently untreatable.

Convincing evidence continues to emerge that gene therapy is effective in patients suffering from other hereditary diseases besides the congenital immune deficiencies (e.g., hemophilia, epidermolysis bullosa, and β-thalassemia) but also from more common disorders like cancer, neurodegenerative or cardiovascular disorders. Even patients (including children) suffering from an inborn genetic disease that is not life-threatening but causes blindness can finally start to see following gene therapy. These few selected recent examples of clinical advances in gene therapy clearly indicate that the momentum in this field is building up. In the absence of effective drugs or alternative therapies, the advances in gene therapy technology clearly represent the best hope for the many patients and families that are blighted by these various diseases. Given the current global economic challenges it is even more important than ever to find sustainable solutions to

[a]Department of Gene Therapy & Regenerative Medicine, Free University of Brussels, Center for Molecular & Vascular Biology, University of Leuven, Belgium
Email: thierry.vandendriessche@vub.ac.be

treat diseases of high unmet medical need. The potential for a one-time curative treatment by gene therapy should be offset against the high costs of continuous therapeutic interventions for the treatment of chronic diseases. The current demographic trends will only worsen the overall burden on our already overstretched health care system. To reduce the economic burden it is absolutely essential to further consolidate this vision and to take into account the mid- and long-term sustainable benefits of gene therapy relative to the initial investments made.

Despite the advances in gene therapy on the clinical and preclinical fronts, clinical progress has been slower than originally anticipated. As detailed in Chapter 3 dedicated to the history of gene therapy, technical obstacles have been compounded with some safety concerns related primarily to unexpected immune complications or insertional oncogenesis due to integrating vectors. However, significant progress has recently been made to improve the safety profiles and risk:benefit ratios in gene therapy. It is particularly reassuring that the gene therapy vectors used today are much safer and/or efficient than some of the vectors used in the first gene therapy trials. Indeed, some of the recent progress in gene and cell therapy could be ascribed to the continuous improvement in gene transfer technologies.

Whereas the earlier gene therapy trials based on the use of genetically modified hematopoietic stem cells (HSCs) relied on γ-retroviral vectors, subsequent trials involved lentiviral vectors instead. It is truly ironic that the human immunodeficiency virus that devastates the immune system during AIDS progression has now been converted into a relatively safe vector for the potential treatment of hereditary diseases by gene therapy, including congenital immunodeficiencies and even AIDS itself. *Virus non semper maleficum est.* Some of the most compelling clinical trial data thus far obtained using lentiviral vectors were based on a phase I/II study for adrenoleukodystrophy (ALD) and β-thalassemia. In addition, adeno-associated viral (AAV) vectors have shown promise for many *in vivo* gene therapy approaches targeting a wide variety of different organs (including liver, heart, skeletal muscle, and brain) and diseases. The emergence of new gene-transfer technologies based on non-viral vectors that rely on chemical and physical gene delivery paradigms also show remarkable progress both in preclinical studies and in clinical trials. For instance, the oligonucleotide-based exon skipping approach for the treatment of Duchenne muscular dystrophy underscores this vision. In addition, the prevailing assumption that non-viral gene delivery could not be used to stably deliver genes into HSCs has recently been challenged through the use emerging transposon-based gene delivery technologies.

Despite the progress in clinical trials and preclinical studies, there is still no gene therapy product on the market that has been approved by the regulatory authorities. Nevertheless, many experts agree that this is likely to happen in the near future. Some of the challenges faced by gene therapists are not unique to the field but are inherent to translational research at the forefront of medical innovation. Though the high hopes for gene therapy in the early 1990s did not immediately translate into clinical success, it is important to recognize that these initial expectations were not realistic, and created the false impression of slow progress.

In many ways, the development of gene therapy mirrors that of therapeutic monoclonal antibodies or clinical bone marrow transplantation. These biomedical and biotechnological innovations took over 25 years to perfect and have now become life-saving therapies for hundreds of thousands of patients suffering from many different diseases, thanks largely to the perseverance and commitments of many academic, medical, and industrial stakeholders. The road to clinical trials and product registration takes 10–15 years in conventional drug development, gene- and cell-therapeutics being no exception. More basic science was required; now that some emerging technologies are becoming real options, the time has come for clinical translation making use of cutting edge technology. It is important to take these realistic timelines into consideration when assessing the overall progress in the field. Sustained funding of both high-quality science and clinical trials is required to guarantee successful clinical developments and the expansion of the pharmaceutical sector, based on genuine innovation and technology transfer securing further phases of clinical development. The continued interactions of gene therapy stakeholders from academia, industry, patient organizations, and regulatory authorities are thereby essential to move this field forward.

While gene therapists continue to perfect the gene delivery vectors from the clinical standpoint of safety and efficacy, several biomedical disciplines are harvesting the fruits of these emerging technologies.

Indeed, gene therapy has fueled the fields of biology and medicine with technologies that allow hypothesis-driven research questions to be addressed. Furthermore, there are also key emerging fields where insights and technologies from the gene therapy community are playing increasingly important roles. In particular, the emerging fields of RNA interference, microRNA and antisense therapies benefit from advances in gene therapy, since safe and efficient delivery of these RNA-based therapeutics is once again the key issue. Furthermore, targeted genomic integration using engineered nucleases not only reduces the risk of insertional oncogenesis associated with random integration but also paves the way towards improved gene targeting and functional genomics approaches in various model systems. Moreover, the recent development of induced pluripotent stem cells (iPS) for regenerative medicine by "genetic reprogramming" is intimately linked to the transfer of genes encoding reprogramming factors into somatic cells and can be considered as a *bona fide* spin-off of the gene therapy field.

The remarkable progress made in vector development and manufacturing, preclinical studies, and clinical gene therapy trials is highlighted in this excellent monograph. Thanks to the leadership of Professor Daniel Scherman as editor and the valuable contributions of the authors of the various chapters, each of them leaders in their respective disciplines, it adequately captures the state-of-the-art developments in the field of gene therapy. It serves as a testimony to the dedication and perseverance of gene therapists across the globe that made this remarkable progress possible. I have no doubt it will provide a valuable and inspiring resource to foster future developments in this exciting field at the forefront of biomedical innovation.

2

BASIC DEFINITIONS AND GENERAL PRINCIPLES

Daniel Scherman[a]

2.1 Definitions

Genetic pharmacology refers to the use of short synthetic oligonucleotides to manipulate gene expression. This can be principally achieved by the so-called antisense, antigene, or RNA-interfering strategies, which will be discussed in Chapters 4 and 5. Other applications of short nucleotide sequences for mRNA exon skipping or for gene targeting and repair are described in Chapters 6 and 7.

The specificity of genetic pharmacology, as compared to classical "small-chemical drug" pharmacology, is that the short single-strand oligonucleotide pharmacological agent recognizes its cognate target through base pairing. The length of the oligonucleotide sequence necessary to ensure target specificity is around 20 bases. The single-strand oligonucleotide generally binds to another single-strand DNA or mRNA target through Watson–Crick hydrogen bonds, but it can also form a triple helix with DNA duplexes by Hoogsteen base pairing. Except for this mechanism of action based on base-pair recognition, genetic pharmacology cannot be distinguished from the more classical small-drug pharmacology, which is based on receptor–ligand molecular recognition (the "lock and key" concept of Paul Ehrlich).

[a]Laboratory of Chemical and Genetic Pharmacology and of Biomedical Imaging, Paris Descartes Pharmacy University, CNRS, Inserm, ChimieParisTech, 4, avenue de l'Observatoire Paris Cedex 06, France
Email: daniel.scherman@parisdescartes.fr

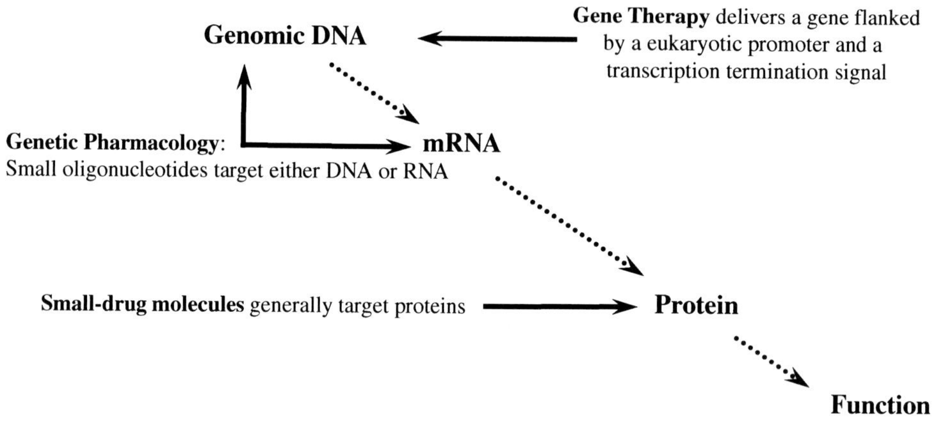
Genomic DNA
Gene Therapy delivers a gene flanked by a eukaryotic promoter and a transcription termination signal
Genetic Pharmacology:
Small oligonucleotides target either DNA or RNA
mRNA
Small-drug molecules generally target proteins
Protein
Function

FIGURE 2.1 ■ While chemical drugs classically target proteins (except in the case of anticancer cytotoxic compounds), small-chemical oligonucleotides used in genetic pharmacology generally target DNA and/or RNA, with the exception of aptamers (Chapter 4). Gene therapy consists of administering a gene-expressing cassette (promoter — gene — polyadenylation signal) to the cells.

Gene therapy more specifically refers to the cell delivery of a gene-expressing cassette, namely a transcribed DNA sequence flanked at its 5′ end by a eukaryotic promoter and at its 3′ end by a polyadenylation signal. This promotes the transcription of an RNA which either by itself displays a therapeutic intracellular effect, or which encodes a missing protein or any protein or peptide allowing a therapeutic or vaccination effect. The distinction between genetic pharmacology and gene therapy (Fig. 2.1) will be preferentially used in the present textbook. However, other authors employ the generic expression "gene therapy" to designate all approaches implying the use of any natural or modified DNA or RNA nucleotide molecule.

Gene therapy might be used, in a non-exhaustive list:

- to compensate for a missing protein in a genetic disease;
- to inhibit the production of a given protein (by generating an antisense mRNA), for instance to render cells resistant to viral infection;
- to express atrophic factor or an anti-inflammatory cytokine;
- to introduce a suicide gene for the treatment of cancer.

Examples of potential applications of gene therapy are just too many to be listed here. Several genetic diseases have attracted large interest for application of a gene-replacement strategy (Table 2.1).

TABLE 2.1 Examples of Genetic Diseases Already Treated or in Clinical Trials Using Gene-Replacement Therapy, and Their Related Clinical Protocols — A Non-Exhaustive List. Several of these Examples are Described in Part III: Therapeutic Applications

Disease	Gene	Clinical Protocol	See Chapter
Immunodeficiency	Adenosine deaminase	*Ex vivo*	
	γ-c IL2 receptor subunit	*Ex vivo*	
	Wiskott–Aldrich syndrome protein	*Ex vivo*	
β-thalassemia	β-globin	*Ex vivo*	23
Hemophilia B	Clotting factor IX	*In vivo*	24
Cystic fibrosis	CFTR	*In vivo*	
α-1 antitrypsin deficiency	α -1 antitrypsin	*In vivo*	
Duchenne muscular dystrophy	Dystrophin	*In vivo*	6, 27
Lipoprotein lipase deficiency	Lipoprotein lipase	*In vivo*	
X-linked adrenoleukodystrophy	ABCD1	*Ex vivo*	
Parkinson's disease	Glutamic acid decarboxylase (GAD)	*In vivo*	26
	Neurturin trophic factor	*In vivo*	26
Lysosomal storage disorders	Lysosomal enzymes		
Canavan leukodystrophy	Aspartoacylase	*In vivo*	
Sanfilippo A (MPSIIIa)	N-sulfoglysosamine sulfohydrolase	*In vivo*	
Retinal degenerative diseases:			
Leber congenital amaurosis	RPE 65 (Type 2 LCA)	*In vivo*	26
Stargardt disease	ABCA4	*In vivo*	

2.2 Critical Advantageous Properties of Gene Therapy

Several fundamental properties render gene therapy essential for the treatment of a large body of diseases, whether genetic or not, and also suggest that it could represent an attractive alternative to protein-substitution therapy.

First, since proteins are generally unable to cross cell membranes, only gene therapy is able to provide for the treatment of any disease which is caused by the lack of expression of an intracellular protein, or which can be cured through the intracellular expression of a therapeutic or antigenic protein. This is the case, for instance, for neuromuscular disorders such as Duchenne muscular dystrophy, for which the missing dystrophin protein has to be expressed in the muscle cytosol. This is also the case for vaccination strategies based on class 1 antigen presentation.

Second, the delivery of a therapeutic protein leads to variations in concentrations. This can induce toxicity when a transient excessive concentration is observed. Conversely, this can be damaging when the tissue concentration of the missing protein is lower than the therapeutic threshold, as a result of the short half-life of the administered protein. This deleterious effect of improper patient coverage is observed, for instance, in hemophilia, where patients suffer from severe chronic joint deterioration due to micro-bleeding events. The repeated administration of a large dose of clotting factors has been proposed to prevent this damage, but this has been shown to induce neutralizing antibodies that inhibit the injected factors (Chapter 24).

Unlike recombinant protein therapy, gene therapy leads to the constant expression of a missing protein, when the gene is delivered under the control of a permanently activated promoter, which represents the most frequent case. This is an important pharmacokinetic advantage of gene therapy, theoretically allowing "constant coverage" of a patient after a single transgene-delivery treatment (Fig. 2.2).

Another advantage of gene therapy results from the spatial control of its effect. Indeed, by combining the local delivery of a therapeutic transgene and the use of a tissue- or cell-type-specific promoter, one can drastically restrict the expression of the desired protein to a precise location in the body of the patient, and even to a specific cell type of a given tissue. Such a cell-specific expression has proven particularly valuable for the treatment of genetic disorders resulting from the complete absence of a protein. Immune response against the missing protein has been observed after gene therapy, because the transgenic protein is then considered to

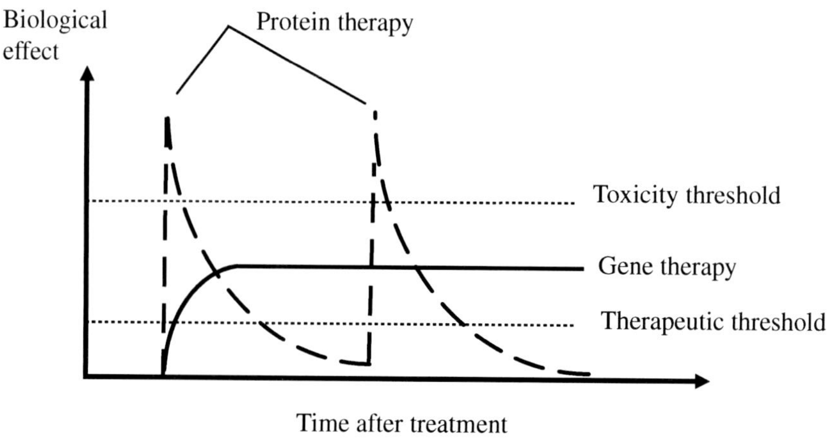

FIGURE 2.2 ■ Comparison of pharmacokinetic profiles of protein therapy and gene therapy.

be exogenous by the patient host. By using a cell-type-specific promoter, one can avoid transgene expression in antigen-presenting cells such as dendritic cells. This results in a decreased immune response against the transgene.

The additional value of cell-type-specific promoters is that they generally lead to a longer and more stable duration of transgene expression, in contrast to very strong promoters such as that of the human cytomegalovirus (hCMV), which induce high but transient expression. This is particularly the case in liver, where use of the liver-specific α-1 antitrypsin gene promoter has proven superior to the hCMV promoter. In muscle also, studies suggest that the restriction of transgene expression to myotubes is an important criterion for the treatment of muscular dystrophies, since it leads to higher and more sustained transgene expression along with the absence of immune response against the transgene, in contrast to what is observed when using the hCMV promoter.

Examples of tissue-specific promoters include those of the following proteins:

- phosphoenolpyruvate carboxykinase (PEPCK) and alpha antitrypsin (AAT) in hepatocytes;
- surfactant protein A (SPA) in epithelial cells (for the treatment of cystic fibrosis);
- desmin (preferentially), skeletal muscle β-actin, or truncated muscle creatin kinase (MCK) in skeletal muscle cells;
- smooth muscle (SM)-specific SM22α promoter or smooth muscle myosin heavy chain for smooth muscle cells;
- CD11c in dendritic cells;
- glial fibrillary acidic protein (GFAP) in glial cells;
- neuron-specific enolase (NSE);
- carcinoembryonic antigen (CEA) for tumor-specific expression.

2.3 Temporal Control of Gene Expression

In the context of *in vivo* gene transfer, the temporal regulation of gene expression would present a broad range of applications. For example, studying gene function in development often requires gene expression at a particular period of time. Knocking-out these genes (knock out animals) or overexpressing them (transgenic animals) might result in embryonic lethality, which requires the capacity to express or down-regulate these genes at a chosen time in development. In addition, particularly in the field of gene therapy, a transcriptional regulation system that would be rapid, reversible, and repeatedly inducible might be required for safety and efficacy reasons. Several inducible systems in mammalian cells have been developed to artificially regulate genes at the transcriptional level by using inducer or repressor small-drug molecules.

An ideal gene regulation system has to meet several requirements, particularly: (1) high inducibility and low basal expression level; (2) no interference with endogenous regulatory networks; and (3) activation by exogenous non-toxic drugs of known pharmacology. Early attempts to develop inducible gene-expression systems made use of cell elements that respond to exogenous signals or stress, such as cytokines, hormones, heat, metal ions, and hypoxia. Heat-inducible promoters derived from stress heat-shock proteins and light-switchable promoters have also been developed. These systems are often subject to pleiotropic effects or show high basal activity in the absence of induction. Several artificial systems regulated by a small-molecule inducer (tetracycline, RU486 or ecdysone for example) have been proposed. The tetracycline-based system, is the most widely used (Fig. 2.3).

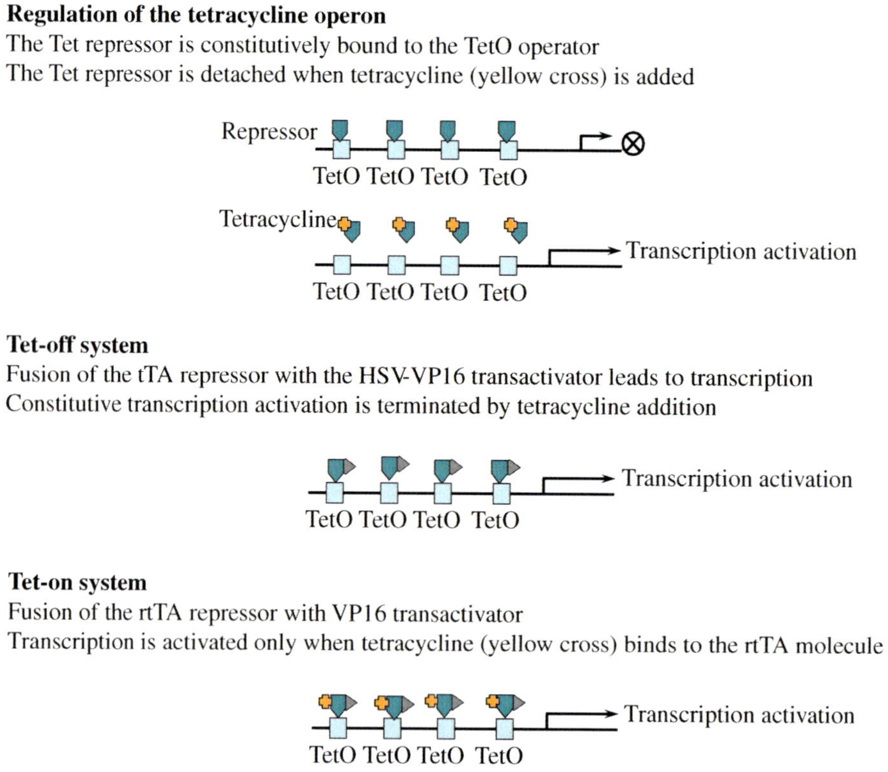
Regulation of the tetracycline operon
The Tet repressor is constitutively bound to the TetO operator
The Tet repressor is detached when tetracycline (yellow cross) is added
Repressor
TetO TetO TetO TetO
Tetracycline
Transcription activation
TetO TetO TetO TetO
Tet-off system
Fusion of the tTA repressor with the HSV-VP16 transactivator leads to transcription
Constitutive transcription activation is terminated by tetracycline addition
Transcription activation
TetO TetO TetO TetO
Tet-on system
Fusion of the rtTA repressor with VP16 transactivator
Transcription is activated only when tetracycline (yellow cross) binds to the rtTA molecule
Transcription activation
TetO TetO TetO TetO

FIGURE 2.3 ■ Different Tet systems for the temporal control of gene expression. The Tet repressor (green arrow) is constitutively bound to the Tet operator. Tetracycline (yellow cross) induces a conformational change leading to the loss of repressor Tet affinity to the Tet operator DNA sequence. In the Tet-off and Tet-on systems, the VP16 transcriptional activator of herpes simplex virus (grey triangle) is fused to the Tet repressor or rtTA mutant (hatched green arrow). When bound to DNA, VP16 activates the transcription of the neighboring gene. In the Tet-off system, the removal of tetracycline is necessary to induce gene expression. In the Tet-on system the addition of tetracycline induces rtTA binding to TetO and thus VP16-mediated gene transcription, whereas repression depends on the clearance of antibiotic.

The original tetracycline system (Tet-off) allows very stringent regulation, but the expression is obtained by removal of the drug. This system thus requires the continuous administration of tetracycline, and induction occurs upon antibiotic removal. Ideally, a candidate gene therapy system with regulated expression should be an "on" system, where transgene expression is switched on by the administration of a small drug. An alternative version of the Tet system based on a mutated form of the tetracycline transactivator (tTA) called reverse tTA (rtTA), has been introduced. The rtTA binds to DNA only when the small-drug inducer is present (Tet-on system).

The major limitations of the Tet-on system in a gene therapy context, which are particularly important for therapeutic applications, are its poor induction kinetics and high basal level of expression. Therefore, several improvements providing a more stringent control of gene expression have been introduced in the form of a mutated transactivator rtTA2^{S}-M2, or by the addition of the tetracycline transcriptional silencer tTS.

2.4 Strategies and Protocols for Gene Therapy and Genetic Pharmacology

In order to define a genetic pharmacology or gene therapy approach, one has to define a large variety of parameters (Fig. 2.4):

- *Definition of the genetic sequence to be used.* Smaller cDNA sequences are the most frequently used, in particular for viral vectors such as adeno-associated virus

Type of therapeutic intervention	Gene complementation Inhibition of gene expression Gene repair
Clinical protocol	*Ex vivo* *In vivo*
Therapeutic object	cDNA, genomic DNA Synthetic oligonucleotide
Delivery mode	Viral vector Non-viral vector Physical delivery technique

FIGURE 2.4 ■ The definition of a gene therapy protocol necessitates the definition of a panel of various parameters.

(AAV) which have a limited encapsidation capacity. On the other hand, genomic sequences or the addition of endogenous introns have been shown to induce a higher and more sustained expression.

- *Duration of transgene expression.* While "burst-type" promoters such as hCMV are suitable for genetic vaccination or short-term cytokine expression, tissue-specific promoters might be optimal for "replacement" gene therapy, where long-term expression is needed.
- *Choice of the cells to be transfected (by non-viral vectors) or transduced (by viral vectors).* Integrating vectors, such as retroviral and lentiviral vectors (Chapter 9) are adapted for transducing stem cells or other dividing cells, since integration allows the maintenance of the transgene in daughter cells. Conversely, adenovirus, AAV, herpes viruses, or plasmid non-viral vectors might be more appropriate for gene transfer to differentiated non-dividing cells (see Part II: Vectors and Gene Delivery Techniques).
- *The required type of genetic intervention.* Gene replacement, gene extinction, or gene repair can be envisioned.
- *The therapeutic protocol.* Should the gene be delivered directly *in vivo* to the patient, or *in vitro* to cells which will be subsequently administered, thus defining an *ex vivo* protocol.

For Further Reading

For more in-depth and specialized information, please refer to the chapters in the present textbook. Several authors of this textbook have also contributed to the following important text:

Cohen-Haguenauer, O. (ed.) (2012). *The CliniBook: Clinical Gene Transfer State of the Art*, Paris: Éditions EDK.

Original Articles on Tissue-Specific Promoters

Chen, X., Wu, J.M., Hornischer K., *et al.* (2006). TiProD: The Tissue-specific promoter database, *Nucleic Acids Res*, **34**, (Database issue):D104–D107.

Edelmann, S.L., Nelson, P.J., Brocker, T. (2011). Comparative promoter analysis *in vivo*: Identification of a dendritic cell specific promoter module, *Blood*, **118**, e40–e49.

Ribault, S., Neuville, P., Méchine-Neuville, A., *et al.* (2001). Chimeric smooth muscle-specific enhancer/promoters valuable tools for adenovirus-mediated cardiovascular gene therapy, *Circ Res*, **88**, 468–475.

Talbot, G.E., Waddington, S.N., Bales, O., *et al.* (2010). Desmin-regulated lentiviral vectors for skeletal muscle gene transfer, *Mol Therapy*, **18**, 601–608.

Original Articles and Reviews on the Regulation of Gene Expression

Agha-Mohammadi, S., O'Malley, M., Etemad, A., *et al.* (2004). Second-generation tetracycline-regulatable promoter: Repositioned Tet operator elements optimize transactivator synergy while shorter minimal promoter offers tight basal leakiness, *J Gene Med*, **6**, 817–828.

Curtin, J.F., Candolfi, M., Puntel, M., *et al.* (2008). Regulated expression of adenoviral vectors-based gene therapies: Therapeutic expression of toxins and immune-modulators, *Methods Mol Biol*, **434**, 239–266.

Goverdhana, S., Puntel, M., Xiong, W., *et al.* (2005). Regulatable gene expression systems for gene therapy applications: Progress and future challenges, *Mol Ther*, **2**, 189–211.

Lewandoski, M. (2001). Conditional control of gene expression in the mouse, *Nat Rev Genet*, **2**, 743–755.

Schönig, K., Bujard, H., Gossen, M. (2010). The power of reversibility: regulating gene activities via tetracycline-controlled transcription, *Methods Enzymol*, **477**, 429–453.

Weber, W., Fussenegger, M. (2006). Pharmacologic transgene control systems for gene therapy, *J Gene Med*, **8**, 535–556.

Weber, W., Fussenegger, M. (2011). Molecular diversity — the toolbox for synthetic gene switches and networks, *Curr Opin Chem Biol*, **15**, 414–420.

3

HISTORY OF GENE THERAPY

Serge Braun [a]

3.1 Historical Evolution

Following the notion of *genetic engineering* that was first presented at the Sixth International Congress of Genetics held in 1932, the concept of *genetic correction* arose after Avery, MacLeod and McCarty in 1944 suggested that "genes could be transferred within nucleic acids" (Avery *et al.*, 1944; Wolff and Lederberg, 1994). Clyde E. Keeler in 1947 was probably the first to use the term *gene therapy*, although the process he was describing (the correction of gene-based deviations in plants and animals) was not envisaged as an effective therapeutic technique to treat genetic diseases in man (Keeler, 1947; Prazeres, 2011). The first real contribution to the field is attributed to Nobel Prize winner Joshua Lederberg, a pioneer in bacterial genetics and plasmid biology, and a visionary in gene therapy (Lederberg, 1963). The "isolation or design, synthesis and introduction of new genes into defective cells or particular organs" was elaborated further in the subsequent years, such that DNA was viewed as "the ultimate drug" (Aposhian, 1970). Originally, the thinking behind this was that by replacing the mutant gene, genetic disorders would no less than "cease to exist".

The 1960s and 1970s set the basis of molecular genetics and gene transfer in bacteria, then into animal and human cells, combined with selection systems for cultured cells and recombinant DNA technology (Graham and van der Eb, 1973).

[a] French Myopathy Association — AFM, Evry, France
Email: sbraun@afm.genethon.fr

The first deliberate transfer of foreign genes into human recipients with a therapeutic purpose, was performed in 1970 by Stanfield Rogers who attempted to treat three German siblings presenting with arginase deficiency, by giving them injections of the native Shope rabbit papilloma virus (Terheggen *et al.*, 1975; Friedmann and Rogers, 2001). In spite of the failure of this clinical trial, this is viewed, before even the establishment of recombinant DNA technology in 1973, as the first anticipation of the therapeutic potential of viruses as carriers of genetic information (Rogers and Pfuderer, 1968).

In July 1980, Martin Cline at UCLA headed another highly controversial human trial designed to treat β-thalassemia. The experiment involved the removal of bone marrow cells from two patients (in Jerusalem and Naples), and their subsequent transformation *in vitro* with plasmids carrying the β-globin gene and the herpes simplex virus thymidine kinase (HSVtk) gene expected to provide a selective proliferative advantage to marrow cells once these were transplanted back into the patients.

In neither case was any real follow-up reported, since both trials were stopped, but apparently neither good nor harm resulted for the patients (Wolff and Lederberg, 1994). Even though Rogers's and Cline's trials were heavily criticized for scientific, procedural, and ethical reasons, they served as catalysts for the development of the field and for the establishment of specific guidelines and regulations on the human use of molecular genetics (Wade, 1980).

In 1983, enzyme-producing-gene-corrected cells were further theorized as a viable approach for treating Lesch–Nyhan disease, a rare neurological disorder (Trubo, 1983).

The first serious approved gene therapy trials happened another decade later in the USA, both relying on retroviruses to transduce the cells. In the first trial, ten patients with advanced melanoma were treated with tumor-infiltrating lymphocytes isolated from solid tumors first marked with the *Escherichia coli* neomycin phosphotransferase gene using a retroviral vector, and then transferred back into the cancer patients. This is also the first example of the use in humans of a marker gene to follow-up the fate and biodistribution of labeled cells (Rosenberg *et al.*, 1990).

The second trial employed enzyme-transduced T-cells for adenosine deaminase (ADA) severe combined immunodeficiency. They turn out to be moderately successful, for reasons we now understand. Nevertheless, a new era of gene therapy began as these first Food and Drug Administration (FDA)-approved experiments have germinated a growing number of trials to refine new gene therapies.

Potentially, gene transfer may provide correction of cellular functions by expressing the deficient gene, or it may add a new function to a cell by transferring an exogenous gene or inhibit the unfavorable action of a cell by introducing a counteracting gene. Non-genetic as well as genetic disorders were therefore targeted. During the 1990s, the pace accelerated — from a handful in the early years, the number of trials had reached more than 2000 by 1999.

This decade was marked by one historical event, the full licensing of the first veterinary gene therapy vaccine that was set up by the French biotech company Transgene SA and licensed to Merial, and registered in 1990 in the form of baits under the name RABORAL V-RG®. This oral vaccine is based on a recombinant vaccinia virus in which the thymidine kinase gene has been replaced by the glycoprotein G of the rabies virus. It has been approved for immunization of wild animals, mainly foxes in Western Europe and raccoons and coyotes, two of the most significant wildlife carriers of rabies in North America. Strategic use of this vaccine by public health officials has been proven to reduce the rate of rabies infection in wildlife populations and therefore potential infection of pets, livestock, and humans. It led to the complete eradication of rabies disease in France (as officially declared on April 30, 2001) and Belgium. This technology was the predecessor of a number of cancer and infectious-disease vaccines now in late-stage clinical trials. In the meantime, more veterinary DNA vaccines have been approved (i.e., one aimed at protecting horses against West Nile virus, another to protect farm-raised salmon against infectious hematopoietic necrosis virus, and one against canine melanoma) (Mackowiak *et al.*, 1999).

According to the Wiley website (The Journal of Gene Medicine, 2012), during the 1990s, for 69% of human patients, the main target disease was cancer. But only 1% of protocols reached phase III. The rather discouraging decade of the 1990s finished badly in September 1999 with the widely publicized death of Jesse Gelsinger, an 18-year-old man undergoing gene therapy for the liver genetic disease ornithine transcarboxylase deficiency (OTCD) — a shadow over the whole field of gene-based therapies. The effects of gene transfer in the other 17 patients who had enrolled in the trial were, in contrast, limited to transient myalgias and fevers, and biochemical abnormalities, suggesting that Gelsinger had predisposing factors to vector toxicity (Raper *et al.*, 2003).

A lesson of this decade is, while the potential of gene therapy is great, many scientific obstacles must be overcome before it becomes a practical form of therapy. The first failures and some (though extremely rare) accidents led unfairly to the perception that the promises of gene therapy were more than what it could deliver. Efficacy, safety, and ethical aspects became focuses of furious debate (Wilson, 2008). Gene therapy involves a whole lot of complicated sets of activities involving tissue targeting, cellular trafficking, delivery of genes to organs, safety of the vector, activity of therapeutic protein, etc. For example, in gene therapy for cystic fibrosis a very early indication is hampered by the complications involved in penetrating the natural barriers that impede viral entry into the obstructed airways.

3.2 Obstacles to the Development of Gene Therapy

3.2.1 Short Life of Treatment

The therapeutic genetic material introduced into target cells must remain functional. Naked DNA or certain viruses (e.g., AAVs) may remain episomal and allow sustained expression in stable tissues (e.g., neurons or skeletal muscles). However, promoters and/or gene sequences different from host codon usage may be subjected to epigenetically induced extinction. In addition, long-term benefit, due to the rapidly dividing nature of many cells, often requires either integration of the therapeutic DNA into the host-cell genome or multiple rounds of gene therapy.

3.2.2 Toxicity and Inflammatory Responses

In the OTCD trial, the patient J. Gelsinger died from fulminant hepatitis four days after beginning treatment with an adenovirus vector. Since then, work using adenovirus vectors has focused on genetically crippled versions of the virus, safer production standards and clinical protocols. It is now being restricted to indications compatible with short-term expression of the transgene (such as vaccination) and more favorable risk:benefit ratio (i.e., cancer or cardiovascular diseases). Developed and manufactured by the Chinese firm Shenzen SiBiono GeneTech and trademarked under the name Gendicine®, a recombinant adenovirus vector 5 expressing the tumor suppressor gene p53 was approved by the State Food and Drug Administration of China for the treatment of head-and-neck squamous cell carcinoma; it became the first ever human gene therapy product to reach the market in April 2004. In November 2005, the Chinese authority approved Oncorine®, a derived product which can selectively kill tumor cells with dysfunctional p53 genes (Räty *et al.*, 2008). Advexin®, the sibling of Gendicine from Introgen, was, however, turned down by the US FDA in 2008.

In Europe, Cerepro®, an adenovirus-mediated gene-based medicine for brain cancer developed by Ark Therapeutics (Finland), is currently being reviewed by the European Medicines Agency (EMA). The therapy combines the adenovirus-mediated local administration of the HSVtk with the intravenous injection of the prodrug ganciclovir. Ganciclovir is converted by the thymidine kinase expressed within the tumor into a substance that specifically kills the dividing tumor cells without affecting the surrounding healthy cells. If successful, Cerepro® could become the first gene therapy medicine of a non-genetic disease to be marketed in the West (Mitchell, 2010).

3.2.3 Insertional Mutagenesis

This has occurred in clinical trials for X-linked severe combined immunodeficiency (SCID-X1) patients, in which hematopoietic stem cells were transduced with the gamma-chain interleukin 2 (γC IL2) receptor gene using a retrovirus, and this led to the development of T-cell leukemia in 5 out of 20 patients (four children in the French trial and one in the British trial). All but one of these children responded well to conventional antileukemia treatment. Clinical trials were halted temporarily in 2002, but resumed after regulatory review of the protocol on both sides of the Atlantic. Gene therapy trials to treat SCID due to deficiency of the ADA enzyme or γC IL2 receptor gene continued with impressive success in Italy (pioneered since 1992 by Claudio Bordignon and Maria-Grazia Roncarolo), the USA, Britain, Ireland, and Japan. Up to 100 immunodeficient children have now been successfully treated (it is noteworthy that the current standard treatment, bone marrow graft, fails in 25% of cases with, however, less media impact) (Aiuti *et al.*, 2009). New immunodeficiency indications including Wiscott–Aldrich syndrome are now being investigated. Gene therapy trials using integrative vectors to treat bone marrow stem cells represent the most successful application of gene therapy to date. Similar approaches have generated breakthroughs in the red blood cell disease β-thalassemia (and sickle cell disease) and adrenoleukodystrotrophy (a neurological disease affecting bone-marrow-derived microglial cells) (Cavazzana-Calvo *et al.*, 2010; Cartier *et al.*, 2009).

3.2.4 Immune Response: Either to the Viral Vector or the Newly Expressed Transgene

Similar to organ transplantation, gene therapy has been plagued by this problem. Not yet fully solved, the issue is being addressed using a selection of patients who are naïve to the virus strain used, adapted concomitant short- or long-term immunosuppressive treatments, switching of viral serotypes, or, as recently proposed, a sophisticated regulation of transgene expression and immunogenes by microRNAs (Brown and Naldini, 2009).

So, through the year 2000, besides transduced stem cells for SCID, adrenoleukodystrophy, and red blood cell disorders, the first anecdotal reports of successes began to provide credible hopes: Intramuscular or subcutaneous delivery of cancer and infectious-disease vaccines, systemic or intratumoral delivery of oncolytic vectors, and recently, AAV-mediated muscle delivery of factor IX for hemophilia B (Nathwani *et al.*, 2011).

Other issues include gene control and targeting issues, especially when fine-tuned therapeutic protein expression is required.

3.3 Significant Recent Progress in Gene Therapy

3.3.1 Deeper Understanding of the Biology of the Transduced Cells

For instance, for treatment of SCID patients, using myeloablative preparative treatment which was not included in the early trials.

3.3.2 Improvements in Our Understanding of the Vectors

For instance: hemophilia and improved AAV vectors and new virus serotypes. The issue of insertional mutagenesis has been addressed by replacing retroviruses with self-complementary lentiviral vectors (first evaluated in 2006 in HIV using genetically modified CD4 T-cells, and now in other immunodeficiencies), or by including certain sequences such as the β-globin locus-control region to direct the site of integration to specific chromosomal sites. Other forms of safer genetic engineering include gene targeting and knocking-out specific genes via engineered nucleases (i.e., zinc-finger nucleases, engineered I-CreI homing endonucleases, or nucleases generated from tumor-associated lymphocyte (TAL) effectors). A recent example of gene-knockout-mediated gene therapy is the ΔΔ32 mutation which disables the CCR5 chemokine receptor in order to control HIV infection (currently tested in HIV patients) (Silva *et al.*, 2011).

Envelope protein pseudotyping of viral vectors (adenoviruses and retroviruses/lentiviruses especially) is being developed to either increase or limit host-cell ranges. These advances allow for the systemic administration of a smaller amount of vector (scAAV9 appears to be a very promising vector in this respect which appears to transduce a wide range of cell types and even to cross the blood–brain barrier which most viral vectors cannot) (Duque *et al.*, 2009). The potential for off-target cell modification could be limited using either chimeric envelope proteins bearing antibody or chemical fragments or by selectively mutating envelope protein genes. The magic bullet gene therapies are not available yet but considerable progress has been made over recent years.

3.3.3 Improved Delivery Procedures

The injection of naked DNA is a perfect example of this issue. Clinical trials of intramuscular injection of a naked DNA plasmid have occurred with some success; however, expression has been very low in comparison to other methods of transfection. Research efforts focusing on improving the efficiency of naked DNA uptake have yielded several novel methods, such as electroporation, sonoporation, and the use of a "gene gun" (ballistic gold, DNA-coated gold particles). The most advanced current indications are DNA vaccines, and angiogenic cardiovascular applications, as they involve only short-term expression and local delivery (Braun, 2008).

Much higher hurdles exist for genetic neuromuscular disorders. The first clinical trial attempted for a neuromuscular disease was for Duchenne muscular dystrophy (DMD) and it consisted of administering the complete coding sequence of the dystrophin gene, carried by a non-viral vector, plasmid DNA. In the phase I trial carried out in 2001–2003, a low dose of vector was used (600 μg) leading to very low and local expression of dystrophin (along the trajectory of the needle). Hydrodynamic limb vein (HLV) administration of the plasmid to the limbs which was developed in primate, rodent, and canine (golden retriever muscular dystrophy (GRMD) dog) models was shown to be efficient and well tolerated. It is possible to deliver large quantities of plasmid (up to several hundred milligrams) to the base of the limb by the intravenous route with the help of an inflatable tourniquet (Wooddell *et al.*, 2011). Efficacy of transfection in primates seems good (40% of the fibers in some muscles), but could be improved. This is possible, thanks to viral vectors such as AAV. These viruses cannot accommodate the complete coding sequence of dystrophin; therefore, shortened versions of the gene have been generated which are capable of correcting the dystrophic phenotype of mouse and dog models. Following a positive safety clinical trial using intramuscular administration of an AAV2.5-minidystrophin (Biodystrophin®) in patients aged 5–11 years (Bowles *et al.*, 2012), locoregional HLV administration of saline buffer in Becker muscular dystrophy (BMD, a mild form of DMD) volunteers has been carried out, setting the optimal conditions for the procedure in humans. Other applications targeting limb-girdle muscular dystrophy, (α or γ sarcoglycanopathies in particular) are already envisaged.

3.3.4 Scale-Up and Good Manufacturing Practice (GMP)

These issues have long been significant obstacles for many of the best vector systems. Even now, producing enough vector for large clinical trials is still problematic. In contrast to plasmids, it remains a challenge to produce viral vectors at a commercial and reproducible scale. Nevertheless, tangible progress is being made, and it is likely that adequate cell-line systems will be available for the manufacture of many vectors by the time clinical efficacy is established in phase III clinical trials.

As the field is learning from its mistakes, and many of the problems are now better understood, another major breakthrough in the field came from another set of rare genetic diseases. Following the announcement in May 2007 by the Moorfields Eye Hospital and University College London's Institute of Ophthalmology of the world's first gene therapy trial for Leber congenital amaurosis (an inherited blinding disease caused by mutations in the RPE65 gene), the results published in 2008 reported safety and increased vision following the subretinal delivery of recombinant AAV carrying the RPE65 gene. More ophthalmic disorders are now being investigated: Optic neuropathy, Stargardt disease, and color blindness (Stieger *et al.*, 2011). It seems obvious now that more therapy products will be approved in the near future both for monogenic and multifactorial disorders. The identification, sequencing and cloning of the estimated 25,000 genes of the human genome will boost the gene therapy field.

Gene therapy comes in forms other than transfer of the complete or partial coding sequence of the healthy gene. Rather than acting at the scale of DNA, one alternative consists of acting downstream, at the level of the gene product. One strategy uses antisense specific to the target gene to disrupt the transcription of the faulty gene. Another uses small interfering RNA (siRNA) to signal the cell to cleave specific sequences in the mRNA transcript of the faulty gene, disrupting its translation, and therefore its expression (first proposed for Huntington's disease in the mid 1990s (Haque *et al.*, 1997)). Another strategy uses double-stranded oligodeoxynucleotides as a decoy for the transcription factors that are required to activate the transcription of the target gene. Oligonucleotides can also mediate gene repair or targeted nucleotide alterations. The most advanced form of this concept is exon skipping (see Chapter 6).

3.4 The Example of Exon Skipping

The principle of exon skipping arose from studies of DMD in 1991 (Matsuo *et al.*, 1991). Mutations in the gene often lead to premature stop codons or destroy the reading frame and prevent the production of any dystrophin at all, producing severe forms of the disease. The more moderate forms result from mutations that preserve the reading frame and are compatible with the production of a shorter dystrophin molecule which may conserve some function.

This concept has been approached experimentally in the *mdx* mouse model, in which it has been possible to study the effect on muscle function of a whole series of deletions of increasing size. These studies indicated that not every part of the dystrophin protein is necessary for its function, leading to a therapeutic approach by modification of RNA splicing. Modifying the pre-messenger RNA splicing by exon skipping in order to restore the reading frame, a natural phenomenon which has been deciphered in the so-called "revertant" fibers of DMD patients, has thus been envisaged. In this situation, antisense nucleotides complementary to the key splicing sequences are chemically modified to make them insensitive to RNases which permit the production of a truncated dystrophin in patients in which there is a complete deficit in dystrophin, i.e., to transform a DMD-type dystrophy into a moderate type.

This method of "splicing therapy" should in theory correct a large number of cases of frameshift or non-sense mutations, by the judicious choice of the exon or exons to eliminate. The persistence of the therapeutic effect is obtained by repeated administration of the antisense oligonucleotides (their half-life in the cell is only a few weeks) or by vectorization of sequences which produce these antisense oligonucleotides in an AAV or a lentivirus (Goyenvalle *et al.*, 2004).

In 2006, the first attempt to do this in humans consisted of the intravenous administration of a low dose of an antisense phosphorothioate oligodeoxynucleotide designed to skip exon 19 in a ten-year-old DMD patient with a deletion of exon 20. Although the level was low, the presence of in-phase mRNA devoid of exons 19 and 20 in blood lymphocytes, and of traces of dystrophin in a muscle biopsy were observed one week after the last injection (Takeshima *et al.*, 2006). Two other molecules (2'O-methyl-phosphorothioate and morpholinos) targeting exon 51 are now in phase II and phase III trials. A phase I/II trial targeting exon 44 is also in progress (van Deutekom *et al.*, 2007; Cirak *et al.*, 2011).

Exon skipping must be adapted to the precise genetic characteristics of each patient. Furthermore, since each antisense nucleotide must be considered as a new medication in itself, it will be indispensable to adapt the regulatory constraints to facilitate their development and their accessibility to rare mutations. Another limitation is represented by the relative incapacity of these products to target the heart, except perhaps AAV-U7 if adequate conditions for administration become available and with new chemistries arising. *Trans*-splicing and skipping of multiple exons are being explored. In parallel, antisense strategies destined to interfere with the expression of RNA are also being developed to interfere with the expression of RNA in other neuromuscular diseases (dystrophinopathies and spinal muscular atrophy, but also myotonic dystrophy and other pathologies characterized by triplet repeats for which the accumulation of "abnormal" RNA has been shown to be toxic). A large proportion of patients are not eligible for exon skipping because some necessary parts of the protein cannot be deleted. This drawback turns out to be of interest for non-genetic disorders, through the skipping of necessary exons of, for instance, oncogenes or viral proteins. Antisense oligonucleotides will most certainly give rise to a whole new class of drugs. Antisense therapy is not strictly a form of gene therapy (except under the form of viral transduction), but is a related, genetically mediated therapy.

More details on exon skipping are given in Chapters 6 and 27.

3.5 Concluding Remarks

Gene therapy has gone through several ten-year cycles. After a first phase of irrational hopes of a revolution in medicine, it experienced ten years of disappointments and skepticism. The two deaths that have been directly attributed to gene therapy have delivered a major setback in terms of public and legal concerns, partly related to the misperception of "genetic manipulation" and the inevitable suspicion of all new cutting-edge therapies. However, the subsequent improvements in vector technology and trial design, more achievable first indications, and ethical standards have benefited the field in the long run. The following decade provided a few real successes in the clinic which helped to regain credibility and attract public support as well as research funds and investments. Suspicion still prevails as revealed by the death of Jolee Mohr: in July 2007, this rheumatoid arthritis patient who received a local knee injection with an anti-inflammatory gene-therapy-based AAV vector carrying the tumor-necrosis-factor-receptor–immunoglobulin Fc fusion gene, died from massive organ failure. The highly publicized hold that was put on this trial, which had enrolled more than 100 patients, was later lifted (but with unfortunately much less media coverage) as the tragic outcome was confirmed to have been caused by a prior fungal infection related to the known side effects of her initial conventional treatment (Wilson, 2008).

We should learn from the history of other medical procedures such as blood transfusion (do we remember that many patients died from non-compatible transfusions, since MHC classes were unknown?), organ transplantation, and monoclonal antibodies, that have all boasted a lot of expectations, then decades of failed clinical trials followed by a much more realistic approach to many of the problems, and by major clinical and commercial successes. We see that gene therapy is following a similar pendulum pattern. A new decade of real progress and success has now started, which will allow gene therapies to strongly affect the practice of medicine.

It is striking that the field which started with a rare disease, moving quickly to a vast majority of large-market indications, which turned out to be more complex to address (this is a reason for the first decade of clinical failures), has now switched to more focused single-gene applications. FDA and EMA rejected in 2011 the application for an AAV-based gene therapy for familial hyperlipidemia, the risk:benefit ratio being far more difficult to address in a very competitive field. However, in 2012, the drug now named "Glybera" was given marketing authorization, following the EMA recommendation earlier that year that the therapy be made available for the most severely ill patients.

Rare diseases are, in fact, often easier targets for cutting-edge therapies, with well-defined, well-understood target genes and very accessible clinical indications. Big pharmaceutical companies are now beginning to reconsider and invest in gene therapy applications. They finally decided to capitalize on successes "in the short term with the easier, but less profitable diseases, rather than failing with the big-market indications" (Scollay, 2009). ADA-SCID, the first indication to be tested in a federally approved gene therapy trial, has been licensed by GSK, and is now under review for drug approval and thus the field has come full circle.

Another lesson from this recent history of gene therapy is the crucial role that patient organizations (along with small biotechnology companies) have played in the advancement of the field. The first logical step was to introduce genes directly into human cells, focusing on diseases caused by single-gene defects, such as cystic fibrosis, hemophilia, muscular dystrophy, and sickle cell anemia. Because this has proven more difficult than modifying bacteria, it is primarily thanks to the long-term, continuous commitment of patient organizations that helped to cross the "death valley" toward proofs of concept in humans that attention from larger companies and stakeholders got finally attracted.

New frontiers in the field can be foreseen. Among them: (1) germ line gene therapy, which is prohibited for application in human beings, at least for the present, for a variety of technical and ethical reasons and (2) preventive gene therapy (for instance anticancer vaccines) which should turn out to be clinically easier to validate but requires much larger human trials. The future of gene therapy will therefore also depend on the ability of private companies to solve technological (i.e., large-scale GMP production) and business (i.e., multiple patents and licenses) issues, and to engage in broader collaborations with patient organizations, regulatory authorities, and the scientific and medical community. As many other diseases remain untouched, patient organizations and other public stakeholders will keep boosting the field.

References

Aiuti, A., Cattaneo, F., Galimberti, S., *et al.* (2009). Gene therapy for immunodeficiency due to adenosine deaminase deficiency, *N Engl J Med*, **360,** 447–458.

Aposhian, H.V. (1970). The use of DNA for gene therapy — the need, experimental approach, and implications, *Perspect Biol Med*, **14,** 98–108.

Avery, O.T., MacLeod, C.M., McCarty, M. (1944). Studies on the chemical nature of the substance inducing transformation of pneumococcal types, *J Exp Med*, **79,** 137–158.

Bowles, D.E., McPhee, S.W., Li, C., *et al.* (2012). Phase 1 gene therapy for Duchenne muscular dystrophy using a translational optimized AAV vector, *Mol Ther*, **20,** 443–455.

Braun, S. (2008). Muscular gene transfer using nonviral vectors, *Curr Gene Ther*, **8,** 391–405.

Brown, B.D., Naldini, L. (2009). Exploiting and antagonizing microRNA regulation for therapeutic and experimental applications, *Nat Rev Genet*, **10,** 578–585.

Cartier, N., Hacein-Bey-Abina, S., Bartholomae, C.C., *et al.* (2009). Hematopoietic stem cell gene therapy with a lentiviral vector in X-linked adrenoleukodystrophy, *Science*, **326,** 818–823.

Cavazzana-Calvo, M., Payen, E., Negre, O., *et al.* (2010). Transfusion independence and HMGA2 activation after gene therapy of human β-thalassaemia, *Nature*, **467,** 318–322.

Cirak, S., Arechavala-Gomeza, V., Guglieri, M., *et al.* (2011). Exon skipping and dystrophin restoration in patients with Duchenne muscular dystrophy after systemic phosphorodiamidate morpholino oligomer treatment: an open-label, phase 2, dose-escalation study, *Lancet*, **378,** 595–605.

Duque, S., Joussemet, B., Riviere C., *et al.* (2009). Intravenous administration of self-complementary AAV9 enables transgene delivery to adult motor neurons, *Mol Ther*, **17,** 1187–1196.

Friedmann, T. (2001). Stanfield Roger: Insights into virus vectors and failure of an early gene therapy model, *Mol Ther*, **4,** 285–288.

Goyenvalle, A., Vulin, A., Fougerousse, F., *et al.* (2004). Rescue of dystrophic muscle through U7 snRNA-mediated exon skipping, *Science*, **306,** 1796–1799.

Graham, F.L., van der Eb, A.J. (1973). A new technique for the assay of infectivity of human adenovirus 5 DNA, *Virology*, **52,** 456–467.

Haque, N.S., Borghesani, P., Isacson, O. (1997). Therapeutic strategies for Huntington's disease based on a molecular understanding of the disorder, *Mol Med Today*, **3,** 175–183.

The Journal of Gene Medicine (2012). *Gene Therapy Clinicial Trials Worldwide* [online]. Available at http://www.abedia.com/wiley/index.html [Accessed 6 February 2013].

Keeler, C.E. (1947). Gene therapy, *J Heredity*, **38,** 294–298.

Lederberg, J. (1963). 'Biological future of man', in Wolstenholme, G. (ed.) *Man and His Future*, London: J & A Churchill, pp. 263–273.

Mackowiak, M., Maki, J., Motes-Kreimeyer, L., *et al.* (1999). Vaccination of wildlife against rabies: Successful use of a vectored vaccine obtained by recombinant technology, *Adv Vet Med*, **41,** 571–583.

Matsuo, M., Masumura, T., Nishio, H., *et al.* (1991). Exon skipping during splicing of dystrophin mRNA precursor due to an intraexon deletion in the dystrophin gene of Duchenne muscular dystrophy Kobe, *J Clin Invest*, **87,** 2127–2131.

Mitchell, P. (2010). Ark's gene therapy stumbles at the finish line, *Nat Biotechnol*, **28,** 183–184.

Nathwani, A.C., Tuddenham, E.G., Rangarajan, S., *et al.* (2011). Adenovirus-associated virus vector-mediated gene transfer in hemophilia B, *N Engl J Med*, **365,** 2357–2365.

Prazeres, D.M.F. (2011). 'Historical perspectives', in Prazeres, D.M.F. *Plasmid Biopharmaceuticals: Basics, Applications, and Manufacturing*, Hoboken NJ: John Wiley & Sons, pp. 1–34.

Raper, S.E., Chirmule, N., Lee, F.S., *et al.* (2003). Fatal systemic inflammatory response syndrome in an ornithine transcarbamylase deficient patient following adenoviral gene transfer, *Mol Genet Metab*, **80,** 148–158.

Räty, J.K., Pikkarainen, J.T., Wirth, T., *et al.* (2008). Gene therapy: The first approved gene-based medicines, molecular mechanisms and clinical indications, *Curr Mol Pharmacol*, **1,** 13–23.

Rogers, S., Pfuderer, P. (1968). Use of viruses as carriers of added genetic information, *Nature*, **219,** 749–751.

Rosenberg, S.A., Aebersold, P., Cornetta, K., *et al.* (1990). Gene transfer into humans — Immunotherapy of patients with advanced melanoma, using tumor-infiltrating lymphocytes modified by retroviral gene transduction, *N Engl J Med*, **323,** 570–578.

Scollay, R. (2009). A brief overview of the past, present, and future gene therapy, *Ann N Y Acad Sci*, **953,** 26–30.

Silva, G., Poirot, L., Galetto, R., *et al.* (2011). Meganucleases and other tools for targeted genome engineering: perspectives and challenges for gene therapy, *Curr Gene Ther*, **11,** 11–27.

Stieger, K., Cronin, T., Bennett, J., *et al.* (2011). Adeno-associated virus mediated gene therapy for retinal degenerative diseases, *Methods Mol Biol*, **807,** 179–218.

Takeshima, Y., Yagi, M., Wada, H., *et al.* (2006). Intravenous infusion of an antisense oligonucleotide results in exon skipping in muscle dystrophin mRNA of Duchenne muscular dystrophy, *Pediatr Res*, **59,** 690–694.

Terheggen, H.G., Lowenthal, A., Lavinha, F., *et al.* (1975). Unsuccessful trial of gene replacement in arginase deficiency, *Z Kinderheilkd*, **119,** 1–3.

Trubo, R. (1983). Genetic manipulation with retroviruses may lead to Lesch-Nyhan treatment, *Med World News*, **24,** 9.

van Deutekom, J.C., Janson, A.A., Ginjaar, I.B., *et al.* (2007). Local dystrophin restoration with antisense oligonucleotide PRO051, *N Engl J Med*, **357,** 2677–2686.

Wade, N. (1980). UCLA gene therapy racked by friendly fire, *Science*, **210,** 509–511.

Wilson, J.M. (2008). Adverse events in gene transfer trials and an agenda for the New Year, *Hum Gene Ther*, **19,** 1–2.

Wolff, J.A., Lederberg, J. (1994). An early history of gene transfer and therapy, *Hum Gene Ther*, **5,** 469–480.

Wooddell, C.I., Hegge, J.O., Zhang, G., *et al.* (2011). Dose response in rodents and nonhuman primates after hydrodynamic limb vein delivery of naked plasmid DNA, *Hum Gene Ther*, **22,** 889–903.

4

GENETIC PHARMACOLOGY USING SYNTHETIC DEOXYRIBONUCLEOTIDES

***Jean-Christophe François*[a] *and Carine Giovannangeli*[a,b]**

4.1 Introduction

Short synthetic oligonucleotides are now used routinely to modulate gene expression in cells and *in vivo*, and oligonucleotide-based drugs are currently being evaluated in clinical trials for many different applications. Most of the approaches are focused on targeting the messenger RNA (mRNA) of a chosen gene. Among the strategies that target mRNA, RNA silencing using RNA interference machinery is at present certainly the most efficient approach to knockdown of gene expression, but some drawbacks restrain use in therapeutics. One promising recent means to modulate gene expression is interference with the splicing machinery using chemically modified antisense oligonucleotides that are unable to induce mRNA cleavage. Such oligonucleotides bind to splice sites and induce production of a novel mRNA and consequently a novel protein. Using combinatorial approaches, catalytic oligodeoxynucleotides have also been developed to control gene expression, even in an inducible manner. Finally, targeting of DNA is also possible through formation of local triple helical structures between the genomic DNA and a synthetic oligonucleotide. This antigene strategy is especially attractive as it can be used to manipulate gene sequences. All these short-oligonucleotide-based approaches require an efficient means of cellular transfection. Transfection can be achieved *in vitro* using specific chemical or physical delivery methods. For *in vivo* delivery, a number of novel formulations and

[a]National Museum of Natural History, Laboratoire de Biophysique, 43 rue Cuvier, 75005 Paris, France
[b]Email: giovanna@mnhn.fr

oligonucleotide conjugates are currently being tested that may achieve sustainable inhibitory effects in the specific cell types of interest. In this chapter, we will describe the general principles of antigene and antisense strategies using synthetic oligonucleotides as artificial modulators of biological functions and as potential drugs. Advantages and limits of each strategy will also be discussed.

4.2 Antigene Strategy

Molecules that specifically recognize a DNA sequence are attractive tools for manipulating gene sequence and expression. These have been termed antigene molecules. Today there are two types of synthetic antigene molecules: Polyamides consisting of *N*-methylpyrrole and *N*-methyimidazole units and oligonucleotides and analogues such as peptide nucleic acids (PNAs). Another class of artificial sequence-specific DNA binding molecules is engineered proteins, such as zinc-finger proteins and transcription activator-like (TAL) effectors. Specific and efficient modulation of gene expression can be achieved by antisense-based strategies (see Section 4.3) and antigene approaches are not really advantageous in this regard. However, antigene molecules are unique tools that provide a means to modify DNA sequences. Here we will focus on their potential for genome engineering.

4.2.1 Antigene Oligonucleotides for DNA-Targeting

DNA-duplex-binding oligonucleotides were first reported in 1987, simultaneously by Peter Dervan and Claude Hélène's laboratories. These pioneering works described triplex-forming oligonucleotides (TFOs) that bind to oligopyrimidine–oligopurine sequences via Hoogsteen hydrogen bonds (Fig. 4.1).

Since their discovery, TFOs have been extensively studied and their cellular activities have been reported in different systems. They have been used to inhibit gene expression at the transcriptional level as well as to induce targeted mutagenesis in model systems. However the major obstacle to the development of TFOs in cellular applications was the limited thermodynamic stability of the triple-helix complex. Considerable progress in specifically stabilizing triplex formation has been achieved using chemically modified oligonucleotides (Fig. 4.2). Today among the most promising TFOs are modified with locked nucleic acids (LNAs) or PNAs; both types of chemically modified oligonucleotides form highly stable and specific complexes.

The biological activity of simple mixed sequences of LNAs and PNAs (i.e., non-exclusively oligopyrimidine) on gene expression, has been reported. These

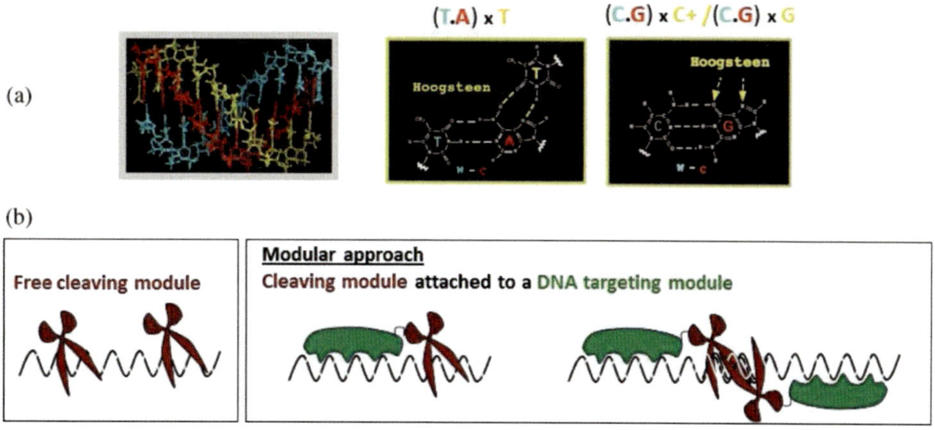

FIGURE 4.1 ■ Antigene molecules. (a) DNA recognition by triplex formation. The triple-helical structure is shown with the oligopurine strand in red, the oligopyrimidine strand in blue and the TFO in yellow. Hoogsteen or reverse Hoogsteen bonding interactions are observed in the (T × A)•T base triplet and in the (C × G)•C+ or (C × G)•G. (b) Sequence-specific nucleases. A DNA-cleaving module attached to a DNA-binding module induces DNA breaks. Conjugation/fusion enables targeting of the cleaving module to the site recognized by the DNA-targeting module and prevents binding to other sites. Sequence-specific nucleases have been described that act as (1) monomers (i.e., synthetic nucleases) with a chemical cleaving agent conjugated to a TFO, a PNA or a polyamide and (2) dimers (i.e., protein-based nucleases) like zinc-finger nucleases (ZFNs), see later, or TALENs with, for example, the FokI catalytic domain fused to zinc-finger or TAL motifs.

molecules should bind by Watson–Crick pairing upon double-stranded DNA invasion and then might theoretically allow targeting of any DNA sequence. Such interaction with the target sequence may occur during biological processes involving local DNA opening, such as transcription or replication. Further studies are needed to evaluate how many genomic sequences are available for such binding and how general this approach could be.

4.2.2 Modifying DNA Sequence in a Controlled Way for Genome Engineering

The induction of stable modification of the cellular genome in a controlled and specific manner is a major challenge for functional studies and gene therapy. Whole-genome sequences are being determined from a growing number of species and DNA sequence variation involved in disease or other biological traits of interest are being identified at an increasing pace. Both exploring and exploiting DNA sequence data therefore creates a huge need for highly efficient methods of genome manipulation. Although RNA interference provides a powerful approach for exploring gene function (see Section 4.3.3), this method is not without caveat

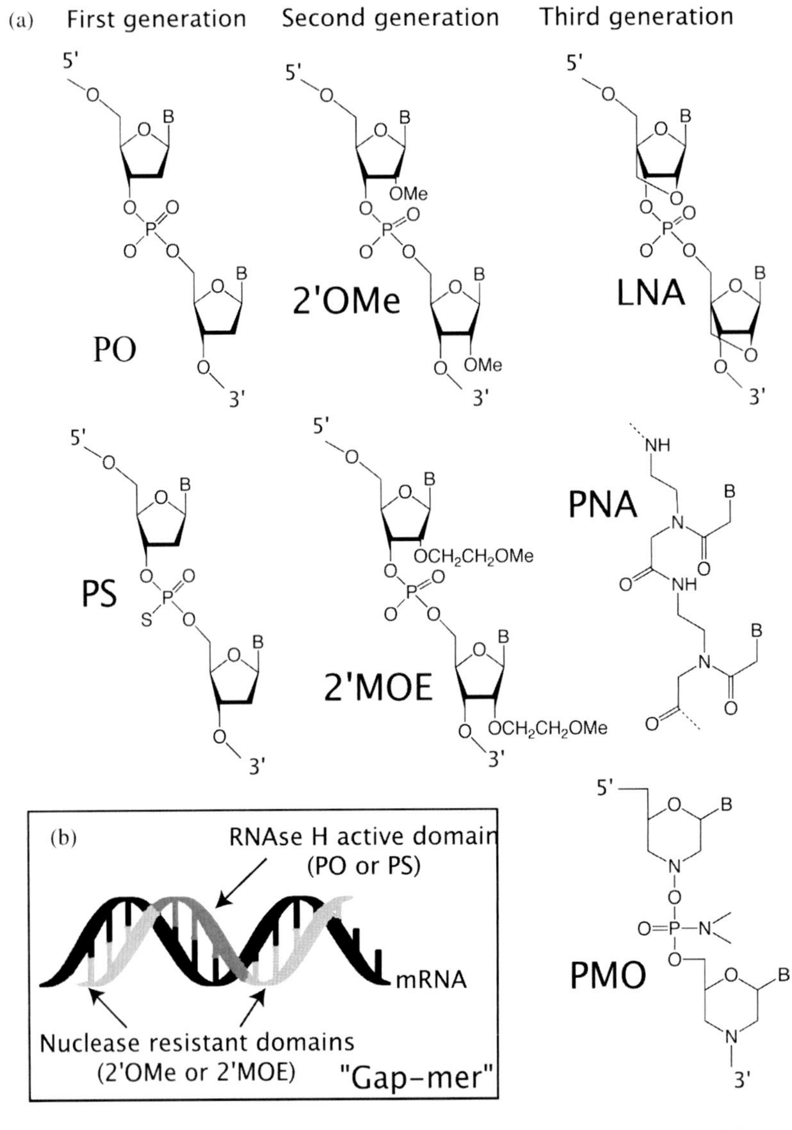

FIGURE 4.2 ■ Chemical structures of some modifications used in nucleic-acid-based strategies. (a) Backbone and sugar residues of oligophosphodiesters were replaced to improve nuclease resistance. Phosphorothioates were among the first modification of RNAse H-dependent antisense molecules (first generation). 2′-O-methyl (2′OMe) and 2′-O-methoxyethyl (2′MOE) oligonucleotides do not induce RNAse H-mediated RNA cleavage (second generation). The third generation of modifications, LNAs, PNAs, and phosphorodiamidate morpholino oligomers (PMOs), are also unable to recruit RNAse H, but form a highly stable complex with their RNA target, allowing their use as a physical block of RNA metabolism. Some of these modifications are also used for small interfering RNAs (siRNAs), ribozymes, DNAzymes and aptamers. (b) Gapmer structure with a central region made of DNA or phosphorothioate, allowing RNAse-H cleavage, and flanks made of 2′OMe, 2′MOE or LNA to protect the oligonucleotide from nuclease degradation and to improve binding affinity for the RNA target.

and cannot substitute for targeted mutagenesis or gene-targeting approaches. Thus, there is a great need to develop gene-modification methods that are applicable for large-scale functional genomics approaches in model organisms.

Seminal studies in the early 1990s demonstrated that a specific double-strand break stimulates homologous recombination by several orders of magnitude in mammalian cells. These studies were performed on a model gene containing the 18-base pair (bp) recognition sequence for the mitochondrial intron *Saccharomyces cerevisiae* I (I-SceI) homing endonuclease. Generalization of this approach therefore requires a sequence-specific endonuclease to create a double-strand break within each endogenous gene of interest. There are currently four types of sequence-specific endonucleases that can potentially be used for controlled genome modification: Zinc-finger nucleases (ZFNs), modified homing endonucleases, TAL effector nucleases (TALENs) and synthetic nucleases. A recent major advance has been made with the use of protein-based nucleases, especially ZFNs, which opens new roads for transgenesis in model organisms and gene therapy.

Synthetic nuclease can also be obtained by conjugation of DNA-damaging agents to synthetic antigene molecules. In fact, TFOs conjugated to a DNA-damaging agent have already been shown to induce gene modification; however, efficiencies obtained so far have been low compared with those of ZFNs. For example, TFOs conjugated to a psoralen moiety were shown to induce targeted interstrand crosslinks after photoactivation and, following lesion repair, to lead to targeted gene-sequence modification. Triplex formation has thus been used for targeted mutagenesis at the triplex sites and also for stimulation of gene modification by homologous recombination. Concerning TFOs linked to cleaving agents, their cleavage activity has been demonstrated *in vitro* and in cells, but it remains to be established how these activities compare with those of protein-based nucleases in stimulating targeted gene modification. TFO-based nucleases have the advantages of simple design and straightforward synthesis and purification and thus might constitute an alternative to protein-based sequence-specific nucleases and an exciting area of development of synthetic DNA-reading moieties.

4.3 Antisense Strategy

Antisense molecules are designed to target complementary sequences in different types of RNAs and then to interfere with RNA metabolism at various levels. This is the so-called antisense approach. We will begin this section with a point of nomenclature. Within a DNA gene, two strands can be defined. One contains the nucleotide information for the amino acid sequence of the protein and is therefore called the sense strand. The second strand is complementary to the sense strand and is referred to as the antisense strand. By extension, all nucleobase-containing molecules that are complementary to messenger or pre-messenger RNAs are called antisense molecules, including — in this chapter — the guide strand of small interfering RNAs (siRNAs). More recently other functional RNAs, like microRNAs (miRNAs), have been targeted with complementary oligonucleotides (Fig. 4.3); such an approach is referred to as antisense and these complementary oligonucleotides as antagomirs.

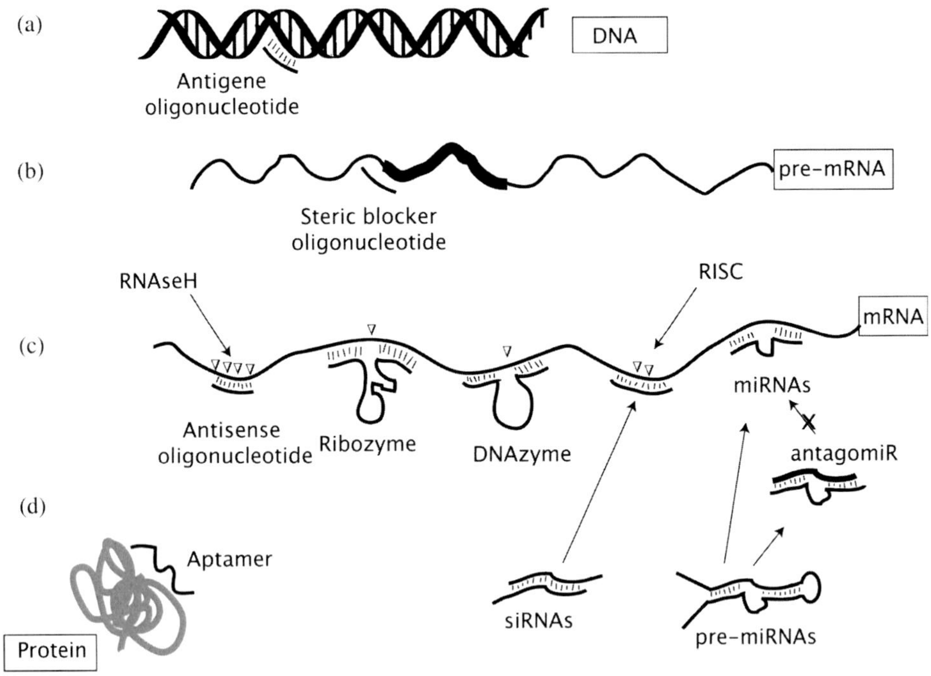

FIGURE 4.3 ■ Targets for synthetic oligonucleotides as artificial modulators of gene expression. (a) In the nucleus, antigene oligonucleotides could target double-stranded DNA to interfere with transcription, by competing with transcription factors or blocking RNA polymerase. When conjugated to DNA-modifying agents, it could be used for controlled genome modification specifically at its binding site. (b) Steric blocker antisense oligonucleotides recognize specific sites on pre-mRNA to interfere with splicing. (c) In the cytoplasm, antisense oligonucleotides that can activate RNAse H upon their binding to mRNA induce mRNA cleavage and therefore result in translation inhibition. Hammerhead ribozymes are RNA molecules that can cleave a complementary RNA strand without the help of proteins. Like ribozymes, DNAzymes obtained by systematic evolution of ligands by exponential enrichment (SELEX) are able to efficiently cleave their RNA targets. siRNAs induce specific cleavage of their target through recruitment of the RNA-induced silencing complex (RISC). miRNAs block translation by forming a double helix with the 3′ untranslated region of an mRNA; the duplex may contain some mismatches. Antagomirs are antisense oligonucleotides designed to bind to endogenous miRNAs and block their gene regulatory function. (d) Aptamer oligonucleotides obtained by SELEX bind specifically to a protein to inhibit activity or block binding to other partners.

Thanks to modern nucleotide chemistry, it is now possible to easily synthesize short oligonucleotide sequences with various sugar, backbone and base modifications. Generally, antisense oligonucleotides are less than 25 nucleotides in length. Such oligonucleotides are commercially available. In contrast, antisense RNAs, which were also used to target a specific mRNA, are often longer and are made of ribonucleotides; they are often produced by expression vectors. Interestingly, natural antisense RNAs have been identified and are now recognized as natural mechanisms for gene regulation in prokaryotes and also in eukaryotes. These natural antisense transcripts regulate gene expression at both transcriptional and post-transcriptional levels and carry out a wide variety of biological roles.

Here we will focus on the advantages and drawbacks of using artificial synthetic antisense oligonucleotides to interfere with mRNA metabolism and intracellular degradation pathways.

4.3.1 Chemically Modified Oligonucleotides — Nature and Properties

4.3.1.1 *Necessity of chemical modification for optimal activity*

More than 35 years ago, it was proposed that short oligodeoxynucleotides may interfere with gene expression through binding to complementary sequences. In the late 1970s, it was demonstrated that short oligodeoxyribonucleotides were able to block *in vitro* RNA translation and to inhibit viral replication in cell culture. Building on these pioneering findings, antisense oligonucleotides have been optimized to improve potency *in vivo*. Usually, antisense oligonucleotides are 15–25 nucleotides in length. This length ensures unique gene recognition, although the optimal length depends on the genome complexity. Because natural oligodeoxyribonucleotides are sensitive to nucleases in biological fluids, chemists have developed modifications that impart nuclease resistance without compromising high affinity and specificity for their RNA target.

Numerous mechanisms are potentially involved in antisense biological activities; the mechanism depends on the nature of the RNA target and the oligonucleotide chemistry (Fig. 4.3). Antisense activities are classified in two main groups. One group involves RNA degradation upon oligonucleotide binding through the recruitment of a specific ribonuclease like RNAse H or the RNA-induced silencing complex (RISC). These nucleases degrade the RNA strand of the hybrid formed between antisense oligonucleotide and target RNA. The other group does not induce RNA cleavage but interferes with cellular RNA metabolism by physical inhibition of splicing or translation machineries. These distinct types of mechanisms of action are invoked by different types of modified oligonucleotides, that are described as follows.

4.3.1.2 *Three generations of oligonucleotides*

Chemical modifications involve substitutions within the sugar, heterocyclic bases or backbone of DNA or RNA (Fig. 4.2). The first generation of modified antisense oligonucleotides includes oligophosphorothioates. In these molecules, one oxygen at each phosphorus in the oligonucleotide chain is replaced by sulfur. These first-generation oligonucleotides were widely used in cellular models and *in vivo*, since this modification confers chemical stability, maintains RNAse H activity, and is easy to synthesize. The second generation of antisense oligonucleotides have modifications to the sugar, particularly to the 2′ position, and include 2′-O-methyl (2′OMe) and 2′-O-methoxyethyl (2′MOE) oligonucleotides. These oligonucleotides have increased affinity for RNA and are nuclease and RNAse H-resistant. Antisense oligonucleotides with third generation chemical modifications include LNAs, PNAs, and phosphorodiamidate morpholino oligomers (PMOs) and also exhibit nuclease and RNAse H resistance and high affinity and specificity. LNAs are RNA analogs where the ribose is locked by a methylene bridge between the 2′O and the 4′C. PNAs are DNA analogs with achiral polyamide backbone consisting of *N*-(2-aminoethyl)glycine units linked to nucleobases. In PMOs, the deoxyribose moiety of DNA is replaced with a six-membered morpholine ring, and the internucleoside linkage is replaced with phosphorodiamidate linkages.

A further improvement incorporates different chemical generations into "gapmers", which are antisense sequences that have their ends modified with 2′MOE, 2′OMe, or LNA chemistry for nuclease protection and increased RNA binding and contain phosphorothioate linkages in their centers to retain RNAse H activity. Four or five phosphorothioate linkages lead to specific RNA cleavage.

These same chemical modifications initially used for "standard" antisense oligonucleotides have been used in siRNAs to confer nuclease resistance or to avoid undesired immunological effects while maintaining the RISC activity for efficient RNA degradation and knockdown efficiency.

The advantages of using a given chemistry for a specific RNA target of interest and a specific mechanism of action will be detailed in the following paragraphs.

4.3.2 RNAse H-Dependent Antisense Oligonucleotides

4.3.2.1 *Choice of the target sequence*

The antisense oligonucleotides can be theoretically targeted to various types of RNAs (Fig. 4.3) and act by various mechanisms. Whether the mechanism involves RNA degradation or not, all regions of an RNA cannot be targeted with equal efficiency, probably due to RNA structure and accessibility. The optimal antisense sequence must be identified and different approaches have been developed to find effective sequences. *In vitro* techniques either based on combinatorial approaches with annealing reactions of the target RNA with oligonucleotide arrays or based on measurement of target accessibility by RNAse H cleavage mapping, have been developed. Other approaches involve *in silico* prediction of accessible regions on mRNA (i.e., unstructured regions). Obviously, *in vitro* strategies cannot take into account the effect of the cellular environment on RNA structure. Experimental screening of numerous oligonucleotides complementary to the RNA sequence of interest thus appears to be a more valuable approach, although it is expensive and time consuming. It should be noted that these approaches have also been used to identify regions of mRNA accessibility to siRNAs.

4.3.2.2 *Specific gene knockdown*

If the antisense oligonucleotide is able to induce RNAse H-dependent cleavage, the mRNA is destroyed, provoking specific gene knockdown. The first-generation phosphorothioate oligonucleotides and later generation gapmers have been extensively used in this way in cell cultures for functional studies and target validation, and for *in vivo* studies and even as therapeutics.

The antisense oligonucleotides that function via an RNAse H-dependent mechanism were the first to be investigated clinically. One of them, called Vitravene™ was approved in 1998 by the US Food and Drug Administration (FDA) for treatment of cytomegalovirus (CMV) retinitis in AIDS patients, demonstrating for the first time the feasibility of antisense oligonucleotides as drugs. This 21-nucleotide long oligophosphorothioate is administered by intraocular injection, making it a highly specialized case.

4.3.2.3 *Non-target-related effects*

Unexpected effects of antisense oligonucleotides and more specifically oligophosphorothioates, unrelated to their RNA binding, have been described. These non-antisense effects can be both non-sequence-specific (resulting from the polyanionic nature of the oligonucleotides) or sequence specific. The non-antisense effects have been mainly described with unmodified oligonucleotides and with oligophosphorothioates and generally occur at high concentrations or with long oligonucleotides. Two motifs that induce sequence-specific, but non-antisense, effects are the CpG motifs and the G-rich sequences.

CpG-containing oligonucleotides, like unmethylated CpG DNA, are recognized by a Toll-like receptor, TLR-9, that triggers protective innate and adaptive immune responses. CpG-containing oligonucleotides stimulate the *in vivo* immune response and have potential for use as vaccine adjuvants in cancer immunotherapy.

G-rich oligonucleotides can also have non-antisense activities, such as cell-cycle arrest and growth inhibitory effects. These G-rich oligonucleotides form G-quadruplex structures that could be specifically recognized by cellular proteins and then exert their biological effect through these interactions.

Despite their possible non-antisense effects, phosphorothioate-containing oligonucleotides are still widely used, especially in the context of gapmers or siRNAs, due to their nuclease resistance. Appropriate controls and experiments must always be done to demonstrate specific antisense activity of oligonucleotides.

4.3.3 RNA Interference — Advantages and Drawbacks for Gene Extinction

Another approach to efficiently knockdown gene expression by targeting mRNA is the use of RNAs that act through the RNA interference (RNAi) pathway (Fig. 4.4). siRNAs are small double-stranded RNAs of 21–23 bp. Like RNAse H-dependent antisense oligonucleotides, this post-transcriptional gene silencing occurs thanks to mRNA degradation carried out by the RISC (Chapter 5).

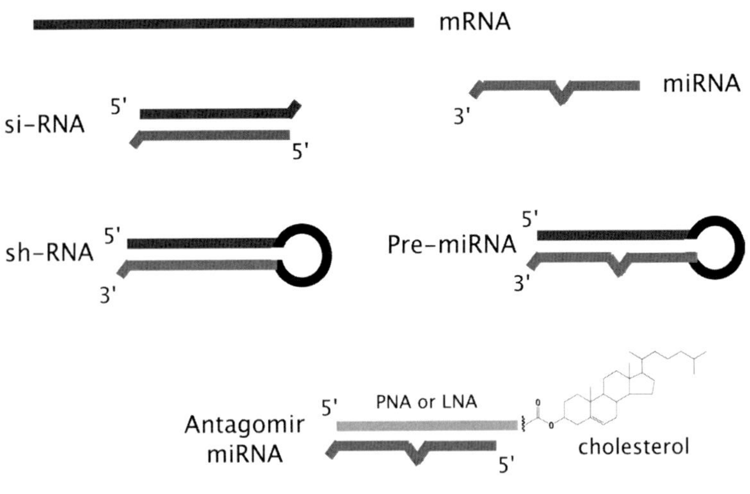

FIGURE 4.4 ■ siRNAs, short hairpin RNAs (shRNAs), antagomirs and their targets. (*Left*) siRNAs are 21-nucleotide-long double-stranded RNAs targeted to mRNA; the synthetic oligonucleotides are often protected at their 3′ ends by two deoxythymidines. On the siRNA scheme, passenger and guide strands are the upper and lower strands, respectively. Delivery of siRNAs is accomplished through the same techniques used for antisense molecules. shRNAs are transcribed intracellularly from plasmids or viral vectors. The loop region is removed by Dicer to yield a siRNA. (*Right*) miRNAs are natural double-stranded RNAs that are matured in the nucleus by Drosha; miRNAs block translation of mRNA upon their binding to the 3′ untranslated region (UTR). Antagomirs can be used to inhibit miRNA function. They generally contain RNAse H-resistant residues and, like siRNAs, have been successfully linked to cholesterol for improved delivery.

RNA interference is a highly conserved mechanism that relies on double-stranded RNA to mediate specific gene silencing. This process occurs in organisms such as plants, fungi and animals. The precursor double-stranded RNA is processed by the cellular enzyme Dicer to generate short RNA duplexes, which are then recognized by RISC. Once loaded with diced double-stranded RNA, RISC chooses the strand that can bind to endogenous messenger RNA, and recognizes the complementary sequence on this mRNA to induce specific cleavage. We will discuss here the advantages and limits of siRNAs compared with "standard" antisense oligonucleotides in gene knockdown applications.

4.3.3.1 *siRNA efficiency*

Some comparisons of antisense oligonucleotides and siRNAs targeted to the same gene sequences have been published. In most cases, the latter are more potent silencers of genes. siRNAs often have activity in the range of 1–100 nM, lower concentrations than the active range of antisense oligonucleotides. RNA interference can be accomplished with synthetic siRNAs or with short hairpin RNAs (shRNAs) that are synthesized through plasmid vectors and processed by the cellular RNAi machinery into siRNAs. The *in vivo* applications of exogenous RNAi, especially using siRNAs, remain difficult due to problems associated with their low chemical stability and their delivery into cells. Researchers are attempting to address these problems, which are shared with antisense molecules, by testing chemical modifications of RNA backbones to enhance nuclease resistance while maintaining RISC activity. As a general rule, limited modifications of siRNA extremities do not significantly impair inhibitory activity, whereas full modification of the guide strand is highly detrimental to silencing activity. Methods for efficient delivery of siRNA have been described; some are applications of previous knowledge on antisense delivery, others are completely innovative such as the conjugation of aptamers or cell-surface-receptor ligands to siRNA for targeted delivery (see Fig. 4.7).

4.3.3.2 *Non-target-related effects*

It was initially claimed that RNA interference was a highly specific strategy to knockdown gene expression. However, recent developments have revealed problems similar to those that were previously found with "standard" antisense oligonucleotides, such as sequence-specific off-target effects and non-sequence-specific ones. Several studies have shown that a moderate concentration of siRNA (100 nM) can silence expression of a range of unrelated genes, referred to as off-target genes, due to partial complementarity with their mRNAs. Microarray analyses have shown that siRNAs with partial complementarity to an mRNA can cause a reduction in expression if there are six to seven consecutive matches between the 5′ end of siRNA guide strand and the off-target mRNA sequence. Some rules now included in web software for design of siRNA sequences attempt to ensure that this partial complementarity is avoided.

The most critical non-sequence-specific effect of siRNAs is the induction of the immune response. The mammalian innate immune system can sense foreign RNA through specific Toll-like receptors (TLR7/8) and is therefore activated by transfected siRNAs, potentially compromising the specificity of silencing. It was found that siRNAs triggering such immune activation contain specific immunostimulatory nucleotide motifs, for example GU-rich, UGGC and UGU sequences. In order to avoid this immune response, siRNA modifications have been used, particularly replacement of uridines in the passenger strand with 2′-O-methyl, 2′-fluoro or 2-deoxy derivatives. The presence of few 2′-O-methyl uridines at the extremities of the passenger strand allows an siRNA to evade immune recognition by Toll receptors without compromising recognition by RISC. Appropriate siRNA controls should be designed to evaluate or exclude potential immune activity.

4.3.3.3 *Exploiting immune response activation*

Rather than avoiding or diminishing the immune response, efforts have also been directed toward synergizing RNAi-mediated silencing with Toll receptor-mediated activation of innate immune response. Certain modifications in the passenger strand of siRNAs can increase their immunostimulatory properties, such as addition of 5′-triphosphate or use of a miRNA-like sequence, such as a non-pairing uridine bulge. Another innovative approach uses an siRNA targeted to an immune suppressor gene signal transducer and activator of transcription 3 (Stat3) conjugated to a TLR9 agonist, a CpG oligonucleotide. This conjugate activates both tumor-associated immune cells by suppression of Stat3 expression and a potent antitumor immune response through the CpG oligonucleotide. These bifunctional siRNA approaches should be further explored.

4.3.4 Steric Blocker Antisense Oligonucleotides

Unlike RNAse H-dependent antisense oligonucleotides and siRNAs that efficiently induce degradation of mRNA and inhibit translation, some chemically modified oligonucleotides are unable to activate RNA cleavage, but can physically interfere with translation or splicing by binding to specific regions of mRNA or pre-mRNA. 2′OMe and 2′MOE-containing oligonucleotides were first used as steric blockers and later LNA-modified oligonucleotides and two types of uncharged oligonucleotides analogs, PNA and PMO, were shown to serve as steric blockers. Some characteristics of such approaches will be described below.

All these modifications confer high affinity for complementary RNA and nuclease and even protease resistance.

4.3.4.1 *Blockers of translation initiation*

Originally, it was postulated that duplex formation between mRNA and antisense oligonucleotides in the 5′ untranslated region (5′ UTR) or on the initiation codon would interfere with ribosome association and sterically inhibit translation initiation. PMOs have been shown to inhibit translation initiation by sterically hindering scanning of the mRNA by the 40S ribosomal subunit. PMOs are now widely used to specifically knockdown gene expression in developmental biology, where they are delivered through microinjection and electroporation into various organisms such as zebrafish, Xenopus, and Drosophila embryos.

4.3.4.2 *Blockers of translation elongation*

Steric blockers may also block translation elongation by binding to the coding region. This blockade can inhibit translation, but may also lead to synthesis of truncated proteins (Fig. 4.5). Truncated proteins were previously observed *in vitro* with oligonucleotides that bind irreversibly to mRNA through covalent attachment to their complementary sequence via transplatin modification. Interestingly, pyrimidine-rich PNAs targeted to coding regions can block *in vitro* translation elongation and generate truncated proteins, probably by formation of a very stable 2:1 PNA:mRNA complex.

Recently, a PNA targeted to insulin-like growth factor I receptor (IGF-IR) mRNA was shown to produce truncated IGF-IR protein *in vitro* and was able to block IGF-IR expression in prostate cancer cells. Despite this promising result, no evidence is at present available to demonstrate that PNA can induce production of truncated proteins that act as dominant-negative proteins and then inhibit downstream signaling. However, using antisense phosphorothioates, it has been shown that certain 3′ cleavage products persist and can be translated to produce an *N*-terminally truncated version of the protein encoded by the target mRNA. This result suggests that if mRNA is bound by a steric blocker, it could be translated and produce truncated proteins.

4.3.4.3 *Splicing modulators*

The alteration of RNA splicing can result from the use of steric-blocker antisense oligonucleotides (Chapter 6). Binding of an antisense RNA to an exon–intron junction can force the splicing machinery to effectively skip an entire exon. If the skipped exon contains a missense or non-sense mutation, activation of gene expression may

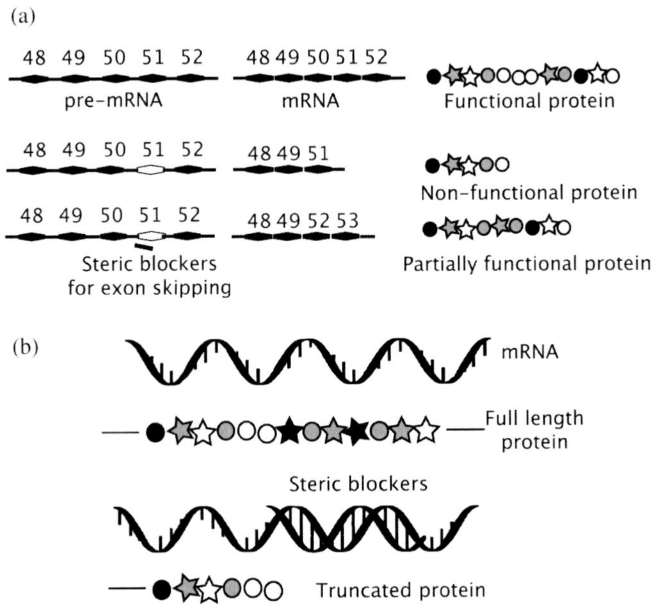

FIGURE 4.5 ■ Splicing and translation blockade by steric blocker oligonucleotides. (a) Steric blockers may be used to block splicing. The example shown is in the Duchenne muscular dystrophy pre-mRNA, where exon 51 of the dystrophin gene contains a cryptic site that leads to aberrant protein translation and loss of exon 50. When an antisense oligonucleotide binds to exon 51, the exon is skipped, producing a partially functional protein (delta 50–51). (b) Steric blockers may also be targeted to mRNA coding regions in order to interfere with translation elongation. Pyrimidine-rich PNAs that form triple helices on mRNA can lead to truncated protein *in vitro*. Other skipped dystrophin, such as delta 51–52, are equally functional (see Chapter 6).

be achieved (Fig. 4.5). Several studies have shown that steric-blocker oligonucleotides can induce exon inclusion, or modulate levels of alternative RNA splicing by masking a splicing site, or block a cryptic splicing site located in an intron.

Recently, the splice-modulation strategy was successfully employed to alter splicing in a Duchenne muscular dystrophy model by targeting the dystrophin mRNA at exon–intron junctions with various steric-blocker antisense oligonucleotides. The injection of these molecules into mouse muscle tissue led to partial recovery of the dystrophin protein functionality by inducing skipping of an exon containing a non-sense mutation. Clinical trials have been initiated with PMO and 2′OMe oligonucleotides in Duchenne muscular dystrophy patients.

4.3.4.4 *Modulators of miRNA functions*

Recently, antisense oligonucleotides were successfully used to target miRNAs, the short non-coding RNAs that regulate gene expression by binding to 3′UTRs of mRNA. miRNAs participate in different cellular functions including pathological ones. For example they are involved in the regulation of the balance between oncogenes and tumor suppressor genes.

Antisense approaches targeting miRNAs have been successfully used to de-repress the expression of the genes targeted by the miRNAs (Fig. 4.4). These oligonucleotides have been called antagomirs. In cell culture, 22–25 nucleotide long antagomirs with 2′ sugar modifications such as 2′OMe, 2′MOE, 2′-fluoro and LNA are effective inhibitors of miRNAs. Antagomirs do not induce RNAse H cleavage and their exact mechanism of action remains to be fully characterized. The first clinical trial with an antagomir, a LNA oligonucleotide targeting miR122, for the treatment of hepatitis C was initiated in 2008. This novel target for antisense oligonucleotides might open new roads in clinical applications.

4.3.4.5 *Specificity*

Whatever the mechanism of action, steric-blocker antisense oligonucleotides are able to modulate gene expression with higher specificity than RNAse H-dependent antisense oligonucleotides or siRNAs and consequently exhibit very few off-targets effects. Indeed, a partially matched duplex formed with inappropriate RNA sequence will be strongly destabilized compared with the fully formed duplex, and therefore have no biological effects, by contrast to what has been observed with oligonucleotides able to activate RNA cleavage. To our knowledge, PNAs and PMOs do not trigger immune responses even when they contain CpG sequences as do antisense oligonucleotides; however, *in vivo* studies must be undertaken to confirm this.

4.3.5 Synthetic Catalytic Nucleic Acids — Ribozymes and DNAzymes

There are also synthetic nucleic acids with RNA-cleaving activity that function without the help of a cellular enzyme. These are the artificial ribozymes and DNAzymes. Both were inspired by natural RNA-cleaving molecules. These molecules with cleavage activity are much less developed than other oligonucleotide-based strategies. Here they will be briefly described to point out the questions that remain to be addressed.

4.3.5.1 *Ribozymes*

Ribozymes are RNA molecules with catalytic activities. These molecules can be designed to recognize and cleave complementary RNA targets. Ribozymes occur naturally and can be divided into two groups based on their catalytic activity: cleaving ribozymes and splicing ribozymes.

Splicing ribozymes exhibit *trans*-splicing activity (i.e., the joining of exons contained on different RNAs). Artificial splicing ribozymes have been designed, but due to their large size they must be transcribed from plasmid vectors in cells. These ribozymes can be engineered to produce novel spliced mRNAs, and the translation products of these mRNAs allow gain of function. For example, *trans*-splicing ribozyme-mediated replacement of cancer-specific RNAs and the generation of new transcripts might exert therapeutic activities in the near future. Repair of defects in mRNAs using these methods is also an interesting road to explore for the therapy of genetic diseases.

Artificial cleaving ribozymes have been used in gene knockdown applications, as they can induce specific mRNA degradation in cells. Cleaving ribozymes range in size from 40 to 200 nucleotides. The most efficient ribozyme developed to date is the hammerhead ribozyme (Fig. 4.6). For cellular and *in vivo* applications, synthetic ribozymes can be chemically modified as is done for antisense oligonucleotides. Alternatively, ribozymes can be produced using appropriate expression vectors. Ribozyme-based anticancer strategies have been evaluated in preclinical models targeting human epidermal growth factor receptor HER-2 and survivin (a well-known antiapoptotic factor), both overexpressed in certain types of cancers. Promising results were obtained from a phase II gene therapy trial with human immunodeficiency virus 1 (HIV-1)-infected adults. The hammerhead ribozyme used in the trial targeted overlapping reading frames of the viral genes viral protein r (vpr) and transactivator of transcription (tat) and was delivered through viral vectors in autologous cluster of differentiation 34 $(CD34)^+$ hematopoietic progenitor cells. This study suggests that ribozymes are promising therapeutic agents that can be employed against HIV-1 infections. However, an advantage compared with other oligonucleotide-based strategies remains to be demonstrated.

4.3.5.2 DNAzymes

In the last decade, oligodeoxynucleotides that exhibit intrinsic RNA cleavage activity like ribozymes have been created. They are called DNAzymes and have been developed using combinatorial techniques referred as *in vitro* selection or systematic evolution of ligands by exponential enrichment (SELEX; see details in Section 4.4.2). Different DNAzymes exhibit specific RNA cleavage *in vitro* when bound to their complementary targets in presence of co-factors such as cationic ions like magnesium. Today the most efficient DNAzyme is the so-called 10–23 DNAzyme that was one of the first used for *in vivo* applications (Fig. 4.6). The 10–23 DNAzyme is composed of a 15-nucleotide-long catalytic central domain

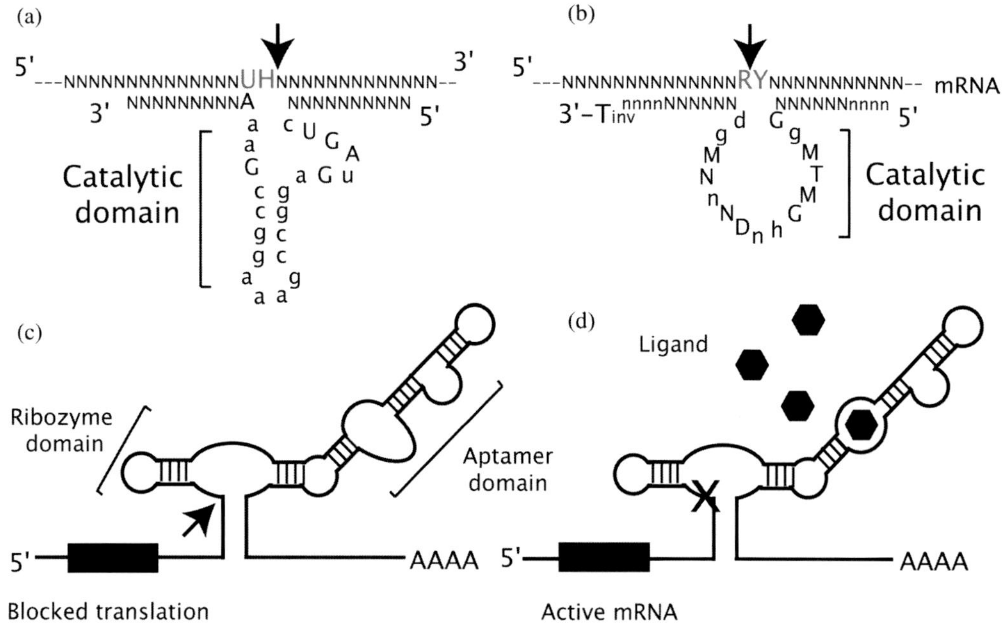

FIGURE 4.6 ■ Ribozymes, DNAzymes and artificial riboswitches. (a) Ribozymes are RNA molecules able to catalytically cleave RNA strands. The most commonly used in artificial regulation of gene expression is the hammerhead ribozyme. Modifications can be incorporated to confer nuclease resistance, while retaining activity of the catalytic core. In positions shown in lower case type, 2′ OMe modifications could be introduced without compromising cleavage activity. Nucleotides indicated in bold type should be unmodified to ensure catalytic activity. Introduction of 2′-C-allyluridine (italicized U) improves activity. Cleavage occurs after the NUH sequence on the mRNA; H can be C, U or A. (b) DNAzymes are DNA molecules selected for their RNA-cleaving ability. The most used DNAzyme for gene inhibition is the 10–23 DNAzyme that can cleave RNA between purine–pyrimidine (RY) dinucleotides. In the catalytic core, some modifications can be introduced to confer nuclease resistance (in lower case type) and some nucleotides facilitate maximal activity (indicated in bold type) (adapted from Zaborowska *et al.*, 2002; Schubert *et al.*, 2003; Wang *et al.*, 2010). DNAzymes are often protected from nucleases by an inverted thymine at the 3′ end and by a few 2′ OMe residues at both ends (in lower case type). N or n indicates any nucleotide complementary to N; R for purine preference; Y for pyrimidine preference; M for A or C; h for A, C, or T; d for G, A, or T. (c, d) Modulation of ribozyme activity can be achieved using an aptameric domain that recognizes a specific ligand (i.e., an artificial riboswitch). Upon ligand binding, ribozyme activity is blocked, allowing the expression of the target gene. This system was recently used in mammals to trigger T-cell proliferation in presence of theophylline, which binds to the aptamer domain.

flanked by two RNA–recognition domains with lengths less than 10 nucleotides to result in high turnover. The cleavage occurs between any unpaired purine–pyrimidine dinucleotide in the targeted mRNA.

For cellular or *in vivo* applications, DNAzymes are chemically modified. 5′ and 3′ ends can be protected using three or four nuclease-resistant nucleotides such as phosphorothioates, 2′OMe residues or LNAs. The catalytic core may also be modified, provided that nucleotides essential for efficient catalytic activity are conserved.

Several DNAzymes have been shown to efficiently destroy specific mRNA targets in cell-based assays. Nevertheless, in most studies, a detailed analysis of RNA cleavage sites or appropriate inactive DNAzyme controls have been lacking so that DNAzyme-mediated mechanisms for gene down-regulation have not been distinguished from a simple antisense phenomenon.

DNAzymes have been designed against infectious or viral diseases such as *Staphylococcus aureus*, HIV, Japanese encephalitis virus and hepatitis, and also against a variety of cancer-related gene products. DNAzymes targeted to vascular endothelial growth factor receptor (VEGF-R), cellular jun proto-oncogene homolog (c-Jun) and early growth response protein 1 (Egr-1) have been tested in relevant mouse models and efficiently reduce tumor growth. DNAzymes are generally delivered locally rather than systemically. Despite their potential promise as therapeutic agents, further studies are clearly needed before DNAzymes can be used in clinical applications. Especially necessary is confirmation of their mechanism of action.

4.3.5.3 *Conditional RNA cleavers based on riboswitches*

Riboswitches are natural regulatory elements mainly located within 5′UTRs of mRNAs that respond to small metabolites such as amino acids or ions. Riboswitches have been found in many prokaryotes and some eukaryotes. The riboswitches sense various stimuli and, via a conformational change, transduce a signal to the gene-expression apparatus, leading to translation regulation. The binding of a specific ligand to the corresponding riboswitch may act positively or negatively on mRNA translation. These natural riboswitches have been reengineered in combination with ribozymes or DNAzymes or oligonucleotides to have new functions, such as tunable cleaving capabilities. Recently for example, ribozymes and riboswitches were associated to produce an RNA-based regulatory system to control T-cell expansion in response to input molecules (Fig. 4.6).

This type of dual system exhibited unique properties and may be exploited to develop new designs of RNA-cleaving oligonucleotides as artificial and conditional regulators of gene expression.

4.4 Delivery Issues — Limitations and Perspectives

4.4.1 Delivery Approaches for Synthetic Oligonucleotides — General Principles and Major Limitations

A critical technical hurdle for all nucleic-acid-based clinical applications is their efficient delivery to their cellular targets that can be RNA, DNA, or protein.

Efficient delivery approaches have been successfully developed for cellular experiments; however, for the clinic, the most critical challenge is to obtain appropriate tissue distribution following systemic administration. Different approaches using viral, chemical and physical delivery have emerged. Plasmid and oligonucleotide delivery methods will be described in detail in Parts II and III of the present textbook. Here, we point out that plasmid DNAs and charged oligonucleotides exhibit many similar behaviors. The major difference is the fact that short oligonucleotides can passively diffuse across the nuclear membrane from the cytoplasm, but plasmids cannot. Uncharged oligonucleotides such as PNA or PMO cannot be delivered using complex formation with cationic agents as can negatively charged oligonucleotides. Here we will focus on the strategy based on the use of molecular conjugates of oligonucleotides and on recent promising advances in the field. Conjugation of various moieties is also being used to improve nanocarriers used for complex formation with nucleic acids (Chapter 15).

We distinguish two main types of delivery: Untargeted and targeted to a specific cell type. Here, we will mainly describe methods for the latter. Briefly, for untargeted delivery, many chemical moieties have been evaluated in complex or directly attached to the cargo oligonucleotide. The approach developed to deliver uncharged PNAs or PMOs consists of conjugation with cell-penetrating peptides (CPP). Various CPPs allow cellular uptake but generally endosomal entrapment is observed and strategies must be developed to allow escape from these vesicles. Concerning targeted delivery, we will focus on the promising use of aptamers.

4.4.2 Targeted Delivery of Oligonucleotides — the Example of Aptamer Conjugates

The objective of drug delivery is to target the right compartment in the right cell. To reach this goal an oligonucleotide can be conjugated to a targeting moiety that binds to a cell-surface receptor that allows internalization. A limited number of receptors have been targeted with oligonucleotide conjugates. They include lipoprotein receptors targeted with cholesterol and integrins targeted with RGD peptides. More recently aptamers were successfully used and will be described in more detail.

Aptamers are DNA or RNA oligonucleotides, usually 15–40 nucleotides long, which can bind with high affinity to a variety of molecules such as small organic or inorganic compounds, peptides, proteins or nucleic acids. Unlike antisense oligonucleotides, aptamers do not recognize their targets through hybridization by Watson–Crick base pairing but through classical three-dimensional ligand–receptor recognition. Aptamers are obtained using the SELEX procedure. With this process, one selects from a combinatorial library of nucleic acid sequences the aptameric sequences that exhibit high binding affinity for the chosen target ligand. Multiple rounds of selection and amplification allow the *in vitro* identification of tight-binding aptamers after 5–15 cycles. Due to their high affinity for their targets, aptamers are often compared to antibodies. Like other nucleic-acid-based strategies, aptamers must be modified for cell-based studies. Using 2′-deoxy, 2′-fluoro, 2′NH_3 or 2′OMe nucleotides during selection steps provides stabilized aptamers that can be used in complex biological media. Several aptamers are currently in clinical trials and one of them, a 2′-fluoropyrimidine-modified RNA aptamer (Macugen™) directed against the 165-amino acid isoform of vascular endothelial growth factor (VEGF), has been approved by US and EU authorities for macular degeneration treatment and is administered by intravitreal injection. Aptamers can be targeted to extracellular, membrane-bound or intracellular proteins. Some aptamers have particular motifs of known structure, such as the G-rich anti-nucleolin aptamer. Indeed all G-rich aptamers discovered in SELEX procedures may adopt a four-stranded structure that probably contributes to the binding affinity towards their targets.

In addition to their use as therapeutic agents for human pathologies, nucleic acid aptamers can be used to direct other molecules such as antisense oligonucleotides, siRNAs or vesicles towards specific tissues (Fig. 4.7). Through recognition of a cell membrane protein, aptamers can be used to deliver other molecules to a specific cell type. For example, aptamers selected to bind to the prostate-specific membrane antigen (PSMA) were successfully used to deliver siRNAs to prostatic cancer cell lines. Covalent or non-covalent links could therefore be used to conjugate siRNAs to the aptamers. Nanoparticles were also delivered through the use of anti-PSMA aptamers to prostatic cells for imaging and drug delivery. Another aptamer selected against a transmembrane receptor, protein tyrosine kinase 7 (PTK7) highly expressed in T-cell acute lymphoblastic leukemia, was successfully used as a targeting agent.

The recently developed cell-SELEX strategy allows identification of aptamer sequences that bind to proteins in native state on certain cell types (Fig. 4.7). In a typical *in vitro* selection for aptamers, a purified protein is often chosen as a target. The cell-SELEX procedure was first applied to a receptor tyrosine kinase RET. A series of positive selections and counter-selections with RNA libraries were performed directly on cells that express or do not express the protein target. One advantage of this technique is that one can generate aptamers that selectively recognize a particular target cell population without knowing specifically their surface protein contents. For example, normal and tumor cells can be differentiated with the help of aptamers obtained using cell-SELEX technology and then further modified for stability in biological fluids. Future studies will determine how effectively the cell-SELEX procedure can be in identification of innovative targeting aptamers.

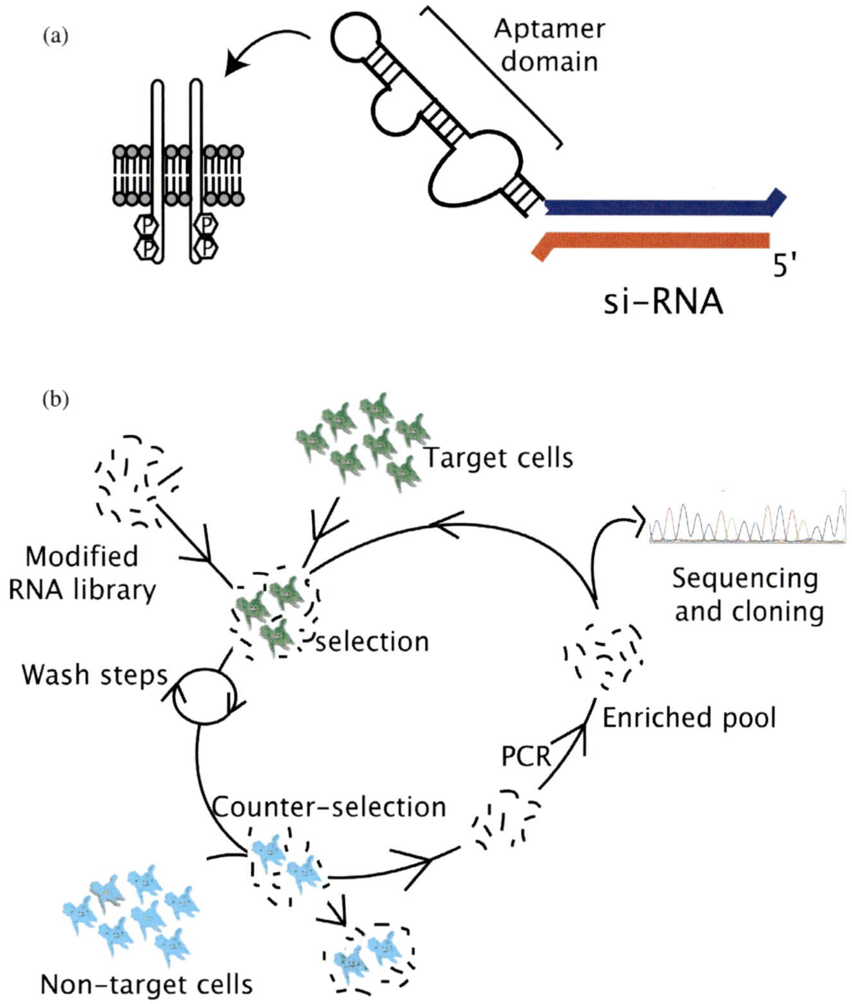

FIGURE 4.7 ■ Aptamer-mediated delivery and cell-SELEX procedure. (a) Aptamers selected for their ability to bind to specific proteins present on cell membranes can be linked covalently to antisense oligonucleotides or siRNAs. These aptamer–oligonucleotide conjugates bind to specific tissues *in vivo*. One limitation of these aptameric approaches is the availability of specific aptamers that target particular cells. (b) If cells are considered as aptamer targets, one can use the recently described cell-SELEX procedure. Aptameric sequences which are able to bind to the target cells are selected from a library. These sequences are then counter-selected against a negative cell line, one that does not express the receptor of interest. After a few rounds of selection and counter-selection, one can expect to isolate aptameric sequences highly specific for the chosen cell line.

4.5 Conclusion

Synthetic oligonucleotides are today commonly used as artificial modulators of gene expression. siRNAs have become very useful tools for knockdown of specific genes for functional studies in basic research. Other chemically modified oligonucleotides can exhibit interesting mechanisms of action that permit gain of functions. The major limitation to use of oligonucleotide-based therapeutics is lack of efficient delivery methods into target cells after systemic administration. Significant advances have been made but more progress is needed.

For Further Reading

Baum, D.A., Silverman, S.K. (2008). Deoxyribozymes: Useful DNA catalysts *in vitro* and *in vivo*, *Cell Mol Life Sci*, **65**, 2156–2174.

Bauman, J., Jearawiriyapaisarn, N., Kole, R. (2009). Therapeutic potential of splice-switching oligonucleotides, *Oligonucleotides*, **19**, 1–13.

Bennett, C.F., Swayze, E.E. (2010). RNA targeting therapeutics: Molecular mechanisms of antisense oligonucleotides as a therapeutic platform, *Annu Rev Pharmacol Toxicol*, **50**, 259–293.

Burnett, J.C., Rossi, J.J. (2012). RNA-based therapeutics: Current progress and future prospects, *Chem Biol*, **19**, 60–71.

Esau, C.C. (2008). Inhibition of microRNA with antisense oligonucleotides, *Methods*, **44**, 55–60.

Faghihi, M.A., Wahlestedt, C. (2009). Regulatory roles of natural antisense transcripts, *Nat Rev Mol Cell Biol*, **10**, 637–643.

Juliano, J., Alam, R., Dixit, V., *et al.* (2008). Mechanisms and strategies for effective delivery of antisense and siRNA oligonucleotides, *Nucleic Acids Res*, **36**, 4158–4171.

Keefe, A.D., Pai, S., Ellignton, A. (2010). Aptamers as therapeutics, *Nat Rev Drug Disc*, **9**, 537–550.

Marlin, F., Simon, P., Saison-Behmoaras, T., *et al.* (2010). Delivery of oligonucleotides and analogues: the oligonucleotide conjugate-based approach, *Chembiochem*, **11**, 1493–1500.

Mulhbacher, J., St-Pierre, P., Lafontaine, D.A. (2010). Therapeutic applications of ribozymes and riboswitches, *Curr Opin Pharmacol*, **10**, 551–556.

Nielsen, P.E. (2010). Peptide nucleic acids (PNA) in chemical biology and drug discovery, *Chem Biodivers*, **7**, 786–804.

Pingoud, A., Wende, W. (2011). Generation of novel nucleases with extended specificity by rational and combinatorial strategies, *Chembiochem*, **12**, 1495–1500.

Rossi, J.J. (2012). Resurrecting DNAzymes as sequence-specific therapeutics, *Sci Transl Med*, 4:139fs20.

Serganov, A., Patel, D.J. (2007). Ribozymes, riboswitches and beyond: Regulation of gene expression without proteins, *Nat Rev Genet*, **8**, 776–790.

Simon, P., Cannata, F., Condorcet, J.P., *et al.* (2008). Targeting DNA with triplex-forming oligonucleotides to modify gene sequence, *Biochimie*, **90**, 1109–1116.

Vollmer, J., Krieg, A.M. (2009). Immunotherapeutic applications of CpG oligodeoxynucleotide TLR9 agonists, *Adv Drug Deliv Rev*, **61**, 195–204.

Wood, M.J., Gait, M.J., Yin, H. (2010). RNA-targeted splice-correction therapy for neuromuscular disease, *Brain*, **133**, 957–972.

Zhou, J., Rossi, J.J. (2009). The therapeutic potential of cell-internalizing aptamers, *Curr Top Med Chem*, **9**, 1144–1157.

5

PRINCIPLES OF RNAi TRIGGER EXPRESSION FOR GENE THERAPY

Lisa J. Scherer[a] and John J. Rossi[a,b]

5.1 Introduction

The field of small RNA therapeutics has seen significant advances in recent years. Small RNAs have a number of advantages for gene therapy applications. RNA synthesis methodologies and attendant cost restrict the size of RNAs for exogenous, short-term applications. Small RNAs can easily be stably expressed alone or in combination even in vectors with size limitations. RNA is less immunostimulatory than protein at the systemic and cellular levels, provided certain criteria are met.

There are a number of different types of small RNA therapeutic molecules, each with their own advantages and disadvantages, including aptamers, decoys, ribozymes and RNA interference (RNAi) triggers (reviewed in Lares *et al.*, 2011; Scherer *et al.*, 2007a). This chapter focuses on the design and expression of RNAi triggers for mRNA knockdown with an emphasis on the use of RNA polymerase III promoters, although many of the principles apply to the expression of other small-RNA gene therapeutics.

An explanation of a series of seemingly unrelated genetic suppression phenomena originally described in worms (Lee *et al.*, 1993; Reinhart *et al.*, 2000) and subsequently in plants and fungi (reviewed in Hannon, 2002; Sen and Blau, 2006) remained elusive until the Nobel-prize winning work of Fire and Mello in *Caenorhabditis elegans* (Fire *et al.*, 1998), which suggested the existence of a common underlying mechanism of post-transcriptional gene silencing (PTGS) referred to

---

[a]Department of Molecular and Cellular Biology, Beckman Research Institute, City of Hope, Duarte, CA 91010, USA
[b]Email: jrossi@coh.org

as RNAi. Subsequent research demonstrated that 21 base pair (bp) synthetic RNA duplexes trigger RNAi in mammalian cells, confirming the conservation of the underlying pathways in higher eukaryotes (Elbashir *et al.*, 2001a, 2001b). This observation opened the door to utilizing short, double-stranded RNAs to mediate specific mRNA knockdown without activating cellular immunity (which occurs in response to ≥30 bp fully double-stranded RNA in the cytoplasm). Shortly thereafter, we and others demonstrated that expressed RNAi triggers could knockdown human immunodeficiency virus 1 (HIV-1) replication, whether the complementary strands of the short hairpin RNA (shRNA) duplex were expressed separately *in trans*, i.e., from different promoters (Lee *et al.*, 2002) or as a single shRNA transcript (Paddison *et al.*, 2002; Paul *et al.*, 2002). Since then, there has been an explosion in RNAi-related research on the molecular and cellular biology of endogenous small RNA regulatory pathways as well as the use of RNAi in reverse genetic studies and gene therapy.

The vast majority of current RNAi gene therapeutic approaches exploit the endogenous microRNA (miRNA) biogenesis pathway of PTGS that is central to the regulation of many basic cellular and developmental programs (reviewed in Li and Liu, 2011; Snead and Rossi, 2010). Figure 5.1 summarizes the basic steps of miRNA biogenesis, which begins with transcription of primary miRNAs (pri-miRNAs), commonly from RNA polymerase II (pol II) promoters. Pri-miRNAs are processed into an approximately 70 nucleotide (nt) pre-miRNA by the microprocessor Drosha/DiGeorge syndrome critical region 8 (DGCR8) complex in the nucleus. Pre-miRNAs have a hairpin structure with stems containing interspersed mismatches. They are then exported to the cytoplasm via exportin5 (exp5) (Gwizdek *et al.*, 2004; Kim, 2004; Lund *et al.*, 2004; Yi *et al.*, 2003) where a cytoplasmic complex containing the enzyme Dicer mediates cleavage of the pre-miRNA at the base of the stem (removing the terminal loop) to form the mature miRNA duplex. Upon association with an Argonaute (Ago) protein within the pre-RNA-induced silencing complex (pre-RISC), the passenger (sense) strand is removed. The remaining guide (antisense) strand forms activated RISC, which binds to the mRNA target and directs either translational repression or mRNA degradation.

Expressed shRNAs have a 19–29 nt stem that is completely base-paired; most formats (discussed below) are designed to bypass the microprocessor step and use triggers that enter the RNAi pathway at the Dicer cleavage step to produce the final small interfering RNA (siRNa). The resulting duplex has symmetric 2 nt 3′ overhangs with a 19–21 base-paired region. Synthetic siRNA duplexes of this form can also bypass Dicer cleavage, but longer versions that are processed by Dicer have greater knockdown efficiency, (Kim *et al.*, 2005) and our expressed shRNA designs takes this into account. Once in the pre-RISC complex, the Ago2 protein cleaves the passenger strand and the guide strand mediates mRNA knockdown.

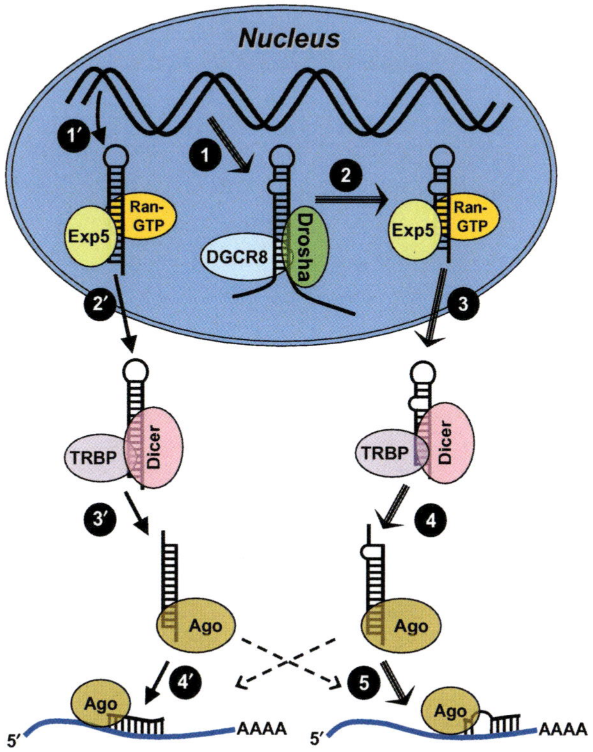

FIGURE 5.1 ■ Overview of cellular RNAi pathways. *miRNA processing (striped arrows)*. Exp5, exportin 5; Ago, Argonaute protein; DGCR8, DiGeorge syndrome critical region 8 gene product; TRBP, HIV-1 transactivating response RNA-binding protein. (1) Endogenous miRNAs and exogenous miRNA mimics are transcribed into pri-miRNAs and first processed by the nuclear microprocessor complex containing Drosha and DGCR8 into pre-miRNAs. (2) Pre-miRNAs, bound by Exp5 and rat sarcoma oncogene-related nuclear protein–guanosine triphosphate (Ran-GTP) are exported to the cytoplasm. (3) The Dicer–TRBP complex processes the pre-miRNA into a miRNA duplex. (4) The miRNA duplex loads into RISC, an Ago-containing complex. (5) The duplex unwinds and the guide strand (which is antisense with respect to the mRNA target) is retained in the miRNA-RISC complex, which binds to the target site, typically in the mRNA 3′ untranslated region (UTR). Nearly all mammalian miRNAs bind with imperfect complementarity, but complementarity of the miRNA guide strand nucleotides 2–8, the "seed sequence", to the mRNA target site is critical to activity. miRNA–RISC binding mediates translation repression and frequently mRNA target degradation. *shRNA processing* (*solid arrows*). (1′) Simple shRNA gene therapeutic primary transcripts bypass Drosha/DGCR8 processing, are (2′) transported to the cytoplasm and (3′) processed by Dicer/TRBP in a similar manner to miRNA precursors. (4′) The resulting siRNA guide strand is incorporated into a siRNA–RISC complex, which after binding to the target site (usually in the coding region of the mRNA) triggers cleavage of the mRNA due to the complete complementarity between siRNA guide strand and the mRNA target site. The dashed lines indicate that a completely complementary siRNA duplex whose guide strand binds to a mRNA target in a miRNA-like manner can potentially trigger miRNA-like repression (a source of off-target effects); conversely, an incompletely complementary miRNA duplex whose guide strand binds to a miRNA target in an siRNA-like manner (complete complementarity) will trigger cleavage. RNAi triggers are typically designed so that one strand of the siRNA or miRNA duplex is highly thermodynamically favored for RISC incorporation as the guide strand, but in some bifunctional trigger designs both strands of the duplex are active.

5.2 RNAi Trigger Types

The mechanism of exogenous-RNAi-mediated mRNA knockdown depends on the degree of guide strand complementarity to the target and on the target site location in the mRNA, which is again based on an understanding of cellular miRNA biogenesis. Endogenous mammalian miRNAs guide strands are not usually fully complementary to their mRNA target; complementarity of nucleotides 2 through 7–8 of the guide strand, the "seed" region is crucial for the majority of targets (although alternate modes of recognition are being discovered (Chi *et al.*, 2012)), with additional complementarity enhancing activity and/or specificity. Natural miRNA target sites tend to cluster in the 3′ untranslated region of mRNAs, although more miRNA sites are being found in other mRNA regions.

shRNAs typically have guide strands fully complementary to the target mRNA which triggers Ago2-mediated cleavage of the target (or "slicer" activity). The target site can theoretically be anywhere in the mRNA but is usually in the coding sequences. In general, miRNA-type guide strands mediate lower levels of target knockdown than fully complementary shRNA-like guide strands due to the cellular mechanisms already described. The important thing to emphasize from a trigger design perspective is that the final mechanism of knockdown depends primarily on the degree of guide strand–mRNA complementarity and target site location, not on the form of the initial trigger. By convention, an expressed miRNA trigger will have the stem mismatches typical of miRNA whether it is expressed in pri-miRNA or pre-miRNA form (synthetic duplexes can mimic the mature form as well). In contrast, shRNAs have completely complementary stems, but it is possible to design a pri-miRNA mimic (Zeng *et al.*, 2002); for example, one that produces a guide strand that is completely complementary to the mRNA target and thus activates Ago2 cleavage of the transcript.

That being said, the most basic shRNA design consists of a 19–29 nt fully complementary base-paired stem and a loop of at least four, and more typically 6–10 nt with a sense (passenger strand)/loop/antisense (guide strand) symmetry. Loop sequences can be divided into "collapsing" and non-collapsing types; the collapsing loop base pairs in such a way as to seamlessly extend the stem. This parameter may be a large part of why some loops are reported to mediate more efficacious RNAi (Schopman *et al.*, 2010a) especially when the stem is only 19–20 bps; the extension of the stem by the collapsing loop may enhance Dicer cleavage efficiency. For these reasons, we typically use a 21 bp stem, where the guide strand is complementary to the target, and a collapsible 8–10 nt loop.

Bifunctional shRNA (bi-shRNA) approaches utilize both cleavage-dependent and -independent PTGS mechanisms. The functions may be triggered by separate units (hairpins) of a multi-cistronic precursor typically built on a pri-miRNA platform (see Section 5.8) (Phadke *et al.*, 2011; Rao *et al.*, 2010). Alternatively, both functionalities can be expressed as part of a single pri-miRNA hairpin, where each strand of the mature miRNA mimic is active against separate target mRNAs for example, one mediating miRNA-based translation repression/degradation and the other the siRNA-based cleavage mechanism (Ehsani *et al.*, 2010).

In summary, there are a number of possible expressed trigger types. In the siRNA format, each strand of the trigger is driven by a separate promoter (Lee *et al.*, 2002, 2008; Miyagishi and Taira, 2002). This is less convenient than the shRNA format and since the mechanism of down-regulation is not completely delineated (Lee *et al.*, 2008) is used less frequently for RNAi-mediated mRNA knock-down. The shRNA format predominates as it is most easily adapted for expression of a single, well-defined RNA. Many free on-line tools are available to assist with target selection and shRNA design. Another version is the miRNA mimic (Amendola *et al.*, 2009; Ehsani *et al.*, 2010; Zeng *et al.*, 2002; reviewed in Arbuthnot, 2011; Manjunath *et al.*, 2009), which can be of the pri-miRNA or pre-miRNA type, and is often modeled on known endogenous miRNA genes. In both cases, the natural guide/passenger strand regions are substituted with sequences directed to the target of interest. Pri-miRNA platforms can be multi-cistronic and easily adapted to expression from pol II promoters (reviewed in Manjunath *et al.*, 2009).

5.3 Toxicity

Off-target effects are a major concern when designing RNAi triggers, as nearly any siRNA has the potential to act as a miRNA through a seed sequence match. Though currently impractical, it should be theoretically possible to predict the likelihood of such an event by searching for target sequences in the relevant cellular transcriptomes. However, it is relatively straightforward to screen the entire genome for high-homology matches that could be problematic. Off-target effects can be exacerbated by poor strand selectivity (i.e., relatively high, unintended passenger strand incorporation into RISC); therefore, testing strand selectivity with the use of dual luciferase assays for both strands under conditions of substrate excess is recommended. Poor efficacy is rarely due to insufficient expression, and is more often the result of low guide strand entrance into or efficacy in RISC (which can be due to poor strand selectivity or other factors), poor target site availability or both. Therefore, it is important to test target knockdown in a biologically relevant assay as well. With cellular targets, it may be worthwhile to make two RNAi triggers against the same target and compare phenotypes for concordance.

Another cause of toxicity is unwanted activation of the cellular immune response. For instance, retinoic acid-inducible protein 1 (RIG-1) helicase recognizes a terminal 5′ triphosphate (5′-pppN) moiety in the context of short blunt, double-stranded cytoplasmic RNAs and triggers a potent interferon-α response. This pathway functions as part of the normal cellular immune response to foreign RNA (Hornung *et al.*, 2006; Schlee *et al.*, 2006). We will return to this issue later.

Finally, toxicity can also arise if the expressed trigger reaches levels sufficient to compete with endogenous miRNA precursors for access to the RNAi pathway, perturbing normal cellular miRNA production and function. The magnitude of this effect may depend on the relative amount of RNAi pathway components, which can differ among cell types and animal strains (Grimm *et al.*, 2010; Martin *et al.*, 2011; Oberdoerffer *et al.*, 2005). In addition, both the trigger type and promoter choice is important in this regard; the latter is a major focus of this discussion (Beer *et al.*, 2010; Boudreau *et al.*, 2008; Grimm *et al.*, 2006, 2010; Manjunath *et al.*, 2009; Martin *et al.*, 2011; McBride *et al.*, 2008).

5.4 RNA Polymerase III (Pol III) Promoter Types

RNA polymerase III (pol III) promoters are widely used to express shRNAs and small therapeutic RNAs. Pol III transcription (reviewed in Orioli *et al.*, 2012; Teichmann *et al.*, 2010; White, 2011) is divided into four categories based on promoter architecture (Fig. 5.2) and transcription complex assembly. We begin with a brief description of each type of promoter followed by a more extensive discussion of those characteristics affecting selection of specific promoters for therapeutic applications.

Type 1 pol III promoters are internal rather than external. Specifically, the internal control region (ICR) recruits the transcription complex which then directs transcription initiation at an upstream, rather than downstream site (as in RNA polymerase II transcription) so the ICR region is actually transcribed. The only example of this type of pol III promoter is the ribosomal 5S gene.

Type 2 promoters are also internal with a bipartite structure, consisting of a proximal A box 10–20 bp downstream of the transcription start site and a distal B box. Transfer RNA (tRNA) genes as well as the adenovirus virus associated VA1 gene are examples; the A and B boxes correspond to portions of the D- and TΨC-loop regions of mature tRNAs, respectively.

Type 3 promoters are external and located primarily in the 5′ flanking region upstream of the transcription start site. The U6 snRNA gene, H1 and 7SK genes are examples. Both U6 and H1 are widely used for expressing shRNAs as well as other small RNA therapeutics.

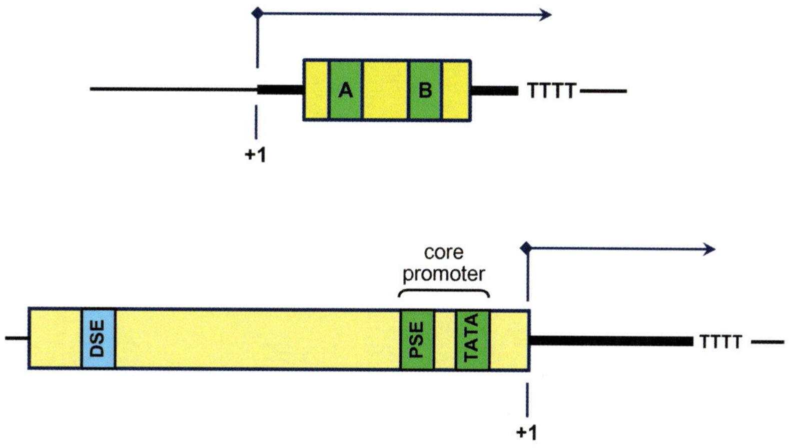

FIGURE 5.2 ■ RNA polymerase III type 2 and type 3 promoter architecture. Diamonds, start site of transcription (+1); arrow, transcription termination site (shown here as TTTT). Type 2, (*upper*). A diagrammatic transfer RNA (tRNA) gene is shown. The internal promoter elements consist of the proximal (to the start site of transcription) A box and distal B box. Open box indicates the mature tRNA sequence; heavy lines, the trimmed 5′ leader and 3′ trailer sequences. Type 3, (*lower*). The basic elements of the U6 promoter are shown. The core promoter consists of a TATA-box from −24 to −29 and the proximal sequence element (PSE) from −47 to −66. The distal sequence element (DSE) activates transcription from the core promoter. Heavy line indicates transcribed sequence.

Finally, type 4 promoters utilize both external and internal promoter elements. Examples include the genes encoding vault RNAs (van Zon *et al.*, 2001), Epstein-Barr encoded small RNAs (EBER) (Howe and Shu, 1989) and the 7SL RNA component of the signal recognition particle (Englert *et al.*, 2004; Ullu *et al.*, 1982; Ullu and Weiner, 1984).

All pol III transcripts (or primary transcripts in the case of tRNAs) begin with a triphosphorylated nucleotide (5′-pppNp). The canonical termination signal for all pol III genes is a run of four or more consecutive dT residues (Richard and Manley, 2009), although other variants have been recently described (Orioli *et al.*, 2011).

5.5 Specific Pol III Promoters

We now consider the attributes of specific pol III promoters for the expression of RNAi triggers, summarized in Table 5.1, which may also be informative for the expression of other therapeutic RNA modalities. Expressing RNAi triggers using the pol III internal promoter types produces a chimeric primary transcript (e.g.,

TABLE 5.1 RNA Polymerase III Promoter Characteristics

Pol III Type	ID	Size	Copies Per Cell	Inducible Versions	Commercially Available?	First Base of shRNA	Multicistronic Transcripts	Comments
2	tRNA	<100	variable	No	No	monophosphate (processed)	Potentially limited transcript length	Graded trigger production possible
2	VAI	~150	1.00E+08	No	No	triphosphate (5′pppN)	No	
3	U6	~275	5.00E+05	Yes	Invitrogen	triphosphate (5′pppN)	Potentially limited transcript length	Toxicity reported
3	H1	~100	5.00E+04	Yes	Invitrogen	triphosphate (5′pppN)	Potentially limited transcript length	Less toxicity than U6
3	7SK	~300	2.00E+05	No	Invivogen Biogenova	triphosphate (5′pppN)	Potentially limited transcript length	Relative toxicity unknown

tRNA-shRNA); the necessity for trigger moiety release adds an extra level of design complexity. For this and other reasons, type 1 and 4 pol III promoters have not been used to express RNAi triggers, so we will continue with applications of type 3 and 2 promoters in that order.

The human U6 promoter (He *et al.*, 1998) has been used extensively to express many different small RNA therapeutic molecules, and was the first used to express both siRNAs and shRNAs (Czauderna *et al.*, 2003; Kwak *et al.*, 2003; Lee *et al.*, 2002; Paul *et al.*, 2002, 2003). The endogenous U6 small nuclear RNA (snRNA) is a critical part of the pre-mRNA splicing machinery (reviewed in Will and Luhrmann, 2011) and is expressed at high levels. The U6 promoter requires that transcription begins with a G nucleotide, which can be somewhat limiting for shRNA expression in particular. In addition, there have been reports of toxicity resulting specifically from expressing shRNAs from U6 (An *et al.*, 2006a; Grimm *et al.*, 2006). The human genome contains five active U6 snRNA genes that differ in transcriptional efficiency (Domitrovich and Kunkel, 2003); moreover, U6 transcription is sensitive to genomic context (Listerman *et al.*, 2007). Consequently, transcription-level-associated toxicity observed with U6-driven shRNA expression may rely on several factors.

H1 RNA is part of the multi-subunit RNAse P complex responsible for 5′ processing of the tRNA primary transcripts (see Section 5.6) (Jarrous and Gopalan, 2010; Puranam and Attardi, 2001). Unlike U6, the H1 promoter can reputedly drive transcription from any +1 nucleotide, although a purine is recommended, and is a less active promoter than U6 as well. This may explain why a shRNA caused liver toxicity when expressed from a U6, but not H1 promoter (An *et al.*, 2006b; Grimm *et al.*, 2006). In addition, the H1 promoter is considerably smaller than U6 which can be useful in multiplexing applications where vector size is limited, as in adeno-associated virus (AAV) vectors.

The endogenous 7SK gene expresses an abundant, highly conserved 330 nt RNA which is part of a complex regulating RNA pol II elongation (Egloff *et al.*, 2006; Wassarman and Steitz, 1991; Zieve and Penman, 1976). There are only a few reports of using the human 7SK promoter (Murphy *et al.*, 1987) to express shRNAs (Cagnon and Rossi, 2000; Chumakov *et al.* 2010; Czauderna *et al.*, 2003; Koper-Emde *et al.*, 2004) which mediated more efficient RNAi than U6 (Koper-Emde *et al.*, 2004). More studies are needed to characterize its use in this application.

RNA pol III type 2 promoters have been used in a variety of small RNA gene therapeutic applications. For example, the VA1 promoter (Bhat and Thimmappaya, 1983) has been used to express an anti-chemokine receptor 5 (CCR5) ribozyme as part of a chimeric VA1–CCR5 ribozyme transcript, to inhibit HIV-1 infection (Cagnon and Rossi, 2000) as well as anti-HIV-1 shRNAs. Inefficient processing of the chimeric VA1-shRNA precursor reduced efficacy (Lee *et al.*, 2008), although adjusting the design to use a pri-mRNA mimic could potentially improve siRNA production. A conceivable disadvantage to this approach is that other regions of the VA1 RNA are processed into miRNAs themselves (Aparicio *et al.*, 2010). Coupled with extremely high innate transcription activity of VA1, the potential for toxicity is a concern. However, it may be possible to alter the structure of the miRNA-producing arms of VA1 RNA to abrogate unwanted VA1 backbone processing without unduly affecting transcriptional activity or alternatively, to express heterologous therapeutic siRNAs.

shRNAs can also be expressed from tRNA promoters as pre-tRNA–shRNA chimeras, but unlike VAI chimeras, can be designed so that the shRNA trigger is released by the cellular pre-tRNA processing machinery (Phizicky and Hopper, 2010) to enter the RNAi pathway (Dyer *et al.*, 2010; Scherer *et al.*, 2007b). Moreover, the degree of shRNA release can be modulated by mutations in the tRNA acceptor stem known to reduce processing efficiency while retaining site specificity, allowing for graded RNAi knockdown. This approach can also be used to express pre-miRNAs and is in fact utilized by some viruses (Bogerd *et al.*, 2010; Diebel *et al.*, 2010).

To summarize, type 3 RNA pol III promoters are still commonly used to express RNAi triggers since transcription start and termination signals are well defined. Enhanced toxicity of U6 versus H1 has been attributed to the higher level of U6 expression and concomitant saturation of the endogenous miRNA pathway. We note that while shRNAs expressed from type 3 promoters typically have a 3′ overhang of two to four uracil nt from the termination signal, which should prevent RIG-1 recognition of the 5′-pppNp, subsequent exonucleolytic trimming of the 3′ overhang could potentially produce blunt 5′pppN hairpins. It is possible that the higher expression levels mediated by the U6 promoter increase the likelihood of accumulating blunt 5′pppN hairpins above the threshold level of RIG-1 activation, accounting for some of the toxicity observed with U6 promoters; indeed, we observed low levels of interferon-related responses to expression of U6 shRNA in 293 EcR cells (Lee *et al.*, 2008). This effect would not be expected with tRNA-mediated expression of shRNA triggers as the processing step leaves a 5′ monophosphate on the shRNA. Toxicity can also be limited by the use of conditional promoters (see Section 5.7).

5.6 RNA Polymerase II (Pol II) Promoters

Pol II promoters have the potential advantages of restricting expression of RNAi triggers to specific cell types and metabolic conditions, which is particularly important since most current delivery methodologies are not cell-type-specific. Pol II promoters can be used to express standard shRNAs, but the 5′ leader sequences of primary transcripts can severely reduce the efficiency and specificity of Dicer processing to siRNA triggers. This difficulty can be circumvented by using pri-miRNA mimics, where the shRNA sequences are inserted into a larger stem–loop context designed to use the microprocessor complex as the first step in trigger processing (see Section 5.8). A number of pol II promoters have been used to express RNAi triggers (Allen *et al.*, 2007; Ely *et al.*, 2009; Giering *et al.*, 2008; Hakamata and Kobayashi, 2010; Kesireddy *et al.*, 2010; Lebbink *et al.*, 2011; Nielsen *et al.*, 2009; Rao and Wilkinson, 2006; Zhou *et al.*, 2005).

5.7 Conditional RNA Interference

Lentivirus and AAV backbones are most commonly employed to introduce stably-expressed RNAi triggers into cells. Current versions of these vector systems transduce a broad range of cell types. Conditional, or regulated RNAi, can be used to limit the circumstances of RNAi trigger production for genetic studies or to avoid toxicity (reviewed in Kleinhammer *et al.*, 2011a; Lee and Kumar, 2009; Wiznerowicz, Szulc and Trono, 2006). Tetracycline-regulated versions of both pol II and pol III promoters exist — some are available commercially — and respond in a concentration-dependent manner for variable trigger expression (Aagaard *et al.*, 2007; Berger *et al.*, 2010; Czauderna *et al.*, 2003; Fei *et al.*, 2007; Kesireddy *et al.*, 2010; Matsukura *et al.*, 2003; Matthess *et al.*, 2005; Raymond *et al.*, 2010). Although they tend to be leaky and inappropriate for many gene therapeutic applications, they remain very useful in basic genetic and preclinical studies. Other systems are: controlled by causes recombination–locus of X over P1 (cre–lox) recombination (Beronja *et al.*, 2010; Coumoul and Deng, 2006; Coumoul *et al.*, 2005; Goncalves *et al.*, 2010; Hakamata and Kobayashi, 2010; Kasim *et al.*, 2003; Kleinhammer *et al.*, 2011b; Shukla *et al.*, 2007); stress via a heat shock promoter (Yang and Paschen, 2008); ligands such as isopropyl β-D-1-thiogalactopyranoside (IPTG) (Wu *et al.*, 2007), ecdysone (Palli *et al.*, 2003; Shea and Tzertzinis, 2010) and theophylline (An *et al.*, 2006a); HIV-1 transactivator of transcription (*tat*) protein (Unwalla *et al.*, 2004); and ribozymes (Allen *et al.*, 2007; Kumar *et al.*, 2009). Many conditional RNAi strategies employ more than one of these modalities.

5.8 Multiplexing Strategies

Certain gene therapeutic applications may require the expression of multiple RNA effectors, referred to as a combinatorial approach or multiplexing, to achieve significant efficacy levels. For example, HIV-1 therapies, whether conventional highly active anti-retroviral therapy (HAART) or genetic-based, employ at least three antivirals, due to the ability of the virus to rapidly evolve drug resistant strains. Both direct experiments and *in silico* modeling support this approach (McIntyre *et al.*, 2011b; Schopman *et al.*, 2010b; ter Brake *et al.*, 2008; reviewed in Grimm *et al.*, 2010; Scherer and Rossi, 2011).

The most common multiplexing approach uses separate promoters for each therapeutic. Promoter duplication should be avoided as it can lead to cassette deletion through recombination or during reverse transcription of retroviral vectors (McIntyre *et al.*, 2009; ter Brake *et al.*, 2008). Typically the entire group of three or four cassettes will be expressed in a single vector backbone (e.g., lentiviral) to ensure co-delivery to each target cell. Recently, a lentiviral triple anti-HIV-1 therapeutic vector expressing a U6 *tat*/regulator of virion expression (*rev*) shRNA, U6/U16 transactivation response (TAR) decoy and a VA1-CCR5 ribozyme was introduced into four HIV-1 patients requiring a bone marrow transplant to treat relapsed or poorly responsive AIDS-related lymphoma. In this phase I clinical trial, expression of antiviral RNAs has persisted for up to 24 months with no adverse effects (DiGiusto *et al.*, 2010). While only one of the three RNA modalities in this combination expressed an RNAi trigger, it demonstrates the feasibility of RNAi-based approaches including multiplexing RNAi strategies, provided care is taken to avoid adverse effects from overexpression of RNAi triggers, as already discussed.

Another approach is the multi-cistronic expression strategy, where multiple triggers are part of a single transcript. This includes long-stem shRNAs or poly-stem shRNAs (Ely *et al.*, 2009; Liu *et al.*, 2007, 2009; McIntyre *et al.*, 2011c; Sano *et al.*, 2008) although the multiple-promoter approaches were more successful than single-promoter configurations in a recent direct comparison (McIntyre *et al.*, 2011a), primarily due to inefficient and inaccurate processing of multi-cistronic precursors. However, we have adapted a naturally occurring human miRNA cluster within exon 13 of the minichromosome maintenance protein 7 (MCM7) gene as a scaffold to express multiple RNAi triggers (Aagaard *et al.*, 2007, 2008; Zhang and Rossi, 2010). The entire region is less than 800 bp even with the inclusion of approximate 200 bp of surrounding native exon sequence that improves miRNA expression levels. In each miRNA unit, the natural pre-miRNA sequences are maintained at the base of the hairpin up to the region of the mature miRNA sequence. The apex of each natural pri-miRNA stem/loop is replaced with a hairpin targeting a conserved HIV-1 sequence; the entire polycistron is expressed from the U1 snRNA pol II promoter. All three units mediate RNAi knockdown against their specific targets, and function is dependent on the presence of flanking pri-miRNA sequences. Some modifications of secondary structure and context significantly improve processing and subsequent RNAi knockdown efficiency. A separate U6–U16 TAR aptamer inserted downstream of the third unit, but still within the intronic sequences, is fully functional in either orientation.

Using a single promoter to express multiple RNAi triggers, rather than separate promoters for each trigger, can reduce overall levels of trigger production and subsequent interference with endogenous RNAi pathways. In addition, the multi-cistronic miRNA platform can use a wide variety of tissue-specific and conditional pol II promoters to regulate expression. These experiments, and others (Ely *et al.*, 2009), show the potential of multi-cistronic approaches and delineate important parameters for their implementation.

5.9 Conclusion

RNAi continues to be employed for genetic studies, especially in cell types not amenable to traditional genetic mutational approaches. A number of ongoing clinical trials are demonstrating the safety of small RNA gene therapeutics as well (reviewed in Lares *et al.*, 2011). These initial studies provide the foundation for subsequent trials of gene therapy efficacy and support the further development of RNAi-based therapeutics.

References

Aagaard, L.A., Amarzguioui, M., Sun, G., *et al.* (2007). A facile lentiviral vector system for expression of doxycycline-inducible shRNAs: Knockdown of the pre-miRNA processing enzyme Drosha, *Mol Ther*, **15**, 938–945.

Aagaard, L.A., Zhang, J., von Eije, K.J., *et al.* (2008). Engineering and optimization of the miR-106b cluster for ectopic expression of multiplexed anti-HIV RNAs, *Gene Ther*, **15**, 1536–1549.

Allen, D., Kenna, P.F., Palfi, A., *et al.* (2007). Development of strategies for conditional RNA interference, *J Gene Med*, **9**, 287–298.

Amendola, M., Passerini, L., Pucci, F., *et al.* (2009). Regulated and multiple miRNA and siRNA delivery into primary cells by a lentiviral platform, *Mol Ther*, **17**, 1039–1052.

An, D.S., Qin, F.X., Auyeung, V.C., *et al.* (2006a). Optimization and functional effects of stable short hairpin RNA expression in primary human lymphocytes via lentiviral vectors, *Mol Ther*, **14**, 494–504.

An, C.I., Trinh, V.B., Yokobayashi, Y. (2006b). Artificial control of gene expression in mammalian cells by modulating RNA interference through aptamer-small molecule interaction, *RNA*, **12**, 710–716.

Aparicio, O., Carnero, E., Abad, X., *et al.* (2010). Adenovirus VA RNA-derived miRNAs target cellular genes involved in cell growth, gene expression and DNA repair, *Nucleic Acids Res*, **38**, 750–763.

Arbuthnot, P. (2011). MicroRNA-like antivirals, *Biochim Biophys Acta*, **1809**, 746–755.

Beer, S., Bellovin, D.I., Lee, J.S., *et al.* (2010). Low-level shRNA cytotoxicity can contribute to MYC-induced hepatocellular carcinoma in adult mice, *Mol Ther*, **18**, 161–170.

Berger, S.M., Pesold, B., Reber, S., *et al.* (2010). Quantitative analysis of conditional gene inactivation using rationally designed, tetracycline-controlled miRNAs, *Nucleic Acids Res*, **38**, e168.

Beronja, S., Livshits, G., Williams, S., *et al.* (2010). Rapid functional dissection of genetic networks via tissue-specific transduction and RNAi in mouse embryos, *Nat Med*, **16**, 821–827.

Bhat, R.A., Thimmappaya, B. (1983). Two small RNAs encoded by Epstein-Barr virus can functionally substitute for the virus-associated RNAs in the lytic growth of adenovirus 5, *Proc Natl Acad Sci USA*, **80**, 4789–4793.

Bogerd, H.P., Karnowski, H.W., Cai, X., *et al.* (2010). A mammalian herpesvirus uses noncanonical expression and processing mechanisms to generate viral MicroRNAs, *Mol Cell*, **37**, 135–142.

Boudreau, R.L., Monteys, A.M., Davidson, B.L. (2008). Minimizing variables among hairpin-based RNAi vectors reveals the potency of shRNAs, *RNA*, **14**, 1834–1844.

Cagnon, L., Rossi, J.J. (2000). Downregulation of the CCRS β-chemokine receptor and inhibition of HIV-1 infection by stable VA1-ribozyme chimeric transcripts, *Antisense Nucleic Acid Drug Dev*, **10**, 251–261.

Chi, S.W., Hannon, G.J., Darnell, R.B. (2012). An alternative mode of microRNA target recognition, *Nat Struct Mol Biol*, **19**, 321–327.

Chumakov, S.P., Kravchenko, J.E., Prassolov, V.S., *et al.* (2010). Efficient downregulation of multiple mRNA targets with a single shRNA-expressing lentiviral vector, *Plasmid*, **63**, 143–149.

Coumoul, X., Deng, C.X. (2006). RNAi in mice: A promising approach to decipher gene functions *in vivo*, *Biochimie*, **88**, 637–643.

Coumoul, X., Shukla, V., Li, C., *et al.* (2005). Conditional knockdown of Fgfr2 in mice using Cre-LoxP induced RNA interference, *Nucleic Acids Res*, **33**, e102.

Czauderna, F., Santel, A., Hinz, M., *et al.* (2003). Inducible shRNA expression for application in a prostate cancer mouse model, *Nucleic Acids Res*, **31**, e127.

Diebel, K.W., Smith, A.L., van Dyk, L.F. (2010). Mature and functional viral miRNAs transcribed from novel RNA polymerase III promoters, *RNA*, **16**, 170–185.

DiGiusto, D.L., Krishnan, A., Li, L., *et al.* (2010). RNA-based gene therapy for HIV with lentiviral vector-modified $CD34^+$ cells in patients undergoing transplantation for AIDS-related lymphoma, *Sci Transl Med*, **2**, 36ra43.

Domitrovich, A.M., Kunkel, G.R. (2003). Multiple, dispersed human U6 small nuclear RNA genes with varied transcriptional efficiencies, *Nucleic Acids Res*, **31**, 2344–2352.

Dyer, V., Ely, A., Bloom, K., *et al.* (2010). tRNA Lys3 promoter cassettes that efficiently express RNAi-activating antihepatitis B virus short hairpin RNAs, *Biochem Biophys Res Commun*, **398**, 640–646.

Egloff, S., Van Herreweghe, E., Kiss, T. (2006). Regulation of polymerase II transcription by 7SK snRNA: two distinct RNA elements direct P-TEFb and HEXIM1 binding, *Mol Cell Biol*, **26**, 630–642.

Ehsani, A., Saetrom, P., Zhang, J., *et al.* (2010). Rational design of micro-RNA-like bifunctional siRNAs targeting HIV and the HIV coreceptor CCR5, *Mol Ther*, **18**, 796–802.

Elbashir, S.M., Harborth, J., Lendeckel, W., *et al.* (2001a). Duplexes of 21-nucleotide RNAs mediate RNA interference in cultured mammalian cells, *Nature*, **411**, 494–498.

Elbashir, S.M., Lendeckel, W., Tuschl, T. (2001b). RNA interference is mediated by 21- and 22-nucleotide RNAs, *Genes Dev*, **15**, 188–200.

Ely, A., Naidoo, T., Arbuthnot, P. (2009). Efficient silencing of gene expression with modular trimeric Pol II expression cassettes comprising microRNA shuttles, *Nucleic Acids Res*, **37**, e91.

Englert, M., Felis, M., Junker, V., *et al.* (2004). Novel upstream and intragenic control elements for the RNA polymerase III-dependent transcription of human 7SL RNA genes, *Biochimie*, **86**, 867–874.

Fei, Z., Chen, Z., Wang, Z., *et al.* (2007). Conditional RNA interference achieved by Oct-1 POU/rtTA fusion protein activator and a modified TRE-mouse U6 promoter, *Biochem Biophys Res Commun*, **354**, 906–912.

Fire, A., Xu, S., Montgomery, M.K., *et al.* (1998). Potent and specific genetic interference by double-stranded RNA in *Caenorhabditis elegans*, *Nature*, 391, 806–811.

Giering, J.C., Grimm, D., Storm, T.A., *et al.* (2008). Expression of shRNA from a tissue-specific pol II promoter is an effective and safe RNAi therapeutic, *Mol Ther*, **16**, 1630–1636.

Goncalves, M.A., Janssen, J.M., Holkers, M., *et al.* (2010). Rapid and sensitive lentivirus vector-based conditional gene expression assay to monitor and quantify cell fusion activity, *PLoS One*, **5**, e10954.

Grimm, D., Streetz, K.L., Jopling, C.L., *et al.* (2006). Fatality in mice due to oversaturation of cellular microRNA/short hairpin RNA pathways, *Nature*, **441**, 537–541.

Grimm, D., Wang, L., Lee, J.S., *et al.* (2010). Argonaute proteins are key determinants of RNAi efficacy, toxicity, and persistence in the adult mouse liver, *J Clin Invest*, **120**, 3106–3119.

Gwizdek, C., Ossareh-Nazari, B., Brownawell, A.M., *et al.* (2004). Minihelix-containing RNAs mediate exportin-5-dependent nuclear export of the double-stranded RNA-binding protein ILF3, *J Biol Chem*, **279**, 884–891.

Hakamata, Y., Kobayashi, E. (2010). Inducible and conditional promoter systems to generate transgenic animals, *Methods Mol Biol*, **597**, 71–79.

Hannon, G.J. (2002). RNA interference, *Nature*, **418**, 244–251.

He, Y., Zeng, Q., Drenning, SD., *et al.* (1998). Inhibition of human squamous cell carcinoma growth *in vivo* by epidermal growth factor receptor antisense RNA transcribed from the U6 promoter, *J Natl Cancer Inst*, **90**, 1080–1087.

Hornung, V., Ellegast, J., Kim, S., *et al.* (2006). 5′-Triphosphate RNA is the ligand for RIG-I, *Science*, **314**, 994–997.

Howe, J.G., Shu, M.D. (1989). Epstein-Barr virus small RNA (EBER) genes: unique transcription units that combine RNA polymerase II and III promoter elements, *Cell*, **57**, 825–834.

Jarrous, N., Gopalan, V. (2010). Archaeal/eukaryal RNase P: Subunits, functions and RNA diversification, *Nucleic Acids Res*, **38**, 7885–7894.

Kasim, V., Miyagishi, M., Taira, K. (2003). Control of siRNA expression utilizing Cre-loxP recombination system, *Nucleic Acids Res Suppl*, **3**, 255–256.

Kesireddy, V., van der Ven, P.F., Furst, D.O. (2010). Multipurpose modular lentiviral vectors for RNA interference and transgene expression, *Mol Biol Rep*, **37**, 2863–2870.

Kim, D.H., Behlke, M.A., Rose, S.D., *et al.* (2005). Synthetic dsRNA Dicer substrates enhance RNAi potency and efficacy, *Nat Biotechnol*, **23**, 222–226.

Kim, V.N. (2004). MicroRNA precursors in motion: Exportin-5 mediates their nuclear export, *Trends Cell Biol*, **14**, 156–159.

Kleinhammer, A., Deussing, J., Wurst, W., *et al.* (2011a). Conditional RNAi in mice, *Methods*, **53**, 142–150.

Kleinhammer, A., Wurst, W., Kuhn, R. (2011b). Constitutive and conditional RNAi transgenesis in mice, *Methods*, **53**, 430–436.

Koper-Emde, D., Herrmann, L., Sandrock, B., *et al.* (2004). RNA interference by small hairpin RNAs synthesised under control of the human 7S K RNA promoter, *Biol Chem*, **385**, 791–794.

Kumar, D., An, C.I., Yokobayashi, Y. (2009). Conditional RNA interference mediated by allosteric ribozyme, *J Am Chem Soc*, **131**, 13906–13907.

Kwak, Y.D., Koike, H., Sugaya, K. (2003). RNA interference with small hairpin RNAs transcribed from a human U6 promoter-driven DNA vector, *J Pharmacol Sci*, **93**, 214–217.

Lares, M.R., Rossi, J.J., Ouellet, D.L. (2011). RNAi and small interfering RNAs in human disease therapeutic applications, *Trends Biotechnol*, **28**, 570–579.

Lebbink, R.J., Lowe, M., Chan, T., *et al.* (2011). Polymerase II promoter strength determines efficacy of microRNA Adapted shRNAs, *PLoS One*, **6**, e26213.

Lee, N.S., Dohjima, T., Bauer, G., *et al.* (2002). Expression of small interfering RNAs targeted against HIV-1 rev transcripts in human cells, *Nat Biotechnol*, **20**, 500–505.

Lee, N.S., Kim, D.H., Alluin, J., *et al.* (2008). Functional and intracellular localization properties of U6 promoter-expressed siRNAs, shRNAs, and chimeric VA1 shRNAs in mammalian cells, *RNA*, **14**, 1823–1833.

Lee, R.C., Feinbaum, R.L., Ambros, V. (1993). The *C. elegans* heterochronic gene lin-4 encodes small RNAs with antisense complementarity to lin-14, *Cell*, **75**, 843–854.

Lee, S.K., Kumar, P. (2009). Conditional RNAi: towards a silent gene therapy, *Adv Drug Deliv Rev*, **61**, 650–664.

Li, L., Liu, Y. (2011). Diverse small non-coding RNAs in RNA interference pathways, *Methods Mol Biol*, **764**, 169–182.

Listerman, I., Bledau, A.S., Grishina, I., *et al.* (2007). Extragenic accumulation of RNA polymerase II enhances transcription by RNA polymerase III, *PLoS Genet*, **3**, e212.

Liu, Y.P., Haasnoot, J., Berkhout, B. (2007). Design of extended short hairpin RNAs for HIV-1 inhibition, *Nucleic Acids Res*, **35**, 5683–5693.

Liu, Y.P., von Eije, K.J., Schopman, N.C., *et al.* (2009). Combinatorial RNAi against HIV-1 using extended short hairpin RNAs, *Mol Ther*, **17**, 1712–1723.

Lund, E., Guttinger, S., Calado, A., *et al.* (2004). Nuclear export of microRNA precursors, *Science*, **303**, 95–98.

Manjunath, N., Wu, H., Subramanya, S., *et al.* (2009). Lentiviral delivery of short hairpin RNAs, *Adv Drug Deliv Rev*, **61**, 732–745.

Martin, J.N., Wolken, N., Brown, T., *et al.* (2011). Lethal toxicity caused by expression of shRNA in the mouse striatum: implications for therapeutic design, *Gene Ther*, **18**, 666–673.

Matsukura, S., Jones, P.A., Takai, D. (2003). Establishment of conditional vectors for hairpin siRNA knockdowns, *Nucleic Acids Res*, **31**, e77.

Matthess, Y., Kappel, S., Spankuch, B., *et al.* (2005). Conditional inhibition of cancer cell proliferation by tetracycline-responsive, H1 promoter-driven silencing of PLK1, *Oncogene*, **24**, 2973–2980.

McBride, J.L., Boudreau, R.L., Harper, S.Q., *et al.* (2008). Artificial miRNAs mitigate shRNA-mediated toxicity in the brain: implications for the therapeutic development of RNAi, *Proc Natl Acad Sci USA*, **105**, 5868–5873.

McIntyre, G.J., Arndt, A.J., Gillespie, K.M., *et al.* (2011a). A comparison of multiple shRNA expression methods for combinatorial RNAi, *Genet Vaccines Ther*, **9**, 9.

McIntyre, G.J., Groneman, J.L., Yu, Y.H., *et al.* (2011b). Multiple shRNA combinations for near-complete coverage of all HIV-1 strains, *AIDS Res Ther*, **8**, 1.

McIntyre, G.J., Yu, Y.H., Lomas, M., *et al.* (2011c). The effects of stem length and core placement on shRNA activity, *BMC Mol Biol*, **12**, 34.

McIntyre, G.J., Yu, Y.H., Tran, A., *et al.* (2009). Cassette deletion in multiple shRNA lentiviral vectors for HIV-1 and its impact on treatment success, *Virol J*, **6**, 184.
Miyagishi, M., Taira, K. (2002). U6 promoter driven siRNAs with four uridine 3′ overhangs efficiently suppress targeted gene expression in mammalian cells, *Nat Biotechnol*, **20**, 497–500.
Murphy, S., Di Liegro, C., Melli, M. (1987). The *in vitro* transcription of the 7SK RNA gene by RNA polymerase III is dependent only on the presence of an upstream promoter, *Cell*, **51**, 81–87.
Nielsen, T.T., Marion, I., Hasholt, L., *et al.* (2009). Neuron-specific RNA interference using lentiviral vectors, *J Gene Med*, **11**, 559–569.
Oberdoerffer, P., Kanellopoulou, C., Heissmeyer, V., *et al.* (2005). Efficiency of RNA interference in the mouse hematopoietic system varies between cell types and developmental stages, *Mol Cell Biol*, **25**, 3896–3905.
Orioli, A., Pascali, C., Pagano, A., *et al.* (2012). RNA polymerase III transcription control elements: Themes and variations, *Gene*, **493**, 185–194.
Orioli, A., Pascali, C., Quartararo, J., *et al.* (2011). Widespread occurrence of non-canonical transcription termination by human RNA polymerase III, *Nucleic Acids Res*, **39**, 5499–5512.
Paddison, P.J., Caudy, A.A., Bernstein, E., *et al.* (2002). Short hairpin RNAs (shRNAs) induce sequence-specific silencing in mammalian cells, *Genes Dev*, **16**, 948–958.
Palli, S.R., Kapitskaya, M.Z., Kumar, M.B., *et al.* (2003). Improved ecdysone receptor-based inducible gene regulation system, *Eur J Biochem*, **270**, 1308–1315.
Paul, C.P., Good, P.D., Li, S.X., *et al.* (2003). Localized expression of small RNA inhibitors in human cells, *Mol Ther*, **7**, 237–247.
Paul, C.P., Good, P.D., Winer, I., *et al.* (2002). Effective expression of small interfering RNA in human cells, *Nat Biotechnol*, **20**, 505–508.
Phadke, A.P., Jay, C.M., Wang, Z., *et al.* (2011). *In vivo* safety and antitumor efficacy of bifunctional small hairpin RNAs specific for the human Stathmin 1 oncoprotein, *DNA Cell, Biol* **30**, 715–726.
Phizicky, E.M., Hopper, A.K. (2010). tRNA biology charges to the front, *Genes Dev*, **24**, 1832–1860.
Puranam, R.S., Attardi, G. (2001). The RNase P associated with HeLa cell mitochondria contains an essential RNA component identical in sequence to that of the nuclear RNase P, *Mol Cell Biol*, **21**, 548–561.
Rao, D.D., Maples, P.B., Senzer, N., *et al.* (2010). Enhanced target gene knockdown by a bifunctional shRNA: a novel approach of RNA interference, *Cancer Gene Ther*, **17**, 780–791.
Rao, M.K., Wilkinson, M.F. (2006). Tissue-specific and cell type-specific RNA interference *in vivo*, *Nat Protoc*, **1**, 1494–1501.
Raymond, C.S., Zhu, L., Vogt, T.F., *et al.* (2010). *In vivo* analysis of gene knockdown in tetracycline-inducible shRNA mice, *Methods Enzymol*, **477**, 415–427.
Reinhart, B.J., Slack, F.J., Basson, M., *et al.* (2000). The 21-nucleotide let-7 RNA regulates developmental timing in *Caenorhabditis elegans*, *Nature*, **403**, 901–906.
Richard, P., Manley, J.L. (2009). Transcription termination by nuclear RNA polymerases, *Genes Dev*, **23**, 1247–1269.
Sano, M., Li, H., Nakanishi, M., *et al.* (2008). Expression of long anti-HIV-1 hairpin RNAs for the generation of multiple siRNAs: Advantages and limitations, *Mol Ther*, **16**, 170–177.
Scherer, L.J., Rossi, J.J., Weinberg, M.S. (2007a). Progress and prospects: RNA-based therapies for treatment of HIV infection, *Gene Ther*, **14**, 1057–1064.

Scherer, L.J., Frank, R., Rossi, J.J. (2007b). Optimization and characterization of tRNA-shRNA expression constructs, *Nucleic Acids Res*, **35**, 2620–2628.
Scherer, L.J., Rossi, J.J. (2011). *Ex vivo* gene therapy for HIV-1 treatment, *Hum Mol Genet*, **20**, R100–107.
Schlee, M., Hornung, V., Hartmann, G. (2006). siRNA and isRNA: Two edges of one sword, *Mol Ther*, **14**, 463–470.
Schopman, N.C., Liu, Y.P., Konstantinova, P., *et al.* (2010a). Optimization of shRNA inhibitors by variation of the terminal loop sequence, *Antiviral Res*, **86**, 204–211.
Schopman, N.C., ter Brake, O., Berkhout, B. (2010b). Anticipating and blocking HIV-1 escape by second generation antiviral shRNAs, *Retrovirology*, **7**, 52.
Sen, G.L., Blau, H.M. (2006). A brief history of RNAi: The silence of the genes, *Faseb J*, **20**, 1293–1299.
Shea, C.M., Tzertzinis, G. (2010). Controlled expression of functional miR-122 with a ligand inducible expression system, *BMC Biotechnol*, **10**, 76.
Shukla, V., Coumoul, X., Deng, C.X. (2007). RNAi-based conditional gene knockdown in mice using a U6 promoter driven vector, *Int J Biol Sci*, **3**, 91–99.
Snead, N.M., Rossi, J.J. (2010). Biogenesis and function of endogenous and exogenous siRNAs, *Wiley Interdiscip Rev RNA*, **1**, 117–131.
Teichmann, M., Dieci, G., Pascali, C., *et al.* (2010). General transcription factors and subunits of RNA polymerase III: Paralogs for promoter- and cell type-specific transcription in multicellular eukaryotes, *Transcr*, **1**, 130–135.
ter Brake, O., t Hooft, K., Liu, Y.P., *et al.* (2008). Lentiviral vector design for multiple shRNA expression and durable HIV-1 inhibition, *Mol Ther*, **16**, 557–564.
Ullu, E., Murphy, S., Melli, M. (1982). Human 7SL RNA consists of a 140 nucleotide middle-repetitive sequence inserted in an alu sequence, *Cell*, **29**, 195–202.
Ullu, E., Weiner, A.M. (1984). Human genes and pseudogenes for the 7SL RNA component of signal recognition particle, *Embo J*, **3**, 3303–3310.
Unwalla, H.J., Li, M.J., Kim, J.D., *et al.* (2004). Negative feedback inhibition of HIV-1 by TAT-inducible expression of siRNA, *Nat Biotechnol*, **22**, 1573–1578.
van Zon, A., Mossink, M.H., Schoester, M., *et al.* (2001). Multiple human vault RNAs. Expression and association with the vault complex, *J Biol Chem*, **276**, 37715–37721.
Wassarman, D.A., Steitz, J.A. (1991). Structural analyses of the 7SK ribonucleoprotein (RNP), the most abundant human small RNP of unknown function, *Mol Cell Biol*, **11**, 3432–3445.
White, R.J. (2011). Transcription by RNA polymerase III: More complex than we thought, *Nat Rev Genet*, **12**, 459–463.
Will, C.L., Luhrmann, R. (2011). Spliceosome structure and function, *Cold Spring Harb Perspect Biol*, **3**. pii: a003707
Wiznerowicz, M., Szulc, J., Trono, D. (2006). Tuning silence: Conditional systems for RNA interference, *Nat Methods*, **3**, 682–688.
Wu, R.H., Cheng, T.L., Lo, S.R., *et al.* (2007). A tightly regulated and reversibly inducible siRNA expression system for conditional RNAi-mediated gene silencing in mammalian cells, *J Gene Med*, **9**, 620–634.
Yang, W., Paschen, W. (2008). Conditional gene silencing in mammalian cells mediated by a stress-inducible promoter, *Biochem Biophys Res Commun*, **365**, 521–527.
Yi, R., Qin, Y., Macara, I.G., *et al.* (2003). Exportin-5 mediates the nuclear export of pre-microRNAs and short hairpin RNAs, *Genes Dev*, **17**, 3011–3016.

Zeng, Y., Wagner, E.J., Cullen, B.R. (2002). Both natural and designed micro RNAs can inhibit the expression of cognate mRNAs when expressed in human cells, *Mol Cell*, **9**, 1327–1333.

Zhang, J., Rossi, J.J. (2010). Strategies in designing multigene expression units to down-regulate HIV-1, *Methods Mol Biol*, **623**, 123–136.

Zhou, H., Xia, X.G., Xu, Z. (2005). An RNA polymerase II construct synthesizes short-hairpin RNA with a quantitative indicator and mediates highly efficient RNAi, *Nucleic Acids Res*, **33**, e62.

Zieve, G., Penman, S. (1976). Small RNA species of the HeLa cell: Metabolism and subcellular localization, *Cell*, **8**, 19–31.

6

ON DEMAND ALTERNATIVE SPLICING FOR GENE RESCUE

Stéphanie Lorain[a] and Luis Garcia[a,b]

6.1 Introduction

While elucidating the mechanisms of pre-messenger RNA transformation, progress in molecular biology has recently opened the prospect of new therapeutic strategies aiming to repair mRNAs. Among them, the use of antisense oligonucleotides masking of key splicing determinants by Watson–Crick pairing has made it possible to rehabilitate mutated mRNAs with great selectivity. Moreover, it is possible to either destroy or to rescue mRNAs *à la carte*: skipping of a mutated exon or re-inclusion of an exon excluded by a mutation. More recently, it was also proposed to reintroduce missing sequences into deleted mRNAs or to replace a mutated exon by *trans*-splicing approaches which combine mRNA repair strategies with classical gene transfer therapy.

6.2 Coding Frame Restoration of a Mutated Gene by Exon Skipping

Duchenne muscular dystrophy (DMD) is a textbook case for therapeutic exon skipping. The dystrophinopathies are pathologies caused by anomalies in the DMD gene that encodes a subsarcolemmal protein called dystrophin (427 kDa). This

[a] UPMC Um76, CNRS UMR7215, Inserm U974, Institut de Myologie, 105 bd de l'Hôpital, 75013 Paris, France
[b] Email: luis.garcia@upmc.fr

protein is absent in severe DMD or present but qualitatively and/or quantitatively altered in moderate Becker muscular dystrophy (BMD). This clinical heterogeneity is the result of the genetic heterogeneity of the mutations in the DMD gene (2.4 Mb, 79 exons). With regard to the large deletions, the most frequent genetic alteration, the severity of the phenotype is primarily conditioned by the effect of the mutation on the protein reading frame of the dystrophin transcript. Thus, very large deletions can lead to moderate clinical picture if the protein reading frame is preserved. It is known that the modular structure of the dystrophin (central rod-domain made of 24 spectrin-like repeats) tolerates large internal deletions (Harper *et al.*, 2002). This observation led to the development of two therapeutic strategies: classical gene therapy with transfer of functional mini- or micro-dystrophin complementary DNAs (cDNAs) (Braun, 2004; Mendell *et al.*, 2010) in muscles, and therapeutic exon skipping. Since the capacity of the virus most suitable for efficiently infecting muscle, the adeno-associated virus (AAV), is limited to 4.7 kb, the only way of transferring the dystrophin cDNA was to reduce its coding sequence to the minimum that will produce a specific force restoration in the dystrophic muscle.

Exon skipping strategies aim to restore a coding reading frame, abolished by mutations, by the forced exclusion of one or more exons (Fig. 6.1). This can be achieved during the splicing of pre-mRNA by using antisense sequences annealing motifs that are crucial for the definition of targeted exons, such as splicing donor or acceptor splice sites, exon splice enhancers (ESE) or branching points.

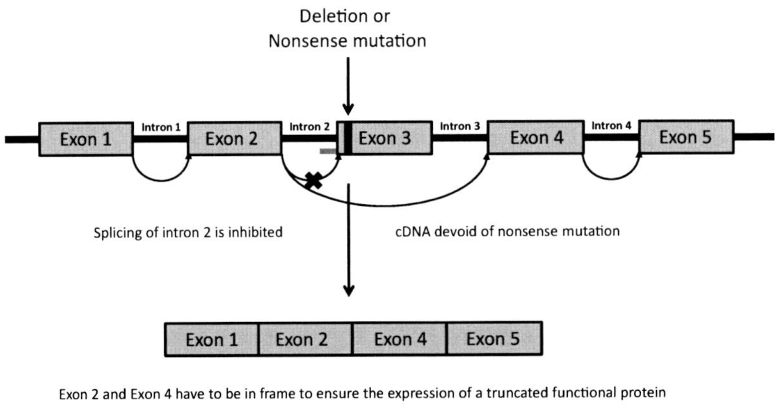

FIGURE 6.1 ■ Exon skipping rationale. Mutations commonly disturb the codon reading frame of affected genes. However, when mutations occur in parts that do not code for essential domains of ensuing proteins, it is occasionally possible to rescue a translational open reading frame by skipping adjacent exons. The goal of exon skipping is to modulate pre-mRNA splicing of dysfunctional genes in order to produce mRNAs retaining the ability to be translated into, at least partly, functional proteins. The figure schematizes the splicing of a hypothetical pre-mRNA made of five exons in the presence of an antisense sequence (gray bar pairing with Intron 2) designed to mask splicing determinants of exon 3.

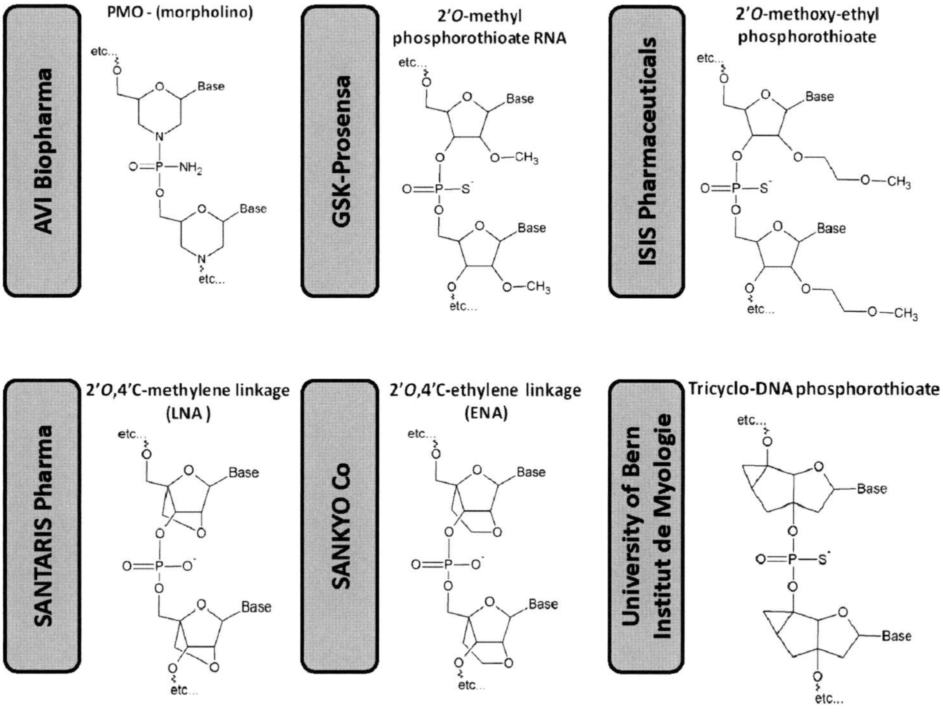

FIGURE 6.2 ■ Most popular chemistries for antisense oligomers. These compounds are designed to resist endogenous nucleases as well as not eliciting RNAse H activity after annealing their pre-mRNA targets. Phosphorodiamidate morpholino oligomer (PMO); 2′O-methyl phosphorothioate RNA; 2′-O-methoxy-ethylene; 2′O,4′C-methylene linkage (locked nucleic acid (LNA)); 2′O,4′C-ethylene linkage (ethylene-bridged nucleic acid (ENA)); Tricyclo-DNA (phosphorothioate).

Two approaches have already been proven to be successful. The first is the use of synthetic oligonucleotide analogues chemically modified to be resistant to endogenous nucleases (Fig. 6.2). Among them, the 2′-O-methyl-phosphorothioates (2′OMePS) and the phosphorodiamidate morpholino oligomers (PMOs) are very popular in the context of DMD and are already under clinical evaluation. They have the enormous advantage of not being immunogenic but have the disadvantage of having to be regularly injected to maintain therapeutic benefit.

In the second approach, the antisense sequences disguised in a small nuclear RNA such as U7 snRNA (Fig. 6.3) or U1 snRNA can be produced from recombinant transgenes (Brun *et al.*, 2003; Goyenvalle *et al.*, 2004; Denti *et al.*, 2006). These therapeutic molecules can be vectorized in AAV vectors as well as in lentiviral vectors (Benchaouir *et al.*, 2007), which ensures permanent production of the therapeutic antisense molecule in transduced cells. Nonetheless, the use of such viral vectors poses the problem of the immunogenicity of the viral capsids, prohibiting recurring treatments (Lorain *et al.*, 2008), to which innate immunity is added, which will eliminate many vectors.

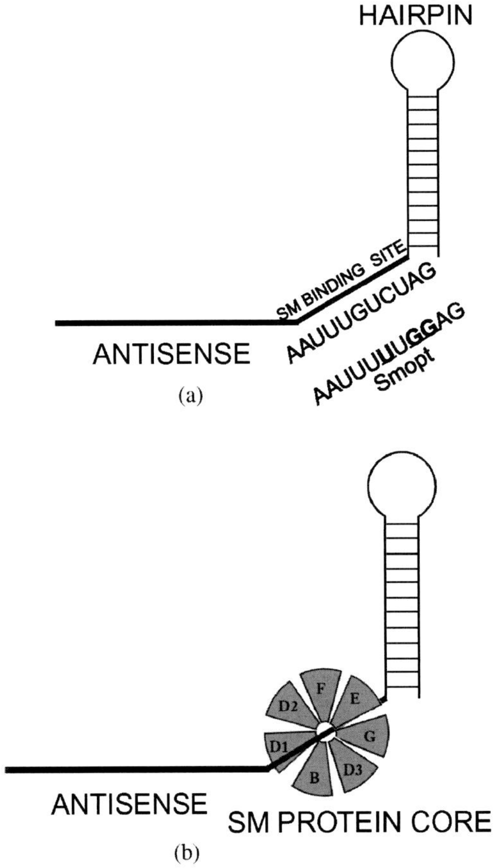

FIGURE 6.3 ■ U7opt snRNA. U7 is a snRNA of about 60 nucleotides involved in 3′ end processing of animal histone mRNAs. The use of U7 as a delivery system for antisense RNA sequences was first pioneered by Daniel Schümperli (University of Berne) in the late 1990s. (a) U7opt snRNA displays three domains transcribed by RNA polymerase II: an antisense sequence (5′ end), a short membrane protein (Sm) binding site optimized to recruit the U2-Sm protein core and a short hairpin (3′ end). (b) U7opt small nuclear ribonucleoprotein assembly takes place in the cytoplasm. Optimization of the Sm binding site allows recruitment of the U2 heptameric Sm protein ring (B, D1, D2, D3, E, F and G).

The effectiveness of these approaches has been validated *ex vivo* (myoblasts of DMD patients in culture) and *in vivo* (intramuscular or intravascular injections) using the *mdx* mouse model (Goyenvalle *et al.*, 2004; Benchaouir *et al.*, 2007; Denti *et al.*, 2006) as well as the canine golden retriever muscular dystrophy (GRMD) model (Yokota *et al.*, 2009). The results of these works are spectacular since stable expression of the expected quasi-dystrophin was obtained, which was associated with a significant improvement of the muscle force.

Clinical trials are ongoing with 2′OMePS (Prosensa/GSK) (van Deutekom *et al.*, 2007; Goemans *et al.*, 2011) and PMO chemistries (AviBioPharma) (Kinali *et al.*,

2009; Cirak *et al.*, 2011). In both cases, the targeted exon is exon 51, because its exclusion in the final transcript restores an operational reading frame in at least six deletion patterns: del 50, del 52, del 49–50, del 48–50, del 47–50, del 45–50, representing nearly 15% of the DMD population, one of the largest proportions of patients that could benefit from targeting a single dystrophin exon (Fig. 6.4). Intramuscular injections (phase I) of these synthetic oligonucleotides have demonstrated very promising dystrophin expression in tibialis anterior and extensor digitorum brevis muscles of DMD patients (van Deutekom *et al.*, 2007; Kinali *et al.*, 2009). Results of phase II trials using systemic administrations, necessary to reach the whole musculature in order to oppose the dystrophic process and obtain clinical effectiveness, have been recently published.

In 2011, a phase I/II study was carried out on 12 patients with weekly injection (subcutaneously) of 0.5–6 mg/kg 2′OMePS PRO051 over 5 weeks (Goemans *et al.*,

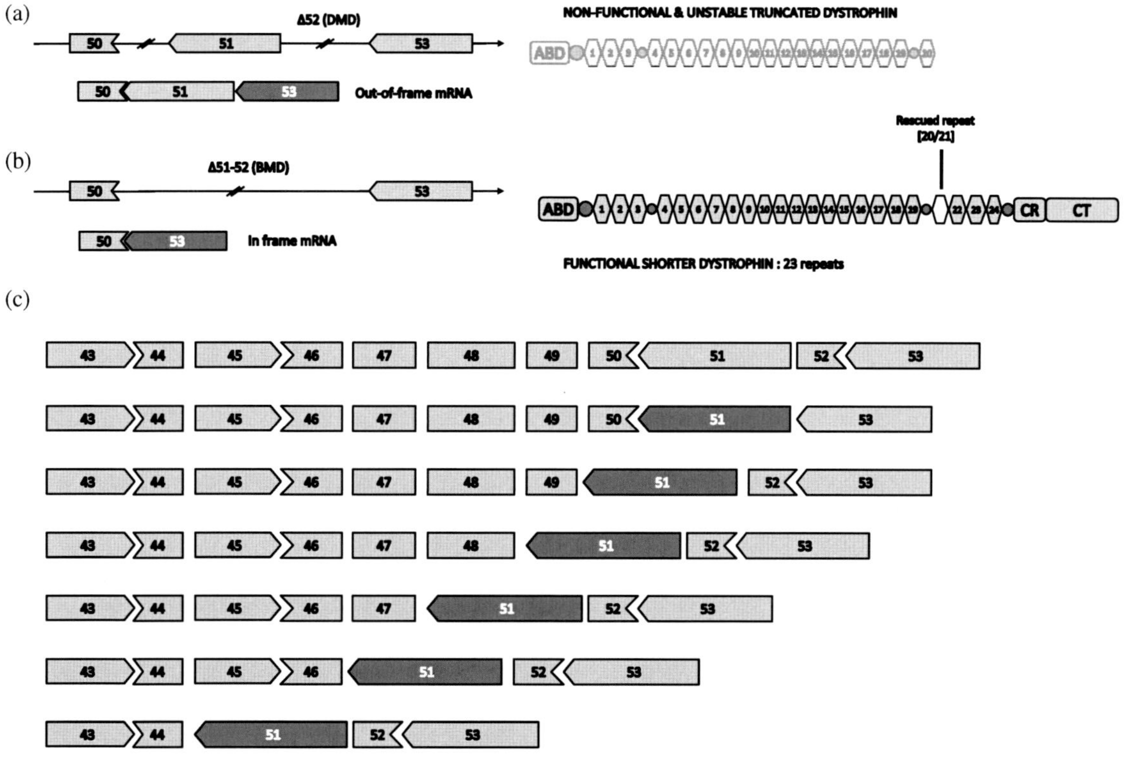

FIGURE 6.4 ■ Rationale for exon skipping in DMD. In DMD, the severity of the phenotype is not correlated to the span of deletions. Instead, it depends on the effect of the deletions on the reading frame of the transcribed mRNA. (a) Loss of exon 52 disrupts the codon reading frame. In this case, dystrophin is not synthesized and the phenotype is severe (DMD). (b) Loss of exons 51 and 52 results in a shorter mRNA, which is still in frame. In this case, a shorter functional protein will be produced and patients will display a milder phenotype such as BMD. (c) Deletion patterns for which skipping exon 51 makes sense. This affects about 15–17% of the DMD population.

2011). The product was well tolerated (followed over 7 weeks). Analyses carried out on muscle biopsies have demonstrated that the muscle fibers (64–97% depending on patients) produced dystrophin, in various amounts that reach 3–12% of the normal level. The extension studies (12, 24 and 48 weeks) showed an improvement of the ambulatory function. A phase III trial for PRO051 is ongoing. On the other hand, AVI BioPharma has completed phase I/II trials with the PMO AVI 4658 (Cirak *et al.*, 2011). Six patients received injections (intravenously) weekly for 12 weeks, with observation over 26 weeks. Significant dystrophin expression was observed at 2 mg/kg. The treatment was well tolerated, with moderate side effects not related to the injected product. A phase III trial is being studied.

However questions still remain concerning the long-term tolerance of these oligonucleotides. One of the limitations of the PMO and the 2′OMePS in this indication is their bad delivery to the cardiac tissue.

6.3 mRNA Repair by *Trans*-Splicing Approaches

Exon skipping strategies would only affect patients for whom forced splicing will generate a shorter but still functional protein. Since many pathological situations escape this prerequisite, additional universal strategies for mRNA repair have been developed in order to replace essential mutated exons by their normal versions or to reintroduce missing exons in the case of large deletions. In this context, the *trans*-splicing approach is hopefully one of the most promising. *Trans*-splicing mechanisms were initially observed in inferior eukaryotes where mature mRNAs can be produced from two distinct pre-messengers (Hastings, 2005). More recently, a similar mechanism has also been described in mammalian cells and interpreted as a mechanism contributing to the molecular diversity of proteins (Gingeras, 2009).

A number of *trans*-splicing devices or spliceosome-mediated RNA *trans*-splicing (SMaRT; Intronn proprietary technology) using engineered RNA molecules that could interfere with the natural splicing of targeted mRNAs have been developed during the last decade (Puttaraju *et al.*, 1999). Schematically, these RNA encoding *trans*-splicing molecules carry the wild-type exon or a series of exons flanked by artificial intronic sequence and a binding site complementary to the mutated pre-mRNA to be repaired (Fig. 6.5). Therefore, the *trans*-spliced end product is a chimeric mRNA that can produce the full-length functional protein. Three construct classes have been designed depending on whether the mutation is localized in the 5′ exons of the messenger (5′-replacement), the 3′ (3′-replacement) or in the middle (exchange by double *trans*-splicing).

The majority of *trans*-splicing studies have developed therapeutic RNAs replacing the 3′ part of the transcript to be repaired. They have been applied to

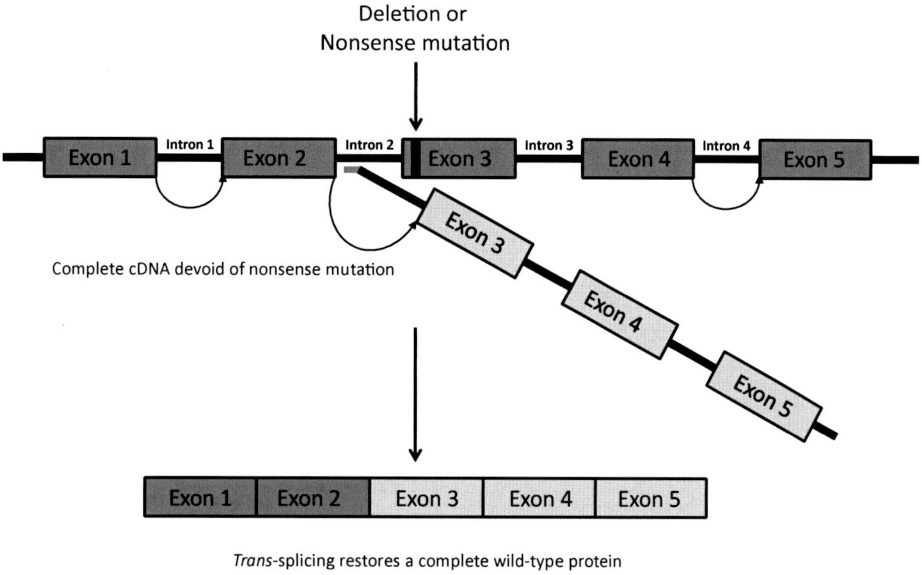

FIGURE 6.5 ■ *Trans*-splicing rationale. While exon skipping allows one sometimes to rescue a mutated gene product, *trans*-splicing is potentially more universal since it allows one to fully repair gene products. By using *trans*-splicing, mutated exons can be replaced by their normal form. This approach also makes it possible to introduce missing exons. The figure schematizes the splicing of a hypothetical pre-mRNA made of five exons in the presence of a *trans*-splicing molecule (TSM). The TSM is made of an antisense sequence (gray bar pairing with Intron 2), targeting an intronic sequence upstream of exons to exchange, an artificial hemi-intron harboring an acceptor splice site to react with the upstream exon, and a coding sequence corresponding to the string of exons to exchange (Exons 3, 4 and 5 in gray).

correct a number of mutations in preclinical models of hemophilia A (Chao *et al.*, 2003), spinal muscular atrophy (Shababi *et al.*, 2010), X-linked immunodeficiency (Tahara *et al.*, 2004) and cystic fibrosis where the widespread mutation CFTRΔF508 was replaced efficiently *in vivo* by the normal sequence via a *trans*-splicing reaction (Liu *et al.*, 2005). In contrast, only a few attempts at 5′ replacement (Mansfield *et al.*, 2003; Kierlin-Duncan and Sullenger, 2007) and internal exon exchange (Fig. 6.6) (Lorain *et al.*, 2010; Koller *et al.*, 2011) have been reported to be successful on minigene transcripts.

Interestingly, in the spinal muscular atrophy (SMA) mouse models, a combination of both *trans*-splicing and antisense approaches was found to increase mRNA repair by *trans*-splicing. The original *trans*-splicing molecule developed to repair survival motor neuron 2 (SMN2) transcript replaced SMN2 mutated exon 7 by SMN1 exon 7 (Coady *et al.*, 2007). Using an AAV vector of serotype 2, the delivery of this construct in severe SMA patients' fibroblasts significantly increased SMN protein levels, which was validated by small nuclear ribonucleoprotein (snRNP) assembly assay for functionality. To improve the *in vivo* efficiency,

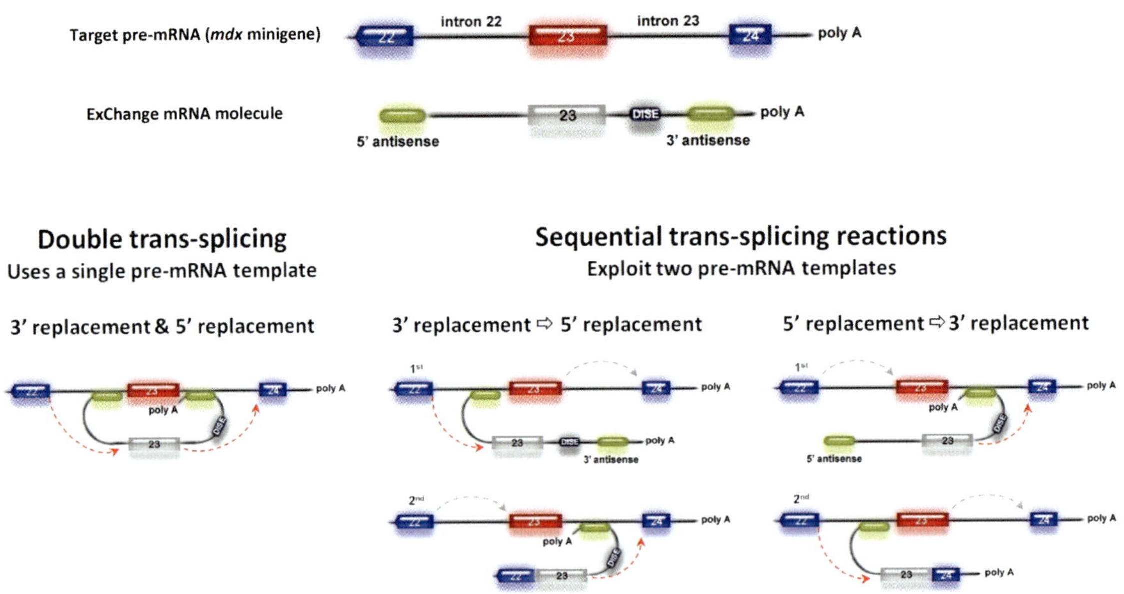

FIGURE 6.6 ■ *Trans*-splicing molecules for exchange of internal exons. *Upper section*: Composition of the ExChange molecule. It is made of two antisense domains (in yellow) designed to anneal intronic regions at both edges of the targeted exon and to replace the exon (in gray) surrounded by two hemi-introns with acceptor and donor splice sites. In some cases a downstream intronic splice enhancer (DISE) might be added. *Lower section*: Possible mechanisms involved during the double *trans*-splicing reaction.

the *trans*-splicing molecule was co-expressed with a short antisense molecule designed to block the downstream splicing at exon 8 and therefore promoting *trans*-splicing events (Coady *et al.*, 2008). This vector was highly effective, as demonstrated after intracerebroventricular injections of a severe model of SMA by extending the lifespan by nearly 70% (Coady and Lorson, 2010).

The majority of *trans*-splicing studies to date have focused on restoring function by replacing the portion of the mRNA transcript containing the disease-causing mutation. However, *trans*-splicing also has potential application in treating tumor cells by suicide gene therapy. Using this approach, several authors have successfully replaced a tumor-specific transcript with one encoding a cell-death-inducing toxin to provide tumor-restricted expression and induce their death (Puttaraju *et al.*, 1999; Gruber *et al.*, 2011).

Trans-splicing approaches have several advantages over conventional replacement gene therapy. First, they could address disorders caused by dominant mutations such as toxic mRNAs, which can be converted via *trans*-splicing to normal transcripts. Moreover, repair will only take place where the target transcript is expressed, preserving both levels and tissue specificity of the expression of the repaired transcript. In addition, because the coding domain can consist of one or more exons, a single *trans*-splicing molecule, in most cases, would be able to address diverse mutations spread over several exons. Improvements in the design of *trans*-splicing molecules have increased their specificity and efficiency and designate *trans*-splicing as a promising approach as a therapy.

6.4 Conclusion

These therapeutic approaches belong to a new class of therapies which are truly allele-specific. Analysis *in silico* suggests that nearly three quarters of the DMD population would be eligible for the exon skipping approach, and even more if we consider the possibility of skipping several exons. Although very promising, the development of these pharmacogenetic molecules is hindered by the elevated cost due to the target multiplicity and the design of clinical trials for very small patient cohorts. Supporting innovation in the field of such personalized medications for rare diseased subpopulations would require strong financial investments, in addition to requiring particular ethical attention, and specific regulatory accompaniment for clinical trials.

Acknowledgments

The authors are supported by the Association Française contre les Myopathies, Duchenne Parent Project France and Association Monégasque contre les Myopathies. We are grateful to Romain Garcia for graphical art.

References

Benchaouir, R., Meregalli, M., Farini, A., *et al.* (2007). Restoration of human dystrophin following transplantation of exon-skipping-engineered DMD patient stem cells into dystrophic mice, *Cell Stem Cell*, **1**, 646–657.

Braun, S. (2004). Naked plasmid DNA for the treatment of muscular dystrophy, *Curr Opin Mol Ther*, **6**, 499–505.

Brun, C., Suter, D., Pauli, C., *et al.* (2003). U7 snRNAs induce correction of mutated dystrophin pre-mRNA by exon skipping, *Cell Mol Life Sci*, **60**, 557–566.

Chao, H., Mansfield, S.G., Bartel, R.C., *et al.* (2003). Phenotype correction of hemophilia A mice by spliceosome-mediated RNA *trans*-splicing, *Nat Med*, **9**, 1015–1019.

Cirak, S., Arechavala-Gomeza, V., Guglieri, M., *et al.* (2011). Exon skipping and dystrophin restoration in patients with Duchenne muscular dystrophy after systemic phosphorodiamidate morpholino oligomer treatment: an open-label, phase 2, dose-escalation study, *Lancet*, **378**, 595–605.

Coady, T.H., Baughan, T.D., Shababi, M., *et al.* (2008). Development of a single vector system that enhances *trans*-splicing of SMN2 transcripts, *PLoS One*, **3**, e3468.

Coady, T.H., Lorson, C.L. (2010). *Trans*-splicing-mediated improvement in a severe mouse model of spinal muscular atrophy, *J Neurosci*, **30**, 126–130.

Coady, T.H., Shababi, M., Tullis, G.E., *et al.* (2007). Restoration of SMN function: delivery of a *trans*-splicing RNA re-directs SMN2 pre-mRNA splicing, *Mol Ther*, **15**, 1471–1478.

Denti, M.A., Rosa, A., D'Antona, G., *et al.* (2006). Chimeric adeno-associated virus/antisense U1 small nuclear RNA effectively rescues dystrophin synthesis and muscle function by local treatment of mdx mice, *Hum Gene Ther*, **17**, 565–574.

Gingeras, T.R. (2009). Implications of chimaeric non-co-linear transcripts, *Nature*, **461**, 206–211.

Goemans, N.M., Tulinius, M., van den Akker, J.T., *et al.* (2011). Systemic administration of PRO051 in Duchenne's muscular dystrophy, *N Engl J Med*, **64**, 1513–1522.

Goyenvalle, A., Vulin, A., Fougerousse, F., *et al.* (2004). Rescue of dystrophic muscle through U7 snRNA-mediated exon skipping, *Science*, **306**, 1796–1799.

Gruber, C., Gratz, I.K., Murauer, E.M., *et al.* (2011). Spliceosome-mediated RNA *trans*-splicing facilitates targeted delivery of suicide genes to cancer cells, *Mol Cancer Ther*, **10**, 233–241.

Harper, S.Q., Hauser M.A., DelloRusso, C., *et al.* (2002). Modular flexibility of dystrophin: implications for gene therapy of Duchenne muscular dystrophy, *Nat Med*, **8**, 253–261.

Hastings, K.E. (2005). SL *trans*-splicing: easy come or easy go? *Trends Genet*, **21**, 240–247.

Kierlin-Duncan, M.N., Sullenger, B.A. (2007). Using 5'-PTMs to repair mutant β-globin transcripts, *RNA*, **13**, 1317–1327.

Kinali, M., Arechavala-Gomeza, V., Feng, L., *et al.* (2009). Local restoration of dystrophin expression with the morpholino oligomer AVI-4658 in Duchenne muscular dystrophy: a single-blind, placebo-controlled, dose-escalation, proof-of-concept study, *Lancet Neurol*, **8**, 918–928.

Koller, U., Wally, V., Mitchell, L.G., *et al.* (2011). A novel screening system improves genetic correction by internal exon replacement, *Nucleic Acids Res*, **39**, e108.

Liu, X., Luo, M., Zhang, L.N., *et al.* (2005). Spliceosome-mediated RNA *trans*-splicing with recombinant adeno-associated virus partially restores cystic fibrosis transmembrane conductance regulator function to polarized human cystic fibrosis airway epithelial cells, *Hum Gene Ther*, **16**, 1116–1123.

Lorain, S., Gross, D.A., Goyenvalle, A., *et al.* (2008). Transient immunomodulation allows repeated injections of AAV1 and correction of muscular dystrophy in multiple muscles, *Mol Ther*, **16**, 541–547.

Lorain, S., Peccate, C., Le Hir, M., *et al.* (2010). Exon exchange approach to repair Duchenne dystrophin transcripts, *PLoS One*, **5**, e10894.

Mansfield, S.G., Clark, R.H., Puttaraju, M., *et al.* (2003). 5′ exon replacement and repair by spliceosome-mediated RNA *trans*-splicing, *RNA*, **9**, 1290–1297.

Mendell, J.R., Campbell, K., Rodino-Klapac, L., *et al.* (2010). Dystrophin immunity in Duchenne's muscular dystrophy, *N Engl J Med*, **363**, 1429–1437.

Puttaraju, M., Jamison, S.F., Mansfield, S.G., *et al.* (1999). Spliceosome-mediated RNA trans-splicing as a tool for gene therapy, *Nat Biotechnol*, **17**, 246–252.

Shababi, M., Glascock, J., Lorson, C.L. (2010). Combination of SMN *trans*-splicing and a neurotrophic factor increases the life span and body mass in a severe model of spinal muscular atrophy, *Hum Gene Ther*, **22**, 135–144.

Tahara, M., Pergolizzi, R.G., Kobayashi, H., *et al.* (2004). *Trans*-splicing repair of CD40 ligand deficiency results in naturally regulated correction of a mouse model of hyper-IgM X-linked immunodeficiency, *Nat Med*, **10**, 835–841.

van Deutekom, J.C., Janson, A.A., Ginjaar, I.B., *et al.* (2007). Local dystrophin restoration with antisense oligonucleotide PRO051, *N Engl J Med*, **357**, 2677–2686.

Yokota, T., Lu, Q.L., Partridge, T., *et al.* (2009). Efficacy of systemic morpholino exon-skipping in Duchenne dystrophy dogs, *Ann Neurol*, **65**, 667–676.

7

NUCLEASE-MEDIATED TARGETED GENETIC CORRECTION

Dieter C. Gruenert[a,b,d], Hamid Emamekhoo[c] and R. Geoffrey Sargent[a]

7.1 Introduction

The term "gene targeting" has been applied in numerous ways over the last few decades. Gene targeting has taken on a new meaning in recent years that goes beyond homologous recombination (HR) and extends to the sequence-specific directed integration of therapeutic complementary DNA (cDNA). Classical gene targeting in mammalian cells (Fig. 7.1) stems from the studies carried out in the early 1980s in mouse embryonic stem (mES) cells (Capecchi, 1989; Kuehn *et al.*, 1987; Smithies *et al.*, 1984). While the prospect of specifically modifying the genome of human cells is therapeutically appealing (Capecchi, 2000), the efficiency of classical HR has been relatively low and requires enrichment through a series of selection protocols to generate clonal populations of cells that have undergone HR (Capecchi, 1994; Mansour *et al.*, 1988).

While classical HR has been demonstrated in numerous mammalian, including human, cell systems (Capecchi, 1989; Kucherlapati *et al.*, 1985; Smithies *et al.*, 1985; Thomas and Capecci, 1986a, 1986b), there has only been limited success in the generation of cells that have been modified by classical HR and have therapeutic potential. The development of oligo/polynucleotide (OPN)-based approaches to specifically modify genomic DNA has provided alternative strategies to the more classical HR approaches for sequence-specific genomic

[a]Departments of Otolaryngology — Head and Neck Surgery, Institute for Human Genetics, Cardiovascular Research Institute, University of California, San Francisco, CA, USA
[b]Laboratory Medicine, Center for Regenerative Medicine and Stem Cell Research, Institute for Human Genetics, Cardiovascular Research Institute, University of California, San Francisco, CA, USA
[c]Department of Internal Medicine, Tri-Health-Good Samaritan Hospital, Cincinnati, OH, USA
[d]Email: dgruenert@ohns.ucsf.edu

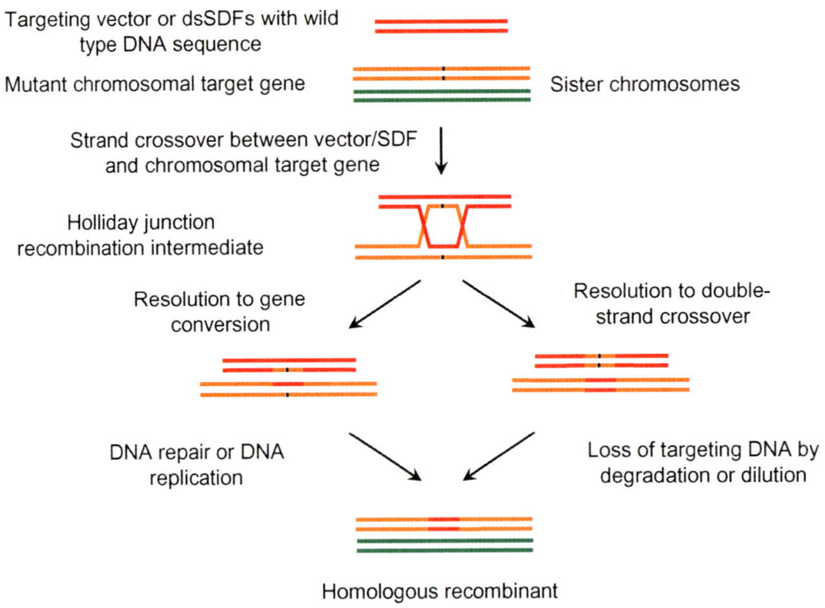

FIGURE 7.1 ■ Gene-targeting by double-stranded targeting vectors: Cellular proteins catalyze HR between classical targeting vectors, small DNA fragments (SDFs) (double-stranded (ds)DNA donor polynucleotides; for single-stranded (ss)DNA polynucleotides see Fig. 7.2), or dsDNA virus genomes (red) to exchange information between the DNA fragments and recipient chromosomal target loci. The mutant target gene is depicted as a black line on one of two homologous sister chromosomes (orange and green). In this model, strand exchange between donor and target DNA sequences generates a Holliday junction that can be resolved into two potential products: (1) a reciprocal crossover with complete exchange of information between donor and recipient, or (2) a gene conversion with exchange of one strand that results in a heteroduplex intermediate product. Gene conversion heteroduplexes require DNA replication or DNA repair to generate chromosome homology after the targeted gene modification. Not pictured in this figure is the targeting of the homologous sister chromosome that is predicted to account for approximately 50% of the gene targeted events.

modification (Campbell *et al.*, 1989; Hunger-Bertling *et al.*, 1990; Igoucheva *et al.*, 2001; Knauert and Glazer, 2001; Kunzelmann *et al.*, 1996a; Radecke *et al.*, 2004; Tsuchiya *et al.*, 2005; Vasquez and Wilson, 1998). These OPN approaches have, in some cases, achieved efficiencies that approach therapeutic significance thereby opening the door to the potential of clinical application (Giovannangeli and Helene, 1997; Gruenert, 2003; Knauert and Glazer, 2001; Porter, 2001; Richardson *et al.*, 2001; Sargent *et al.*, 2011; Seidman and Glazer, 2004; Thomas, 1994; Vasquez *et al.*, 2001b; Yanez and Porter, 1998). Moreover, the development of sequence-specific double-strand break (DSB)-inducing nuclease technologies, has further enhanced the therapeutic possibilities of HR-mediated, sequence-specific modification/correction of genomic DNA (Bibikova *et al.*, 2003; Porteus and Baltimore, 2003; Rouet *et al.*, 1994; Smith *et al.*, 2000).

In light of these advances and the development of virally based, adeno-associated virus (AAV) (Inoue *et al.*, 1999; Miller *et al.*, 2006; Russell and Hirata, 1998) and helper-dependent adenovirus (HDAV) (Kim and Svendsen, 2011; Liu *et al.*, 2011; Oka and Chan, 2005) gene-targeting systems, there is now an impressive repertoire of strategies that can make sequence-specific genomic modification a therapeutic and clinical reality.

7.2 Cell and Gene Therapy

The treatment of inherited diseases through the introduction of nucleic acids to complement or repair the disease-causing mutation has long since been the foundation of gene therapy. Given that most diseases are also associated with tissue/organ damage due to the pathology of the disease state, a comprehensive therapy will involve not only the correction of the disease phenotype through gene complementation or repair, but also the repair and regeneration of the damaged tissue/organ with corrected tissue-specific precursor cells. Ultimately, a comprehensive cell and gene therapy will entail: (1) correction of the disease phenotype by gene complementation or repair of the genomic mutant sequences with wild-type DNA, (2) restoration of normal cell/tissue/organ function, (3) repair of tissue damage with non-immunogenic cells, (4) delivery/transplant of corrected cells and DNA by a non-toxic/non-immunogenic system, (5) use of a rapid and efficient delivery system, (6) optimization of long-term efficacy with minimal applications or short-term efficacy with multiple equally effective applications (Table 7.1).

7.2.1 Cell Therapy

Cell-based therapies can be traced back to the development of hematopoietic and bone marrow stem cell transplantation strategies initiated by Mathé and colleagues in 1959 (Mathé *et al.*, 1959) and further developed by Thomas and colleagues in 1970 (Buckner *et al.*, 1970). These studies were focused on the restoration of the hematopoietic system after radiation therapy for leukemia and relied on the cells from donors that were not from the patient. As a result, the studies also led to a better understanding of histocompatibility and immune rejection during transplantation of the allogenic tissue. It was clear from these studies that the ultimate goal of any transplantation is the mitigation of an immune response to the transplanted cells. One strategy to overcome this limitation is through the transplantion of cells from the patient, i.e., autologous transplantation. While this may be possible for some disease states that involve the hematopoietic system, it becomes more complex when dealing with inherited diseases or non-hematopoietic tissues/organs.

TABLE 7.1 Specific Goals of Cell and Gene Therapy

1.	Correct mutant genetic sequences with normal (wild-type) sequences and eliminate disease pathology
2.	Express normal cell, tissue and/or organ function
3.	Repair damaged tissue with human leukocyte antigen (HLA)-matched/autologous cells
4.	Non-toxic/non-immunogenic delivery/transplantation system
5.	Rapid and efficient
6.	Long-term correction with minimal applications or short-term correction with multiple applications

The development of an autologous cellular therapy for tissue and organs outside of the hematopoietic system has been difficult. In this context, the generation of human pluripotent stem (PS) cells from the mature somatic cells of any patient has opened the door to the possibility of developing a comprehensive autologous multi-organ cellular therapy. The human induced pluripotent stem (hiPS) cell systems are clearly an important advance in the development of autologous cell-based therapies for tissue/organ regeneration (Takahashi *et al.*, 2007; Yamanaka, 2007; Yu *et al.*, 2007). Through directed differentiation of patient-derived hiPS cells to generate tissue-specific precursor cells it should be possible to overcome the immune rejection limitations of allogeneic-cell-based therapies.

Such a cellular therapy would be ideally suited to interface with genetic therapies for inherited diseases. The ability to generate clonal populations of corrected patient-derived cells that are pluripotent and have the potential of differentiating into any cell lineage would be of tremendous clinical value. It will also be important to maintain the cell-type-dependent expression of the corrected gene to ensure maximal therapeutic benefit. Most genetic therapies have focused on the correction of a pathological phenotype through complementation of the defective gene by cDNA transgenes (Bushman, 2007; Cavazzana-Calvo *et al.*, 2001; Friedmann and Roblin, 1972; Thomas, 1994). There have been other, alternative approaches focusing on sequence-specific gene modification that have been developing over the last two decades and have recently received greater attention as potential therapeutic interventions (Chapdelaine *et al.*, 2010; Gruenert *et al.*, 2003; Knauert and Glazer, 2001; McNeer *et al.*, 2011; Richardson *et al.*, 2002; Suzuki, 2008; Thomas, 1994; Vasquez and Wilson, 1998; Yanez and Porter, 1998). These are discussed in greater detail below.

7.2.2 Gene Therapy

Most gene therapy have involved cDNA-based complementation systems strategies. These strategies have had limited success in ameliorating the disease phenotype and little long-term efficacy. Notable exceptions are the clinical studies by Cavazzana-Calvo and Fischer on X-linked severe combined immune deficiency (SCID-X1), which have demonstrated long-term efficacy (Fischer *et al.*, 2011). However, these studies have also been plagued by the occurrence of retroviral insertional-mutagenesis-linked leukemia in several patients (Cattoglio *et al.*, 2007; Cavazzana-Calvo and Fischer, 2007; Pike-Overzet *et al.*, 2007). Another recent study using AAV, a virus that generally remains episomal within the nucleus, showed long-term correction of the factor IX gene hemophilia B phenotype (Nathwani *et al.*, 2011). These are significant advances in the application of gene therapy; however, there are still concerns about insertional-mutagenesis as well as limitations, such as the size of the inserted cDNA (AAV has a maximum packaging size of approximately 4.7 kb and will therefore not accommodate many disease-causing genes). Moreover, the inherent dissociation between the coding sequences and their endogenous regulatory elements in all cDNA-based gene therapies leads to a breakdown of cell-type-specific regulation of gene expression, and results in cell-independent expression of the therapeutic cDNA. While cell-independent gene expression might not be a factor for many genes, the expression of the transgene could result in a change of the cellular microenvironment such that it elicits a tissue-specific pathophysiological response. Furthermore, unregulated expression of the therapeutic cDNA could interfere with the ability of a cell to differentiate along a relevant lineage-specific pathway, as well as maintain long-term phenotypic correction.

Sequence-specific gene modification or gene editing strategies provide a viable alternative to cDNA-based systems and ensure that the relationship between the gene-specific regulatory elements and the coding sequences is retained upon correction of the mutant sequences. Several gene editing strategies have already been shown to have therapeutic potential: (1) OPN-mediated, (2) AAV-mediated and (3) HDAV-mediated.

The OPN strategies employ single-stranded oligonucleotides (SSOs), triplex-forming oligonucleotides (TFOs) and polynucleotide SDFs to facilitate homologous exchange (Sargent *et al.*, 2011). Virus-based strategies using AAV (Miller, 2011; Vasileva *et al.*, 2006) and HDAV (Li *et al.*, 2011; Liu *et al.*, 2011) have also been shown to facilitate homologous exchange at potential therapeutic targets and have the advantage of virus mediated delivery of therapeutic donor DNA.

7.3 Oligo/Polynucleotide-Mediated Sequence-Specific Modification

OPNs are essentially homologous to the target sequences with the exception of a few (generally between one and five) modifying bases within the segment of DNA.

More mismatches may be used; however, their number and placement within the OPN can influence efficacy and must be tested empirically for individual OPNs. As indicated above, OPN-mediated homologous exchange engages elements of endogenous DNA repair and replication pathways in order to facilitate the exchange between the incoming nucleotide sequences and the genomic target sequences (Sargent *et al.*, 2011). Such site-specific 'genetic surgery' maintains gene integrity and minimizes permanent regional chromatin remodeling that could change gene expression patterns (Cavazzana-Calvo *et al.*, 2007; Cereseto and Giacca, 2004; Hackett *et al.*, 2007; Pike-Overzet *et al.*, 2007). There has been a great deal of speculation about the cellular machinery facilitating OPN-mediated homologous exchange in human cells; however, the pathways have not been well defined and require further analysis (Gruenert *et al.*, 2003; Igoucheva *et al.*, 2006; Sargent *et al.*, 2011; Vasquez *et al.*, 2002; Yanez and Porter, 1999). Understanding the role these pathways play in modulating OPN-based DNA sequence editing is essential for maximizing therapeutic efficacy and assessing the potential adverse effects.

7.3.1 Single-Stranded Oligodeoxynucleotides-Mediated Gene Modification

Single-stranded oligodeoxynucleotides (SSOs/ssODNs, SSOs will be used from this point on) can be as much as 200 nucleotides (nt) long, but are generally less than 100 nt and contain a single mismatch, generally in the middle of the molecule (Fig. 7.2). SSOs were initially evaluated in human cells targeting plasmids carrying a mutant neomycin gene (Campbell *et al.*, 1989) or in lymphoblasts targeting the hypoxanthine–guanine phosphoribosyl transferase 1 (HPRT1) gene (Hegele *et al.*, 2008; Hunger-Bertling *et al.*, 1990; Kenner *et al.*, 2004; Wuepping *et al.*, 2009). In addition to unmodified SSOs, those with phosphorothioate (Gamper *et al.*, 2004) and 2′-O-methyl uracil backbones (Igoucheva *et al.*, 2001) or with 5′ or 3′ thymidine clamps (Hegele *et al.*, 2008) that inhibit degradation have also been used for effecting homologous exchange.

Microarray gene expression analysis comparing transfection with SSOs and plasmids, indicated that more DNA repair, cell-cycle arrest and apoptosis genes were upregulated after transfection with a plasmid than the SSOs (Igoucheva *et al.*, 2006). As a result, the frequency of SSO-mediated genomic correction was approximately five-fold higher when the SSOs were co-transfected with an unrelated plasmid rather than the SSO alone. These studies support the notion that DNA repair influences SSO-mediated homologous exchange in genomic DNA. Additional study analyses indicated that the mismatch repair pathway is an important component of SSO-induced gene modification (Aarts and te Riele, 2010; Igoucheva *et al.*, 2008; Pichierri *et al.*, 2001; Surtees *et al.*, 2004).

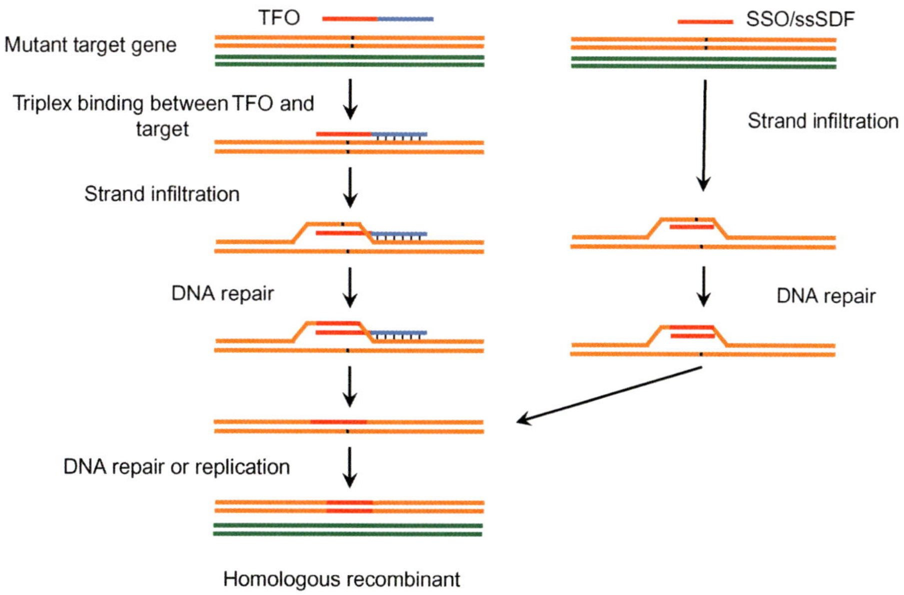

FIGURE 7.2 ■ Gene-targeting by TFOs, SSOs and ssSDFs: TFOs have a bipartite structure typically comprised of a polypurine DNA sequence and a DNA sequence homologous to the chromosomal target sequence. The polypurine sequence is designed to have high binding affinity for the DNA duplex adjacent to the target and forms a three-stranded DNA structure that allows strand infiltration of the unbound end of the oligonucleotide. Cellular DNA repair pathways catalyze transfer of sequence information to one strand of the recipient gene target forming a heteroduplex intermediate structure. SSOs and single-stranded SDFs (ssSDFs) may be introduced directly into cells or can result from intracellular denaturation or degradation of double-stranded oligonucleotides. In this model, the SSO/ssSDF strands infiltrate the chromosomal duplex target and the DNA repair machinery catalyzes the transfer of donor sequence information into the targeted strand. Chromosomes with homologous gene correction would be generated by another round of DNA repair or replication.

Some studies evaluating the correction of reporter genes indicate higher correction efficiencies with the non-transcribed, antisense SSO rather than with the sense strand (Igoucheva *et al.*, 2003; Nickerson and Colledge, 2003; Yin *et al.*, 2005). One study with lacZ suggests that transcription may be a factor in the correction process (Igoucheva *et al.*, 2003); however, these observations could also indicate differences in chromatin accessibility, because regional differences in genomic integration of the reporter transgene between individual clones. Thus, the chromosomal context and the expression of resident genes are different. There is also conflicting evidence that there are no strand-associated differences in SSO-mediated homologous exchange (Dekker *et al.*, 2003).

Higher efficiencies of SSO-mediated modification were also observed when cells were inhibited in S-phase (Brachman and Kmiec, 2005; Olsen *et al.*, 2005a; Wu *et al.*, 2005), indicating that DNA replication played a substantive role in SSO-induced homologous exchange. Mechanistically, one could envision incorporation of an SSO

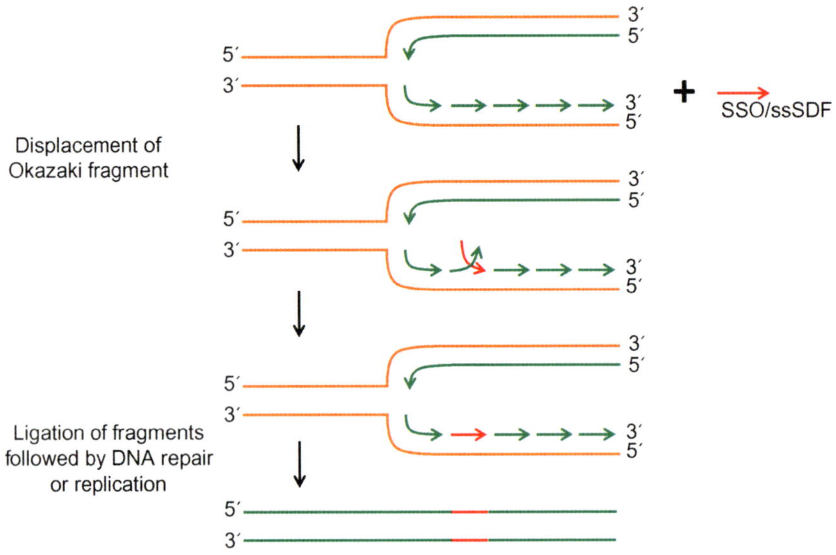

FIGURE 7.3 ■ Recombination of oligonucleotides at replication forks: SSOs/ssSDFs (red) approximately the size of Okazaki fragments (approximately 200 nt) may form as a consequence of intracellular denaturation or degradation of double-stranded oligonucleotides or may be introduced directly into cells. SSOs/ssSDFs could displace a homologous Okazaki fragment and be subsequently ligated into the mature replicated DNA strand to generate a heteroduplex intermediate structure. The gene correction is fixed by DNA repair or another round of DNA replication.

into a replication fork displacing an Okazaki fragment (Gruenert, 1998; Radecke *et al.*, 2006b; Wu *et al.*, 2005) (Fig. 7.3). SSOs can incorporate into the genome, but appear to act as a template rather than incorporate in the presence of double strand breaks (DSBs) (Hegele *et al.*, 2008; Radecke *et al.*, 2006b). Therefore the mechanism for incorporating the sequence information carried by the SSO into the target gene may be independent of the HR pathway activated by DSBs (Sargent *et al.*, 2011).

SSO-mediated modification *in vitro*, *in vivo* and in mES cells suggest that they can be used to target disease-related genes and create transgenic animal models (Aarts and te Riele, 2010; Alexeev *et al.*, 2002; Andrieu-Soler *et al.*, 2007; Lu *et al.*, 2003). However, it will be important to further characterize the mechanisms that underlie the homologous exchange process to provide insight into the potential long-term risk/benefit of the gene modification.

7.3.2 SFHR/SDF-Mediated Gene Modification

Small fragment homologous replacement (SFHR) is the process whereby SDFs potentiate sequence-specific homologous exchange. While SFHR still requires further mechanistic elucidation, it is clear that: (1) an SDF with defined base alterations finds its genomic or episomal sequence homologue, and (2) cellular enzymatic machinery facilitates sequence-specific homologous exchange between the SDF and target sequences (Fig. 7.1). More in-depth analyses will be necessary to define the cellular factors that modulate SFHR, even though a number of studies have implicated certain SFHR-associated mechanistic pathways (Goncz *et al.*, 2006; Gruenert *et al.*, 2003; Luchetti *et al.*, 2012; Sargent *et al.*, 2011).

SDF polynucleotides are distinct from SSOs in that they are larger than SSOs (generally between 200 and 2000 bp/nt) and are individual ssDNA, complementary ssDNA or dsDNA (Sargent *et al.*, 2011). Because of their size/length and strandedness, the term polynucleotide is used to describe and distinguish SDFs more accurately from other multinucleotide constructs.

SFHR has been demonstrated *in vitro* and *in vivo*, targeting a number of genes associated with different inherited diseases (Sargent *et al.*, 2011). Studies in human hematopoietic stem cells (HSCs) have demonstrated SDF-mediated modification of the human β-globin gene, targeting both the sickle cell disease (SCD) locus (Goncz *et al.*, 2006) and a β-thalassemia codon 39 (C>T) mutant locus (Colosimo *et al.*, 2007). Analysis of a clonal population of modified HSCs indicated that cell division, thereby implicating DNA replication, played a substantive role in SFHR (Goncz *et al.*, 2006). Studies were also carried out targeting the most common mutation at codon 508 (delF508 or ΔF508) in the cystic fibrosis (CF) transmembrane conductance regulator (CFTR) gene (Goncz *et al.*, 1998, 2001; Kunzelmann *et al.*, 1996b; Sangiuolo *et al.*, 2002; Sangiuolo *et al.*, 2008). These studies were carried out *in vitro* and *in vivo* and modified human and mouse airway epithelial cells and mES cells. Mutations in the dystrophin gene, responsible for Duchenne muscular dystrophy, were also modified *in vitro* and *in vivo* (Kapsa *et al.*, 2001, 2002; Todaro *et al.*, 2007). One of these studies suggested that DNA replication played a key role in SFHR and that its modulation could enhance the efficiency of SFHR (Todaro *et al.*, 2007). In addition, studies in cells from a mouse model of SCID with mutations in the DNA-dependent protein kinase catalytic subunit (DNA-PKCS) gene showed a restoration of resistance to X-ray damage after exposure to targeting SDFs (Zayed *et al.*, 2006). This study is significant because it demonstrated that mutations in the DNA-PKCS gene, a component of the non-homologous end joining (NHEJ) DNA repair pathway, were not required for SFHR-mediated homologous exchange, implying that the NHEJ pathway is not involved in SFHR. Other studies, targeting mutations in the α-1 antitrypsin gene (McNab *et al.*, 2007), the survival motor neuron-1 (SMN-1) gene (Sangiuolo *et al.*, 2005) or the HPRT gene (Bedayat *et al.*, 2010), showed correction of the phenotypic features associated with α-1 antitrypsin deficiency, spinal muscular atrophy or Lesch–Nyhan syndrome, respectively.

Efficiencies within the range of 0.1–15% were observed when evaluating SDF-induced homologous exchange (Gruenert *et al.*, 2003; Sargent *et al.*, 2011). This range of sequence-specific modification efficiencies was observed when targeting both episomal and genomic DNA, and was highest when the SDFs were microinjected directly into human hematopoietic stem cell nucleus (Goncz *et al.*, 2006) or nucleofected a mES cell (Sangiuolo *et al.*, 2008). There are still numerous elements of SFHR that still require elucidation to determine whether the SDF-mediated homologous exchange will be stable or whether there are cellular apoptotic pathways that are activated and might compromise the viability of the transfected cells, as has been observed with SSOs (Liu *et al.*, 2009a).

7.3.3 TFO-Mediated Gene Modification

The ability of DNA to form triple helices was originally described more than 50 years ago (Felsenfeld and Rich, 1957). Subsequently it was demonstrated that polypurine or polypyrimidine regions of the DNA can form triple-stranded structures with TFOs via Hoogsteen hydrogen bonds (Kallenbach *et al.*, 1976). These single-stranded TFOs are 10–30 nt in length and can bind to specific duplex DNA sequences to form a triple helix. TFOs have been used to modulate gene function by inducing sequence-specific mutations in plasmids (Wang *et al.*, 1995), mice genomic DNA (Vasquez *et al.*, 2001b) and to facilitate intrachromosomal recombination (Datta and Glazer, 2001; Luo *et al.*, 2000) (Fig. 7.2). TFOs tethered or untethered to an SDF, directed to a sequence near the homology, as well as protein nucleic acid (PNA) TFOs have also been successfully used for gene targeting (Chan *et al.*, 1999; Maurisse *et al.*, 2002; Schleifman *et al.*, 2008).

Mechanistically, there have been numerous studies evaluating the enzymatic pathways that underlie TFO-mediated homologous exchange. One DNA repair pathway, nucleotide excision repair (NER), has been shown to be involved (Faruqi *et al.*, 2000; Vasquez *et al.*, 2002; Wang *et al.*, 1996). Sequence-specific mutagenesis or intramolecular episomal HR was not observed in the Xeroderma pigmentosum (XP) group A cells defective in the XP, complementation group (XPA) gene, a component of NER. However, cells complemented with the wild-type XPA gene cDNA or wild-type cells showed HR. Studies in Cockayne syndrome group B (CSB) cells that are defective in transcription coupled repair (TCR), a component of NER showed that TFOs were unable to induce sequence-specific mutagenesis in these cells (Wang *et al.*, 1996). In addition to NER, NHEJ and HR repair pathways involved in repair of DSBs also appear to influence the effectiveness of TFO-mediated gene targeting (Chin and Glazer, 2009; Vasquez and Glazer, 2002; Villemure *et al.*, 2003; Zhang *et al.*, 2007).

The limitations imposed by the requirement of polypurine and polypyrimidine motifs in the region of the desired target sequence have been overcome to some extent by the modification of TFOs with locked nucleic acid (LNA) nucleotides or PNA residues in order to increase TFO target sequence affinity (Simon *et al.*, 2008). TFOs have also been linked to psoralens that form covalent bonds with DNA upon irradiation with ultraviolet light, thereby anchoring the TFO at the target site and stimulating DNA repair (Liu *et al.*, 2009b; Pathak *et al.*, 1959; Varganov *et al.*, 2007; Vasquez *et al.*, 2001a). Even though psoralens are useful in the laboratory, safety standards and their potential for carcinogens make them impractical for clinical applications (Gruenert and Cleaver, 1985; Pathak *et al.*, 1959; Tamaro *et al.*, 1986).

7.4 Enhancement of Gene-Targeting Efficacy

Therapeutic efficacy depends largely on the efficiency of sequence-specific gene correction. Since sequence-specific gene editing employs components of the DNA repair and replication pathways (Sargent *et al.*, 2011), there are numerous potential enzymatic targets that can be modulated to enhance homologous exchange.

There are a number of gene and cell systems that have been used to phenotypically assay the efficiency of homologous exchange. These include HPRT1 (Bedayat *et al.*, 2010; Doetschman *et al.*, 1987; Hendrie *et al.*, 2003; Hunger-Bertling *et al.*, 1990; Kenner *et al.*, 2002), green fluorescent protein (GFP) (Kamiya *et al.*, 2008; Olsen *et al.*, 2005b; Radecke *et al.*, 2004; Thorpe *et al.*, 2002; Vasileva *et al.*, 2006), neomycin resistance (G418R) (Campbell *et al.*, 1989; Song *et al.*, 1985), zeocin resistance (ZeoR) (Colosimo *et al.*, 2001) and β-galactosidase (β-gal) (Nickerson and Colledge, 2003). While the reporter/selectable marker gene systems used are cDNA-based and do not have the same intron/exon structure as mammalian genes, the HPRT1 gene is in its natural genomic context and may more accurately reflect homologous exchange at other genomic loci.

Some studies have sought to gain insight into the mechanisms underlying OPN-mediated homologous exchange and have used cell systems with a genetic defect in an element of the DNA repair or replication pathways. However, because of the complexity and the redundancy in DNA repair and replication pathways, it is difficult to determine which pathways predominate for any given OPN gene-targeting strategy. As a result, the modulation of gene-targeting efficiency and its enhancement is more challenging in its implementation.

7.5 Targeted Cleavage of Genomic DNA

One approach for enhancing the efficiency of homologous exchange has been the introduction of DSBs (Golding *et al.*, 2004; Nickoloff and Brenneman, 2004; Thompson and Schild, 2002; Zhang *et al.*, 2007). Cells have the ability to repair DSBs and mitigate the potential cytotoxic and mutagenic effects through NHEJ or

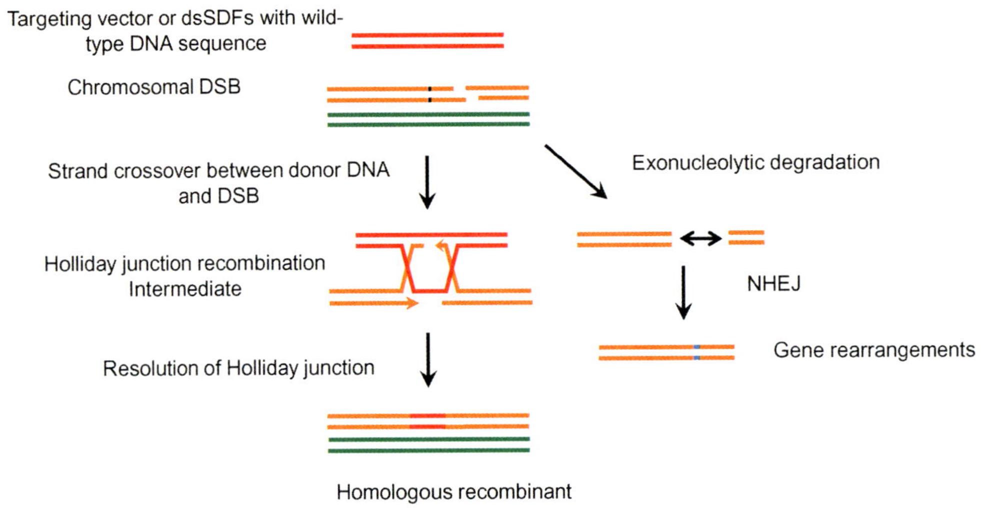

FIGURE 7.4 ■ DSB repair enhancement of gene editing. Chromosomal DNA DSBs can be induced by sequence-specific endonucleases and dramatically stimulate HR between donor DNA and the chromosomal target. The cellular DSB repair enzymes use the homologous donor DNA (e.g., SSOs, SDFs, TFOs, recombinant viruses, classical plasmid targeting vectors) to repair DSBs by a pathway similar to that shown in Fig. 7.1. Chromosomal breaks can also be repaired by NHEJ, a process that may result in mutations (e.g., deletions, insertions, translocations, etc.). Donor DNA used in DSB enhancement of HR may contain positive and negative drug-selectable markers, as indicated in Fig. 7.7.

HR (Jeggo, 1998; Johnson and Jasin, 2001) (Fig. 7.4). While both NHEJ and HR pathways are invoked by DSBs, it is the HR pathway that is thought to be error-free and therefore a prime candidate for therapeutic enhancement.

There have been a number of strategies developed for generating sequence-specific DBSs; however, it is rare-cutting, sequence-specific endonucleases that appear to have the best potential for mitigating random DSBs throughout the genome (Choulika *et al.*, 1995; Elliott *et al.*, 1998; Grizot *et al.*, 2010; Johnson and Jasin, 2001; Rouet *et al.*, 1994; Smih *et al.*, 1995). Homing/meganucleases (Cabaniols and Paques, 2008; Chevalier *et al.*, 2002; Grizot *et al.*, 2010; Smith *et al.*, 2006; Stoddard, 2005) (Fig. 7.5) and chimeric endonucleases in which the *Flavobacterium okeanokoites* I (FokI) nuclease is linked to sequence-specific zinc-finger binding (Bibikova *et al.*, 2003; Kim and Chandrasegaran, 1994; Porteus and Baltimore, 2003) (Fig. 7.6) or transcription activator-like effector (TALE) (Cermak *et al.*, 2011; Christian *et al.*, 2010; Hockemeyer *et al.*, 2011; Miller *et al.*, 2011; Romer *et al.*, 2009) (Fig. 7.6) DNA binding motifs. The chimeric zinc-finger nuclease (ZFNs) and TALE nucleases (TALENs) can then be directed to produce DSBs at specific target sequences (Fig. 7.4).

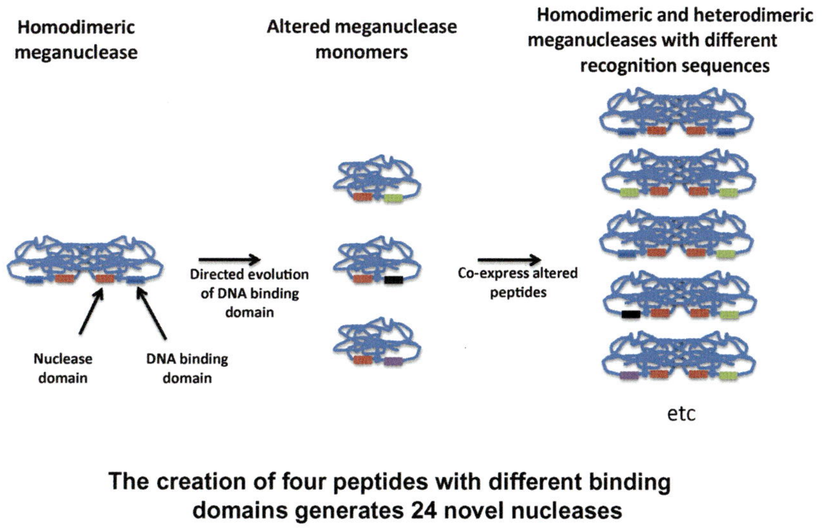

FIGURE 7.5 ■ Structure and evolution of meganucleases: The sequence-specific mutation of polypeptide motifs involved in DNA binding and specification of the meganuclease recognition sequence can generate new enzyme monomers. Co-expression of novel peptide monomers *in vivo* and their dimerization into an active enzyme generates meganucleases that can recognize new DNA recognition sequences to cleave at specific target sites.

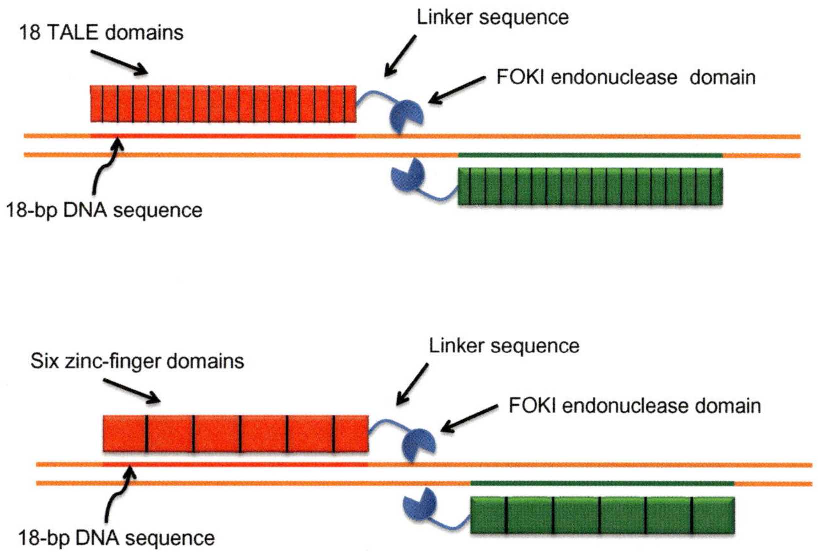

FIGURE 7.6 ■ ZFNs and TALENs: ZFNs and TALENs are designed for sequence-specific cleavage of chromosomal target DNA. These proteins have a sequence recognition domain (red and green) coupled by a linker peptide sequence (blue line) to a FokI nuclease domain that is active only as a homodimer. When the FokI peptides are juxtaposed to dimerize, DNA cleavage will occur.

7.5.1 Meganucleases/Homing Endonucleases

The homing endonucleases or meganucleases (MNs) are based on rare-cutting endonucleases that recognize specific DNA sequences and then bind genomic DNA and cut it to produce a DSB (Figs. 7.4 and 7.5). The most-studied rare-cutting MN, mitochondrial intron *Saccharomyces cerevisiae* I (I-SceI), is derived from yeast and has an 18-bp recognition site not found in mammalian cells. The recognition site must therefore be introduced into a plasmid or genomic target region. Studies in mammalian cells using the I-SceI enzyme/18-bp recognition sequence system demonstrated that DSBs, whether they were introduced into an episomal or genomic target, enhance HR (Choulika *et al.*, 1995; Cohen-Tannoudji *et al.*, 1998; Johnson and Jasin, 2001; Liang *et al.*, 1998; Nickoloff and Brenneman., 2004; Radecke *et al.*, 2006a; Rouet *et al.*, 1994; Sargent *et al.*, 1997). Unfortunately, there are no I-SceI recognition sequences in unmodified human DNA, making this system impractical for therapeutic enhancement.

Another DSB-generating gene-targeting system, based on the intronic *Chlamydomonas reinhardtii* I (I-CreI) MN, has been shown to be more amenable than the I-SceI to protein engineering and in the introduction of DSBs at specific genomic loci in mammalian cells (Chevalier *et al.*, 2002; Fonfara *et al.*, 2012; Stoddard, 2011). The I-CreI MN functions as a homodimeric protein with a 22-bp pseudo-palindromic recognition sequence (Heath *et al.*, 1997) and is a member of the LAGLIDAG family of MNs. All members of this family have the conserved LADLIDAG amino acid motif found at the homodimer protein–protein interface that contributes amino acid residues in order to facilitate the enzyme cleavage. Crystal structure analysis showed that many of the LADLIDAG family of MNs have a compact modular structure. Their recognition sequences have been identified and the DNA–amino acid interactions have been mapped, thereby providing information essential for directed gene engineering with which to create MNs with novel recognition sequences (Grizot *et al.*, 2010). The new MNs have already been used *in vivo* to correct genes at specific chromosomal targets (Cabaniols *et al.*, 2010; Chapdelaine *et al.*, 2010; Munoz *et al.*, 2011). These studies suggest that sequence-specific I-CreI MNs can be engineered to both induce DSBs in the genomic DNA, and to facilitate homologous exchange at specific disease loci (Fig. 7.5).

7.5.2 Zinc-Finger Nucleases

Chimeric ZFNs complement the homing MNs in their ability to generate sequence-specific DSBs (Carroll *et al.*, 2006; Porteus and Carroll, 2005; Wu *et al.*, 2007). ZFNs are comprised of a customized sequence-specific ZF domain linked to a non-specific FokI nuclease for cleavage at a ZF-defined sequence domain (Fig. 7.6). These chimeric nucleases have been successfully used in conjunction with a plasmid that carries the donor DNA to enhance HR in Drosophila (Carroll, 2008) and human cells (Alwin *et al.*, 2005; Lombardo *et al.*, 2007; Mittelman *et al.*, 2009; Porteus and Baltimore, 2003; Urnov *et al.*, 2005; Zou *et al.*, 2009).

ZFN/donor plasmid studies showed correction of a single-base frameshift mutation in the IL2RG gene of K562 human erythroleukemia cells and expression of wild-type IL2RG mRNA and protein (Urnov *et al.*, 2005). Studies with ZFNs and SSO (Olsen *et al.*, 2009) or SDF (H Parsi and DC Gruenert, unpublished data) donor DNA have shown enhanced correction of mutant enhanced green fluorescent protein (EGFP or eGFP) in HEK293 cells. One recent study also showed high levels of gene targeting in the adeno-associated virus integration site 1 (AAVS1) locus and in the ribosomal S6 kinase 2 (RSK2) gene that is associated with Coffin–Lowry syndrome, a disease that results in mental retardation, psychomotor and developmental abnormalities and cancer (Chen *et al.*, 2011).

7.5.3 Transcription Activator-Like Effector Nucleases (TALENs)

TALE motifs were originally described for a family of genes associated with the genus *Xanthomonas* bacteria (Gonzalez *et al.*, 2007; Romer *et al.*, 2009; Sugio *et al.*, 2007). The *Xanthomonas* TALE mimic host cell transcription factors and are the primary bacterial virulence elements that alter the cells and interfere with development.

TALEs are comprised of a nuclear localization signal (NLS), a central region of tandem repeats coding for 34–35 amino acids and a transcriptional activation domain (Boch and Bonas, 2010). It is the sequence and number of the tandem repeats and in particular, the 12th and 13th amino acids of the repeats that determine TALE sequence-selectivity and the genes whose transcription is modulated (Morbitzer *et al.*, 2010; Romer *et al.*, 2010; Scholze and Boch, 2011). Therefore, the sequence-specificity of DNA binding is effectively determined by the 12th and 13th amino acids of the tandem repeat elements (Boch *et al.*, 2009; Rusk, 2011). Given the TALE DNA binding specificity, it is an obvious extrapolation to envision how the TALEs might be further developed to generate sequence-specific targeting nucleases (TALENs) that are similar to the ZFNs (Cermak *et al.*, 2011; Christian *et al.*, 2010; Wood *et al.*, 2011) (Fig. 7.6). The advantage of the TALEN system is that it is technically more accessible than the ZFNs in that the code for the TALENs is relatively straightforward, making it easier to generate sequence-specific motifs. Thus, the technical challenges associated with TALEN synthesis and the apparent greater offsite cutting by MNs and ZFNs, makes them a more likely candidate for therapeutic application (Carroll, 2008; Clark *et al.*, 2011; Cradick *et al.*, 2011; Fonfara *et al.*, 2012; Mussolino *et al.*, 2011; Radecke *et al.*, 2010; Sargent *et al.*, 2011).

7.5.4 Limitations

While endonuclease systems have a distinct appeal because of their ability to enhance homologous exchange, they also have their limitations. The potential for eliciting an immune response to the foreign ZFN protein is clearly worth evaluating *in vivo* and *ex vivo*. In addition, the possibility of offsite cutting by the ZFNs will increase the cytotoxic and mutagenic potential of these nuclease systems (Carroll, 2008; Mussolino *et al.*, 2011; Porteus and Carroll, 2005; Radecke *et al.*, 2010). Moreover, if offsite cutting were excessive, particularly in the modification of pluripotent stem cells, it would undermine karyotypic and genetic integrity and would likely compromise the viability and genomic stability of the target cells. These are potential limitations that could have a profound effect on the therapeutic potential of any approach that relies on the introduction of DSBs. However, these limitations can be addressed prior to clinical implementation by evaluating immunogenicity and by measuring the generation of offsite DSBs in cell systems directly relevant to the cells requiring therapeutic intervention and not surrogate cell lines. One preclinical validation, whole-genome sequencing, will reveal any offsite mutations due to ZFN, TALEN or MN activity and will likely be required to assay genomic DNA of therapeutic cells that have been modified by gene editing.

7.6 Recombinant-Virus-Mediated Gene Targeting

Other strategies facilitating sequence-specific homologous exchange have relied on recombinant viruses to deliver the therapeutic DNA homologue. Virus-based DNA delivery systems present distinct advantages for gene targeting when compared with physical (electroporation, microinjection, etc.) or chemical (liposome, polyamidoamine dendrimer, polyethylenimine, etc.) transfection of DNA (Colosimo *et al.*, 2000). While microinjection can deliver the DNA directly to the nucleus, it is time consuming, technically challenging and can be tedious. On the other hand, electroporation and chemical delivery requires the DNA to traverse the cytoplasm before entering the nucleus. Consequently, exposure to nucleases, ligases and other proteins with an affinity for DNA can disrupt the progress of the modifying DNA to its genomic target. With a few exceptions, electroporation and chemical transfection are typically associated with enhanced cell death and often require additional culture manipulations that can cause further cell loss.

Viruses in general, AAVs and adenoviruses (AdVs) more specifically, have evolved to infect/transduce cells in order to deliver their DNA to the nucleus in a protected virus capsid. Thus, they have a significant advantage over electroporation and chemical methods in that they can deliver an intact nucleic acid payload directly to the cell nucleus. The AAV and the AdV gene delivery systems have a high efficiency of DNA delivery with little apparent cytotoxicity. In addition, they avoid DNA degradation and catenation often associated with electroporation and chemically-based DNA delivery (Colosimo *et al.*, 2000). For all practical purposes, viruses can be considered as nature's equivalent of DNA microinjection into the nucleus without the technical and human limitations. Both the AAV and AdV-derivative systems have recently been adapted as targeting vectors for HR-mediated sequence-specific genetic modification of chromosomal loci (Aizawa *et al.*, 2011; Li *et al.*, 2011; Liu *et al.*, 2011; Miller, 2011; Russell and Hirata, 1998).

AAV is a single-stranded DNA virus with a linear genome of approximately 4.7 kb, while AdV is a double-stranded virus with a genome of approximately 26–45 kbp (Fig. 7.7). AAVs are not known to cause disease in humans, whereas AdV is responsible for approximately 20% of respiratory infections in children and adults (Ahmed *et al.*, 2012). Each virus has several serotypes that define

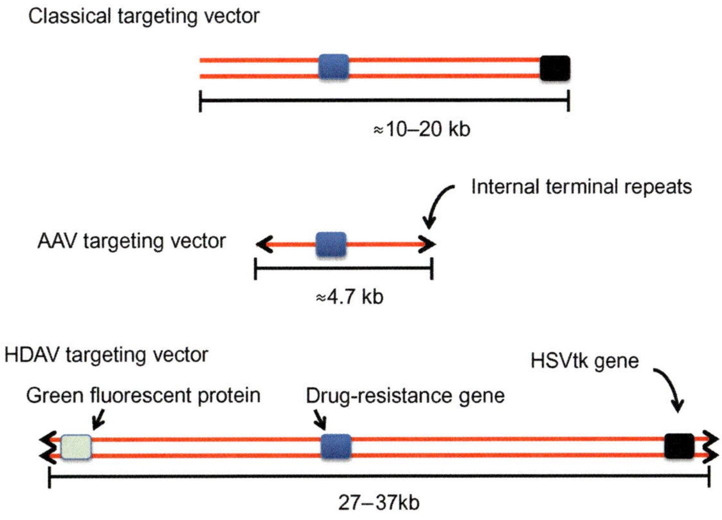

FIGURE 7.7 ■ Structure of classical, AAV and AdV-derived HDAV vectors: Classical gene-targeting vectors typically contain between 10 and 20 kbp dsDNA homologous to the target, and may contain a positive drug-selectable marker (neomycin, hygromycin, puromycin, etc. in blue) gene, and a negative drug-selectable marker, the herpes simplex virus thymidine kinase (HSVtk) gene (black). Not shown is the plasmid backbone associated with the targeting vector. AAV targeting vectors are ssDNA molecules, <4.7 kbp and may contain a drug-selectable marker. Internal terminal repeat sequences (black arrowheads) are required for DNA packaging and replication. HDAV targeting vectors are dsDNA between 27 and 37 kbp in size. In addition to almost 30 kbp of DNA sequence homologous to the chromosomal target, HDAV vectors may also contain positive drug-selectable markers (blue) and a negative selectable marker (black), in addition to a reporter gene to monitor the number of infected cells (green).

cell- and tissue-specificity, and recombinant vectors derived from AAV and AdV have been used to infect various human cells and tissues *in vivo* and *ex vivo*. In recombinant AAV and AdV-derived HDAV gene-targeting vectors, the genes encoding the viral genome are replaced with donor DNA that is homologous to the chromosomal target to be modified, such that the only viral DNA sequences remaining in the recombinant vector are those required for viral replication and packaging into the virus capsids. Hence, these vectors are sometimes called "gutted" viruses. The proteins required for viral replication and capsid production are provided by a second, "helper" virus vector that is co-transfected with the recombinant viral vector into the host, human cell line (e.g., HEK293) for the generation of "packaged" recombinant virus (Fig. 7.7).

7.6.1 AAV

Due to their small genome size, recombinant AAV vectors are limited to donor fragments of approximately 3–4.5 kb. Thus, the engineering of AAV vectors becomes more challenging because selectable marker genes, used to isolate cell lines that have undergone homologous exchange, are typically 0.5–1.5 kb of the recombinant virus gene-targeting vector genome, leaving only 2.5–3 kb for the homologous donor DNA fragment. The frequency of HR with classical gene-targeting vectors (Figs. 7.1 and 7.8) is thought to be directly related to the length of the donor fragment homology and vectors with small donor fragments would be expected to undergo recombination less frequently with target genomic sites (Hasty *et al.*, 1991).

However, despite the relatively small size of donor DNA present in AAV vectors, gene targeting is remarkably efficient. Typical experiments targeting chromosomal genes with the AAV system varies from about 1×10^{-8}/cell to almost 1% at very high multiplicities of infection (Khan *et al.*, 2011). Gene-targeting efficiencies appear to be position-dependent and vary from gene to gene (Cornea and Russell, 2010). However, there does not seem to be a consistent trend in efficiency based on cell type or the targeted modification (Hendrie and Russell, 2005). Gene-targeting efficiencies do appear, however, to depend on the amount of donor DNA homology present in the AAV targeting vector. AAV vectors have been used to correct and introduce mutations including point mutations, insertions and deletions (Hendrie and Russell, 2005; Khan *et al.*, 2011).

AAV vectors have also been successfully used to target chromosomal genes in human embryonic stem (hES) and human iPS (hiPS) cells, both with and without ZFN-induced DSBs (Asuri *et al.*, 2011; Khan *et al.*, 2011). In the presence of ZFNs, AAV gene targeting was enhanced approximately ten-fold to efficiencies of about 1% of treated cells (Asuri *et al.*, 2011). This efficiency of gene targeting would make it possible to screen individual colonies and identify recombinant clones without drug selection, and would avoid the need to include selectable markers with flanking sequence-specific recognition sequences (e.g., locus of X-over P1 (loxP) or flippase recognition target (FRT)) for the eventual deletion of the marker.

7.6.2 HDAV

Unlike the AAV vectors, HDAV gene-targeting vectors can accept donor DNA fragments in the range of 25–35 kbp. Similar to AAV vectors, the HDAV vectors have the potential of circumventing one of the issues associated with endonuclease-enhanced gene targeting by minimizing the potential for non-specific disruption of the genomic DNA. The HDAV system relies on the more classical positive–negative selection markers (Mansour *et al.*, 1988). G418 (neomycin resistance, G418R) is used as the positive selection agent for cells that have incorporated the recombinant targeting vector into their genomic DNA and gancyclovir is used for negative selection (through loss of the HSVtk gene), in order to enrich for homologous recombinants that are gancyclovir-resistant (GANCR) (Figs. 7.7 and 7.8).

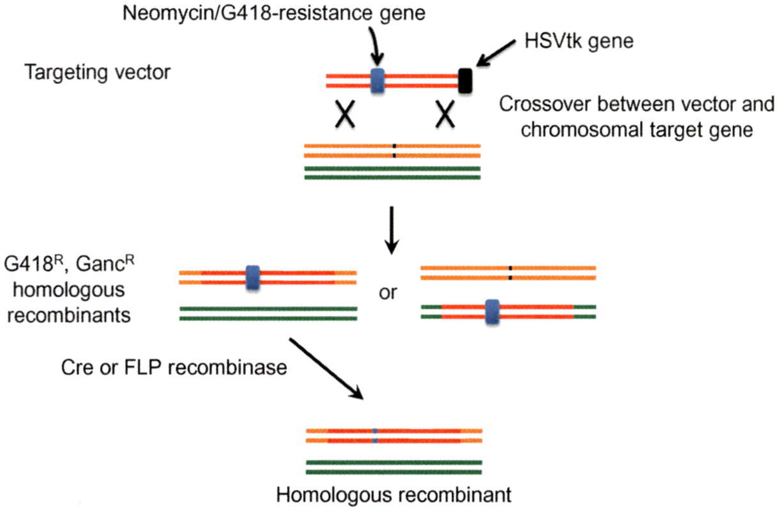

FIGURE 7.8 ■ Drug selection strategies used for gene-targeted cell lines: Positive–negative drug selection is performed by selecting for cells that have integrated the AAV, HDAV or classical targeting vector into genomic DNA (e.g., with G418), and then selecting for cells that have lost the negative drug-selectable marker (e.g., HSVtk) due to recombination or damage. Cells that survive in the presence of G418 and gancyclovir (Ganc) are presumptive homologous recombinants and are screened by PCR or Southern analysis for the predicted recombinant gene structure. The residual positive drug-selectable markers are often flanked by sequence-specific locus of X-over P1 (loxP) or FRT recombinase recognition sequences, and excised by intracellular expression of Cre or flippase (Flp) recombinase, leaving behind a loxP or FRT DNA sequence footprint.

HDAV vectors have been used in a number of cell systems; however, the recent application to pluripotent stem cells has been very intriguing and exciting (Gruenert and Sargent., 2012). In one study, over 106 hES or hiPS cells were infected with individual HDAV vectors targeting five different native genes (HPRT, DNA ligase 1 (LIG1), LIG3, Ku antigen 80 kDA subunit (KU80), and homeobox 9 (HB9)) (Aizawa *et al.*, 2011). The gene-targeting efficiency ranged from 5.6×10^{-5} to 2×10^{-7} homologous recombinants per cell, with relative HR efficiencies per G418R/GANCR enrichment ranging from 7% to a remarkable 81% without apparent bias for the gene, hES/hiPS cell line or transcriptional status. While no random integration of the HDAV vectors was detected in cells that had undergone HR, rare recombinants with unexplained gene structures were detected by Southern analysis. Although they were unexplained by the authors, these structures resemble alternative recombination products described in other systems. Similar HDAV gene editing results were also reported by other groups at the untranscribed laminin A (LMNA) gene locus in hES and hiPS cells (Liu *et al.*, 2011) and the human β-globin gene (Li *et al.*, 2011). In all the studies, genetically modified cell lines maintained the expression of markers diagnostic for hES/hiPS cells, retained a normal karyotype and were capable of differentiation.

7.6.3 Limitations

As is the case with most nucleic acid delivery and gene modification systems there are limitations inherent to AAV and HDAV gene-targeting strategies. *In vivo* administration of the gene-targeting virus could elicit an immune response to HDAV capsid proteins; however, this possibility is less likely in *ex vivo* gene modification. It appears that the residual capsid proteins are lost over several cell divisions. Even though random integrants were not detected in these studies, random integration of intact targeting vectors is always a concern. Ultimately, cells used for human therapies will require exhaustive characterization for potential ectopic vector fragment integration.

Another factor, currently required for clinically destined gene-corrected cells generated by the HDAV system, is the removal of the positive drug-selectable maker (e.g., neomycin) gene. In the studies described above the marker gene was flanked by loxP or FRT sequences that required intracellular expression of the Cre or FLP recombinase, respectively, for the excision of the gene. Whether the short loxP or FRT DNA "footprint" left behind will have any detectable biological activity that could compromise targeted cell lines for human therapeutics, remains to be determined (Fig. 7.8).

7.7 Conclusion

Overall, the systems described here have opened a new chapter in the quest for developing comprehensive cell and gene therapy. The effective sequence-specific modification and editing of the genomic DNA is the future of clinical medicine. While their therapeutic potential remains to be tested and the systems need to be refined, the OPN, AAV and HDAV homologous exchange systems as well as the MN, ZFN and TALEN nuclease enhancements of homologous exchange have established an exciting new element of the gene and cell therapy toolbox. The potential of gene-targeting sequence-specific modification of genomic DNA for the treatment of inherited diseases, envisioned over 20 years ago (Capecchi, 1989; Friedmann and Roblin., 1972; Smithies *et al.*, 1985), may finally be realized.

Acknowledgments

The authors would like to thank all those in the gene and cell therapy fields whose scientific efforts and rigor have contributed to and inspired this chapter. We would also like to acknowledge support from National Institutes of Health (NIH) grants P01DK88760, R01GM75111, R01 GM 075111 — 04 S1, Cystic Fibrosis Research, Inc and Pennsylvania Cystic Fibrosis, Inc.

References

Aarts, M., te Riele, H. (2010). Parameters of oligonucleotide-mediated gene modification in mouse ES cells, *J Cell Mol Med*, **14**, 1657–1667.

Ahmed, J.A., Katz, M.A., Auko, E., *et al.* (2012). Epidemiology of respiratory viral infections in two long-term refugee camps in Kenya, 2007–2010, *BMC Infect Dis*, **12**, 7.

Aizawa, E., Hirabayashi, Y., Iwanaga, Y., *et al.* (2011). Efficient and accurate homologous recombination in hESCs and hiPSCs using helper-dependent adenoviral vectors, *Mol Ther*, **20**, 424–431.

Alexeev, V., Igoucheva, O., Yoon, K. (2002). Simultaneous targeted alteration of the tyrosinase and c-kit genes by single-stranded oligonucleotides, *Gene Ther*, **9**, 1667–1675.

Alwin, S., Gere, M.B., Guhl, E., *et al.* (2005). Custom zinc-finger nucleases for use in human cells, *Mol Ther*, **12**, 610–617.

Andrieu-Soler, C., Halhal, M., Boatright, J.H., *et al.* (2007). Single-stranded oligonucleotide-mediated *in vivo* gene repair in the *rd1* retina, *Mol Vis*, **13**, 692–706.

Asuri, P., Bartel, M.A., Vazin, T., *et al.* (2011). Directed evolution of adeno-associated virus for enhanced gene delivery and gene targeting in human pluripotent stem cells, *Mol Ther*, **20**, 329–338.

Bedayat, B., Abdolmohamadi, A., Ye, L., *et al.* (2010). Sequence-specific correction of genomic hypoxanthine-guanine phosphoribosyl transferase mutations in lymphoblasts by small fragment homologous replacement, *Oligonucleotides*, **20**, 7–16.

Bibikova, M., Beumer, K., Trautman, J.K., *et al.* (2003). Enhancing gene targeting with designed zinc finger nucleases, *Science*, **300**, 764.

Boch, J., Bonas, U. (2010). *Xanthomonas* AvrBs3 family-type III effectors: discovery and function, *Annu Rev Phytopathol*, **48**, 419–436.

Boch, J., Scholze, H., Schornack, S., *et al.* (2009). Breaking the code of DNA binding specificity of TAL-type III effectors, *Science*, **326**, 1509–1512.

Brachman, E.E., Kmiec, E.B. (2005). Gene repair in mammalian cells is stimulated by the elongation of S phase and transient stalling of replication forks, *DNA Repair* (*Amst*), **4**, 445–457.

Buckner, C.D., Epstein, R.B., Rudolph, R.H., *et al.* (1970). Allogeneic marrow engraftment following whole body irradiation in a patient with leukemia, *Blood*, **35**, 741–750.

Bushman, F.D. (2007). Retroviral integration and human gene therapy, *J Clin Invest*, **117**, 2083–2086.

Cabaniols, J.P., Paques, F. (2008). Robust cell line development using meganucleases, *Methods Mol Biol*, **435**, 31–45.

Cabaniols, J.P., Ouvry, C., Lamamy, V., *et al.* (2010). Meganuclease-driven targeted integration in CHO-K1 cells for the fast generation of HTS-compatible cell-based assays, *J Biomol Screen*, **15**, 956–967.

Campbell, C.R., Keown, W., Lowe, L., *et al.* (1989). Homologous recombination involving small single-stranded oligonucleotides in human cells, *New Biol*, **1**, 223–227.

Capecchi, M.R. (1989). Altering the genome by homologous recombination, *Science*, **244**, 1288–1292.

Capecchi, M.R. (1994). Targeted gene replacement, *Scientific American*, March, 52–59.

Capecchi, M.R. (2000). How close are we to implementing gene targeting in animals other than the mouse? *Proc Natl Acad Sci USA*, **97**, 956–957.

Carroll, D. (2008). Progress and prospects: zinc-finger nucleases as gene therapy agents, *Gene Ther*, **15**, 1463–1468.

Carroll, D., Morton, J.J., Beumer, K.J., *et al.* (2006). Design, construction and *in vitro* testing of zinc finger nucleases, *Nat Protoc*, **1**, 1329–1341.

Cattoglio, C., Facchini, G., Sartori, D., *et al.* (2007). Hot spots of retroviral integration in human $CD34^+$ hematopoietic cells, *Blood*, **110**, 1770–1778.

Cavazzana-Calvo, M., Fischer, A. (2007). Gene therapy for severe combined immunodeficiency: Are we there yet? *J Clin Invest*, **117**, 1456–1465.

Cavazzana-Calvo, M., Hacein-Bey, S., Yates, F., *et al.* (2001). Gene therapy of severe combined immunodeficiencies, *J Gene Med*, **3**, 201–206.

Cereseto, A., Giacca, M. (2004). Integration site selection by retroviruses, *AIDS Rev*, **6**, 13–21.

Cermak, T., Doyle, E.L., Christian, M., *et al.* (2011). Efficient design and assembly of custom TALEN and other TAL effector-based constructs for DNA targeting, *Nucleic Acids Res*, **39**, e82.

Chan, P.P., Lin, M., Faruqi, A.F., *et al.* (1999). Targeted correction of an episomal gene in mammalian cells by a short DNA fragment tethered to a triplex-forming oligonucleotide, *J Biol Chem*, **274**, 11541–11548.

Chapdelaine, P., Pichavant, C., Rousseau, J., *et al.* (2010). Meganucleases can restore the reading frame of a mutated dystrophin, *Gene Ther*, **17**, 846–858.

Chen, F., Pruett-Miller, S.M., Huang, Y., *et al.* (2011). High-frequency genome editing using ssDNA oligonucleotides with zinc-finger nucleases, *Nat Methods*, **8**, 753–755.

Chevalier, B.S., Kortemme, T., Chadsey, M.S., *et al.* (2002). Design, activity, and structure of a highly specific artificial endonuclease, *Mol Cell*, **10**, 895–905.

Chin, J.Y., Glazer, P.M. (2009). Repair of DNA lesions associated with triplex-forming oligonucleotides, *Mol Carcinog*, **48**, 389–399.

Choulika, A., Perrin, A., Dujon, B., *et al.* (1995). Induction of homologous recombination in mammalian chromosomes by using the I-*Sce*I system of *Saccharomyces cerevisiae*, *Mol Cell Biol*, **15**, 1968–1973.

Christian, M., Cermak, T., Doyle, E.L., *et al.* (2010). Targeting DNA double-strand breaks with TAL effector nucleases, *Genetics*, **186**, 757–761.

Clark, K.J., Voytas, D.F., Ekker, S.C. (2011). A TALE of two nucleases: Gene targeting for the masses? *Zebrafish*, **8**, 147–149.

Cohen-Tannoudji, M., Robine, S., Choulika, A., *et al.* (1998). I-*Sce*I-induced gene replacement at a natural locus in embryonic stem cells, *Mol Cell Biol*, **18**, 1444–1448.

Colosimo, A., Goncz, K.K., Holmes, A.R., *et al.* (2000). Transfer and expression of foreign genes in mammalian cells, *Biotechniques*, **29**, 314–318, 320–322, 324 passim.

Colosimo, A., Goncz, K.K., Novelli, G., *et al.* (2001). Targeted correction of a defective selectable marker gene in human epithelial cells by small DNA fragments, *Mol Ther*, **3**, 178–185.

Colosimo, A., Guida, V., Antonucci, I., *et al.* (2007). Sequence-specific modification of a β-thalassemia locus by small DNA fragments in human erythroid progenitor cells, *Haematologica*, **92**, 129–130.

Cornea, A.M., Russell, D.W. (2010). Chromosomal position effects on AAV-mediated gene targeting, *Nucleic Acids Res*, **38**, 3582–3594.

Cradick, T.J., Ambrosini, G., Iseli, C., *et al.* (2011). ZFN-site searches genomes for zinc finger nuclease target sites and off-target sites, *BMC Bioinformatics*, **12**, 152.

Datta, H.J., Glazer, P.M. (2001). Intracellular generation of single-stranded DNA for chromosomal triplex formation and induced recombination, *Nucleic Acids Res*, **29**, 5140–5147.

Dekker, M., Brouwers, C., te Riele, H. (2003). Targeted gene modification in mismatch-repair-deficient embryonic stem cells by single-stranded DNA oligonucleotides, *Nucleic Acids Res*, **31**, e27.

Doetschman, T., Gregg, R.G., Maeda, N., *et al.* (1987). Targetted correction of a mutant HPRT gene in mouse embryonic stem cells, *Nature*, **330**, 576–578.

Elliott, B., Richardson, C., Winderbaum, J., *et al.* (1998). Gene conversion tracts from double-strand break repair in mammalian cells, *Mol Cell Biol*, **18**, 93–101.

Faruqi, A.F., Datta, H.J., Carroll, D., *et al.* (2000). Triple-helix formation induces recombination in mammalian cells via a nucleotide excision repair-dependent pathway, *Mol Cell Biol*, **20**, 990–1000.

Felsenfeld, G., Rich, A. (1957). Studies on the formation of two- and three-stranded polyribonucleotides, *Biochim Biophys Acta*, **26**, 457–468.

Fischer, A., Hacein-Bey-Abina, S., Cavazzana-Calvo, M. (2011). Gene therapy for primary adaptive immune deficiencies, *J Allergy Clin Immunol*, **127**, 1356–1359.

Fonfara, I., Curth, U., Pingoud, A., *et al.* (2012). Creating highly specific nucleases by fusion of active restriction endonucleases and catalytically inactive homing endonucleases, *Nucleic Acids Res*, **40**(2), 847–860. doi: 10.1093/nar/gkr788. Epub 2011 Sep 29. PMID: 21965534.

Friedmann, T., Roblin, R. (1972). Gene therapy for human genetic disease? *Science*, **175**, 949–955.

Gamper, H.B., Jr., Gewirtz, A., Edwards, J., *et al.* (2004). Modified bases in RNA reduce secondary structure and enhance hybridization, *Biochemistry*, **43**, 10224–10236.

Giovannangeli, C., Helene, C. (1997). Progress in developments of triplex-based strategies, *Antisense Nucleic Acid Drug Dev*, **7**, 413–421.

Golding, S.E., Rosenberg, E., Khalil, A., *et al.* (2004). Double strand break repair by homologous recombination is regulated by cell cycle-independent signaling via ATM in human glioma cells, *J Biol Chem*, **279**, 15402–15410.

Goncz, K.K., Colosimo, A., Dallapiccola, B., *et al.* (2001). Expression of ΔF508 CFTR in normal mouse lung after site-specific modification of CFTR sequences by SFHR, *Gene Ther*, **8**, 961–965.

Goncz, K.K., Kunzelmann, K., Xu, Z., *et al.* (1998). Targeted replacement of normal and mutant CFTR sequences in human airway epithelial cells using DNA fragments, *Hum Mol Genet*, **7**, 1913–1919.

Goncz, K.K., Prokopishyn, N.L., Abdolmohammadi, A., *et al.* (2006). Small fragment homologous replacement-mediated modification of genomic β-globin sequences in human hematopoietic stem/progenitor cells, *Oligonucleotides*, **16**, 213–224.

Gonzalez, C., Szurek, B., Manceau, C., *et al.* (2007). Molecular and pathotypic characterization of new *Xanthomonas oryzae* strains from West Africa, *Mol Plant Microbe Interact*, **20**, 534–546.

Grizot, S., Epinat, J.C., Thomas, S., *et al.* (2010). Generation of redesigned homing endonucleases comprising DNA-binding domains derived from two different scaffolds, *Nucleic Acids Res*, **38**, 2006–2018.

Gruenert, D.C. (1998). Gene correction with small DNA fragments, *Curr Res Mol Ther*, **1**, 607–613.

Gruenert, D.C. (2003). Genomic medicine: development of DNA as a therapeutic drug for sequence-specific modification of genomic DNA, *Discovery Medicine*, **3**, 58–60.

Gruenert, D.C., Cleaver, JE. (1985). Repair of psoralen-induced cross-links and monoadducts in normal and repair-deficient human fibroblasts, *Cancer Res*, **45**, 5399–5404.

Gruenert, D.C., Sargent, RG. (2012). Virus-mediated genetic surgery: homologous recombination with a little "helper" from my friends, *Mol Ther Nucleic Acids*, **1**, e2.

Gruenert, D.C., Bruscia, E., Novelli, G., *et al.* (2003). Sequence-specific modification of genomic DNA by small DNA fragments, *J Clin Invest*, **112**, 637–641.

Hackett, C.S., Geurts, AM., Hackett, PB. (2007). Predicting preferential DNA vector insertion sites: implications for functional genomics and gene therapy, *Genome Biol*, **8 Suppl 1**, S12.

Hasty, P., Rivera-Perez, J., Bradley, A. (1991). The length of homology required for gene targeting in embryonic stem cells, *Mol Cell Biol*, **11**, 5586–5591.

Heath, P.J., Stephens, K.M., Monnat, R.J., Jr., *et al.* (1997). The structure of I-Crel, a group I intron-encoded homing endonuclease, *Nat Struct Biol*, **4**, 468–476.

Hegele, H., Wuepping, M., Ref, C., *et al.* (2008). Simultaneous targeted exchange of two nucleotides by single-stranded oligonucleotides clusters within a region of about fourteen nucleotides, *BMC Mol Biol*, **9**, 14.

Hendrie, P.C., Russell, D.W. (2005). Gene targeting with viral vectors, *Mol Ther*, **12**, 9–17.

Hendrie, P.C., Hirata, R.K., Russell, D.W. (2003). Chromosomal integration and homologous gene targeting by replication-incompetent vectors based on the autonomous parvovirus minute virus of mice, *J Virol*, **77**, 13136–13145.

Hockemeyer, D., Wang, H., Kiani, S., *et al.* (2011). Genetic engineering of human pluripotent cells using TALE nucleases, *Nat Biotechnol*, **29**, 731–734.
Hunger-Bertling, K., Harrer, P., Bertling, W. (1990). Short DNA fragments induce site specific recombination in mammalian cells, *Mol Cell Biochem*, **92**, 107–116.
Igoucheva, O., Alexeev, V., Yoon, K. (2001). Targeted gene correction by small single-stranded oligonucleotides in mammalian cells, *Gene Ther*, **8**, 391–399.
Igoucheva, O., Alexeev, V., Yoon, K. (2006). Differential cellular responses to exogenous DNA in mammalian cells and its effect on oligonucleotide-directed gene modification, *Gene Ther*, **13**, 266–275.
Igoucheva, O., Alexeev, V., Anni, H., *et al.* (2008). Oligonucleotide-mediated gene targeting in human hepatocytes: implications of mismatch repair, *Oligonucleotides*, **18**, 111–122.
Igoucheva, O., Alexeev, V., Pryce, M., *et al.* (2003). Transcription affects formation and processing of intermediates in oligonucleotide-mediated gene alteration, *Nucleic Acids Res*, **31**, 2659–2670.
Inoue, N., Hirata, R.K., Russell, D.W. (1999). High-fidelity correction of mutations at multiple chromosomal positions by adeno-associated virus vectors, *J Virol*, **73**, 7376–7380.
Jeggo, P.A. (1998). Identification of genes involved in repair of DNA double-strand breaks in mammalian cells, *Radiat Res*, **150**, S80–S91.
Johnson, R.D., Jasin, M. (2001). Double-strand-break-induced homologous recombination in mammalian cells, *Biochem Soc Trans*, **29**, 196–201.
Kallenbach, N.R., Daniel, W.E., Jr., Kaminker, M.A. (1976). Nuclear magnetic resonance study of hydrogen-bonded ring protons in oligonucleotide helices involving classical and nonclassical base pairs, *Biochemistry*, **15**, 1218–1224.
Kamiya, H., Uchiyama, M., Nakatsu, Y., *et al.* (2008). Effects of target sequence and sense versus antisense strands on gene correction with single-stranded DNA fragments, *J Biochem*, **144**, 431–436.
Kapsa, R.M., Quigley, A.F., Lynch, G.S., *et al.* (2001). *In vivo* and *in vitro* correction of the mdx dystrophin gene nonsense mutation by short-fragment homologous replacement, *Hum Gene Ther*, **12**, 629–642.
Kapsa, R.M., Quigley, A.F., Vadolas, J., *et al.* (2002). Targeted gene correction in the mdx mouse using short DNA fragments: towards application with bone marrow-derived cells for autologous remodeling of dystrophic muscle, *Gene Ther*, **9**, 695–699.
Kenner, O., Kneisel, A., Klingler, J., *et al.* (2002). Targeted gene correction of hprt mutations by 45 base single-stranded oligonucleotides, *Biochem Biophys Res Commun*, **299**, 787–792.
Kenner, O., Lutomska, A., Speit, G., *et al.* (2004). Concurrent targeted exchange of three bases in mammalian *hprt* by oligonucleotides, *Biochem Biophys Res Commun*, **321**, 1017–1023.
Khan, I.F., Hirata, R.K., Russell, D.W. (2011). AAV-mediated gene targeting methods for human cells, *Nature Protocols*, **6**, 482–501.
Kim, H.W., Svendsen, C.N. (2011). Gene editing in stem cells hits the target, *Cell Stem Cell*, **9**, 93–94.
Kim, Y.G., Chandrasegaran, S. (1994). Chimeric restriction endonuclease, *Proc Natl Acad Sci USA*, **91**, 883–887.
Knauert, M.P., Glazer, P.M. (2001). Triplex forming oligonucleotides: sequence-specific tools for gene targeting, *Hum Mol Genet*, **10**, 2243–2251.
Kucherlapati, R., Spencer, J., Moore, P. (1985). Homologous recombination catalyzed by human cell extracts, *Mol Cell Biol*, **5**, 714–720.
Kuehn, M.R., Bradley, A., Robertson, E.J., *et al.* (1987). A potential animal model for Lesch-Nyhan syndrome through introduction of HPRT mutations into mice, *Nature*, **326**, 295–298.

Kunzelmann, K., Kathofer, S., Hipper, A., *et al.* (1996a). Culture-dependent expression of Na^+ conductances in airway epithelial cells, *Pflugers Arch*, **431**, 578–586.

Kunzelmann, K., Legendre, J.Y., Knoell, D.L., *et al.* (1996b). Gene targeting of CFTR DNA in CF epithelial cells, *Gene Ther*, **3**, 859–867.

Li, M., Suzuki, K., Qu, J., *et al.* (2011). Efficient correction of hemoglobinopathy-causing mutations by homologous recombination in integration-free patient iPSCs, *Cell Res*, **21**, 1740–1744.

Liang, F., Han, M., Romanienko, P.J., *et al.* (1998). Homology-directed repair is a major double-strand break repair pathway in mammalian cells, *Proc Natl Acad Sci USA*, **95**, 5172–5177.

Liu, C., Wang, Z., Huen, M.S., *et al.* (2009a). Cell death caused by single-stranded oligodeoxynucleotides-mediated targeted genomic sequence modification, *Oligonucleotides*, **19**, 281–286.

Liu, Y., Nairn, R.S., Vasquez, K.M. (2009b). Targeted gene conversion induced by triplex-directed psoralen interstrand crosslinks in mammalian cells, *Nucleic Acids Res*, **37**, 6378–6388.

Liu, G.H., Suzuki, K., Qu, J., *et al.* (2011). Targeted gene correction of laminopathy-associated LMNA mutations in patient-specific iPSCs, *Cell Stem Cell*, **8**, 688–694.

Lombardo, A., Genovese, P., Beausejour, C.M., *et al.* (2007). Gene editing in human stem cells using zinc finger nucleases and integrase-defective lentiviral vector delivery, *Nat Biotechnol*, **25**, 1298–1306.

Lu, I.L., Lin, C.Y., Lin, S.B., *et al.* (2003). Correction/mutation of acid α-D-glucosidase gene by modified single-stranded oligonucleotides: *in vitro* and *in vivo* studies, *Gene Ther*, **10**, 1910–1916.

Luchetti, A., Filareto, A., Sanchez, M., *et al.* (2012). Small fragment homologous replacement: evaluation of factors influencing modification efficiency in an eukaryotic assay system, *PLoS One*, **7**, e30851.

Luo, Z., Macris, M.A., Faruqi, A.F., *et al.* (2000). High-frequency intrachromosomal gene conversion induced by triplex-forming oligonucleotides microinjected into mouse cells, *Proc Natl Acad Sci USA*, **97**, 9003–9008.

Mansour, S.L., Thomas, K.R., Capecchi, M.R. (1988). Disruption of the proto-oncogene *int-2* in mouse embryo-derived stem cells: a general strategy for targeting mutations to non-selectable genes, *Nature*, **336**, 348–352.

Mathé, G., Bernard, J., Schwarzenberg, L., *et al.* (1959). Trial treatment of patients afflicted with acute leukemia in remission with total irradiation followed by homologous bone marrow transfusion, *Rev Fr Etud Clin Biol*, **4**, 675–704.

Maurisse, R., Feugeas, J.P., Biet, E., *et al.* (2002). A new method (GOREC) for directed mutagenesis and gene repair by homologous recombination, *Gene Ther*, **9**, 703–707.

McNab, G.L., Ahmad, A., Mistry, D., *et al.* (2007). Modification of gene expression and increase in α1-antitrypsin (α1-AT) secretion after homologous recombination in α1-AT-deficient monocytes, *Hum Gene Ther*, **18**, 1171–1177.

McNeer, N.A., Chin, J.Y., Schleifman, E.B., *et al.* (2011). Nanoparticles deliver triplex-forming PNAs for site-specific genomic recombination in $CD34^+$ human hematopoietic progenitors, *Mol Ther*, **19**, 172–180.

Miller, D.G. (2011). AAV-mediated gene targeting, *Methods Mol Biol*, **807**, 301–315.

Miller, D.G., Wang, P.R., Petek, L.M., *et al.* (2006). Gene targeting *in vivo* by adeno-associated virus vectors, *Nat Biotechnol*, **24**, 1022–1026.

Miller, J.C., Tan, S., Qiao, G., *et al.* (2011). A TALE nuclease architecture for efficient genome editing, *Nat Biotechnol*, **29**, 143–148.

Mittelman, D., Moye, C., Morton, J., *et al.* (2009). Zinc-finger directed double-strand breaks within CAG repeat tracts promote repeat instability in human cells, *Proc Natl Acad Sci USA*, **106**, 9607–9612.

Morbitzer, R., Romer, P., Boch, J., *et al.* (2010). Regulation of selected genome loci using de novo-engineered transcription activator-like effector (TALE)-type transcription factors, *Proc Natl Acad Sci USA*, **107**, 21617–21622.

Munoz, I.G., Prieto, J., Subramanian, S., *et al.* (2011). Molecular basis of engineered meganuclease targeting of the endogenous human RAG1 locus, *Nucleic Acids Res*, **39**, 729–743.

Mussolino, C., Morbitzer, R., Lutge, F., *et al.* (2011). A novel TALE nuclease scaffold enables high genome editing activity in combination with low toxicity, *Nucleic Acids Res*, **39**, 9283–9293.

Nathwani, A.C., Tuddenham, E.G., Rangarajan, S., *et al.* (2011). Adenovirus-associated virus vector-mediated gene transfer in hemophilia B, *N Engl J Med*, **365**, 2357–2365.

Nickerson, H.D., Colledge, W.H. (2003). A comparison of gene repair strategies in cell culture using a lacZ reporter system, *Gene Ther*, **10**, 1584–1591.

Nickoloff, J.A., Brenneman, M.A. (2004). Analysis of recombinational repair of DNA double-strand breaks in mammalian cells with I-*Sce*I nuclease, *Methods Mol Biol*, **262**, 35–52.

Oka, K., Chan, L. (2005). Construction and characterization of helper-dependent adenoviral vectors for sustained *in vivo* gene therapy, *Methods Mol Med*, **108**, 329–350.

Olsen, P.A., Randol, M., Krauss, S. (2005a). Implications of cell cycle progression on functional sequence correction by short single-stranded DNA oligonucleotides, *Gene Ther*, **12**, 546–551.

Olsen, P.A., Randol, M., Luna, L., *et al.* (2005b). Genomic sequence correction by single-stranded DNA oligonucleotides: role of DNA synthesis and chemical modifications of the oligonucleotide ends, *J Gene Med*, **7**, 1534–1544.

Olsen, P.A., Solhaug, A., Booth, J.A., *et al.* (2009). Cellular responses to targeted genomic sequence modification using single-stranded oligonucleotides and zinc-finger nucleases, *DNA Repair (Amst)*, **8**, 298–308.

Pathak, M.A., Daniels, F., Hopkins, C.E., *et al.* (1959). Ultra-violet carcinogenesis in albino and pigmented mice receiving furocoumarins: Psoralen and 8-methoxypsoralen, *Nature*, **183**, 728–730.

Pichierri, P., Franchitto, A., Piergentili, R., *et al.* (2001). Hypersensitivity to camptothecin in MSH2 deficient cells is correlated with a role for MSH2 protein in recombinational repair, *Carcinogenesis*, **22**, 1781–1787.

Pike-Overzet, K., van der Burg, M., Wagemaker, G., *et al.* (2007). New insights and unresolved issues regarding insertional mutagenesis in X-linked SCID gene therapy, *Mol Ther*, **15**, 1910–1916.

Porter, A.C. (2001). Correcting a deficiency, *Mol Ther*, **3**, 423–424.

Porteus, M.H., Baltimore, D. (2003). Chimeric nucleases stimulate gene targeting in human cells, *Science*, **300**, 763.

Porteus, M.H., Carroll, D. (2005). Gene targeting using zinc finger nucleases, *Nat Biotechnol*, **23**, 967–973.

Radecke, F., Peter, I., Radecke, S., *et al.* (2006a). Targeted chromosomal gene modification in human cells by single-stranded oligodeoxynucleotides in the presence of a DNA double-strand break, *Mol Ther*, **14**, 798–808.

Radecke, S., Radecke, F., Peter, I., *et al.* (2006b). Physical incorporation of a single-stranded oligodeoxynucleotide during targeted repair of a human chromosomal locus, *J Gene Med*, **8**, 217–228.

Radecke, F., Radecke, S., Schwarz, K. (2004). Unmodified oligodeoxynucleotides require single-strandedness to induce targeted repair of a chromosomal EGFP gene, *J Gene Med*, **6**, 1257–1271.

Radecke, S., Radecke, F., Cathomen, T., *et al.* (2010). Zinc-finger nuclease-induced gene repair with oligodeoxynucleotides: Wanted and unwanted target locus modifications, *Mol Ther*, **18**, 743–753.

Richardson, P.D., Kren, B.T., Steer, C.J. (2001). Targeted gene correction strategies, *Curr Opin Mol Ther*, **3**, 327–337.

Richardson, P.D., Kren, B.T., Steer, C.J. (2002). Gene repair in the new age of gene therapy, *Hepatology*, **35**, 512–518.

Romer, P., Recht, S., Lahaye, T. (2009). A single plant resistance gene promoter engineered to recognize multiple TAL effectors from disparate pathogens, *Proc Natl Acad Sci USA*, **106**, 20526–20531.

Romer, P., Recht, S., Strauss, T., *et al.* (2010). Promoter elements of rice susceptibility genes are bound and activated by specific TAL effectors from the bacterial blight pathogen, *Xanthomonas oryzae* pv. *Oryzae*, *New Phytol*, **187**, 1048–1057.

Rouet, P., Smih, F., Jasin, M. (1994). Expression of a site-specific endonuclease stimulates homologous recombination in mammalian cells, *Proc Natl Acad Sci USA*, **91**, 6064–6068.

Rusk, N. (2011). TALEs for the masses, *Nat Methods*, **8**, 197.

Russell, D.W., Hirata, R.K. (1998). Human gene targeting by viral vectors, *Nat Genet*, **18**, 325–330.

Sangiuolo, F., Bruscia, E., Serafino, A., *et al.* (2002). *In vitro* correction of cystic fibrosis epithelial cell lines by small fragment homologous replacement (SFHR) technique, *BMC Med Genet*, **3**, 8.

Sangiuolo, F., Filareto, A., Spitalieri, P., *et al.* (2005). *In vitro* restoration of functional SMN protein in human trophoblast cells affected by spinal muscular atrophy by small fragment homologous replacement, *Hum Gene Ther*, **16**, 869–880.

Sangiuolo, F., Scaldaferri, M.L., Filareto, A., *et al.* (2008). Cftr gene targeting in mouse embryonic stem cells mediated by small fragment homologous replacement (SFHR), *Front Biosci*, **13**, 2989–2999.

Sargent, R.G., Kim, S., Gruenert, D.C. (2011). Oligo/Polynucleotide-based gene modification: strategies and therapeutic potential, *Oligonucleotides*, **21**, 55–75.

Sargent, R.G., Rolig, R.L., Kilburn, A.E., *et al.* (1997). Recombination-dependent deletion formation in mammalian cells deficient in the nucleotide excision repair gene *ERCC1*, *Proc Natl Acad Sci USA*, **94**, 13122–13127.

Schleifman, E.B., Chin, J.Y., Glazer, P.M. (2008). Triplex-mediated gene modification, *Methods Mol Biol*, **435**, 175–190.

Scholze, H., Boch, J. (2011). TAL effectors are remote controls for gene activation, *Curr Opin Microbiol*, **14**, 47–53.

Seidman, M.M., Glazer, P.M. (2004). Setting standards in gene repair, *Oligonucleotides*, **14**, 79.

Simon, P., Cannata, F., Concordet, J.P., *et al.* (2008). Targeting DNA with triplex-forming oligonucleotides to modify gene sequence, *Biochimie*, **90**, 1109–1116.

Smih, F., Rouet, P., Romanienko, P.J., *et al.* (1995). Double-strand breaks at the target locus stimulate gene targeting in embryonic stem cells, *Nucleic Acids Res*, **23**, 5012–5019.

Smith, J., Bibikova, M., Whitby, F.G., *et al.* (2000). Requirements for double-strand cleavage by chimeric restriction enzymes with zinc finger DNA-recognition domains, *Nucleic Acids Res*, **28**, 3361–3369.

Smith, J., Grizot, S., Arnould, S., *et al.* (2006). A combinatorial approach to create artificial homing endonucleases cleaving chosen sequences, *Nucleic Acids Res*, **34**, e149.

Smithies, O., Gregg, R.G., Boggs, S.S., *et al.* (1985). Insertion of DNA sequences into the human chromosomal β-globin locus by homologous recombination, *Nature*, **317**, 230–234.

Smithies, O., Koralewski, M.A., Song, K.Y., *et al.* (1984). Homologous recombination with DNA introduced into mammalian cells, *Cold Spring Harb Symp Quant Biol*, **49**, 161–170.

Song, K.Y., Chekuri, L., Rauth, S., *et al.* (1985). Effect of double-strand breaks on homologous recombination in mammalian cells and extracts., *Mol Cell Biol*, **5**, 3331–3336.

Stoddard, B.L. (2005). Homing endonuclease structure and function, *Q Rev Biophys*, **38**, 49–95.

Stoddard, B.L. (2011). Homing endonucleases: From microbial genetic invaders to reagents for targeted DNA modification, *Structure*, **19**, 7–15.

Sugio, A., Yang, B., Zhu, T., *et al.* (2007). Two type III effector genes of *Xanthomonas oryzae* pv. *oryzae* control the induction of the host genes *OsTFIIAγ1* and *OsTFX1* during bacterial blight of rice, *Proc Natl Acad Sci USA*, **104**, 10720–10725.

Surtees, J.A., Argueso, J.L., Alani, E. (2004). Mismatch repair proteins: Key regulators of genetic recombination, *Cytogenet Genome Res*, **107**, 146–159.

Suzuki, T. (2008). Targeted gene modification by oligonucleotides and small DNA fragments in eukaryotes, *Front Biosci*, **13**, 737–744.

Takahashi, K., Tanabe, K., Ohnuki, M., *et al.* (2007). Induction of pluripotent stem cells from adult human fibroblasts by defined factors, *Cell*, **131**, 861–872.

Tamaro, M., Gastaldi, S., Carlassare, F., *et al.* (1986). Genotoxic activity of some water-soluble derivatives of 5-methoxypsoralen and 8-methoxypsoralen, *Carcinogenesis*, **7**, 605–609.

Thomas, K.R. (1994). Impact of gene targeting on medicine, *Mol Genet Med*, **4**, 153–178.

Thomas, K.R., Capecchi, M.R. (1986a). Targeting of genes to specific sites in the mammalian genome, *Cold Spring Harb Symp Quant Biol*, **51**, 1101–1113.

Thomas, K.R., Capecchi, M.R. (1986b). Introduction of homologous DNA sequences into mammalian cells induces mutations in the cognate gene, *Nature*, **324**, 34–38.

Thompson, L.H., Schild, D. (2002). Recombinational DNA repair and human disease, *Mutat Res*, **509**, 49–78.

Thorpe, P.H., Stevenson, B.J., Porteous, D.J. (2002). Functional correction of episomal mutations with short DNA fragments and RNA–DNA oligonucleotides, *J Gene Med*, **4**, 195–204.

Todaro, M., Quigley, A., Kita, M., *et al.* (2007). Effective detection of corrected dystrophin loci in mdx mouse myogenic precursors, *Hum Mutat*, **28**, 816–823.

Tsuchiya, H., Sawamura, T., Harashima, H., *et al.* (2005). Correction of frameshift mutations with single-stranded and double-stranded DNA fragments prepared from phagemid/plasmid DNAs, *Biol Pharm Bull*, **28**, 1958–1962.

Urnov, F.D., Miller, J.C., Lee, Y.L., *et al.* (2005). Highly efficient endogenous human gene correction using designed zinc-finger nucleases, *Nature*, **435**, 646–651.

Varganov, Y., Amosova, O., Fresco, J.R. (2007). Third strand-mediated psoralen-induced correction of the sickle cell mutation on a plasmid transfected into COS-7 cells, *Gene Ther*, **14**, 173–179.

Vasileva, A., Linden, R.M., Jessberger, R. (2006). Homologous recombination is required for AAV-mediated gene targeting, *Nucleic Acids Res*, **34**, 3345–3360.

Vasquez, K.M., Glazer, P.M. (2002). Triplex-forming oligonucleotides: Principles and applications, *Q Rev Biophys*, **35**, 89–107.

Vasquez, K.M., Wilson, J.H. (1998). Triplex-directed modification of genes and gene activity, *Trends Biochem Sci*, **23**, 4–9.

Vasquez, K.M., Christensen, J., Li, L., *et al.* (2002). Human XPA and RPA DNA repair proteins participate in specific recognition of triplex-induced helical distortions, *Proc Natl Acad Sci USA*, **99**, 5848–5853.

Vasquez, K.M., Dagle, J.M., Weeks, D.L., *et al.* (2001a). Chromosome targeting at short polypurine sites by cationic triplex-forming oligonucleotides, *J Biol Chem*, **276**, 38536–38541.

Vasquez, K.M., Marburger, K., Intody, Z., *et al.* (2001b). Manipulating the mammalian genome by homologous recombination, *Proc Natl Acad Sci USA*, **98**, 8403–8410.

Villemure, J.F., Abaji, C., Cousineau, I., *et al.* (2003). MSH2-deficient human cells exhibit a defect in the accurate termination of homology-directed repair of DNA double-strand breaks, *Cancer Res*, **63**, 3334–3339.

Wang, G., Seidman, M.M., Glazer, P.M. (1996). Mutagenesis in mammalian cells induced by triple helix formation and transcription-coupled repair, *Science*, **271**, 802–805.

Wang, G., Levy, D.D., Seidman, M.M., *et al.* (1995). Targeted mutagenesis in mammalian cells mediated by intracellular triple helix formation, *Mol Cell Biol*, **15**, 1759–1768.

Wood, AJ., Lo, T.W., Zeitler, B., *et al.* (2011). Targeted genome editing across species using ZFNs and TALENs, *Science*, **333**, 307.

Wu, J., Kandavelou, K., Chandrasegaran, S. (2007). Custom-designed zinc finger nucleases: what is next?, *Cell Mol Life Sci*, **64**, 2933–2944.

Wu, X.S., Xin, L., Yin, W.X., *et al.* (2005). Increased efficiency of oligonucleotide-mediated gene repair through slowing replication fork progression, *Proc Natl Acad Sci USA*, **102**, 2508–2513.

Wuepping, M., Kenner, O., Hegele, H., *et al.* (2009). Higher efficiency of thymine-adenine clamp-modified single-stranded oligonucleotides in targeted nucleotide sequence correction is not correlated with lower intracellular degradation, *Hum Gene Ther*, **20**, 283–287.

Yamanaka, S. (2007). Strategies and new developments in the generation of patient-specific pluripotent stem cells, *Cell Stem Cell*, **1**, 39–49.

Yanez, R.J., Porter, A.C. (1998). Therapeutic gene targeting, *Gene Ther*, **5**, 149–159.

Yanez, R.J., Porter, A.C. (1999). Influence of DNA delivery method on gene targeting frequencies in human cells, *Somat Cell Mol Genet*, **25**, 27–31.

Yin, W., Kren, B.T., Steer, C.J. (2005). Site-specific base changes in the coding or promoter region of the human β- and γ-globin genes by single-stranded oligonucleotides, *Biochem J*, **390**, 253–261.

Yu, J., Vodyanik, M.A., Smuga-Otto, K., *et al.* (2007). Induced pluripotent stem cell lines derived from human somatic cells, *Science*, **318**, 1917–1920.

Zayed, H., McIvor, R.S., Wiest, D.L., *et al.* (2006). *In vitro* functional correction of the mutation responsible for murine severe combined immune deficiency by small fragment homologous replacement, *Hum Gene Ther*, **17**, 158–166.

Zhang, N., Liu, X., Li, L., *et al.* (2007). Double-strand breaks induce homologous recombinational repair of interstrand cross-links via cooperation of MSH2, ERCC1-XPF, REV3, and the Fanconi anemia pathway, *DNA Repair (Amst)*, **6**, 1670–1678.

Zou, J., Maeder, M.L., Mali, P., *et al.* (2009). Gene targeting of a disease-related gene in human induced pluripotent stem and embryonic stem cells, *Cell Stem Cell*, **5**, 97–110.

8

GENOME ENGINEERING AND GENOME EDITING USING CRISPR/CAS9–RNA-GUIDED NUCLEASE

Daniel Scherman[a] and Jean-Louis Mandel[b]

8.1 Introduction

New generation sequencing has opened the way to accessible whole genome determination, thus allowing to envision revolutionary personalized medicine, where genetic mutations might be identified and corrected at the individual level. Targeted gene modification via homologous recombination (HR) described in Chapter 7 represents the key, necessary tool for this exciting personalized genetic correction prospect.

Genome engineering, which means the capacity to delete, insert, and modify the genomic DNA sequences of cells or organisms, encompasses a much wider range of applications than personalized gene therapy, covering fields as large as functional genomics (allowing to study the function of DNA sequences in their endogenous genomic site), cell line modification, generation of knock-out or knock-in transgenic animal models by injection into fertilized eggs, optimization of recombinant microorganisms for biotechnological applications — crop production and fermenter bioproduction, etc. Two classes of techniques have emerged for targeted genome modification, using either protein-based or, more recently, RNA-based recognition of targeted genomic DNA sequences.

[a]Laboratory of Chemical and Biological Technologies for Health, Pharmacy School, Paris Descartes University, CNRS, Inserm, 4 avenue de l'Observatoir, 75006, Paris, France
Email: daniel.scherman@parisdescartes.fr
[b]Institut de Génétique et de Biologie Moléculaire et Cellulaire, IGBMC, Strasbourg
Email: jlmandel@igbmc.fr

8.2 Protein-Based DNA-Targeting Nucleases and Recombinases

Chapter 7, describes non-homologous end-joining (NHEJ), HR, homology-directed repair (HDR), and the different techniques initially used for specifically modifying the genome of living cells. These techniques, which are based on the use of protein nucleases bearing specific DNA-recognition properties, such as meganuclease, zinc-finger nucleases, or TALENs, do allow an efficient modification of various eukaryotic cells. All these early approaches use customized protein domains for sequence-specific DNA binding. The meganucleases integrate their nuclease and DNA-binding domain in the same protein module and are obtained by molecular evolution and selection. The zinc-finger nucleases and TALEN technologies use a building block "lexicon" consisting of protein domains which recognize specific DNA sequences (see Sections 4.2.2 and 7.5). In the Cys_2–His_2 zinc-finger nucleases, each domain recognizes three DNA base-pairs, while, for TALEN, each domain is specific of a single base pair. These protein domains are then assembled linearly in order to target specific 18 bp DNA sequence on the genome, and linked to a Fok1 endonuclease moiety.

An additional customized protein-based DNA-targeting system described in Fig. 8.1 has appeared more recently. It involves a recombinase-mediated step

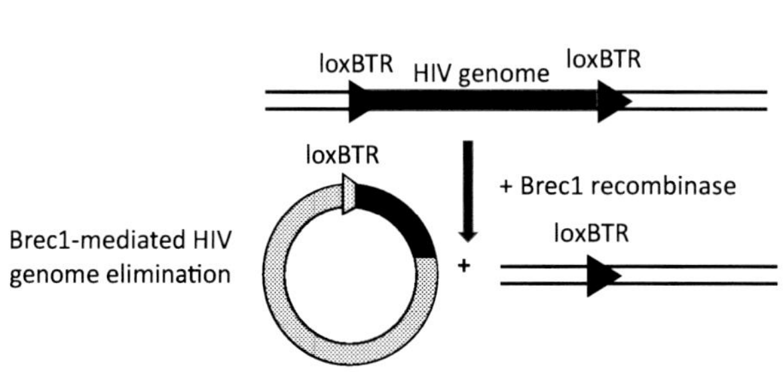

FIGURE 8.1 ■ Mechanism of sequence excision from a prokaryotic genome by the classically used Cre/*loxP* system and by the Brec1 recombinase. The *Cre* recombinase excises the sequence comprised between two *loxP sites*. The Brec1 recombinase has been optimized by molecular evolution and subsequent rationale structure-based modification in order to recognize the *loxBTR* sequences present in the genome of a large number of HIV-1 subtypes and variants.

(Abi-Ghanem *et al.*, 2013; Karpinski *et al.*, 2016) analog to the *Cre/loxP* system. In this system, the *Cre* recombinase enzyme has been optimized by selection-directed molecular evolution and further structure-aided computational optimization in order to bind and recombine a sequence present on a large proportion of HIV-1 strains, which was called *loxBTR*. The obtained *Brec1* recombinase is able to excise an integrated HIV-1 genome from infected cells. Such an elegant genome engineering approach potentially allows to "cure" HIV-1 infected cells by excision of the integrated HIV-1 genomes by gene transfer of the Brec1 recombinase.

8.3 RNA-Guided CRISPR/Cas9-DNA Endonuclease: A Process of Microbial Acquired Immunity

The above protein-based genome-editing technologies are considered to be time-consuming and, in some instances, of low efficacy, and some particular limitations arise in the cases of TALENs and zinc-finger nucleases from undesired intramolecular interaction between the adjacent modular motives used to target genomic sequences. A very promising and fast developing genome-editing technology has been introduced since 2013: CRISPR/Cas9, which makes use of an RNA-guide able to locate the endonuclease Cas9 to a specific genetic genomic locus in both prokaryotic and eukaryotic cells. The CRISPR/*Cas* system has been elucidated through years of fundamental studies and represents a curious and remarkable process of microbial "acquired immunity", which was surprising since adaptive immune response was widely considered as the specific prerogative of multicellular eukaryotic species.

The abbreviation "CRISPR" codes for: Clustered regularly interspaced short palindromic repeats. These CRISPR *direct-repeats* had been found several years before in the genome of a high proportion of bacterial genomes, but their signification had remained mysterious. It was then discovered that the *SPACERS* genomic sequences which are inserted between the CRISPR *direct-repeats* originate from bacteriophage genomes, and that when a bacteriophage sequence is inserted into the CRISPR array, the bacteria then becomes resistant to the bacteriophage. Thus, the bacterial CRISPR array is used as an immune memory and defense system by a wide range of bacteria.

Figure 8.2 describes how a short 20–40 nucleotide (nt) sequence from an invading bacteriophage is cut from the phage genome and inserted into the multiple CRISPR *direct-repeat* array. The bacteriophage DNA sequences (called "protospacer" sequences) are cut and inserted as *SPACERS* into the bacterial CRISPR locus through the action of CRISPR-associated nucleases (*Cas* nuclease), this conferring to the bacteria the capacity to become resistant to the phage.

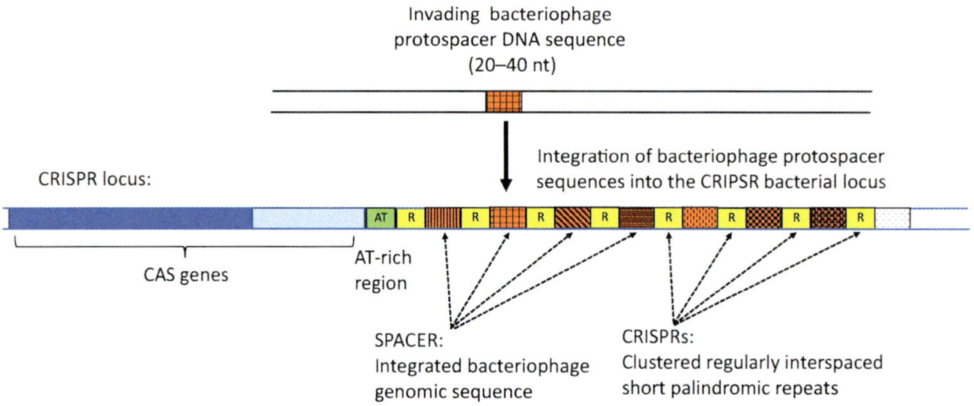

FIGURE 8.2 ■ Protospacer acquisition and insertion as a *SPACER* into the CRISPR array.

In the following step of the CRISPR defense mechanism, the CRISPR locus is transcribed into a non-coding RNA array, the pre-CRISPR-RNA (*pre-crRNA*). This long *pre-crRNA* is maturated in a process which varies between the three different types of CRISPR systems: types I, II, and III. Each CRISPR type has a specific cleavage and process mechanism, but all three types lead to the generation of small *crRNAs* containing an individual *SPACER* RNA sequence and a *direct-repeat* sequence (or part of it). The small *crRNAs* possess a double function:

- the *direct-repeat* sequence is able to bind and modulate the properties of a set of *Cas* nucleases specific for each CRISPR type;
- the *SPACER* RNA sequence is used as an invading RNA strand which base pairs according to Watson–Crick with the bacteriophage DNA targeted sequence.

According to this dual function, the *SPACER* RNA "guides" the *Cas* nuclease for specific cleavage at a given site of an invading bacteriophage genome. This exquisite targeting property is the cornerstone of the gene-editing and gene therapy use of the CRISPR defense system.

While the types I and III CRISPR systems recruit a complex assembly of several *Cas* proteins, on the opposite, type II CRISPR system only necessitates a single endonuclease called *Cas9*. This simpler type II system is the one used for genome editing in eukaryotic cells, and it will be further detailed in this chapter. Most progress has been obtained with the CRISPR type II system of *Streptococcus pyogenes* which uses the SpCas9 nuclease, and of *Streptococcus thermophilus* (StCas9).

In the type II CRISPR, a bacterial *trans*-activating CRISPR RNA (*tracrRNA*) hybridizes to the *direct-repeat* sequence (Fig. 8.3). Through a non-completely elucidated mechanism involving the RNAse III enzyme, a set of RNA duplexes is generated by Watson–Crick base-pairing between a *crRNA* and *tracrRNA*, as shown in Fig. 8.3.

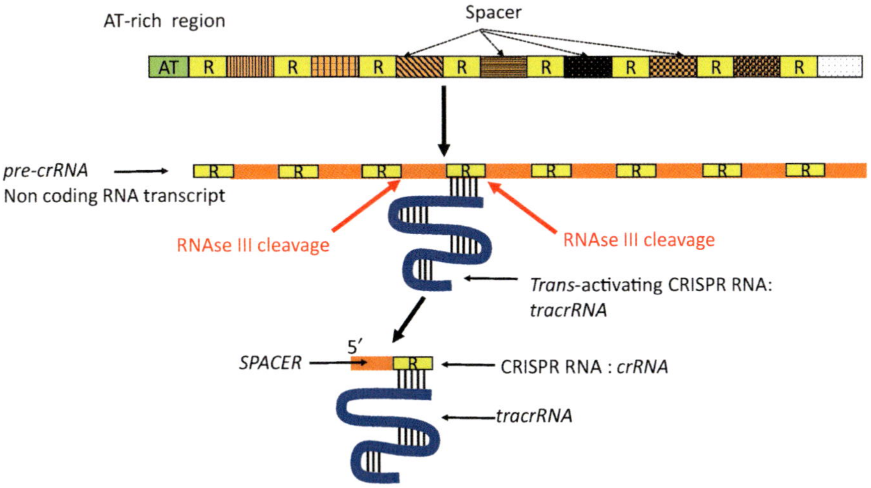

FIGURE 8.3 ■ Annealing of tracrRNA and RNAse III cleavage releases a set of RNA duplexes with bacteriophage-targeting capacity.

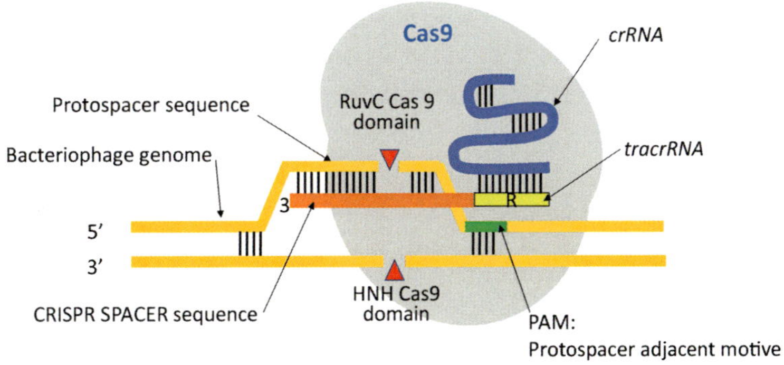

FIGURE 8.4 ■ Loading of the *crRNA/tracrRNA* duplex onto the Cas9 nuclease is followed by recognition of the PAM motive on the bacteriophage genome, then by invasion and hybridization of the phage DNA by the SPACER RNA, and finally by phage genome nucleolytic cleavage by the cas9 active sites RuvC and HNH (red arrows).

The final steps of the type II CRISPR defense mechanism are schematized in Fig. 8.4. The *crRNA/tracrRNA* duplex associates to the Cas9 nuclease by binding to a Cas9 positively charged groove. The resulting RNA/RNA/protein ternary ribonucleoprotein complex then scouts the bacteriophage DNA and binds weakly to a genetic motive adjacent to the protospacer sequence: the "PAM" *protospacer adjacent motive*. This weak binding allows the 20 nt SPACER sequence of the *crRNA/tracrRNA* duplex to invade the phage DNA and to hybridize to the bacteriophage protospacer. This annealing then leads to a Cas9 conformational change, allowing nuclease activity and double-strand cleavage by Cas9 of the bacteriophage DNA.

The double-strand break occurs through the action of two Cas9 enzymatic active sites: RuvC and HNH. This last step ultimately leads to phage elimination and confers bacterial resistance. The RuvC Cas9 catalytic site cleaves the DNA strand bearing the PAM, while the HNH site cuts the phage DNA strand complementary to the *crRNA* guide. The cleavage occurs about 3–4 nt upstream of the PAM sequence.

It is worth mentioning that once the ribonucleoprotein CRISPR complex binds to the PAM site, the protospacer strand invasion begins through a seed sequence located 8–10 bases at the 3′ end of the *gRNA* SPACER sequence. Only if the seed annealing shows a perfect match, the *gRNA* will continue to anneal in the 3′ to 5′ direction in a zipper-like cooperative Watson–Crick hybridization. In contrary to the first "seed" annealing process, the latter cooperative process might occur even in the presence of a slight mismatch, which thus induces an *off-target* double-strand break. As will be described later, several strategies have been proposed to decrease *off-target* cleavage in order to improve the biosafety of the CRSPR/Cas9 system.

Several key features are necessary to the efficiency and safety of the CRISPR immune defense process within the bacteria. First, the Cas9 nuclease alone (apoCas9) is naturally inactive, and is only activated when bound as shown in Fig. 8.4 to both the PAM motive and the targeted protospacer sequence on the bacteriophage DNA. Second, the direct repeats of the CRISPR locus are devoid of PAM motives, which prevents the CRISPR locus from being "self"-attacked by the ribonucleoprotein complex.

8.4 Improvements of CRSPR/Cas9 System for Genome Editing and Gene Therapy

As soon as it was discovered that the CRISPR/Cas9 ribonucleoprotein complex could be efficiently transplanted into and remain active in eukaryotic cells, the gene editing capacity of this simple and versatile system appeared tremendously promising. Indeed, instead of having to reconstruct *de novo* an entire protein for every targeted sequence, the CRISPR/Cas9 systems only requires an appropriate RNA guide containing a SPACER sequence complementary to the targeted genomic site, the main remaining condition to fulfill being that a convenient PAM is located at the vicinity of this targeted genomic site. In contrary to other gene editing systems (TALEN, zinc-finger nucleases, etc.), what is only required is a proper RNA guide to be synthesized or constructed within a plasmid, and associated with a generic Cas9 nuclease.

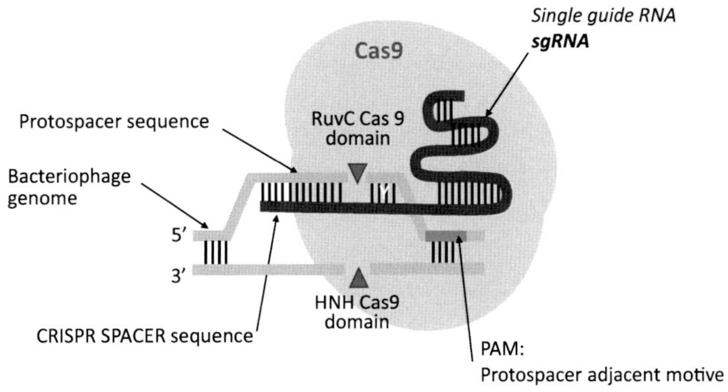

FIGURE 8.5 ■ The CRISPR/Cas9 *sgRNA* system.

In order to further simplify this platform, the *crRNA/tracrRNA* duplex described in Fig. 8.4 is now most frequently replaced by a single-stranded guiding RNA (single guide RNA, *sgRNA*) covering the dual function of the duplex, as shown in Fig. 8.5. The targeting moiety of the *sgRNA* is about 20 nt long. The use of several RNA guides in the same cell opens the way to parallel multiplex editing in several genomic loci, thus expanding the perspectives of the technology for genome functional screening and multiple genetic correction. In consequence, the CRISPR/Cas9 system has rapidly become the major technique presently used for any gene-targeting application.

The PAM motive necessary for the initial binding of the ribonucleoprotein complex is specific of every *Cas* nuclease ortholog from different bacterial strains. For instance, the *Streptococcus pyogenes* SpCas9 PAM sequence is 5′ NGG, a degenerated recognition codon which allows the CRISPR ribonucleoprotein complex to bind on average to every 8 bp in a human genome. This confers a large but incomplete targeting capacity for gene editing, and thus restricts the versatility of the technology. Such a limitation represents an important concern for HR, which requires very precise double-strand break location. To circumvent this problem, other Cas9 nucleases requiring a different PAM sequence have been made available, such as:

- *Streptococcus thermophilus* StCas9 (PAM: 3′ NNAGAAW);
- *Acidaminococcus sp.* AsCpf1 (PAM: 5′ TTTV);
- *Lachnospiraceae bacterium* LbCpf1 (PAM: 5′ TTTV); or
- *Staphylococcus aureus* (PAM: 3′ NNGRRT or NNGRR(N)), etc.

In addition, mutated variants of SpCas9, of AsCpf1, and of LbCpf1 have been identified with different PAM sequences. All in all, the diversity of PAMs recognized by different *Cas9* or other *Cas* orthologs greatly expands the targeting capacity of the CRISPR/*Cas* system to virtually any targeted site in eukaryotic genomes.

Most applications make use of the nuclease activity of the Cas enzymes to generate double-strand breakage for either gene inactivation as a result of NHEJ, or for gene repair or gene insertion by HR (also called HDR). However, other applications have been developed which only make use of the *genome-targeting* capacity of the ribonucleoprotein system. Adequate domains of the Cas nucleases have been modified, for instance, by eliminating the Mg^{++} binding sites necessary for the nuclease activity, and it has been possible to generate mutant ribonucleoprotein complexes with maintained DNA-targeting efficiency but devoid of nuclease activity.

The RuvC catalytic domain is composed of three parts of the Cas9 protein: RuvC I near the *N*-terminal region, and RuvC II and III adjacent to the second HNH catalytic sequence near the middle of the protein. The D10A (aspartate ⇒ Ala) mutation in RuvC I inactivates the RuvC catalytic activity, and the H840A (histidine ⇒ Ala) substitution inactivates the HNH second nuclease domain of Cas9. The double mutant, whose nuclease activity is completely inactivated (called "nuclease dead" dCas9), has been fused to a transcriptional activator (such as VP64 or VPR) or repressor (such as KRAB) in order to obtain targeted gene activation or repression, respectively. Other cell biology applications include the fusion of *dCas9* with a fluorescent protein such as green fluorescent protein (GFP), which allows to monitor *in cellulo* the spatial dynamics of a given genomic locus.

The CRISPR single guide sgRNA is introduced into the cells either within a plasmid under a polymerase III promoter such as U6, or as an RNA produced by *in vitro* transcription. The nuclease Cas9 can also be introduced as a gene, under the dependency of a ubiquitous constitutive promoter such as CMV. When the Cas and *sgRNA* genes are inserted in a plasmid, they are delivered by classical transfection techniques using chemical vectors (cationic lipids or polymers) or physical delivery techniques (electrotransfer, nucleofection, sonoporation, hydrodynamic delivery, etc.), which are described in Chapters 14–18 of this Handbook. For cells difficult to transfect or for some *in vivo* application, recombinant delivery viruses have been developed, mainly lentivirus and adeno-associated virus (AAV). However, the size cDNA of the commonly used SpCas9 (about 4 kb) is too large for proper encapsidation into an AAV particle. For this specific application, *Staphylococcus aureus Cas9*, whose gene is about 3 kb, represents a suitable alternative.

8.5 Reducing Off-Target Effects

The negative potential of *off-target* double-strand break of the CRSPR has received immediate consideration. *Off-target* cleavage might result from the risk of redundancy of the 20 nt targeting sequence and from the fact that the annealing guide

RNA accepts a certain degree of mismatch. Several strategies have been proposed to circumvent this limitation.

The first approach is based on the decrease of nt size of the annealing *sgRNA*. It has been noted that a truncated *sgRNA* significantly increases SpCas9 specificity, presumably because mismatches have a stronger negative effect of the binding and subsequent nuclease efficiency of the ribonucleoprotein complex.

Second, rationally based mutations have been introduced in SpCas9 in order to generate high fidelity Cas9 mutants which either possess a lower affinity for the targeted DNA strands or have improved proof-reading capacity (Hsu *et al.*, 2014).

The third strategy aims to limit the presence time of the Cas nuclease in the cell. For this purpose, Cas enzymes have been inserted under inducible promoters, such as Tet-on.

The fourth option to limit *off-target cleavage* is to use already assembled CRISPR ribonucleoprotein obtained from *in vitro* transcribed *sgRNA* and recombinant *Cas9*. Such complexes can be directly delivered into the cells by *ex vivo* electroporation or by association with cationic lipids. This leads to a very transient ribonucleoprotein presence, thus allowing to decrease off-target double-strand breaks.

Finally, a very elegant last approach uses Cas9 variants which have been inactivated in either the RuvC (D10A mutation) or the HNH (H840A mutation) catalytic domains. These mutants have lost their capacity to generate double-strand breaks, but still retain a nickase activity (i.e., cutting a single strand). Figure 8.6 displays how two Cas9 derived nickases within a CRISPR complex can be used to

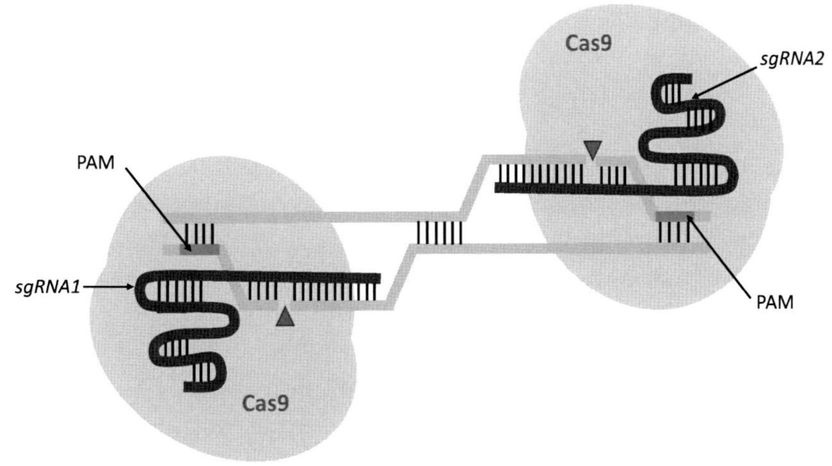

FIGURE 8.6 ■ Use of two Cas9-derived nickases for highly selective double-strand break targeting with limited *off-target*. In the shown configuration, a D10A mutated RuvC domain has been introduced in the SpCas9 nuclease.

increase selectivity. In this configuration, a double-strand break will only occur if the two Cas9 nickases are properly located at proximal genomic sites by two different *sgRNA* targeting different genomic sequences. The probability of two mismatches occurring simultaneously on these two targeted 20 nt sequences is greatly reduced.

8.6 Therapeutic Applications of CRISPR/Cas9

A very large panel of potential gene therapy therapeutic applications of CRISPR/*Cas9* are being investigated, such as, in a non-limiting list, several genetic or infectious diseases affecting liver, CNS, retina, or muscle.

8.6.1 CRISPR/Cas9 for HIV Gene Therapy Through Genome Disruption

Viral diseases and, in particular, AIDS, represent one of the most promising therapeutic perspective of CRISPR/Cas9. While antiretroviral tritherapy effectively controls viremia in HIV-1 patients and partially restores $CD4^+$ T-cells normal level, this poly-chemotherapy fails to eliminate integrated HIV-1 from latently infected T-cells or from other reservoir tissues such as the central nervous system. The integrated proviral DNA copies persist in a dormant state, and can be reactivated to produce replication-competent virus, thus reinfecting the patient when the drug treatment is stopped. This explains the necessity to develop strategies to effectively "cure" chronically-infected T-cells and other cell types, by a treatment which realizes the excision or disruption of integrated HIV-1 proviral genomes.

There are two intracellular steps where CRISPR/Cas9 can target the HIV-1 genome and can be used as an intracellular defense against HIV-1: (i) after reverse transcription and (ii) when the provirus is integrated in the cell chromosomes. In both cases, the double-strand HIV genome is not anymore protected by the viral envelope and capsid, and it is thus prone to nuclease attack. Indeed, Kaminski *et al.* (2016a) and Liao *et al.* (2016b) have reported an efficient CRISPR/Cas9 directed disruption of both infecting HIV-1 viruses and integrated provirus in the human embryonic kidney cell line HEK293, in several T-cell lines, and in human $CD4^+$ T-cells obtained from HIV-1^+ patients. In the case of virus disruption, the hypothesis is that intracellular exonucleases further degrade the retro-transcribed HIV genome after CRISPR/Cas9-induced double-strand break. In the case of integrated provirus, mutations, insertions, or deletions (*indels*) are considered to be the cause of provirus disruption and of the excision of segments of integrated provirus. Targeting HIV-1 coding regions and the LTR-R region, which contain relatively conserved TAR sequences, is considered as a favorable strategy.

Noticeably, this excision/disruption process is much more efficient in a *multiplex* approach, meaning that multiple *gRNA* directed against proviral sequences are necessary in order to achieve significant disruption/excision of the integrated provirus. One of the advantages of the multiplex approach is that it allows to circumvent possible resistance resulting from a mutation occurring in a single sequence targeted by the *sgRNA*.

These pioneering studies have demonstrated the possibility to generate cells which are self-protected against HIV-1, and even able to "cure" themselves after HIV-1 infection. Importantly, it was also shown that engineered human-induced pluripotent stem cells stably expressing HIV-targeted CRISPR/Cas9 could be efficiently differentiated into cell types which are normally HIV reservoirs, and that these engineered cells maintain their resistance to HIV-1 infection. Hence, these primary results have opened the way to the generation of engineered cells possessing acquired protection against HIV-1, and which can be self-grafted into patients after modification for personalized gene/cell therapy of HIV-1.

In a subsequent work, Kaminski *et al.* (2016b) described a proof of concept of *in vivo* eradication of the HIV-1 in transgenic mice and rats encompassing the HIV-1 genome. These exciting results were obtained by delivering RNA-guides/Cas9 through the use of a recombinant adeno-associated virus 9 (rAAV9) vector. Interestingly, AAV9 has been shown to be able to cross the blood–brain barrier (BBB), which is an important clue for treating HIV-1 reservoir in the central nervous system. As already mentioned, the smaller Cas9 from *Staphylococcus aureus* (3.3 kDa) was used here because of the AAV's limited encapsidation capacity. A multiplex approach was employed, with two *sgRNAs* targeting the LTR1 and Gag D sequences. After tail-vein IV administration into mice or after retro-orbital administration into rats, the excision of the of a 940 bp DNA fragment spanning between the LTR and Gag gene were observed in all tissues (spleen, liver, heart, lung, brain, kidney, and circulating lymphocytes). In addition, the level of viral RNA was drastically decreased (about 80%) in the circulating blood cells and lymph nodes of the treated rats. Thus, the multiplex excision with two *sgRNA*/SaCas9 resulted in a significant decrease in viral transcripts. These encouraging results of *in vivo* HIV-1 partial eradication should pave the way to clinical trials using CRISPR *sgRNA*/Cas9 in association with classical poly-chemotherapy.

Finally, an indirect way to treat HIV is based on using T lymphocyte genome editing in order to render these lymphocytes resistant to HIV infection. The C–C chemokine receptor type 5 (CCR5) plays a major role as a co-receptor in the infection process. A naturally occurring CCR5 inactivating mutation present in the European population is protective against AIDS in homozygous individuals. Gene editing tools to knock-out CCR5 in the genome of autologous cells using targeted CRISPR-Cas9 are thus actively investigated. In 2017, several reports have described CRISPR/Cas9-mediated CCR5 ablating system in cells such as hematopoietic stem cells, HSCs, which confers HIV-1 resistance *in vivo* (Xu *et al.*, 2017).

8.6.2 CRISPR/Cas9 Gene Therapy of Non-Infectious Diseases Through Genome Disruption

Multiple applications other than viral diseases can potentially benefit from CRISPR-Cas9-induced targeted genome disruption. For instance, it has been noted that naturally occurring loss-of-function of the PCSK9 gene induced a reduction of low-density lipoprotein and cholesterol levels in blood. It can be expected from this that PCSK9 gene disruption could confer protection against cardiovascular disease. By using an adenoviral vector to deliver a PCSK9-targeted CRISPR/Cas9 system, a recent study in mouse has reported a mutagenesis rate of PCSK9 in the liver, superior to 50%. This resulted in decreased blood plasma cholesterol levels by 35–40% (Ding *et al.*, 2014).

A second example concerns a subtype of Leber congenital amaurosis, LCA10, which is a severe retinal dystrophy caused by mutations in the CEP290 gene. The most frequent CEP290 mutation found in human patients is an intronic mutation that generates a cryptic splice donor site. The large size of the CEP290 gene precludes using an AAV vector for subretinal gene delivery of CEP290, thus eliminating the possibility of gene augmentation therapy which has been successful in other Leber congenital amaurosis subtypes (see Chapter 25). Hence, other gene therapy solutions have been proposed, such as the use of an antisense oligonucleotide to mask and thus inactivate the cryptic splice donor site (Gerard *et al.*, 2012).

A recent study (Ruan *et al.*, 2017) has shown that targeted genomic deletion using CRISPR)/Cas9 can suppress this cryptic splice donor in a CEP290 cellular model. Moreover, the study showed the feasibility of introducing *in vivo* a precise deletion into retinal cells by subretinal administration of two AAVs, one carrying the *sgRNA*, the second one a SpCAS9 protein.

Finally, another interesting technique proposed in this study is the use of a self-limiting CRISPR/Cas9 system to reduce the duration of expression of SpCas9: in this improved system, recognition sites for the *sgRNA* are introduced into the SpCas9 sequence itself, thus allowing the SpCas9 gene to be excised and eliminated following the intracellular expression of SpCas9 protein. As a result, the continuous expression of SpCas9 is abolished and its cellular residency time is limited by its cellular turnover, which increases biosecurity by decreasing the occurrence of off-target double-strand breaks.

8.6.3 Targeted Transcriptional Activation

Independently of the potency to correct mutation through HR or to induce indels, the capacity of single-guide RNA to target a precise genomic localization can be used to "reactivate" a silent gene or increase a given protein expression (Gersbach and Perez-Pinera, 2014). As already mentioned, the system comprises a *sgRNA* allowing to target a specific genomic site associated to an inactivated Cas9 endonuclease. This deadCas9 (dCas9) is fused to a transcriptional activation domain. For instance, the Cas9 endonuclease catalytic residues were mutated (D10A, H840A) to create a dCas9 which was genetically fused with a C-terminal VP64 acidic transactivation domain (Perez-Pinera *et al.*, 2013).

Multiple therapeutic applications can be envisioned. Utrophin is a fetal muscle protein localized at the muscle fiber (myotube) membrane, whose expression is lost after birth while being replaced by dystrophin. The dystrophin gene is among the largest human genes, and in consequence its mutation or deletions are responsible for a high occurrence genetic disease, Duchenne Muscular Dystrophy. Attempts to reactivate utrophin expression has been proposed as a treatment for Duchenne Muscular Dystrophy, and this has been obtained at the preclinical level in a cellular model by using zinc-finger proteins transcription activator-like effectors and CRISPR/Cas9 fused to a transcription activator.

Another application could allow to compensate for β-globin mutations observed in β-thalassemia and sickle-cell anemia by reactivation of the fetal γ-globin chain.

Friedrich Ataxia is another disease where compensatory reactivation could prove useful, since frataxin expression is reduced in cells from Friedreich's ataxia patients due to a trinucleotide repeat expansion in the intron 1.

A last example is that of age-related muscular degeneration. In its "wet" form, this disease causes loss of vision in a large proportion of elderly humans and represents the main cause of blindness in aging population. About 30% of wet-AMD patients are non-responders to classical anti-VEGF therapy (anti-VEGF monoclonal antibody, VEGF soluble receptor, or VEGF "trap"). For these patients, the use of pigmented-epithelial derived factor PEDF has been proposed: PEDF represents the most potent identified anti-angiogenic factor, and its reactivation at the vicinity of a retinal macula represents an interesting application of CRISPR-targeted transactivation.

Finally, CRISPR/Cas9-mediated gene induction has been envisioned for improving the characteristics of cell therapy products in regenerative medicine, for instance, for inducing the required reprogrammation (differentiation and specialization) of induced pluripotent stem cells (iPSCs) (see Chapter 29).

8.6.4 Programmable Editing of a Target Base Without DNA Cleavage

A last very exciting application of the CRISPR-Cas9 system concerns single-base edition. Before the introduction of CRISPR/Cas9, the correction of single-base mutations required double-strand DNA breaks and the use of donor templates in order to repair the DNA (Chapter 7). Frequent undesired insertions/deletions (*indels*) result from this approach, and CRISPR/Cas9 base editors represent an exciting potentially superior alternative.

Base editing enables the direct conversion of one single mutation base pair, thus allowing to restore a wild-type gene. The most-used base editors are now based on CRISPR-Cas9 system. In brief, these base editors comprise a CRISPR-Cas9 inactivated mutant that cannot make double-strand breaks, but has maintained a "nickase" single-strand breakage capacity. The Cas9 protein is fused to an enzyme which allows single-base conversion in the single-stranded DNA bubble created by Cas9. The opposite strand is then cleaved by the nickase, and DNA repair enzymes insert the corresponding base in front of the edited base, thus leading to complete and irreversible conversion of a single-mutation sequence into a "wild-type" corrected sequence.

Several single-base editing enzymes have been introduced (Gaudelli *et al.*, 2017). In the first version, a single-strand specific cytidine deaminase has been used to convert C to uracil within the DNA bubble. The system is completed by an uracil glycosylase inhibitor that impedes uracil excision. The nickase activity then nicks the non-edited DNA strand and mobilizes cellular DNA repair processes which replace the G-containing DNA strand. This combination leads to efficient and permanent C•G to T•A base pair conversion in several prokaryotic and eukaryotic cell types, including human embryonic cells. More recently, the genomic editing toolkit has been completed by the introduction of an adenosine deaminase selected by molecular evolution, which allows the A•T to G•C conversion (Gaudelli *et al.*, 2017).

An application of CRISPR/Cas9 gene editing was proposed for preventing cardiovascular risk (Chadwick *et al.*, 2017, 2018). As for the PCSK9 gene (see Section 8.6.2), it was found that naturally occurring loss-of-function mutations in ANGPTL3 (angiopoietin-like 3) had a beneficial cardiovascular effect, being associated with reduced blood triglycerides and low-density lipoprotein cholesterol, and thus decreasing the risk of coronary heart disease. The two studies reported that CRISPR/Cas9 base editing of both PCSK9 and ANGPTL3 were inducing an inactivation of each of these the two proteins, and that this effect could be obtained *in vivo* in mice. Adenoviruses vectors carrying the CRISPR/Cas9 editing system were IV administered and up to 35% editing rate was observed in the liver of treated mice. A markedly reduced level of triglycerides and cholesterol was observed in hyperlipidemic mice (up to 50%). These precursor studies establish the proof of concept of *in vivo* base editing to treat patients with atherogenic dyslipidemia.

For Further Reading

Abi-Ghanem, J., Chusainow, J., Karimova, M., *et al.* (2013). Engineering of a target site-specific recombinase by a combined evolution and structure-guided approach, *Nucleic Acids Res*, **41**(4), 2394–2403.

Addgene: CRISPR guide, http://www.addgene.org/crispr/guide/.

Chadwick, A.C., Evitt, N.H., Lv, W., *et al.* (2018). Reduced blood lipid Levels with *in vivo* CRISPR-Cas9 base editing of ANGPTL3, *Circulation*, **137**, 975–977.

Chadwick, A.C., Wang, X., Munsuru, K. (2017). *In vivo* editing of PCSK9 (proprotein convertase subtilisine/kexin type 9) as a therapeutic alternative to genome editing, *Arterioscler Thromb Vasc Biol*, **37**, 1741–1747.

Charpentier, E., Marraffini, L. (2014). Harnessing CRISPR-Cas9 immunity for genetic engineering, *Curr Opin Microbiol*, 114–119.

Cornu, *et al.* (2015). Editing CCR5: a novel approach to HIV gene therapy, *Adv Exp Med Biol*, **848**, 117–30.

Ding, Q., Strong, A., Patel, K.M., *et al.* (2014). Permanent alteration of PCSK9 with *in vivo* CRISPR-Cas9 genome editing, *Circ Res*, **115**(5), 488–492.

Doudna, J.A., Charpentier, E. (2014). Genome editing. The new frontier of genome engineering with CRISPR-Cas9, *Science*, **346**(6213), 1258096, doi: 10.1126/science.158096.

Gaj, T., Gersbach, C., Barbas, C.F. (2013). ZFN, TALEN, and CRISPR/Cas-based methods for genome engineering (2013), *Trends Biotechnol*, **31**(7), 397405.

Gaudelli, N.M., Komor, A., Holly, A., *et al.* (2017). Programmable base editing of A•T to G•C in genomic DNA without DNA cleavage, *Nature*, **551**(7681), 464–471, doi: 10.1038/nature24644.

Gerard, X., Perrault, I., Hanein, S., *et al.* (2012). AON-mediated exon skipping restores ciliation in fibroblasts harboring the common Leber congenital amaurosis CEP290 mutation, *Mol Ther Nucleic Acids*, **1**, e29, doi: 10.1038/mtna.2012.21.

Gersbach, C.A., Perez-Pinera, P. (2014). Activating human genes with zinc finger proteins, transcription activator-like effectors and CRISPR/Cas9 for gene therapy and regenerative medicine, *Expert Opin Ther Targets*, **18**(8), 835–839, doi: 10.1517/14728222.2014.913572, https://doi.org/10.1517/14728222.2014.913572.

Giannelli, S.G., Luoni, M., Castoldi, V., *et al.* (2017). Cas9/sgRNA selective targeting of the P23H Rhodopsin mutant allele for treating Retinitis Pigmentosa by intravitreal AAV9. PHP.B-based delivery, *Hum Mol Genet*, doi: 10.1093/hmg/ddx438.

Hsu, P., Lander, E.S., Zhang, F. (2014). Development and applications of CRISPR-Cas9 for genome engineering, *Cell*, **157**, 1262–1278.

Kaminski, R., Chen, Y., Fischer, T., *et al.* (2016a). Elimination of HIV-1 genomes from human T-lymphoid cells by CRISPR/Cas9 gene editing, *Sci Rep*, **6**, 22555, doi: 10.1038/srep22555.

Kaminski, R., Bella, R., Yin, C., *et al.* (2016b). Excision of HIV-1 DNA by gene editing: A proof-of-concept *in vivo* study, *Gene Ther*, **23**(8–9), 690–695.

Karpinski, J., Hauber, I., Chemnitz, J., *et al.* (2016). Directed evolution of a recombinase that excises the provirus of most HiV-1 primary isolates with high specificity, *Nat Biotechnol*, **34**(4), 401409.

Liao, H.K., Gu, Y., Diaz, A., *et al.* (2015). Use of the CRISPR/Cas9 system as an intracellular defense against HIV-1 infection in human cells, *Nat Commun*, **6**, 6413, doi: 10.1038/ncomms7413.

Perez-Pinera, P., Kocak, D.D., Vockley, C.M., *et al.* (2013). RNA-guided gene activation by CRISPR-Cas9-based transcription factors, *Nat Meth*, **10**, 973–976.

Ruan, G.X., Barry, E., Yu, D., *et al.* (2017). CRISPR/Cas9-mediated genome editing as a therapeutic approach for leber congenital amaurosis 10, *Mol Ther*, **25**, 331–341.

Xu, L., Yang, H., Gao, Y., *et al.* (2017). CRISPR/Cas9-Mediated CCR5 ablation in human hematopoietic stem/progenitor cells confers hiv-1 resistance *in vivo*, *Mol Ther*, **25**, 1782–1789.

PART II

VECTORS AND GENE DELIVERY TECHNIQUES

9

γ-RETROVIRUS- AND LENTIVIRUS-DERIVED VECTORS FOR GENE TRANSFER AND THERAPY

Caroline Duros[a] *and Odile Cohen-Haguenauer*[a,b,c]

9.1 Introduction

Viral vectors can be manipulated *in vitro* to modify their genomes in order to insert a gene of interest. Intrinsic viral cycle properties, such as DNA importation into the nucleus and its expression, are therefore used to deliver the transgene to target cells. The major advantage of viral vectors derived from retroviruses is the integration of the transferred transgene into the chromosome of the host cell, which may enable long-lasting transgene expression, not only in the transduced target cell, but also in all offspring cells after cell division.

9.2 The Concept: Designing Retrovirus-Based Vectors

9.2.1 Starting From the Knowledge of Helper Retrovirus Biology

Retroviruses are composed of oncoviruses, lentiviruses and spumaviruses. Oncoretroviruses can only infect dividing cells, because they require the disruption of the nuclear membrane for the viral genome to get into the nucleus, whereas lentiviruses can also infect non-dividing cells. Oncoretroviruses and lentiviruses

[a]Laboratory of Biotechnology and Applied Pharmacogenetics, CNRS UMR8 H3, Ecole Normale Supérieure de Cachan, France
[b]Oncogenetics, Department of Clinical Oncology, Hôpital Saint-Louis, Unversité Paris7-Paris Diderot, Sorbonne Paris-Cité Paris, France
[c]Email: odile.cohen-haguenauer@sls.aphp.fr

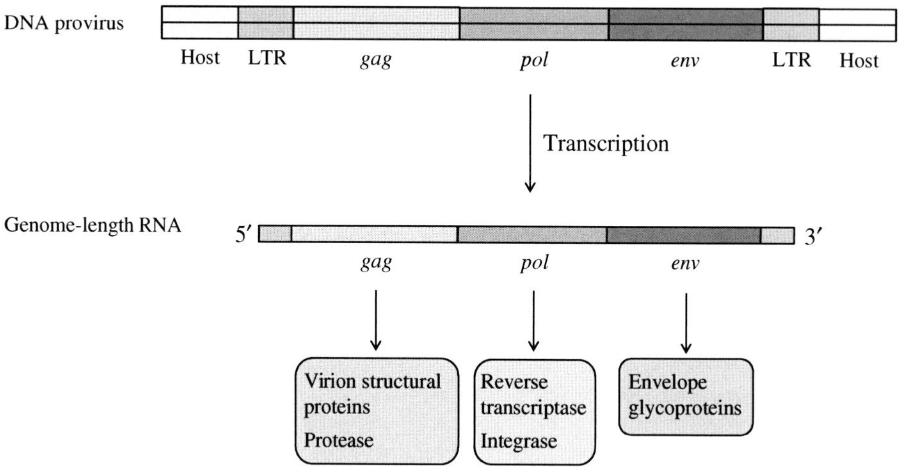

FIGURE 9.1 ■ Organization of the γ-retrovirus genome.

are also first-choice vectors because they can be manipulated to obtain defective, non-replication-competent viruses in the target cells. Retroviruses are enveloped viruses with positive RNA, two identical RNAs being encapsidated in the viral particle. Their genome is composed, on one hand, of 5′ and 3′ long terminal repeat (LTR) regulatory sequences, and on the other hand of coding sequences for structural and enzymatic proteins: group-specific antigen (gag), polymerase (pol), envelope (env) and accessory proteins (Fig. 9.1).

The LTR consists of replication-initiation sequences, transcription- and translation-regulatory sequences (U3 and U5), as well as sequences necessary for transgene integration (R: repeated sequences). The encapsidation/packaging signal (φ) is located just downstream of the 5′ LTR encompassing the 5′ part of the gag sequences. Gag encodes nucleocapsid proteins and internal peptides. Pol encodes proteins required for replication and viral genes' expression (protease, reverse transcriptase and integrase). Env encodes envelope proteins (Fig. 9.2).

During a natural viral cycle (Fig. 9.3), the viral envelope glycoprotein interacts with its receptor at the target cell surface, which leads to the fusion of the two membranes, the virus thus injecting its RNA genome into the cytoplasm. The reverse transcriptase, a unique and essential enzyme, reverse transcribes RNA into double-stranded DNA in the cytoplasm, thanks to a complex mechanism involving several steps (Fig. 9.4). This leads to a DNA flanked by two LTRs, in contrast to the RNA that has only one of each sequence, apart from the repeat that flanks the RNA at both its 5′ and 3′ ends (Fig. 9.5). Viral DNA obtained is then imported into the nucleus and will integrate into the host cell genome: A process which is mediated by the integrase. This DNA will later be transcribed by the cellular machinery, and can be used in two forms: the non-spliced RNA can be packaged (two per virion) and also serve as a matrix for translation of gag and pol proteins, and the spliced RNA will be the matrix for the translation of the envelope proteins (Fig. 9.2).

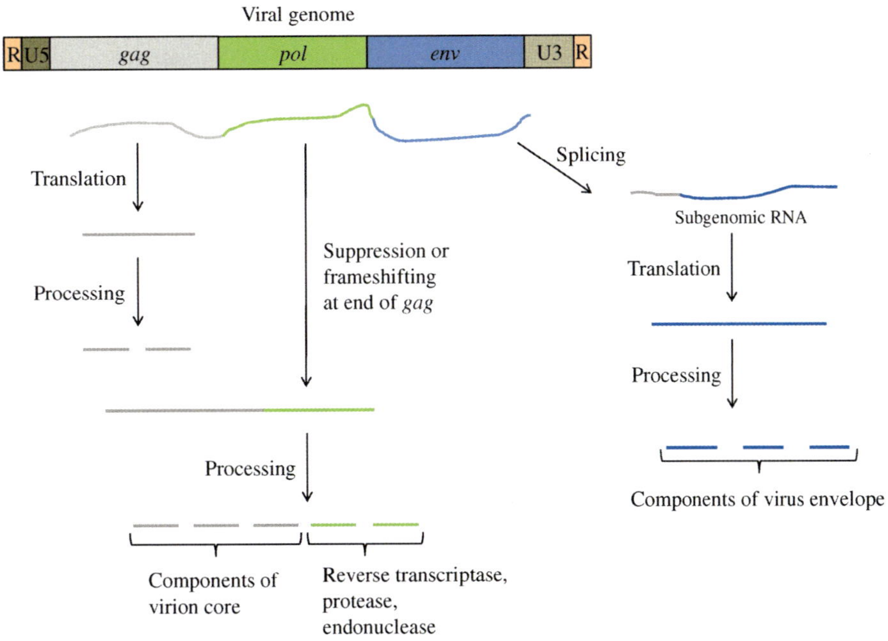

FIGURE 9.2 ■ Synthesis of viral proteins from the γ-retrovirus genome.

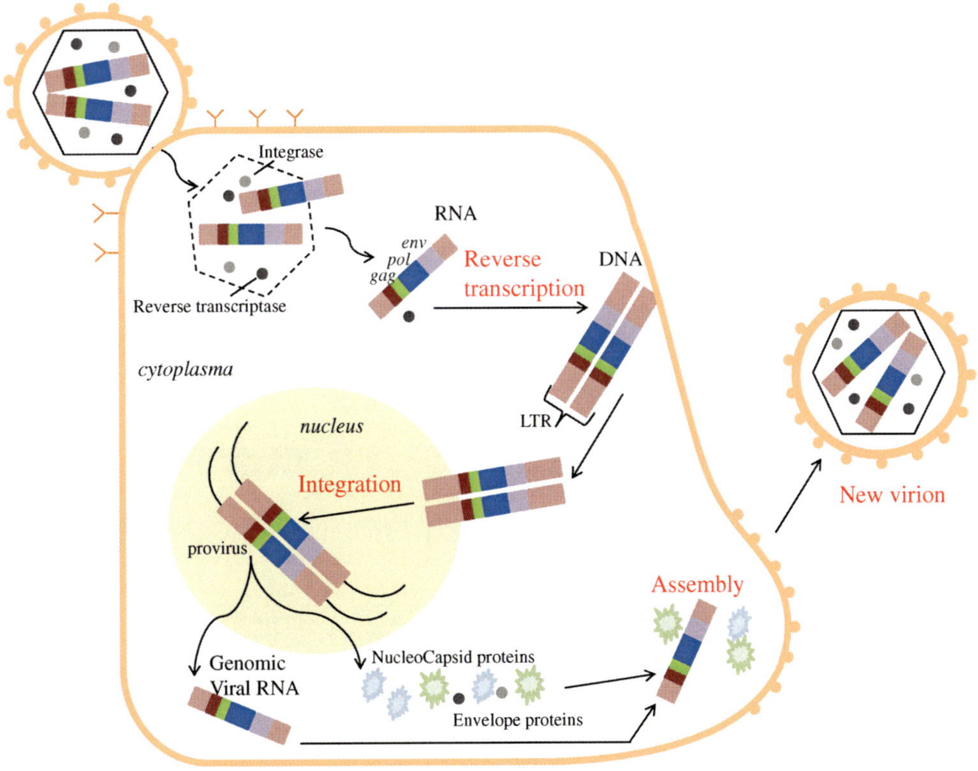

FIGURE 9.3 ■ Schematic representation of the sequential steps involved in the retrovirus viral cycle.

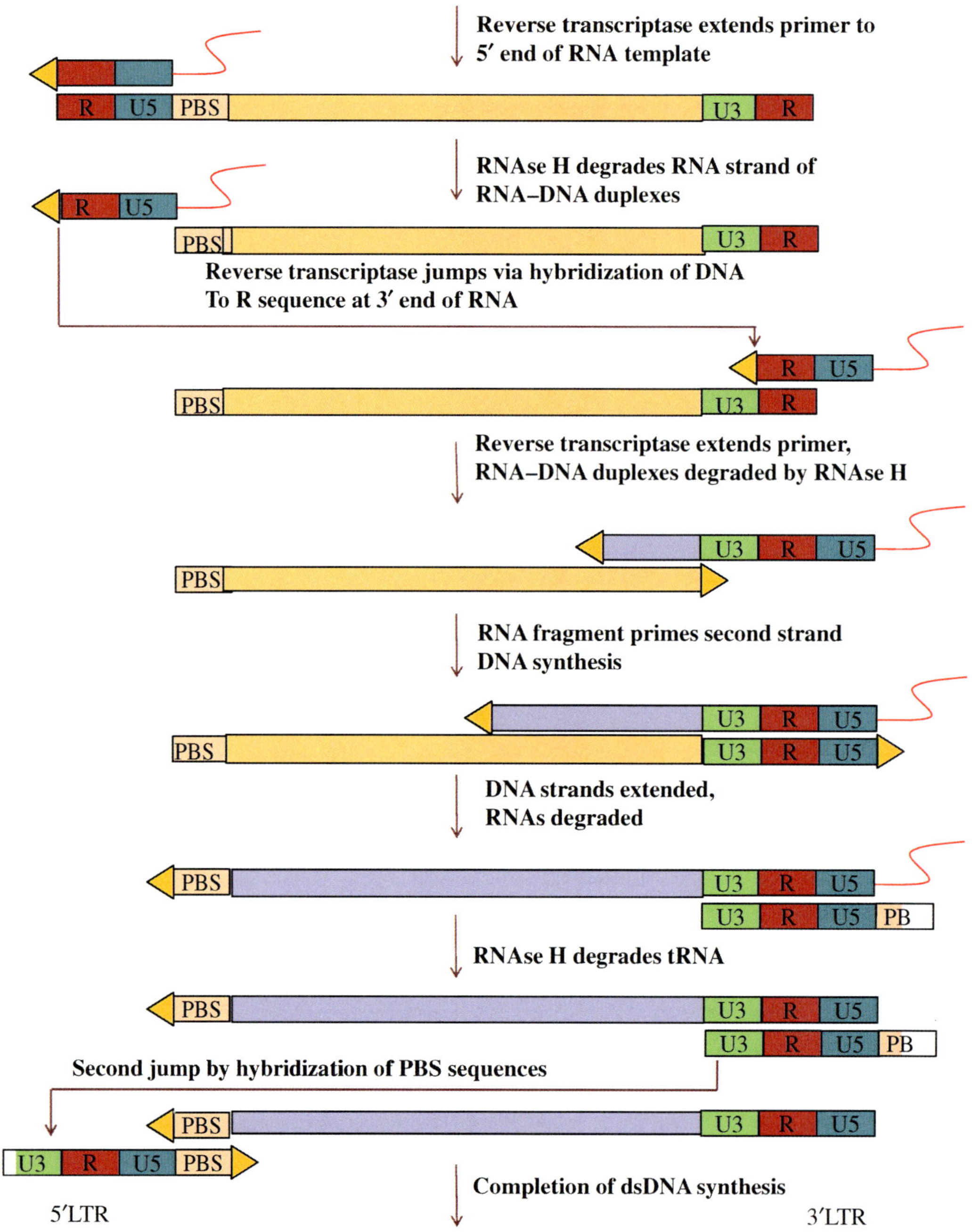

FIGURE 9.4 ■ Schematic representation of retrovirus replication mediated by the key enzyme: Reverse transcriptase.

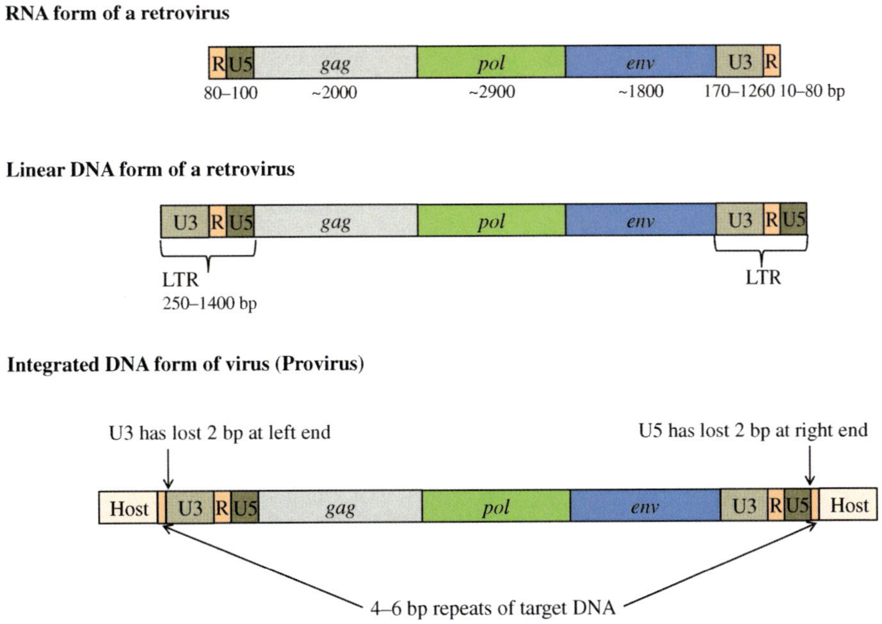

FIGURE 9.5 ■ Structure of the γ-retrovirus genome as its RNA form (*top*); linear DNA intermediate synthesized by the reverse transcriptase (*middle*) and integrated provirus, inserted into the host cell chromosomal DNA (*bottom*).

Indeed, the retroviral genome structure is an overlapping one and can produce different proteins depending on splicing (Fig. 9.6). This special feature allows saving of a lot of space in the viral genome but makes the latter more complex, particularly in lentiviruses. It is of note that for lentiviruses, the efficient expression of gag and pol requires a virally-encoded post-transcriptional activator called rev (Fig. 9.7). Following capsid proteins assembly and packaging of two non-spliced RNAs, viral particles bud at the cell membrane of the infected cell, taking along with them a tiny part of the cellular membrane into which envelope proteins are anchored through their transmembrane domains while the surface glycoproteins confer their tropism to the newly formed virions.

9.2.2 The Basic Principle of Deriving Vectors from Retrovirus

The strategy for viral vector construction is based on the dissociation of the viral cycle events. In particular, viral replication mechanisms are split into two parts: One will be the defective gene-transfer vector (whose destination is the target cell nucleus) and the other, a separate genome in charge of synthesis of enzymatic and structural viral proteins. Altogether, this system enables an abortive viral cycle called transduction. In the vector part, after deletion of viral protein-coding

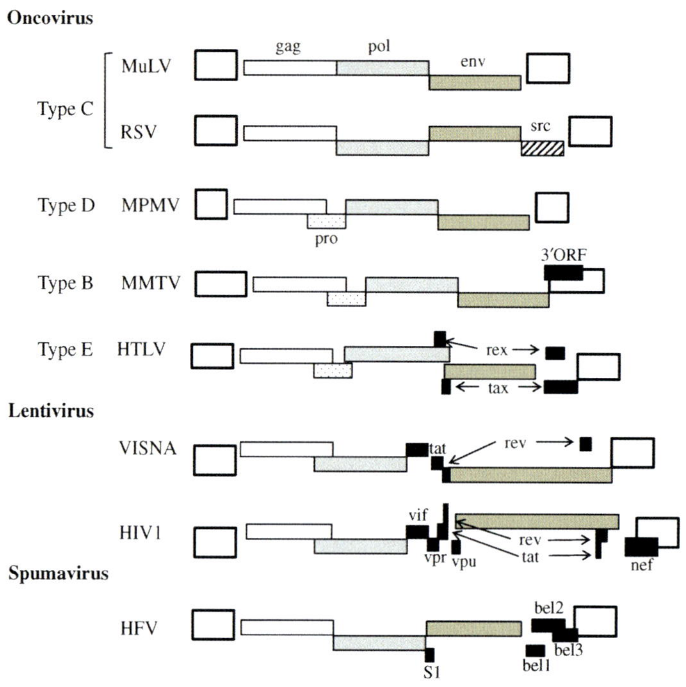

FIGURE 9.6 ■ Genomic organization of the various retrovirus species.

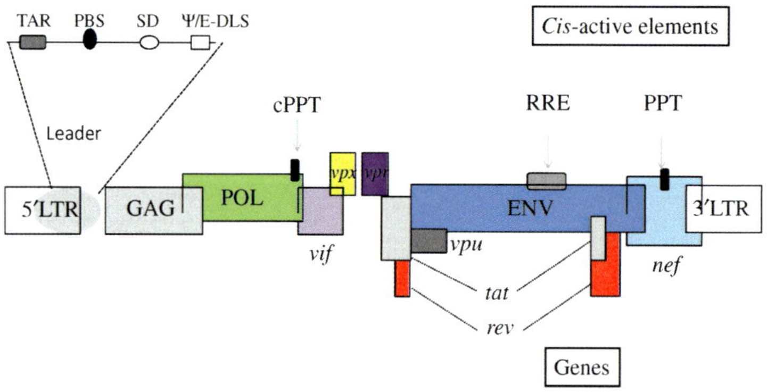

FIGURE 9.7 ■ Genomic organization of lentiviruses in e.g., human immunodeficiency virus-1 (HIV-1); cPPT, central polypurine tract; E/DLS, encapsidation and dimerization sequence; PBS, primer binding site; PPT, polypurine tract; SD, splice donor; TAR, transactivating response region.

sequences, the remaining space is used to introduce either reporter genes or genes of therapeutic interest, including the sequences necessary for their integration and expression. To insure production of viral particles, it is thus necessary to bring the missing viral proteins *in trans* to the system, through engineered cells called packaging or transcomplementing cell lines. The genomic RNA contains all the *cis*-acting sequences, whereas the packaging plasmids contain all the *trans*-acting proteins, necessary for adequate transcription, packaging, reverse transcription and integration. A diagram of the human immunodeficiency virus-1 (HIV-1)-based third-generation system is presented in Fig. 9.8. The U3 region of the 3′ LTR is essential for the replication of a wild-type retrovirus, since it contains the viral promoter in its RNA genome. It is dispensable for replication-defective vectors and has been deleted to remove all transcriptionally active sequences, creating the so-called self-inactivating (SIN) LTR. SIN vectors are thus unable to reconstitute their promoter and are safer than their counterparts with full length LTRs (Yu *et al.*, 1986).

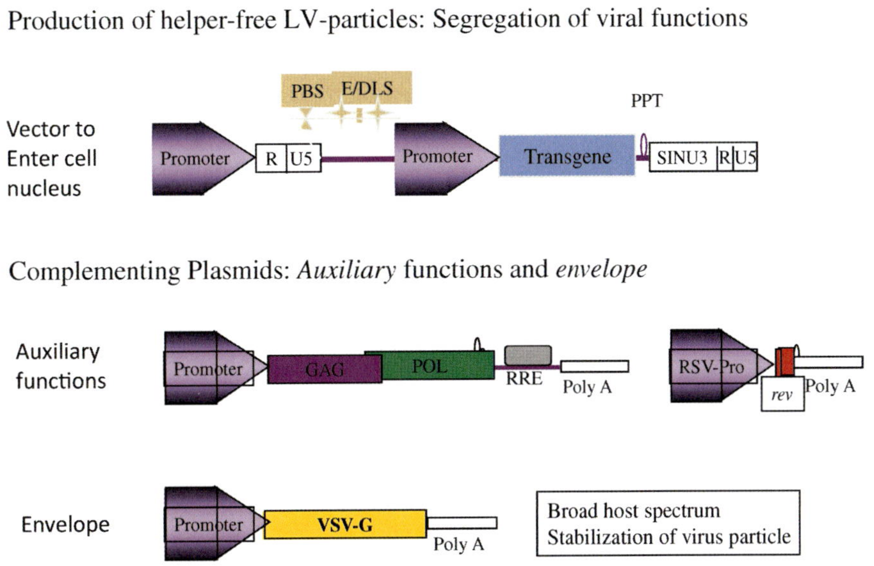

FIGURE 9.8 ■ Transient production of lentivirus vectors (LVs) based on three-plasmids transfection: Recombinant SIN vector shuttling the transgene and viral complementing proteins split on two additional plasmids (or three when rev is separate). The genetic information contained in the vector genome is the only information transferred to the target cells. Early genomic vectors were composed of the following components. The 5′ LTR, the major splice donor, the packaging signal (encompassing the 5′ part of the gag gene), the rev-responsive element (RRE), the envelope splice acceptor, the internal expression cassette containing the transgene and the 3′ LTR. In the latest generations, several improvements have been introduced. The central polypurine tract of HIV has also been added back in the central portion of the genome of the transgene RNA. This increases titers, at least in some targets, and decreases pseudotransduction. The U3 region of the 3′ LTR has been deleted to remove all transcriptionally active sequences, creating the so-called SIN LTR. Finally, chimeric 5′ LTRs have been constructed, in order to render the LV promoter transactivator (Tat)-independent. This has been achieved by replacing the U3 region of the 5′ LTR with either the cytomegalovirus (CMV) enhancer (CCL LTR) or the corresponding Rous sarcoma virus (RSV) U3 sequence (RRL LTR). Vectors containing such promoters can be produced at high titers in the absence of the Tat HIV transactivator. However, the rev-dependence of these third-generation LV has been maintained, in order to maximize the number of recombination events that would be necessary to generate a replication-competent retrovirus. This latest generation represents the system of choice for future therapeutic projects.

Two distinct strategies are used for vector production: either sustained from optimized stable transcomplementing cell lines, which up until now are available for γ-retroviruses only or transient production through multiple-plasmids cotransfection. For the production of γ-retroviruses, a stable cell line such as TEFLY GA (Cosset *et al.*, 1995) can be used which has been engineered from a TE671 medulloblastoma cell line in order to stably integrate two different plasmids at two distinct loci and express at a high level, on one hand sequences encoding gal and pol, and env protein on the other hand. Each sequence of interest is linked by an internal ribosome entry site to one antibiotic-resistance gene consisting of either phleoblastin or blasticydin, respectively. Antibiotic selection actually ensures that the cell synthesizes viral proteins. In order to produce vector particles, packaging cells are then merely transfected with the vector-containing plasmid.

A transient production system is so far the main available option for lentivirus production, where, for example, 293T cells, which are immortalized cells from the human kidney, are transfected with four plasmids simultaneously: First, the plasmid containing the gene of interest; second, the plasmid containing a packaging plasmid (Pack) and including mostly gag and pol sequences; third, the separate plasmid encoding rev protein, and, finally, the fourth plasmid coding the envelope protein, for example the amphotropic G protein of vesicular stomatitis virus (VSV-G). VSV-G is widely used to pseudotype both lentivirus-derived as well as γ-retroviral vector particles, given its robust stability, which allows for the concentration of the vector by ultracentrifugation, and because its phospholipid receptor is ubiquitously expressed in mammalian cells (Fig. 9.7). The separation of viral sequences allows a reasonable level of safety in considerably decreasing the risk of generating helper viral particles, which would replicate once in the host cell. Such a scenario would require several recombination events, the probability of which is substantially decreased with the dissociation of the retroviral genome and the sequence overlap reduced to a minimum. Vector-containing virions are produced by transcomplementing/packaging cells and particles are released into the culture medium from where they can be collected, purified and concentrated. The particle produced is replication-defective since it lacks most viral sequences driving this function and will be able to achieve just one single abortive viral cycle when inside the target cell. Making use of both the reverse transcriptase and the integrase provided *in trans* and co-encapsidated into the virion, the viral genome will be retro-transcribed and further inserted into the host cell genome. The reporter gene or the gene of interest can then be transcribed and translated by the cellular machinery, like any other cellular gene. The absence of structural viral proteins ensures that the integrated transgenic genome will not form new infectious virus particles. The pathogenic potential of HIV stems from the presence of nine genes which each encode important virulence factors. In fact, with the multiply-attenuated design of HIV vectors, six of these genes (namely env, viral infectivity factor (Vif), viral protein r (Vpr), viral protein u (Vpu), negative regulatory factor (Nef) and transactivator (Tat)) (Figs. 9.7 and 9.8) can be deleted from the system without altering its gene-transfer ability. This safety lock ensures that the parental virus cannot be reconstituted.

9.2.3 Advantages of Lentivectors Over γ-Retrovectors

These vectors are replication-defective since regions encoding the proteins necessary for additional rounds of virus replication and packaging are deleted from the viral genome. Defective γ-retrovirus genomes have a cloning capacity up to 6–8 kb, and are able to transduce target cells according to the envelope pseudotype under use. In actively dividing cells the efficiency of transgene delivery can reach up to 90%. A major limitation of this technology is that slowly- or non-dividing cells, such as neurons, are resistant to γ-retrovirus-mediated transduction. It has been long identified that retrovirally-shuttled transgenes are silenced in embryonic stem (ES) cells, as well as in induced pluripotent stem cells (iPS) through mechanisms involving methylation. Despite practical advantages, γ-retroviruses have been associated with major drawbacks, in particular in clinical trials where insertional mutagenesis resulted in the development of malignancies. It thus became obvious that alternative approaches to γ-retrovirus-mediated gene transfer should be considered.

Unlike γ-retrovectors, so far no malignancy resulting from insertional mutagenesis has been reported with lentivectors. Vectors have been successfully derived from HIV-1, HIV-2, simian immunodeficiency virus (SIV) or equine infectious anemia virus (EIAV). A unique feature of lentiviruses is that they are able to transduce both non-dividing (slowly dividing or quiescent but metabolically active cells) and dividing cells. In addition, their cloning capacity is broader than that of γ-retrovectors and they exhibit higher transduction efficiency, of human cells in particular. It has been shown that the integration of murine leukemia virus (MLV)-based vectors is not random as integration occurs mainly within, or close to, 5′ regulatory regions of transcriptionally active genes (Wu *et al.*, 2003); LVs do not show such preference and mostly integrate at random in the open reading frame (Schroder *et al.*, 2002).

9.2.4 Biosafety and Assessment of Genotoxicity

Despite great strides made in developing animal models and *in vitro* assays to predict the safety of new vectors, there are limits to our ability to accurately predict the risk of genotoxicity with a new vector (Corrigan-Curay *et al.*, 2012). This may be a function of the assays or models, or it may indicate that clinical data will always remain the true test. Nevertheless, continued refinement of preclinical models is mandatory. Ideally, standard assays or platforms might be developed for use across trials. One of the challenges highlighted is that some animal models, which best recapitulate what is seen in the clinical trials, require long-term follow-up and may transpire to be outrageously expensive as a standard model. Whether such models or others based on cancer-prone animals can be adapted by combining them with validated *in vitro* assays and biological markers is an important research question with encouraging preliminary evidence in studies which involve high-throughput integration-sites analysis. Further, the validity of the assay or model may be dependent upon the experience of the lab; interoperability across labs will be key in establishing validity. With the new generations of coming vectors, comparative studies can be performed which will progressively help establish reference systems harboring significantly improved safety and efficacy features.

9.3 Technology Developments to Overcome Bottlenecks and Improve Safety

9.3.1 Genotoxicity and Counteracting Insertional Mutagenesis

9.3.1.1 *Genotoxicity*

A number of bottlenecks have partially hampered further use of retrovirus vectors in the clinic. In particular, safety issues related to integration into the host cell genome and manufacturing limitations related to vector titer and stability have been addressed during the past few years. Several groups focused on improving vector design; others have concentrated their efforts on developing new and safer producer cells and improved production and purification processes. In fact, MLV LTR-driven retroviral vectors have been shown to activate proto-oncogenes. This risk is a major concern that hampers potential therapeutic applications of retroviral vectors because the severe adverse events are directly attributable to gene-transfer-vector integration. In otherwise successful gene therapy trials in patients with X-linked severe combined immunodeficiency (SCID-X1), these types of effects have resulted in five reported cases of vector-induced leukemia (Hacein-Bey-Albina *et al.*, 2003, 2008; Thrasher *et al.*, 2006). Similar effects have also been seen in several other gene therapy trials, for instance *X*-linked chronic granulomatous disease (Ott *et al.*, 2006; Stein *et al.*, 2010) and more recently in Wiskott–Aldrich syndrome (WAS) (Botzug *et al.*, 2010). In all of these cases, insertion of the γ-retroviral vector near known proto-oncogenes led to enhancer-mediated expression of these proto-oncogenes. These observed side effects of integrative-vector-mediated gene therapy reveal current limitations of integrative vectors for gene transfer used in clinical studies, as SIN vectors with strong promoters have also been shown to partially retain oncogenic potential (Modlich *et al.*, 2009). This issue was recently considered and discussed at an international level and a report issued (Corrigan-Curay *et al.*, 2012).

In response to insertional mutagenesis leading to enhancer-mediated expression of proto-oncogenes, a number of new vector designs are being explored, including the use of LVs, which have a propensity to insert into genes rather than near transcriptional start sites, as well as modified γ-retroviral vectors.

Indeed, over the past few years, several approaches have been pursued along distinct and potentially synergistic rationales, as follows:

- The development of SIN retroviral vectors. In particular, all lentivectors are SIN which lack enhancer/promoter elements at viral LTRs, a feature established to significantly reduce the oncogenic potential. This data confirms the importance of retroviral vector design in vector safety (Modlich, 2006). Further improvements on SIN vectors include the use of physiological promoters and led to significant decrease in the ability to induce clonal dominance in hematopoietic stem cells (HSCs) (Zychlinski *et al.*, 2008).
- Insulating the transgenic cassette in order to prevent cross-talk *in cis* from and towards the integrated exogenous sequences: While in keeping with random integration, this strategy has the potential to operate should integration occur at any given locus. This is providing the genetic insulator elements (GIEs) under use are robust enough to act as both enhancer-blockers and boundaries against potential silencing.
- Inhibiting the natural integration process by knocking down lens epithelium-derived growth factor (LEDGF), the cell-partner of HIV integrase in target cells. This may allow redirection of integration towards defined genetic regions.
- Further along this line, the possibility to select *ad hoc* genetic regions for integration to take place in these safe harbors has been investigated, in particular in taking advantage of zinc-finger nucleases (ZFNs) with predetermined specificity; more recently, technologies have been developed based on MNs or transcription activator-like effector nucleases (TALENs) which hold the potential to specifically recognize a mutation at a given gene locus (see Section 9.3.1.3).
- Investigating alternative integrative systems with a more random integration profile in human cells, such as foamy-virus vectors or improving γ-retroviral vectors with a novel α-retroviral SIN vector platform based on Rous sarcoma virus (RSV). RSV-derived vectors indeed show a more neutral integration spectrum, which is likely to increase the safety profile. Recently, codon optimization of the gag/pol expression construct contributed to overcoming the otherwise poor titers commonly observed; combined with further optimization steps, titers of over 10^7 infectious particles per mL were obtained in human cells (Suerth *et al.*, 2010).

9.3.1.2 Genetic insulators to prevent cis-regulatory cross-talk

GIEs are *cis*-elements that act as boundaries, inhibiting enhancer or silencer interactions between the integrated transgene and the surrounding chromatin. We have established new species of GIE which have been shown in a plasmid-based quantitative assay to effectively insulate an exogenous gene from genomic regulatory interference, with both enhancer-blocking and boundary effect (Gaussin *et al.*, 2011). In order to evaluate the potential of these GIEs in integrating gene-transfer vectors intended for therapy, SIN-insulated retrovectors have been constructed and compared with non-insulated SIN-LTRs counterparts in both γ-retro and lentivectors. Taking advantage of the deletions of enhancer–promoter sequences from the 3′U3 region in SIN viral vectors mentioned above, these elements could be introduced to further improve vector performance. We have identified a specific combination of synthetic GIE repeats, which is genetically stable when placed in the 3′SIN-LTR of both γ-retro and lentivectors and allows vector production at high titers. In target cells a far less scattered expression profile is observed as compared with controls, the level of which is determined by the nature and strength of the internal promoter only; in fact, a strong cell-type-specific promoter was constrained by the insulator in non-target cells. This data shows that the synthetic GIE efficiently insulates the transgenic expression cassette, which as a result, operates autonomously from its *cis*-environment (Duros *et al.*, 2011). The implementation of GIEs in SIN retroviral vectors contributed to the reduction of genotoxicity potential, as shown in cell-culture-based immortalization assays. The development of these powerful insulators with established efficacy holds interesting promise with broad application potential.

9.3.1.3 Targeting integration — Safe harbor

The most successfully used protein-based nucleases are ZFNs, recombinant homing endonuclease (rHEs) and TALENs. ZFNs are chimeric enzymes that consist of two major functional domains, the DNA binding domain and the catalytic domain originating from the type II restriction endonuclease *Flavobacterium okeanokoites* I (FokI) (Wah *et al.*, 1997). ZF-based DNA binding domains are frequently found in nature, especially in transcription factors. ZFNs currently used in gene therapy consist of three to six ZF modules per subunit (Pruett-Miller, 2008). A heterodimeric ZFN pair therefore duly recognizes a DNA stretch of 18–36 bp, which is estimated to be long enough to define a specific and statistically single target site in the human genome. Natural FokI must form a dimer before going through a conformational change which activates the enzyme, resulting in a staggered cut which is 9–13 bp distant from the cognate sequence. Based on this technology, the specific bi-allelic disruption of the HIV-1 co-receptor chemokine receptor 5 (*CCR5*) in human cluster of differentiation (CD)4$^+$ T-cells or CD34$^+$ hematopoietic stem and progenitor cells (HSPCs) and the successful engraftment of these genetically-modified cells in an immunodeficient HIV mouse model resulted in the significant reduction in HIV-1 levels and sustained CD4$^+$ T-cell counts (Perez *et al.*, 2008; Holt *et al.*, 2010). Altogether, these data bring along a novel genetic anti-HIV approach based on ZFN-mediated genome engineering. A clinical trial was initiated in 2010 with genetically modified autologous T-cells at the *CCR5* locus (NCT00842634; NCT01044654).

Similarly, Naldini and co-workers have taken advantage of ZFN technology to direct integration into safe genomic sites, which may allow robust transgene expression without disrupting endogenous transcription. Initially *CCR5* and adeno-associated virus integration site (AAVS) loci were selected since ZFNs for these targets have been developed and extensively characterized (Lombardo *et al.*, 2011). Targeting integration to the *CCR5* locus and sorting enhanced green fluorescent protein (eGFP)-positive cells enriched for 90% of cells with a site-specific integration, up-regulation of the *CCR5* transcript was observed when phosphoglycerate kinase (PGK) or spleen focus-forming virus (SFFV) promoters were used and some flanking genes were also up-regulated. In contrast, targeting into the AAVS1 site did not result in deregulation of flanking genes. Therefore, locus- and promoter-dependent effects do not correlate with the promoter strength.

9.3.1.4 *Targeting integration — Site-specific integration*

Rare-cutting endonucleases have emerged as powerful tools for precise genome engineering. Different types of proteins have been used as designer scaffolds to generate artificial endonucleases with highly specific cleaving sites, which include ZFNs, natural MNs from the mitochondrial intron *Saccharomyces cerevisiae* I (I-SceI) family and, more recently, TALENs. Natural MNs, also called homing endonucleases, are the most specific endonucleases in nature and, thus, should provide ideal scaffolds for the creation of new genome engineering tools. At Cellectis Inc., a genome-wide study to characterize the potential of the MN platform and demonstrate that: (1) efficacy of MN-induced genome editing is locus-dependent, as epigenetic modifications heavily affect the process; (2) cleavage by MNs can induce targeted mutagenesis or homologous gene targeting with a relative ratio which remains markedly stable throughout the genome (Arnould *et al.*, 2006; Smith *et al.*, 2006). Therapeutic applications are extremely challenging and depend on several important criteria: Intrinsic properties such as specific activity, together with the set up of appropriate vector systems in order to streamline targeted recombination efficacy (Takeuchi *et al.*, 2011). So far, depending on the vector system, targeted modifications frequency could reach over 1% in primary cells, including stem cells.

9.3.1.5 *Non-integrating retrovirus vectors: The integration deficient lentivirus vector platform*

As persistent expression of transgenes of interest is not always necessary, transient expression based on non-integrating vectors could help circumvent putative insertional mutagenesis. Along this line, integration-defective retrovectors have been engineered, taking advantage of inactivating mutations introduced in the viral integrase. Integration-deficient γ-retroviral vectors have been described, which translate into very low titers. In addition to this bottleneck, their inability to transduce non-dividing cells makes it unlikely to satisfy the demands of most experiments. In contrast, the so-called integration deficient lentivirus vector (IDLV)-platform has attracted a lot of attention, including with a view to clinical translation in gene therapy settings. IDLVs can be produced through the use of classical integrase mutations affecting the active site. IDLVs generate normal amounts of viral DNA, which is then converted into circular episomes, mostly by cellular DNA repair pathways. The circular molecules are metabolically stable but lack replication signals; they are progressively diluted out in dividing cells but stable in quiescent cells. IDLVs support a variety of genetic modifications, including gene expression, site-specific recombination, transposition and homologous-recombination-mediated gene targeting. Efficient gene expression from IDLVs results in transient transduction of dividing cells and stable gene expression in quiescent cells *in vitro* and *in vivo* (Philippe *et al.*, 2006; Yanez-Munoz *et al.*, 2006).

9.3.2 Targeting Expression

Constraining unwanted expression of the transgene in non-target cells is believed to add to both gene-transfer efficacy and safety, mostly when considering potential toxicity linked to the accumulation of the transgene product and the requirement to regulate expression in time and space from a designed expression cassette. In some cases, it might also be instrumental in preventing insertional mutagenesis in the event where overstimulation of the regulatory sequences driving the transgene might be at stake; however, in target cells, like T-cells in X-SCID patients, this strategy would be unlikely to have been of any added value.

9.3.2.1 *Promoters*

Improvement in the therapeutic index of vectors is a key issue in gene therapy since off-target unwanted expression in cells or tissues might result in toxicity and may also relate to the immune responses directed at the transduced cells and/or the therapeutic proteins themselves, which can curtail long-term gene expression.

Improvement of expression specificity in time and space can be obtained, at least in part, by placing the transgene under the control of a specific promoter which will be turned off and on according to identified functional characteristics. Muscle- or pancreas-specific promoters have been shown to perform. In some cases knowledge generated from transgenic mice has been successfully implemented in order to engineer tissue-specific gene-transfer vectors. In other instances, more complex structures have been used, as in the case of the locus control region when tight control of hemoglobin β-chain synthesis has been sought. Recently, synthetic promoters were designed by VandenDriessche and collaborators (American Society of Gene & Cell Therapy (ASGCT) meeting, Philadelphia, May 2012), which successfully target the liver (Cantore *et al.*, 2012). As mentioned above, the control of specificity can be improved by making use of insulator elements placed inside the retro- or lentivirus SIN-LTR, which prevent variegation due to random integration (Duros *et al.*, 2011 and in preparation). These elements also improve the potential of inducible promoters and the stringency of control, thereby improving both efficacy and safety.

9.3.2.2 MicroRNAs to target expression from retrovirus vectors

MicroRNAs (miRNAs) guide a complex of proteins towards specific target mRNAs, thereby inducing mRNA decay and/or inhibiting translation of the corresponding protein. With this novel approach to gene regulation, third-generation miRNA-regulated expression vectors can be designed and constructed to improve target cell specificity and/or inducibility of viral expression vectors. Indeed, Naldini and co-workers have investigated this novel approach in lentiviral vectors which include a miRNA targeting sequence in the 3′ untranslated region so that the transgenic mRNA is specifically degraded in a cell type that expresses the cognate miRNA. Based on a bi-cistronic vector system to screen candidates, miRNA-126 was defined as specific for HSC expression in murine and human models. Because overexpression of galactocerebrosidase (GALC) in differentiated cells has been proposed for correction of globoid cell leukodystrophy (GLD) and limited by the toxicity of GALC in human HSCs, the target sequence for miRNA-126 was incorporated into a GALC-expressing lentiviral vector. While allowing expression in differentiated progeny, as required, GALC expression was successfully prevented in HSC. This approach resulted in significant survival in a mouse model (Gentner *et al.*, 2010). Moving to neurobiology, studies in Huntington's disease (HD) indicate that the selective dysfunction or death of target neurons type which are most at risk is not solely mediated by damage from the mutant protein. One of the main drivers of the disease and brain damage might be mutant protein toxicity within the glial cells of the central nervous system, especially astrocytes and microglia, and this may even be a prominent contributor to disease initiation in some instances. Deglon and co-workers recently developed a new lentiviral vector, which specifically targets astrocytes with the aim of studying their contribution to HD pathogenesis. They show that pseudotyping with Mokola-virus envelope leads to a partial shift in the tropism of lentiviruses toward astrocytes and when combined with a detargeting strategy with the neuron-specific miRNA124, residual expression is eliminated in neuronal cells. Using this approach, targeted expression of a gene of interest was achieved in astrocytes *in vivo*, and allowed the *in vivo* assessment of the contribution to HD pathogenesis of mutant Huntington when expressed in astrocytes (Faideau *et al.*, 2010). These results also illustrate the potency of the miRNA detargeting approach for cell-type-specific expression in the brain.

9.3.3 Targeting Cell Entry

Further along the way of increasing safety is targeting the cell entry of gene-transfer vectors.

LVs are mainly derived from HIV-1, but also from non-human lentiviruses, such as SIV, feline immunodeficiency virus (FIV) or EIAV. The overall vector design is very similar for all lentiviral vector types. The env protein mediates the entry of the vector particle into its target. HIV-1 env specifically recognizes CD4, a molecule present on the surface of helper T-cells, macrophages and some glial cells. Interestingly, in all retroviruses, including HIV-1, the envelope protein can be substituted by the corresponding protein of another virus, though with somewhat distinct yields. This process, which alters the tropism of the virion, is called pseudotyping. In particular, the envelope glycoprotein of the VSV is predominantly used for pseudotyping of vector particles, which enables transduction of a broad range of target cells. More selective cell entry was achieved in taking advantage of the natural tropism of glycoproteins from other membrane-enveloped viruses.

Several strategies have, in fact, been considered in order to target vector cell-entry to cells of particular interest and with a view to addressing *in vivo* administration routes as the technology and related safety progressively advance. One obvious strategy is to take advantage of established tropism of other virus envelopes. Screening of a large panel of pseudotyped vectors established the superiority of the gibbon ape leukemia virus (GALV) and the cat endogenous retroviral glycoproteins (RD114) for efficient transduction of human hematopoietic cells. In addition, with the replacement of the cytoplasmic tail of RD114 and GALV glycoproteins with that of MLV-A, incorporation of these chimeric glycoproteins was significantly improved and infectious titers increased. Measles virus (MV) glycoproteins also require a modification of their cytoplasmic tails to allow efficient incorporation as part of lentiviral particles. Importantly, Verhoyen, Cosset

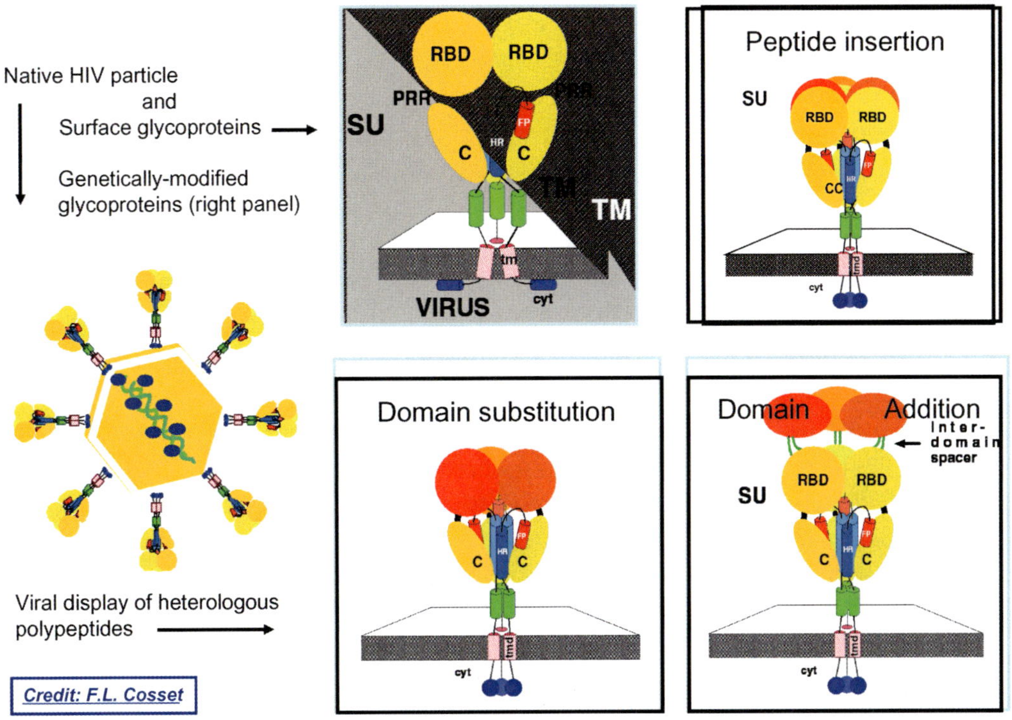

FIGURE 9.9 ■ Surface engineering of LVs by two strategies: (1) genetically-modified glycoproteins; (2) viral display of heterologous peptides.

and co-workers (Frecha *et al.*, 2008) showed that lentivectors pseudotyped with such modified MV glycoproteins can transduce quiescent T- and B-cells more efficiently than VSV-G-pseudotyped LVs. In addition, envelope glycoproteins can be genetically engineered into chimeric glycoproteins (Fig. 9.9) through peptide insertion, domain substitution or polypeptide display.

The targeted *in vivo* transduction of HSCs would represent a major advance, especially in some diseases where these cells are particularly fragile. In addition, vectors intended for *in vivo* transduction must be specific to the target cell, in order to avoid vector spreading while enhancing transduction efficiency. Specific pseudotypes engineered by Verhoyen and Cosset hold major promises in that regard since they have recently reported on successful *in vivo* transduction of human haematopoietic progenitors with a novel lentiral vector displaying stem cell factor (SCF) and the mutant cat endogenous retroviral glycoprotein (RDTR) derived from the RD114 glycoprotein (Frecha *et al.*, 2012). The rationale for using SCF-display is that, *in vivo*, the majority of HSCs are residing in the G0 phase of the cell cycle and are not quite permissive to classical VSV-G pseudotyped LVs. In contrast, vector particles displaying early-activating-cytokine (SCF and/or thrombopoietin) allow a slight and transient stimulation of human

(h) $CD34^+$ cells which results in efficient gene transfer to these target cells. First, these RDTR/SCF-LVs outperformed RDTR-LVs for transduction of $hCD34^+$ cells *in vitro* since 30–40% of these cells could be readily transduced from cord blood mononuclear cells and in the unfractionated bone marrow of healthy and Fanconi anemic donors; subsequent correction of patients' cells was achieved. Then, and most interestingly, *in vivo*, these novel RDTR/SCF-displaying LVs were then shown to be able to distinguish between the target $hCD34^+$ cells of interest and non-target cells, when the vector was directly injected into the bone marrow cavity of humanized BALB/c recombinase activating gene 2 (Rag2)-null/interleukin2 receptor γC (IL2rgc)-null (BALB/c RAGA) mice. This resulted in the highly selective transduction of candidate $hCD34^+$ Lineage $(Lin)^-$ HSCs. Indeed, in the future, these RDTR/SCF LVs might completely alleviate *ex vivo* handling and target cells loss related to *ex vivo* manipulations, with the combined risk of cell death, differentiation, loss of stemness characteristics and homing/engraftment potential along with the *ex vivo* procedures for culture, transduction and/or expansion. In summary, with surface-engineering of LVs the goal is to confer specific properties to vectors intended in particular for *in vivo* gene delivery, since it is meant to: (1) enhance vector stability; (2) decrease sensitivity to human complement; (3) increase gene transfer to specific cell types; (4) mediate targeted gene delivery. This is likely to simplify gene therapy for hematopoietic defects as well as for other diseases using specific targeting strategies, should direct *in vivo* inoculation of the vector off the shell become a clinical reality.

9.3.4 Overcoming Restrictions

Although LVs can transduce metabolically active non-dividing cells, the transduction of some primary human cells is restricted. In particular, primary human myeloid cells can hardly be transduced by LVs: Whereas dendritic cells or macrophages can be transduced, albeit at a comparably low efficiency, primary human monocytes are not transducible with common LVs without previous stimulation. The development of LVs derived from specific LVs which allow the genetic modification of monocyte transduction has been pursued as it can be helpful in both gene therapy approaches targeting myeloid cells, and also for investigating the basis for HIV-1 restriction in these cells. To improve the target cell spectrum of LVs, Schweitzer and collaborators (Schule *et al.*, 2009) derived vectors from SIV of sooty mangabeys (SIVsmmPBj) which are capable of efficient transduction of myeloid cells (monocytes, dendritic cells and macrophages). The viral accessory gene product Vpx was identified as the factor responsible for this interesting feature. Investigation of Vpx-binding proteins led to identification of cellular HIV-1 restriction factors present in myeloid cells (Laguette *et al.*, 2011).

9.3.5 Production of Retroviral Vectors and Scaling Up Towards Manufacture of Clinical-Grade Products

Improvement of retroviral vectors production involves different levels: producer cell design and engineering, optimization of culture medium and biotech-compatible production including process development, bioreaction, downstream processing and storage.

One of the most successful strategies for the generation of optimized transcomplementing cell lines is targeted genome integration, which has been used to successfully derive well-defined, so-called "modular producer cells" engineered to produce high-titer γ-retroviral vectors. Two cell lines were created; Flp293A and 293 FLEX, both derived from 293 cells, which produce particles pseudotyped with either amphotropic or GALV envelopes, respectively (Schucht *et al.*, 2006). The initial step in cell-engineering is based on the identification and tagging of a favorable genetic locus, which is permissive to stable and high retroviral vector production. In order to make this tool compatible with any vector shuttling any gene of interest, two heterologous non-compatible flippase recognition target (FRT) sites flank the tagged retroviral genome at the defined chromosomal site, allowing for recombinase-mediated cassette exchange and a predictable structure of the packaged RNA as a result. Indeed, targeting of gene therapy vectors into this defined genomic site can exclude the generation of dangerous RNA species which might arise from viral vector readthrough up into cellular genes with oncogenic potential. All producer cell clones are similar and high yields of viruses are reproducibly collected whatever the transgene under test: e.g., up to 2.5×10^7 infectious particles per 10^6 cells in 24 h can be obtained within three weeks. The single, defined integration site in a modular producer cell line thus provides the high level of safety required for the clinical implementation of γ-retrovectors.

In addition, during the past years and within the scope of the CliniGene European Union Network of Excellence (EU-NoE), several production steps were optimized, including culture conditions, times of harvesting and the end-steps with purification process and storage conditions (Cruz *et al.*, 2011). Culture conditions and medium composition have a significant effect on viral titers due to their effect upon producer cell metabolism. In particular, elevated medium osmolality, created by addition of osmolytes or high sugar concentrations to the culture medium, can increase specific retrovirus productivity and stability, also related to lipid metabolism in the producer cell. The use of serum-free media is an important step which helps both development of purification strategies and prevention of immunological response to viral preparation, in accordance with quality requirements recommended by regulatory agencies.

Currently, LV-production is mostly based on transient transfection protocols in 293T cells, a simian virus 40 (SV40) T-antigen transfected HEK293 cell line. For gene therapy applications, this transient production protocol has two limitations: (1) it cannot exclude carry-over of plasmid-derived sequences and accidental recombination events with the putative formation of vectors with altered properties which cannot easily be detected; (2) the nature of the producer cell line with e.g., SV40 T antigen raises safety concerns. A stable and safe producer cell line intended for biotech-based manufacture of LVs would both represent a key safety improvement and broaden the potential for clinical applications. Nevertheless, the experience of the last decade shows that the construction of a stable lentiviral packaging cell-line is not straightforward. Due to well-identified cytotoxic properties of helper proteins such as protease, rev and VSV-G envelope, one of the major obstacles to the successful generation of a stable packaging cell line is the need for controlled expression systems that allow accurate adjustment of protein expression. Strategies that circumvent or reduce the need for gene regulation and the exploitation of novel gene switcher systems (Brousseau *et al.*, 2008) resulted in the generation of lentiviral helper cell lines with reasonable stability and titers. So far, the copy number of the therapeutic vector has not been considered since in all published reports the number of lentiviral genomes in stable producer cell lines has not been determined. Protocols used in these studies commonly allow integration of increased numbers of integrated vector copies. For example, satisfactory titers were obtained upon introduction of hundreds of concatemerized lentiviral vector copies into the cellular genome (Wirth *et al.*, 2007). These data point towards the potential need for multiple-copy integration of LVs in order to bypass a still undefined bottleneck in their biotech-compatible manufacture. The observation that both T-antigen (T-Ag) and rat sarcoma virus proto-oncogene (ras) expression improve this blockade potentially opens a way to overcome current limitations. However, strategies should be pursued which preferably disregard the use of elements with oncogenic potential such as T-Ag and ras in order to achieve significant improvement in lentivector production.

9.4 Applications

9.4.1 Clinical Gene Transfer and Therapy

Gene therapy with HSCs is an attractive therapeutic strategy for several forms of primary immunodeficiencies. Current approaches are based on *ex vivo* gene transfer of the therapeutic gene into autologous HSC by vector-mediated gene transfer. Results of the adenosine deaminase deficiency-severe combine immunodeficiency (ADA-SCID) gene therapy trial conducted at the H. San Raffaele-Telethon Institute for Gene Therapy (HSR-TIGET, Milan) have demonstrated long-term restoration of immune competence and clinical benefit when infusion of bone marrow $CD34^+$ cells transduced with a retroviral vector encoding ADA was combined with a reduced conditioning (Aiuti *et al.*, 2009). The multi-lineage contribution to haematopoiesis of progenitor clones was demonstrated, showing fluctuating lineage outputs for several years. Despite the occurrence of insertions near potentially oncogenic genomic sites, data show that transplantation of ADA-transduced HSC does not result in skewing or expansion of malignant clones *in vivo* (Biasco *et al.*, 2011). In contrast to the SCID-X1 trials, no vector-mediated leukemias developed in trials for ADA-SCID. The ADA-SCID trials used similar vectors to those used in the SCID-X1 trials and enrolled as many subjects. This difference raised questions as to whether the disease itself was a factor in the development of these cases of leukemia.

Moving to the treatment of other primary immunodeficiencies, extensive non-clinical studies have demonstrated that a SIN-lentiviral vector encoding for human WAS protein under the control of a homologous 1.6 kb promoter efficiently transduced human $CD34^+$ cells corrected the human and mouse phenotype and was not associated with toxicity. In 2010, a phase I/II gene therapy protocol was started at HSR-TIGET based on LV-gene transfer into WAS patient's $CD34^+$ cells, combined to reduced intensity conditioning. Long-term assessment will provide key information on the safety, biological activity and efficacy of LV-based HSC gene therapy for WAS.

In X-linked adrenoleukodystrophy (X-ALD) the first clinical success has been reported with lentivirus-vector transduced haematopoietic cells (Cartier *et al.*, 2009). A gene therapy strategy based on the reinfusion of autologous HSC corrected with a lentiviral vector was developed, since in X-ALD cerebral demyelination can be stopped or reversed within 12–18 months by allogeneic HSC transplantation (HSCT). Stable correction of peripheral leukocytes was demonstrated and stabilization of the demyelinating lesions for up to nearly five years. Integration sites analysis revealed a common pattern in both myeloid and lymphoid lineages, suggesting that true stem cells have been genetically-modified, resulting in clinical benefit. In the first two patients treated, HSC-based gene therapy resulted in neurological improvement similar to that for HSCT.

9.4.2 Induction of iPS and Stem Cell Engineering

LVs have been successfully used to generate iPS cells. A unique feature of lentiviruses is that they are able to transduce both non-dividing (slowly dividing or quiescent but metabolically active cells) and dividing cells, allowing the generation of iPS from most cell types. As a next step towards safety improvement, excisable integrating vectors have been engineered in order to generate transgene-free iPS and help prevent the above-mentioned drawbacks as well as the following. In addition to being placed under the control of viral promoters, the stable integration of transgenes encoding transcription factors or oncogenes involved in cell proliferation such as the avian myelocytomatosis viral oncogene homolog (*c-MYC*) harbors a substantial risk of malignant transformation should reprogramming factors not be fully silenced or incidentally be reactivated during differentiation. Moreover, viral promoter reactivation could lead to the deregulation of *cis*-neighboring genes; the latter represents an additional mechanism which might compromise cell-cycle integrity. Excisable LVs have been engineered which include both a locus of X-over P1 (loxP) site in the 3′ LTR and an inducible promoter driving transgene expression. During virus replication, the loxP site is duplicated in the 5′ LTR so that the integrated transgenic cassette is flanked with a loxP site at both ends. The excision of the reprogramming factors follows the targeted and transient expression of the "causes recombination" (Cre) recombinase in transduced cells which induces a recombination event between loxP sites. Using this system, Jaenisch and his group (Soldner *et al.*, 2009) were able to generate transgene-free human iPS cells which are able to maintain their pluripotent state and display a global gene expression profile similar to that of human ES cells. These iPS cells could further differentiate into dopaminergic neurons. The major limitation of this study is that reprogramming factors were primarily integrated at different independent sites, which resulted in multiple transgene excisions upon Cre recombinase expression. In fact, multiple and simultaneous recombination reactions could lead to genome rearrangement and genomic instability. In order to overcome this drawback, Chang *et al.* (2009) designed a polycistronic lentiviral vector encoding defined reprogramming factors separated by 2A sequences, resulting in the integration of a single reprogramming cassette flanked by two loxP sites. Following Cre-recombinase-mediated excision, the iPS cells lines generated only three lentiviral LTR signatures which consist of a single loxP site that does not interrupt coding sequences, promoters or regulatory elements. Although conceptually elegant, this system holds a risk of non-specific recombination events and genomic instability should Cre recombinase expression not be tightly enough controlled.

Another commonly used heterologous recombination system is the flippase (Flp)/FRT recombinase/targets system from *Saccharomyces cerevisiae*. While it is supposedly less efficient than the Cre/loxP system it conversely exhibits far less toxicity, a feature which is essential when working with primary cells. To date, there has been no report of generation of human iPS cells, while murine iPS cells using this system with a polycistronic lentivector in which the reprogramming cassette was flanked with two FRT sites. These mouse iPS cells were further transduced with empty MLV retrovirus-like particles which shuttle the Flp recombinase fused to the gag–pol polyprotein. This process resulted in the complete removal of the reprogramming cassette (Voelkel *et al.*, 2010). Factor-free iPS resulting from heterologous recombination systems thus represents a more suitable source of cells for human disease modeling. However, these iPS cells still harbor scars of insertion sites and are not "genetically clean" pluripotent stem cells, a feature which might still alleviate translation to cell-based therapies.

9.4.3 Gene Ablation: Short Hairpin RNAs

In parallel with the development of non-replicative vectors, important developments were achieved using replication-competent retroviruses to deliver interfering RNA to tumor cells *in vitro* and *in vivo*. The primary goal was to study how gene function may affect tumor biology but efficacy was also evaluated. *In vitro* studies were performed to determine both delivery and knockdown efficiencies of a miRNA-adapted short hairpin RNA (shRNA) expression cassette inserted between env and 3′ LTR by silencing luciferase and eGFP *in vitro*. In a next step, proof of concept was established *in vivo* by silencing Polo-like kinase 1 (PLK1) by intratumoral application in a HT1080-derived tumor. Significant reduction of tumor volume was observed after 25 days (Schaser *et al.*, 2011). Tumor specificity and long-term down-regulation of the desired target gene were confirmed.

There are too many other reports to mention where lentivectors have been used to efficiently deliver shRNA and successfully knockdown a gene function, including with conditional ablation, making use of a regulatory switch such as the tetracycline (Tet)-inducible system. This leads us to mention that lentivectors have also been used in transgenesis and for this purpose, have been directly injected into oocytes to generate transgenic mice.

9.5 Prospects

One key bottleneck of translating innovative technologies from bench to bedside is related to the production of the retargeted cell-type-specific vectors or alternative serotypes. As high-titer vector preparation strongly depends on the capsid or envelope, the production efficiency substantially varies between different vector types and according to the design of the transgene cassette. The ability to produce high-titer vectors that are capable of achieving increased gene-transfer efficiency and expression in the desired target cells while preventing expression in non-target cells and genotoxicity are measurable outcomes and addresses a current unmet need in the field of gene therapy. In addition, challenges in the use of retrovirus vectors in future clinical trials, include determining whether the data from preclinical models are sufficient to move into the clinic, addressing those patients for whom the risk:benefit ratio is reasonably appropriate and ensuring that the informed consent process adequately communicates the complex challenges and risks that the use of integrating vectors for long-term gene correction presents. While the design of better vectors and models for testing will likely lead to decreased risk of oncogenic disease, it will likely take many years of clinical experience to definitively establish this safety. In addition, combining tight transcription regulation, targeting of cell entry and reducing the injected dose of vectors will likely facilitate the clinical implementation and expedite first-in-man clinical trials, including via direct *in vivo* administration routes.

References

Aiuti, A., Cattaneo, F., Galimberti, S., *et al.* (2009). Gene therapy for immunodeficiency due to adenosine deaminase deficiency, *N Engl J Med*, **360**, 447–458.

Arnould, S., Chames, P., Perez, C., *et al.* (2006). Engineering of large numbers of highly specific homing endonucleases that induce recombination on novel DNA targets, *J Mol Biol*, **355**, 443–458.

Biasco, L., Ambrosi, A., Pellin, D., *et al.* (2011). Integration profile of retroviral vector in gene therapy treated patients is cell-specific according to gene expression and chromatin conformation of target cell, *EMBO Mol Med*, **3**, 89–101.

Brousseau, S., Jabbour, N., Lachapelle, G., *et al.* (2008). Inducible packaging cells for large-scale production of lentiviral vectors in serum-free suspension culture, *Mol Ther*, **16**, 500–507.

Cantore, A., Nair, N., Della Valle, P., *et al.* (2012). Hyperfunctional coagulation factor IX improves the efficacy of gene therapy in hemophilic mice, *Blood*, **120(23)**, 4517–4520.

Cartier, N., Hacein-Bey-Abina, S., Bartholomae, C.C., *et al.* (2009). Hematopoietic stem cell gene therapy with a lentiviral vector in X-linked adrenoleukodystrophy, *Science*, **326**, 818–823.

Chang, C.W., Lai, Y.S., Pawlik, K.M., *et al.* (2009). Polycistronic lentiviral vector for "hit and run" reprogramming of adult skin fibroblasts to induced pluripotent stem cells, *Stem Cells*, **27**, 1042–1049.

Corrigan-Curay, J., Cohen-Haguenauer, O., O'Reilly, M. (2012). Challenges in vector and trial design using retroviral vectors for long-term gene correction in hematopoietic stem cell gene therapy: Summary of a symposium sponsored by the NIH Office of Biotechnology Activities and the EC DG-research NoE for the Advancement of Clinical Gene Transfer and Therapy, *Mol Ther*, **20**, 1084–1089.

Cosset, F.L., Takeuchi, Y., Battini, J.L., *et al.* (1995). High-titer packaging cells producing recombinant retroviruses resistant to human serum, *J Virol*, **69**, 7430–7436.

Cruz, P.E., Rodrigues, T., Carmo, M., *et al.* (2011). Manufacturing of retroviruses, *Methods Mol Biol*, **737**, 157–182.

Duros, C., Artus, A., Gaussin, A., *et al.* (2011). Insulated lentiviral vectors towards safer gene transfer to stem cells. In 2011 Annual Meeting of the Society of Gene and Cell Therapy, Seattle Washington, *Mol Ther*, **19**, Supplement 1, S149.

Faideau, M., Kim, J., Cormier, K., *et al.* (2010). *In vivo* expression of polyglutamine-expanded huntingtin by mouse striatal astrocytes impairs glutamate transport: A correlation with Huntington's disease subjects, *Hum Mol Genet*, **19**, 3053–3067.

Frecha, C., Costa, C., Lévy, C., *et al.* (2008). Stable transduction of quiescent T cells without induction of cycle progression by a novel lentiviral vector pseudotyped with measles virus glycoproteins, *Blood*, **112**, 4843–4852.

Frecha, C., Costa, C., Negre, D., *et al.* (2012). A novel lentiviral vector targets gene transfer into human hematopoietic stem cells in marrow from patients with bone marrow failure syndrome and *in vivo* in humanized mice, *Blood*, **119**, 1139–1150.

Gaussin, A., Modlich, U., Bauche, C., *et al.* (2011). CTF/NF1 transcription factors act as potent genetic insulators for integrating gene transfer vectors, *Gene Ther*, 1–10.

Gentner, B., Visigalli, I., Hiramatsu, H., *et al.* (2010). Identification of hematopoietic stem cell-specific miRNAs Enables gene therapy of globoid cell leukodystrophy, *Sci Transl Med*, **2**, 58ra84.

Hacein-Bey-Abina, S., Garrigue, A., Wang, G.P., *et al.* (2008). Insertional oncogenesis in 4 patients after retrovirus-mediated gene therapy of SCID-X1, *J Clin Invest*, **118**, 3132–3142.

Hacein-Bey-Abina, S., Von Kalle, C., Schmidt, M., *et al.* (2003). LMO_2-associated clonal T cell proliferation in two patients after gene therapy for SCID-X1, *Science*, **302**, 415–419.

Holt, N., Wang, J.B., Kim, K., *et al.* (2010). Human hematopoietic stem/progenitor cells modified by zinc-finger nucleases targeted to CCR5 control HIV-1 *in vivo*, *Nat Biotechnol*, **28**, 839–847.

Laguette, N., Sobhian, B., Casartelli, N., *et al.* (2011). SAMHD1 is the dendritic- and myeloid-cell-specific HIV-1 restriction factor counteracted by Vpx, *Nature*, **474**, 654–657.

Lombardo, A., Cesana, D., Genovese, P., *et al.* (2011). Site-specific integration and tailoring of cassette design for sustainable gene transfer, *Nat Meth*, **8**, 861–869.

Modlich, U., Bohne, J., Schmidt, M., *et al.* (2006). Cell-culture assays reveal the importance of retroviral vector design for insertional genotoxicity, *Blood*, **108**, 2545–2553.

Modlich, U., Navarro, S., Zychlinski, D., *et al.* (2009). Insertional transformation of hematopoietic cells by self-inactivating lentiviral and gammaretroviral vectors, *Mol Ther*, **17**, 1919–1928.

Ott, M.G., Schmidt, M., Schwarzwaelder, K., *et al.* (2006). Correction of X-linked chronic granulomatous disease by gene therapy, augmented by insertional activation of MDS1-EVI1, PRDM16 or SETBP1, *Nat Med*, **12(4)**, 401–409.

Perez, E.E., Wang, J.B., Miller, J.C., *et al.* (2008). Establishment of HIV-1 resistance in $CD4^+$ T cells by genome editing using zinc-finger nucleases, *Nat Biotechnol*, **26**, 808–816.

Philippe, S., Sarkis, C., Barkats, M., *et al.* (2006). Lentiviral vectors with a defective integrase allow efficient and sustained transgene expression *in vitro* and *in vivo*, *Proc Natl Acad Sci USA*, **103**, 17684–17689.

Pruett-Miller, S.M., Connelly, J.P., Maeder M.L, *et al.* (2008). Comparison of zinc finger nucleases for use in gene targeting in mammalian cells, *Mol Ther*, **16**, 707–717.

Schaser, T., Wrede, C., Duerner, L., *et al.* (2011). RNAi-mediated gene silencing in tumour tissue using replication-competent retroviral vectors, *Gene Ther*, **18**, 953–960.

Schroder, A.R., Shinn, P., Chen, H., *et al.* (2002). HIV-1 integration in the human genome favors active genes and local hotspots, *Cell*, **110**, 521–529.

Schucht, R., Coroadinha, A.S., Zanta-Boussif, M.A., *et al.* (2006). A new generation of retroviral producer cells: predictable and stable virus production by Flp-mediated site-specific integration of retroviral vectors, *Mol Ther*, **14**, 285–292.

Schule, S., Kloke, B.P., Kaiser, J.K., *et al.* (2009). Restriction of HIV-1 replication in monocytes is abolished by Vpx of SIVsmmPBj, *PLoS One*, **4**, e7098.

Smith, J., Grizot, S., Arnould, S., *et al.* (2006). A combinatorial approach to create artificial homing endonucleases cleaving chosen sequences, *Nucleic Acids Res*, **34**, e149.
Soldner, F., Hockemeyer, D., Beard, C., *et al.* (2009). Parkinson's disease patient-derived induced pluripotent stem cells free of viral reprogramming factors, *Cell*, **136**, 964–977.
Stein, S., Ott, M.G., Schultze-Strasser, S., *et al.* (2010). Genomic instability and myelodysplasia with monosomy 7 consequent to EVI1 activation after gene therapy for chronic granulomatous disease, *Nat Med*, **16**, 198–204.
Suerth, J.D., Maetzig, T., Galla, M., *et al.* (2010). Self-inactivating alpharetroviral vectors with a split-packaging design, *J Virol*, **84**, 6626–6635.
Takeuchi, R., Lambert, A.R., Mak, A.N., *et al.* (2011). Tapping natural reservoirs of homing endonucleases for targeted gene modification, *Proc Natl Acad Sci USA*, **108**, 13077–13082.
Thrasher, A.J., Gaspar, H.B., Baum, C., *et al.* (2006). Gene therapy: X-SCID transgene leukaemogenicity, *Nature*, **443**, E5–6, discussion E6–7.
Voelkel, C., Galla, M., Maetzig, T., *et al.* (2010). Protein transduction from retroviral Gag precursors, *Proc Natl Acad Sci USA*, **107**, 7805–7810.
Wah, D.A., Hirsch, J.A., Dorner, L.F., *et al.* (1997). Structure of the multimodular endonuclease FokI bound to DNA, *Nature*, **388**, 97–100.
Wirth, D., Gama-Norton, L., Riemer, P., *et al.* (2007). Road to precision: recombinase-based targeting technologies for genome engineering, *Curr Opin Biotechnol*, **18**, 411–419.
Wu, X., Li, Y., Crise, B., *et al.* (2003). Transcription start regions in the human genome are favored targets for MLV integration, *Science*, **300**, 1749–1751.
Yanez-Munoz, R.J., Balaggan, K.S., MacNeil, A., *et al.* (2006). Effective gene therapy with nonintegrating lentiviral vectors, *Nat Med*, **12**, 348–353.
Yu, S.F., von Rüden, T., Kantoff, P.W., *et al.* (1986). Self-inactivating retroviral vectors designed for transfer of whole genes into mammalian cells, *Proc Natl Acad Sci USA*, **83**, 3194–3198.
Zychlinski, D., Schambach, A., Modlich, U., *et al.* (2008). Physiological promoters reduce the genotoxic risk of integrating gene vectors, *Mol Ther*, **16**, 718–725.

10

ADENOVIRUS VECTORS

Stefan Kochanek [a]

10.1 Introduction

Adenovirus was first isolated in 1953 from the adenoids of a patient (resulting in the name of the virus) and is known as a pathogen which frequently causes relatively mild diseases of the respiratory tract, the eye or other organs. However, in some cases, in particular in immunocompromised patients, the disease can be very serious, resulting in disseminated adenovirus infection, and may even follow a fatal clinical course.

On the other hand, adenoviruses have been very valuable models for studying transformation, oncogenesis, replication, transcription, translation, cell-cycle control and other important principles in cell and molecular biology. The discovery of RNA splicing, for example, is based on work with adenoviruses and the Nobel Prize in Physiology/Medicine 1993 has been awarded to Drs. Roberts and Sharp on this topic. Therefore, the biology of adenovirus is understood in great molecular detail, although many things are still to be learned.

To date, more than 400 clinical gene therapy trials have been based on adenovirus vectors. Thus, among all viral and non-viral vectors this vector type has been used most frequently in the clinic.

10.2 General Aspects of Adenovirus Biology

Adenoviruses have been isolated from many vertebrates, including mammals, reptiles, birds, amphibia and fish. The virus family of *Adenoviridae* is divided

[a] Department of Gene Therapy, Ulm University Helmholtz Str. 8/1, D-89081 Ulm, Germany
Email: stefan.kochanek@uni-ulm.de

into five genera (Mastadenovirus, Atadenovirus, Aviaadenovirus, Siadadenovirus and Ichtadenovirus). Within the genus of Mastadenovirus there are currently seven human species (A to G), encompassing altogether 52 human adenovirus types. Adenoviruses are non-enveloped viruses with a molecular weight of approximately 150 MDa. The shape of the particle is that of an icosahedron, meaning that the particle has 20 faces and 12 vertices. The diameter of the particles is approximately 100 nm. The virus particle carries a double-stranded DNA genome with a size of approximately 30–40 kb, depending on the specific adenovirus type. Human adenovirus particles consist of 11 structural proteins, seven to form the capsid and four that are packaged with the DNA in the core (Fig. 10.1).

A regular infectious cycle starts with the uptake of the virus particle by receptor-mediated endocytosis. Here the tip of the fiber protein (called the fiber knob) binds to the Coxsackie adenovirus receptor (CAR). This is true for most human adenovirus types from species A, C, D, E and F. However, some human adenovirus types have been shown to bind to other cellular receptors including, for example, cluster of differentiation 46 (CD46) or CD80/86. Interestingly, recent research indicates that while for virus uptake *in vitro* the above-mentioned CAR is used in general, the mechanism for cell entry *in vivo* might be more complicated. For example, it has recently been shown that serum proteins, after binding to the virus particle, may influence tropism.

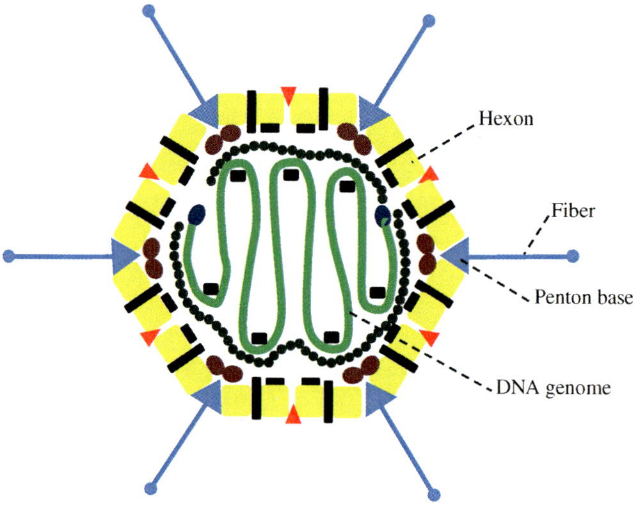

FIGURE 10.1 ■ Scheme of an adenovirus particle as a non-enveloped virus carrying a double-stranded DNA genome in its core. The major capsid protein hexon and two proteins, involved in entry (fiber and penton basis) are indicated. (Figure modified from Stewart and Bumett, 1993).

After receptor-mediated endocytosis the virus particle is disassembled in endosomes. The virus has invented mechanisms to penetrate from the endosome to the cytoplasm. The particle is then transported on microtubules to the nucleus and, following further disassembly, the adenoviral genome enters the nucleus through the nucleopore.

In the nucleus a transcriptional program is initiated starting with expression of the so-called early (E) genes. Expression of the early genes has three main functions: First, to induce the host cell to enter S-phase of the cell cycle as an optimal environment for virus replication. Second, synthesis of viral proteins that are needed for replication of the viral DNA genome. Third, strategies to protect the infected cell from antiviral strategies of the host.

Expression of the so-called late (L) genes is initiated once the viral genomic DNA has started to replicate. Now the structural proteins are synthesized leading to the assembly of new capsids and followed by the encapsidation of the newly synthesized viral DNA. This leads to the assembly of new capsids and also encapsidation of the newly synthesized viral DNA.

The infectious cycle ends with the death of the cell and the release of the newly formed virus particles.

10.3 First-Generation Adenovirus Vectors

Most first-generation adenovirus vectors have been based on human adenovirus type 5 (hAd5). However, more recently adenoviral vectors have been developed based on other human adenovirus types including hAd26 and hAd35, or on adenoviruses derived from other species, including from the chimpanzee. First-generation vectors retain most of the adenoviral genes, except the E1A and E1B genes, the first genes expressed after the viral genome enters the nucleus, and which are essential both for tuning the cell into a factory for adenovirus production and for the activation of all other viral genes during the natural course of an infection. As a result, adenoviral vectors not expressing E1 genes are able to transduce replicating or non-replicating target cells however, they will not replicate in these cells. This is the reason that this vector type is also called replication-deficient or ΔE1 adenoviral vector. Since the E1 genes are removed from the vector genome, the vector has to be produced in a cell line that stably expresses the E1 genes. A commonly used cell line for production of adenoviral vectors is the well-known 293 cell line. Replacing the E1 genes at the left part of the viral genome, the vector carries an expression cassette coding, for example, for a therapeutic protein. The capacity of first-generation vectors for the transport of foreign DNA is limited to about 7 or 8 kb.

A distinct disadvantage of this vector for classical gene therapy applications, in which a missing gene function in a host is to be replaced by the introduced transgene, is the fact that despite removal of the E1 genes there is some residual expression of other adenoviral genes. In an immunocompetent host, this results in an anti-adenoviral immune response against vector-transduced cells and, therefore, elimination of these cells from the organism. Thus, transient gene expression is the consequence. This is the reason why this vector is commonly used for genetic vaccination or tumor therapy, i.e., applications that do not depend on long-term gene expression after gene transfer.

10.4 High-Capacity Adenovirus Vectors

High-capacity adenovirus (HC-Ad) vectors, which are also called helper-dependent and also "gutless" adenoviral vectors, address both the limitation in capacity for uptake of foreign DNA and the frequently observed immunogenicity of first-generation adenoviral vectors, discussed in the previous sections. In this vector type, all viral genes are removed, resulting in a total capacity for the uptake of foreign DNA of approximately 35 kb (Fig. 10.2). For stability reasons during production of this

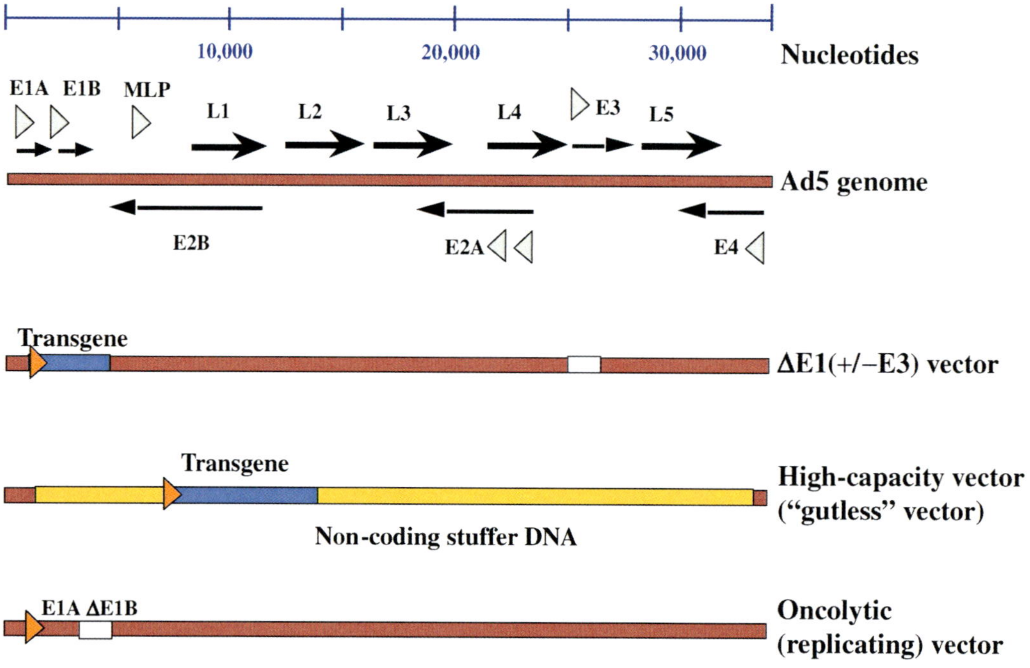

FIGURE 10.2 ■ Scheme of transcription map of hAd5 (upper part) and the main adenoviral vector types (lower part). Arrows correspond to transcripts, triangles to promoters.

vector type (vectors with very small genomes of less then 20 kb size tend to rearrange during production), "stuffer DNA" is included in this vector to bring the total size of the vector (including the therapeutic gene) to about 30 kb. Several studies performed in small and large animals have indicated that toxicity and immunogenicity of this vector type is considerably reduced compared with first-generation vectors, resulting in long-term gene expression following transduction of non-replicating cell types.

The production system of this vector type is based on the causes recombination–locus of X-over P1 (Cre–loxP) recombination system of bacteriophage P1. For production, a helper virus is used that has a deletion of the E1 genes and a packaging signal that is flanked by two loxP sites. The vector carrying the therapeutic gene is produced in the 293 cell line that, in addition to E1A and E1B, expresses the Cre recombinase of bacteriophage P1. The production cell is infected with the vector and the helper virus at the same time. Only the vector genome is packed into viral capsids while the packaging signal of the helper virus is removed via Cre-mediated excision and thus cannot be packaged into the viral capsid. While the biological properties of this vector type are clearly advantageous compared with first-generation vectors, production is very cumbersome and difficult to standardize. This is the reason why currently nearly all clinical trials have been performed with first-generation adenoviral vectors.

10.5 Conditionally Replicating Adenoviral Vectors

There is a strong interest in the development of adenoviral vectors that are suitable for cancer treatment. First-generation adenoviral vectors are efficient in transduction of primary cells, including tumor cells, but are very limited in their ability to spread within a tumor. Therefore, adenoviral vectors have been developed that replicate in tumor cells but not in normal cells, thereby resulting in amplification and improved spreading within a tumor. Different principles have been pursued to achieve this aim. The best known examples are based on adenoviral vectors that carry mutations either in the E1B region or in the E1A region, thus preventing interaction with cellular binding partners such p53 and retinoblastoma protein (Rb). Due to mechanisms that are not yet totally clear, these viruses replicate more efficiently in transformed cells compared with normal untransformed cells, resulting in selective amplification in tumor cells with sparing of normal untransformed cells. A second type of conditionally replicating adenoviruses (also called CRAds) takes advantage of using a tumor- or cell-specific promoter to control the expression of the E1A gene, which is the first viral gene that is expressed after the virus genome reaches the nucleus of a cell following infection, thereby restricting replication of the virus to cells in which that specific promoter is active (Fig. 10.2).

10.6 Application of Adenoviral Vectors

There are three main areas in which adenoviral vectors are preferentially used.

As a research tool and for analysis of gene function under normal or pathological conditions adenoviral vectors are very frequently used to express RNAs coding either for specific proteins or inhibitory principles such as short hairpin RNAs (shRNAs). This can be done very efficiently *in vitro*, since adenoviral vectors allow for efficient transduction of replicating and non-replicating cells. Very frequently adenoviral vectors are used to address scientific questions by *in vivo* application in small or large animals, either through systemic administration by intravenous injection or by local administration by direct injection (for example into the muscle).

A second area of interest, already mentioned above, relates to the use of adenoviral vectors for tumor therapy. A large number of clinical trials have been performed with first-generation adenoviral vectors or with oncolytic adenoviral vectors, respectively, with anecdotal evidence of antitumor activity in some cases. In fact, an adenovirus with a mutation in the E1B region has been approved as a drug in China. Overall results, however, have been rather disappointing, pointing to a need for improvements in the activity of these vectors and/or also attempts to use these vectors in combination with chemotherapy or radiotherapy.

The use of adenoviral vectors as a genetic vaccine is a third area of current significant interest. Adenoviral vectors are very potent genetic vaccines. This is likely to do with an adjuvant effect brought about by the viral capsid and the ability of this vector type to transduce antigen-presenting cells, resulting in strong immune responses both at T- and B-cell levels.

10.7 Recent Developments in Adenoviral Vector Technology

Introducing specific peptide ligands into the capsid of adenoviral vectors by genetic engineering at the same time as preventing interaction with the natural receptors is a current research topic of many investigators. The aim is to change the tropism of adenoviral vectors towards one that is more favorable for a specific application.

Another activity of interest in the genetic vaccine field relates to the use of adenovirus serotypes other than those based on human hAd5. About 60% of adults have pre-existing immunity against hAd5 due to previous contact with hAd5 through natural infection. In principle, pre-existing cellular and/or humoral immunity directed towards hAd5 may influence immune responses against the vector-expressed protein of interest. Although there are also indications from large-animal studies and clinical trials that pre-existing immunity against hAd5 does not prevent vaccination with hAd5-based vectors, many research groups are currently investigating the use of adenoviral vectors based on other serotypes from humans or other species or by generating chimeric adenoviral vectors, in which part of the capsid comes from one adenovirus serotype, the rest from another; in both cases with the aim of further improving vaccination efficiency with adenoviral vectors.

Again another current research activity addresses the finding that adenoviral vectors interact *in vivo* with many cellular or non-cellular components. Here, the vector particle is furnished with a shield consisting of an inert polymer such as polyethylenglycol (PEG) or poly(N-2-hydroxypropyl)methacrylamide (HPMA). For example, it has been recently shown that the natural receptor CAR is not only present on nuclei-containing cells but also on human erythrocytes, which likely will result in binding of adenoviral vectors to erythrocytes after intravenous injection, thereby functionally inactivating the adenoviral vector for delivery to target cells.

Taken together, improving vector activity and *in vivo* performance will remain an important research area in the future.

For Further Reading

Original Articles

Carlisle, R.C., Di, Y, Cerny, A.M., *et al.* (2009). Human erythrocytes bind and inactivate type 5 adenovirus by presenting Coxsackie virus-adenovirus receptor and complement receptor 1, *Blood*, **113**, 1909–1918.

Stewart, P.L., Burnett, R.M. (1993). Adenovirus structure as revealed by X-ray crystallography, electron microscopy, and difference imaging, *Jpn J Appl Phys*, **32**, 1342–1347.

Waddington, S.N., McVey, J.H., Bhella, D., *et al.* (2008). Adenovirus serotype 5 hexon mediates liver gene transfer, *Cell*, **132**, 397–409.

Review Articles

Berk, A.J. (2007). 'Adenoviridae: the viruses and their replication', in Knipe, D.M., Howley, P.M. (eds.) *Fields Virology*, Philadelphia PA: Lippincott Williams & Wilkins, pp. 2355–2394.

Fisher, K.D., Seymour, L.W. (2010). HPMA copolymers for masking and retargeting of therapeutic viruses, *Adv Drug Deliv Rev*, **62**, 240–245.

Imperiale, M.J., Kochanek, S. (2004). Adenovirus vectors: biology, design, and production, *Curr Top Microbiol Immunol*, **273**, 335–357.

Kreppel, F., Kochanek, S. (2007). Modification of adenovirus gene transfer vectors with synthetic polymers: a scientific review and technical guide, *Mol Ther*, **16**, 16–29.

Wold, W.S.M., Horwitz, M.S. (2007). 'Adenoviruses', in Knipe, D.M., Howley, P.M. (eds.) *Fields Virology*, Philadelphia PA: Lippincott Williams & Wilkins, pp. 2395–2436.

Yamamoto, M., Curiel, D.T. (2010). Current issues and future directions of oncolytic adenoviruses, *Mol Ther*, **18**, 243–250.

11

ADENO-ASSOCIATED VIRUS (AAV) VECTORS

Aurélie Ploquin[a], Hildegard Büning[b] and Anna Salvetti[a,c]

11.1 Introduction

Among the variety of viral vectors, those derived from the human parvovirus adeno-associated virus (AAV) have emerged as a very efficient tool for *in vivo* gene transfer into a variety of tissues and animal species during the last two decades. The relative simplicity of the organization of the AAV genome and the non-pathogenic properties of the parental AAV have greatly contributed to the use of this viral vector among the gene transfer community. However, the limited knowledge of the wild-type virus compared with other viral vectors has required considerable effort to gain insight into wild-type AAV biology in order to be able to improve the AAV vector system for therapy. This review will summarize the most important features of both wild-type and recombinant AAV (rAAV) to show how the increased understanding of the biology of the virus and vector have enabled rAAV to lead the *in vivo* gene transfer field.

11.2 Wild-Type AAV Biology

AAV was discovered in the early 1960s as a contaminant of adenovirus preparations and soon thereafter classified as a member of the parvovirus family.

[a] INSERM U758-Human Virology Unit, Ecole Normale Supérieure de Lyon, 46 allée d'Italie, 69007 Lyon, France
[b] Department I of Internal Medicine and Center for Molecular Medicine, Cologne, University of Cologne, Robert-Koch Str. 21, 50931 Cologne, Germany
[c] Email: anna.salvetti@ens-lyon.fr

Owing to its replication-defective phenotype, this virus, initially named "satellite", defined a new genus within the *Parvoviridae*, the *Dependoviruses*. Indeed, further studies showed that this virus was unable to replicate alone but required the assistance of a helper virus. Besides adenovirus, the first helper virus identified and responsible for the term "adeno-associated", other viruses have also been shown to be able to mediate the helper effect including members of the herpes virus family, herpes simplex virus (HSV) or *Cytomegalovirus*, and human papillomavirus (HPV) (Daya and Berns, 2008). This section will review the major properties of wild-type AAV derived from *in vitro* studies and *in vivo* analyses.

11.2.1 The AAV Genome and Life Cycle

AAV particles are composed of a naked 20 nm capsid that contains a 4.7 kb single-stranded (ss) DNA genome. The viral genome contains two genes, replication (*rep*) and capsid (*cap*), flanked by 145 bp inverted terminal repeats (ITRs) that serve as origins of DNA replication (Fig. 11.1). Four Rep proteins are produced from the *rep* gene using two different promoters and splicing patterns. The two major proteins,

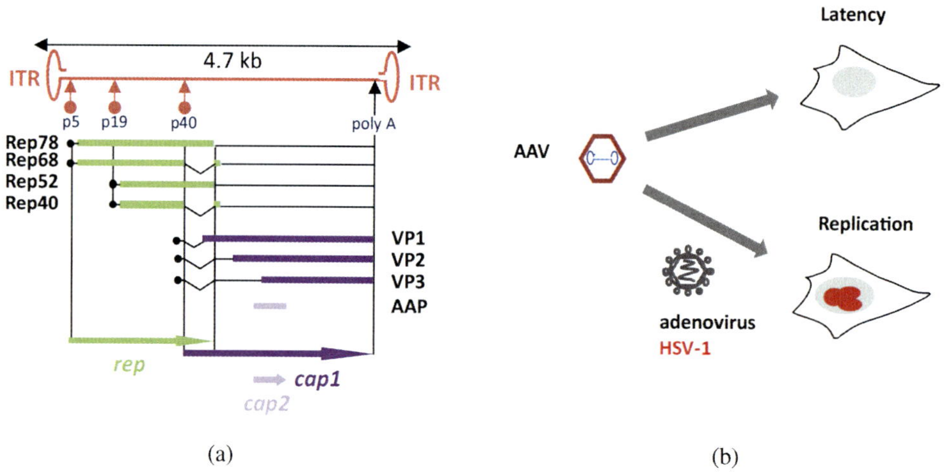

FIGURE 11.1 ■ (a) Organization of the wild-type AAV2 genome. The viral genome is composed by the *rep* and *cap* open reading frames flanked by the ITRs. Four rep proteins are produced from messenger RNAs (mRNAs) initiated at the p5 and p19 promoters, whereas the three viral proteins (VP) and the assembly activating protein (AAP) factors are produced from a mRNA transcribed from the p40 promoter. The ITRs are palindromic sequences that can form T-shaped structures at the end of the ssDNA molecule. (b) AAV life cycle. In the absence of a helper virus, AAV enters a latent state characterized by the persistence of the viral genome in an episomal or integrated form. In the presence of a helper virus, here adenovirus or HSV-1, AAV enters a replicative and productive phase that results in the production of infectious particles.

Rep78 and Rep68, display DNA-binding, endonuclease and helicase activities that are essential for AAV genome replication. The smaller Rep proteins, Rep52 and Rep40, are involved in packaging viral DNA but they are dispensable for AAV replication (Daya and Berns, 2008). The *cap* gene encodes three structural proteins VP1/2/3 and one AAP (Sonntag *et al.*, 2010).

AAV has a biphasic life cycle that depends upon the presence or not of a helper virus (Fig. 11.1). When AAV infects a cell in the absence of a helper virus it enters a latent state that is characterized by the persistence of viral DNA in an integrated or episomal form and by the absence of expression of the viral genes. In contrast, in the presence of a helper virus AAV enters a productive phase that is characterized by transcription of the *rep* and *cap* genes, replication of viral DNA, assembly of empty capsids and, finally, packaging of the newly replicated ssDNA genome into the particles (Flotte and Berns, 2005). Importantly, these steps, including particle assembly and packaging, occur in the nucleus of the cells concomitantly with the replication of the helper virus that is responsible for the cytopathic effect. It is currently unknown if the AAV particles found outside of the cell are an indirect consequence of the cell lysis or they are released through an active process.

11.2.2 AAV Serotypes and Variants

Of the initially isolated six AAV serotypes, five were contaminants of adenovirus stocks, while AAV5 was isolated from human condylomatous warts. Among these serotypes, AAV2, 3 and 5 are thought to be of human origin whereas AAV1 and 4 appeared to derive from monkeys, based on the prevalence of neutralizing antibodies against these particles in these animal species. Interestingly AAV6 was characterized as a hybrid serotype resulting from a recombination between AAV1 and 2. Subsequently, the capsid sequences of five additional serotypes and more than 100 variants have been isolated from human and non-human primate tissues using a polymerase chain reaction (PCR) strategy allowing the amplification of a variable *cap* gene region (Gao *et al.*, 2002, 2004). Therefore, the complete AAV genome is not available for these new serotypes and variants.

In addition to these primate serotypes, virus isolates derived from bovine, avian and caprine cells were also cloned (Wu *et al.*, 2006). Beside their application for rAAV vector production (see Section 11.3), these studies highlighted two important aspects of wild-type AAV biology. First, the AAV sequences can be found at relatively high levels in a large proportion of species, including humans, further confirming epidemiological data showing that a majority of the human population is infected with this virus. Second, AAV sequences can be found in a wide variety of tissues and several variants co-exist within the same individual

(Gao *et al.*, 2003, 2004; Schnepp *et al.*, 2005, 2009). This diversity in terms of tropism and sequence likely reflects the ability of AAVs to infect animals through different entry pathways, probably related to the helper virus that accompanies AAVs, to spread and recombine within a single individual and to persist in these tissues. However, it must be noted that it is unclear at the moment if the AAV sequences found *in vivo* truly represent latent AAV genomes that can be rescued upon infection with a helper virus, or just a signature of an older infection that cannot be reactivated. As previously noted by other authors, all these features make AAV a non-pathogenic ubiquitous commensal of mammals (Flotte and Berns, 2005).

11.2.3 Interaction with the Host Cell

Information on the initial steps of AAV infection, including receptor attachment and trafficking to the nucleus has been obtained mostly from studies conducted on AAV2, the best studied AAV serotype. Initial studies identified heparan sulfate proteoglycan (HSPG) as the primary receptor for rAAV2. Several other molecules including the $\alpha v\beta 5$ and $\alpha 5\beta 1$ integrins, fibroblast growth factor receptor-1 (FGFR1), the hepatocyte growth factor receptor, the 36/67-kDa laminin receptor (LamR) and cluster of differentiation 9 (CD9) were further shown to serve as co-receptor molecules. Less information is available for other AAV serotypes except for platelet-derived growth factor receptor (PDGFR) for AAV5, LamR for AAV8 and 9 and N- or O-linked sialic acid residues for AAV 4, 5, and 6 (Büning *et al.*, 2008). More recently, AAV9, one of the most promising AAV serotypes for gene transfer, was also shown to bind terminal N-linked galactose residues. The large variety of receptors and/or co-attachment molecules reflects the widespread tropism of most AAV serotypes that, like AAV2, are able to transduce nearly any tissue *in vivo* (Bell *et al.*, 2011; Shen *et al.*, 2011). This indicates the use of specific AAV serotypes for tissue-restricted disease treatment.

Upon engagement with receptor/co-receptor molecules, AAV particles enter the cells using, in most cases, a clathrin-dependent endocytosis pathway and move via a cytoskeleton-driven mechanism toward the nucleus. Endosomal release of the particles into the cytoplasm occurs following a conformational modification of the capsids and results in their accumulation around the nuclear membrane. It is still unclear if entire capsids can enter the nucleus intact or if a partial uncoating is required before nuclear entry (Lux *et al.*, 2005; Sonntag *et al.*, 2006).

After release into the nucleus, the fate of the ssDNA AAV genome, whether it is that of wild-type or rAAV, includes an essential step converting the ssDNA into a transcriptionally competent double-stranded (ds) DNA molecule. Many studies conducted with rAAV vectors have identified this conversion as a major rate-limiting step that can prevent efficient rAAV transduction *in vitro* and *in vivo* (see Section 11.3) (McCarty, 2008). Further analyses have shown that *in vivo* rAAV genomes are also rapidly circularized, forming monomeric or concatemeric episomes that can persist in the nucleus of the cell (Duan *et al.*, 1998, 1999; Nakai *et al.*, 2001; Schnepp *et al.*, 2003). However, in contrast to other DNA viruses, rAAV episomes are rapidly lost upon cell division, indicating the absence of any mechanism to promote their replication and segregation concomitantly with cellular DNA.

Most of these observations that were made using rAAV are likely to be transposable to wild-type AAV with two notable differences: (1) wild-type AAV genes are rapidly repressed in the absence of a helper virus co-infection. This leads to the persistence of transcriptionally silent viral genomes; (2) wild-type AAV has developed the capacity to site-specifically integrate into a locus of human chromosome 19 (Smith, 2008). This property requires functional Rep proteins and is probably responsible for maintenance of the wild-type AAV genome in dividing cells. Both episomal and integrated wild-type AAV genomes are transcriptionally silent, but can be re-activated upon infection of the cell with a helper virus, leading to a new viral cycle and to the spread of AAV particles.

11.2.4 Interaction with the Helper Viruses

To successfully replicate, AAV exploits a helper virus. For historical reasons, most of the studies conducted on the relationship between AAV and its helper viruses initially focused on adenoviruses and the identification of the adenoviral genes required for the helper effects. Use of diverse adenovirus strains and individual adenoviral genes led to the conclusion that only five adenoviral factors are necessary to induce the helper effect. These are the early protein E1A, required to activate *rep* gene expression, the E1B (55 kDa) and E4 (open reading frame 6 (orf6)) proteins and the viral-associated 1 (VA1) RNAs involved in AAV2 RNA transport, stability and translation, and finally the E2A ssDNA-binding protein (ssDBP) that intervenes at the level of AAV DNA replication, gene expression and translation (Geoffroy and Salvetti, 2005). More recently, new functions for these helper factors have emerged, including the capacity of the E1B(55 kDa)/E4 (orf6) complex to stimulate AAV2 DNA replication by inactivating some cellular DNA repair factors (Stracker *et al.*, 2002; Schwartz *et al.*, 2007). Altogether these studies have contributed to the understanding of how adenovirus stimulates the AAV life cycle but also, and most notably, they have led to the development of essential tools for rAAV vector production (see Section 11.3.1).

Other studies have focused on the helper activities provided by HSV-1. In particular, recent reports have identified at least nine HSV-1 factors contributing to AAV replication. These include factors like infected cell protein (ICP)0, ICP4 and ICP22 that stimulate the synthesis of the AAV Rep proteins, and six herpes replication proteins, the upstream long (UL)5/8/52 helicase/primase complex, the ICP8 ssDBP, and the UL30/UL42 polymerase complex that cooperate to induce AAV DNA replication (Alazard-Dany *et al.*, 2009).

Altogether these studies indicate that the helper virus helps AAV not only by directly providing some specific factors required for critical steps of the AAV life cycle, but also indirectly by rendering the cell permissive for AAV replication. Accordingly, two recent proteomic analyses of the factors recruited by AAV Rep proteins in cells co-infected with adenovirus or HSV-1 have confirmed that a common pathway of cellular factors contributes to the AAV life cycle (Nash *et al.*, 2009; Nicolas *et al.*, 2010).

11.3 Recombinant AAV Vectors

Recombinant AAV vectors were developed in the late 1980s and were rapidly demonstrated to possess unique properties for *in vivo* gene transfer. Since their initial development, the field has very rapidly evolved to integrate the progressive knowledge on wild-type AAV biology towards the development of more efficient vectors. This section will summarize the main achievements in the field of rAAV vector technology and use for *in vivo* gene transfer. It will also highlight some recent findings on the interaction of rAAV particles with the immune system.

11.3.1 Design, Production and Purification of rAAV Particles

Because of its simple genome organization, the design of rAAV vectors is relatively straightforward where the AAV vector genome replaces the *rep* and *cap* genes with a transgene cassette containing the cDNA of choice surrounded by transcriptional regulation elements, i.e., a promoter and a polyadenylation signal (Fig. 11.2). The 145-bp long ITRs are the sole viral sequences that are retained *in cis* in the vector genome. Because of probable space constraints imposed by the capsid structure, the size of the transgene cassette is restricted to approximately 4.5 kb. Although possible solutions have been proposed, this property still represents one of the major limitations for the use of these vectors (Daya and Berns, 2008).

In contrast to their design, the production of rAAV vectors has proven more difficult to optimize. Indeed, as for the wild-type AAV, the production of infectious rAAV particles requires not only the presence of the *rep* and *cap* gene products but also of the helper virus activities. In addition, because the newly formed particles do not seem to be efficiently secreted by the cells, they have to be purified

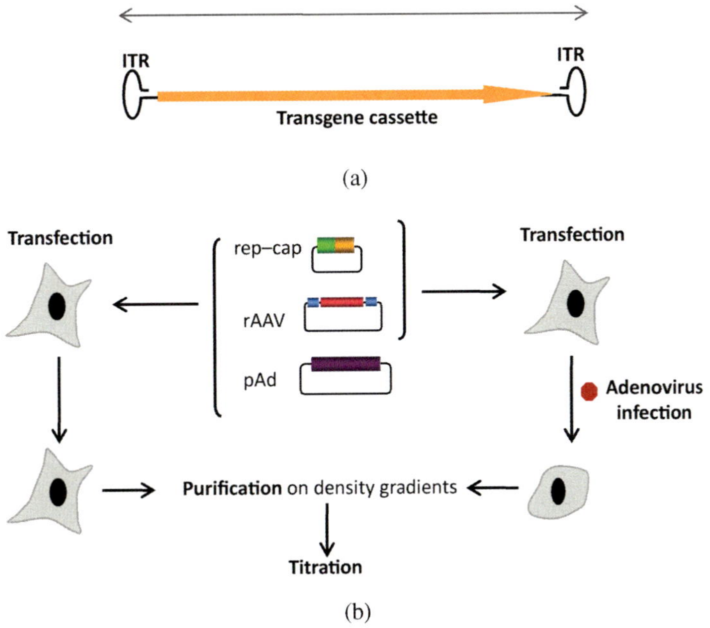

FIGURE 11.2 ■ (a) Structure of a rAAV vector genome. Recombinant vectors are obtained by replacing the *rep* and *cap* genes by a transgene cassette that includes a promoter, a cDNA, and a polyadenylation signal. The ITRs are the only viral sequences retained within the vector. (b) Production of rAAV vectors using adenovirus helper functions. rAAV vectors can be produced using two different procedures that involve the use of either infectious adenovirus particles (*right*) or adenoviral plasmids containing the minimal adenoviral helper genes (*left*). In both situations, the rAAV vector and the *rep* and *cap* open reading frames are introduced into the cells by transient transfection. rAAV particles are then purified from cell lysates e.g., by density-gradient centrifugation (Salvetti *et al.*, 1998).

from cell extracts. Initial studies on the production of rAAV particles involved the use of transformed human kidney cells (293) or the cervical carcinoma cell line HeLa that were transfected with two plasmids containing the rAAV vector backbone and the *rep* and *cap* genes followed by infection of the cells with adenovirus particles (Fig. 11.2). The rAAV particles were then obtained after purification of the cell extracts on density gradients. However, the helper virus can also replicate, and because the density of AAV and adenovirus are relatively close, this procedure resulted in rAAV stocks that were highly contaminated with adenovirus particles. Heat inactivation allowed inactivation of infectious adenovirus particles but did not prevent contamination with adenovirus-derived proteins or even capsids, thus preventing a clear evaluation of the potential of rAAV vector *in vivo*. This major hurdle was overcome a few years later by exploiting information available on the helper activities of adenoviruses, resulting in the generation of adenovirus-derived plasmids that contained the adenoviral genes required for the helper function, but lacked information encoding structural adenoviral proteins and, thus, prevented generation of infectious adenovirus particles.

Current methods for small-scale rAAV production are based on the co-transfection of cells with two or three plasmids that provide the sequences of the rAAV vector, the *rep* and *cap* genes, and the adenovirus helper functions (Fig. 11.2). Further developments of or variations from this standard protocol use helper activities from other viruses (Ayuso *et al.*, 2010). This includes for example the use of replication-defective HSV or baculovirus vectors, which encode all required information for the production of rAAV vectors and introduction into the cells by infection. These latter methods that circumvent the transfection step greatly facilitate the establishment of large-scale rAAV production. Additional improvements include using stable packaging cell lines that can provide *in trans* the *rep* and *cap* gene products (Thorne *et al.*, 2009). Last, but not least, several improvements and/or modifications have been introduced in the purification procedure. These include the use of chromatography procedures that are easily scalable and result in a high level of purity (Ayuso *et al.*, 2010).

11.3.2 Use of rAAV Vector for Gene Transfer and Gene Therapy in Animal Models

Initial *in vivo* evaluation of the efficiency of rAAV vectors were mostly performed in mice using vectors derived from AAV2 and encoding a reporter gene such as β-galactosidase or green fluorescent protein (GFP). Surprisingly, these studies demonstrated that a single injection of rAAV vectors into tissues such as muscle resulted in a high and sustained transgene expression that could last life-long. This observation contrasted strikingly with previous experience of *in vivo* gene transfer using adenovirus-derived vectors, which resulted in a strong but transient expression. Loss of adenovirus-mediated transgene expression is caused by an immune response against the transduced cells. Further analyses confirmed the ability of rAAV vectors to sustain stable transgene expression *in vivo* and, also, to efficiently transduce a variety of highly relevant tissues, such as liver, central nervous system (CNS) and retina. These initial studies rapidly prompted the use of rAAV vectors in animal models of genetic diseases to validate their potential for gene therapy. Indeed, several studies were performed not only in mice but also in larger animal models such as dogs and non-human primates.

Evidence for partial or complete therapeutic efficiency in animal models were obtained for many diseases: hemophilia B, lysosomal storage diseases, congenital inherited forms of blindness and acquired neurological disorders are some examples of clinical applications of rAAV vectors (Daya and Berns, 2008).

The detailed analysis of rAAV-mediated *in vivo* gene transfer has also fostered the improvement of vector design. Notably, efforts have focused on the improvement and/or modification of the tropism of the AAV particles. This was achieved using two non-exclusive strategies: The first involved the replacement of the original AAV2 capsid with that derived from other available AAV serotypes. The evaluation of these new serotypes has rapidly demonstrated that although all serotypes can transduce nearly all tissues they individually possess specific preferences for some cell types (Wu *et al.*, 2006). This has rapidly led to the identification of the optimal serotype for use in a given tissue. Production of rAAV vectors from alternative serotypes can be easily achieved by replacing the AAV2 *cap* gene present in the rep–cap plasmid construct with the homologous sequence derived from any other AAV serotype leading to the production of pseudotyped particles containing an AAV2-derived vector genome packaged into a capsid derived from different serotypes. The second strategy employed to modify the vector tropism consists of the genetic modification of some capsid by insertion of peptide ligands to redirect the particles to a specific receptor or cell type (Büning *et al.*, 2008).

The *in vivo* evaluation of rAAV vectors has also led to the development of modified vector backbones to improve the kinetics and/or the efficiency of transgene expression. The most impressive improvement was brought about by the development of ds rAAV vectors. As indicated above, several observations had led to the conclusion that conversion of the ssDNA vector molecule, delivered into the nucleus by the AAV particle, into a double stranded form was a major limiting step for efficient transduction of several tissues (see Section 11.2.3). To overcome this barrier, a strategy was developed to produce AAV capsids containing dsDNA vector molecules. *In vivo* evaluation of these dsAAV vectors, also called self-complementary AAV (scAAV) vectors, confirmed that the delivery of a dsDNA molecule can generally improve both the kinetics and the efficiency of transduction (McCarty, 2008). However, the major limitation is that the size of the vector is half that of ssAAV genomes, thus limiting the application of these vectors to small transgene cassettes (<2.5 kb).

11.3.3 Use of rAAV Vectors in Clinical Trials

Experience with rAAV vectors rapidly demonstrated that these vectors are very efficient tools for *in vivo* transduction of post-mitotic tissues, in particular liver, muscle and the CNS. Translation into clinical trials emerged from studies conducted in small- and large-animal models (Mingozzi and High, 2011). Up until now, approximately 80 clinical trials have been conducted or are still underway using rAAV vectors (http://www.wiley.com/legacy/wileychi/genmed/clinical/) in both inherited and acquired diseases. Three major observations can be made on the basis of this list: the first is that in nearly all these trials rAAV vectors were directly injected *in vivo*; they are only rarely used in *ex vivo* protocols in contrast to other vectors such as lentiviral vectors. The reason for this is that rAAV vectors remain mainly in an episomal form that is stable in differentiated post-mitotic cells but is lost upon cell division. The second observation is the wide variety of tissues that were targeted for transduction. Finally, even though the initial trials were conducted with AAV2-derived vectors, other more efficient AAV serotypes, such as AAV8 or AAV1, are now replacing them.

The rAAV vectors so far have demonstrated an excellent safety profile and therapeutic efficacy was reported following local *in vivo* application of rAAV vectors in particular for the treatment of Leber congenital amaurosis (LCA), lipoproteinlipase (LPL) deficiency and hemophilia B (Mingozzi and High, 2011). For patients suffering from LCA, a retinal degenerative disorder due to a defect in the retinal pigment epithelium-specific protein 65 kDa (*RPE65*) gene that leads to the early onset of blindness, three clinical trials were simultaneously initiated in the USA and in the United Kingdom. In all three trials, rAAV2 vectors encoding a functional version of the RPE65 protein were injected subretinally into one eye. Increasing brightness in the treated eye was reported to begin two weeks after treatment (High, 2009). In addition, psychophysical and measurable improvements in visual function were observed in all three trials. Based on these promising results, further studies are being initiated that will include younger patients in whom the disease is less advanced. The second major success was recently obtained in hemophilia B patients. The treatment of this disease using rAAV vectors first was initiated more than ten years ago with a clinical trial developed in the USA and involving the intra-muscular injection of rAAV2 vectors encoding the blood coagulation factor IX (Matrai *et al.*, 2010). In those studies, encouraging results were reported in terms of vector persistence and absence of toxicity, although there was no evidence of therapeutic effect due to a limitation in the vector doses that could be injected into the patients. In animal studies, a 10–100-fold higher therapeutic efficacy was observed following liver-injection compared with muscle-injection, most likely due to higher protein secretion capacity.

Furthermore, rAAV-mediated expression of transgenes in liver seemed to induce tolerance towards the transgene product (High, 2009). Consequently, further studies focused on the intrahepatic-injection of the rAAV2 vectors. Initially, therapeutic levels of factor IX expression were reported in a patient of the high-dose cohort. However, the circulating levels of factor IX of 10–12% declined to baseline levels by week 10. This decline was accompanied by an asymptomatic rise in liver transaminases first measured four weeks post-vector injection. To decipher the cause that initiated the loss of transgene expression, detailed immunological investigations were initiated. Currently, the most likely explanation is a re-activation of memory T-cells directed against the viral capsid. More recently a positive outcome was finally reached by introducing substantial modifications in the protocol including a change of serotype (from rAAV2 to rAAV8) and vector genome conformation (from single-stranded to self-complementary) in order to increase the transduction efficacy and, thereby reduce the vector dose required.

This latter trial, which is still ongoing, has reported thus far a stable therapeutic level of factor IX (Nathwani *et al.*, 2011).

In contrast to the factor IX experience, the clinical trial on inherited LPL deficiency was begun with a non-serotype-2 vector (Gaudet *et al.*, 2010). Patients with LPL deficiency, a rare autosomal-recessive disorder that causes accumulation of triglyceride-rich lipoproteins in the blood, received a rAAV1 vector encoding a gain-of-function variant of LPL (Glybera®) by intramuscular injection. Clinical benefit, as indicated by a decrease in triglyceride level and frequency of pancreatitis, was reported. However, in this trial, activation of T-cell response towards the AAV capsid was observed, revealing that the use of an alternative serotype may not be sufficient to avoid induction of a cytotoxic T-cell response.

11.3.4 Major Limitations and Future Directions

Altogether, preclinical and clinical trials in patients have clearly demonstrated the great potential of rAAV vectors. Nevertheless, there is still room for improvement. Currently, researchers focus on the problem of pre-existing and *de novo*-induced immune response against the viral vector particles. Humans are frequently infected by AAV during childhood. The induced adaptive immune response is a considerable obstacle that hampers the efficiency of rAAV transduction *in vivo*. In particular, pre-existing neutralizing antibodies against AAV particles prevent initial cell transduction. In addition, as already discussed, T cell responses towards the capsid are believed to be responsible for the decline of therapeutic efficacy due to destruction of transduced cells. Therefore, besides including an immune-suppressive regime, an impressive amount of research is ongoing to identify the origin of the capsid-specific T-cell-epitopes that are presented on major histocompatibility complex (MHC) molecules on transduced cells. In addition, at least three strategies have already been developed to deal with the problem of pre-existing neutralizing antibodies: (1) serotype switch; (2) genetic modifications of epitopes; and (3) mosaic capsids (Büning *et al.*, 2008). The serotype switch strategy uses capsids of an alternative serotype not, or significantly less efficiently, neutralized by a given serum. While this strategy worked well in animal studies, the value for clinical applications remains to be proven since existence of cross-reactive antibodies has been observed in patient sera. Furthermore, this strategy is restricted by the serotype-specific variations in tissue tropism. Modification of antibody epitopes by rational design or by *in vitro* evolution, the second strategy that has been successfully exploited, does not affect vector tropism. However, the immune-escape phenotype of currently available mutants is still notably less efficient than that achieved by a serotype switch. The third strategy is also based on a combined approach. In particular, the capsid open reading frames of a set of AAV serotypes are fragmented by enzymatic digestion and combined e.g., by PCR. Thereby, a library of virions with mosaic capsids is generated (following packaging), which is screened for infectivity in the presence of neutralizing antibodies. Mutants selected by this approach showed an impressive immune-escape phenotype, but their tropism remains to be fully characterized.

Off-target transduction is a further limitation caused by the broad tropism of rAAV vectors and by unspecific uptake of vectors, for example in the liver. Cell-surface, transcriptional as well as post-transcriptional targeting strategies have been developed to circumvent this limitation. Cell-surface-targeting strategies aim to restrict the vector tropism to a receptor of choice by genetic or non-genetic modification of the viral capsid. While this strategy targets the vector entry step, transcriptional and post-transcriptional targeting restricts transgene expression and production, respectively. Briefly, by using cell-type-specific promoters, transgene expression can be limited to the tissue of choice while incorporation of microRNA target sites avoids transgene production in off-target cells. Combination of the different targeting strategies is possible and is likely to increase the specificity of *in vivo* gene transfer.

A further limitation concerns the high doses of vector that have to be administered to achieve a therapeutic effect. Indeed, doses of approximately 10^{12} vector particles per kg were required in the initial hemophilia B trials using rAAV2 vectors. Even if lower doses were used in the later AAV8 trials, the amount of particles delivered as a single injection in patients still remains impressively high. This is also the case for subretinal injection if one considers that the vector particles remains sequestered in a very small zone of the retina. Besides the challenge of producing high particle numbers, application of high vector doses increases the risk of inducing immune responses and of off-target transduction. Hence, efforts are focused on improving transduction efficiencies by identifying barriers to AAV transduction followed by introduction of specific modifications. Successful examples are the development of self-complementary vector genomes (see Section 11.3.2), mutation of exposed tyrosine residues on the capsid surface to avoid capsid degradation and introduction of peptide ligands (cell-surface targeting) to enable transduction of cell types that lack AAV receptors.

All the efforts discussed above will likely result in a considerable refinement of the AAV vector tool and in the precise identification of the most suitable conditions to achieve efficient gene transfer and a successful therapeutic outcome.

References

Alazard-Dany, N., Nicolas, A., Ploquin, A., *et al.* (2009). Definition of herpes simplex virus type 1 helper activities for adeno-associated virus early replication events, *PLoS Pathog*, **5**, e1000340.

Ayuso, E., Mingozzi, F., Bosch, F. (2010). Production, purification and characterization of adeno-associated vectors, *Curr Gene Ther*, **10**, 423–436.

Bell, C.L., Vandenberghe, L.H., Bell, P., *et al.* (2011). The AAV9 receptor and its modification to improve *in vivo* lung gene transfer in mice, *J Clin Invest*, **121**, 2427–2435.

Büning, H., Perabo, L., Coutelle, O., *et al.* (2008). Recent developments in adeno-associated virus vector technology, *J Gene Med*, **10**, 717–733.

Daya, S., Berns, K.I. (2008). Gene therapy using adeno-associated virus vectors, *Clin Microbiol Rev*, **21**, 583–593.

Duan, D., Sharma, P., Dudus, L. *et al.* (1999). Formation of adeno-associated virus circular genomes is differentially regulated by adenovirus E4 ORF6 and E2a gene expression, *J Virol*, **73**, 161–169.

Duan, D., Sharma, P., Yang, J. *et al.* (1998). Circular intermediates of recombinant adeno-associated virus have defined structural characteristics responsible for long-term episomal persistence in muscle tissue, *J Virol*, **72**, 8568–8577.

Flotte, T.R., Berns, K.I. (2005). Adeno-associated virus, a ubiquitous commensal of mammals, *Hum Gene Ther*, **16**, 401–407.

Gao, G., Alvira, M.R., Somanathan, S., *et al.* (2003). Adeno-associated viruses undergo substantial evolution in primates during natural infection, *Proc Natl Acad Sci USA*, **100**, 6081–6086.

Gao, G., Alvira, M.R., Wang, L., *et al.* (2002). Novel adeno-associated viruses from rhesus monkeys as vectors for human gene therapy, *Proc Natl Acad Sci USA*, **99**, 11854–11859.

Gao, G., Vandenberghe, L.H., Alvira, M.R., *et al.* (2004). Clades of adeno-associated viruses are widely disseminated in human tissues, *J Virol*, **78**, 6381–6388.

Gaudet, D., de Wal, J., Tremblay, K., *et al.* (2010). Review of the clinical development of alipogene tiparvovec gene therapy for lipoprotein lipase deficiency, *Atheroscler Suppl*, **11**, 55–60.

Geoffroy, M.C., Salvetti, A. (2005). Helper functions required for wild type and recombinant adeno-associated virus growth, *Curr Gene Ther*, **5**, 265–271.

High, K.A. (2009). The Jeremiah Metzger Lecture: gene therapy for inherited disorders: from Christmas disease to Leber's amaurosis, *Trans Am Clin Climatol Assoc*, **120**, 331–359.

Lux, K., Goerlitz, N., Schlemminger, S., *et al.* (2005). Green fluorescent protein-tagged adeno-associated virus particles allow the study of cytosolic and nuclear trafficking, *J Virol*, **79**, 11776–11787.

Matrai, J., Chuah, M.K., VandenDriessche, T. (2010). Preclinical and clinical progress in hemophilia gene therapy, *Curr Opin Hematol*, **17**, 387–392.

McCarty, D.M. (2008). Self-complementary AAV vectors; advances and applications, *Mol Ther*, **16**, 1648–1656.

Mingozzi, F., High, K.A. (2011). Therapeutic *in vivo* gene transfer for genetic disease using AAV: progress and challenges, *Nat Rev Genet*, **12**, 341–355.

Nakai, H., Yant, S.R., Stor, T.A., *et al.* (2001). Extrachromosomal recombinant adeno-associated virus vector genomes are primarily responsible for stable liver transduction *in vivo*, *J Virol*, **75**, 6969–6976.

Nash, K., Chen, W., Salganik, M., *et al.* (2009). Identification of cellular proteins that interact with the adeno-associated virus rep protein, *J Virol*, **83**, 454–469.

Nathwani, A.C., Tuddenham, A.G.D., Rosales, C., *et al.* (2011). A phase I/II clinical trial entailing peripheral vein administration of a novel self complementary adeno-associated viral vector encoding human FIX for haemophilia B gene therapy. (Abstract) 10th Annual Congress of the Société Française de Thérapie Cellulaire et Génique, June 6–8, 2011, Nantes, France, *Hum Gene Ther*, **22**, A-7–A-8.

Nicolas, A., Alazard-Dany, N., Biollay, C., *et al.* (2010). Identification of rep-associated factors in herpes simplex virus type 1-induced adeno-associated virus type 2 replication compartments, *J Virol*, **84**, 8871–8887.

Salvetti, A., Orève, S., Chadeuf, G., *et al.* (1998). Factors influencing recombinant adeno-associated virus production, *Hum Gene Ther*, **9**, 695–706.

Schnepp, B.C., Clark, K.R., Klemanski, D.L., *et al.* (2003). Genetic fate of recombinant adeno-associated virus vector genomes in muscle, *J Virol*, **77**, 3495–3504.

Schnepp, B.C., Jensen, R.L., Chen, C.L., *et al.* (2005). Characterization of adeno-associated virus genomes isolated from human tissues, *J Virol*, **79**, 14793–14803.

Schnepp, B.C., Jensen, R.L., Clark, K.R., *et al.* (2009). Infectious molecular clones of adeno-associated virus isolated directly from human tissues, *J Virol*, **83**, 1456–1464.

Schwartz, R.A., Palacios, J.A., Cassell, G.D., *et al.* (2007). The Mre11/Rad50/Nbs1 complex limits adeno-associated virus transduction and replication, *J Virol*, **81**, 12936–12945.

Shen, S., Bryant, K.D., Brown, S.M., *et al.* (2011). Terminal N-linked galactose is the primary receptor for adeno-associated virus 9, *J Biol Chem*, **286**, 13532–13540.

Smith, R.H. (2008). Adeno-associated virus integration: virus versus vector, *Gene Ther*, **15**, 817–822.

Sonntag, F., Bleker, S., Leuchs, B., *et al.* (2006). Adeno-associated virus type 2 capsids with externalized VP1/VP2 trafficking domains are generated prior to passage through the cytoplasm and are maintained until uncoating occurs in the nucleus, *J Virol*, **80**, 11040–11054.

Sonntag, F., Schmidt, K., Kleinschmidt, J.A. (2010). A viral assembly factor promotes AAV2 capsid formation in the nucleolus, *Proc Natl Acad Sci USA*, **107**, 10220–10225.

Stracker, T.H., Carson, C.T., Weitzman, M.D. (2002). Adenovirus oncoproteins inactivate the Mre11-Rad50-NBS1 DNA repair complex, *Nature*, **418**, 348–352.

Thorne, B.A., Takeya, R.K., Peluso, R.W. (2009). Manufacturing recombinant adeno-associated viral vectors from producer cell clones, *Hum Gene Ther*, **20**, 707–714.

Wu, Z., Asokan A., Samulski, R.J. (2006). Adeno-associated virus serotypes: Vector toolkit for human gene therapy, *Mol Ther*, **14**, 316–327.

12

HERPES SIMPLEX VIRUS (HSV-1)-BASED VECTORS: APPLICATIONS FOR GENE TRANSFER, GENE THERAPY, CANCER VIROTHERAPY AND VACCINATION

Matias E. Melendez[a], *Aldo Pourchet*[a], *Anna Greco*[a] *and Alberto L. Epstein*[a,b]

12.1 Introduction

The improvement of methods for efficient delivery and regulated expression of genetic material into mammalian cells has been a major objective of molecular and cellular biology, gene therapy and vaccine development over the last 30 years and is still an area of intensive research. Virus-derived vectors are one of the most promising gene transfer tools due to the fact that viruses are naturally occurring molecular devices that have evolved to ensure targeted gene delivery and efficient expression in most cell types. This review focuses on vectors derived from herpes simplex virus type 1 (HSV-1), one of the most powerful and versatile gene transfer tools.

12.2 Short Introduction to HSV-1 Biology

HSV-1 is a widespread human pathogen, infecting 40–80% of people worldwide, whose lifestyle is based on a long-term dual interaction with the host. After initial infection and lytic replication at the body periphery, generally at the oral mucosa

[a]Centre de Génétique et Physiologie Moléculaire et Cellulaire, CNRS UMR5534, Université Lyon I-16, rue Raphaël Dubois 69100 Villeurbanne, France
[b]Email: alberto.epstein@univ-lyon1.fr

(the primary infection), the virus particles enter the sensory neurons innervating the infected region and travel in a retrograde direction along the axons to reach the nucleus of the neurons in the sensory ganglia. Following delivery into the sensory neuron nucleus, the virus genome will generally remain in a silent, latent state, for long periods. Periodic reactivation from latency usually leads to the return of the virus to epithelial cells, where it produces secondary lytic infections (recurrences) resulting in mild illness symptoms, such as cold sores. A short introduction to selected molecular aspects of the biological cycle of HSV-1 will help with understanding how HSV-1 derived vectors are generated and used (Roizman *et al.*, 2007).

12.3 The Virus Particle

The mature, extracellular HSV-1 particle (diameter approximately 220 nm), is made of four concentric layers: (1) a core of double-stranded DNA, (2) an icosadeltahedral capsid composed of 162 capsomers, (3) the tegument, which is an amorphous layer of proteins located between the envelope and the capsid and (4) the lipid envelope in which are embedded viral proteins and glycoproteins involved in several functions, including receptor-mediated cellular entry (Fig. 12.1). Proteomic analyses of purified extracellular HSV-1 particles identified more than 40 virus-encoded proteins as constituents of the virus particle, including eight capsid-associated polypeptides, 23 tegument proteins and 13 glycoproteins or membrane-associated proteins (Loret *et al.*, 2008).

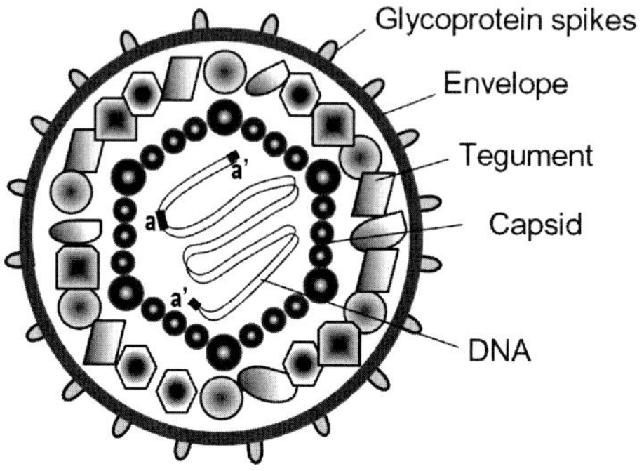

FIGURE 12.1 ■ Schematic representation of the HSV-1 particle. The viral particle is made of a linear 152 kilobase pairs (kbp) DNA genome, which is contained within an icosadeltahedral capsid, a lipid envelope, in which are embedded a dozen viral-encoded proteins and glycoproteins, and the layer of tegument proteins, which are located between the capsid and the envelope.

12.4 The HSV-1 Genome

The 153 kbp HSV-1 linear DNA genome packaged within the capsid is devoid of histones. This long molecule is composed of two parts, designated as L (long) and S (short). Each part consists of unique sequences (UL and US) bracketed by inverted repeats. The repeats surrounding the L component are designated *ab* and *b′a′*, while those surrounding the S component are designated *a′c′* and *ca*. The structure of the virus genome and its replication cycle is shown in Fig. 12.2.

The UL component contains at least 56 genes (UL1–UL56) while the US component contains at least 13 genes (US1–US13). The *b* repeats flanking UL contain three genes: Two of them encode the immediate early protein infected cell polypeptide 0 (ICP0) and the late neurovirulence protein ICP34.5, while the third one encodes a family of transcripts known as latency-associated transcripts (LATs). The LATs, which encode no proteins, are expressed antisense respective to the genes encoding ICP0 and ICP34.5. The *c* repeats flanking US contain a single gene, encoding the immediate early protein ICP4. The *a* sequences contain no *trans*-acting genes (Fig. 12.3). Thus, the HSV-1 genome contains at least 77 canonical genes of which four are duplicated. A few other overlapping genes or genes entirely contained within larger open reading frames have recently been identified, increasing the number of protein-encoding genes to more than 80. In addition, the virus genome expresses several microRNAs (miRNAs), most of which are encoded in the repeated *b* sequences (Umbach *et al.*, 2008). Lastly, the viral genome contains three origins of DNA synthesis (ori), one located in UL (oriL) and two in the repeated *c* sequences surrounding US (oriS), and the cleavage/packaging (*pac*) sequences, which are contained in the repeated *a* sequences. Although linear within the particle, the virus genome circularizes after infection (Fig. 12.2). The virus genome replicates using a combination of homologous recombination and rolling-circle amplification (Ingvarsdottir and Blaho, 2009).

12.5 Virus Gene Expression

Expression of HSV-1 lytic genes is temporarily regulated in a cascade fashion, giving rise to three phases of gene expression (Fig. 12.3).

The lytic expression program begins with the transcription of the immediate-early (IE) genes. The resulting IE proteins, referred to as ICP0, ICP4, ICP22, ICP27 and ICP47, are mostly regulatory proteins that control gene expression during subsequent, early (E) and late (L) phases of the lytic cycle. Transcription of IE genes is highly stimulated by viral protein 16 (VP16), a virion tegument protein

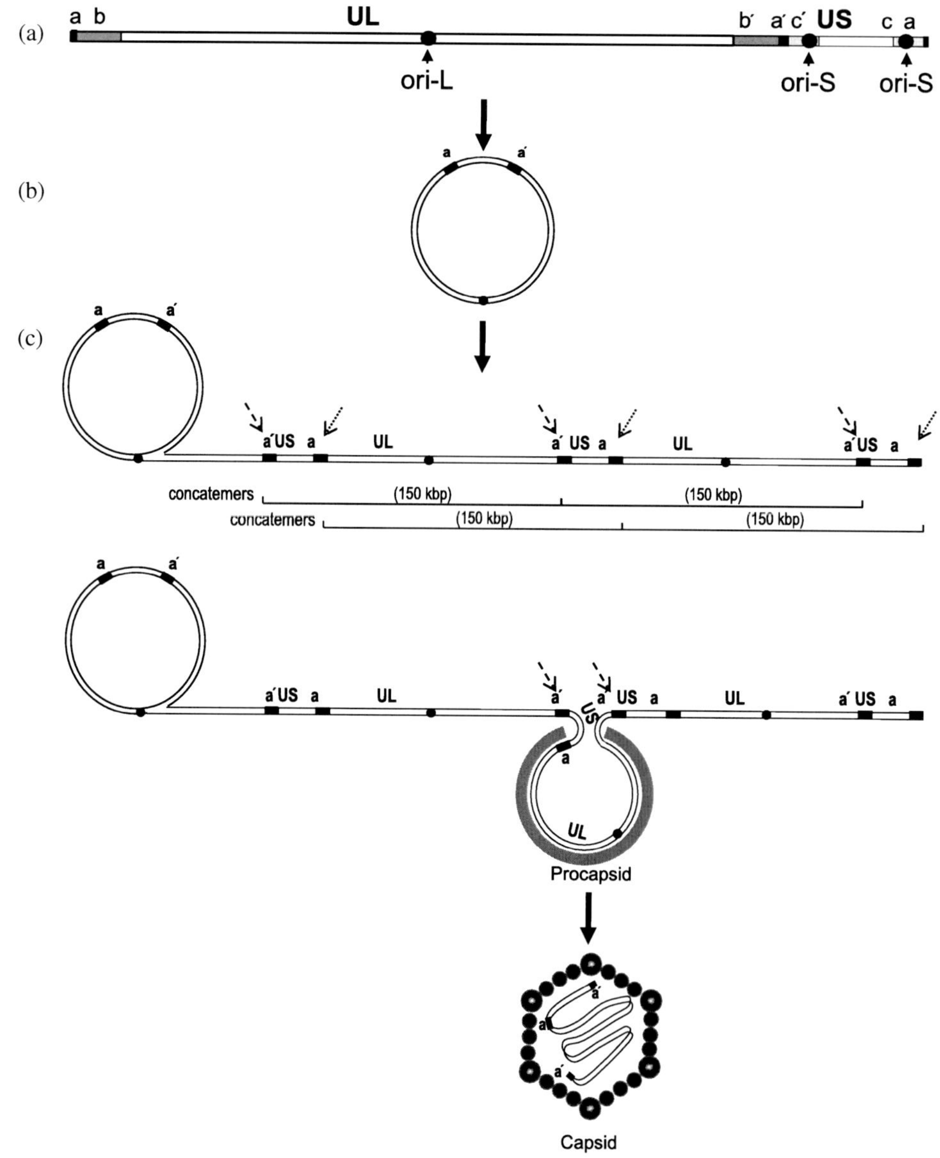

FIGURE 12.2 ■ Structure and replication of the HSV-1 genome. (a) The viral DNA is composed of two unique sequences, designated as unique long (UL) and unique short (US), each bracketed by inverted repeated sequences: *ab* and *b′a′* surround UL while *a′c′* and *ca*, surround US. The *a* and *a′* sequences (black squares) contain the cleavage/packaging (*pac*) signals. The genome also contains three origins of viral DNA synthesis (black circles), one located in UL (oriL) and two in the repeated sequences surrounding US (oriS). (b) The DNA molecule is linear within the capsid but it becomes circular after entry into the cell nucleus. (c) The circularized viral DNA replicates through a complex rolling circle mechanism, generating a DNA concatemer composed of tandem genomic units. The genomic units are then cleaved at two successive *pac* signals in the same orientation (i.e., within two *a* sequences or within two *a′* sequences) and are packaged into pre-capsids, which then mature into capsids.

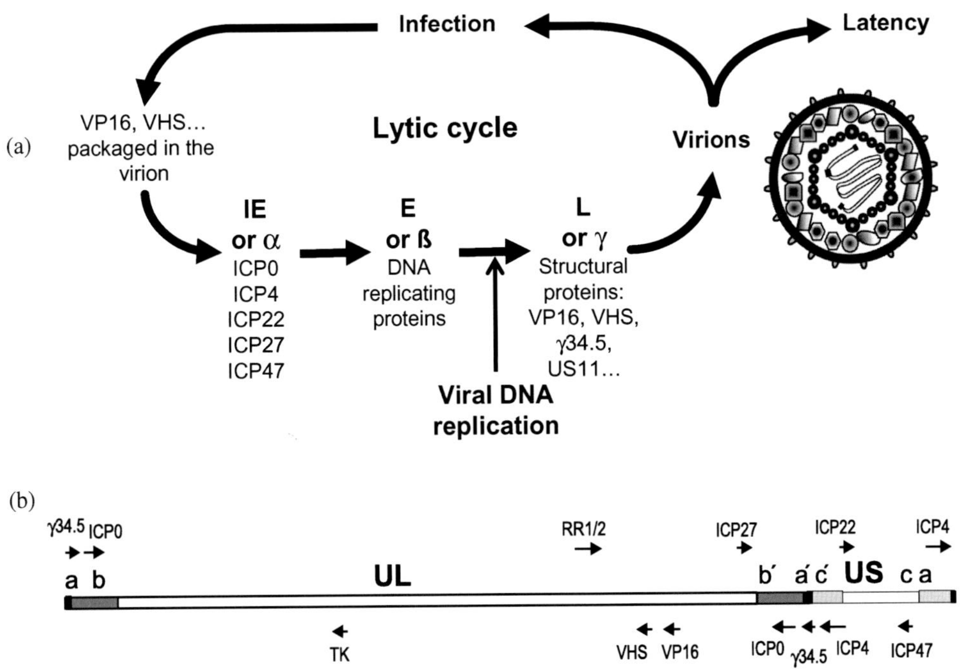

FIGURE 12.3 ■ Expression of the lytic program of the viral genome. (a) During productive, lytic infection, the viral genome express its genes following three sequential steps, which are regulated mainly at the transcription level, leading to the synthesis of immediate-early (IE or α), early (E or β) and late (L or γ) proteins. (b) Scheme of the virus genome showing the location of the five IE genes, as well as of some of the E and L genes referred in the text.

that is a powerful transcription factor. The early (E) gene products comprise (1) several enzymes that increase the pool of deoxynucleotides of the infected cells, such as a thymidine kinase (TK), and (2) replication proteins involved in viral DNA synthesis, including a DNA polymerase. Then the late (L) genes are expressed, which encode the structural proteins involved in the assembly of the capsid, the tegument and the envelope of the virus particles, including some proteins that play important roles during the next infectious cycle. Lytic viral replication results in the impairment of host macromolecular synthesis, the release of progeny particles, and the death of the host cell (Roizman *et al.*, 2007). About 50% of the proteins expressed during the lytic cycle are not essential in cell culture and can be deleted without significantly perturbing virus production. These proteins, however, play important roles *in vivo*, such as control of neurovirulence or escape from host antiviral responses (Ingvarsdottir and Blaho, 2009). During latency, the viral genome remains as a circular episome within the neuron cell nucleus and the lytic genes are completely silenced. Only the latency locus (LAT) is transcribed during latency, giving rise to non-coding LATs, driven by the latency-associated promoter (LAP), whose function is not yet completely elucidated, though it is thought that they protect neurons from apoptosis (Bloom *et al.*, 2010). The LATs also encode several miRNA that can down-regulate expression of key viral genes and seem to control the switch between lytic and latent infection (Umbach *et al.*, 2008).

12.6 HSV-1-Based Vectors

The interest in HSV-1-based vectors stems from three properties of HSV-1, not shared with other viral systems. The first is the very large capacity of the virus particle, which allows efficient delivery of up to 150 kbp of DNA. This DNA will remain as a nuclear episome from where the transgenes are expressed. The non-integration of the vector genome into host DNA decreases the risk of mutagenesis. The second feature is the complexity of the virus genome, which contains some 40 genes that are non-essential for virus replication in cell culture and can be deleted without disturbing virus production. The third one is the remarkable adaptations of HSV-1 to the nervous system, which includes the control of neurovirulence, the ability to *trans*-synaptically spread between neurons in anterograde and retrograde directions, and the capacity to establish latency in neurons, a non-toxic condition that allows transgenic transcription under the control of appropriate promoters (Marconi *et al.*, 2009).

Three different types of vectors have been derived from HSV-1, in order to exploit one or more of the above-mentioned properties: attenuated recombinant vectors, defective recombinant vectors and amplicon vectors. Although these vectors are used mainly for gene delivery to neural cells, they can also transduce genes to other cell types, including epithelial cells, fibroblasts, muscle cells, hepatocytes, mesenchymal stem cells and cell lines derived from gliomas, hepatocellular carcinomas, osteosarcomas, epidermoid carcinomas and many other human and murine malignancies (Marconi *et al.*, 2009).

12.6.1 Attenuated Recombinant Vectors

Attenuated vectors are replication-competent viruses that carry mutations in genes that are not essential for virus multiplication in cultured cells but are important for virulence in the inoculated hosts. Such genes are generally involved in multiple interactions with cellular proteins that optimize the ability of the virus to grow within cells or to evade host responses. Understanding such interactions has permitted the deletion of these genes, alone or in combination, to create HSV-1 mutants that can replicate only within specific tissues or within actively dividing cells, but displaying no or low toxicity, as they cannot spread within the nervous system.

12.6.1.1 *Oncolytic HSV-1 vectors*

The most common of these attenuated viruses are the oncolytic vectors, which are used to treat tumors. HSV-1 has indeed many attractive features for development as an oncolytic vector. Firstly, it is a weakly pathogenic virus endemic to the human population. Secondly, several non-essential genes can be deleted from the HSV-1 genome without affecting its capacity to be produced in cultured cells and to replicate in cancer cells, while being non-virulent for most normal quiescent primary cells. In addition, efficient anti-herpetic drugs provide a safety mechanism in case undesired virus spread occurs. These viruses can serve, in addition, as gene delivery vehicles, to express therapeutic transgenes that are detrimental to cancer cells (Friedman *et al.*, 2009).

Several generations of oncolytic HSV-1 vectors have already been developed. The first generation vectors contain deletions in a single gene that restrict their replication to dividing cells, such as the genes encoding TK or ribonucleotide reductase (RR) or, alternatively, in genes that enable the virus to fight against cellular innate antiviral responses, such as those encoding the proteins ICP34.5, ICP0 or US3. Both TK and RR are involved in optimizing nucleic acid metabolism required for virus growth and are necessary for efficient replication in quiescent cells, such as neurons. Viral mutants lacking these proteins are highly neuroattenuated. In contrast, when used in mouse models of brain cancers they multiply, inducing slower tumor growth and prolonged host survival. However, clinical trials with these vectors were not pursued because of their high-level toxicity at high titers.

The virus proteins ICP0, US3 or ICP34.5, although non-essential for virus growth in most types of cultured cells, are essential for HSV-1 virulence *in vivo*. These proteins inhibit cellular innate responses, both at the transcription (ICP0) and translation (US3, ICP34.5) levels, thus supporting efficient virus gene expression in normal cells. Tumor cells, however, often display impaired antiviral responses, allowing replication of HSV-1 carrying deletions in these genes. These HSV-1 mutants have shown highly attenuated neurovirulence, considerable antitumor activity and good safety levels in rodent models and in non-human primates. They have consequently reached the clinics and are being used in human trials, mainly to address tumors of the central nervous system (CNS). Nevertheless, though safe, these mutants replicate with reduced efficiency even in tumor cells and, despite the promising experimental results obtained, the results from clinical trials showed that they did not significantly affect tumor growth.

To further augment their inherent antitumor efficacy, second-generation oncolytic HSV-1 vectors have been developed by further engineering the virus genome or by incorporating transgenes that express antitumor proteins that act synergistically with the inherent antitumor effect of lytic replication. For instance, an ICP34.5-deleted HSV-1 with secondary mutations in the genes expressing US11 and US12 was found to give improved antitumor activity *in vivo*, without compromising safety (Mohr and Gluzman, 1996). In other cases, genes encoding immune-modulator proteins or anti-angiogenic proteins were introduced into the genome of an oncolytic HSV-1 to increase immunity or to modify the environment of the tumors. Still other oncolytic HSV-1 were engineered to express toxic prodrug-activating systems (Nakamura *et al.*, 2001), or to increase the cell membrane fusion capability (Fu *et al.*, 2003). Overall, incorporating suicide, fusogenic, immune-modulator or anti-angiogenic-expressing transgenes have been shown to increase antitumor efficacy. Oncolytic viruses expressing granulocyte–macrophage colony-stimulating factor (GM-CSF) are particularly interesting since they induce immune responses that also reduce metastasis, and are being used in clinical protocols with quite encouraging results (Hu *et al.*, 2006).

A third-generation of oncolytic HSV-1 vectors is currently being developed. These are targeted viruses that keep their wild-type virulence features but are engineered to express them only in cancer cells, thereby specifically increasing their antitumor capabilities while preventing damage of healthy tissues. Tumor cell targeting can be obtained at three levels: (1) by redirecting viral entry through engineering the envelope glycoproteins, (2) by using cancer-specific promoters to target gene transcription, or (3) by exploiting differences in miRNA expression between cancer cells and their normal counterparts. While efforts in these directions are being carried out, no targeted oncolytic HSV-1 has so far reached the clinic.

12.6.1.2 *Attenuated HSV-1 vectors for gene delivery to the peripheral nervous system*

Inoculation of HSV-1 vectors by peripheral routes takes advantage of the natural life cycle of the virus, which usually infects nerve ends at peripheral sites before retrograde transport to neuronal cell bodies where latency is generally established. In the peripheral nervous system (PNS) there are a number of potential applications for vectors capable of peripheral replication and axonal transport. These include the regrowth of damaged nerves, the treatment of pain and neuropathies, or the protection of neurons in neurodegenerative diseases, amongst others. Viruses mutated in either TK or RR have been used to this end. These viruses can multiply in the epithelial cells but cannot do it in neurons, where they will remain in latency. These vectors were shown to efficiently express nerve growth factor beta subunit (β-NGF) in latently infected dorsal root ganglia (DRG), or to deliver genes into monkey eyes or rodent visual systems, or to drive preproenkephalin expression in DRG, for instance (Braz *et al.*, 2001; Manservigi *et al.*, 2010).

12.6.2 Defective Recombinant Vectors

These are disabled, replication-incompetent, non-pathogenic vectors that lack one or more essential IE genes (Fig. 12.4). These vectors cannot spread within the host but retain many advantageous features of the wild-type virus, particularly the ability to express transgenes after establishing latent infections in central and peripheral neurons following local administration.

The interesting features of defective recombinant HSV-1 vectors are the ability to efficiently transduce non-dividing cells, the large capacity of their capsids, and the ability to establish latent infections, which may be exploited for long-term expression of therapeutic transgenes in neurons. The main problems of these vectors are: (1) the elimination of lytic gene expression and the control of innate and immune host responses; (2) the identification of strategies to target transgene expression to specific tissues; and (3) simultaneous expression of multiple genes. In recent years, novel technologies have allowed deeper study of these problems (Manservigi *et al.*, 2010).

As quoted above, the IE genes ICP0, 4, 22, 27 and 47 are required for expression of early and late viral genes. Both ICP4 and ICP27 are essential for virus replication and the deletion of one of these genes requires adequate complementing cell lines to provide *in trans* the missing viral proteins. ICP0 is a multi-functional transactivator protein, able to inhibit the innate host responses and to promote viral gene expression. ICP47 inhibits major histocompatibility complex (MHC) class I antigen presentation, thus contributing to virus escape from immune surveillance. The role of ICP22 is less well understood (Roizman *et al.*, 2007). Several

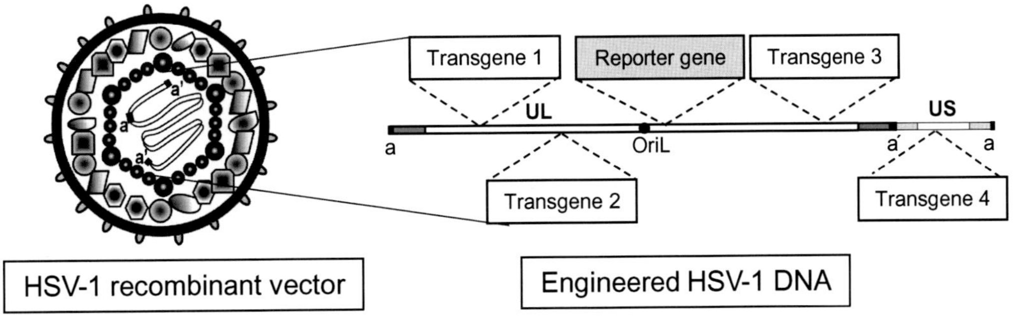

FIGURE 12.4 ■ Recombinant vectors are HSV-1 particles carrying an engineered HSV-1 genome. They usually carry one (or several) transgene(s) of interest, and a reporter gene (such as luciferase, β-galactosidase (*LacZ*), or green fluorescent protein (GFP)) to facilitate the pinpointing of the infected cells. Recombinant vectors can be either defective or attenuated.

replication-defective vectors have been constructed in which most or all IE genes were deleted in various combinations. These vectors do not display E and L gene expression and provide enough space to introduce expression cassettes for different transgenes. Deletion of all IE genes prevents toxicity for cells at high multiplicity of infection, allowing the vector genome to persist in cells for long periods of time. However, these vectors grow poorly in culture and express transgenes at very low levels in the absence of ICP0.

Defective HSV-1 can be used for: (1) delivery and expression of genes to CNS and PNS neurons, (2) selective destruction and immunotherapy of tumors and (3) prophylaxis against HSV-1 and other infectious diseases. Each application requires a different type of genetic engineering. A major advantage of these vectors is the possibility to achieve a synergistic therapeutic effect in gene therapy of multifactorial diseases (Fig. 12.4).

12.6.2.1 *Defective HSV-1 vectors for gene transfer in the nervous system*

The fact that HSV-1 can persist in neurons in a latent state, without interference with host-cell function, makes replication-defective viruses an attractive vehicle for gene therapy of neural disorders. However, the genetic modifications introduced to reduce their pathogenicity and increase their safety often result also in the loss of viral activities required for life-long persistence within the host. As for long-term transgene expression, it is controversial which promoter would be better suited for this purpose *in vivo*. It seems likely that the LAP promoter may be the most effective in long-term transgene expression in neurons. Hybrid promoters, constituted by LAP elements linked to viral strong promoters, have also been reported to sustain long-term transgenic expression (Manservigi *et al.*, 2010).

Defective recombinant HSV-1 vectors have been tested in gene therapy models of neurodegenerative, autoimmune and neuropathy diseases, like Parkinson's disease, multiple sclerosis (MS), chronic pain or spinal cord injury pain (Marconi *et al.*, 2010). Vectors expressing multiple trophic factors seem to be very promising as a side treatment for neurodegenerative diseases, including motor neuron diseases (MND). The advantage of these vectors is the possibility to achieve therapeutic effects in applications requiring simultaneous and synergistic expression of multiple genes. For instance, HSV-1 vectors were engineered to simultaneously express multiple neurotrophic factors, and experimental evidence indicates that treatment with these vectors can significantly increase motor neuron survival in comparison with the delivery of a single factor alone. One demyelinating disease that might benefit from anti-inflammatory therapy is MS, an autoimmune-mediated disease of the CNS. Defective HSV-1 vectors expressing immune-modulators are being used to treat experimental autoimmune encephalomyelitis (EAE), a mouse model for MS. This approach has established the therapeutic efficacy of defective HSV-1 vectors expressing anti-inflammatory genes, such as interleukin (IL4) or interleukin-1 receptor antagonist (IL1Ra). Defective HSV-1 vectors expressing anti-inflammatory cytokines, such as IL-4 or IL-10, have also been used to examine the involvement of cytokines in inflammatory pain. In a rat model of inflammatory pain, expression of IL10 by a HSV-1 vector prevents activation of p38 mitogen-activated protein kinase (p38 MAPK) and expression of full-length membrane-spanning tumor necrosis factor-α (TNF-α) in the DRG, suggesting the involvement of these molecules in the development of inflammatory and neuropathic pain. Other groups have used vectors to deliver genes that encode antisense or miRNA sequences or genes that antagonize ion channel function whose activities are essential to the development of chronic pain. Preclinical studies on animal models of pain demonstrated the capacity of these vectors to transfer genes into the DRG neurons following subcutaneous inoculation, and to efficiently express and release inhibitory neurotransmitters or anti-inflammatory peptides that modulate pain-related behaviors and provide a therapeutic effect in models of poly-neuropathy and chronic regional pain (Manservigi *et al.*, 2010).

Based on these results, a phase I human clinical trial, conceived to treat chronic pain of cancer patients using defective HSV-1 vectors expressing preproenkephalin, started in 2008 (Wolfe *et al.*, 2009). The entry of defective recombinant HSV-1 vectors into clinical trials will generate important information regarding the toxicity, safety, therapeutic potential and behavior of these vectors (Manservigi *et al.*, 2010; Marconi *et al.*, 2010).

12.6.2.2 *Defective HSV-1 vectors for vaccination*

The use of HSV-1 vectors for prophylaxis against viral infections requires mutants incapable of replicating in CNS and of spreading in immune-compromised individuals, unable to reactivate and not transmissible from vaccinated individuals to contacts. Although these viruses will make proteins only in the initially infected cells, the presentation of viral and/or transgene antigens to the host's immune system should be able to elicit long-lived and potent cellular and humoral immune responses. Various defective HSV-1 vectors were tested as potential anti-herpes vaccines in murine model or as vaccine vectors in murine and simian models. For instance, two out of seven rhesus monkeys vaccinated with defective HSV-1 expressing human immunodeficiency virus (HIV) antigens were protected, while a third showed a sustained reduction in viral load, following rectal challenge with pathogenic simian immunodeficiency virus mac 239 strain (SIVmac239). In another study, a single vaccination with a vector expressing chicken ovalbumin (OVA) was sufficient to elicit a strong immune response characterized by high frequency of primary and memory antigen-specific cytotoxic T-lymphocyte (CTL) response. Poor induction of cluster of differentiation 4 (CD4) helper T-cell responses and a weak induction of T-dependent antibodies, however, were observed. HSV-1 recombinants have been reported to elicit antigen-specific responses despite pre-existing immunity against viral antigens, which is important in the prospect of their potential therapeutic use in human populations where the virus is widely distributed. No genetic vaccine relying upon defective HSV-1 vector has so far been released (Marconi *et al.*, 2009).

12.6.2.3 *Defective HSV-1 vectors for cancer therapy*

Defective HSV-1 vectors were used to deliver anticancer transgenes to tumors such as melanoma or glioblastoma. These vectors express, in association with the autologous HSV-1 TK gene acting as a suicide gene when accompanied by its prodrug gancyclovir, further transgenes chosen for their potential to synergize in tumor cell killing and induction of antitumor immunity. In particular, soluble human cytokines IL-2, GM-CSF, interferon gamma (IFN-γ), and human TNF-α genes have been studied. Human umbilical cord vein endothelial cells HUVECs infected with HSV-1 vectors expressing anti-angiogenic fusion proteins, such as endostatin::angiostatin, were shown to induce cytostatic effects in proliferation assays *in vitro*. However, most groups willing to develop antitumor therapy focus today on oncolytic HSV-1 (Manservigi *et al.*, 2010; Marconi *et al.*, 2009).

12.6.3 Amplicon Vectors

Amplicons (Spaete and Frenkel, 1982) are HSV-1 particles identical to wild-type HSV-1 but which carry a concatemeric form of a DNA plasmid (the amplicon plasmid) instead of the viral genome. An amplicon plasmid is a standard *Escherichia coli* plasmid carrying one origin of replication (oriS) and one packaging signal (*pac*) of HSV-1, in addition to the transgenic sequences of interest (Fig. 12.5).

As amplicons carry no virus genes, they do not induce synthesis of virus proteins, making these vectors fully non-toxic for the infected cells. The lack of virus genes allows that most of the 153 kbp capacity of the HSV-1 particle can be used to accommodate large foreign pieces of DNA. This is the most outstanding property of amplicons, as there is no other viral vector able to deliver such a large amount of foreign DNA. In addition, amplicons are safer than defective recombinant HSV-1 vectors because the absence of virus genes reduces the risk of reactivation, complementation or recombination with latent HSV-1 genomes. In contrast, amplicons require a helper HSV-1 genome to be produced and high-titer helper-free amplicon vectors are difficult to prepare (Epstein, 2009). It is however, now possible to generate relatively large amounts of non-toxic amplicon vector stocks, essentially free of contaminant helper particles (reviewed in Epstein, 2009).

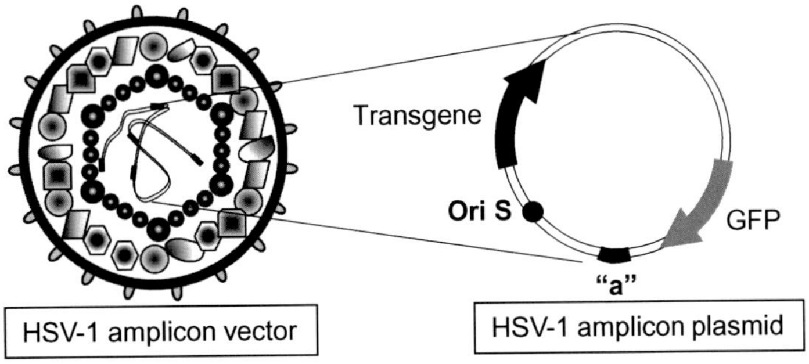

FIGURE 12.5 ■ Amplicons are HSV-1 particles carrying a head-to-tail concatemer of a DNA derived from the amplicon plasmid. This is a standard *E. coli* plasmid carrying one origin of DNA replication (oriS) and one *pac* sequence from HSV-1, thus allowing amplification and packaging of the plasmid into HSV-1 particles. In addition, it carries the reporter (here GFP) and the transgenic sequences of interest (represented by arrows). The amplicon vector carries around 150 kbp of a head-to-tail DNA concatemer derived from the amplicon plasmid.

12.6.3.1 *Production of helper-free amplicon vectors*

Two different helper systems are used today to produce amplicon stocks not or only barely contaminated with helper particles. One system is based on the cotransfection of amplicon plasmids with bacterial artificial chromosomes (BACs) supplying the full set of transacting HSV-1 functions (BAC-HSV). The HSV-1 helper genome carried by the BAC lacks the viral packaging signals (*pac*) and its size greatly exceeds that of the wild-type HSV-1 genome. As a consequence, these helper genomes cannot be packaged into newly assembled HSV-1 particles, producing helper-free amplicons (Saeki *et al.*, 2001). However, since the amount of amplicon vectors produced is limited, this method is well suited to producing amplicons for fundamental research and small-scale gene transfer, but it appears hardly suitable for large-scale production, as required for human vaccination or gene therapy approaches.

The alternative helper system is based on the deletion, through Cre/loxP1-based site-specific recombination, of the packaging signals of the helper virus in the cells that are producing the amplicons. This system uses as helper a recombinant HSV-1 that carries a unique and ectopic *pac* signal flanked by two parallel loxP sites. This helper virus genome is thus Cre-sensitive and will not be packaged in Cre-expressing cells due to deletion of the *pac* signal. As some helper genomes can, however, escape recombination, the vector stocks can be marginally contaminated (usually less than 0.1%) with helper particles, which are defective and non-pathogenic due to the engineering of further mutations in the helper genome (Zaupa *et al.*, 2003). Although this system is well suited to large-scale production of amplicons it still needs further improvements, as the presence of contaminant helper virus, even if marginal, could be a problem for gene therapy applications.

12.6.3.2 *Infectious transfer of very large foreign DNA sequences*

Amplicons can be used to deliver complete genomic DNA loci, multiple minigenes or DNA sequences that regulate chromatin structure and function, including vegetative replication and segregation. This may prove useful, for instance, to design improved gene therapy vectors displaying stable tissue-specific expression in proliferating cells, for generation of multiple splice variants, or for construction of polyvalent vaccines expressing several antigens.

For example, amplicons were used to deliver a 135 kbp fragment carrying the human low-density lipoprotein receptor (LDLR) into human fibroblasts derived from patients with familial hypercholesterolemia. The transduced LDLR locus was expressed at physiological levels for three months following infectious delivery to proliferating Chinese hamster ovary (CHO) cells (Hibbitt and Wade-Martins, 2006). This and other studies were possible thanks to the ability to avoid dilution of the episomal amplicon genomes in proliferating cells. As the amplicon genome does not contain HSV-1 replication functions and does not integrate into the host chromosomes, it will be diluted upon cell proliferation. However, it has been possible to generate amplicons that persist in proliferating cells by introducing the sequences of oriP and Epstein–Barr nuclear antigen-1 (EBNA-1) genes from Epstein–Barr virus (EBV) into the amplicon genome (Wang and Vos, 1996), thus allowing replication and segregation of episomic amplicon genomes. An alternative, more recent, method is based on the introduction of alphoid DNA. The alphoid-containing amplicon genome is converted in the infected cells into a human artificial chromosome (HAC), which remains transcriptionally active and is stably transmitted, at least in some cell types (Moralli *et al.*, 2006).

12.6.3.3 *Applications of amplicon vectors*

Amplicons have been successfully used to transduce genes of neurobiological, immunological or therapeutic interest in cultured cells or living organisms. They have been used to deliver toxic, proapoptotic or immune-stimulatory genes into experimental gliomas, showing potential to kill cancer cells. They were also used to deliver neurotrophins, antiapoptotic genes, heat-shock proteins or antioxidant enzymes, in studies addressing protection of neurons against neurological insults (reviewed in Epstein, 2009). Amplicons have been used to express frataxin, a mitochondrial protein whose dysfunction causes Friedreich ataxia, with encouraging results in a mouse model of this disease (Gomez-Sebastian *et al.*, 2007).

Other amplicons, delivering genes affecting neurotransmitter expression or neuroreceptor synthesis, have been used to study neuroplasticity and behavioral features, as well as to study aspects of Alzheimer's disease (Jerusalinsky *et al.*, 2012). Amplicons were used to deliver tyrosine hydroxylase and other genes of the dopamine pathway to the nigro-striatal system or to cultured striatal cells, in studies aimed at treating Parkinson's disease. Amplicons were used to deliver genes to the pigment epithelial cells of the rat retina (Epstein, 2009; Marconi *et al.*, 2010).

Amplicons can deliver genes to dendritic cells, primary hepatocytes, skeletal muscle, myoblasts and myotubes, cultured cardiomyocytes and heart slices. They have been used to express the structural proteins of Moloney murine leukemia virus (MoMLV) retrovirus, thus rescuing integrated retrovirus vectors, as well as the non-structural or structural proteins of hepatitis C virus (HCV). They were also used as experimental vaccines against HIV or intracellular bacteria (Epstein, 2009). Most interestingly, amplicons that carry the inverted terminal repeat sequences from adeno-associated virus (AAV) genome or from sleeping beauty transposons were constructed in order to increase the stability of transduced genes, by amplification and/or integration into host chromosomes, in the presence of replication (*rep*) or the transposase proteins respectively (de Oliveira and Fraefel, 2010; de Silva *et al.*, 2010). Lastly, amplicons encoding antisense, short hairpin RNA (shRNA) or miRNA molecules are being used to efficiently generate RNA interference (RNAi) molecules that induce specific silencing of cellular genes. However, in spite of the great potential of amplicons as gene delivery tools, they have not yet been used for gene therapy, likely due to the difficulties in obtaining large amounts of helper-free amplicon stocks.

12.7 Concluding Remarks

The different types of HSV-1-based vectors attempt to exploit different biological properties of the virus. Recombinant HSV-1 vectors, either attenuated or defective, try to exploit the adaptations of HSV-1 to the nervous system, such as its natural neurotropism or its ability to establish latency. Promising results have been obtained in treatment of cancer, in experimental models of PNS and CNS diseases and in treatment of pain. Although these vectors have been used mainly for gene transfer to neurons or glial cells, they can efficiently deliver genes to other cell types or tissues, and hold a big potential as vector vaccines, both against infectious diseases and against cancer. In particular, attenuated recombinant vectors are being used to treat malignancies, including brain, melanoma, breast and gastrointestinal cancers. In all cases these vectors are well tolerated and, furthermore, evidence of therapeutic effects have been often observed, although the first-generation vectors used seem to be insufficiently aggressive to eradicate the tumors. A phase I clinical trial using defective HSV-1 recombinant vectors expressing opioid peptides to treat chronic pain is ongoing and a second clinical trial, with vectors expressing glutamic acid decarboxylase (GAD) should start soon. The entry of defective recombinant HSV-1 vectors into clinical trials will generate critical information regarding the toxicity, safety, therapeutic potential and behavior of these vectors, and is expected to lead to new insights that will enable the further improvement of these vectors and accelerate their development for other neurological diseases.

Amplicon vectors attempt to exploit the capacity of the virus capsid to accommodate and deliver more than 100 kbp of foreign DNA. These vectors have been used only in experimental systems, and have proven to be versatile and useful tools for gene delivery to different cell types, both in culture conditions and in living organisms, including for the investigation of complex behavioral traits. Amplicons possess the unique feature of being able to deliver entire genomic loci, including large regulatory elements and introns and to convert them into HAC. A major area of amplicon research that should be developed in the future addresses the possibility to produce still larger amounts of purified vectors. There is also a large place for better controlling transgenic expression and for avoiding transgenic silencing of amplicon vectors.

Encouraging results have been obtained in experimental systems, and hopeful data are coming from clinical trials. However, more work remains to be carried out with HSV-1 vectors, especially if we intend to prolong and control transgene expression and to allow cell targeting, in order to increase their efficacy, to restrict transgene expression to predefined subsets of cells, and to decrease their undesired effects, such as infection of healthy cells. No doubt that many of these improvements will be achieved in the next few years.

For Further Reading

Original Articles

Braz, J., Beaufour, C., Coutaux, A., *et al.* (2001). Therapeutic efficacy in experimental polyarthritis of viral driven enkephalin overproduction in sensory neurons, *J Neurosci*, **21**, 7881–7888.

de Silva, S., Mastrangelo, M.A., Lotta, L.T., Jr., *et al.* (2010). Extending the transposable payload limit of Sleeping Beauty (SB) using the Herpes Simplex Virus (HSV)/SB amplicon-vector platform, *Gene Ther*, **17**, 424–431.

Fu, X., Tao, L., Jin, A., *et al.* (2003). Expression of a fusogenic membrane glycoprotein by an oncolytic herpes simplex virus potentiates the viral antitumour effect, *Mol Ther*, **7**, 748–754.

Gomez-Sebastian, S., Gimenez-Cassina, A., Diaz-Nido, J., *et al.* (2007). Infectious delivery and expression of a 135 kb human FRDA genomic DNA locus complements Friedreich's ataxia deficiency in human cells, *Mol Ther*, **15**, 248–254.

Hu, J.C., Coffin, R.S., Davis, C.J., *et al.* (2006). A phase I study of OncoVEXGM-CSF, a second-generation oncolytic herpes simplex virus expressing granulocyte macrophage colony-stimulating factor, *Clin Cancer Res*, **12**, 6737–6747.

Loret, S., Guay, G., Lippé, R. (2008). Comprehensive characterization of extracellular herpes simplex virus type 1 virions, *J Virol*, **82**, 8605–8618.

Mohr, I., Gluzman, Y. (1996). A herpesvirus genetic element which affects translation in the absence of the viral GADD34 function, *Embo J*, **15**, 4759–4766.

Moralli, D., Simpson, K.M., Wade-Martins, R., *et al.* (2006). A novel human artificial chromosome gene expression system using herpes simplex virus type 1 vectors, *EMBO Rep*, **7**, 911–918.

Nakamura, H., Mullen, J.T., Chandrasekhar, S., *et al.* (2001). Multimodality therapy with a replication-conditional herpes simplex virus 1 mutant that expresses yeast cytosine deaminase for intratumoural conversion of 5-fluorocytosine to 5-fluorouracil, *Cancer Res*, **61**, 5447–5452.

Saeki, Y., Fraefel, C., Ichikawa T., *et al.* (2001). Improved helper virus-free packaging system for HSV amplicon vectors using an ICP27-deleted, oversized HSV-1 DNA in a bacterial artificial chromosome, *Mol Ther*, **3**, 591–601.

Spaete, R.R., Frenkel, N. (1982). The herpes simplex virus amplicon: A new eucaryotic defective-virus cloning-amplifying vector, *Cell*, **3**, 295–304.

Umbach, J.L., Kramer, M.F., Jurak, I., *et al.* (2008). MicroRNAs expressed by herpes simplex virus 1 during latent infection regulate viral mRNAs, *Nature*, **45**, 780–783.

Wang, S., Vos, J.M. (1996). A hybrid herpesvirus infectious vector based on Epstein–Barr virus and herpes simplex virus type 1 for gene transfer into human cells *in vitro* and *in vivo*, *J Virol*, **70**, 8422–8430.

Wolfe, D., Wechuck, J., Krisky, D., *et al.* (2009). A clinical trial of gene therapy for chronic pain, *Pain Medicine*, **10**, 1325–1330.

Zaupa, C., Revol-Guyot, V., Epstein, A.L. (2003). Improved packaging system for generation of high level non-cytotoxic HSV-1 amplicon vectors using Cre–loxP site-specific recombination to delete the packaging signals of defective helper genomes, *Human Gene Ther*, **14**, 1049–1063.

Reviews

Bloom, D.C., Giordani, N.V., Kwiatkowski D.L. (2010). Epigenetic regulation of latent HSV-1 gene expression, *Biochim Biophys Acta*, **1799**, 246–256.

de Oliveira, A.P., Fraefel, C. (2010). Herpes simplex virus type 1/adeno-associated virus hybrid vectors, *Open Virol J*, **18**, 109–122.

Epstein, A.L. (2009). HSV-1-derived amplicon vectors: recent technological improvements and remaining difficulties — A review, *Mem Inst Oswaldo Cruz*, **104**, 399–410.

Friedman, G.K., Pressey, J.G., Reddy, A.T., *et al.* (2009). Herpes simplex virus oncolytic therapy for pediatric malignancies, *Mol Ther*, **7**, 1125–1135.

Hibbitt, O.C., Wade-Martins, R. (2006). Delivery of large genomic DNA inserts >100 kb using HSV-1 amplicons, *Curr Gene Ther*, **6**, 325–336.

Ingvarsdottir K., Blaho J.A. (2009). Role of viral chromatin structure in the regulation of herpes simplex virus 1 gene expression and replication, *Future Microbiol*, **4**, 703–712.
Jerusalinsky, D., Baez M.V., Epstein A.L. (2012). Herpes simplex virus type 1-based amplicon vectors for fundamental research in neurosciences and gene therapy of neurological diseases, *J Physiol Paris*, **106**, 2–11.
Manservigi, R., Argnani, R., Marconi, P. (2010). HSV recombinant vectors for gene therapy, *Open Virol J*, **4**, 123–156.
Marconi, P., Manservigi, R., Epstein, A.L. (2010). HSV-derived helper-independent defective vectors, replicating vectors and amplicon vectors for the treatment of brain diseases, *Curr Opin Drug Discov Devel* **13**, 169–183.
Marconi, P., Argnani, R., Epstein, A.L., *et al.* (2009). HSV as a vector in vaccine development and gene therapy, *Adv Exp Med Biol*, **655**, 118–144.
Roizman, B., Knipe, D.M., Whitley, R. (2007). 'Herpes simplex viruses', in Knipe, D.M., Howley, P.M. (eds.) *Fields Virology*, Philadelphia PA: Lippincott, Williams & Wilkins, pp. 2501–2601.

13

NON-VIRAL DNA VECTORS

Martin Schleef [a]

13.1 Plasmids — Tools in Molecular Biology

The first plasmids were identified by Joshua Lederberg (1952). The "wild-type" plasmids (to distinguish them from synthetic derivatives made later) were either circular (e.g., *Bacillus subtilis, Escherichia coli*) or linear (e.g., *Streptomyces coelicolor*) and present within the host cell at copy numbers of 1–50.

When the value of these epigenetic DNA molecules became obvious, an up-to-now growing list of vectors for cloning and gene expression was developed. Those plasmids are circular duplex molecules, which may be stably maintained as episomal genetic information within bacteria (Helinski, 1979; Summers, 1996; Schumann, 2001). Plasmid size ranges from 1.5 to approximately 120 kilo base pairs (kbp), and the plasmid copy number per bacterial cell may vary considerably (Davis *et al.*, 1980). In the case of small plasmids, plasmid copy numbers as high as 1000 (copies per cell) have been reported. The replication (amplification) does not depend on any plasmid-encoded protein and is not synchronized with the replication of the bacterial host chromosome (Davis *et al.*, 1980).

The plasmid's dimensions depend on its form. A linearized plasmid of 3 kbp has a molecular mass of 2×106 Da and a length of 1 μm (Davis *et al.*, 1980). The exact form of a plasmid molecule depends on its integrity. While the covalently closed circular ccc plasmid is in a supercoiled plasmid (sc plasmid) state, the open circular plasmid (oc plasmid) form is in a relaxed or nicked state. In addition, monomeric and multimeric forms are distinguished (Schmidt *et al.*, 1999).

[a] PlasmidFactory GmbH & Co. KG, Meisenstrasse 96, D-33739 Bielefeld, Germany
Email: martin.schleef@plasmidfactory.com

Figure 13.1 shows the schematic supercoiled (ccc-)form, the nicked (oc-)form and dimers of both topologies as well as linear forms. A potential influence of the plasmid form on the efficacy of DNA vaccines, transfection, cotransfection and virus production has been published (Maucksch *et al.*, 2009). The effect of

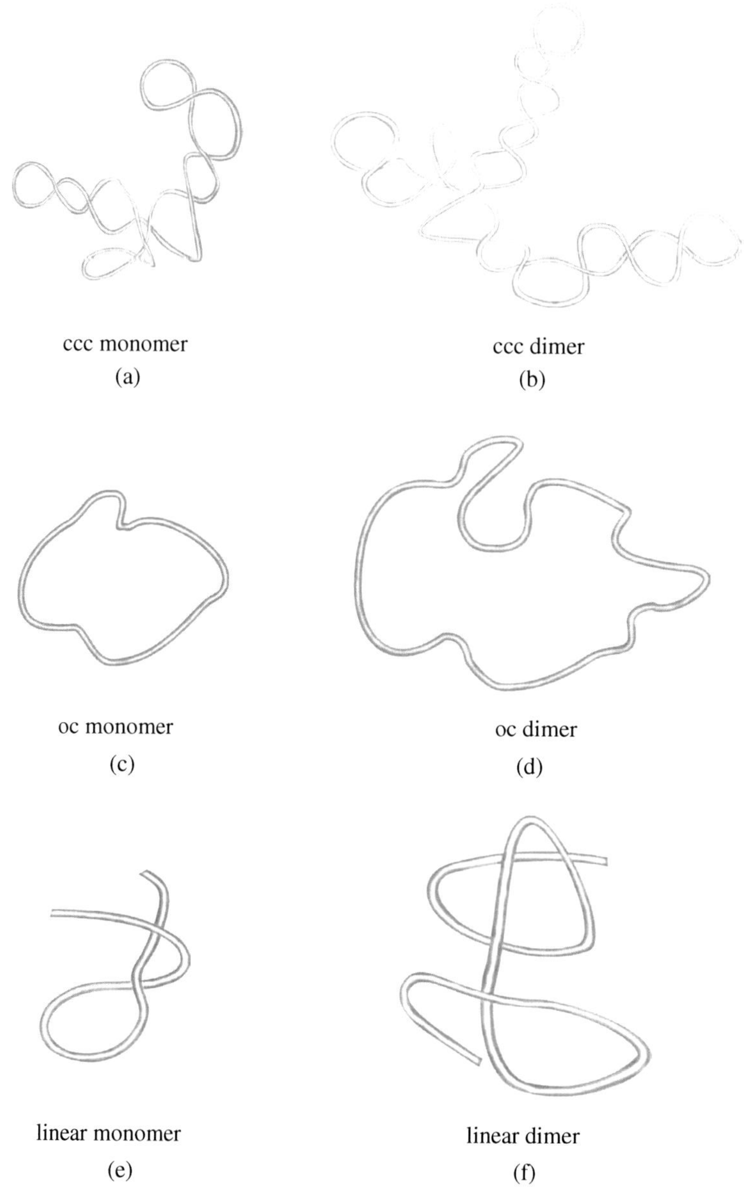

FIGURE 13.1 ■ Typical plasmid topologies present within bacterial cells used to amplify plasmid cloning vectors.

plasmid size was analyzed with different plasmids carrying the same transgene, with the comparison of monomeric and dimeric plasmids (Maucksch *et al.*, 2009; Voss *et al.*, 2006) and most recently with plasmids and so-called "minicircle" DNA (see Section 13.4 and Darquet *et al.*, 1997; Kreiss *et al.*, 1999; Schleef *et al.* 2010) containing identical expression cassettes.

Manufacturing plasmids within *E. coli* cells requires them to have a specific sequence that is used to replicate the plasmids, which are then able to distribute some of them to both daughter cells after cell division (Del Solar *et al.*, 1998). Most cloning vectors have the ColE1 origin of replication (*ori*). This ColE1 sequence — in the wild type encoding a protein (called RNA polymerase I modulator "Rom") — is bound by a small RNA (RNAII) expressed by RNA polymerase I, resulting in a DNA–RNAII-hybrid. If such RNAII is subject to cleavage by RNaseH, the resulting 3′-OH end works as a primer for the prolongation by the bacterial DNA polymerase I. This process is regulated via a second RNA: RNAI. This RNAI is complementary to parts of RNAII and forming this RNAI–RNAII-hybrid results in less frequent DNA–RNAII-association and — further down the line — less initiation of replication. This regulation and other types of plasmid replication are nicely summarized in Hayes (2003).

During plasmid replication, a certain portion of plasmids ends up in a larger supercoiled but not monomeric form: The plasmid dimer. Such a dimer is visible on an agarose gel by its slower migration speed (in other words a second band "above" the intended monomeric supercoiled plasmid DNA band, see Fig. 13.2). The *E. coli* cell has an interest in having an equally distributed set of plasmids between the daughter cells in cell division. Bacterial partitioning systems help to do so (ParA, ParB). In addition, other systems help to transfer a dimer plasmid (which is generated by homologous recombination of two identical plasmid sequences in a head-to-tail fusion resulting in a double-size large plasmid with absolutely identical restriction pattern; see Fig. 13.3 and Maucksch *et al.* (2009)) into two monomers by site specific resolution.

A special type of plasmid is the "cloning vector". This is not naturally occurring but was designed in the course of development of molecular biology and genetics to handle gene sequences. A plasmid cloning vector has a structure that is typically circular, has an *ori* sequence, exists in high copy numbers within *E. coli* cells and can be re-identified within bacterial cells by transferring a so-called selectable marker, such as antibiotic resistance genes located on the plasmid. Such a cloning vector allows the integration of a variety of different genes of interest (GOI) at a certain position. This position is — if capable of taking up different (multiple) gene sequences in an area with a high number of nucleic acid restriction sites in one narrow position of a plasmid — a "multiple cloning site" (MCS) as described earlier e.g., by Preston (2003). This part of the plasmid carries a cargo that is integrated by a cloning step, being either an expression cassette for a protein or an RNA or it is just a DNA stretch with a certain functional structure.

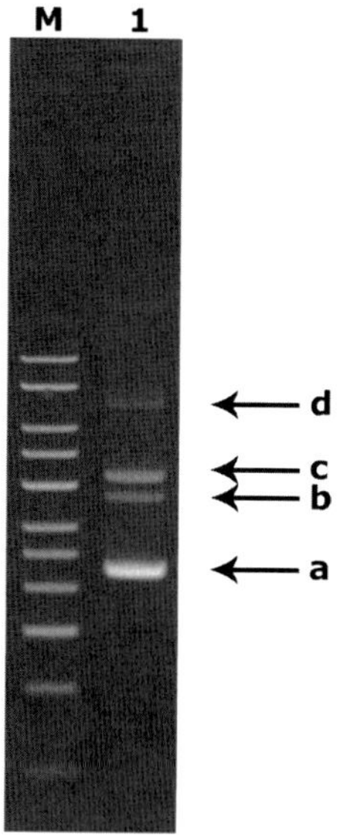

FIGURE 13.2 ■ Agarose gel electrophoresis of different undigested plasmids (0.8% agarose, 1 V/cm, staining after electrophoresis). The 1 kbp DNA ladder (lane M) is a positive staining control — the size of the contained linear DNA fragments does not indicate the sizes of the (undigested) plasmids. The band labeled with "a" is the fastest migrating plasmid form (ccc = covalently closed circular, the supercoiled monomer). The bands at "b" and "c" are either the dimer of ccc or the oc (= open circular or "nicked plasmid") form, that shows one broken strand of two. "d" is the oc form of the dimer.

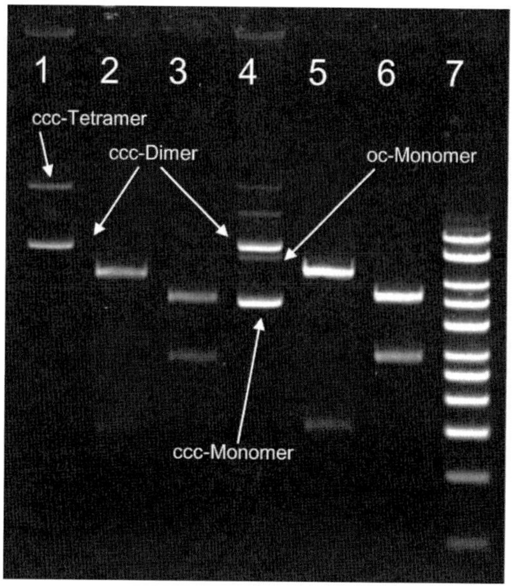

FIGURE 13.3 ■ Comparison of monomeric (lanes 4 to 6 together with dimer) versus dimeric (lanes 1 to 3 together with a tetramer) plasmid DNA, separated by agarose gel electrophoresis. The restriction digest pattern (lanes 2 and 3 or lanes 5 and 6) indicates, that both DNAs being visibly different in the undigested form (lanes 1 and 4) show identical bands after digestion.

This sequence can either be functional within the *E. coli* cell (e.g., for the production of recombinant proteins within such prokaryotic cells) or the functionality is reduced to a minimum in *E. coli* cells but is required to be of high relevance in the heterologous system of an eukaryotic target cell when transfected with such plasmid DNA. This is the case for the production of antibodies, viral particles, antigens in DNA vaccination or — *in vitro* — mRNA on the basis of the respective blueprint on the plasmid. A typical plasmid with different sequence elements is shown as an example in Fig. 13.4.

The number of plasmids per cell is of relevance for cloning vectors. The intended goal is to obtain a large amount of plasmid DNA to be able to purify this for subsequent work. The high copy number of ColE1 *ori*-containing cloning vectors (e.g., pUC21) was achieved by protecting this from the RNAI-hybridization through a point mutation within the RNAII-sequence of its *ori* (see Section 13.1) (Vieira and Messing 1982).

An overview is given in Chapter 31 on manufacturing and quality control aspects of plasmid DNA used in clinical and research applications.

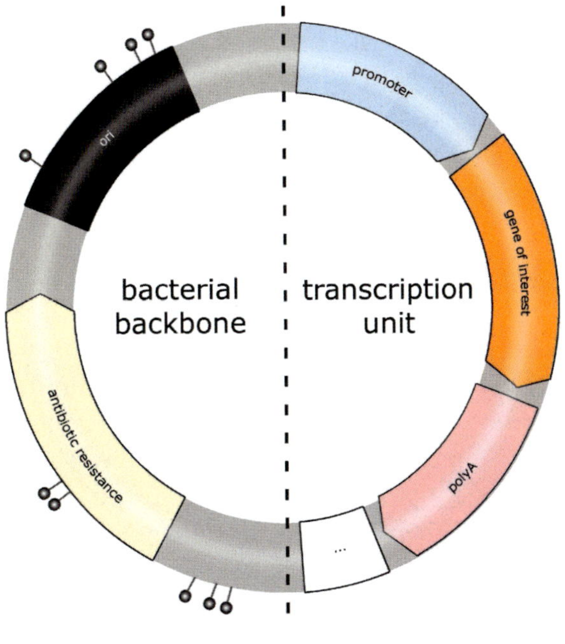

FIGURE 13.4 ■ Typical structure of a plasmid vector. The different structural elements are indicated: ori, origin of replication, bla, β-lactamase ampicillin resistance gene, GOI, gene of interest — a gene expression cassette including promoter, coding region and polyA-signal, MCS — multiple cloning site for the insertion of additional sequences. The left part representing sequences of bacterial origin may contain CpG motifs (needles).

13.2 DNA Gene Transfer and Application

13.2.1 Use of Plasmid DNA in Gene and Cell Therapy as well as in DNA Vaccination

Approximately 25% of all gene-therapy protocols performed so far in clinical studies were directly based on plasmid DNA vectors (Edelstein *et al.*, 2007). The market share of plasmid DNA vectors with respect to all vaccines was initially expected to rise to about 60% (Jain, 1996). This also includes the plasmid DNA used to produce viral vectors — e.g., by transient transfection of producer cells for adeno-associated virus (AAV) vectors or lentiviral vectors. The two-plasmid AAV packaging/helper system from the laboratory of Jürgen Kleinschmidt (Grimm *et al.*, 1998) was initially developed for serotypes 1 to 6. Heparin-binding site deficient mutants (pDG(R484E/R585E)) (Kern *et al.*, 2003) and further serotypes are available for cotransfection with just the transfer plasmid (containing the inverted terminal repeats (ITRs)) on one side and the helper/packaging plasmid (both functions on one other plasmid > 20 kbp in size). Descriptions of further versions of such a system have been published (Lock *et al.*, 2010) and two international reference standards were applied to ensure proper clinical manufacturing of this AAV by use of the pDG plasmid system (Moullier and Snyder 2008). In such cases, optimization of transfection is relatively easy if only the ratio of the quantities of two plasmids have to be evaluated when starting the work with a new batch of material of the one or the other plasmid. This is significantly more difficult if three (or more) plasmids have to be triple-transfected and their individual relative amounts need to be (re-)optimized each time a new batch is enclosed. An overview is provided by Ayuso *et al.* (2010).

After Chardeuf *et al.* (2005) demonstrated that the structural elements of plasmid vectors for the production of AAV particles — namely the antibiotic resistance gene elements carried on the transfer plasmid — were detectable within virus preparations, the regulatory authorities strongly requested avoiding such sequences in AAV production strategies, leading us to consider the development of a minicircle system to avoid this in the future. A further insight into AAV vectors is presented in Chapter 11 of this textbook.

13.2.2 Transferring Genes into Cells

Transferring genes into eukaryotic cells is called transfection, while the transfer of a plasmid, with the purpose of performing amplification of the plasmid, into, for example, an *E. coli* cell transforms this cell — and is called transformation.

The transfer of genes either in the form of a plasmid or a fragment of a plasmid or an RNA or any other genetic material is achieved by bringing the nucleic acid into close contact with the target cell, accessing the cell for the nucleic acid and trying to obtain expression or function within the target cell. To reach this goal, different approaches were developed over the years, involving either chemical or physical uptake of the nucleic acid by a cell.

The following summary should help in understanding the huge differences in transferring nucleic acids and how it depends on the type of target cell or organism but also the type of nucleic acid (DNA or RNA) and finally the intracellular target (nucleus, cytoplasm, organelles or other cellular structures) to be addressed.

13.2.2.1 *Lipofection and complexing agents for gene transfer*

So-called lipoplexes (Barron *et al.*, 1999), micelles of positively charged lipids surrounding the DNA are able to enter into the cells (Koltover *et al.*, 1998) and to release their DNA cargo there. In addition, cationic lipids can potentially be targeted to specific cell types by incorporating certain ligands into their structure. The effect of such complexing agents hence is protection, changes in features influence the access (in)to the cell and targeting. Most transfection kits contain such substances. Further details are presented in Chapters 14 and 15 of this textbook.

13.2.2.2 *Electro gene transfer (EGT)*

Application of short electric field pulses to cells leads to transiently porous areas in the cell membranes; therewith, they become permeable to substances that cannot otherwise enter. The various applications include the direct (electrophoretic) transfer of genes, proteins, ionic dyes and drugs into the cell interior. To get effective transmembrane transport the electric field strength has to overcome a certain threshold value, but has to be lower than the value that causes massive irreversible cell damage (Schmeer *et al.*, 2004). In tissue, these damaged cells may cause necrosis, possible ulceration and appearance of wounds. In gene therapy especially this effect has to be avoided because a loss of viability causes reduced efficacy of the DNA transfer (Miklavcic *et al.*, 2000). Another important parameter is the pulse duration, because the electric field, besides the above mentioned permeabilization, has to electrophoretically draw the DNA across the cell membrane and through the cytosol. Most of all the method is limited by the accessibility of the target tissue to the electrodes for field pulse application (Schmeer, 2009; Trollet *et al.*, 2005; Mir, 1999, 2005; Bureau *et al.*, 2002; for reviews: Trollet *et al.*, 2006; review Favard *et al.*, 2007). In addition, the type of tissue is an important parameter affecting access to the cell membrane. Clinical applications were reported for skin tumors and in gene therapy. The combination of EGT and needle injection was used in DNA vaccination, including human clinical trials. Further information is provided in Chapter 17 of this textbook.

13.2.2.3 *Hydrodynamic gene transfer*

Injecting large volumes rapidly by intravenous infusion results in the highest transgene expression detectable in the liver (Andrianaivo *et al.*, 2004). The applied DNA probably enters the liver tissue through the portal vein enabled by the increased venous pressure after the high-volume bolus injection. The leakage of the liver capillary system may be another factor enabling the DNA to enter the liver tissue. However, this method only leads to a sufficient amount of transfected cells if the efflux of the DNA solution is obstructed by clamping the vein. Please also refer to Chapter 16 of this textbook.

13.2.2.4 *Aerosolization for gene transfer*

The delivery of DNA to the lung can be applied by simple injection to certain distinct parts of the lung. A more widespread delivery is not easy and several approaches have been tested so far (reviewed by Davies *et al.*, 2005) including systemic administration — ending up with certain portions of the DNA in other tissues (e.g., liver). The DNA was complexed in order for it to be protected against degradation on its way. Later approaches to ensure topical application of DNA made use of an instillation (Davies *et al.*, 2005) and further success was obtained by use of aerosols with complexed DNA — e.g., to treat cystic fibrosis (Davies *et al.*, 2005; Yang *et al.*, 1995; Ziady *et al.*, 2003).

13.2.2.5 *Ultrasound-driven gene transfer*

Sonoporation uses ultrasound to permeabilize the cell membrane in order to allow uptake of large molecules such as DNA into the cell. Microbubbles, similar to those used as contrast agents, significantly increase transfection efficacy when exposed to the low-frequency field. Hence, sonoporation is used in targeted gene transfer *in vitro* as well as *in vivo* (Newman and Bettinger, 2007; Li *et al.*, 2008).

13.2.2.6 *Laser beam gene transfer*

A further approach for the transfer of naked DNA into tissue was presented by Zeira *et al.* (2003), where a femtosecond infrared laser beam was used to transfer plasmid DNA into muscle tissue.

13.2.2.7 *Particle bombardment gene transfer*

In this technique, the DNA is precipitated on small particles, usually made of gold. Other substances, such as proteins and peptides, may be added. The DNA-coated particles are shot through the skin of the recipient tissue using a gas stream. Due to this bombardment, parts of the skin get damaged, depending on the velocity of the particles, but nevertheless this method has been proven to be highly effective for gene delivery into the epidermis. This method is only suitable for gene delivery into regions near the surface (Fuller *et al.*, 2006; Steele *et al.*, 2001).

13.2.2.8 *Needle or needle-free jet injection for gene transfer*

Wolff and co-workers were the first to show that simple application of naked DNA by needle injection into muscle tissue is sufficient for gene transfer leading to the expression of the transgene (Wolff *et al.*, 1990, 1991). However, although needle injection can transduce naked DNA into muscle tissues, this technique was largely inefficient for other tissues, including tumors. A combination of injection and EGT was mentioned above.

Meanwhile, numerous studies are dealing with the optimization of this procedure (Furth *et al.*, 1995a, 1995b). This led to the development of the hydrodynamic-based procedure to deliver large volumes (>1 ml) of naked DNA by injection into the tissue within short times (Liu *et al.*, 1999; Zhang *et al.*, 2004).

Jet-injection technology is based on jets of high velocity possessing the force to penetrate skin and underlying tissues, leading to transfection of the affected areas. *In vivo* application of this technology does not induce tissue damage or significant inflammatory reactions at jet-injection sites (Cartier *et al.*, 2000; Walther *et al.*, 2001). The use of such a device requires the optimization of injection pressure to avoid DNA damage by shear forces.

13.2.2.9 *Magnetofection for gene transfer*

In this method, DNA is associated with magnetic nanoparticles, resulting in molecular complexes. Those are transferred towards, or even into, the target cells by an appropriate magnetic field (Scherer *et al.*, 2002; Huettinger *et al.*, 2008; Jahnke *et al.*, 2007) and most recently are used for the transfection of primary neural stem cells (Sapet *et al.*, 2011).

13.2.3 Applications of Plasmid DNA

In addition to the use of plasmid DNA for, e.g., AAV vector, lentiviral vector or adenoviral vector production, the DNA can be used as a DNA vaccine in human or animal applications. Promising results were obtained from veterinary clinical studies in collaboration with Plank *et al.* (Huettinger *et al.*, 2008; Jahnke *et al.*, 2007), where we produced a DNA vaccine against the cat fibrosarcoma.

Also, the use of various plasmids or a single plasmid with a polycistronic expression cassette (a stretch of DNA where various expression units or an expression unit with various sequence motifs is located) is an approach to modify transfected cells with the goal of modulating their expression profile. This modulation may add a certain feature (e.g., a surface antigen of the cell) or the expression of an interesting protein to be produced, or just to make the cell visible (e.g., green fluorescent protein (GFP)). Recent findings demonstrate that such modification may even end up in the reprogramming of such cells, leading to a strong requirement for minicircle DNA (see Section 13.4) in induced pluripotent stem cell (iPS) research (Jia *et al.*, 2010).

Recently, a strong requirement for highly purified DNA was observed in the field of using plasmids for the *in vitro* mRNA production (Kormann *et al.*, 2011).

Other applications for plasmid DNA are those where a transient transfection is applied to obtain initial amounts of the expressed protein (e.g., antibodies), even if manufacture by a stably expressing cell line (not transiently but constantly expressing the protein) seems to be more economical. This is clearly not the case as Backliwal and Wurm (2009) explain, since the time and effort to generate, test, validate and use such cell lines is significantly higher than that for performing all initial and, to a certain extent, later expression studies with plasmids.

Part III of this textbook gives an overview of recent applications in gene therapy and genetic vaccination.

13.3 Next Generation Plasmids and Transposons

13.3.1 Future Non-Viral DNA Systems — Antibiotic-Free and Miniplasmids

Usually, non-viral vectors are applied as episomal vectors with the strong intention of avoiding any integration into the target cell, except for cases as outlined below (transposons) to exclude any risk of insertional mutagenesis (Somia and Verma, 2000; Li *et al.*, 2002; Flotte 2004; Zaiss and Muruve 2008).

Initially, non-viral vectors were mainly represented by plasmid DNA that mostly consists of the following elements introduced earlier: (1) an origin of replication, (2) an eukaryotic expression cassette made of the GOI controlled by regulatory elements, and (3) a selection marker (typically an antibiotic resistance element like a kanamycin resistance (neomycin phosphotransferase (*nptI*/*nptII*)) or ampicillin resistance β-lactamase (*bla*) gene) that ensures plasmid maintenance and propagation in dividing bacterial cells. Due to the fact that not all of these sequence elements are finally required (see Fig. 13.4, *left*) and — even worse — could be disadvantageous, a strong need for avoiding them made us and other researchers search for approaches to reduce the size and sequence content of plasmids. In particular, the observation that bacterial sequences injected into muscle cause unintended immune responses and that bacterial DNA contains unmethylated CpG motifs that bind to the Toll-like receptor 9 (TLR9) of antigen-presenting cells, leading to the activation of the immune system of the host (Mir *et al.*, 1999; Krieg *et al.*, 1995; Hyde *et al.*, 2008) as well as an up-regulated expression of cytokines and chemokines, requires reduction of their content within such vectors. Even if such effects may support DNA vaccination approaches, from a therapeutic perspective and with respect to a pharmaceutical view, they are not helpful.

A few years ago certain approaches were presented, where in some cases the size of a plasmid was reduced by replacing the antibiotic resistance marker sequences by other expression elements to perform selection for the presence of a plasmid within the bacterial cell. Either an essential gene was deleted from the genome of the cell and placed on the plasmid backbone to select for the presence of the plasmid within a cell or the presence of the plasmid caused a titration effect influencing the expression regulation of a genome-based gene to influence growth or life of the bacterial strain — all with the goal of selecting for the presence of plasmids.

In the operator-repressor-titration (ORT) (Cranenburgh *et al.*, 2001) system the special *E. coli* production strain contains the essential chromosomal gene *dapD* under control of the lactose (*lac*) operator/repressor system of the *lac* operon. The *lac* promoter system can be induced by adding isopropyl β-D-1-thiogalactopyranoside (IPTG) due to inactivation of the Lac repressor (LacI), thus enabling the bacteria culture to grow. In the absence of IPTG, the essential tetrahydrodipicolinate succinylase subunit (*dapD*) gene is not expressed, thus leading to cell lysis. Transformation of the bacteria with a high-copy-number plasmid carrying the *lac* operator site, *lacOS*, leads to a large excess of the plasmid-encoded compared with the chromosomally-encoded *lacOS*. Hence, most of the few chromosomally encoded inhibitors LacI bind to the plasmids rather than to the chromosomal sites. This "titration" leads to the expression of the essential gene to a level similar to that obtained in the presence of IPTG.

If such special production strain (e.g., DH1/lacdapD) is transformed with a pORT plasmid containing no antibiotic resistance gene but a *lacOS*, the cells carrying the plasmid can survive without the presence of IPTG, which represents a selection for plasmid-containing cells. However, the pORT plasmids like all the biosafe miniplasmids, although lacking the antibiotic resistance markers, still contain bacterial backbone sequences like the origin of replication (*ori*) only needed for amplification in bacteria and not for the intended application in the target cells.

Another approach has been described by the group of R. Grabherr (Mairhofer *et al.*, 2008, 2010). Here, the strategy for the antibiotic-free selection and propagation of plasmids involves the origin of replication carried by pUC-derived expression vectors (ColE1 *ori*). The origin-encoded RNAI silences a host-encoded repressor by RNA–RNA antisense interaction. The inhibition of the repressor allows the expression of an essential gene, which enables growth of plasmid-carrying cells. For this purpose, an operator was introduced upstream of an essential gene in the chromosome of the *E. coli* production strain and the repressor gene corresponding to this operator was fused to an RNAII-like sequence. RNAI (plasmid replication inhibitor) and RNAII (primer for plasmid replication), are naturally encoded by the *ori* sequence. If the cells have been successfully transformed with the plasmid, the RNAI is produced from the plasmid replication control region and translation of the repressor is inhibited with high efficiency by antisense hybridization of the RNAI to the RNAII-like sequence fused to the mRNA of the repressor. Now expression of the essential gene is possible and the cells can survive.

A recent technology made use of an approach earlier presented as a bacterial plasmid vector system, free of an antibiotic resistance marker and published by Seed (1983). The system for pFAR (free of antibiotic resistance) plasmid vectors relies — as the initial system — on the use of a suppressor transfer RNA (tRNA) — here the suppression of an amber mutation introduced into a chromosomal essential gene of the bacterial producer strain. This nonsense mutation leads to the incomplete translation of the essential protein and no cell multiplication. The introduction into those strains of plasmid encoding a suppressor tRNA expressed from prokaryotic regulatory elements restores a full synthesis of the protein and bacterial growth. These two antibiotic-free resistance systems present the following differences:

The pCOR plasmids are produced from a bacterial strain that contains a mutation introduced into the *argE* gene that encodes a protein involved in the arginine biosynthetic pathway. Therefore, their production requires the utilization of a medium completely devoid of proteins. The pCOR plasmids were entirely *de novo* synthesized. They contain a suppressor tRNA gene, a eukaryotic expression cassette and a R6K conditional origin of replication that requires a specific initiator protein encoded by the plasmid initiator of replication (*pir*) gene, thus limiting plasmid host range to bacterial strains that encode this *trans*-acting protein.

The pFAR plasmids are produced from an *E. coli* derivative which contains an amber mutation in the *thyA* gene that encodes a thymidylate synthase required for DNA synthesis. This mutation leads to thymidine auxotrophy. The introduction of pFAR plasmid into the *E. coli* mutant restored normal growth to the auxotrophic strain.

pFAR4 is a new plasmid DNA vector that was entirely synthesized. It carries an allelic form of *hisR*, the expression of which is under the control of prokaryotic regulatory elements, and encoding a histidine suppressor tRNA that allows the insertion of the expected amino acid with a high efficiency (up to 100%) (Kleina *et al.*, 1990; Michaels *et al.*, 1990). Indeed, the introduction of pFAR plasmids into the optimized *thyA* mutant allowed the selection of plasmid-containing bacteria, the restoration of normal growth to thymidine auxotrophic strains and a high-yield production of monomeric supercoiled expression vectors, as required for pharmaceutical products.

Initial injection and electrotransfer of luciferase-encoded pFAR4 led to higher expression levels in skin and transplanted tumor (of almost an order of magnitude), thus further showing the importance of reducing plasmid size by eliminating redundant elements.

The major advantage of these host/vector systems is that no antibiotic resistance gene is present on the plasmids. However, sequence motifs (e.g., the origin of replication) redundant for gene expression in the target cells are still present on these plasmids, which leads to a strong requirement for even further reduced DNA vectors, such as *minicircles* (see Section 13.4).

13.3.2 Jumping Genes — Transposons as Non-Episomal Carriers for Genetic Information

Transgene expression in liver after hydrodynamic delivery is of short duration, because a hepatocyte regeneration process occurs and because non-integrated plasmids are lost during cell division. Using the phage phiC31 integrase, Olivares *et al.* (2002) have demonstrated that transgene integration greatly increases the level and duration of expression after hydrodynamic delivery. In such an integrating strategy, the sleeping beauty (SB) transposon is particularly attractive as a gene delivery tool since it has the distinctive ability to integrate into the host genome through a conservative cut-and-paste mechanism with no integration bias into genes. The pioneering work of Izsvak and colleagues has shown that plasmid-based SB system provides long-term transgene expression in vertebrates (Izsvak *et al.*, 2009). In addition, SB shows interesting properties in the context of hydrodynamic liver delivery, leading to partial biochemical correction in mucopolysaccharidosis Type I (MPSI) mice in both peripheral organs and the brain.

Integrating plasmid-based gene transfer systems are composed of: (1) an expression plasmid that encodes the integrase or transposase which can act *in trans* and (2) a donor plasmid containing the DNA to be integrated, which is flanked by specific sequences recognized by the integrase or transposase for the integration process.

Recently, a dramatic improvement has been obtained by the generation of novel engineered SB transposases using molecular evolution. The hyperactive SB100X is 100-fold more potent *in vitro* compared with the originally resurrected SB. Figure 13.5 shows the two plasmids — one carrying the enzyme and the other with the expression cassette (here expressing GFP) between the IR sites of the transposon.

The human application of the SB system was recently presented by Hackett *et al.* (2010), where a chimeric antigen receptor was applied to redirect the specificity of human T-cells — an approach performed *ex vivo* for clinical phase I/II.

The overview of an alternative approach using mobile DNA elements initially observed in baculoviruses known as "piggyBac" (PB) has been presented by Alan Bradley (Yusa *et al.*, 2011). The initial work on the mammalian application was presented by Ding *et al.* (2005). Recent applications have demonstrated the use of PB for the induction of pluripotent stem cells through reprogramming fibroblasts (Woltjen *et al.*, 2009).

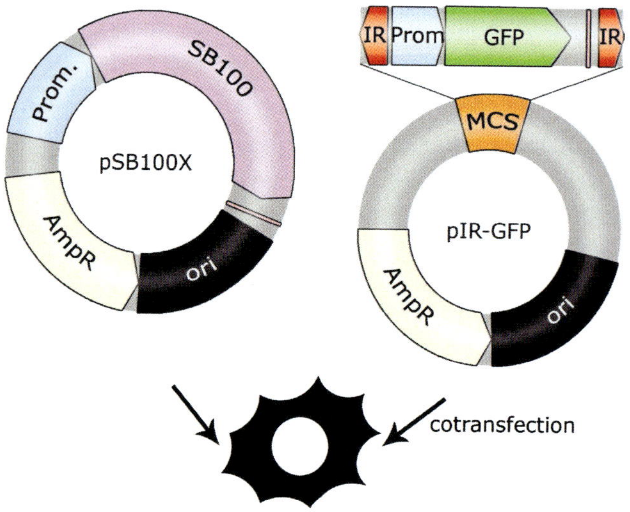

FIGURE 13.5 ■ Co-transfection of a plasmid (pSB100X) carrying the gene for an SB enzyme (SB100) and a second plasmid, carrying the expression cassette for a GFP gene between the inverted repeat (IR) sequences of the transposase, located within the MCS of plasmid pIR-GFP. Both are co transfected into the target cell.

13.4 Minicircle DNA — "Plasmids" Without "Bits and Pieces"

Recent developments show that a new type of vector may be the system of choice with which to overcome the disadvantages of plasmid DNA: The "minicircle" DNA. This circular, non-viral DNA molecule derives from plasmid DNA but lacks those elements of plasmids not useful or even detrimental for (pre-)clinical applications.

These vectors derive from parental plasmids with antibiotic resistance markers, the GOI and *ori*, as well as two special recombination sequences right and left of the GOI. Through an intramolecular recombination process the GOI (plus one of the signal sequence elements) is cut out of that parental plasmid, circularized and finally results in only the GOI and the signal sequence in a circular molecule. The residual part of the parental plasmid, the miniplasmid, contains the *ori* and the selection marker, and by definition is still a plasmid (Fig. 13.6).

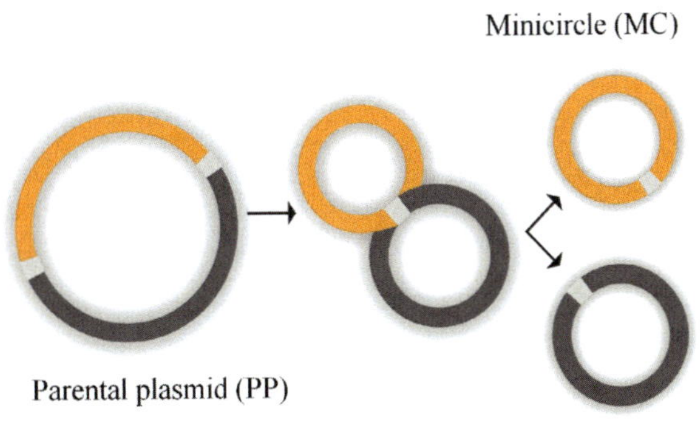

(a)

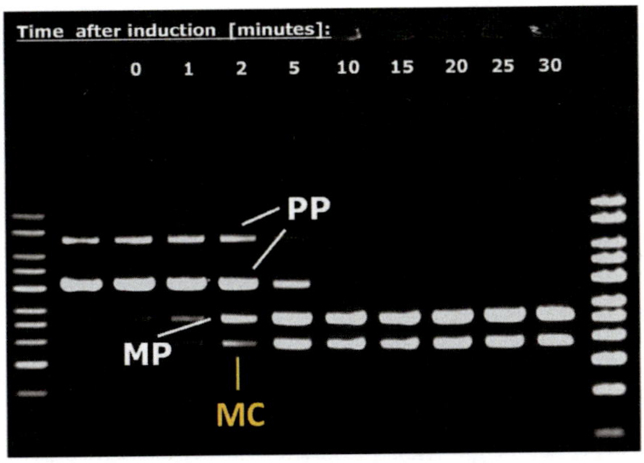

(b)

FIGURE 13.6 ■ (a) Scheme of *cis*-recombinaton of a parental plasmid (PP) resulting in two independent circular DNA molecules being the *minicircle* (MC) and the miniplasmid (MP). (b) time course of the minicircle building process before (0) and throughout the recombination (sampling times indicated). Data kindly provided by A. Herzig, PlasmidFactory, Bielefeld.

13.4.1 Initial Approaches to Remove the Bacterial Backbone Sequences

Different approaches have been proposed over the last few years to obtain a DNA gene therapy vector containing only the gene expression cassette. The initial idea of purifying a restriction fragment cut from a plasmid and protecting this against the exonuclease activity within the target cell leads to a unique approach known as MIDGE (Schakowski *et al.*, 2001). This molecule is still a relaxed and linear DNA, although protected at its ends from being degraded.

The idea of also obtaining a circular, if possible supercoiled, DNA molecule with as little bacterial sequences as possible was approached by Darquet *et al.* (1997) and Kreiss *et al.* (1999), who made use of the fact that recombinases may also perform a recombination between two recombination sites positioned within the same molecule (*cis*) and result in two smaller intact circular DNA molecules, both containing one of the two recombination sites. This approach was initially patented and is until today the basis of all other minicircle approaches ever performed. To date, all further patent applications and patents depend on these initial findings. The lambda integrase used by Darquet *et al.* (1997) is known to favor unidirectional recombination events between attachment sites attB and attP in the absence of the excisionase (Xis) protein (Sadowski, 1986). The resulting molecules carry attL and attR sites. However, about 30% of the minicircles produced with this system were still present as dimers in a Xis-negative environment (Kreiss *et al.*, 1998). Thus, these structures had to be removed in the approach described here using the multimer resolution system of the broad host range plasmid RK4 (Kreiss *et al.*, 1998).

The enzymes used so far to achieve the (intramolecular) recombination process were derived from two enzyme families: (1) tyrosine recombinases, such as the integrase of bacteriophage lambda, the causes recombination (Cre) recombinase from bacteriophage P1 and the flippase (FLP) recombinase of the yeast plasmid 2 μm circle, or (2) serine recombinases, e.g., the integrase of Streptomyces bacteriophage PhiC31 or the ParA resolvase from the multimer resolution system of the broad host range plasmid RK2 or RP4 (Bigger *et al.*, 2001; Chen *et al.*, 2003; Darquet *et al.*, 1997; Jechlinger *et al.*, 2004; Nehlsen *et al.*, 2006). The recombination events of the Cre recombinase (locus of X-over P1 (loxP) sites) and of the FLP recombinase (flippase recognition target (FRT) sites) result in identical or highly similar sites and thus the recombination is bidirectional and fully reversible, finally resulting in several multimer structures due to intramolecular and intermolecular recombination (Gilbertson, 2003). Thus, Bigger and co-workers constructed a parental plasmid with one lox site being mutated to prefer unidirectional recombination (Bigger *et al.*, 2001). Therewith, the generation of monomeric minicircle molecules was improved but a high amount of concatemers or catenates was still observed (Bigger *et al.*, 2001). Also the miniplasmid still present within the same cell as the minicircle made it difficult to obtain a pure minicircle fraction. The idea of digesting the

TABLE 13.1 Overview of Experimental *Minicircle* Systems (Modified from Schleef and Schmeer, 2011)

Reference	Recombination System	Recognition Sequence	Induction
Bigger *et al.* (2001)	*Cre* recombinase (bacteriophage P1)	loxP recognition sites	Ara induction
Nehlsen *et al.* (2006) Broll *et al.* (2009)	Flp recombinase (*Saccharomyces cerevisiae*)	FRT	Heat induction
Darquet *et al.* (1997)	λ Integrase (bacteriophage λ)	attB and attP (bacteriophage λ)	Heat induction
Chen *et al.* (2003)	φC31 Integrase (bacteriophage φC31)	attB and attP (bacteriophage φC31)	Ara induction
Mayrhofer *et al.* (2008)	ParA resolvase of RP4 and RK2	res	Ara induction

non-required miniplasmid (but not the minicircle) was initially proposed (within the producer cell or after total DNA preparation) by Bigger and co-workers (Bigger *et al.*, 2001) and later fine-tuned by Chen and colleagues (Chen *et al.*, 2003) using a specific type of restriction enzyme only cutting extremely specifically — a meganuclease. An overview of different recombination systems is shown in Table 13.1.

13.4.2 Recent Minicircle Technology

Based on the earliest patented approach to obtain a functional minicircle (Cameron *et al.*, 1996, with a priority date of 1995; and Crouzet *et al.*, 2000), the idea was to overcome the low level of recombination and the low level of purity by applying different recombianeses (see Table 13.1). In addition, the approach to selectively bind a sequence motif with the purpose of separating this from a mixture of different DNAs (Gossen *et al.*, 1993) led to the approach we initially published in 2008 (Mayrhofer *et al.*, 2008). The resulting minicircle DNA consists almost only of the gene of interest, leading to significant size reduction and improved performance. Comparison of plasmid and minicircle-mediated gene expression shows improved performance of the minicircle in different cell lines, i.e., minicircle vectors improve transfection efficiency and transgene expression in different cell lines, mainly due to size reduction, but also due to the high purity of such DNA products (Mayrhofer *et al.*, 2008, 2009). Availability of minicircles at reproducible quality (for details see Schleef and Schmeer, 2011) and in sufficient amounts makes this system an applicable and effective alternative to conventional plasmid vectors.

A comprehensive overview is available in Schleef (2013).

The following minicircle-based expression systems exist and have been tested already: Cytomegalovirus–luciferase (CMV–luc), CMV–GFP, CMV–β-galactosidase (lacZ) as well as a DNA vaccine against hepatitis B virus (HBV)-S2S surface antigen based on minicircle or a minicircle expressing enhanced GFP (eGFP) under the control of the SV40 promoter containing an scaffold/matrix attachment region (S/MAR) element for months-long GFP expression.

The advantage of a minicircle is not only the reduced size, resulting in a far better transfer rate, but also the fact that certain sequences are not present within the vector molecule — e.g., CpG motifs. Recent publications also show that, using S/MAR elements, they are able to be maintained and expressed for a significantly longer time compared with a plasmid backbone (Broll *et al.*, 2009) and other versions reduced in their CpG content, with optimized sequence/promoter elements staying active through the lifetime of the animal tested (Haase *et al.*, 2010; Argyros *et al.*, 2011). Finally, the use of a minicircle DNA allows the successful induction of iPS cells (Jia *et al.*, 2010). First results of the comparison of regular plasmid DNA with minicircle DNA are summarized in Fig. 13.7.

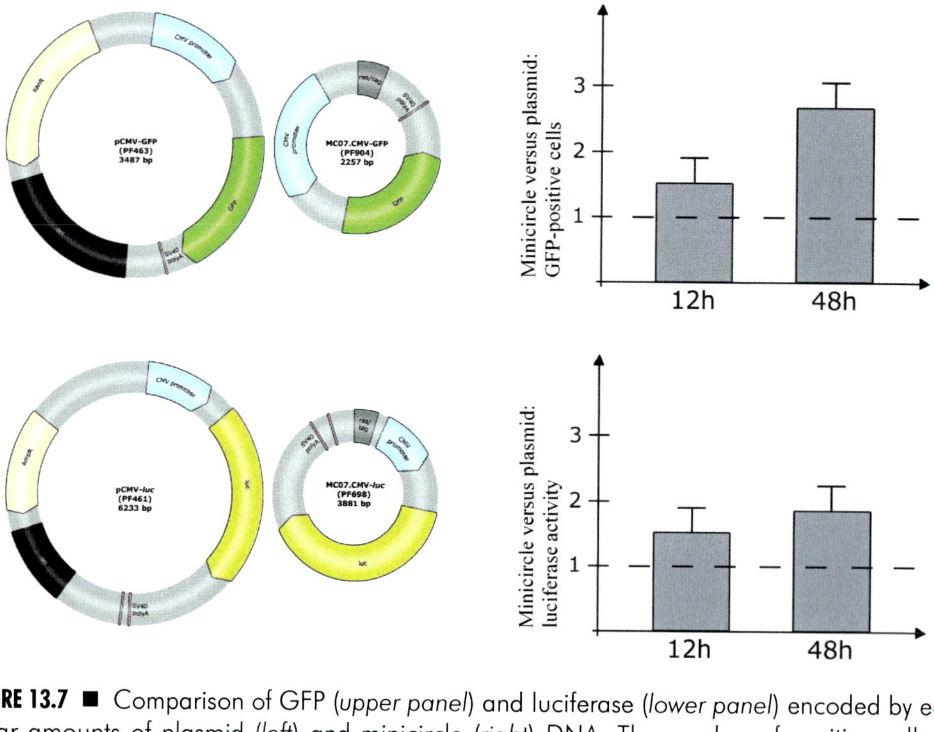

FIGURE 13.7 ■ Comparison of GFP (*upper panel*) and luciferase (*lower panel*) encoded by equimolar amounts of plasmid (*left*) and minicircle (*right*) DNA. The number of positive cells (for GFP) and luciferase activity (for luc) was tested at 12 h and 48 h. The bar size is the level of the minicircle relative to the plasmid (set to 1 — dashed line). Data kindly provided by W. Walther, Max Delbrück Center MDC, Berlin.

Acknowledgment

The author thanks Marco Schmeer for critical discussion, Anja Herzig (PlasmidFactory, Bielefeld, Germany) for kindly providing Fig. 13.6(b), Wolfgang Walther (MDC, Berlin, Germany) for kindly providing Fig. 13.7, Janine Conde-Lopez for support with figures and Emilie Coupriannoff for editorial support. The author acknowledges the support of the German Federal Ministry of Education and Research (BMBF) for grants BioChancePLUS (0313749) and Nano-4-Life (13N9063), of MOLEDA STREP, of the research team of PlasmidFactory, Bielefeld, Germany for contributing to the work and the whole manufacturing team of PlasmidFactory for their discussion. Part of this work has also been supported by the CliniGene Network of Excellence funded by the European Commission FP6 Research Programme under contract LSHB-CT-2006–018933.

References

Andrianaivo, F., Lecocq, M., Wattiaux-De Coninck, S., *et al.* (2004). Hydrodynamics-based transfection of the liver, entrance into hepatocytes of DNA that causes expression takes place very early after injection, *J Gen Med*, **6**, 877–883.

Argyros, O., Wong, S.-P., Fedonidis, C. *et al.* (2011). Development of S/MAR minicircles for enhanced and persistent transgene expression in the mouse liver, *J Mol Med*, **89**, 515–529.

Ayuso, E., Mingozzi, F., Bosch, F. (2010). Production, purification and characterization of adeno-associated vectors, *Curr Gene Ther*, **10**, 423–436.

Backliwal, G., Wurm, F.M. (2009). Large scale transient gene expression, *European Biopharmaceutical Review*, 88–94.

Barron, L.G., Gagne, L., Szoka, F.C. (1999). Lipoplex-mediated gene delivery to the lung occurs within 60 minutes of intravenous administration, *Hum Gene Ther*, **10**, 1683–1694.

Bigger, B.W., Tolmachov, O., Collomber, J.M., *et al.* (2001). An araC-controlled bacterial cre expression system to produce DNA minicircle vectors for nuclear and mitochondrial gene therapy, *J Biol Chem*, **276**, 23018–23027.

Broll, S., Oumard A., Hahn K., *et al.* (2009). Minicircle performance depending on S/MAR-nuclear matrix interactions, *J Mol Biol*, **395**, 950–965.

Bureau, M.F., Gehl, J., Deleuze, V., *et al.* (2002). Importance of association between permeabilization and electrophoretic forces for intramuscular DNA electrotransfer, *Biochim Biophys Acta*, **1474**, 353–359.

Cameron. B., Crouzet, J., Darquet, A.-M., *et al.* (1996). DNA molecules, preparation thereof and use thereof in gene therapy, World Intellectual Property Organization patent WO 1996/026270.

Cartier, R., Ren, S.V., Walther, W., *et al.* (2000). *In vivo* gene transfer by low volume jet injection, *Anal Biochem*, **282**, 262–265.

Chadeuf, G., Ciron, C., Moullier, P., *et al.* (2005). Evidence for encapsidation of prokaryotic sequences during recombinant adeno-associated virus production and their *in vivo* persistence after vector delivery, *Mol Ther*, **12**, 744–753.

Chen, Z.Y., He, C.Y., Ehrhardt, A., *et al.* (2003). Minicircle DNA vectors devoid of bacterial DNA result in persistent and high-level transgene expression *in vivo*, *Mol Ther*, **8**, 495–500.

Cranenburgh, R.M., Hanak, J.A., Williams, S.G., *et al.* (2001). *Escherichia coli* strains that allow antibiotic-free plasmid selection and maintenance by repressor titration, *Nucleic Acids Res*, **29**, 2120–2124.

Crouzet, J., Scherman, D., Cameron, B., *et al.* (2000). Circular DNA expression cassettes for *in vivo* gene transfer, US patent 6143530.

Darquet, A.M., Cameron, B., Wils, P., *et al.* (1997). A new DNA vehicle for nonviral gene delivery: supercoiled minicircle, *Gene Ther*, **4**, 1341–1349.

Davies, L.A., Hyde, S.C., Gill, D.R. (2005). 'Plasmid inhalation: delivery to the airways', in Schleef, M. (ed.) *DNA Pharmaceuticals — Formulation and Delivery in Gene Therapy, DNA Vaccination and Immunotherapy*, Weinheim: Wiley-VCH, pp. 145–164.

Davis, B.D., Dulbecco, R., Eisen, H.H., *et al.* (1980). *Microbiology*, Philadelphia PA: Harper and Row.

Del Solar, G., Giraldo, R., Ruiz-Echevarria, M.J., *et al.* (1998). Replication and control of circular bacterial plasmids, *Microbiol Mol Biol Rev*, **62**, 434–464.

Ding, S., Wu, X., Li, G., *et al.* (2005). Efficient transposition of the piggyBac (PB) transposon in mammalian cells and mice, *Cell*, **122**, 473–483.

Edelstein, M.L., Abedi, M.R., Wixon, J. (2007). Gene therapy clinical trials worldwide to 2007 — an update, *J Gene Med*, **9**, 833–842.

Favard, C., Dean, D.S., Rols, M.P. (2007). Electrotransfer as a non viral method of gene delivery, *Curr Gene Ther*, **7**, 67–77.

Flotte, T.R. (2004). Gene therapy progress and prospects: Recombinant adeno-associated virus (rAAV) vectors, *Gene Ther*, **11**, 805–811.

Fuller, D.H., Loudon, P., Schmaljohn, C. (2006). Preclinical and clinical progress of particle-mediated DNA vaccines for infectious diseases, *Methods*, **40**, 86–97.

Furth, P.A., Kerr, D., Wall, R. (1995a). Gene transfer by jet injection into differentiated tissues of living animals and in organ culture, *Mol Biotechnol*, **4**, 121–127.

Furth, P.A., Shamay, A., Hennighausen, L. (1995b). Gene transfer into mammalian cells by jet injection, *Hybridoma*, **14**, 149–152.

Gilbertson, L. (2003). Cre–lox recombination: creative tools for plant biotechnology, *Trends Biotechnol*, **21**, 550–555.

Gossen, J.A., de Leeuw, W.J.F., Molijn, A.C., *et al.* (1993). Plasmid rescue from transgenic mouse DNA using lacI repressor protein conjugated to magnetic beads, *BioTechniques*, **14**, 624–629.

Grimm, D., Kern, A., Rittner, K., *et al.* (1998). Novel tools for production and purification of recombinant adenoassociated virus vectors, *Hum Gene Therapy*, **9**, 2745–2760.

Haase, R., Argyros, O., Wong, S.P., *et al.* (2010). pEPito: A significantly improved non-viral episomal expression vector for mammalian cells, *BMC Biotechnol*, **10**, 20.

Hackett, P.B., Largaespada, D.A., Cooper, L.J.N. (2010). A transposon and transposase system for human application, *Mol Therapy*, **18**, 674–683.

Hayes, F. (2003). 'The function and organization of plasmids', in Casali, N., Preston, A. (eds.) *E. coli Plasmid Vectors*, Totowa NJ: Humana Press, pp. 1–17.

Helinski, D. (1979). Bacterial plasmids: Autonomous replication and vehicles for gene cloning, *Crit Rev Biochem*, **7**, 83–101.

Huettinger, C., Hirschberger, J., Jahnke, A., *et al.* (2008). Neoadjuvant gene delivery of feline granulocyte-macrophage colony-stimulating factor using magnetofection for the treatment of feline fibrosarcomas: a phase I trial, *J Gene Med*, **10**, 655–667.

Hyde, S.C., Pringle, I.A., Abdullah, S., *et al.* (2008). CpG-free plasmids confer reduced inflammation and sustained pulmonary gene expression, *Nature Biotech*, **26**, 549–551.

Izsvak, Z., Li, M.A., Mátés, L., *et al.* (2009). Transposon-mediated genome manipulations in vertebrates, *Nat Meth*, **6**, 415–422.

Jahnke, A., Hirschberger, J., Fischer, C., *et al.* (2007). Intra-tumoral gene delivery of feIL-2, feIFN-γ and feGM-CSF using magnetofection as a neoadjuvant treatment option for feline fibrosarcomas: A Phase-I study, *J Vet Med A Physiol Pathol Clin Med*, **10**, 599–606.

Jain, K.K. (1996). *Vectors for Gene Therapy: Current status and future prospects*, London: PJB Publications Ltd.

Jechlinger, W., Azimpour Tabrizi, T., Lubitz, W. (2004). Minicircle DNA immobilized in bacterial ghosts: *in vivo* production of safe non-viral DNA delivery vehicles, *J Mol Microbiol Biotechnol*, **8**, 222–231.

Jia, F., Wilson, K.D., Sun, N., *et al.* (2010). A nonviral minicircle vector for deriving human iPs cells, *Nature Methods*, **3**, 197–199.

Kern, A., Schmidt, K., Leder, C., *et al.* (2003). Identification of a heparin-binding motif on adeno-associated virus type 2 capsids, *J Virol*, **77**, 11072–11081.

Kleina, L.G., Masson, J.-M., Normanly, J., *et al.* (1990). Construction of *Escherichia coli* amber suppressor tRNA genes. II, Synthesis of additional tRNA genes and improvement of suppressor efficiency, *J Mol Biol*, **213**, 705–717.

Koltover, I., Salditt, T., Rädler, J.O., *et al.* (1998). An inverted hexagonal phase of cationic liposome–DNA complexes related to DNA release and delivery, *Science*, **281**, 78–81.

Kormann, M.S.D., Hasenpusch, G., Aneja, M.K., *et al.* (2011). Expression of therapeutic proteins after delivery of chemically modified mRNA in mice, *Nature Biotchnol*, **2**, 154–157.

Kreiss, P., Cameron, B., Darquet, A.M., *et al.* (1998). Production of a new DNA vehicle for gene transfer using site-specific recombination, *Appl Microbiol Biotechnol*, **49**, 560–567.

Kreiss, P., Cameron, B., Rangara, R., *et al.* (1999). Plasmid DNA size does not affect the physiological properties of lipoplexes but modulates gene transfer efficiency, *Nucleic Acids Res*, **27**, 3792–3798.

Krieg, A.M., Yi, A.K., Matson, S., *et al.* (1995). CpG motifs in bacterial DNA trigger direct B-cell activation, *Nature*, **374**, 546–549.

Lederberg, J. (1952). Cell genetics and hereditary symbiosis, *Physiol Rev*, **32**, 403–430.

Li, Y.S., Davidson, E., Reid, C.N., *et al.* (2008). Optimising ultrasound-mediated gene transfer (sonoporation) *in vitro* and prolonged expression of a transgene *in vivo*: Potential applications for gene therapy of cancer, *Cancer Lett*, **273**, 156–162.

Li, Z., Düllmann, J., Schiedlmeier, B., *et al.* (2002). Murine leukemia induced by retroviral gene making, *Science*, **296**, 497.

Liu, F., Song, Y.K., and Liu, D. (1999). Hydrodynamics-based transfection in animals by systemic administration of plasmid DNA, *Gene Ther*, **6**, 1258–1266.

Lock, M., McGorray, S., Auricchio, A., *et al.* (2010). Characterization of a recombinant adeno-associated virus type 2 reference standard material, *Hum Gene Ther*, **21**, 1273–1285.

Mairhofer, J., Cserjan-Puschmann, M., Striedner, G., *et al.* (2010). Marker-free plasmids for gene therapeutic applications – Lack of antibiotic resistance gene substantially improves the manufacturing process, *J Biotechnol*, **146**, 130–137.

Mairhofer, J., Pfaffenzeller, I., Merz, D., *et al.* (2008). A novel antibiotic free plasmid selection system: Advances in safe and efficient DNA therapy, *Biotechnol J*, **3**, 83–89.

Maucksch, C., Hoffmann, F., Schleef, M., *et al.* (2009). Transgene expression of transfected supercoiled plasmid DNA concatemers in mammalian cells, *J Gene Med*, **11**, 444–453.
Mayrhofer, P., Blaesen, M., Schleef, M. *et al.* (2008). Minicircle-DNA production by site specific recombination and protein-DNA interaction chromatography, *J Gene Med*, **10**, 1253–1269.
Mayrhofer, M., Schleef, M., Jechlinger, W. (2009). 'Use of minicircle plasmids for gene therapy', in Walther, W., Stein, U.S. (eds.) *Methods in Molecular Biology, Gene Therapy of Cancer*, Totowa NJ: Humana Press, pp. 157–165.
Michaels, M.L., Kim, C.W., Mattews, D.A., *et al.* (1990). *Escherichia coli* thymidylate synthase: Amino acid substitutions by suppression of amber nonsense mutations, *Proc Natl Acad Sci USA*, **87**, 3957–3961.
Miklavcic, D., Semrov, D., Mekid, H., *et al.* (2000). A validated model of *in vivo* electric field distribution in tissues for electrochemotherapy and for DNA electrotransfer for gene therapy, *Biochim et Biophys Acta*, **1523**, 73–83.
Mir, L.M., Bureau, M.F., Gehl, J., *et al.* (1999). High-efficiency gene transfer into skeletal muscle mediated by electric pulses, *Proc Natl Acad Sci USA*, **96**, 4262–4267.
Mir, L.M. (2005). 'Electrogenetransfer in clinical applications', in Schleef, M. (ed.) *DNA Pharmaceuticals: Formulation and Delivery in Gene Therapy, DNA Vaccination and Immunotherapy*, Weinheim: Wiley-VCH, pp. 219–226.
Moullier, P., Snyder, R.O. (2008). International efforts for recombinant adeno-associated viral vector reference standards, *Mol Ther*, **16**, 1185–1188.
Nehlsen, K., Broll, S., Bode, J. (2006). Replicating minicircles: generation of nonviral episomes for the efficient modification of dividing cells, *Gene Ther Mol Biol*, **10**, 233–244.
Newman, C.M.H., Bettinger, T. (2007). Gene therapy progress and prospects: Ultrasound for gene transfer, *Gene Ther*, **14**, 465–475.
Olivares, E.C., Hollis, R.P., Chalberg, T.W., *et al.* (2002). Site-specific genomic integration produces therapeutic Factor IX levels in mice, *Nat Biotechnol*, **20**, 1124–1128.
Preston, A. (2003). 'Choosing a cloning vector', in Casali N., Preston A. (eds.), *E. coli Plasmid Vectors*. Totowa NJ: Humana Press, pp. 19–26.
Sadowski, P. (1986). Site-specific recombinases: changing partners and doing the twist, *J Bacteriol*, **165**, 341–347.
Sapet, C., Laurent, N., de Chevigny, A., *et al.* (2011). High transfection efficiency of neural stem cells with magnetofection, *BioTechniques*, **50**, 187–189.
Schakowski, F., Gorschlüter, M., Junghans, C., *et al.* (2001). A novel minimal-size vector (MIDGE) improves transgene expression in colon carcinoma cells and avoids transfection of undesired DNA, *Mol Ther*, **3**, 793–800.
Scherer, F., Anton, M., Schillinger, U., *et al.* (2002). Magnetofection: enhancing and targeting gene delivery by magnetic force *in vitro* and *in vivo*, *Gene Ther*, **9**, 102–109.
Schleef, M., Blaesen, M., Schmeer, M., *et al.* (2010). Production of non viral vectors, *Curr Gene Ther*, **10**, 487–507.
Schleef, M., Schmeer, M. (2011). Minicircle — Die nächste Generation nicht-viraler Gentherapie-Vektoren, *Pharm Unserer Zeit*, **40**, 220–224.
Schleef, M. (ed.) (2013). *Minicircle and Miniplasmid DNA Vectors: The Future of Nonviral and Viral Gene Transfer*, Weinheim: Wiley-VCH.
Schmeer, M., Seipp, T., Pliquett, U., *et al.* (2004). Mechanism for the conductivity changes caused by membrane electroporation of CHO cell-pellets, *PCCP*, 6, 5564–5574.
Schmeer, M. (2009). 'Electroporative gene transfer', in Walther, W., Stein, U.S. (eds.) *Methods in Molecular Biology, Gene Therapy of Cancer*, Totowa NJ: Humana Press, pp. 157–165.

Schmidt, T., Friehs, K., Schleef, M., *et al.* (1999). Quantitative analysis of plasmid forms by agarose and capillary gel electrophoresis, *Analyt Biochem*, **274**, 235–240.

Schumann, W. (2001). 'The biology of plasmids', in Schleef M (ed.) Plasmids for therapy and vaccination, Weinheim: Wiley-VCH, pp. 1–43.

Seed, B. (1983). Purification of genomic sequences from bacteriophage libraries by recombination and selection *in vivo*, *Nucleic Acid Res*, **11**, 2427–2445.

Somia, N., Verma, I.M. (2000). Gene therapy: Trials and tribulations, *Nat Rev Genet*, **1**, 91–99.

Steele, K.E., Stabler, K., VanderZanden, L. (2001). Cutaneous DNA vaccination against Ebola virus by particle bombardment: Histopathology and alteration of CD3-positive dendritic epidermal cells, *Vet Pathol*, **38**, 203–215.

Summers, D.K. (1996). *The Biology of Plasmids*, Oxford: Blackwell Science.

Trollet, C., Bigey, P., Scherman, D. (2005). 'Electrotransfection – an overview', in Schleef, M. (ed.), *DNA Pharmaceuticals: Formulation and Delivery in Gene Therapy, DNA Vaccination and Immunotherapy*, Weinheim: Wiley-VCH, pp. 189–218.

Trollet, C., Bloquel, C., Scherman, D., *et al.* (2006). Electrotransfer into skeletal muscle for protein expression, *Curr Gene Ther*, **5**, 561–578.

Vieira, J., Messing, J. (1982). The pUC plasmids, an M13 mp7-derived system for insertion mutagenesis and sequencing with synthetic universal primers, *Gene*, **19**, 259–268.

Voss, C., Schmidt, T., Schleef, M., *et al.* (2006). Verwendung von isolierten homogenen Nukleinsäure-Multimeren für die nicht-virale Gentherapie oder genetische Impfung, DE10101761.

Walther, W., Stein, U., Fichtner, I., *et al.* (2001). Non-viral in vivo gene delivery into tumors using a novel low volume jet-injection technology, *Gene Ther*, **8**, 173–180.

Wolff, J.A., Malone, R.W., Williams, P., *et al.* (1990). Direct gene transfer into mouse muscle *in vivo*, *Science*, **247**, 1465–1458.

Wolff, J.A., Williams, P., Acsadi, G., *et al.* (1991). Conditions affecting direct gene transfer into rodent muscle *in vivo*, *Biotechniques*, **11**, 474–485.

Woltjen, K., Michael, I.P., Mohseni, P., *et al.* (2009). piggyBac transposition reprograms fibroblasts to induced pluripotent stem cells, *Nature*, **458**, 766–770.

Yang, Y., Li, Q., Ertl, H.C., *et al.* (1995). Cellular and humoral immune responses to viral antigens create barriers to lung-directed gene therapy with recombinant adenoviruses, *J Virol*, **69**, 2004–2015.

Yusa, K., Zhou, L., Li, M.A., *et al.* (2011). A hyperactive piggyBac transposase for mammalian applications, *Proc Natl Acad Sci USA*, **108**, 1531–1536.

Zaiss, A.K., Muruve, D.A. (2008). Immunity to adeno-associated virus vectors in animals and humans: a continued challenge, *Gene Ther*, **15**, 808–816.

Zeira, E., Manevitch, A., Khatchatouriants, A., *et al.* (2003). Femtosecond infrared laser-an efficient and safe in vivo gene delivery system for prolonged expression, *Mol Ther*, **8(2)**, 342–350.

Zhang, X., Dong, X., Sawyer, G.J., *et al.* (2004). Regional hydrodynamic gene delivery to the rat liver with physiological volumes of DNA solution, *J Gen Med*, **6**, 693–703.

Ziady, A.G., Davis, P.B., Konstan, M.W. (2003). Non-viral gene transfer therapy for cystic fibrosis, *Expert Opin Biol Ther*, **3**, 449–458.

14

MACROMOLECULAR CONJUGATES FOR NON-VIRAL NUCLEIC ACID DELIVERY

Mark Ericson[a], Kevin Rice[a,c] and Guy Zuber[b,d]

14.1 Multi-Functional Nucleic Acid Delivery Systems

The aim of this chapter is to introduce the reader to the principles governing the current design and preparation of synthetic nucleic acid delivery systems using macromolecular conjugates. The plasma membrane lipid bilayer is impermeable to nucleic acids and delivery of an exogenous therapeutic nucleic acid into the cell requires detailed knowledge of cell-entry mechanisms. To do so, systems organized around the nucleic acids and equipped with multiple components should be conceived in such ways that individual elements can perform specific tasks at the right time and location (Fig. 14.1). Not surprisingly, the ideal system is highly reminiscent of a viral particle and should: (1) pack and protect the nucleic acids from naturally occurring enzymatic degradation, (2) facilitate circulation and biodistribution to the target tissue, (3) bind to cell-surface receptors that are endocytosed, (4) mediate particle escape from the endosome, (5) promote intracellular trafficking and release the nucleic acids from the delivery agent, and (6) facilitate nuclear uptake to allow transcription and translation. The following sections discuss the basic principles of why and how current synthetic nucleic acid delivery systems are designed and synthesized in order to overcome the various physiological and cellular barriers.

[a]Division of Medicinal and Natural Products Chemistry, College of Pharmacy, The University of Iowa, 115 S. Grand Avenue , #300 PHAR, Iowa City, IA, USA
[b]CNRS, Laboratoire de Chimie Génétique, Faculté de Pharmacie, Université Louis Pasteur, Strasbourg I, 67401 Illkirch Cedex, Strasbourg, France
[c]Email: kevin-rice@uiowa.edu
[d]Email: zuber@unistra.fr

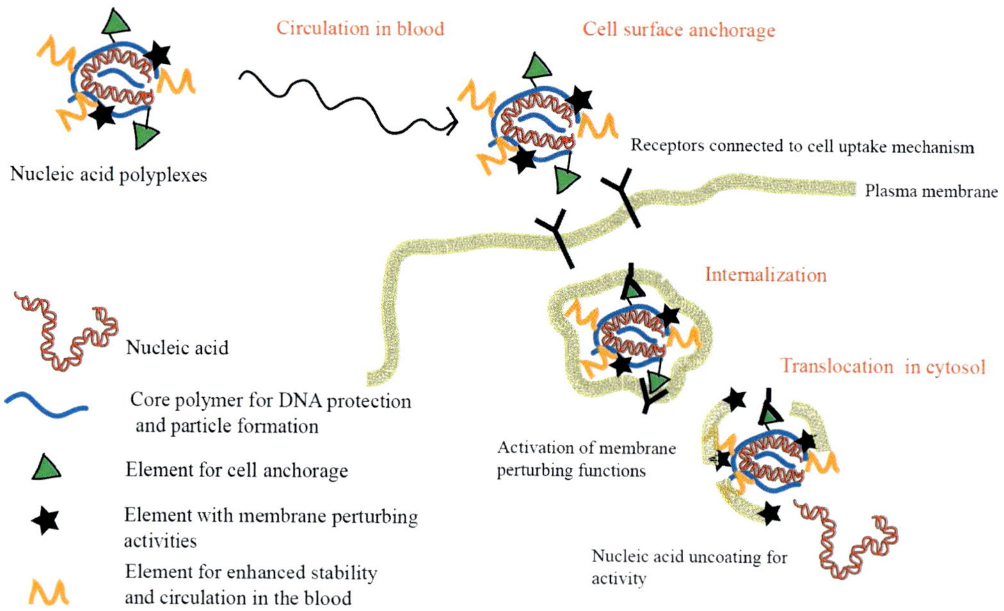

FIGURE 14.1 ■ Schematic representation of how elements of macromolecular nucleic acid delivery systems function to achieve effective delivery into eukaryotic cells.

14.2 Role of the Carrier in the Delivery System

The role of the carrier in a synthetic nucleic acid delivery system is to allow formation of a macromolecular assembly around DNA or small interfering RNA (siRNA), resulting in a polyplex, which is cohesive enough to protect the oligonucleic acid from degradation during circulation. The carrier is most often a cationic polymer that binds to DNA or siRNA by simple electrostatic interaction. However, other modes of binding, such as intercalation, are possible and provide additional stability to the polyplexes. Regardless of the binding mode, polyplex stability relies on achieving multiple interactions between the DNA or siRNA and the carrier. Not unexpectedly, the interaction between binding partners governs key properties of the polyplexes. The stability of the polyplex depends on the structure and molecular weight of the cationic carrier, the molecular weight of the anionic nucleic acid, the incubation medium and the cation (from the carrier) to anion (from DNA) stoichiometry. Stoichiometry is an important parameter and is usually given as the nitrogen of carrier to phosphate of DNA ratio (N:P). This ratio is derived from most polymeric carriers consisting of positively charged protonated amines, and DNA or siRNA bearing one negative charge per phosphate. The most typical polyplexes are described in the following sections.

14.2.1 Kinetically Driven Cationic DNA Polyplexes

At the nanoscale, double-stranded plasmid DNA is 5,000 bp long and occupies a large hydrodynamic volume due to electrostatic repulsion between anionic phosphates. The addition of polycations induces a cooperative binding leading to DNA condensation and protection from degradation. Examples of well-studied cationic polymers and oligocations used for nucleic acid condensation are shown in Fig. 14.2.

Polymers are generally composed of 20–200 primary amines that are protonated at physiological pH and rapidly ion-pair with DNA to form polyelectrolyte aggregates. The speed of condensation does not allow thermodynamic equilibration. When the cationic carrier is in excess (N:P > 2), the rapid, kinetically driven process results in the formation of cationic polyplexes in which several plasmids are compacted into a single polyplex that is surrounded by surface-bound polycations (Fig. 14.3).

Being thermodynamically unstable, these systems tend to evolve toward lower energy states. Equilibration and eventual expulsion of the cationic carriers from the polyplex is rendered difficult by ionic repulsion between particles. In the absence of buffer salts, ionic repulsion ensures an almost infinite colloidal stability of the polyplexes (Goula *et al.*, 1998). The picture is different in the presence of normal saline or serum. Indeed, ions allow the system to equilibrate again. At a macroscopic level, fast intra-particle aggregation occurs and particles may interact with any anionic molecule encountered, including anionic cell surfaces. As we will see in the other sections, this fact is undesirable for *in vivo* administration but can be advantageously exploited for binding polyplexes to cell surfaces in culture.

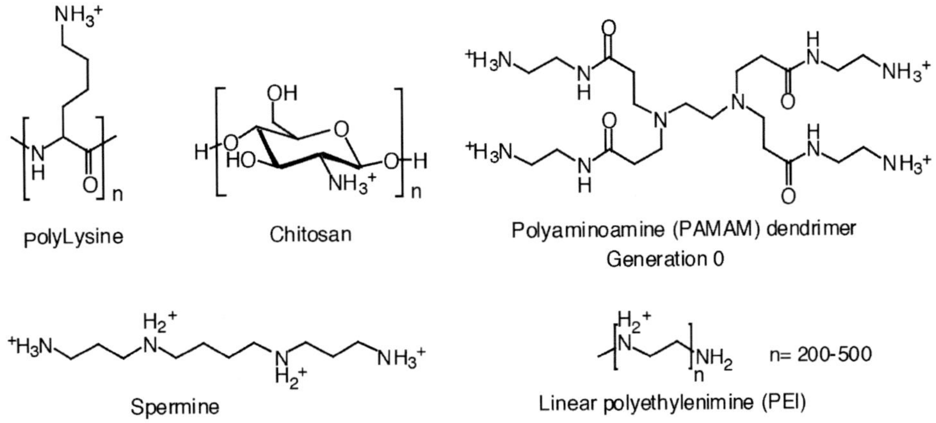

FIGURE 14.2 ■ Chemical structures of polyLysine, chitosan, polyaminoamine (PAMAM) dendrimer, spermine and linear polyethylenimine (PEI).

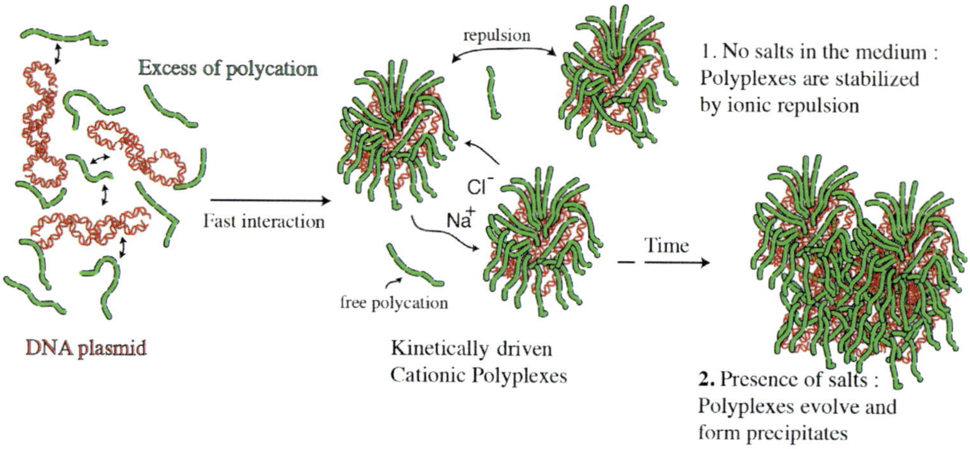

FIGURE 14.3 ■ Schematic representation of DNA polyplex formation from polycationic partners and aggregation behaviors in the presence of ions.

14.2.2 Caged Polyplexes

Unlike larger polycations, oligocations such as spermine, cationic lipids or cationic peptides interact with DNA reversibly. Equilibration tends to direct the system towards its most stable state in which smaller lipoplexes and polyplexes of precise stoichiometry (N:P = 1) are formed that may contain a single plasmid. A significant design advantage afforded by oligocations over polycations is their small size, allowing more controlled chemical modification and greater homogeneity. This advantage is multiplied when considering the number of carrier modifications which may be required to incorporate all necessary elements for successful *in vivo* delivery. However, the smaller size also results in lower affinity for DNA and siRNA. Spontaneous reversibility of the polyplex is deleterious for *in vivo* delivery applications due to premature dissociation in the blood, resulting in rapid metabolism of DNA by DNAses. This limitation has led to the development of cross-linking strategies to improve the stability of polyplexes produced by oligocationic carriers (Fig. 14.4). Theoretically, caged polyplexes can be designed to stabilize the polyplex and trigger the intracellular release of DNA.

The design of caged polyplexes must establish that the cross-linking method is inert to DNA to maintain the integrity of the genetic information. Several homobifunctional cross-linking agents, such as glutaraldehyde or dimethyl-3, 3′-dithiobispropionimidate, which react with residual primary amines on the polyplex, have been used to increase polyplex stability. Even though caged polyplexes produced a dramatic increase in the metabolic stability of DNA *in vivo*, these formulations fail to produce appreciable gene expression, perhaps due to the excessively slow release of DNA or other missing intracellular delivery elements.

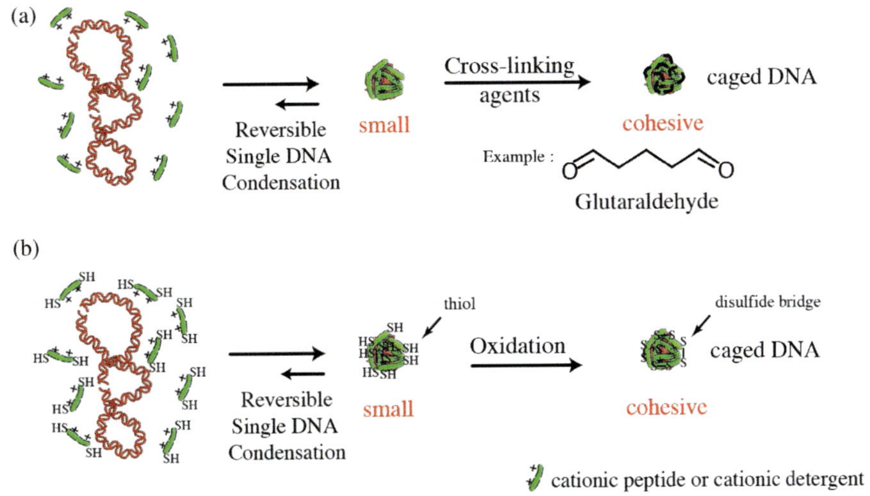

FIGURE 14.4 ■ Schematic representation of stable, caged polyplexes formed using chemical cross-linking of oligocations nucleic acid polyplexes with (a) amine cross-linking agents or (b) oxidation of thiol-containing oligocations. At first, the plasmid DNA is condensed with oligocations at a stoichiometry required for electrostatic neutrality. After equilibration, the system is stabilized by caging.

Another strategy is to integrate thiol groups within the oligocation and to use oxidation of two thiols to form intermolecular disulfide bonds (McKenzie *et al.*, 2000). Disulfide bonds have the advantage of being reduced inside cells, facilitating the triggered intracellular release of nucleic acids. While elegant in concept, sulfhydryl cross-linked polyplexes possessing targeting and stealthing elements are overly complex to prepare and only weakly active at producing gene transfer *in vivo*, most likely due to the premature release of DNA following polyplex reduction.

These two examples emphasize the importance of controlling the metabolic stability and release of the nucleic acids. Improving caged polyplexes requires flexibly designed macromolecular conjugates that provide much better control over triggered release of DNA.

14.2.3 Oligonucleotide Polyplexes

siRNA duplexes and DNA share similar anionic charge density; however the smaller size of siRNA significantly reduces the electrostatic cohesion of the polyplexes and dissociation readily occurs outside the cell. To increase the stability of siRNA polyplexes, the caging strategies that have already been described for DNA can be applied (Taratula *et al.*, 2009), as seen in Fig. 14.5. siRNA duplexes

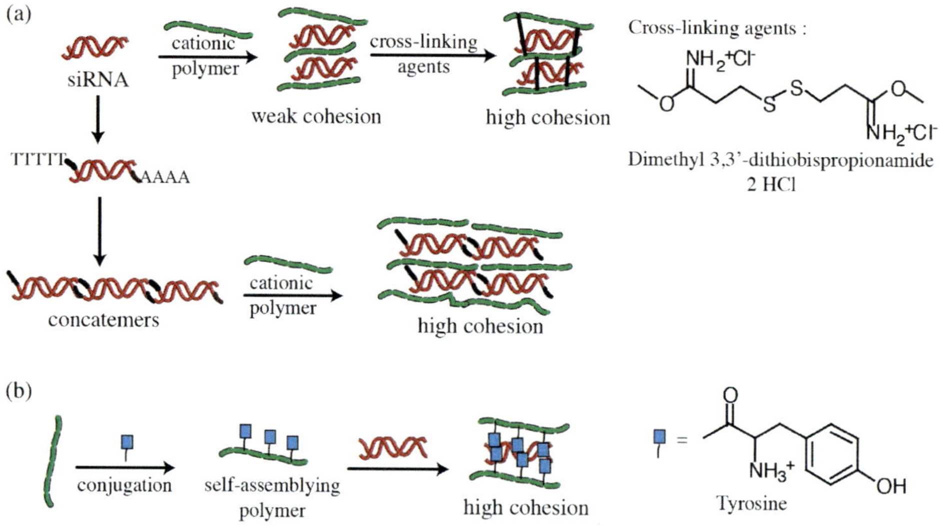

FIGURE 14.5 ■ Strategies for formation of stable siRNA complexes. (a) Caged polyplexes via polymer cross-linking or via formation of concatemers of oligonucleotides by base pairing. (b) Caged polyplexes via non-covalent interactions by modification of the vectors with hydrophobic domains.

can also be oligomerized to resemble DNA by incorporating self-complementary overhanging nucleotides (Bolcato-Bellemin *et al.*, 2007). An alternative to the chemically reversible caged polyplexes is to modify the polycations with hydrophobic domains. Non-covalent interactions between these domains can maintain the cohesion of the polyplexes extracellularly and are abolished upon sensing intracellular pH changes (Creusat *et al.*, 2010).

14.2.4 Intercalating Core Systems

In addition to the negatively charged phosphate backbone, other properties of nucleic acids can be utilized to efficiently bind and protect DNA or siRNA. Molecules which contain planar aromatic domains can be inserted between base pairs of nucleic acids, interacting with the aromatic nucleobases. Core systems anchored by intercalation protect the nucleic acids from enzymatic degradation without relying on ionic interactions. As a result, these systems do not dissociate or form aggregates when placed in a biological medium. Intercalator-based vectors with minimal cationic character can also be used, allowing for the formation of neutral or anionic particles when complexed with nucleic acids. While these advantages may be useful for *in vivo* applications, concerns about potential toxicity and carcinogenicity must be addressed for these vectors.

14.3 Transport of Polyplexes Across the Plasma Membrane

Certain cationic polyplexes demonstrate an intrinsic activity to transfect cells grown in culture. Their mechanism of action is likely due to the presence of proteoglycans on the plasma membrane. Most adherent cell lines have large quantities of cell-surface sulfated proteoglycans that bind to cationic polyplexes. These highly negatively charged polysaccharides are normally involved in cellular adhesion but also promote the uptake of polyplexes by electrostatic binding and zippering to form intracellular vacuoles named endosomes. The endosomes are then transported to perinuclear lysosomal compartments for degradation (Fig. 14.6). During intracellular trafficking, the endosomal environment undergoes progressive acidification. Within 30 min, the pH drops from 7.4 to as low as 4.5 due to an active influx of protons, water and chloride ions (Sonawane *et al.*, 2003).

To rupture the endosomal membrane, fusogenic peptides can be incorporated within the polyplexes. These relatively short 20–30 amino acid peptides are structurally arranged as amphipathic α-helixes in their membrane lytic form (Fig. 14.7). Ideally, fusogenic peptides should remain inactive when bound to DNA at a high pH of 7.4 and become active at the lower pH of 4.5 as the endosome is transported intracellularly. This design feature affords triggered intracellular endosomal lysis and avoids indiscriminate cell lysis during systemic circulation. The amphipathic nature of the helix allows the peptide to insert into the lipid bilayer, creating pores that disrupt the membrane and allowing for endosomal escape. Cationic fusogenic peptides, such as KALA (a peptide rich in lysine (K), alanine (A) and leucine (L)) or melittin, are lytic across the pH range of 7.4–4.5. Anionic fusogenic peptides, such as hemagglutinin 2 (HA-2) or JST-1, become α-helical and

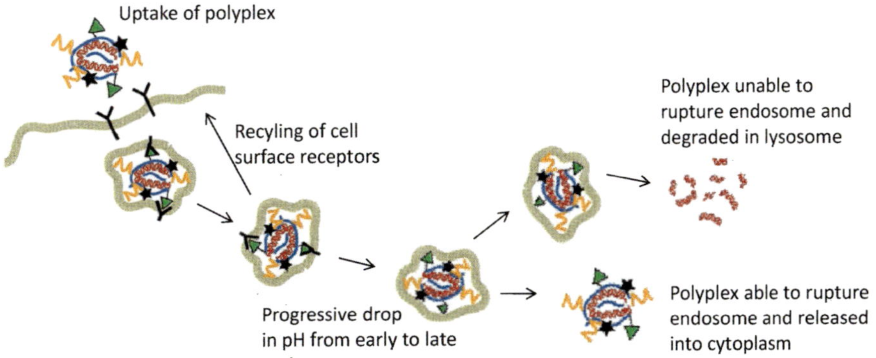

FIGURE 14.6 ■ Schematic representation of a polyplex being internalized and processed by the endosomal pathway.

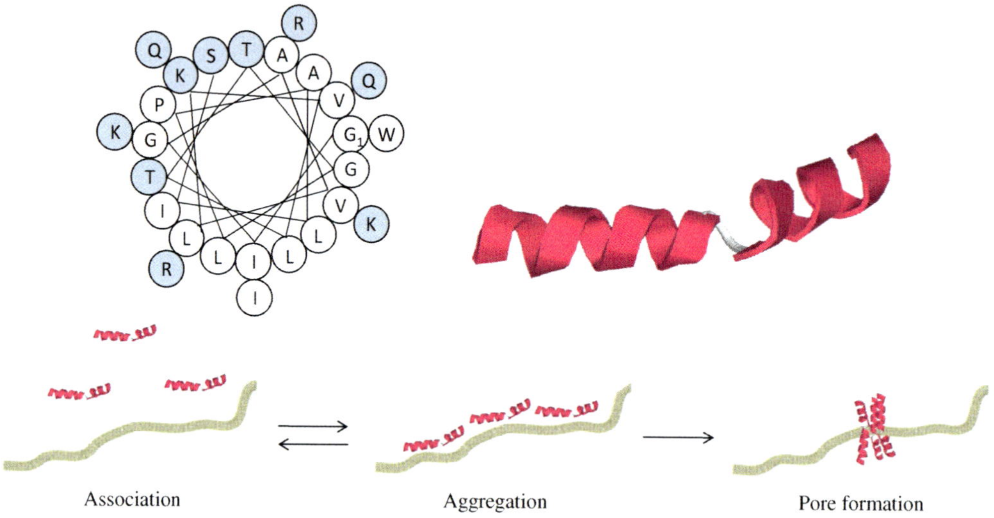

FIGURE 14.7 ■ A helical wheel model of melittin is presented with polar and ionic amino acids indicated in blue, demonstrating the amphipathic nature of the peptide. A ribbon structure of melittin is illustrated during association, aggregation and pore formation.

membrane lytic as the pH approaches 4.5–5. Both anionic and cationic fusogenic peptides have been incorporated into DNA delivery systems.

An alternative to fusogenic peptides is to use endosomal buffering agents. It has been shown that the addition of chloroquine facilitates disruption of the endosomal membrane. Chloroquine is a membrane-permeable molecule that overrides the adenosine triphosphatase (ATPase)-dependent proton pump by buffering endosomes. Overactivity of the proton pump causes accumulation of water and ions in the endosomes, resulting in osmotic pressure strong enough to facilitate rupture of the lipid membrane and partial release of the complex into the cytoplasm.

Modification of cationic polymers to incorporate buffering capacities in the pH range of 7.4–5.0 has led to successful DNA transfection reagents (Fig. 14.8). One method of adding buffering capacity is to incorporate histidine residues to polyLysine or polyarginine carriers. The imidazole side chain of histidine (H) has an acid dissociation constant (pKa) of 6 versus pKas of 10.6 and 12.5 for amine and guanidine side chains of lysine (K) and arginine (R), respectively. At physiological pH 7.4, the side chain of histidine exists primarily in an unprotonated form and is not involved in DNA binding. As the pH drops from the early to late endosome, the histidine residues are protonated, serving to buffer the endosome and override the proton pump. This "proton sponge" effect may, in part, also explain why polyethylenimine (PEI) is such a successful *in vitro* transfection reagent, as its

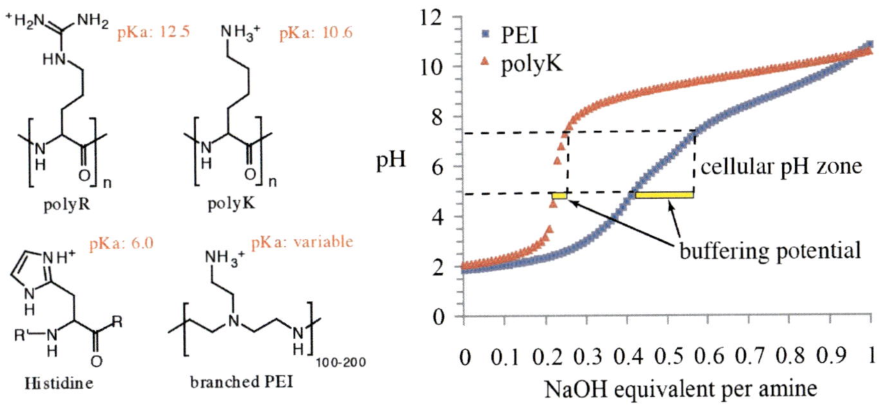

FIGURE 14.8 ■ Chemical structures and pKas of polyarginine (polyR), polyLysine (polyK), histidine and PEI. Acidobasic titration (graph at *right*) shows branched PEI with an average degree of polymerization (DP) of 500 to have significant buffering abilities in the pH 7.5–5.0 range in comparison with polyK (DP 100).

chemical structure allows DNA binding but diminishes the pKas of some secondary amines in the effective range to buffer the endosome (Boussif *et al.*, 1995; Kichler *et al.*, 2001).

While the discovery of the endosomal-buffering carrier PEI is among the most important advances in non-viral gene delivery, the main utility of PEI is for nucleic acid delivery in cell culture, with very little activity *in vivo*. This is best understood by appreciating that PEI is most active *in vitro* at an N:P ratio of 9:1. At this stoichiometry, most of the PEI is free in solution, since an N:P of 2:1 is sufficient to completely ion-pair DNA to form polyplexes. At an N:P of 9:1, the unbound free PEI is co-endocytosed along with the polyplex, leading to endosomal buffering. The successful application of PEI *in vivo* would necessitate avoiding polyplex aggregation in serum and delivery at excessively high concentrations and stoichiometry, such that free PEI would co-endocytose into the same cells as the polyplex.

As with endosomal lytic agents such as chloroquine or PEI, the membrane lytic activity of fusogenic peptides is also both pH- and concentration-dependent. The most potent fusogenic peptides cause membrane lysis at a minimum concentration of 1 μM. This concentration is easily achieved for hours or days during *in vitro* gene transfer experiments, but is nearly impossible to achieve *in vivo* without serious toxicity consequences. Thereby, it is possible to conclude that both endosomal buffering and endosomal lytic agents are currently insufficiently potent to cause significant endosomal lysis when incorporated into macromolecular conjugates and administered in a typical intravenous dose of 1 μg of DNA polyplex *in vivo*.

14.4 Design of Polyplexes for *in vivo* Delivery

For successful *in vivo* delivery, polyplexes must be stable enough to reach their distant target cell surfaces. Due to their gene transfer activity *in vitro*, cationic polyplexes have been extensively evaluated *in vivo*, but provided poor results. The salt concentration in serum is sufficiently high (150 mM) to immediately begin dissociation of the ionic interactions of a carrier from DNA upon dosing *in vivo*. Alternatively, if not dissociated, cationic polyplexes immediately bind to anionic proteins and erythrocytes in the blood (Chollet *et al.*, 2002). These undesirable interactions significantly diminish the ability of cationic polyplexes to reach the targeted tissue, as large aggregates are formed that become trapped in the lung vasculature.

To increase their biocompatibility, the charge on cationic polyplexes is usually masked. This is done by conjugation of inert polymers including polyacrylamide, polyvinylpyrrolidone and polyethyleneglycol (PEG) to the cationic polymer. PEGylated carriers can shield polyplexes to prevent protein binding, thereby maintaining a small particle size and increasing circulatory half-life to increase distribution to tissues and cells.

Another important obstacle for molecular conjugate delivery *in vivo* is the ability to target certain tissues or organs for increased and selective uptake of the complexed conjugate. This can be achieved by either actively targeting the desired cell type, or by relying on physical properties of the polyplex that result in a passive accumulation within a given tissue. Many approaches have been explored for active delivery; perhaps the most studied involves receptor-mediated internalization. Often a ligand for a particular cell-surface receptor is attached to the vector, enabling the vector to selectively bind and be taken up into the desired tissue. Numerous ligand–receptor systems have been examined for their ability to mediate cell-specific uptake. Protein and peptide ligands such as transferrin, epidermal growth factor (EGF) and arginine–glycine–aspartate (RGD) increase delivery to select tumors. Carbohydrate ligands such as triantennary N-glycan or other galactose-bearing ligands target the asialoglycoprotein receptor found on hepatocytes. Mannose-bearing ligands have been developed to target mannose receptors on dendritic cells. In addition to known ligand–receptor interactions, such as those described above, novel ligands can be discovered using techniques, such as phage display, to identify new targeting modalities.

Other targeting methods beyond ligand–receptor interactions have been employed to achieve specific uptake of nucleic acids by molecular conjugates. Antibodies which recognize specific antigens present on select cell types have been conjugated to delivery systems to improve specificity. While antibodies have had some success in targeted delivery, the cost of production, unreliable chemical modification and lack of long-term stability limit the prospects of antibody-mediated targeting. Conversely, the development of short oligonucleotide aptamers is emerging as a promising field of targeted therapy. These single-stranded RNA or DNA molecules achieve the high affinity and specificity of antibodies by multiple rounds of selection and amplification yet possess many advantages, including potential cheaper cost of a synthetic process and ease of synthesis and chemical modification for prolonged stability. While aptamers are often developed as agonists for various enzymes, chimeric aptamer–siRNA constructs have successfully delivered siRNA in a cell-specific fashion with the aptamer serving as the targeting ligand.

The incorporation of a targeting ligand into a gene-delivery system continues to be a challenging aspect of design. While incorporation of multiple copies of the ligand on a polyplex tends to improve the binding affinity to the receptor on cells through the multivalent effect, the ligand must be presented to allow receptor recognition. The attachment of the ligand directly to DNA in the polyplex can result in ligand-masking when a shielding agent is incorporated, such as high-molecular-weight (5000 Da) PEG. Alternatively, the ligand can be extended away from the DNA by covalent attachment to the unbound terminus of the PEG, but without guarantees that PEG folding will still mask the ligand. Perhaps the most assured approach is to choose a larger ligand that is able to project its binding site above the stealth layer of PEG on a polyplex. There are further design considerations regarding how to control the conjugation chemistry and the number of ligands bound to the polyplex. These design considerations must also work in concert with the incorporation and release of endosome-disrupting agents and other subcellular targeting agents.

Without a doubt, the most difficult aspect of polyplex design has been to target its delivery to the nucleus. This is partly because the nuclear envelope restricts foreign DNA entry. Viruses have evolved to overcome this barrier, but do so using mechanisms or disassembly strategies that are not yet fully understood. It is also difficult to study nuclear targeting in isolation, without an otherwise fully functional delivery system. Studies utilizing cells in culture are often inadequate due to rapid cell division, at which time foreign DNA enters the nucleus when the nuclear envelope is dissolved. Nuclear targeting studies in slowly-dividing or non-dividing cells in culture are poor surrogates for cells *in vivo*, in part because of the unrealistic concentrations of gene-transfer agents afforded *in vitro* that are not achievable *in vivo*. Despite these limitations, there are many reports of improved gene transfer *in vitro* by incorporating nuclear-localizing peptides into polyplexes. However, none of these strategies have worked to a degree that improves *in vivo* gene delivery to a level remotely comparable to that of a viral delivery system or to the most efficient physical methods.

14.5 Assembly, Disassembly and Testing of Macromolecular Conjugates

As the field continues to search for breakthroughs, macromolecular conjugates continue to evolve. The current focus is on the overall design of assembled elements into a flexible polyplex that allows the triggering of each component to act at the appropriate time.

Newer designs favor a more controlled assembly versus a random assembly of elements. Lower molecular weight polymers allow for more controlled derivatization resulting in greater homogeneity and the opportunity for systematic optimization. Individual elements can, in principle, be assembled into well-defined polymers. The linkages between elements must also be dissociable to insure their disassembly at the appropriate time in the cell.

Not only is the design of macromolecular nucleic acid delivery systems fundamental to success, but it is equally important to design experiments that test the efficacy of individual elements of the delivery system. While this is possible using *in vitro* gene transfer, it is increasingly clear that these studies do not predict efficacy *in vivo*. The use of modern gene transfer reporter gene readouts provides the opportunity for optimization of individual elements *in vivo*. However, unlike *in vitro* gene transfer, this strategy relies on having a fully integrated and functionally active gene-transfer system in place. What is typically observed is no expression when dosing 1 μg of DNA polyplex, a dose used to achieve high levels of expression in muscle by intramuscular-electroporation (im-EP) or in liver by hydrodynamic dosing. This conundrum has left most in the field little option but to increase the dose of DNA polyplexes during *in vivo* studies to 25 μg or higher, just to achieve detectable levels of transgene expression that is still many orders of magnitude lower than that achievable by a 1 μg dose delivered by im-EP or hydrodynamic dosing. Clearly a breakthrough is needed to find the missing element that will increase efficiency to the level of expression achieved by physical methods and viral delivery.

Given the nearly limitless chemical space in which to design macromolecular conjugates, and the increased understanding of the various intracellular trafficking pathways, it is only a matter of time before combinations of elements are assembled that function to deliver DNA *in vivo* as efficiently as viruses. As these systems evolve to function *in vivo*, increasing their potency will be the primary concern, since this will lower the effective dose and reduce the risk of potential toxicities.

Acknowledgments

Mark Ericson is supported by a National Institutes for Health (NIH) Predoctoral Training Grant in Biotechnology and an American Foundation for Pharmaceutical Education (AFPE) Predoctoral Fellowship in the Pharmaceutical Sciences.

For Further Reading

Bolcato-Bellemin, A.L., Bonnet, M.E., Creusat G., *et al.* (2007). Sticky overhangs enhance siRNA-mediated gene silencing, *Proc Natl Acad Sci USA*, **104**, 16050–16055.

Boussif, O., Lezoualc'h, F., Zanta, M.A. *et al.* (1995). A versatile vector for gene and oligonucleotide transfer into cells in culture and *in vivo*: Polyethylenimine, *Proc Natl Acad Sci USA*, **92**, 7297–7301.

Chollet, P., Favrot, M.C., Hurbin, A., *et al.* (2002). Side-effects of a systemic injection of linear polyethylenimine-DNA complexes, *J Gene Med*, **4**, 84–91.

Creusat, G., Rinaldi, A.S., Weiss, E., *et al.* (2010). Proton sponge trick for pH-sensitive disassembly of polyethylenimine-based siRNA delivery systems, *Bioconjugate Chem*, **21**, 994–1002.

Goula, D., Remy, J.S., Erbacher, P., *et al.* (1998). Size, diffusibility and transfection performance of linear PEI/DNA complexes in the mouse central nervous system, *Gene Therapy*, **5**, 712–717.

Kichler, A., Leborgne, C., Coeytaux, E., *et al.* (2001). Polyethylenimine-mediated gene delivery: A mechanistic study, *J Gene Med*, **3**, 135–144.

McKenzie, D.L., Smiley, E., Kwok, K.Y., *et al.* (2000). Low molecular weight disulfide cross-linking peptides as nonviral gene delivery carriers, *Bioconjugate Chem*, **11**, 901–909.

Sonawane, N.D., Szoka, F.C., Jr., Verkman, A.S. (2003). Chloride accumulation and swelling in endosomes enhances DNA transfer by polyamine-DNA polyplexes, *J Biol Chem*, **278**, 44826–44831.

Taratula, O., Garbuzenko, O.B., Kirkpatrick P., *et al.* (2009). Surface-engineered targeted PPI dendrimer for efficient intracellular and intratumoral siRNA delivery, *J Control Release*, **140**, 284–293.

15

AUTO-ASSOCIATIVE LIPID-BASED SYSTEMS FOR NON-VIRAL NUCLEIC ACID DELIVERY

Virginie Escriou[a,d], Nathalie Mignet[a] and Andrew Miller[b,c]

15.1 Lipid Self-Association and Self-Assembly

The unrivalled power of lipids to form macromolecular lipid assemblies and the shear three-dimensional plasticity of these assemblies have suggested a number of very important applications. Most significantly, varieties of macromolecular lipid-based assemblies can be adapted to mediate delivery of nucleic acids to cells *in vitro* and even *in vivo*. However, before we address how this has been achieved, we need to look at basic lipid structures in biology and some of the key macromolecular lipid assemblies that can be generated by self-assembly using a variety of naturally available lipids (Miller and Tanner, 2008).

15.1.1 Monomeric Lipid Structures

Monomeric lipids may be broadly defined as molecules of intermediate molecular weight (MWt. 100–5000 Da) that contain a substantial portion of aliphatic or

[a]Laboratory of Chemical and Genetic Pharmacology and Laboratory of Imaging, Paris Descartes Pharmacy University, CNRS, Inserm, Chimie ParisTech, 4, avenue de l'Observatoire Paris Cedex 06, France
[b]Institute of Pharmaceutical Science, King's College London, Franklin-Wilkins Building, Waterloo Campus, 150 Stamford Street, London SE1 9NH, UK
[c]GlobalAcorn Limited, London, UK
[d]Email: virginie.escriou@parisdescartes.fr

aromatic hydrocarbons. Most biologically important lipids belong to a subset of this broad class of molecules known as *complex lipids*. Major members of this subset are the *acylglycerols*, *glycerophospholipids* and *sphingolipids*. Acylglycerols are storage lipids and play no obvious structural roles in biology. Glycerophospholipids may be thought of as derivatives of triacylglycerols, wherein the carbon atom C-3, carboxylate ester has been replaced by a phosphate ester. Whilst the number of possible glycerophospholipids is vast, some of the most widely studied and arguably most biologically important are illustrated in Fig. 15.1. These glycerophospholipids are important predominantly as constituents of biological membranes.

Sphingolipids are a combination of the base sphingosine and long-chain fatty acids. Acylation of the carbon atom C-2 amine group by a fatty acid gives rise to the *ceramides* from which *phosphoceramides* are derived by phosphate ester derivatization of the C-1 hydroxyl group, as illustrated in Fig 15.2. Ceramides and phosphoceramides are important constituents of human skin lipids and neural membranes. The phosphoceramides, together with the glycerophospholipids, are collectively known as *phospholipids*. They have an unparalleled capacity for self-association.

15.1.2 Lyotropic Mesophases of Phospholipids

Macromolecular lipid assemblies arise from the amphiphilic character of phospholipids (Fig. 15.3). Broadly speaking, all phospholipids contain a compact *polar region*, which is hydrophilic in character, and an extended *chain region*, which is hydrophobic in character. In the presence of water, the tendency of the hydrophobic chain regions to self-associate and simultaneously exclude water leads to macromolecular lipid assemblies that may be described as extended non-covalent structures held together by van der Waals interactions and the hydrophobic effect. These assemblies adopt different mesophase states depending upon the character of the phospholipid involved and the local conditions.

These phospholipid mesophase states are also known as *lyotropic mesophases*. Some of the most well-established lyotropic mesophases have been summarized (Table 15.1) using the *Luzzati nomenclature*. Structures and transitions between mesophases are very relevant to the biological behavior of lipids and so these will be discussed. However, given the current state of knowledge on lipids and their macromolecular assemblies, there can be no absolute certainty concerning precisely which of the many, perhaps all, lyotropic mesophases play a role in biology.

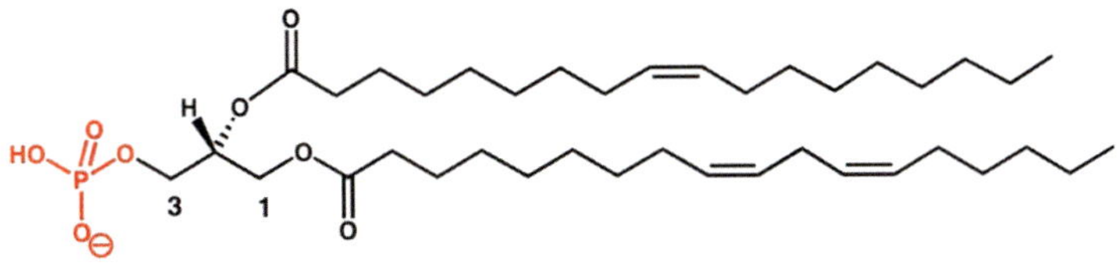

1-Linoleoyl-oleoyl-*sn*-glycero-3-phosphatidic acid (LOPA)

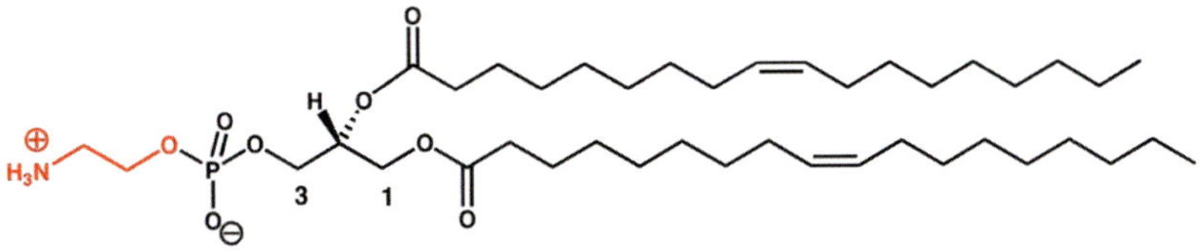

1,2-Dioleoyl-*sn*-glycero-3-phosphatidylethanolamine (DOPE)

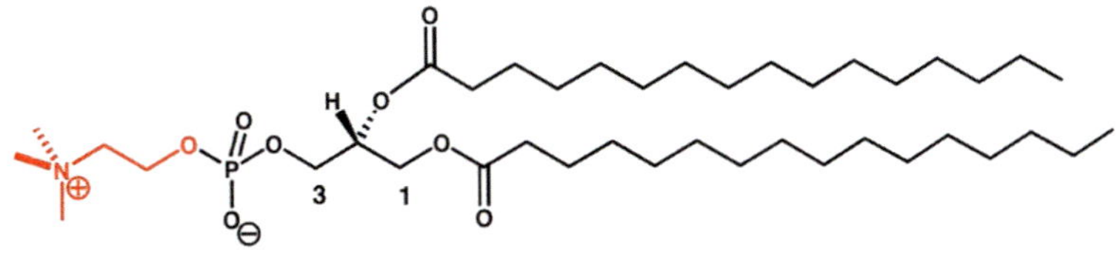

1,2-Dipalmitoyl-*sn*-glycero-3-phosphatidylcholine (DPPC)

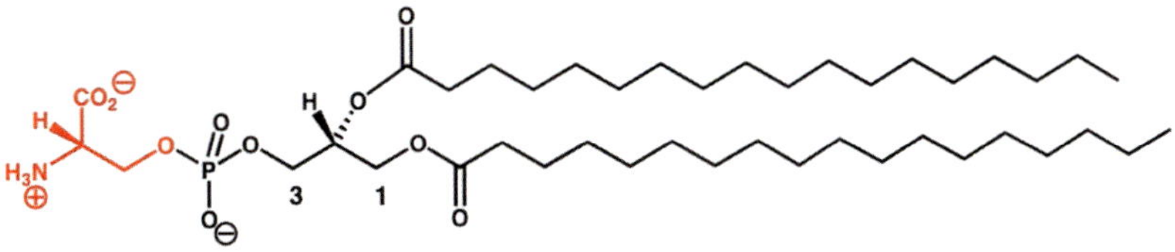

1,2-Distearoyl-*sn*-glycero-3-phosphatidylserine (DSPS)

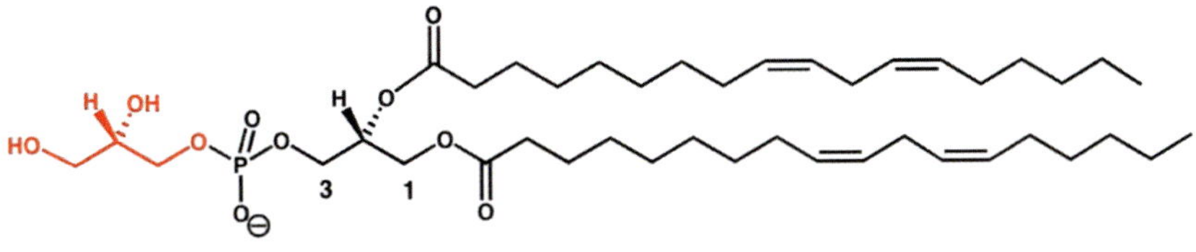

1,2-Dilinoleoyl-*sn*-glycero-3-phosphatidylglycerol (DLPG)

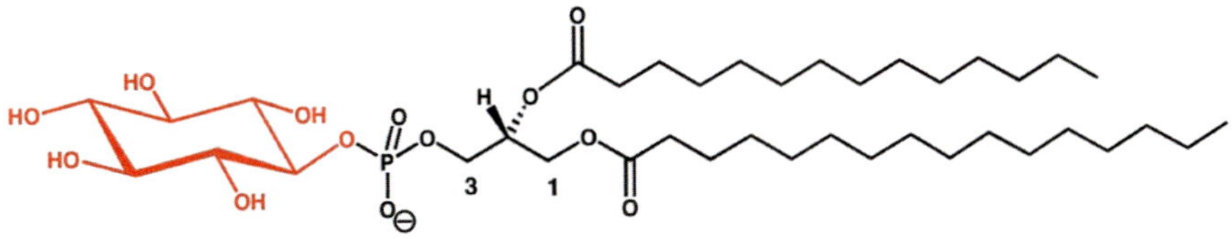

1-Palmitoyl-2-myristoyl-*sn*-glycero-3-phosphatidylinositol (PMPI)

FIGURE 15.1 ■ Representative structures of first-rank glycerophospholipids that are major and integral components of macromolecular lipid assemblies.

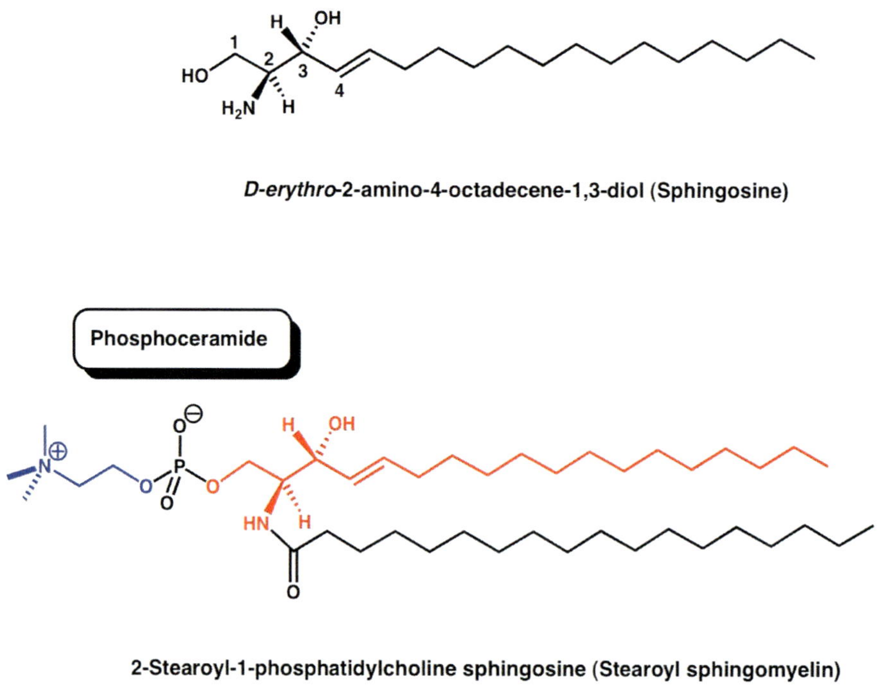

FIGURE 15.2 ■ Representative structure of phosphoceramides that are formed from the amine diol sphingosine.

15.1.3 Solid-Like Mesophases

One or more *crystalline lamellar* L_c *soild-like mesophases* may be formed by self-assembly from phospholipids at low temperatures and/or low levels of hydration. When long- and short-range order is found in three dimensions then the result is a *three-dimensional lamellar crystal*, which is a true crystal. The three-dimensional crystalline order results from the close packing of two-dimensional phospholipid crystalline sheets. In all crystalline and other ordered states, phospholipid close packing and molecular configuration is defined in terms of a number of parameters. These parameters are σ, the *mean cross sectional area of a fatty acid alkyl chain perpendicular to the chain axis*; ϕ, the *tilt angle of the chain* with respect to bilayer plane; d_p, the *thickness of the head group region* and S, the *surface area at the bilayer plane occupied by the individual phospholipid*. When the two-dimensional phospholipid crystalline sheets cease to maintain regular stacking arrangements with respect to each other, three-dimensional crystalline order breaks down leading to series of two-dimensional crystalline sheets, each irregularly stacked with respect to the next. Such mesophases are known as *two-dimensional lamellar crystals* since they still maintain a good deal of crystalline order.

One-dimensional lamellar solid-like mesophases are also known as gel states and occur when phospholipids are arranged into bilayers that then stack into a

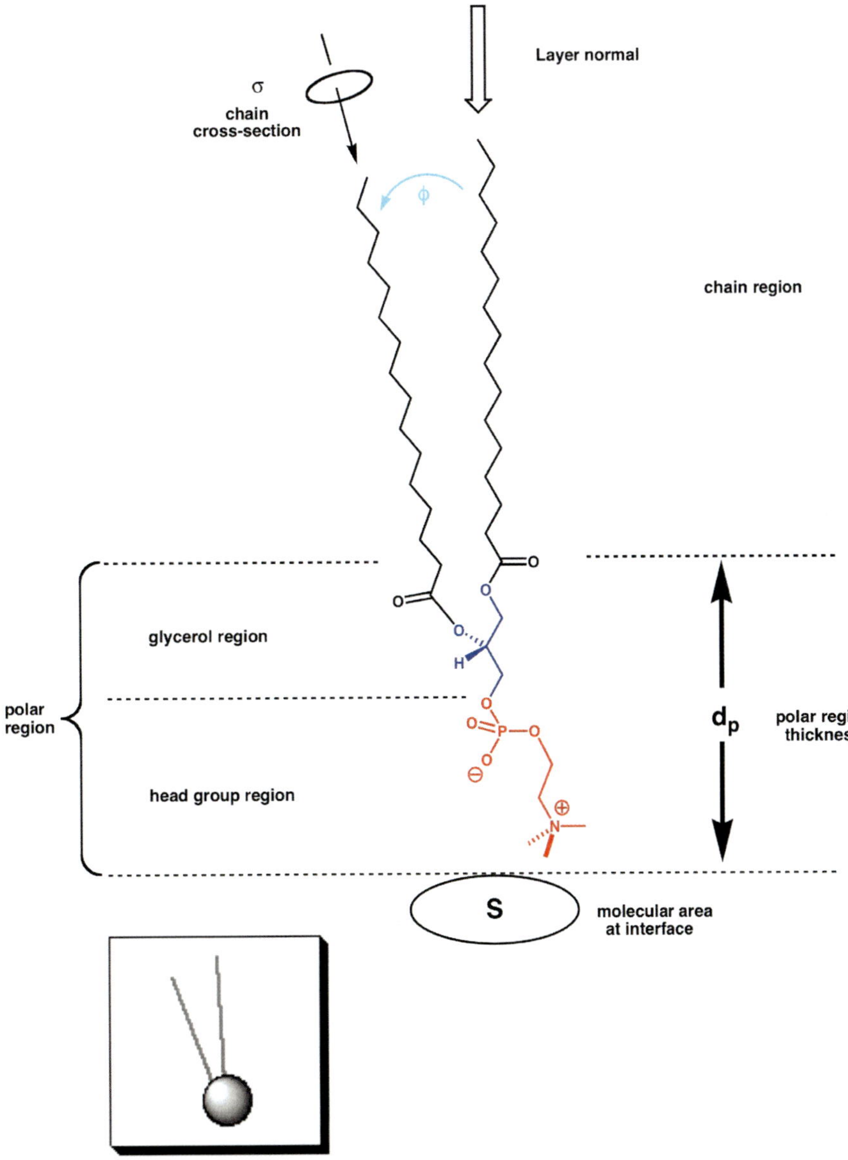

FIGURE 15.3 ■ Structure of dipalmitoyl L-α-phosphatidylcholine (DPPC) to illustrate the main molecular parameters and structural features that dictate the formation of crystalline lamellar solid-like mesophases of macromolecular lipid assemblies. Schematic representation of a phospholipid molecule is also shown (*inset*). Reproduced from Miller and Tanner (2008).

multilayer with each bilayer separated by water. The fatty acid alkyl side chains are stiff and extended, is in three-dimensional and two-dimensional lamellar crystals, but may undergo hindered rotations about their chain axes. There are a number of types of one-dimensional lamellar solid-like mesophases depending upon the tilt angle ϕ; these are L_β ($\phi = 0$), L_β' ($\phi > 0$) and an interdigitated mesophase $L_{\beta I}$

TABLE 15.1 Summary of Known Structural Assemblies from Self-Assembly of Lipids

Mesophase Type	Name	Mesophase Structure
Lamellar solid-like		
Three-dimensional	L_c	Three-dimensional crystal
Two-dimensional	L_c^{2D}	Two-dimensional crystal
One-dimensional	$P_{\beta,}$	Rippled gel
	L_β	Untilted gel
	$L_{\beta,}$	Tilted gel
	$L_{\beta I}$	Interdigitated gel
	$L_{\alpha\beta}$	Partial gel
Fluid mesophases		
One-dimensional	L_α	Lamellar fluid
Two-dimensional	H	Hexagonal
	H^C	Complex hexagonal
	R	Rectangular
	M	Oblique
Three-dimensional	Q	Cubic
	T	Tetragonal
	R	Rhombohedral
	O	Orthorhombic

($\phi = 0$), where fatty acid alkyl side chains from different monolayers overlap with each other (Fig. 15.4).

One other one-dimensional lamellar solid-like mesophase, $L\delta$, is known, in which the fatty acid alkyl side chains cease to be linear but adopt a helical conformation.

15.1.4 Fluid Mesophases

Under certain conditions of temperature and hydration, solid-like mesophases will undergo a transition into fluid mesophases. The physical conditions under which transitions of this type occur are very important in biological terms. The vast majority of *three-dimensional fluid mesophases* so far identified have *cubic* (Q) symmetry (Table 15.1). Six cubic fluid mesophases have been characterized so far and these appear to fall into two distinct families, one family based upon periodic

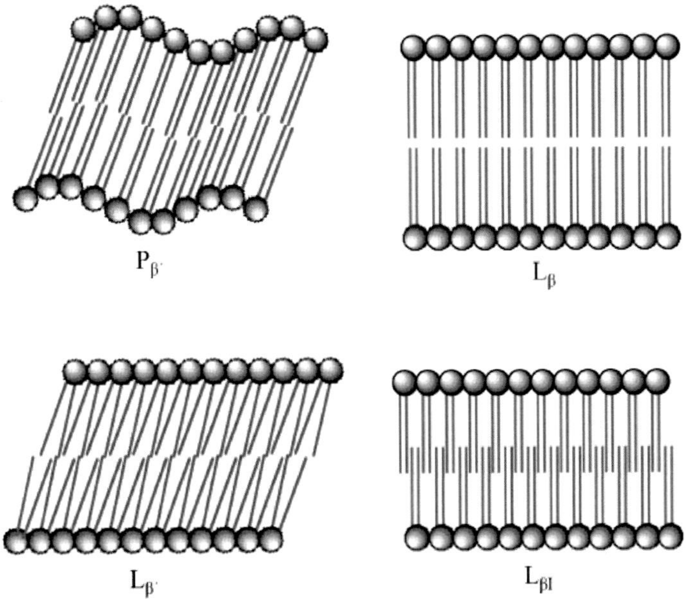

FIGURE 15.4 ■ Selection of one-dimensional lamellar solid-like mesophases that are partially disordered, but exhibit translational ordering in two dimensions. These gel states exist half way between crystalline L_c states and completely fluid mesophases such as the lamellar $L_{\alpha 1}$ fluid mesophases and hexagonal H_{II} fluid mesophase. In lipid assembly terms, these represent the equivalent of secondary/tertiary structure formation. Adapted from Miller and Tanner (2008).

minimal surfaces (bicontinuous) and the other upon discrete lipid aggregates (micellar). Three-dimensional fluid mesophases may exist as either *Type I* (*normal topology*, oil-in-water) or *Type II* (*inverse topology*, water-in-oil) structures. Frequently, cubic mesophases of both Type I and Type II exist. For instance, the main bicontinuous mesophases (*Q230*, *Q224* and *Q229*) all form Type I and Type II structures consisting of two separate interwoven but unconnected networks of channels or rods which are formed from either fatty acid side chains or water, respectively (Fig. 15.5).

Type II rhombohedral (R) or *tetragonal* (T) three-dimensional fluid mesophases are also known. For example, The R_{II} mesophase consists of planar two-dimensional hexagonal arrays, formed from aqueous channels, which are then regularly stacked to form a three-dimensional lattice. In a similar way, the T_{II} mesophase is comprised of planar two-dimensional square arrays that are once again stacked to from the three-dimensional lattice. A debate is now raging amongst some chemical biology researchers concerning the potential existence of cubic mesophases in cells, indeed there is a proposition that the membranes of cellular organelles may in fact adopt the normal topology cubic Q_I fluid mesophase form of organization.

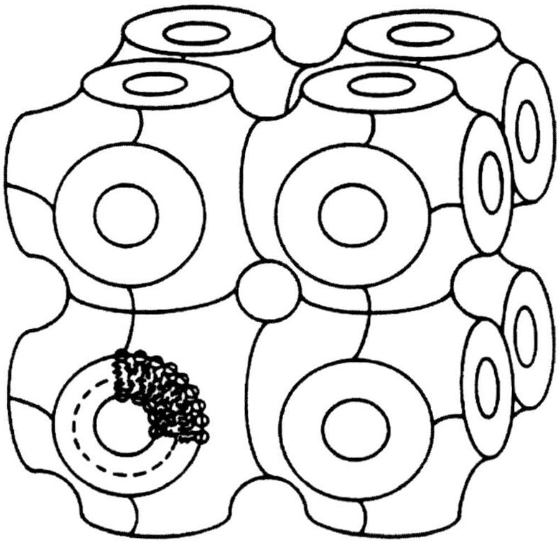

FIGURE 15.5 ■ Structural representation of a principal cubic fluid mesophase, Im3m (Q^{229}), that is formed by lipid self-assembly and which is also known as a cubic Q_I fluid mesophase. Three-dimensional lipid self-assembly supports a regular network of aqueous channels or pores. Reproduced from Miller and Tanner (2008).

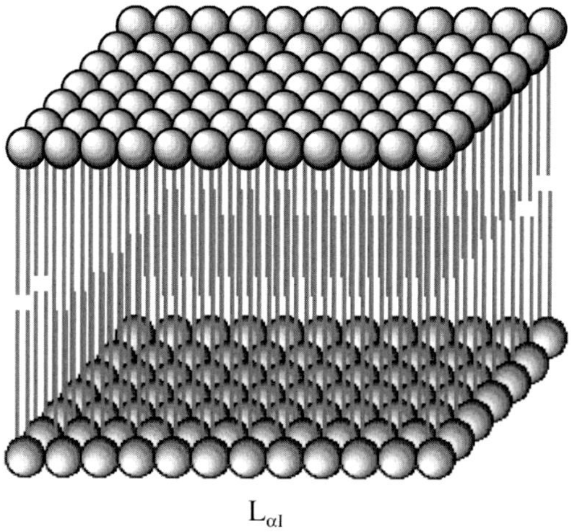

FIGURE 15.6 ■ Structural representation of $L_{\alpha I}$, the main lamellar fluid mesophase formed by lipid assemblies and the primary fluid mesophase adopted by biological membranes (also called bilayer membranes). Reproduced from Miller and Tanner (2008).

Having said the above, membranes in cells are usually considered to adopt *one-dimensional fluid mesophases* forms of organization, in particular the *lamellar L_α fluid mesophase* in normal topology (Fig. 15.6, also called *bilayer*). This lamellar $L_{\alpha I}$ fluid mesophase is widely considered to represent the default state of all biological membranes under normal physiological conditions. Cellular membranes are also

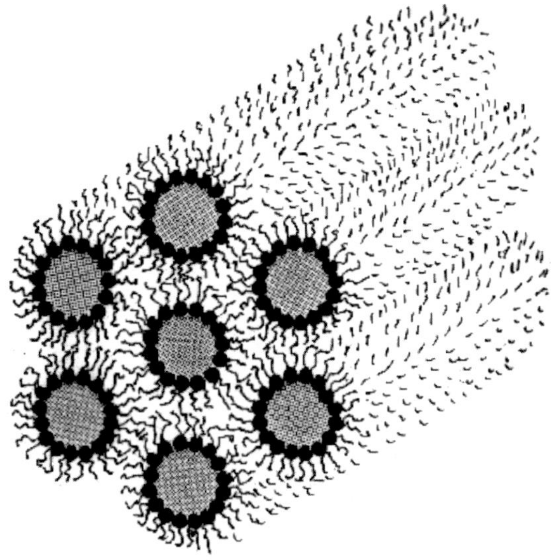

FIGURE 15.7 ■ Structural representation of the H_{II}, the main hexagonal fluid mesophase formed by lipid assemblies. Reproduced from Miller and Tanner (2008).

thought to be able to adopt certain two-dimensional fluid mesophases under certain circumstances, in particular the *hexagonal fluid mesophase* H (Fig. 15.7), such as the inverse topology Type II structure, H_{II}.

Both normal- and inverse-topology hexagonal fluid mesophases, H_I and H_{II}, have the appearance of stacked cylinders. In the H_I fluid mesophase, fatty acid alkyl side chains are contained within the cylinders whilst polar regions make up cylinder surfaces; in the H_{II} fluid mesophase, cylinders of water are bordered by polar regions and the spaces between cylinders are occupied by fatty acid alkyl side chains. The interconversion within biological membranes between the $L_{\alpha I}$ fluid mesophase and H_{II} fluid mesophase is currently considered to be central to the dynamic behavior of all manner of biological membranes, in particular biological membranes become temporarily more porous in the H_{II} fluid mesophase prior to returning to the much less porous $L_{\alpha I}$ fluid mesophase. In addition, mesophase interconversion appears to be important to facilitate fusion and trafficking events involving biological membranes.

15.2 Cationic Lipids to Cationic Liposomes/Micelles

Cationic lipids are synthetic and as such must be prepared by synthetic organic chemistry. Nevertheless, like phospholipids, cationic lipids are amphiphiles and the same physical parameters can be used to define their structures and behaviors

(Fig. 15.3). As amphiphiles, they are capable of interacting with other lipids to form macromolecular lipid assemblies. Different lyotropic mesophase states are formed according to the nature of lipids involved and the ratio between hydrophilic polar regions and corresponding hydrophobic chain regions. The ability of the lipid polar regions to interact with water will also influence the mesophases formed by the lipid (Tranchant *et al.*, 2004). As mentioned above, the $L_{\alpha I}$ mesophase is widely considered to represent the default mesophase state of all biological membranes under normal physiological conditions. Importantly, although this mesophase is described as a one-dimensional lamellar fluid mesophase, this mesophase can extend in three dimensions indefinitely and with curvature leading to the formation of lipid bilayer spheres consisting typically of a single lipid bilayer membrane that encapsulates a spherical cavity in which aqueous buffer is entrapped. These are known as unilamellar vesicles (ULVs). ULVs are known to vary in diameter from about 50 nm in diameter up to, and including, several hundred nanometres. Multilamellar vesicles (MLVs) involving multiple lipid bilayers are also known. For biological utility, ULVs with diameters of 100–150 nm are ideal for several mainstream applications where solubility and stability are an issue. In particular there has been a substantial interest over the past 20 years in the subject of *cationic liposomes* and their main role to enable functional delivery and expression (*transfection*) of therapeutic nucleic acids such as plasmid DNA (pDNA) to target cells *in vivo*.

15.2.1 Formulation of Cationic Liposomes/Micelles

Simple cationic liposomes (or even micelles) are formed by self-assembly from cationic lipids (also known as cytofectins; *cyto-* for cell and *-fectin* for transfection) alone or more commonly from the combination of a cationic lipid and a neutral lipid such as dioleoyl L-α-phosphatidylethanolamine (or 1,2-dioleoyl-*sn*-glycero-3-phosphothanolamine) (DOPE), dioleoyl L-α-phosphatidylcholine (or 1,2-dioleoyl-*sn*-glycero-3-phosphocholine) (DOPC) or cholesterol. The molecular structures of representative cationic lipids are shown alongside potential neutral lipid partners in Fig. 15.8.

Typically, cationic lipid and neutral lipid components are mixed together in an appropriate molar ratio and then induced or formulated into ULVs by any one of a number of methods including *reverse-phase evaporation* (REV), dehydration-rehydration and extrusion (Miller, 1998, 2008; Kostarelos and Miller, 2005). Alternatively, cationic lipids may be assembled into cationic micellar structures after being dispersed in water or aqueous organic solvents. The dehydration-rehydration scheme is represented schematically in Fig. 15.9.

Lipid molecular shape is an important determinant of both cationic liposome size and also the possibility that micelles are formed in preference to cationic

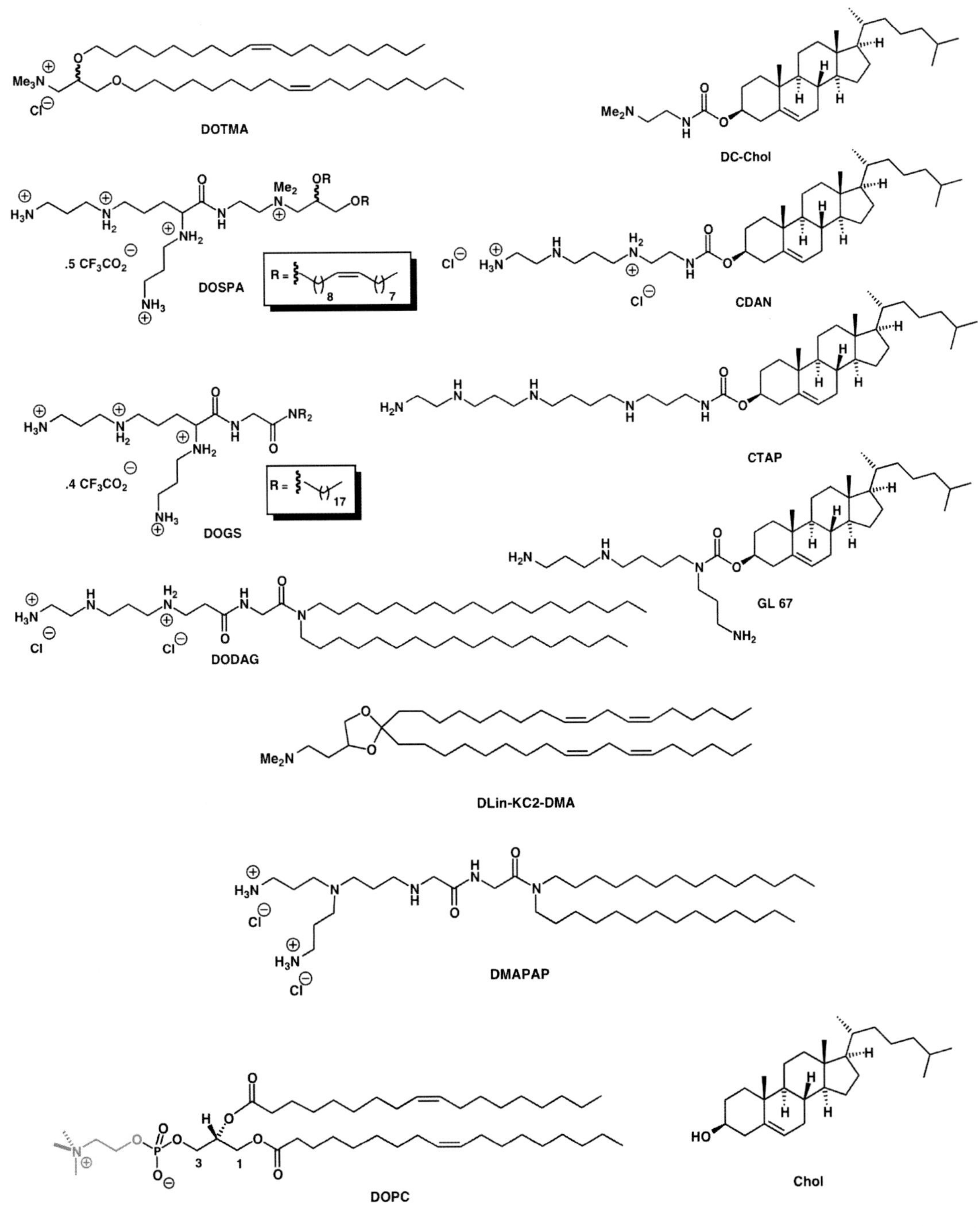

FIGURE 15.8 ■ Set of typical, well-established cationic lipids used in cationic liposome/micelle formulations; *N*-[1-(2,3-dioleyloxy)propyl]-*N,N,N*-trimethyl ammonium chloride (DOTMA); 2,3-dioleyloxy-*N*-[2 (sperminecarboxamido)ethyl]-*N,N*-dimethyl-1-propanaminium trifluoro-acetate (DOSPA); dioctadecylamidoglycylspermine (DOGS): *N′,N′*-dioctadecyl-*N*-4,8-diaza-10-aminode-canoylglycine amide (DODAG): 3β-[*N*-(*N′,N′*-dimethylaminoethane)carbamoyl]cholesterol (DC-Chol): N^1-cholesteryl-oxycarbonyl-3, 7-diazanonane-1,9-diamine (CDAN): N^{15}-cholesteryloxycarbonyl-3,7,12-triaza-pentade-cane-1,15-diamine (CTAP): 2-{3-[bis-(3-aminopropyl)amino]propylamino}-*N*-ditetradecylcarbamoylmethylacetamide (DMAPAP) (RPR 209120): DOPC and Chol abbreviations are explained in the text.

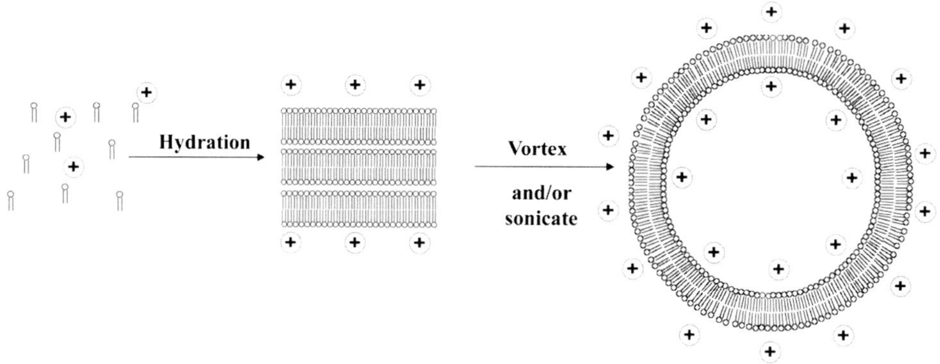

FIGURE 15.9 ■ Two stage dehydration–rehydration procedure for formation of cationic liposomes: A thin lipid film on the walls of a glass vessel are hydrated in the presence of buffer yielding multilamellar aggregates that are then converted to unilamellar vesicles by vortex mixing and/or sonication procedures.

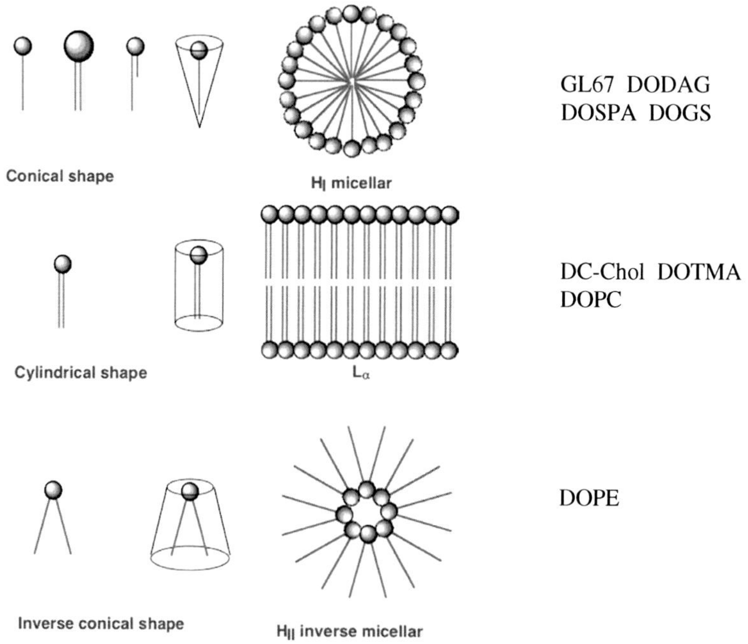

FIGURE 15.10 ■ The relationship between the molecular shape of lipid molecules (see Figs. 15.1, 15.2 and 15.8) and their preferred lyotropic mesophase structures (see Table 15.1).

liposomes. According to structure, lipids can encourage the formation of different mesophases (Fig. 15.10).

15.2.2 Syntheses of Cationic Lipids

Syntheses of cationic lipids appear simple but are often complex to implement. One of the greatest contributors to complexity is the difficulty of purifying such amphiphilic molecules free of minor impurities and contaminants and of achieving efficient syntheses when nitrogen is involved in amine and amide functional group formation. Both difficulties can be resolved by using at least one of two cardinal rules during synthetic procedures. These are:

- Assemble polar cationic headgroups separately using solid-phase chemistry.
- Protect amine functional groups throughout a given synthesis and then reveal all these functional groups at the end of the synthesis by global deprotection.

The following three synthetic schemes illustrate these principles in action. The first is the synthesis of *N*[15]-cholesteryloxycarbonyl-3,7,12-triazapentadecane-1,15-diamine (CTAP) (Fig. 15.11) which was prepared using a hyperflexible synthetic organic chemistry approach involving the coupling of amine-protected fragments leading to a final global deprotection step to unmask amine functional groups.

The second synthesis is that of cationic lipid *N*[1]-cholesteryloxycarbonyl-3,7-diazaonane-1,9-diamine (CDAN) (Fig. 15.12) which embodies the potential of solid phase synthesis for high-yielding polyamine synthesis.

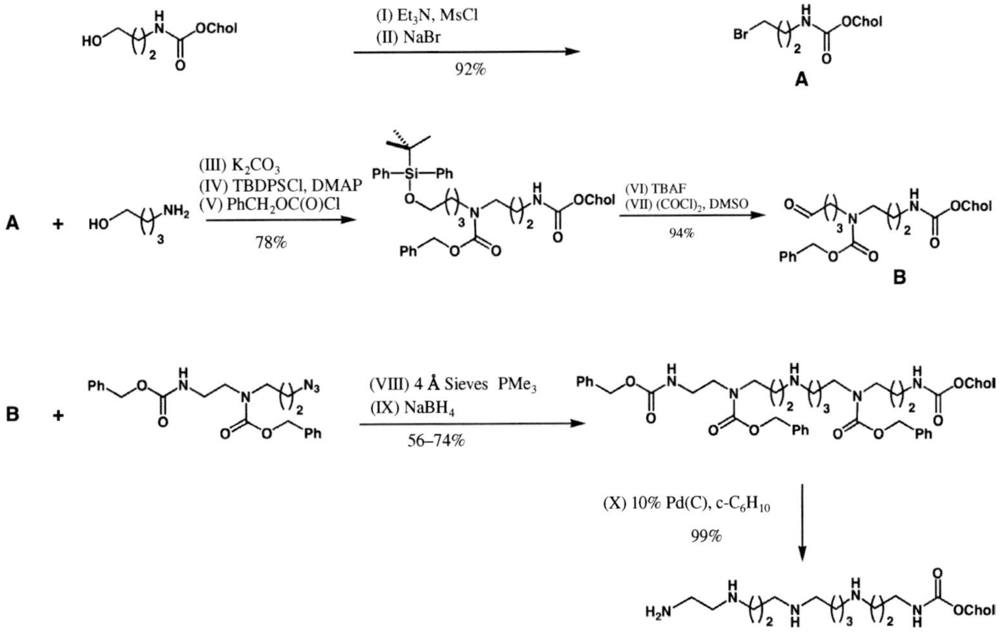

FIGURE 15.11 ■ The synthesis of CTAP demonstrating the solution-phase fragment-based approach to polyamine synthesis (see Cooper *et al.*, 1998; Keller *et al.*, 2003a).

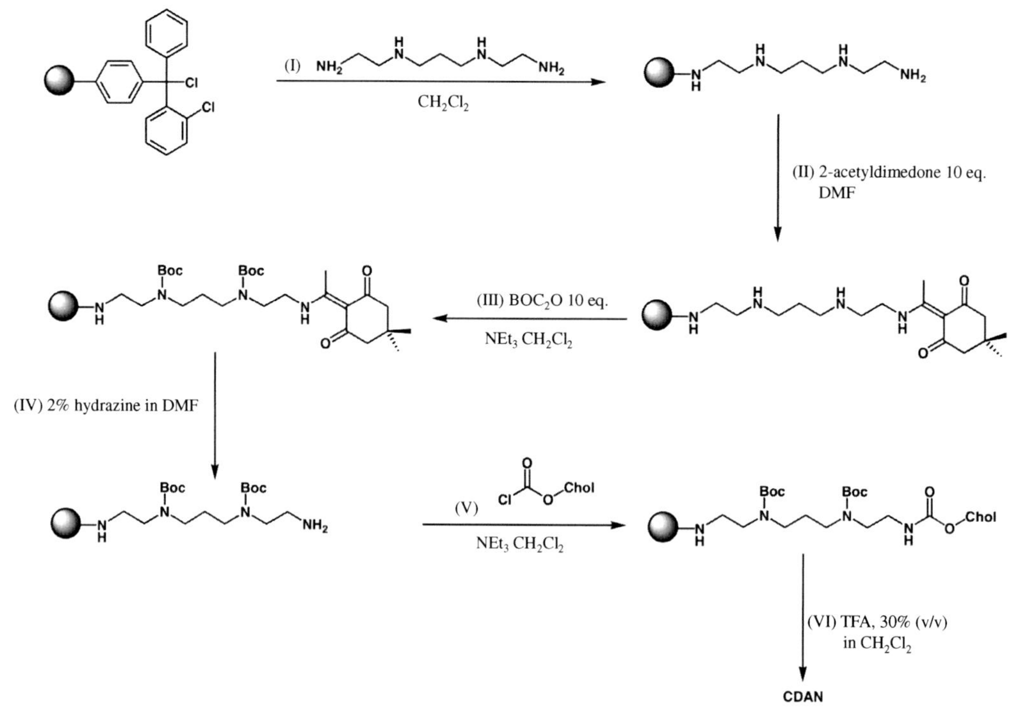

FIGURE 15.12 ■ The synthesis of CDAN demonstrating the solid-phase fragment-based approach to polyamine synthesis (see Oliver *et al.*, 2004).

Finally, the synthesis of *N*′,*N*′-dioctadecyl-*N*-4,8-diaza-10-aminodecanoylglycine amide (DODAG) (Fig. 15.13) is shown to exemplify a convergent synthesis design which employs fragment synthesis of a key protected polyamine synthon (A) that is coupled to a hydrophobic moiety in a penultimate step prior to global amine functional group deprotection.

15.3 Characterization of the Interactions between Lipids and DNA

Cationic unilamellar vesicles or micelles have an impressive ability to interact with and encapsulate complex and large nucleic acids such as pDNA. Combination under kinetic mixing conditions results in the formation of nanometric liposome/ micelle–pDNA (lipoplex; LD) particles. Nowadays, these are frequently described as pDNA–lipoplex nanoparticles. The population of nanoparticles is typically

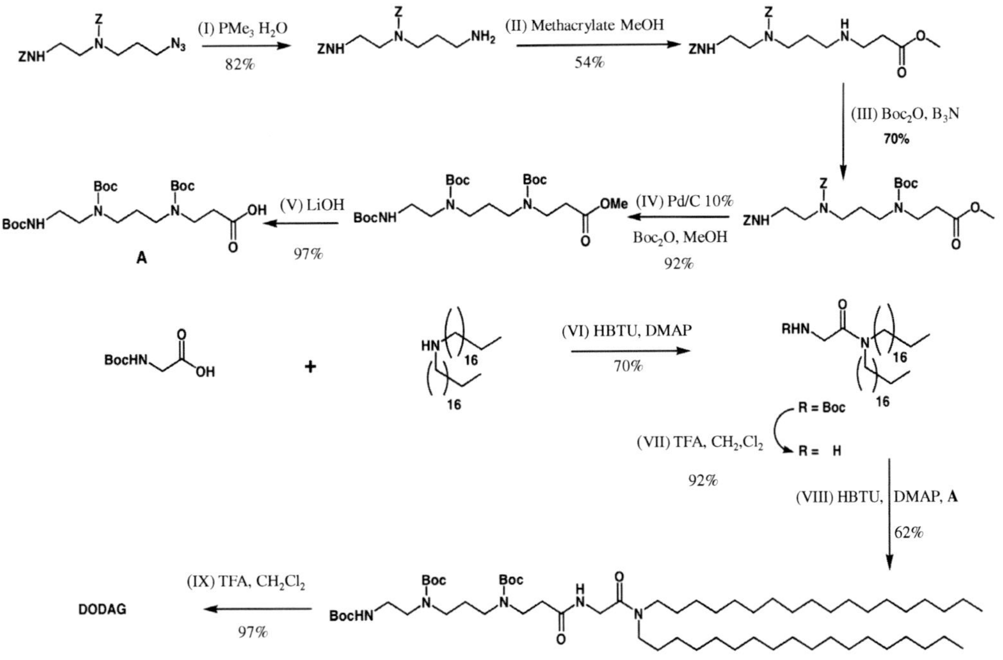

FIGURE 15.13 ■ The highly convergent synthesis of DODAG demonstrating the convergent solution-phase fragment-based approach to polyamine synthesis (see Cooper *et al.*, 1998; Mével *et al.*, 2010).

quite heterogeneous in size but pDNA–lipoplex nanoparticles have been known for at least 20 years to mediate functional delivery of pDNA to cells *in vitro*, in tissue *ex vivo* and have even been used *in vivo* (although with limited success). There have been numerous cationic liposome/micelle systems reported to mediate successful pDNA delivery to cells *in vitro*, of which a growing number have been commercialized (Miller, 1998, 2003, 2008; Nicolazzi *et al.*, 2003; Kostarelos and Miller, 2005, 2008). More are being reported all the time, but in each case the key ingredient is the cationic lipid used.

Given that the $L_{\alpha I}$ fluid mesophase and inverse topology hexagonal H_{II} fluid mesophase are the two fluid mesophases significant for biological membrane structure and membrane fusion events, respectively, there should be little surprise that these same two fluid mesophases dominate interactions between DNA and cationic liposomes or micelles. DNA in solution has a substantial negative charge associated with DNA phosphodiester links. These negative charge centers are surrounded by cationic counterions that can themselves become displaced by the introduction of three-dimensional arrays of positively charged head group regions associated with cationic lipids embedded in cationic liposome or cationic micelles. Such substantial, cooperative charge neutralization of DNA phosphodiester backbone renders DNA hydrophobic and subject to hydrophobic collapse. Over a period of several minutes, there takes place a process of kinetic mixing and structural negotiation until DNA–lipoplex nanoparticles are eventually formed (typically 60–250 nm in diameter). There remain disagreements about the nature of these lipoplex nanoparticles. However, a growing consensus view is that these consist of a multilamellar core comprising multiple $L_{\alpha I}$ fluid mesophase cationic bilayer membranes separated by layers of equally spaced DNA helices (Fig. 15.14). The core is then thought to be surrounded by a bilamellar liposome retaining layer.

Many characterization techniques are used to define the nature of the DNA–lipoplex nanoparticles. Dynamic light scattering measures the hydrodynamic diameter of the lipoplex nanoparticles; electron microscopy, static and neutron light scattering give non-hydrated diameters, the molecular mass and nanoparticle three-dimensional shape; otherwise small-angle X-ray scattering (SAXS) has been used to probe internal structure. Quantitation of DNA condensation and encapsulation has been demonstrated typically by gel retardation (or delayed migration) experiments which make use of gel electrophoresis or alternatively by DNA fluorescence binding exclusion assays (Stewart *et al.*, 2001; Tranchant *et al.*, 2004).

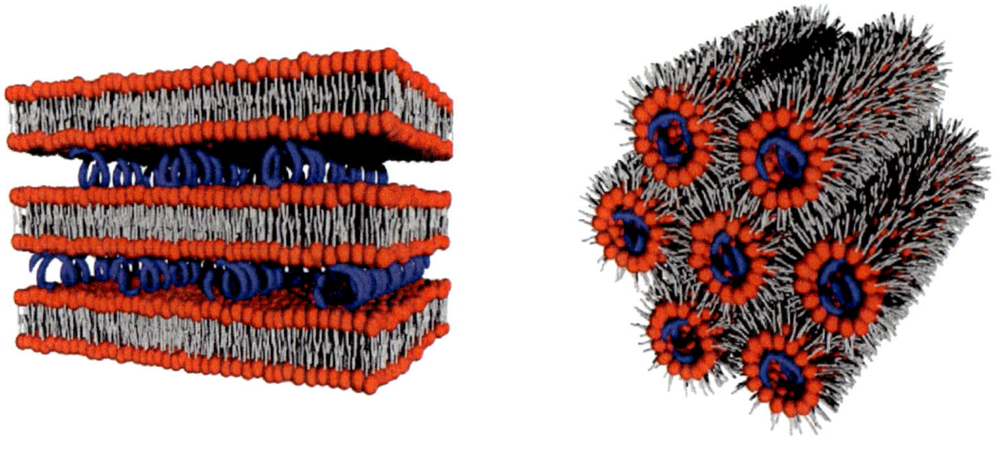

(a) (b)

FIGURE 15.14 ■ DNA interactions with macromolecular lipid assemblies. (a) Representation of multilamellar core of DNA–lipoplex nanoparticle. DNA helices are shown as ribbons (blue), lipid polar regions (red) and chains (grey); (b) Representation of metastable lipoplex nanoparticle hexagonal mesophase thought to play a role in functional DNA delivery to cells (see Section 15.4 on mechanism).

15.4 DNA Delivery to Cells

Typically, pDNA–lipoplex nanoparticles are prepared through the combination of cationic liposome/micelle and pDNA in a lipid:pDNA ratio around 12:1 by weight. This ratio could also be replaced by a lipid:pDNA molar ratio or more commonly by a corresponding cationic lipid:deoxynucleotide [cat]:[nt] mol ratio. From the latter mol ratio, a lipid–pDNA charge ratio (N/P) can be derived where N defines the number of protonated amine functional groups presented by each cationic lipid present and P is the negative charge (always 1) presented by the attached phosphate of each deoxynucleotide. For the most part, lipoplex nanoparticles are prepared such that N/P is greater than 1, under which circumstances nanoparticle formulations are at their most reproducible and the lipoplex nanoparticles so prepared are at their smallest size and least heterogeneous. Such cationic lipoplex nanoparticles are commonly used for pDNA delivery as well. Having said this, there may be cases where lipoplex nanoparticles prepared such that N/P less than 1 are known to be required for certain target cells such as lung epithelial cells (Stewart *et al.*, 2001).

15.4.1 Cellular Entry

Cellular entry of lipoplex nanoparticles is generally achieved by *endocytosis* triggered by contact between cationic lipids and negatively charged proteoglycans attached at the surfaces of plasma membranes. Endocytosis is a cellular uptake mechanism in which a cell membrane invaginates and pinches off to engulf extracellular substances into the cell (Fig. 15.15) (Miller, 1999; Medina-Kauwe *et al.*, 2005). Clathrin is a cytosolic coat protein which plays a critical role in the formation of membrane-bounded vesicles for endocytosis. After being internalized, clathrin-coated vesicles become uncoated (depolymerization), resulting in early endosomes. The pH of these endosome compartments is decreased over at least 30 min to pH 5–6 due to proton import activities of H^+-adenosine triphosphatase (ATPase) proton pumps located in endosome membranes. Endosome contents are destined typically for digestion by various hydrolytic enzymes following fusion with lysosomal vesicles. Therefore, endosomal disruption of the endocytosed nanoparticles is essential to avoid lysosomal degradation and potential transportation back to the cell surface.

15.4.2 Intracellular Trafficking of DNA

Cationic lipids not only facilitate nanoparticle formation and trigger cellular entry (see Section 15.4.1), but may also be able to assist in the process of endosomolysis

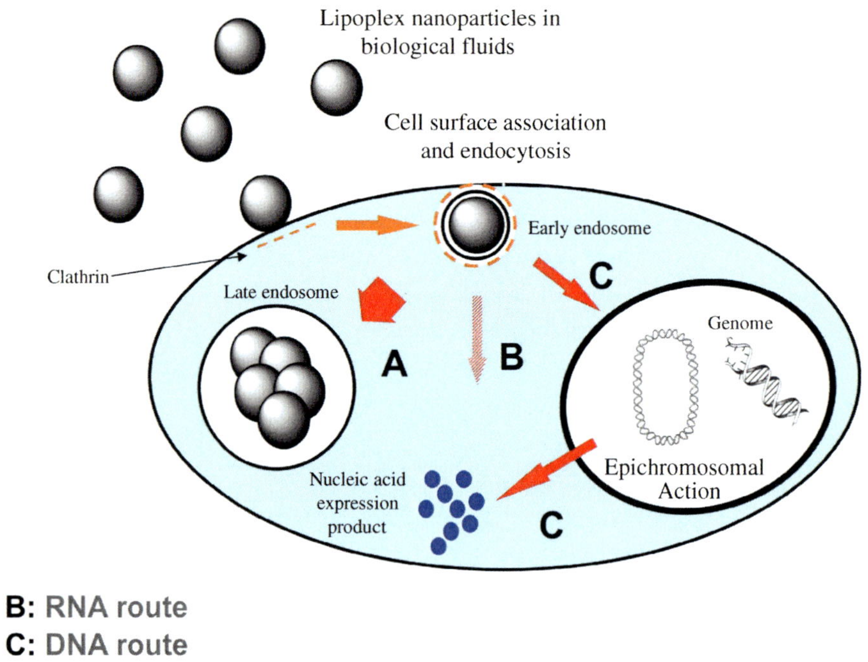

FIGURE 15.15 ■ Schematic representation of intracellular hurdles encountered by lipoplex nanoparticles. Lipoplex nanoparticles first interact with anionic glycoproteins, such as proteoglycans, found on the cell surface. This leads to the accumulation of clathrin proteins at the binding site for endocytosis. After internalization, the clathrin coat depolymerizes, resulting in early endosomes. The plasma proteins then recycle back to the cell surface. Endocytosed lipoplex nanoparticles must escape from endosomes to be able to release their contents into the cytoplasm or else enter large late endosomes (Path A). The released nucleic acid must then traffic through to the cytoplasm (Path B), if RNA, or else traverse the nuclear membrane to enable transfection effects in the nucleus (Path C). Adapted from Miller (1999).

(endosome escape) or the internal storage vesicle escape of nucleic acids to the cytoplasms of target cells (Miller, 2008; Martin *et al.*, 2005). In particular, where cationic lipids are designed with multivalent head groups, such as DODAG or CDAN, amine functional groups can adopt a range of pKa values from approximately six to ten; these amine functional groups are not fully protonated during lipoplex nanoparticle formulation at neutral pH but will become fully protonated in early endosomes after acidification. Such a characteristic could change the osmotic pressure in endosome compartments leading to endosomolysis by the *osmotic shock mechanism* (Fig. 15.16) (Stewart *et al.*, 2001).

Delivered nucleic acids also need to escape nanoparticle associations. This may arise through the *flip-flop mechanism* (Fig. 15.16). According to this mechanism, internalized lipoplex nanoparticles induce electrostatically lateral diffusion of anionic lipids, predominantly located on the cytoplasmic side, to form charge-neutralized ion pairs with cationic lipids. The ion pairing is proposed to destabilize

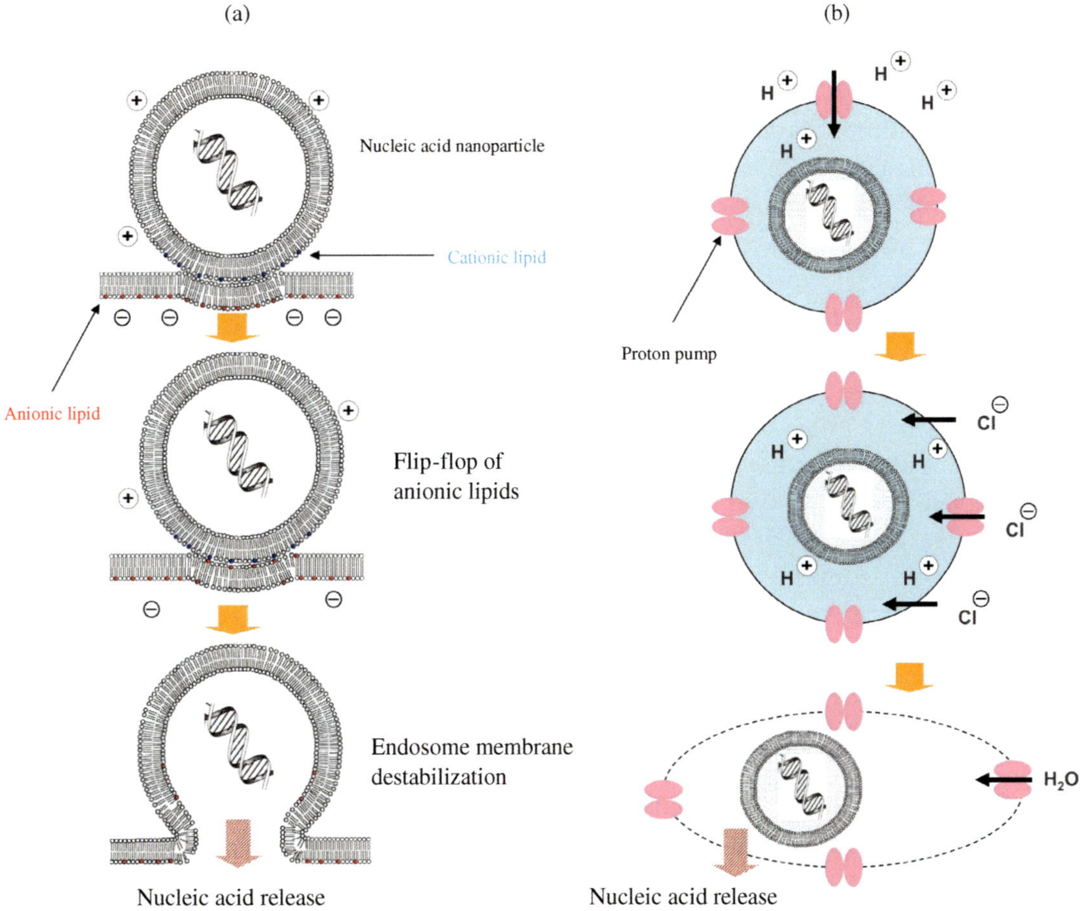

FIGURE 15.16 ■ Representation of endosmolysis escape hypotheses: (a) the flip-flop mechanism and (b) the osmotic shock mechanism.

endosomal and nanoparticle membranes, allowing for mobilization of pDNA into the cell cytoplasm via membrane channels, presumably established in regions of inverse topology hexagonal H_{II} fluid mesophase. Accordingly, the inclusion of neutral co-lipids, such as dioleoyl L-α-phosphatidylethanolamine (DOPE) in a lipoplex nanoparticle formulation, is thought highly beneficial. DOPE is a natural fusogenic lipid with a tendency to adopt the hexagonal H_{II} fluid mesophase, over a wide range of temperatures (from 10 to over 37°C), a characteristic typically expected to improve intracellular trafficking of nucleic acids post lipoplex nanoparticle internalization (Koynova *et al.*, 2006).

There may well be other vesicular membrane barriers and membrane structures that need lysis in target cells in order to maximize efficient intracellular trafficking of delivered nucleic acids. Other physical properties of lipids or conjugates may yet need to be harnessed to deal with these supplementary problems.

Once in cytoplasm, nucleic acids such as pDNA must traffic to the nucleus. In the process, pDNA must run the gauntlet of cytosolic nuclease enzymes, other proteins and organelles that digest and/or restrict the mobility of pDNA, constituting additional intracellular barriers to functional pDNA delivery. Unfortunately, the nuclear membrane itself can represent a substantial barrier. Nuclear entry is governed by nuclear pores (~55 Å) in the nuclear envelope which only permit the passive transport of molecules up to a molecular weight of 70 kDa or a diameter of ~10 nm. This size limit is much smaller than the size of DNA (e.g, a 5 kb DNA has a molecular weight of approximately 3.3 million Da), hence pDNA cannot enter the nucleus without assistance. Accordingly, nuclear entry of pDNA will only occur during cell division during mitosis, when the nuclear membrane loses its integrity (allowing facile entry of pDNA), or via an active transport mechanism that makes use of nuclear localization sequences (NLS) to facilitate nuclear import by association with specific cytosolic receptors — importins (Medina-Kauwe *et al.*, 2005). Once present in the nucleus, pDNA needs to undergo epichromosomal expression giving rise to a therapeutic protein product. Unfortunately, even this process is troubled by gene silencing that occurs typically within 1 week of delivery to target cells. Thankfully, the means are now being developed to overcome this particular problem (Kostarelos and Miller, 2005; Argyros *et al.*, 2008).

Ultimately, efficient functional delivery of pDNA to target cells *in vitro* has not been achieved. Barriers are recognized but mechanisms are incomplete and functional pDNA delivery to cells remains deeply inefficient with potentially severe consequences for effective *in vivo* delivery of pDNA too. Structure–activity relationships relating cationic lipid and lipoplex nanoparticle structure to transfection outcomes are emerging (Stewart *et al.*, 2001; Koynova *et al.*, 2006), but a great deal of work really remains to be done.

15.5 Delivery of pDNA *in vivo*

Over the past 15–20 years there have been some useful results obtained for the delivery of pDNA *in vivo*. For instance, in cystic fibrosis and rheumatoid arthritis gene therapy in which pDNA–lipoplex nanoparticles were used to deliver therapeutic genes to established animal models of both diseases, leading to temporary corrections of disease phenotypes in both cases (Alton *et al.*, 1993; Fellowes *et al.*, 2000). However, in general, in spite of considerable efforts, pDNA–lipoplex nanoparticles have proved unable to mediate efficient delivery of pDNA to target cells in many animal models *in vivo* and even more so in the clinic. The primary reasons for this are that pDNA–lipoplex nanoparticles are fundamentally difficult to formulate reproducibly, difficult to store long-term, and unstable with respect to colloidal aggregation and structural breakdown in biological fluids such as serum and lung mucus. Therefore, pDNA–lipoplex nanoparticles are inadequate for clinical use (Miller, 2003).

15.5.1 Extracellular Barriers to Functional Delivery

In deciding how to improve pDNA–lipoplex nanoparticles for routine *in vivo* use, a proper understanding of biological barriers to successful nanoparticle-mediated delivery of nucleic acids *in vivo* is essential. For instance, in the case of lung delivery, the main biological barrier is represented by the consequences of non-specific, destructive/inhibitory interactions between nanoparticles and mucus components. In the case of intravenous delivery, the main biological barrier is defined by the consequences of non-specific, destructive/inhibitory interactions between nanoparticles and serum components. Furthermore, there will be barriers which stand in the way of access to desired target cells in tissue. At the very least, pDNA–lipoplex nanoparticles need to be upgraded for stability in biological fluids. In addition, there are also other problems of *immune-activation* involving immune cells and *complement*. In particular, the lipoplex association of pDNA and lipids is now known to be potentially immunogenic, a problem that can, fortunately, be abrogated by creating genuinely CpG-motif-free pDNA that is reportedly much less immunogenic than wild-type pDNA and has since been used in clinical trials (Zhiady and Davis, 2006). In addition, whole pDNA–lipoplex nanoparticles can interact with opsonins and other plasma proteins, thereby provoking either the complement cascade or the reticulo-endothelial system (RES) including Kupffer cells in the liver and mononuclear phagocytic cells. Typically, nanoparticles >150 nm in dimensions are thought to be more vulnerable. Accordingly, bearing all this in mind, pDNA–lipoplex nanoparticles must be upgraded to overcome these limitations. In addition, problems brought by biological fluid and immunological barriers will also be supplemented by other structural/physical barriers to nanoparticle-mediated delivery where access to target cells in organs of interest is required. These must be resolved on a case-by-case basis.

15.5.2 Overcoming Extracellular Barriers

Fortunately, many of the extracellular biological barriers described above can be solved with reference to the *Derjaguin–Landau–Verwey–Overbeek* (DLVO) theory (Verwey and Overbeek, 1948). According to this theory, the stability of nanoparticles in solution is explained by two types of interparticle interaction

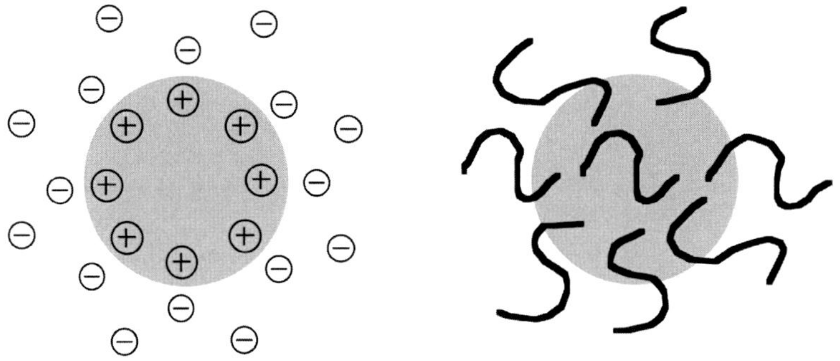

FIGURE 15.17 ■ According to DLVO theory, colloidal stability of nanoparticles is maintained when repulsive force is greater than attractive force. Stabilization of nanoparticles in solution can be achieved by electrostatic repulsion, the nanoparticles are surrounded by a double charged (Stern) layer, (*left*) or by steric repulsion, the surface of the nanoparticles is functionalized with polymer, (*right*).

force: Repulsive forces generated by electrostatic and/or steric repulsion, and van der Waals attractive forces. To maintain dispersion of particles, the repulsive forces must be greater than the attractive forces in order to prevent nanoparticles from coming into contact and aggregating them (Fig. 15.17).

In nanoparticles, there is now a common appreciation that the DVLO theory can be realized using a suitable hydrophilic, stealth/biocompatibility layer that covers nanoparticle surfaces. The most versatile and widely tested hydrophilic and stealth/biocompatible surface cover to date appears to be provided by the polymer polyethylene glycol (PEG); PEG^{2000} for systemic use and approximately PEG^{5000} for lung use. A PEG layer provides stability to nanoparticles by increasing steric repulsive forces between nanoparticles, and thus preventing them from coming into close contact with one another. In addition, PEG creates a hydrophilic layer on the nanoparticle surface, which repels plasma proteins and slows down RES uptake.

Fortunately, there are delivery problems where biology actually appears to help! For instance, nanoparticle-mediated delivery to liver or tumors can benefit from the fact that both liver and tumors can have leaky vasculature. Gaps or fenestrations between vascular endothelial cells in liver and in tumor are easily sufficient to allow nanoparticles of 100 nm or less in diameter access to cells beyond. In the case of tumor tissue, nanoparticle accumulation is said to be under the influence of the *enhanced permeability and retention* (EPR) mechanism (Maeda *et al.*, 2000). This mechanism involves a process of leaving the blood stream (*extravasation*) and inner tissue migration, assuming that nanoparticle properties are such as to allow this to happen (Fig. 15.18).

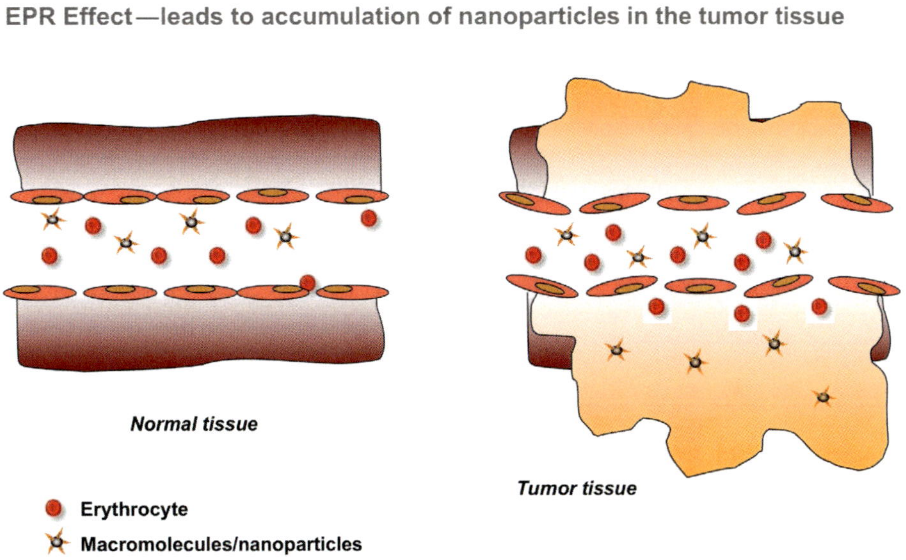

FIGURE 15.18 ■ A schematic representation of nanoparticle localization to tumor tissues by the enhanced permeability and retention (EPR) effect. Nanoparticles are unable to diffuse through normal vasculature (*left*) due to the tight junction of endothelial cells, while leaky vasculature in tumors (*right*) enhances extravasation and increases accumulation of nanoparticles, contributing to the EPR effect.

15.5.3 ABCD Nanoparticle Structural Paradigm

As noted in Section 15.5.1, pDNA–lipoplex nanoparticles are fundamentally unstable for use *in vivo* with respect to colloidal aggregation in biological fluids and immune system surveillance. The DVLO theory suggests the need for surface attachment of PEG moieties. Nevertheless there has been significant effort to determine the means to stabilize lipoplex nanoparticles for *in vivo* use without recourse to surface attachment of PEG moieties. For instance, two simple methods developed relatively recently to achieve greater nanoparticle stability *in vivo* were:

(1) Substitution of lipids comprising double-bond-containing fatty acids with lipids comprising triple-bond-containing fatty acids. The result is that lamellar fluid mesophases can be stabilized relative to the fusogenic, inverse topology hexagonal H_{II} fluid mesophase. Such a simple change in macromolecular lipid physical property does appear to stabilize pDNA–lipoplex nanoparticles to the extent that *in vivo* applications can be improved (Fletcher *et al.*, 2008).
(2) Precondense pDNA with a cationic entity, such as the cationic μ (mu) peptide of the adenovirus core, prior to mixing with cationic liposome/micelles. In this instance, the resulting liposome/μ/pDNA (LMD) nanoparticles are stable enough for lung instillation and functional delivery of pDNA *in vivo* (Tagawa *et al.*, 2002).

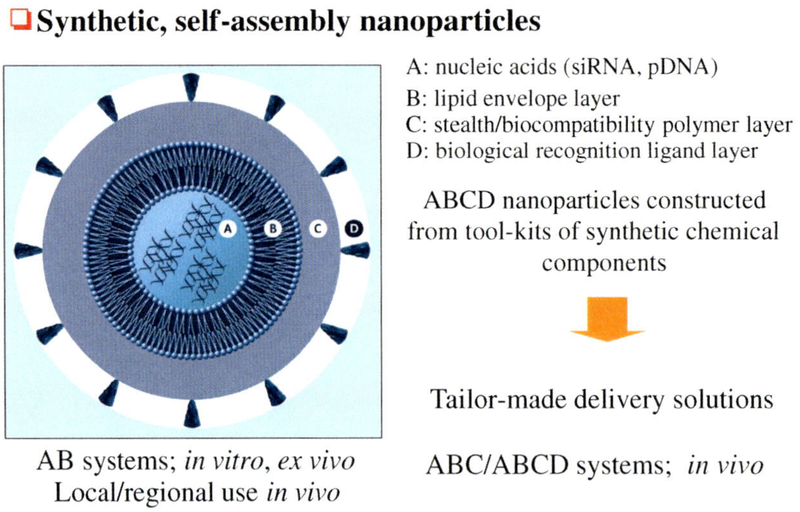

FIGURE 15.19 ■ In these nanoparticles, therapeutic nucleic acids of interest — such as pDNA (*A*) are condensed within functional concentric layers of chemical components designed for compaction/association, delivery into cells and intracellular trafficking; (*B*-components — typically lipids), biological stability; (*C*-stealth/biocompatibility components — typically polyethylene glycol [PEG]) and biological receptor targeting to target cells; (*D*-biological targeting ligand components).

However, these solutions are unlikely to be clinically applicable. Hence, there has been a growing need for a structural paradigm that defines all features necessary to ensure successful functional delivery of pDNA, or any other such nucleic acid, *in vivo*. For this reason, the *ABCD* nanoparticle structural paradigm was introduced. The essential features of lipid-based *ABCD* nanoparticles are illustrated (Fig. 15.19). The principal idea is that nanoparticles should be built up through condensation and entrapment of a therapeutic nucleic acid of interest (*A*-component) within regular, functional concentric layers of chemical components designed to overcome barriers to successful functional delivery of therapeutic nucleic acids *in vivo*. Implicit in the structural paradigm is that *AB* core nanoparticles, equivalent to lipoplex nanoparticles, are surrounded by a PEG stealth-biocompatibility polymer layer (*C*-component). Appropriate biological targeting ligands may also be added (*D*-components) to enhance further the functional delivery of therapeutic nucleic acids *in vivo*. Implicit to the *ABCD* nanoparticle structural paradigm is the high degree of nanoparticle regularity and the use of bespoke tool-kits of chemical components to set up the modular ("lego-model") self-assembly of tailor-made, *ABC* and *ABCD* nanoparticles (<100 nm in diameter, monodisperse) designed for the purpose of functional *in vivo* delivery of any therapeutic nucleic acids to any particular target cells in organs of interest. Every inch a chemical set of solutions looking for biological problems.

Examples of such *ABC* nanoparticles have now been employed for successful, functional systemic delivery of pDNA to lung and to tumors in animal models of human disease *in vivo* and have even been evaluated in clinical trials (Thanou *et al.*, 2007).

15.6 Fundamentals of siRNA Delivery

Persistent failures to reach the clinic using synthetic cationic lipid-based pDNA–nanoparticles have created some disquiet over the future of such systems (Zhu and Mahato, 2010). However, the recent discovery of non-coding RNAs (ncRNAs) and the revelations of the RNA interference (RNAi) field have pointed to an alternative, brighter future for cationic lipid-based RNAi-effector nanoparticles as a primary means of implementing RNAi-based therapeutic approaches to disease treatment and therapy.

15.6.1 siRNA, Structure and Function

Central to RNAi are small interfering RNAs (siRNAs). These are short RNA duplexes comprised of two short RNA strands assembled into A-form double helices (typically 19- to 21-nucleotide base pairs in length) with two nucleotide-base overhangs at the 3′ end of each strand. These are central actors in the RNAi pathway found in most cells of most eukaryotic organisms. RNAi is realized by sequence-specific gene silencing, in a process which offers a powerful platform for functional genomics, *in vivo* target validation and gene-specific medicines. Unlike conventional drugs or small molecules, which have a limited range of protein targets, siRNAs can be designed by sophisticated bioinformatics methods with precise nucleotide sequences complementary to the most appropriate nucleotide sequence regions of any selected target mRNA of choice (Kostarelos and Miller, 2005; Miller 2008; Kurreck, 2009). Provided that synthetic siRNAs can then be introduced into target cells in organs of interest, then sense and antisense strands become separated in the presence of Argonaute (Ago) proteins and individual antisense strand–Ago complexes become integrated into a multi-protein complex known as the RNA-induced silencing complex (RISC) (Fig. 15.20). Activated RISC uses bound antisense strand as a complementary Watson–Crick base pair recognition motif with which to sequester the target mRNA of choice that is then catalytically destroyed by phosphodiester link hydrolysis. Over the past few years, there has been considerable work done to identify those siRNAs capable of

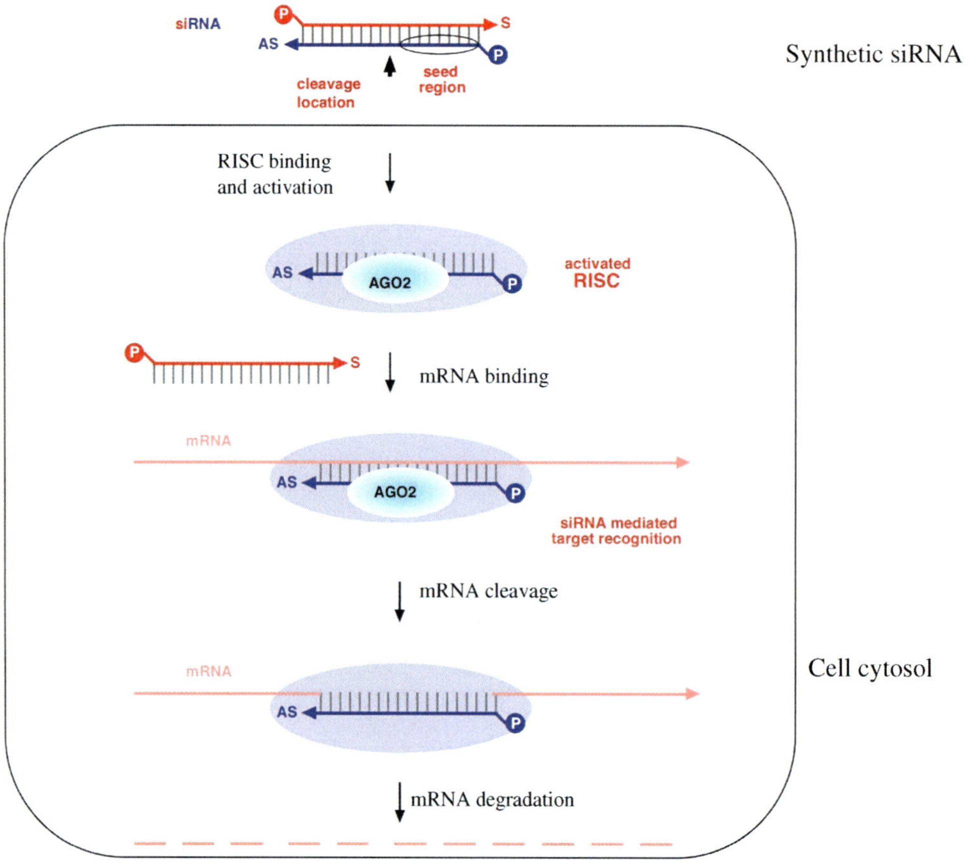

FIGURE 15.20 ■ Schematic representation of a siRNA molecule and its interfering function. An siRNA (top) has a well-defined structure: A short (usually 21 nt) double-strand RNA with 2 nt 3′ overhangs on either end (often DNA), each strand has a 5′ phosphate group and a 3′ hydroxyl group. Once introduced inside the cytoplasm of a cell, siRNA is incorporated into the RNA-inducing silencing complex (RISC) that includes an Argonaute (Ago) protein as one of its main components. Ago cleaves and discards the passenger (sense, in red) strand of the siRNA duplex leading to activation of the RISC. The remaining guide (antisense, in blue) strand of the siRNA guides RISC to its homologous mRNA, resulting in the endo-nucleolytic cleavage of the target mRNA (Kurreck, 2009). (Figure adapted from Kostarelos and Miller, 2005.)

the highest levels of specific mRNA selection and destruction, minimizing the likelihood of off-target effects (unintentional mRNA selection and destruction) and other unintended consequences such as ncRNA interference and RNA-mediated immune effects. RNAi effects take place in target cell cytoplasm. Therefore, in principle successful, functional delivery of siRNA to a target cell involves far fewer intracellular barriers than the corresponding delivery of pDNA. In particular there is no nuclear barrier problem or problems of "gene silencing". See Chapter 5 for more details.

15.6.2 Cationic Liposomes/Micelles and siRNA Delivery

Due to its relatively large molecular weight (14 kDa) and phosphodiester backbone, naked siRNA does not freely cross cell membranes. Unmodified, naked siRNAs are relatively unstable in blood and serum, are rapidly degraded by endonucleases and exonucleases. Typically, chemical modifications can be introduced into RNA duplex structures so as to enhance biological stability without adversely affecting gene-silencing activities but the problem of access to target cells remains. Given that siRNA is not pDNA, while functional delivery of siRNA has to face all the same extracellular problems as functional delivery of DNA, target cell entry and intracellular trafficking are less demanding. In fact, the basic requirement is only to deliver siRNA to RISC at levels set by the maximum reaction velocity (ν_{max}) of the Ago protein and the catalytic turnover of the RISC protein system in totality. Consequently, estimates have suggested that only 30–100 molecules of siRNA are required per cell to achieve the most efficient siRNA-mediated gene-knockdown and corresponding protein down-regulation (Mescalchin *et al.*, 2007). Another key factor concerning functional siRNA delivery is that siRNA is much smaller then pDNA and can be placed in the same class as synthetic drug molecules or smaller biopharmaceutical agents. Therefore, when it comes to siRNA delivery, viral delivery systems that so often out-perform cationic lipid-based nanoparticle delivery systems in the context of gene delivery have no direct role. Viral delivery can mediate the delivery of non-coding pDNA constructs that express ncRNAs such as short hairpin RNAi (shRNAi) effectors, but cannot deliver synthetic siRNAs. Hence, synthetic nanoparticle systems have been in great demand over the past few years for RNAi-effector delivery to cells.

Several years ago, kinetic mixing of siRNA with a cationic liposome system adapted for the purpose, resulted in the formation of siRNA–lipoplex nanoparticles (lipid:siRNA ratio 12:1, by weight) that retained many of the characteristics and properties of pDNA–lipoplex nanoparticles, including the capacity for functional delivery of the entrapped nucleic acid to cells *in vitro* (Spagnou *et al.*, 2004). Bouxsein *et al.* (2007) also demonstrated that siRNA–lipoplex nanoparticles can exhibit combinations of lamellar and inverted topology hexagonal fluid mesophases, qualitatively similar to pDNA–lipoplex nanoparticles. In addition, they showed that siRNA–lipoplexes also comprised of some other non-lamellar structures. These studies also suggested that functional siRNA–lipoplex nanoparticle formation required cationic lipid:nucleic acid molar ratios nearly an order of magnitude larger than those required to form functional pDNA–lipoplex nanoparticles (Bouxsein *et al.*, 2007). One interesting hypothesis has been evaluated that suggests that pre association of siRNA with pDNA would help siRNA form more robust lipoplex nanoparticles. A combination of pDNA (minus transgene) and siRNA admixed with cationic liposomes can be used to derive mixed siRNA/pDNA–lipoplex nanoparticles with the capacity for focused *in vivo* applications of siRNA-mediated gene knockdown in therapeutic models of arthritis and cancer (Rhinn *et al.*, 2009; Khoury *et al.*, 2006). Nevertheless, studies on siRNA–lipoplex nanoparticles have all indicated that they have relatively similar properties to pDNA-based lipoplex nanoparticles and so, in the same way, are likely to have only relatively limited use *in vivo* from the physico–chemical point of view (see Section 15.5).

Given the massive potential for successful functional delivery of siRNA, a variety of other positively charged delivery systems are also being intensively scrutinized for delivery. Protamine–antibody fusion proteins, cyclodextrin-containing polycation nanoparticles, oligoethylenimine or polyethylenimine (OEI or PEI) and atellocollagen have all been tested for their capacities to mediate siRNA delivery *in vivo*. Of these, chitosan looks promising as a delivery system in topical lung applications and atellocollagen in direct intratumoral applications. However, there has been a general preference from amongst industry leaders in this field to spearhead the drive towards the clinic starting with siRNA–lipoplex nanoparticles then upgrading into more robust cationic lipid-based siRNA–nanoparticle delivery systems that can be circumscribed by the *ABCD* nanoparticle paradigm described above (Section 15.5.3).

15.7 Delivery of siRNA *in vivo*

Currently, more than 100 reports have been published that describe *in vivo* administration of siRNA in mammals involving various routes of administration, delivery systems or experimental models (reviewed in Kurreck, 2009). Of all the carriers, formulations that fit an siRNA–*ABC* nanoparticle description are currently the most widely validated means for systemic delivery of siRNA to various organs (Table 15.2). For instance, siRNA–*ABC* nanoparticle variants, described in the literature as stabilized nucleic acid-lipid particles (SNALPs), have been used to mediate efficient delivery of anti-hepatitis B Virus (HBV) synthetic siRNAs to liver in a murine model of transient HBV replication and have also been used successfully to silence expression of an endogenous hepatic gene in non-human primates. SNALPs have also been used to inhibit apolipoprotein B (ApoB) expression for at least 11 days and reduce serum cholesterol levels in monkeys. Refined SNALPs have also been reported very recently to lower factor VII gene expression at very low doses (Semple *et al.*, 2010).

The number of active RNAi therapeutic clinical trials is also rapidly growing, in the treatment of important diseases such as age-related macular degeneration (AMD), cancer and respiratory diseases. Most of these trials involve the administration of naked siRNA, modified or unmodified. However, these are being superseded by trials involving siRNA delivery mediated by siRNA–*ABC* nanoparticle delivery systems for systemic applications (Table 15.3).

TABLE 15.2 Examples of Therapeutic Application for *in vivo* Lipid-based Nanoparticle-Mediated Delivery of siRNA via Systemic Administration

Applications	Delivery Vehicle	Target	Reference
Cancer (prostate and pancreatic models)	AtuPLEX (cationic lipid + neutral lipid + pegylated lipid)	Protein kinase N3	Aleku *et al.* (2008)
Cancer (ovarian)	LEsiRNA (Gd-lipid + cationic lipid + neutral lipid + pegylated lipid + fluorescent lipid)	Survivin	Kenny *et al.* (2011)
Viral infections (mouse model of HBV replication)	SNALP (cationic lipid + neutral lipid + pegylated lipid)	HBV RNA	Morrissey *et al.* (2005)
Viral infections (mouse model of HBV infection)	siFECTplus (cationic lipid + neutral lipid + aminoxylipid + PEG^{2000}-$[CHO]_2$)	HBV RNA	Carmona *et al.* (2009)
Inflammation (mouse model of induced arthritis)	DMAPAP + DOPE (cationic + neutral lipid)	TNF-α	Khoury *et al.* (2006)

Notes: HBV: Hepatitis B virus; TNF: tumor necrosis factor; DMAPAP: see Fig. 15.8.

TABLE 15.3 Current Clinical Trials for siRNA Therapeutics Mediated by siRNA–*ABC* Nanoparticles

Disease	siRNA Product	Target	Formulation	Delivery Mode	Company
Hypercholesterolemia	TKM-ApoB	ApoB	SNALP	IV	Tekmira
Cancer (solid tumor)	Atu27	Protein kinase N3	Atu27	IV	Silence Therapeutics
Cancer (liver)	ALN-VSP	Kinesin Spindle Protein/VEGF	SNALP	IV	Alnylam

Notes: IV: intravenous; VEGF: vascular endothelial growth factor; SNALP: stabilized nucleic acid lipid particles.

SNALP systems in particular are being used by Alnylam Pharmaceuticals, for instance, in two different siRNA therapeutic applications aimed at targeting the kinesin spindle protein or vascular endothelial growth factor (VEGF) respectively for their potential against liver tumor activities. Tekmira are also looking at SNALP systems to mediate their RNAi therapeutic approach against hypercholesterolemia. Finally, Silence Therapeutics have introduced Atu27, an alternative siRNA-ABC nanoparticle delivery system that invokes the use of a highly chemically modified variation of the typical siRNA RNAi-effector designed against protein kinase 3, a novel anti-angiogenesis target gene.

15.8 Conclusion: Next Steps in Therapeutic Nucleic Acid Delivery

The state of the field for pDNA and siRNA delivery using lipid-based *ABC/ABCD* nanoparticle-mediated delivery is poised. Proofs of concept in animal models of human disease abound. However, today's nanoparticles are unlikely to be first-choice agents for clinical use. In the development of any scientific field, proof of concept is only the first stage from developing the confidence of success until general utility can be established. Think of the simplicity of Alexander Graham Bell's first prototype telephone compared with the first commercial telephones and then the modern day mobile telephone. Currently, the field of delivery of therapeutic nucleic acids is at the working prototype stage. There should be no doubt that we will get there; it is just a matter of "when" and not "if". The next steps must be to introduce further designer chemical innovations based on currently successful prototype nanoparticles in order to achieve more efficient delivery and more widespread applications. These innovations could come with new cationic lipid designs such as those formed by pDNA-mediated detergent dimerization creating low-charge *AB* core nanoparticles from monomolecular pDNA (Chittimalla *et al.*, 2005). Alternatively, cationic lipid–pDNA ionic interactions may be replaced by non-ionic interactions making use of other bio-inspired molecular recognition systems. For instance, varieties of nucleolipids, polysaccharide glycoclusters and lipothiourea-based pDNA-nanoparticles have been proposed and in some cases investigated for future applications (Zhu and Mahato, 2010).

There is now the need for better rational frameworks of understanding to be created as the means to guide future research work. Therefore, based on the *ABCD* nanoparticle structural paradigm, future nanoparticle developments could be guided as follows. There are four main influences upon nanotechnology, namely nanoparticle *size, shape, surface* and *structure*. All of these four "S's" are important for cationic lipid-based pDNA- and siRNA-nanoparticle delivery systems. Recent work with gadolinium (Gd)-*ABC* (passive targeted) and Gd-*ABCD* (active targeted) imaging nanoparticles has shown how important all four "S's" are for the cellular labeling of tumor cells post intravenous, systemic administration of nanoparticles (Kamaly *et al.*, 2008, 2009, 2010). Size should be <100 nm and formulations need to be as monodisperse as possible (and formulations scalable). In addition, nanoparticle surfaces should be as close to neutral in charge as possible and should present a sufficient hydrophilic barrier (such as PEG) for colloidal stability and for the avoidance of immune surveillance. Finally, the *ABCD* structural paradigm appears to matter although the capacity to establish organized nanoparticles of this type depends very heavily on the molecular tool-kits of chemical components on which organized and disciplined self-assembly can take place. Almost certainly the same four "S" words *ABC*/*ABCD*, although differently parameterized, will determine the outcomes of lipid-based nanoparticle-mediated therapeutic nucleic acid delivery to other organs of interest.

At the same time, the lipid-based imaging nanoparticle work mentioned above has also suggested four other main factors of influence on lipid-based–nanoparticle-mediated therapeutic nucleic acid delivery which may well transpire to be central to success or failure in the clinic. These are the four "T's", namely *targeting*, intracellular *trafficking*, nanoparticle *triggerability* and *timing*. The results of two studies involving the cellular labeling effects of two targeted *ABCD* imaging nanoparticles in comparison to the effects of corresponding *ABC* imaging nanoparticle controls, are revealing (Kamaly *et al.*, 2009; Song *et al.*, 2009). Targeting ligands do not appear to influence nanoparticle biodistribution and nucleic acid pharmacokinetics. Instead the primary influence of targeting ligands appears to be to assist specific, highly efficient functional delivery of imaging agents and/or entrapped therapeutic nucleic acids once target cells are reached. Another issue to bear in mind is that receptor-mediated cell entry can be compromized by many non-specific effects that can be completely overwhelming (Waterhouse *et al.*, 2005; Andreu *et al.*, 2008), an observation that has led to ongoing studies intended to determine and characterize those factors that might influence non-specific enhanced cell uptake mechanisms such as nanoparticle size and heterogeneity, surface hydrophilicity and charge, not to forget other factors such as ligand conformation, points of attachment and ligand surface density.

On the other hand, collective experience with *in vivo* therapeutic nucleic delivery studies suggests that the best way to reduce the effective doses of nucleic acids and retain functional delivery is to substantially improve the efficiency of intracellular trafficking of delivered nucleic acids. This is true of both pDNA and siRNA delivery. Furthermore, efficiency of functional delivery should benefit hugely from nanoparticle triggerability (i.e., where nanoparticles are designed for stability in biological fluids but are enabled for controlled release of the particular therapeutic drug cargo at the target site(s) in response to endogenous changes in local conditions, such as pH, redox state, enzyme, temperature etc., or exogenous stimuli (Carmona *et al.*, 2009; Drake *et al.*, 2010). The primary need for triggerability arises from the fact that while PEG layers are indeed potent structural features to promote nanoparticle stability and immune system avoidance, the presence of PEG can be exceedingly refractory to efficient functional delivery of entrapped nucleic acids once delivery nanoparticles reach their target cells (Keller *et al.*, 2003b). These negative PEG effects may be mitigated in whole or in part by targeting ligands, but this fact remains to be properly established. Finally, there is the importance of timing as a contributor towards successful functional delivery of nucleic acids mediated by lipid-based *ABC* nanoparticles. Timing of delivery both in terms of disease progression and dosing regimes are certain to have substantial impacts on the efficacy of lipid-based nanoparticle delivery as well, a reality that can only be solved by extensive clinical trials in the future. The results of ongoing current clinical trials with lipid-based nanoparticles for siRNA delivery (Table 15.3) will provide important information regarding the translatability of delivery systems developed in rodents and primates. However, further technical developments taking into account the four "S's" of nanotechnology and the four "T's" of delivery will almost certainly be needed.

Overall, the rate of progress in the delivery challenge involving siRNA has been remarkably rapid and the future for RNAi therapeutics founded on lipid-based *ABC/ABCD* nanoparticle-mediated delivery looks very bright indeed. We hope that future research and development will allow the same to be said about gene therapy before very long, based upon lipid-based-nanoparticle-mediated pDNA delivery. However, although there has been excellent progress, lipid-based *ABC/ABCD* nanoparticle-mediated delivery for functional delivery of nucleic acids probably requires substantial further technical development in order to offer real clinical benefit.

For Further Reading

Original Articles

Aleku, M., Schulz, P., Keil, O., *et al.* (2008). Atu027, a liposomal small interfering RNA formulation targeting protein kinase N3, inhibits cancer progression, *Cancer Res*, **68**, 9788–9798.

Alton, E.W.F.W., Middleton, P.G., Caplen, N.J., *et al.* (1993). Non-invasive liposome-mediated gene delivery can correct the ion transport defect in cystic fibrosis mutant mice, *Nat Genet*, **5**, 135–142.

Andreu, A., Fairweather, N., Miller, A.D. (2008). Clostridium neurotoxin fragments as potential targeting moieties for liposomal gene delivery to the CNS, *ChemBioChem*, **9**, 219–231.

Argyros, O., Wong, S.P., Niceta, M., *et al.* (2008). Persistent episomal transgene expression in liver following delivery of a scaffold/matrix attachment region containing non-viral vector, *Gene Ther*, **15**, 1593–1605.

Bouxsein, N.F., McAllister, C.S., Ewert K.K., *et al.* (2007). Structure and gene silencing activities of monovalent and pentavalent cationic lipid vectors complexed with siRNA, *Biochemistry*, **46**, 4785–4792.

Carmona, S., Jorgensen, M.R., Kolli, S., *et al.* (2009). Controlling HBV replication *in vivo* by intravenous administration of triggered PEGylated siRNA-nanoparticles, *Mol Pharmaceutics*, **6**, 706–717.

Chittimalla, C., Zammut-Italiano, L., Zuber, G., *et al.* (2005). Monomolecular DNA nanoparticles for intravenous delivery of genes, *J Am Chem Soc*, **127**, 11436–11441.

Cooper, R.G., Etheridge, C.J., Stewart L., *et al.* (1998). Polyamine analogues of 3β-[*N*-(*N*', *N*'-dimethylaminoethane)carbamoyl]cholesterol (DC-Chol) as agents for gene delivery, *Chem Eur*, **4**, 137–152.

Drake, C., Aissaoui, A., Argyros, O., *et al.* (2010). Bioresponsive small molecule polyamines as non-cytotoxic alternative to polyethylenimine, *Mol Pharmaceutics*, **7**, 2040–2055.

Fellowes, R., Etheridge, C.J., Coade, S., *et al.* (2000). Amelioration of established collagen induced arthritis by systemic IL-10 gene delivery, *Gene Ther*, **7**, 967–977.

Fletcher, S., Ahmad, A., Price, W.S., *et al.* (2008). Biophysical properties of CDAN/DOPE-analogue lipoplexes account for enhanced gene delivery, *ChemBioChem*, **9**, 455–463.

Kamaly, N., Kalber, T., Ahmad, A., *et al.* (2008). Bimodal paramagnetic and fluorescent liposomes for cellular and tumor magnetic resonance imaging, *Bioconjugate Chem*, **19**, 118–129.

Kamaly, N., Kalber, T., Kenny, G., *et al.* (2010). A novel bimodal lipidic contrast agent for cellular labelling and tumour MRI, *Org Biomol Chem*, **8**, 201–211.

Kamaly, N., Kalber, T., Thanou, M., *et al.* (2009). Folate receptor targeted bimodal liposomes for tumor magnetic resonance imaging, *Bioconjugate Chem*, **20**, 648–655.

Keller, M., Jorgensen, M.R., Perouzel, E., *et al.* (2003a). Thermodynamic aspects and biological profile of CDAN/DOPE and DC-Chol/DOPE lipoplexes, *Biochemistry*, **42**, 6067–6077.

Keller, M., Harbottle, R.P., Perouzel, E., *et al.* (2003b). Nuclear localisation sequence templated nonviral gene delivery vectors: Investigation of intracellular trafficking events of LMD and LD vector systems, *ChemBioChem*, **4**, 286–298.

Kenny, G.D., Kamaly N., Kalber T.L., *et al.* (2011). Novel multifunctional nanoparticle mediates siRNA tumour delivery, visualisation and therapeutic tumour reduction *in vivo*, *J Control Rel*, **149**, 111–116.

Khoury, M., Louis-Plence, P., Escriou, V., *et al.* (2006). Efficient new cationic liposome formulation for systemic delivery of small interfering RNA silencing tumor necrosis factor α in experimental arthritis, *Arthritis Rheum*, **54**, 1867–1877.

Koynova, R., Li, W., McDonald, R. (2006). An intracellular lamellar–nonlamellar phase transition rationalizes the superior performance of some cationic lipid transfection agents *Proc Natl Acad Sci USA*, **103**, 14373–14378.

Medina-Kauwe, L.K., Xie, J., Hamm-Alvarez, S. (2005). Intracellular trafficking of nonviral vectors, *Gene Ther*, **12**, 1734–1751.

Mével, M., Kamaly, N., Carmona, S., *et al.* (2010). DODAG; a versatile new cationic lipid that mediates efficient delivery of pDNA and siRNA, *J Control Rel*, **143**, 222–232.

Morrissey, D.V., Lockridge, J.A., Shaw, L., *et al.* (2005). Potent and persistent *in vivo* anti-HBV activity of chemically modified siRNAs, *Nature Biotech*, **23**, 1002–1007.

Oliver, M., Jorgensen, M.R., Miller, A.D. (2004). The facile solid-phase synthesis of cholesterol-based polyamine lipids, *Tetrahedron Lett*, **45**, 3105–3108.

Rhinn, H., Largeau, C., Bigey, P., *et al.* (2009). How to make siRNA lipoplexes efficient? Add a DNA cargo, *Biochim Biophys Acta*, **1790**, 219–230.

Semple, S.C., Akinc, A., Chen, J., *et al.* (2010). Rational design of cationic lipids for siRNA delivery, *Nat Biotechnol*, **28**, 172–176.

Song, S., Liu, D., Peng, J., *et al.* (2009). Novel peptide ligand directs liposomes toward EGF-R high-expressing cancer cells *in vitro* and *in vivo*, *FASEB J*, **23**, 1396–1404.

Spagnou, S., Miller, A.D., Keller, M. (2004). Lipidic carriers of siRNA: Differences in the formulation, cellular uptake, and delivery with plasmid DNA, *Biochemistry*, **43**, 13348–13356.

Stewart, L., Manvell, M., Hillery, E., *et al.* (2001). Physico-chemical analysis of cationic liposome-DNA complexes (lipoplexes) with respect to *in vitro* and *in vivo* gene delivery efficiency, *J Chem Soc Perkin Trans*, **2**, 624–632.

Tagawa, T., Manvell, M., Brown, N., *et al.* (2002). Characterisation of LMD virus-like nanoparticles self-assembled from cationic liposomes, adenovirus core peptide μ (mu) and plasmid DNA, *Gene Ther*, **9**, 564–576.

Verwey, E.J.W., Overbeek, J.T.G. (1948). *Theory of the Stability of Lyophobic Colloids*, Amsterdam: Elsevier.

Waterhouse, J.E., Harbottle, R.P., Keller, M., *et al.* (2005). Synthesis and application of integrin targeting lipopeptides in targeted gene delivery, *ChemBioChem*, **6**, 1212–1223.

Reviews

Kostarelos, K., Miller, A.D. (2005). Synthetic, self-assembly ABCD nanoparticles; a structural paradigm for viable synthetic non-viral vectors, *Chem Soc Rev*, **34**, 970–994.

Kurreck, J. (2009). RNA interference: From basic research to therapeutic applications, *Angew Chem Int Ed*, **48**, 1378–1398.

Maeda, H., Wu, J., Sawa T., *et al.* (2000). Tumor vascular permeability and the EPR effect in macromolecular therapeutics: A review, *J Control Rel*, **65**, 271–284.

Martin, B., Sainlos, M., Aissaoui, A., *et al.* (2005). The design of cationic lipids for gene therapy, *Curr Pharm Des*, **11**, 375–394.

Mescalchin, A., Detzer, A., Wecke, M., *et al.* (2007). Cellular uptake and intracellular release are major obstacles to the therapeutic application of siRNA: Novel options by phosphorothioate-stimulated delivery, *Expert Opin Biol Ther*, **7**, 1531–1538.

Miller, A.D. (1998). Cationic liposomes for gene therapy, *Angew Chem Int Ed*, **37**, 1768–1785.

Miller, A.D. (1999). 'Nonviral delivery systems for gene therapy', in Lemoine, N.R. (ed.) *Understanding Gene Therapy*, Oxford: Bios Scientific Publishers, pp. 43–69.

Miller, A.D. (2003). The problem with cationic liposome/micelle-based non-viral vector systems for gene therapy, *Current Med Chem*, **10**, 1195–1211.

Miller, A.D. (2008). Towards safe nanoparticle technologies for nucleic acid therapeutics, *Tumori*, **94**, 234–245.

Miller, A.D., Tanner J. (2008). *The Essentials of Chemical Biology*, Chichester: Wiley-Blackwell.

Nicolazzi, C., Garinot, M., Mignet, N., *et al.* (2003). Cationic lipids for transfection, *Current Med Chem*, **10**, 1263–1277.

Thanou, M., Waddington, S., Miller, A.D. (2007). 'Gene Therapy', in Taylor, J.B., Triggle, D.J. (eds.), *Comprehensive Medicinal Chemistry II*, Oxford: Elsevier, pp. 297–320.

Tranchant, I., Thompson, B., Nicolazzi C., *et al.* (2004). Physicochemical optimisation of plasmid delivery by cationic lipids, *J Gene Med*, **6**, S24–S35.

Zhiady, A., Davis P. (2006). Current prospects for gene therapy of cystic fibrosis, *Curr Opin Pharmacol*, **6**, 515–521.

Zhu, L., Mahato, R. (2010). Lipid and polymeric carrier-mediated nucleic acid delivery, *Expert Opin Drug Deliv*, **7**, 1209–1226.

16

HYDRODYNAMIC-PRESSURE-BASED NON-VIRAL NUCLEIC ACID DELIVERY

Takeshi Suda[a], *Kenya Kamimura*[a], *Guisheng Zhang*[b] *and Dexi Liu*[b,c]

16.1 Definition of Hydrodynamic-Pressure-Based Gene Delivery

Hydrodynamic-pressure-based gene delivery is a method for intracellular gene delivery accomplished by a rapid injection of a large volume of DNA solution into a vasculature. It is usually used as a non-viral method to introduce genes into hepatocytes in rodents by a tail vein injection of plasmid DNA solution. Due to its physical nature, hydrodynamic delivery can also be used for intracellular delivery of RNAs, proteins and other membrane-impermeable substances.

The name of hydrodynamic delivery was chosen to reflect the combined effect of the large injection volume and high injection speed. The elevated pressure in the vasculature is usually created by a pressurized injection into a vasculature with an input volume far exceeding the exit volume from the target tissue. Hydrodynamic delivery can be performed on any vasculature as long as a catheter or needle can be placed.

[a]Division of Gastroenterology and Hepatology, Graduate School of Medical and Dental Sciences, Niigata University, Niigata 951–8510, Japan
[b]Department of Pharmaceutical and Biomedical Sciences, University of Georgia College of Pharmacy Athens, GA 30602, USA
[c]Email: dliu@uga.edu

16.2. Fundamentals of Hydrodynamic Gene Delivery

The concept of hydrodynamic gene delivery using hydrodynamic pressure as the driving force to facilitate intracellular gene delivery was established in 1999 (Liu *et al.*, 1999; Zhang *et al.*, 1999). The procedure, initially called "hydrodynamics-based transfection" (Liu *et al.*, 1999), involves a tail vein injection into a mouse of plasmid DNA solution in a volume equal to 8–10% of body weight in 5–7 s, resulting in significant level of reporter gene expression in all internal organs including lung, heart, spleen, kidney and liver. The highest level of gene expression was seen in the liver with 40% of hepatocytes transfected compared with an estimate of less than 0.1% in other organs.

Mechanistically, a bolus injection of a large volume of DNA solution into the tail vein induces cardiac congestion, accumulation of injected solution in the inferior vena cava (IVC), and an increase of intravascular pressure in this venous section (Fig. 16.1, central portion). Consequently, retrograde flow of injected DNA solution from IVC to hepatic vein pushes the pre-existing blood in liver vasculature towards the portal side, avoiding immediate mixing of DNA with blood components, and the hydrodynamic force carried by the moving solution in the vasculature enlarges the endothelial fenestrae and allows the direct access of plasmid DNA to hepatocytes. Because of the tight junctions holding the hepatocytes together and preventing pericellular passage of DNA solution, the hydrodynamic pressure works directly onto the basolateral surface of the hepatocytes and/or forces invagination of plasma membrane to allow DNA into the cytoplasm. With increase of the interior pressure of the hepatocytes due to the entry of DNA solution, and decrease of the intravascular pressure due to cardiac activity and tissue distribution of the injected DNA solution, the broken plasma membrane reseals, trapping DNA molecules inside. With time, regular blood circulation resumes, the body-wide systemic homeostasis reestablishes, and transgene products appear in transfected cells (Fig. 16.1, *right*). The unique structures of the liver including the fenestrated sinusoids, no basal membrane, its location adjacent to the IVC, low blood pressure, a large surface area of hepatocytes facing the lumen and high capacity of hepatocytes for gene expression are the primary reasons that a significantly higher transgene expression was seen in the liver, compared with that for other internal organs such as the lung, heart, spleen and kidney.

Hydrodynamic tail vein injection is well tolerated by rodents. However, effects of the procedure on animals are observed. Immediately after the injection, mice become immobile and manifest labored breathing for about 5 min. Rats react more severely and may stop breathing after the injection. Irregularity of cardiac activity, significant liver expansion and increase of serum concentration of liver

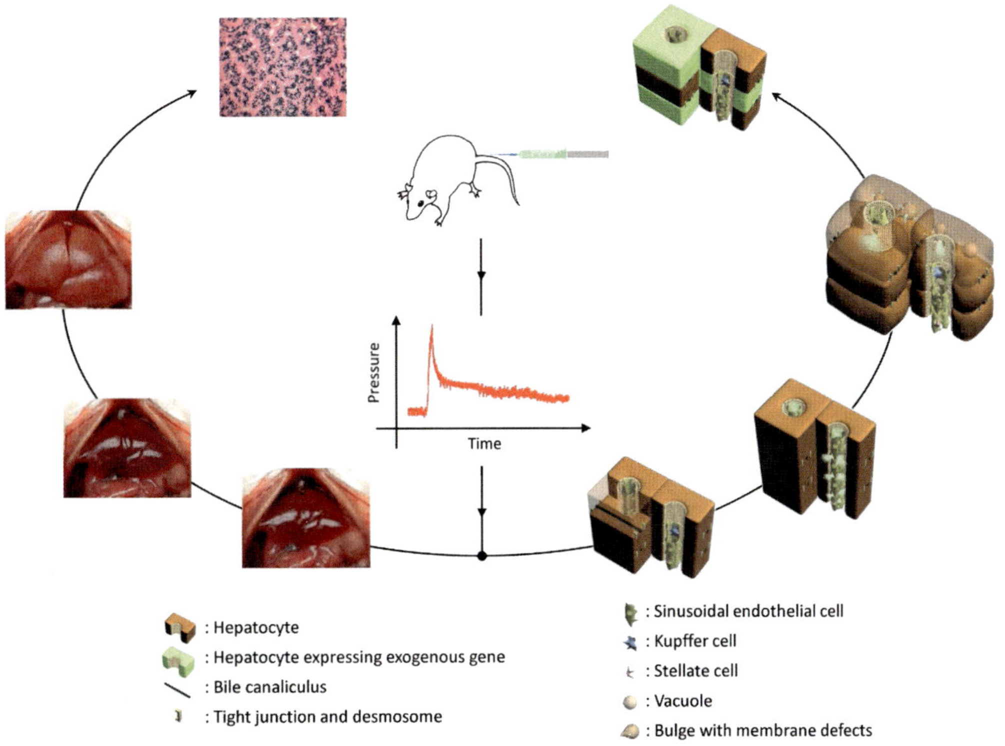

FIGURE 16.1 ■ Mechanisms of hydrodynamic gene transfer to hepatocytes. Hydrodynamic tail vein injection of DNA solution generates rapid increase in IVC pressure (*center*). Cartoon illustration (*right*) depicts the sinusoidal structure with hepatocytes surrounding the sinusoid. With hydrodynamic injection, the injected solution is forced out of liver fenestrae due to an elevated hydrodynamic pressure, generating membrane defects on hepatocytes, and resulting in DNA entry and swelling of hepatocytes. With increasing time, the solution in the cells is eliminated from the hepatocytes and gene expression occurs. The images (*left*) depict the global effect of hydrodynamic injection on the liver. Significant liver expansion was visible upon hydrodynamic injection. β-galactosidase (LacZ) staining of the liver section from a mouse hydrodynamically transfected is shown in the photograph of histochemical staining.

enzymes have been reported (Liu *et al.*, 1999; Zhang *et al.*, 2004; Suda *et al.*, 2007, Zhou *et al.*, 2010). These acute signs of the hydrodynamic impact are transient and disappear quickly. The electrocardiogram becomes normal in 90 s (Zhang *et al.*, 2004). The liver, which expanded to almost 240% of original volume, returns to its original size in 30 min (Suda *et al.*, 2007). The blood concentrations of aspartate aminotransferase (AST) and alanine aminotransferase (ALT) drop into the normal range 72 h post injection (Liu *et al.*, 1999). No apparent long-term complication was identifiable, even in animals receiving eight weekly hydrodynamic injections over a period of two months (Yang *et al.*, 2001).

Experimentally, a 27-gauge needle is commonly used for tail vein injection into mice (Liu *et al.*, 1999), whereas 20–24-gauge needles are appropriate for tail vein injection into rats, depending on body weight (Zhou *et al.*, 2010). Other sizes of needles can also be used depending on the injection volume and speed, and the type of blood vessel receiving the injection. The volume administered through the tail vein to rodents is usually 80–100 ml/kg (Liu *et al.*, 1999). However, a volume of approximately 15 ml/kg has been used for hydrodynamic gene delivery to an isolated rabbit liver (Eastman *et al.*, 2002). For maximal delivery efficiency, injections into mice should be performed as rapidly as possible, typically in 3–5 s (Liu *et al.*, 1999). Instead, rats are typically given injections at a rate of 2 ml/s using a power injector (Maruyama *et al.*, 2002), although a hand-held syringe can be used (Zhou *et al.*, 2010). The injection into isolated rabbit liver was performed at a rate of 15–20 ml/s, which can only be accomplished by a power injector. DNA amount employed ranges from 0.1 to 10 mg/kg. An optimal dose is 3 mg/kg for rats, and 7 mg/kg for rabbits. For mice, the plasmid DNA dose range is 0.5–2.5 mg/kg. While saline is the most commonly used vehicle for hydrodynamic gene delivery, other solutions have also been employed, including Ringer's solution and phosphate-buffered saline.

The transfected cells in the liver are hepatocytes located at the transition area of zones 2 and 3 to the central vein (Fig. 16.2) where the hydrodynamic injection has the highest physical effects (Zhang *et al*, 2004; Budker *et al.*, 2006). It appears that there is an anatomical watershed along zones 2 and 3 toward the central vein that is most sensitive to the mechanical force and/or pressure. The highest level of gene expression reported for a single hydrodynamic injection into the mouse tail vein was 500–1000 μg per ml of serum, and 45 μg of cellular protein per gram of the liver. It was shown that such level of gene expression was sufficient for restoring the function of blood coagulation in hemophilia mice, establishing viral infection of human virus in mice, elimination of established tumors and analysis of DNA sequences for their function in transcription regulation (for recent reviews, see Bonamassa *et al.*, 2010; Herweijer and Wolf, 2007; Suda and Liu, 2007).

Because of its physical nature, the method of hydrodynamic delivery has been utilized for intrahepatic delivery of a variety of molecules including circular DNA, DNA fragments, bacterial artificial chromosomes, RNA (single or double stranded, synthetic or genomic), oligonucleotides, proteins (antibodies, enzymes), polymers and small compounds. Modified procedures for gene delivery to muscle, kidney, heart and spleen have been reported for hydrodynamic gene delivery in different animal species (Suda and Liu, 2007; Herweijer and Wolf, 2007).

The major advantages of hydrodynamic delivery are its simplicity and reproducibility. The procedure does not require special equipment, trained skills or laborious preparations. Compared with the methods of viral and non-viral vector-based gene delivery, hydrodynamic delivery has the least probability of inducing adverse biological effects or immunogenic reactions. Neither histological nor biochemical studies revealed prolonged tissue damage to animals when proper hydrodynamic parameters were employed.

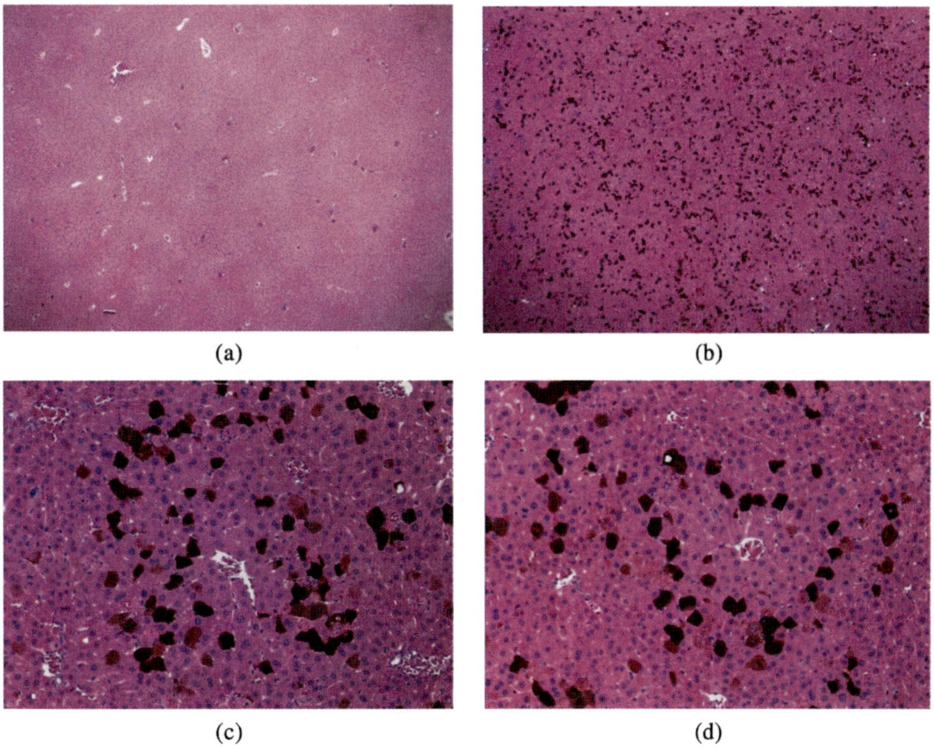

FIGURE 16.2 ■ Distribution of hydrodynamically transfected hepatocytes in mouse liver. Saline equal to 10% of body weight containing cytomegalovirus promoter driving luciferase reporter gene (pCMV-luc) plasmids was injected through the mouse tail vein either slowly (60 s) or hydrodynamically (10 s). The liver was harvested 24 h post injection and subjected to immunohistochemistry using anti-luciferase antibody followed by hematoxylin–eosin staining. (a) Slow injection resulting in no luciferase-positive cells but areas from zone 3 (the third of the acinus closest to the central vein) to zone 2 (between zones 1 and 3) exhibited light staining in comparison to zone 1 (the third of the acinus closest to the portal tract). (b) Hydrodynamic injection resulting in luciferase-positive hepatocytes locating primarily along the boundaries between light and dense hematoxylin–eosin staining. (c) and (d) Higher magnifications of (b) revealing that positive hepatocytes are surrounding the central vein. Original magnifications are 40× in (a) and (b), and 200× in (c) and (d).

16.3 Image-Guided, Computer-Assisted Hydrodynamic Delivery for Site-Specific Gene Transfer in Large Animals

Despite hydrodynamic tail vein injection being well tolerated in rodents, application of the same procedure to humans was considered less likely because, for a man with body weight of 60 kg, an injection of 6000 mL of DNA solution in less than 10 s into the blood circulation is medically unacceptable. When a large volume of solution is rapidly infused into non-compressive blood the infused solution exudes into the so-called third space, the interstitial space between cells, and forms tissue edema. If this happens in the lung, the solution accumulates in the interstitial air spaces, alveoli, and interferes with effective gas exchange, leading to pulmonary failure. Too much of an extension of the heart muscle due to volume overload causes reduction of the contractility and the insufficient coronary circulation, leading to relative ischemia of the myocardium and malfunction of cardiac conduction system.

To develop a clinically applicable procedure for hydrodynamic gene delivery, we (Suda *et al.*, 2008; Kamimura *et al.*, 2009, 2010) and others (Eastman *et al.*, 2002) have attempted to perform localized hydrodynamic gene delivery. The procedure we have been working on involves an insertion of a balloon catheter through the jugular vein to the hepatic vein of the selected liver lobe under a fluoroscope. Upon inflation of the balloon to prevent the backflow of the injected DNA solution, a large volume of DNA solution is hydrodynamically injected into the targeted liver lobe by a computer-controlled injection device employing the pressure of a CO_2 gas tank as the driving force. The injection volume and the intravascular effects are controlled by duration of injection and by the intravascular pressure detected by a pressure transducer placed through the balloon catheter near the tip of the inserted catheter. Using a reporter construct, Kamimura *et al.* (2009) have demonstrated, in pig liver, a significant level of reporter gene expression only in the targeted liver lobe with limited spillover to the non-targeted lobes. Sequential injections to two, three or four lobes of the liver in the same animals were safely performed with the same level of gene delivery efficiency as that of the injection to a single lobe. Importantly, due to direct injection to the hepatic vein, the volume required for optimal gene delivery has been reduced from about 10% body weight in rodents to less than 5% in pigs (Suda *et al.*, 2008). It was shown that the diameter of catheter used, complete blockade of backflow of injected solution and injection pressure are critical for the success of gene delivery.

In comparison to the approach where the occlusion of portal vein and/or the IVC above and below its junction to the hepatic vein is in place before hydrodynamic gene delivery to the entire liver, the procedure of image-guided, computer-assisted hydrodynamic gene delivery to the liver has a great advantage in gene delivery efficiency, minimal effect on cardio functions, better controllability and low tissue damage (Kamimura *et al.*, 2009). Our recent work in assessing the safety of this procedure on dogs and baboons confirms that the image-guided, computer-assisted hydrodynamic delivery for site-specific gene delivery to the liver is safe.

Kamimura *et al.* (2009) have also shown that the additional occlusion of blood flow in IVC or IVC plus portal vein was effective in elevating hydrodynamic pressure in the targeted vasculature but did not enhance gene delivery efficiency.

Physiological examination of pigs with IVC occlusion revealed transient decreases of blood pressure and respiration rate (Kamimura *et al.*, 2009). Removal of occlusion from IVC resulted in a rapid and transient increase in heart rate. Occlusion of the portal vein and hepatic vein showed no effect on physiological and cardiac activities. No major changes in serum composition were observed. These results suggest that blockade in IVC should be avoided for hydrodynamic gene delivery to the liver.

Regional hydrodynamic gene delivery employing the image-guided, computer-assisted approach has also been assessed for gene delivery to skeletal muscle in large animals (Kamimura *et al.*, 2010). Similar to intrahepatic gene delivery, image-guided insertion of a balloon catheter from the jugular vein to the femoral vein was performed followed by hydrodynamic injection of a reporter construct. A high level of reporter gene expression was obtained with minimal tissue damage (Kamimura *et al.*, 2010). Among the factors affecting the outcome of the hydrodynamic gene delivery to skeletal muscle in pigs, the size of balloon catheter, the injection pressure and injection volume are most critical. Immunohistochemical staining of the muscles hydrodynamically transfected revealed a 70–90% muscle cells expressing the reporter gene 60 days after a single hydrodynamic gene delivery (Kamimura *et al.*, 2010).

16.4 Future Perspectives

Hydrodynamic gene delivery using hydrodynamic pressure as the driving force has been well established as a tool for gene delivery and gene therapy studies in small animals (Suda and Liu, 2007; Herweijer and Wolf, 2007). More recent results in large animals have indicated that it is possible to apply hydrodynamic gene delivery to clinical situations. Development of image-guided, computer-assisted hydrodynamic delivery for site-specific gene transfer represents a conceptual and technological breakthrough in overcoming the cardiopulmonary problems of systemic hydrodynamic gene delivery. However, more efforts are needed to address some issues toward successful clinical application of hydrodynamic gene delivery. The most fundamentally important issue for now is a direct demonstration of the effectiveness and safety of hydrodynamic gene delivery in humans, or at least in animals that resemble the size, anatomy and physiology of humans.

The liver represents an important organ for hydrodynamic gene delivery because of its involvement in numerous genetic and metabolic diseases. The route of hydrodynamic gene delivery to the liver is, on the other hand, an important issue. Four different vasculatures can reach hepatocytes including the hepatic artery, portal vein, bile duct and hepatic vein. Injection via the hepatic artery or portal vein generates a flow of injected solution in the same direction of natural blood flow and requires blockade at the conjunction of IVC and hepatic vein in order to generate sufficient hydrodynamic pressure. In addition, a surgical intervention is needed to expose the portal vein for needle insertion at an extrahepatic site. Rapid injection and/or balloon occlusion at the hepatic artery could cause endothelial damage, leading to a permanent vascular obstruction. Taking these problems together, hydrodynamic gene delivery through portal vein or hepatic artery may not be the best choice. The injection from the bile duct, successfully performed on rats, requires endoscopic retrograde cholangiography in humans, which is a procedure known to have a high risk of a complication including life-threatening pancreatitis. It has been reported that the complication rate of endoscopic retrograde cholangiography is approximately 5%.

An ideal path for hydrodynamic gene delivery to the liver appears to be from the hepatic vein in retrograde. Firstly, it is rare to induce severe vascular damage in venous systems, based on interventional radiology techniques. The natural blood flow from the portal vein and hepatic artery to hepatic vein provides resistance to injected DNA solution and helps in building hydrodynamic pressure. In addition, the approach of lobe-specific and regional hydrodynamic injection through a hepatic vein affects part of the liver leaving the non-injected part to sustain the regular liver functions or for liver regeneration if a complication occurs. Furthermore, the well-established interventional radiology technique provides the essential accuracy for localizing the balloon catheter for hydrodynamic gene delivery. Thus, regional hydrodynamic gene delivery through hepatic vein is the most promising route to target the liver.

Regional hydrodynamic gene delivery can be used for gene delivery to organs other than the liver. It was shown that hydrodynamic gene delivery to kidney and muscle worked well in pigs (Suda *et al.*, 2008; Kamimura *et al.*, 2010). Due to their unique endothelial structure (continuous endothelium versus fenestrae of the liver), a higher hydrodynamic pressure is needed for successful gene delivery. Applications of hydrodynamic gene delivery to other organs/tissues are also feasible once the optimal hydrodynamic parameters are established.

One of the significant findings from large-animal studies is that the level of gene expression obtained is at least 100-fold lower than that of mice using a standard hydrodynamic tail vein injection (Kamimura *et al.*, 2009). While efforts are ongoing to improve the efficiency, different hypotheses have been proposed to explain the underlying mechanisms. The overall effect of the hydrodynamic flow on the hepatocytes is of major interest. When performed on rodents, the whole liver is affected more homogeneously, with only the portal vein as the leaky site. Since the liver is an isotropic elastic body, the physical effect that is evoked by the hydrodynamic pressure is likely to be more homogenous with maximal gene delivery efficiency. As expected, the delivery efficiency is a function of the injection volume and speed. In contrast, when localized hydrodynamic injection is performed targeting part of the liver in large animals, the injected solution readily flows out from the target area through physiological vascular connections of the meshwork of veins, arteries and portal veins, resulting in lower impact on the liver cells and lower gene delivery efficiency. Practically, it is difficult to target the entire cell population for maximal gene delivery efficiency in organs such as the liver with enriched vasculature. Obstruction of the portal vein or portal vein plus IVC did not seem to help (Kamimura *et al.*, 2009), suggesting that the interlobular leakage is likely to be responsible for the reduced hydrodynamic impact at the distant sites from the site of injection. In addition, different gene delivery efficiency could be due to species differences. For example, liver compliance, as measured by surface indentation, is much lower in the pig when compared with that of mouse, rat, dog and man (Carter *et al.*, 2001). The higher content of fiber proteins in pig liver (Kamimura *et al.*, 2009) and rodents with liver fibrosis (Yeikilis *et al.*, 2006) reduced the effectiveness of hydrodynamic gene delivery. Other species-related differences such as sinusoidal structure, endothelial strength against hydrodynamic pressure and the structure of plasma membrane of hepatocytes could also play an important role in determining the efficiency of hydrodynamic delivery. Evidently, more studies are needed in large animals to identify factors that affect the hydrodynamic effects in favor of high gene delivery efficiency. Hopefully, the new information generated by additional work will be able to guide the development of improved technology for hydrodynamic gene delivery. Ultimately, it is likely that precise parameters for optimal hydrodynamic gene delivery in man can be defined only in clinical trials.

Although it was proven that the principles of hydrodynamic gene delivery can feasibly be adapted to gene delivery to various organs in large animals, future studies need to focus on illustration of the processes of endothelial and membrane permeabilization and their functional and structural recovery. For liver gene delivery, more work is necessary to study the mechanism underlying the heterogeneous effects of the procedure on the liver sinusoids and hepatocytes. Towards clinical applications, it is required to establish a proper interventional radiology technique for regional application. It is also desirable to establish an ideal time–pressure curve for target organ, and to develop a real-time injection control system compliant with good clinical practice. Evidently, further endeavor has to involve a wide variety of sciences and technology including engineering, computer science, medicine and molecular biology in order to convert an effective tool for basic research into a clinically applicable procedure for treatment of human diseases.

References

Bonamassa, B., Hai, L., Liu, D. (2010). Hydrodynamic gene delivery and its applications in pharmaceutical research, *Pharm Res*, **28**, 694–701.

Budker, V.G., Subbotin, V.M., Budker, T., *et al.* (2006). Mechanism of plasmid delivery by hydrodynamic tail vein injection. II. Morphological studies, *J Gene Med*, **8**, 874–888.

Carter, F.J., Frank, T.G., Davies, P.J., *et al.* (2001). Measurements and modeling of the compliance of human and porcine organs, *Med Image Anal*, **5**, 231–236.

Eastman, S.J., Baskin, K.M., Hodges, B.L., *et al.* (2002). Development of catheter-based procedures for transducing the isolated rabbit liver with plasmid DNA, *Hum Gene Ther*, **13**, 2065–2077.

Herweijer, H., Wolff, J.A. (2007). Gene therapy progress and prospects: hydrodynamic gene delivery, *Gene Ther*, **14**, 99–107.

Kamimura, K., Suda, T., Xu, W., *et al.* (2009). Image-guided, lobe-specific hydrodynamic gene delivery to swine liver, *Mol Ther*, **17**, 491–499.

Kamimura, K., Zhang, G., Liu, D. (2010). Image-guided, intravascular hydrodynamic gene delivery to skeletal muscle in pigs, *Mol Ther*, **18**, 93–100.

Liu, F., Song, Y., Liu, D. (1999). Hydrodynamics-based transfection in animals by systemic administration of plasmid DNA, *Gene Ther*, **6**, 1258–1266.

Maruyama, H., Higuchi, N., Nishikawa, Y., *et al.* (2002). High-level expression of naked DNA delivered to rat liver via tail vein injection, *J Gene Med*, **4**, 333–341.

Suda, T., Gao, X., Stolz, D.B., *et al.* (2007). Structural impact of hydrodynamic injection on mouse liver, *Gene Ther*, **14**, 129–137.

Suda, T., Liu, D. (2007). Hydrodynamic gene delivery: its principles and applications, *Mol Ther*, **15**, 2063–2069.

Suda, T., Suda, K., Liu, D. (2008). Computer-assisted hydrodynamic gene delivery, *Mol Ther*, **16**, 1098–1104.

Yang, J., Chen, S.P., Huang, L., *et al.* (2001). Sustained expression of naked plasmid DNA encoding hepatocyte growth factors in mice promotes liver and overall body weight, *Hepatology*, **33**, 848–859.

Yeikilis, R., Gal, S., Kopeiko, N., *et al.* (2006). Hydrodynamics-based transfection in normal and fibrotic rats, *World J Gastroenterol*, **12**, 6149–6155.

Zhang, G., Budker, V., Wolff, J.A. (1999). High levels of foreign gene expression in hepatocytes after tail vein injections of naked plasmid DNA, *Hum Gene Ther*, **10**, 1735–1737.

Zhang, G., Gao, X., Song, Y.K., *et al.* (2004). Hydroporation as the mechanism of hydrodynamic gene delivery, *Gene Ther*, **11**, 675–682.

Zhou, T., Kamimura, K., Zhang, G., *et al.* (2010). Intracellular gene transfer in rats by tail vein injection of plasmid DNA, *AAPS J*, **12**, 692–698.

17

ELECTROTRANSFER/ELECTROPORATION FOR NON-VIRAL NUCLEIC ACID DELIVERY

Pascal Bigey[a,c]***, Richard Heller***[b,d] ***and Daniel Scherman***[a,e]

17.1 History and Principles of Electrotransfer

17.1.1 History

Efficient and safe *in vivo* DNA delivery is a major requirement for gene therapy or for the study of gene function. *In vitro* gene transfer of plasmid DNA has been solved reasonably by means of cationic lipid or polymer transfection, calcium phosphate precipitation or electroporation. Conversely, efficient *in vivo* gene transfer appears much more difficult to achieve, since each delivery system has specific drawbacks and limitations. While viral vectors present higher cell transduction efficiency, non-viral methods using plasmid DNA remain attractive for a variety of reasons: They are easier and cheaper to produce and handle, tissue-specific in some cases and they have no DNA insert size limitations. In addition, unlike viral vectors, plasmid DNA does not directly stimulate the acquired immune system, allowing non-viral gene therapy treatments to be periodically re-administered.

Among the variety of non-viral delivery methods currently under investigation, *in vivo* DNA electrotransfer has proven to be one of the most effective. Several other terms have been used to designate the same technique: *electrotransfection,*

[a]Laboratory of Chemical and Genetic Pharmacology and of Biomedical Imaging, Paris Descartes Pharmacy University, CNRS, Inserm, Chimie ParisTech, 4, avenue de l'Observatoire Paris Cedex 06, France
[b]College of Health Sciences, Frank Reidy Research Center for Bioelectrics, Old Dominion University, 830 Southampton Avenue, Suite 5100, Norfolk VA 23510, USA
[c]Email: pascal.bigey@parisdescartes.fr; [d]Email: rheller@odu.edu; [e]Email: daniel.scherman@parisdescartes.fr

DNA electroporation, electrogenetherapy. This technology is based on plasmid DNA injection into a tissue, followed by the application of a defined set of electric pulses that induce DNA cellular entry. Since the 1982 publication by Neumann *et al.*, *electropermeabilization* has been used to introduce small molecules or nucleic acids into prokaryotic and eukaryotic cells first *in vitro*, then subsequently *in vivo* (Potter, 1988; Titomirov *et al.*, 1991; Heller, 1996).

Electrotransfer can be applied to almost any tissue of a living animal and potentially human, including skeletal or smooth muscle, tumors, skin, liver, kidney, artery, retina, cornea or even brain. A surprisingly high efficiency of gene transfer *in vivo* was observed in muscle by the groups of Scherman and Mir (Mir *et al.*, 1998, 1999) and by others (Aihara *et al.*, 1998; Mathiesen, 1999). Skeletal muscle has consequently been the major tissue of choice for therapeutic and vaccination purposes, but many investigators have also explored the advantages of delivery to tumors (Rols, 1998). The efficiency of this electroporation approach has led to clinical trials, mainly in the field of DNA immunization or immunotherapy against infectious diseases and cancer.

17.1.2 Principle and Mechanism of Electropermeabilization and Electrotransfer

Figure 17.1 represents the different types of electric pulses that can be delivered by commercial electropulsators. Exponential pulses are often used for *in vitro* electrotransfection, with a time constant dependent on the resistance of the incubation

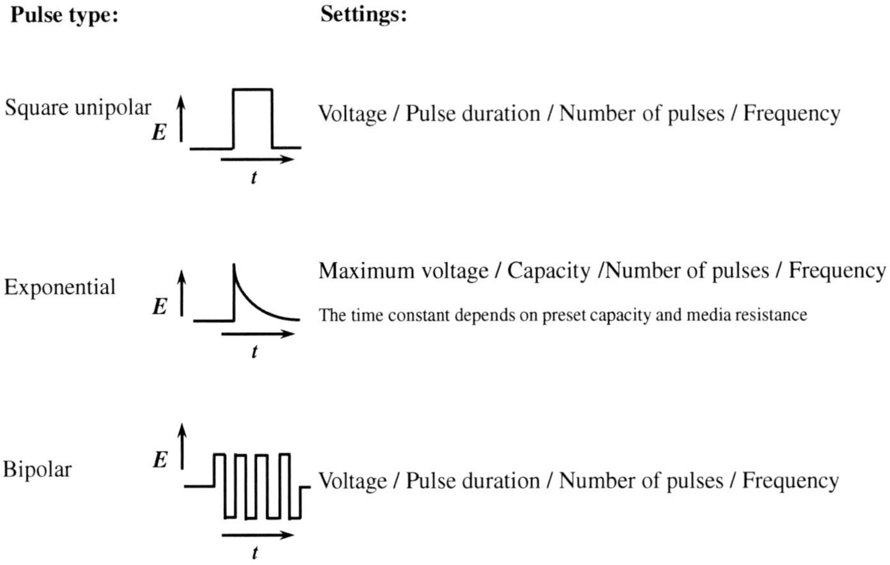

FIGURE 17.1 ■ Pattern of applied electric pulses.

media. Unipolar square wave electric pulses are preferred for *in vivo* experiments, since the voltage and duration of the pulses can be set independently of the electrical resistance of the tissue. Square wave bipolar pulses are used for electrophysiology, even if they have proved efficient for electrotransfer.

When a cell is submitted to an external electric field, and because of the ionic conductance of the intracellular and extracellular medium which are separated by the non-conducting cellular membrane, a surface charge density is observed at the cell membrane (Fig. 17.2). The ionic charge density is negative on the anode side of the cell and positive on the cathode side of the cell. The charge density is of maximal absolute value at sites where the membrane is perpendicular to the external electric field and null where the cell membrane is parallel to this field.

The ionic charge density at the cell membrane induces a transmembrane potential ΔV_m, whose value depends on the strength of the electric field E_{ext} and on the angle "q" between the electric field and the axis perpendicular to the cell membrane:

$$\Delta V_m = E_{ext}\, 1.5\, r \cos q \text{ (Laplace law).}$$

When ΔV_m exceeds about 0.2 V, this transmembrane potential causes a disorganization of the biological phospholipid bilayer and its concomitant permeabilization.

It has to be noted that this transmembrane potential is superimposed onto the constitutive negative transmembrane potential present in all cells in their basal state. Thus, a more significant depolarization and electropermeabilization is observed on the anode side, where the two negative values of the constitutive and electro-induced transmembrane potential are added. This has been confirmed by ultrafast videomicroscopy at the single cell level (Golzio *et al.*, 2002).

Notably, the surface charge density of ionic charges which accumulate at the membrane is proportional to the ratio of the total amount of intracellular ions (which is correlated to the cell volume) divided by the cell surface. Hence, the cell membrane surface charge density and resulting ΔV_m also depend on the cell dimension or radius "r". The crucial consequences of this "radius" effect are numerous and give clues to the efficiency of the technology.

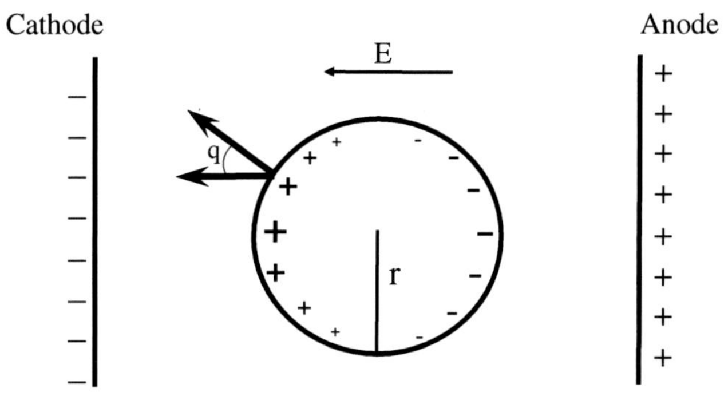

FIGURE 17.2 ■ Ionic charge density inside the cell membrane caused by an external electric field.

- Large cells are permeabilized at lower and potentially less toxic electric field strength, which explains the exceptional efficiency of electrotransfer on myofibers; also, myotubes are permeabilized at lower field strength than their stem cells (the so-called "satellite cells"), which is advantageous for repair. Indeed, skeletal muscle cells are electrotransferred using electric fields of 100–200 V/cm, while 500–1000 V/cm are used for cancer cells, and 5000 V/cm for bacteria.
- Small intracellular compartments are not permealized. This spares mitochondria whose function depends on the internal mitochondrial membrane polarization, and also avoids leakage from lysosomes, which could harm and potentially kill the electrotransferred cells.
- Extracellular matrices are not affected, which also contributes to adequate tissue repair of damage occurring to cells which have undergone excessive permeabilization. The absence of scar and exquisite tissue repair after electrotransfer has been confirmed on various species, including dogs.

It is generally admitted that membrane destabilization and the resulting permeability to hydrophilic external molecules is necessary for DNA electrotransfer to occur. Such permeabilization have been evaluated by measuring the uptake by muscles cells of a small hydrophilic complex of ethylenediaminetetra acetate (EDTA) to chromium[51] (Mir *et al.*, 1999). As shown in Fig. 17.3(a), the cell membrane permeabilization to Cr-EDTA was similar whether it was injected 30 s before or 30 s after delivery of pulses.

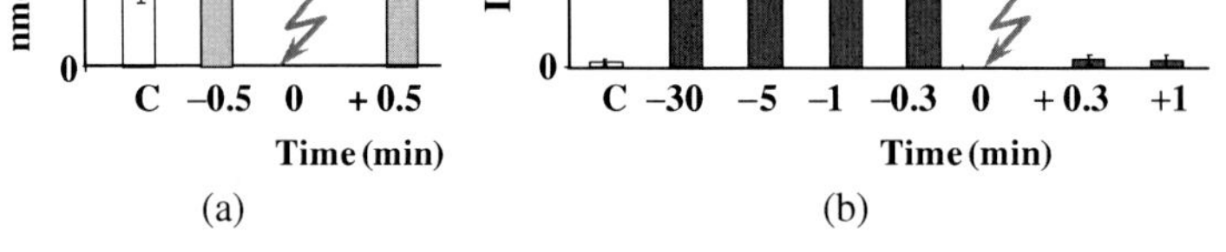

FIGURE 17.3 ■ Effect of DNA injection time. In (a), the uptake of the Cr-51/EDTA complex is identical whether electropermabilization is performed shortly before, or after the administration of Cr-51/EDTA, which indicates that the permeabilization effect is durable for at least 30 s. In (b) the expression of the luciferase transgene is only observed when plasmid DNA is present during the electric field delivery. This means that the electric field needs to exert a direct effect on DNA (i.e., electrophoresis) for gene transfer to occur. In coherence with that result, the efficacy of electrotransfer is strictly proportional to the duration of the pulses (data not shown). C: control.

In contrast, DNA strictly needs to be present during the delivery of electric pulses for transfection to occur (Fig. 17.3(b)). This demonstrates the direct electrophoretic effect of the electric field on the DNA molecule during the electrotransfer process. Thus, DNA transfer by electric pulses appears to be a two-component phenomenon requiring not only cell "permeabilization", but also the active electrophoresis of DNA during the electric pulse stimulation. This is shown schematically in Fig. 17.4.

This electrophoretic component of DNA electrotransfer has been demonstrated *in vitro* by an experiment in which transfection efficiency on a cell in monolayer was found to vary depending on whether the electric field applied had a polarity inducing DNA electrophoresis toward the cells or away from the cells (Klenchin *et al.*, 1991). Electrophoresis of DNA has several possible effects, such as promoting the movement of DNA into the permeabilized cell or favoring the insertion of DNA in a membrane destabilized by an electric field.

The association of permeabilization and electrophoresis of DNA during electrotransfer has been highlighted by studying the combination of a low-voltage non-permeabilizing pulse of long duration (electrophoretic) and of a high-voltage permeabilizing pulse of short duration (permeabilizing pulse) (Fig. 17.5). Only the primary permeabilization pulse followed by the electrophoretic pulse led to highly efficient gene transfer (Bureau *et al.*, 2000). This indicates that the electrophoretic effect of a low-voltage pulse can contribute to an efficient transfection only if the cell membrane has been previously destabilized by a permeabilizing pulse. The importance of cell permeabilization was also studied by magneticresonance imaging with a contrast agent gadolinium complex, Gd-diethyltriaminepentaacetic acid (Gd-DTPA). Results show that the zone of permeabilization to the Gd-DTPA complex is similar to the zone of expression of

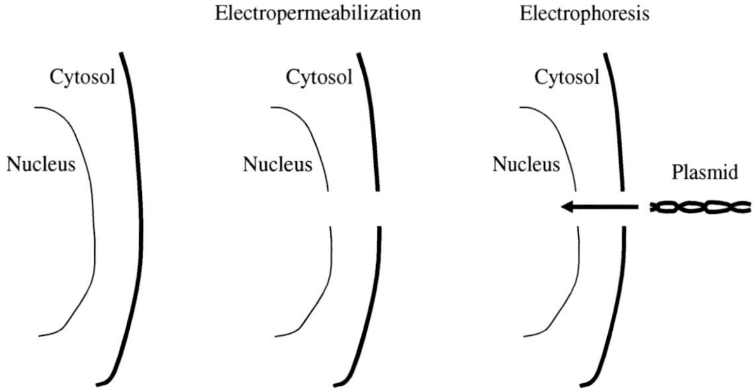

FIGURE 17.4 ■ Two-component mechanism of DNA electrotransfer mediated by the external electric field: electropermealization allows the transmembrane crossing of hydrophilic molecules such as plasmid DNA, and electrophoresis promotes DNA migration towards and within the cell to the nucleus.

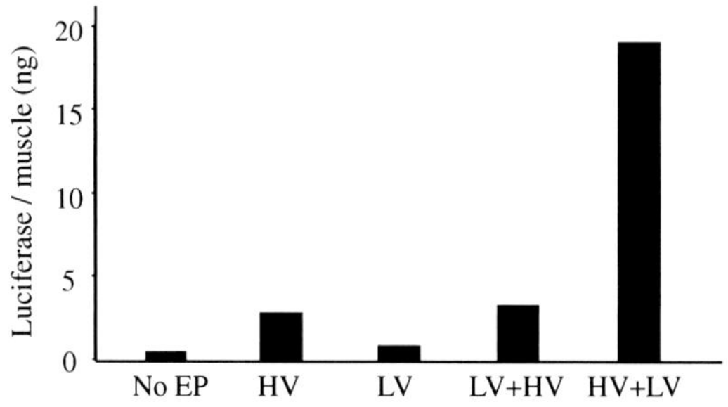

FIGURE 17.5 ■ The two-component mechanism of DNA electrotransfer is revealed by varying the succession of permeabilizing and electrophoretic pulses. HV: Permeabilizing high-voltage pulse of 800 V/cm, 100 ms. LV: Electrophoretic low-voltage pulse of 80 V/cm, 100 ms. No EP: absence of electric pulses. Only HV followed by LV is successful, i.e., permeabilization followed by electrophoresis.

an electrotransferred plasmid encoding β-galactosidase (Paturneau-Jouas *et al.*, 2003). The mechanism of *in vivo* electrotransfer and the association between permeabilization and electrophoresis have been further detailed in Satkauscas *et al.* (2002) and Bigey *et al.* (2002).

17.1.3 Factors of Interest for a Successful Electrotransfer

In addition to the characteristics of the electric pulses, several other factors might be critical for optimizing the efficacy of electrotransfer, thus enabling clinical applications.

17.1.3.1 *Improvement of plasmid biodistribution*

Efficiency of plasmid electrotransfer into muscle is high in young and small rodents, but it is decreased in larger species, possibly due to the level of connective tissue present in the muscle. Indeed, one limitation to plasmid gene transfer is the access of the plasmid to the muscle fiber surface, which is prevented by the extracellular matrix containing hyaluronan and collagen. The pretreatment of mice skeletal muscle with hyaluronidase results in a significant increase both in total transgene expression and in the proportion of transfected fibers. Moreover, this improved protocol allows a reduction in the voltage that needs to be applied for high-efficiency gene transfer, thus leading to a substantial reduction in damage to the muscle.

In addition, injected DNA diffuses in the muscle tissue: Some is lost in blood circulation, and the other part is localized in interstitial spaces where degradation can occur. When a small amount of plasmid DNA is used, the addition of

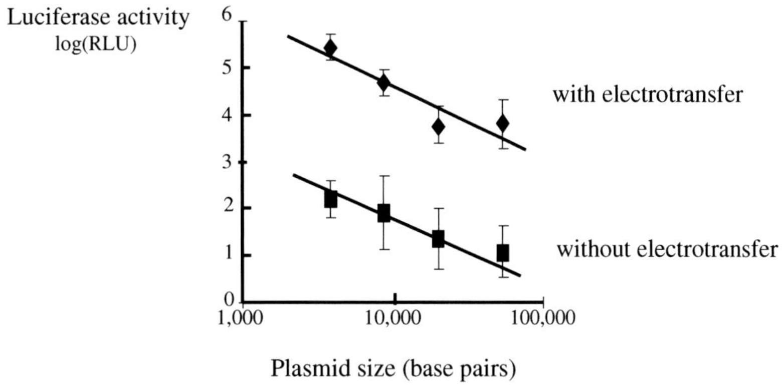

FIGURE 17.6 ■ The level of electrotransfer into tibial cranial skeletal muscle is inversely dependent on plasmid size and strongly increased by electrotransfer (more than 2-log, i.e., 100-fold). RLU, relative log units.

non-coding DNA to the plasmid of interest leads to improved transfection efficiency, presumably by saturating extracellular or intracellular DNAse.

17.1.3.2 *Optimization of plasmid structure*

An alternative form of supercoiled DNA molecule for non-viral gene transfer, named minicircles, has been developed for gene therapy. Minicircles have neither bacterial origins of replication nor antibiotic resistance markers (see Chapter 13). The *in vivo* injection into mouse cranial–tibial muscle, or into human head and neck carcinoma grafted in nude mice, resulted in 13–50 times more reporter gene expression with minicircles than with the unrecombined plasmid or larger plasmids. Histological analysis in muscle showed there were more transfected myofibers with minicircles than with unrecombined plasmid. In order to further investigate the mechanism of the increase in gene transfer efficiency observed with minicircles, we constructed plasmids ranging from 900 to 52,500 bp. These plasmids bore the same luciferase reporter gene cassette but had prokaryotic backbones of various sizes. By intramuscular naked DNA injection, a drastic decrease in transgene expression was observed when the size of the plasmid increased. This was also observed in the context of electrotransfer (Fig. 17.6) (for review see Bloquel *et al.*, 2004b).

17.1.3.3 *Design of molecules for electrotransfer*

Modifications can be carried out on the transgene in order to improve the stability of the produced protein, or its bioactivity. Such a construction was used to produce a dimeric erythropoietin (Epo) protein. In this study, the dimeric form has proven more active *in vitro* on erythroid cells and *in vivo* in electrotransferred mice than the monomeric form. Even if kinetics of monomeric and dimeric proteins were the same with a rapid decrease, this improved effect of the dimer is due to a higher binding affinity because of the presence of two molecules of Epo.

Similar results have been obtained by using a dimer of tumor necrosis factor (TNF)-α soluble receptor, which has a higher affinity for the trimeric TNF-α than the natural occurring monomeric TNF-α soluble receptor. The stability of the produced protein and time of residence in the blood circulation can be increased by fusing an immunoglobulin constant fragment to the gene of interest. Such a construct, made with the TNF-α soluble receptor gene, leads to a fusion protein produced and still secreted by the electrotransferred muscle and, above all, much more stable. This chimeric protein is bioactive, as shown by electrotransfer of plasmid encoding this protein in a model of lipopolysaccharide (LPS) systemic inflammation: A clear reduction in the amount of circulating TNF-α and interleukin (IL)-1β is observed. It has also proven efficient in collagen-induced arthritis on clinical and histological scores of the disease (Kim *et al.*, 2003; Bloquel *et al.*, 2004a). The same kind of protein was constructed by linking the viral IL-10 gene to the non-cytolytic immunoglobulin Fc portion. The use of this fusion cytokine led to a peak serum concentration 100-fold higher than with the non-fusion vIL-10 expression plasmid. Thus, the construction of fusion proteins appears to be a simple way to deliver enhanced levels of secreted proteins without altering their biological activities (Adachi *et al.*, 2002).

17.2 Animal Studies

Electroporation is a physical approach and as such almost any tissue of a living animal could be targeted to deliver DNA by this method. Examples are given in Table 17.1.

The skeletal muscle has been the most utilized target since:

- it constitutes a large and easily accessible volume of tissue in which DNA electroporation is very efficient,
- muscle fibers have a long lifespan as they are post-mitotic, potentially allowing long-term expression (more than a year) in transfected cells (in the absence of regeneration due to injury or cytotoxic immune response). This latter point is of great importance since electroporation does not lead to DNA integration, but rather to an episomal state of the plasmid,
- skeletal muscle is made up of thousands of cylindrical muscle fibers bound together by connective tissue through which run blood vessels and nerves. This

TABLE 17.1 Target Tissues for DNA Electroporation

Target Tissue	Species
Bladder	Rat
Brain	Mouse, chicken, *Xenopus*, bee, ferret
Carotid artery	Rabbit
Ciliary muscle	Rat, rabbit
Cornea	Mouse, rat
Diaphragm	Rat
Embryo	Chick, mouse
Heart	Pig, rat
Joint	Mouse, rat
Kidney	Rat
Liver	Mouse, rat
Lung	Mouse
Ovary	Mouse
Retina	Mouse, rat
Skeletal muscle	Mouse, rat, rabbit, guinea pig, pig, sheep, goat, cattle, non human primate, dog
Skin	Zebrafish, calf
Spinal cord	Mouse, rat
Spleen	Rat
Testis	Rat
Tumor	Mouse, rat

constitutes an abundant blood vascular system and skeletal muscle is therefore able to produce secreted proteins with functional post-translational modifications which can easily reach the blood circulation, including large proteins such as monoclonal antibodies, and

- cotransfection of multiple unlinked genes can be easily performed by electroporation.

Recently, a number of investigators have started to utilize skin as a target for delivery, since:

- it constitutes an easily accessible tissue in which DNA electroporation can be performed in a non-invasive manner,
- skin can be easily monitored,
- delivery can induce localized or systemic expression,
- skin contains a large number of antigen-presenting cells making it an excellent target for DNA vaccines or immunomodulating applications,
- like skeletal muscle, the cotransfection of multiple unlinked genes can be easily performed by electroporation.

17.2.1 Delivery Principle

In vivo DNA electrotransfer is a very simple physical technique for gene delivery into various mammalian tissues, which consists of injecting plasmid DNA into a targeted tissue and applying a series of electric pulses. Practically, a plasmid solution in isotonic saline (NaCl 150 mM) is injected into the targeted tissue through a syringe, and electric pulses are then delivered by means of two or more electrodes placed on each side or around the injection site (electrodes can be either needles or plates). This is illustrated in Fig. 17.7 for the skeletal muscle of an anesthetized mouse. The electroporation efficiency greatly depends on the electrical parameters. The optimal conditions result from a compromise between an efficient plasmid transfer and a minimal toxicity of the electric field. The potential toxicity may involve different parameters, but permeabilization is a main factor, since the external media diffuses into cells and modifies their internal media composition.

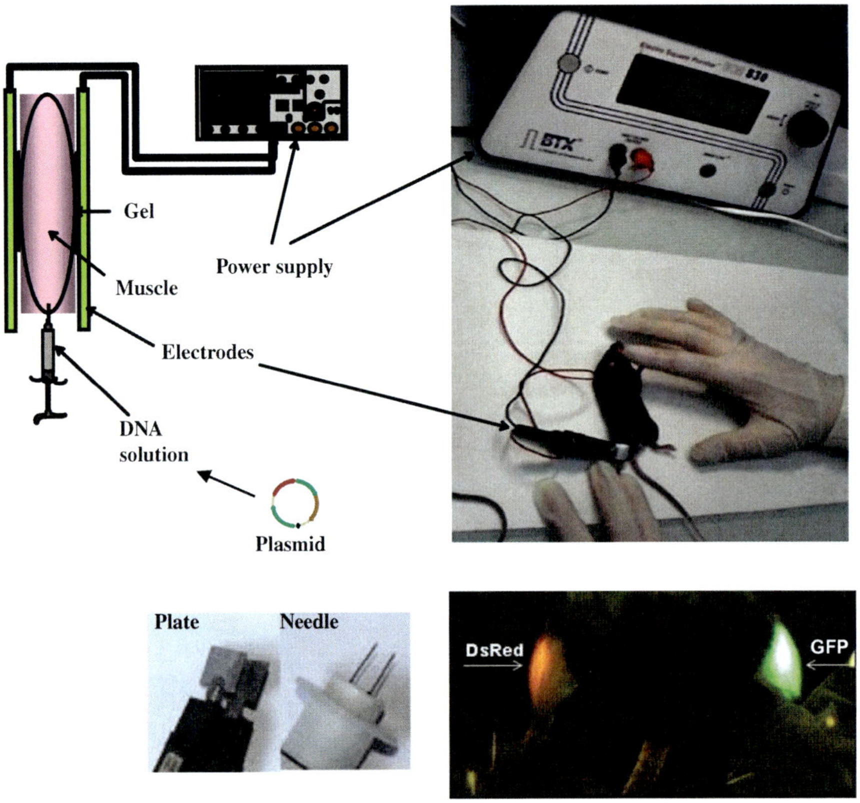

FIGURE 17.7 ■ Practical intramuscular DNA electroporation shown on an anesthezied mouse. Below are shown plate electrodes suitable for mouse and rat, needle electrodes used for larger animals and the expression in a mouse at day seven of the reporter genes green fluorescent protein (GFP) (right) and *Discosoma* red (DsRed) (left).

Internal medium may also leak out of the cell. The electric fields are applied through electrodes configured for the specific tissue in which the delivery is being performed in. The shape of the electrodes will influence the electrical parameters needed for efficient delivery. Plate electrodes have been the electrode configuration of choice for mouse or rat skeletal muscle as the procedure can be done in a non-invasive manner, while needle electrodes are more suitable for larger animals to better target a specific area of a large muscle and avoid the necessity of applying too high a voltage (across a large area of tissue) that could lead to deleterious effects to the tissue. For skin, plate electrodes, arrays of non-penetrating electrodes or needles (short or microneedles) are typically used. To restrict delivery to the skin, it is preferable to use plates or non-penetrating electrodes. For non-penetrating electrodes, the electroporation parameters needed for efficient delivery change when moving between species, particularly when delivery is being attempted to thicker tissue (larger animals).

As specified above, muscle is an excellent target for DNA gene transfer applications. Indeed, the persistence of DNA in an episomal state for months and the ability of skeletal muscle to secrete proteins allow multiple therapeutic approaches such as direct gene transfer for muscle disorders, DNA vaccination or systemic delivery of therapeutic proteins to be considered. Histological studies of muscle after electrotransfer treatment showed that muscle damage is maximal within the first seven days. In central areas affected by electroporation, tissue damage is mostly characterized by muscle lesions containing necrotic myofibers and is heavily populated with inflammatory cells. Shortly after treatment, the muscles start to regenerate from satellite cells and by three weeks after the electrotransfer experiment the muscle has completely recovered (normal structure of muscle fibers, lesions no longer evident). Cell necrosis and decreased gene expression have been reported at excessive electric field intensities, but this can be minimized by determining thresholds for each species.

Currently, a great diversity of protocols has been successfully used in skeletal muscle (Table 17.2).

The easy accessibility of skin makes it an attractive target for potential gene therapy applications. This is particularly the case for applications pertaining to cutaneous diseases, vaccines and some metabolic disorders. Skin makes an excellent target because there are a large number of antigen-presenting cells as well as an abundant blood supply in the dermal layer of skin, which may help transgenic products distribute into distant organs through the circulation. Electroporation of skin is a simple, direct, *in vivo* method to deliver genes for therapy. As with muscle, parameter selection is critical to achieve the appropriate expression level and kinetics. In addition to pulse width and field strength an important parameter is electrode design. There are many protocols that have been successfully used in the skin (Table 17.3). Typically, expression will peak at between 24 and 48 h although there have been some reports of peak expression at one week. Unlike

TABLE 17.2 Electrical Parameters Used for Intramuscular DNA Electroporation for Different Species

Species	Electrodes	Electrical Parameters (Electric Field, Pulses, Duration, Frequency)
Cow	Needles	150–200 V/cm, 1000 pulses, 200 μs, 5–10 trains of pulses
Guinea pig	Six needles	200 V/cm, 6 pulses, 50 ms, 1 Hz
Goat	Array	150–200 V/cm, 1000 pulses, 200 μs, 5–10 trains of pulses
Macaque	Needles	200 V/cm, 6 pulses, 50 ms, 1 Hz
Mouse	Needles	200 V/cm, 8 pulses, 20 ms, 1–2 Hz
	Plate	90 V/cm, 1000 pulses, 10 trains, 1 s
Pig	Needles	200 V/cm, 6 pulses, 60 ms, 1 Hz
	Plates	200 V/cm, 6 pulses, 20 ms, 5 Hz
Sheep	Needles	150–200 V/cm, 1000 pulses, 200 μs, 10 trains of pulses
Rabbit	Needles	200 V/cm, 6 pulses, 50 ms, 1 Hz or 120 V/cm, 8 pulses, 20 ms 5 Hz
Rat	Plates	160 V/cm, 8 pulses, 20 ms, 5 Hz
	Needles	200 V/cm, 8 pulses, 50 ms, 1 Hz

TABLE 17.3 Electrical Parameters Used for DNA Electroporation to the Skin for Different Species

Species	Electrodes	Electrical Parameters (Electric Field, Pulses, Duration, Frequency)
Mouse	Plate	100–300 μs, 400–1750 V/cm; 2–30 ms, 50–1000 V/cm; 150 ms, 100 V/cm, 100 μs+400–ms, 700–1000 V/cm + 80–200 V/cm, 2–8 pulses
	Needles	100 μs, 1750–1800 V/cm; 20 ms, 400–1800 V/cm; 100 μs+400 ms, 700–1000 V/cm + 80–200 V/cm, 2–18 pulses
Rat	Plates	100 μs, 1750–1800 V/cm; 20–150 ms, 75–200 V/cm, 2–10 pulses
	Needles	100 μs, 1750 V/cm, 20 ms, 200–V/cm
Pig	Plates	100 μs, 1750 V/cm, 100 μs+400 ms, 1000 V/cm + 80–200 V/cm, 2–6 pulses
	Needles	100 μs, 1750–2000 V/cm; 20–52 ms, 0.1–0.4 Amps, 100 μs+400 ms, 1000 V/cm + 80–240 V/cm, 2–6 pulses
Guinea pig	Needle array	20–200 ms, 75–200 V/cm, 8–72 pulses

muscle, expression typically goes down after two weeks. Histological evaluation of skin samples following delivery with electroporation revealed low levels of damage that was present for approximately one week. These short-term damage levels increased significantly as the field strength was increased. The primary effect observed was surface burns or scars (particularly at high field strengths (>400 V/cm and long pulses >10 ms) as well as the appearance of localized subepidermal necrosis. Twitching (pain) associated with applied pulses and potential damage can be reduced or eliminated by utilizing pairs or arrays of electrodes in a non-penetrating format with reduced distance between them. By reducing the distance between the electrodes, the area of tissue affected by each electric pulse is reduced as well as the depth of penetration of the field.

17.2.2 Muscle as a Secretory Organ for Therapeutic Proteins

An important issue for gene transfer applications is the level and duration of gene expression. Different kinetics of gene expression have been described after DNA electrotransfer into skeletal muscle. Long-term expression has been observed for a variety of transgenes such as human secreted alkaline phosphatase (hSeAP), the luciferase reporter gene, human factor IX (hFIX) in severe combined immunodeficiency (SCID) mice or murine erythropoietin (mEpo) in immunocompetent mice. By studying luciferase expression with a charge-coupled device (CCD) camera (which allows an *in vivo* kinetic study without sacrificing the animals), it was observed that gene expression increased with time during the first few days, and then decreased a little and stayed at a stable level for several months. This pattern of expression is observed for most of the secreted transgenes. It is not yet clearly established why expression in skeletal muscle lasts so long. The main belief is that it is due to the non-dividing, multi nucleated mature muscle fibers which are transfected during the electrotransfer process.

The ability of skeletal muscle to produce post-translationally modified functional proteins opens up the possibility of using it as an endocrine tissue for secretion of proteins into the bloodstream and therefore its use in a large number of applications. Moreover, the stable expression which can be obtained by DNA electrotransfer makes it very appealing for the treatment of numerous pathologies. The examples of proteins secreted by the skeletal muscle reported in Table 17.4 have all shown improvements of the related pathology in animal models.

Rheumatoid arthritis is a very good example to illustrate the advantages and challenges of DNA electrotransfer. It is a chronic inflammatory autoimmune disease in which the main observed characteristics are joint destruction and systemic inflammation. DNA electrotransfer of anti-inflammatory cytokines like IL-10 or IL-4 has shown promising results. However, elevated levels of circulating cytokines are very difficult to obtain. The most efficient current treatments are based on the use of anti-TNF-α molecules, the best results being obtained with recombinant proteins acting as soluble receptors for TNF-α. At least five recombinant proteins are currently used in clinics (Etanercept, Infliximab, Golimumab, Certolizymab and Adalimumab). However, serious adverse effects have been observed following these treatments: Infections, malignancy, heart failure,

TABLE 17.4 Examples of Therapeutic Proteins Secreted by the Skeletal Muscle after DNA Electroporation in Animal Models

Therapeutic Area	Encoded Protein
Cancer	Cytokines : IL-12, interferon (IFN)-γ, fibroblast growth factor (FGF)-2 ligand metalloprotease inhibitors, plasminogen fragment, endostatin
Cardiovascular diseases	IL-10, IL-18, platelet activating factor (PAF) acetyl hydrolase (atherosclerosis)
	IL-10, IL-18, vascular endothelial growth factor (VEGF)-B, FGF-1 (ischemia)
	IL-10, IL1-Ra (stroke)
Genetic diseases	Coagulation factor VIII or IX (hemophilia)
	Epo (β-thalassemia)
	Follistatin, dystrophin, laminin α2 (myopathies)
Autoimmune disease	TNF-α receptor, IL1-Ra, IL-4, IL-10, proopiomelanocortin (POMC) (rheumatoid arthritis)
Metabolic disorders	Epo (anemia)
	Insulin, pro-insulin, IGF-1, IL-4, VEGF (diabetes)
Organ-specific diseases	TNF-α receptor, plasminogen (ocular diseases)
	Bone morphogenetic protein (BMP)-2, BMP-4 (bone regeneration)
	Hepatocyte growth factor (HGF) (liver regeneration)
	Insulin-like growth factor (IGF)-1 (muscle regeneration)
DNA immunization	Surface antigen of hepatitis B virus (HbsAg)
	Hepatitis C virus envelope glycoprotein
	Influenza virus hemagglutinin
	Botulinum or tetanic toxins fragments
	Plasmodium yoelli parasite proteins

multiple sclerosis or autoimmune responses have been reported, although it is not proven yet that this is a class effect. This might be due to the high doses of recombinant proteins injected. DNA electrotransfer might allow a satisfactory response at low doses and avoid a deleterious peak of protein following injection. This is illustrated in Fig. 17.8, where a single treatment of arthritic mice with a plasmid encoding a soluble TNF-α receptor showed the same effect as three injections per week of the reference recombinant protein treatment.

17.2.3 DNA Immunization

The principle of immunization is to induce the generation of memory T- and B-cells and the presence of neutralizing antibody in the serum following the

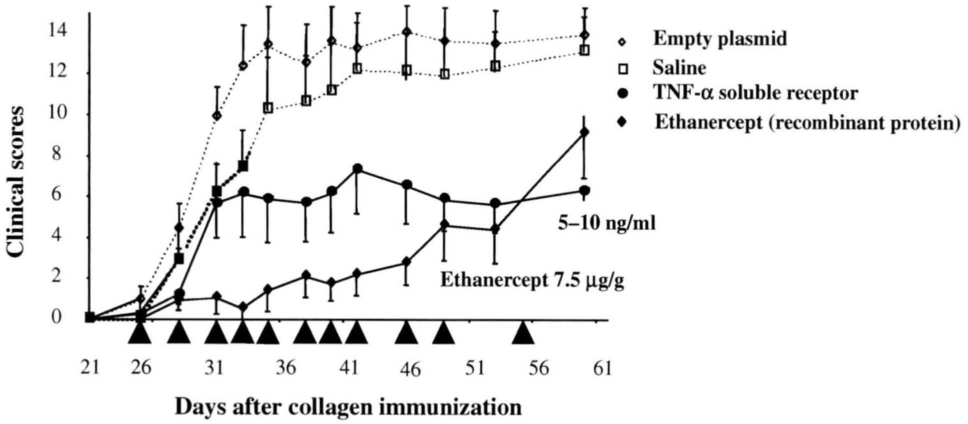

FIGURE 17.8 ■ Example of therapeutic effect on a mouse model of arthritis after intramuscular DNA electroporation of a plasmid encoding a soluble receptor derivative of TNF-α. Mice were treated once at day 23. Arrows indicates a reference recombinant protein treatment that is used in clinics. The clinical score is related to the severity of the disease: the higher the score, the more symptomatic the disease.

injection of a foreign protein. The observation that direct *in vivo* gene transfer of recombinant DNA resulted in expression of protein *in situ*, led to the development of DNA vaccines. Introducing the gene encoding a protein directly into the skin or muscle of an animal elicits an immune response. In fact, this plasmid-based vaccine injection is an attractive approach as it provides several advantages over current vaccines (mainly live-attenuated pathogens, e.g., bacteria and viruses, or recombinant proteins). Plasmid DNA can be manufactured in a very cost-effective manner, can be stored with relative ease (no need for a "cold chain" to maintain the efficacy of the vaccine) and there are none of the safety concerns associated with live-attenuated vaccine or pathogen. The organism will produce the antigen inducing the immune response itself. With "naked" DNA immunization high titers of neutralizing antibodies can be obtained in animals but, because of low or poorly reproducible gene transfer efficiency, multiple immunizations of high DNA doses are often required to achieve modest responses, particularly in primates. One reason for the lack of efficacy of DNA vaccine in large animals and in the first human clinical trials seems to be inefficient uptake of DNA by cells in the muscular tissue, which differs between small and large animal species.

In this context, electrotransfer greatly increases the potential of DNA vaccines since it increases antigen expression level by several orders of magnitude, and it has been suggested that, in this specific case, the level of antibodies produced is related to antigen expression level. While increased antigen expression may explain the increased immune response in animals treated with DNA electrotransfer, damages induced to muscle cells and release of "danger signals" after electroporation may also contribute. Several recent studies have shown the enhanced efficiency of electrotransfer in DNA immunization: after electrotransfer of a plasmid encoding a surface antigen of hepatitis B virus, antibody titers were increased in mice, rabbits and guinea pigs. This was also demonstrated in mice after electrotransfer of a plasmid encoding a tuberculosis protein. High antiserum titers against botulinum toxins could also be raised in mice after DNA electroporation. For clinical application in humans and also for veterinary studies, it was crucial to demonstrate that DNA injection and electrotransfer also induced an immune response in larger animals. This technique has therefore also been applied to pigs, sheep, cattle and rhesus macaque, and in all cases an improved immune response was observed. See Chapter 20 for further details.

17.2.4 DNA Delivery to Tumors

Treatment of cancer remains a viable target for electroporation-based therapies. There have been many studies that have evaluated the effectiveness of delivering plasmids with a variety of effector genes. These have included tumor suppressors, immunomodulating agents, inhibitors of cell growth and pro-apoptotic agents. Several of these studies have also evaluated combination approaches (delivery of both drugs and genes).

Immunogenetherapy has been extensively examined for treatment of tumors in mouse models. IL-12 was demonstrated to induce complete long-term responses with protection from recurrence following electroporation delivery to melanoma tumors. Other cytokines that have been tested include interferon (IFN)α in both melanoma and squamous cell carcinoma, IL-15 for melanoma and IL-21 for mouse rectal carcinoma. These studies have demonstrated the ability to induce significant responses. In some cases, complete long-term regression and protection from recurrence. Combination approaches have also been evaluated. The B7.1 with either granulocyte-macrophage colony-stimulating factor (GM-CSF) or IL-12 was used for treating fibrosarcoma or squamous cell carcinoma, respectively. Electroporation has been used successfully to deliver chemotherapeutics (electrochemotherapy (ECT)) in both preclinical and clinical trials. Several studies have evaluated combining ECT with immunogenetherapy using electroporation. These included delivering bleomycin with IL-2, GM-CSF or IL-12 in a variety of tumor models. Another approach tested was to deliver a suicide gene such as thymidine kinase followed by administration of gancyclovir in a pancreatic model. Studies have also shown that tumor regression may also be induced by interference with tumor signaling pathways. Intratumor delivery to a mouse melanoma of plasmids encoding signal transducer and activator of transcription (Stat) 3 inhibitor or the human immunodeficiency virus (HIV) protein viral protein R (Vpr) induced tumor regression. Delivery of antisense oligonucleotides to polo-like kinase 1 induced tumor regression in a variety of tumor types.

It is clear from the results obtained in these preclinical studies that the number of clinical trials utilizing this gene therapy approach for direct tumor delivery will continue to increase. A popular approach has been to stimulate the immune system by the delivery of immunomodulating agents. Clearly there are distinct advantages to the use of this approach as systemic responses have been seen following localized delivery. As the technology matures and the results from more studies are published, it will be interesting to see how well this approach translates to clinical efficacy.

17.2.5 DNA Electroporation as a Tool

In addition to its potential use in gene therapy, DNA electrotransfer is a powerful laboratory tool to study *in vivo* gene expression and function in any given tissue. Besides skeletal muscle, skin and tumors, a number of other tissues have been shown to express reporter genes after electrotransfer, although each tissue requires that specific optimal electroporation parameters be determined. Electrotransfer can be used to study gene expression and function, in a spatially and temporally restricted manner. A very recent example was the first demonstration that sustained intraocular placental growth factor (PGF) production induces vascular and retinal changes similar to those observed in the early stages of diabetic retinopathy, while the role of PGF was controversial.

Another very interesting use of electrotransfer described the preparation of a monoclonal antibody against chemokine-like factor 1 (CKLF1), a newly cloned human cytokine. As CKLF1 is a highly hydrophobic protein, the purification of native CKLF1 was unsuccessful. Electrotransfer of a CKLF1-encoding plasmid into the skeletal muscle of mice (and therefore *in vivo* secretion of the protein) instead of the conventional protein immunization strategy overcame this problem. Treated mice were checked for an antibody response after three treatments, and the spleen of the most responsive mouse was harvested and used to generate a monoclonal antibody by conventional methods.

Currently, *in vivo* DNA electroporation is a widely used technique. Its efficiency has led to at least one veterinary product which is licensed for pigs (growth hormone releasing hormone (GHRH) treatment in sows). The preclinical successes have allowed this approach to make its way towards clinical use, mainly in the fields of infectious diseases and cancer.

17.3 Clinical Applications

There has been a steady increase in clinical trials utilizing electroporation to deliver plasmid DNA. Currently, there are 23 trials listed on www.clinicaltrials.gov that have utilized *in vivo* electroporation to deliver plasmid DNA. Of these studies, there are eleven studies evaluating cancer therapies, ten studies evaluating vaccines for infectious agents and two that were evaluating safety of the electroporation approach.

17.3.1 Clinical Cancer Studies

Treating cancer via non-viral gene transfer using *in vivo* electroporation has thus far focused on immune stimulation. This has included delivery of cytokines directly to the tumor or delivering tumor antigens or antibodies to the muscle or skin to stimulate a response. Cancer types that have been tested include melanoma, colorectal cancer, prostate, cervical and haematological malignancies. Three of the studies utilized intratumor delivery, six utilized intramuscular delivery and two utilized intradermal delivery. Two of the trials have been completed, one administering immunotherapy via intratumor administration and the other a cancer vaccine for prostate cancer administered via intramuscular delivery. There have been two safety trials, one intramuscular and one intradermal that have been completed as well and reports indicate that delivery of plasmid DNA using electroporation was safe and tolerable.

The first clinical trial utilizing *in vivo* electroporation delivered interleukin-12 (hIL-12) directly to cutaneous lesions of patients with malignant melanoma (www.clinicaltrials.gov identifier NCT00323206). This phase I trial had a primary objective of safety and tolerability and each patient received an intralesional injection of a plasmid encoding hIL-12 directly into up to four surface tumors. The deliveries were performed three times over eight days. The major adverse affect was transient pain during the administration of the electroporation pulses. Increased hIL-12 protein levels were observed in treated tumors and 10% of patients showed complete regression of all metastases, both treated and untreated, and disease stabilization or partial responses were seen in 42% of patients treated (Daud *et al.*, 2008).

The completed cancer vaccine trial evaluated intramuscular delivery of a plasmid encoding a tetanus toxin domain fused to prostate-specific membrane antigen which was delivered to patients with recurrent prostate cancer. The DNA vaccine was administered five times over a 48-week period. Both the DNA and the electric pulses were administered using the same two-needle device. The pulses were a train of five 20-ms pulses at 8.3 Hz delivered with a maximum current of 250 mA using an Elgen Twinjector device (Inovio Pharmaceutical Inc., San Diego, CA). The outcome was based on immune responsiveness following vaccine administration. The investigators have thus far reported that there was increased antibody titers against tetanus toxin that persisted up to 18 months (Low *et al.*, 2009). In addition, preliminary evaluations of the treated patients showed signs of suppression of tumor growth and evidence for clinically manageable concomitant autoimmunity.

17.3.2 Clinical Vaccine Studies for Infectious Agents

There are ten clinical trials listed on www.clinicaltrials.gov that report utilizing *in vivo* electroporation to deliver DNA vaccines against infectious agents. Five of these studies are evaluating DNA vaccines against HIV, two are against influenza (Flu), two are for hepatitis C (Hep C) and one against malaria. Nine of these studies administer the DNA injection and electroporation via the intramuscular route and the other via the intradermal route. Two of these studies are completed, five are still recruiting and three are not recruiting. Results have only been reported for one of these studies. In this study, healthy volunteers received an intramuscular injection of a HIV-1 vaccine encoding the group-specific antigen (gag), envelope (env), polymerase (pol), negative regulatory factor (nef) and transactivator of transcription (tat) antigens either alone or followed by the administration of electroporation. A third group received a saline placebo followed by electroporation. The administration of the prime and boost injections were eight weeks apart. Some volunteers received a third injection at 36 weeks. The procedure was found to be safe and tolerable. When compared to naked DNA intramuscular injection, the group receiving the vaccine with electroporation was observed to have an up to 70-fold increase in HIV-1-specific cell mediated immunity, as indicated by a gamma interferon spot enzyme linked immunosorbent assay (ELISpot). Responders were both cluster of differentiation (CD)4^+ and $CD8^+$. There was also an increase in the number of antigens against which the electroporation group showed a response.

References

Adachi, O., Nakano, A., Sato, O., *et al.* (2002). Gene transfer of Fc-fusion cytokine by *in vivo* electroporation: application to gene therapy for viral myocarditis, *Gene Ther,* **9**, 577–583.

Aihara, H., Miyazaki, J. (1998). Gene transfer into muscle by electroporation *in vivo*, *Nat Biotechnol*, **16**, 867–870.

Bigey, P., Bureau, M.F., Scherman, D. (2002). *In vivo* plasmid DNA electrotransfer, *Curr Opin Biotechnol*, **13**, 443–447.

Bloquel, C., Bessis, N., Boissier, M.C., *et al.* (2004a). Gene therapy of collagen-induced arthritis by electrotransfer of human tumor necrosis factor-α soluble receptor I variants, *Hum Gene Ther*, **15**, 189–201.

Bloquel, C., Fabre, E., Bureau, M.F., *et al.* (2004b). Plasmid DNA electrotransfer for intracellular and secreted proteins expression: new methodological developments and applications, *J Gene Med*, **Suppl 1,** S11–S23.

Bureau, M.F., Gehl, J., Deleuze V., *et al.* (2000). Importance of association between permeabilization and electrophoretic forces for intramuscular DNA electrotransfer, *Biochim Biophys Acta*, **1474**, 353–359.

Daud, A.I., DeConti, R.C., Andrews, S., *et al.* (2008). Phase I trial of interleukin-12 plasmid electroporation in patients with metastatic melanoma, *J Clin Oncol*, **26**, 5896–5903.

Golzio, M., Teissie, J., Rols, M.P. (2002). Direct visualization at the single-cell level of electrically mediated gene delivery, *Proc Natl Acad Sci USA*, **99**, 1292–1297.

Heller, R., Jaroszeski, M., Atkin, A., *et al.* (1996). *In vivo* gene electroinjection and expression in rat liver, *FEBS Letters*, **389**, 225–228.

Kim, J.M., Ho, S.H., Hahn, W., *et al.* (2003). Electro-gene therapy of collagen induced arthritis by using an expression plasmid for the soluble p75 tumor necrosis factor receptor-Fc fusion protein, *Gene Ther*, **10**, 1216–1224.

Klenchin, V.A., Sukharev, S.I., Serov, S.M. *et al.* (1991). Electrically induced DNA uptake by cells is a fast process involving DNA electrophoresis, *Biophys J*, **60**, 804–811.

Low, L., Mander, A., McCann, K., *et al.* (2009). DNA vaccination with electroporation induces increased antibody responses in patients with prostate cancer, *Hum Gene Ther*, **20**, 1269–1278.

Mathiesen, I. (1999). Electropermeabilization of skeletal muscle enhances gene transfer *in vivo*. *Gene Ther*, **6**, 508–514.

Mir, L., Bureau, M., Gehl, J., *et al.* (1999). High efficiency gene transfer into skeletal muscle mediated by electric pulses, *Proc Natl Acad Sci USA*, **96**, 4262–4267.

Mir, L., Bureau, M., Rangara, R., *et al.* (1998). Long-term, high level *in vivo* gene expression after electric pulse-mediated gene transfer into skeletal muscle, *C R Acad Sci III*, **321**, 893–899.

Neumann, E., Schaefer-Ridder, M., Wand, Y., *et al.* (1982). Gene transfer into mouse lyoma cells by electroporation in high electric fields, *EMBO J*, **1**, 841–845.

Paturneau-Jouas, M., Parzy, E., Vidal, G., *et al.* (2003). Electrotransfer at MR imaging: tool for optimization of gene transfer protocols — feasibility study in mice, *Radiology*, **228**, 768–775.

Potter, H. (1988). Electroporation in biology: methods, applications, and instrumentation, *Anal Biochem*, **174**, 361–373.

Rols, M.P., Delteil, C., Golzio, M., *et al.* (1998). *In vivo* electrically mediated protein and gene transfer in murine melanoma, *Nat Biotechnol*, **16**, 168–171.

Satkauskas, S., Bureau, M.F., Puc, A., *et al.* (2002). Mechanism of *in vivo* DNA electrotransfer: respective contribution of cell electropermeabilization and DNA electrophoresis, *Mol Ther*, **5**, 133–140.

Titomirov, A., Sukharev, S., Kristanova, E. (1991). *In vivo* electroporation and stable transformation of skin cells of newborn mice by plasmid DNA, *Biochimt Biophys Acta*, **1088**, 131–134.

For Further Reading

Reviews

Bodles-Brakhop, A.M., Heller R., Draghia-Akli, R. (2009). Electroporation for the delivery of DNA-based vaccines and immunotherapeutics: current clinical developments, *Mol Ther*, **17**, 585–592.

Heller, L.C., Heller, R. (2010). Electroporation gene therapy preclinical and clinical trials for melanoma, *Curr Gene Ther*, **10**, 312–317.

Hojman, P. (2010). Basic principles and clinical advancements of muscle electrotransfer, *Curr Gene Ther*, **10**, 128–138.

Mir, L.M. (2009). Nucleic acids electrotransfer-based gene therapy (electrogenetherapy): past, current, and future, *Mol Biotechnol*, **43**, 167–176.

Reed, S.D., Li, S. (2009). Electroporation advances in large animals, *Curr Gene Ther*, **9**, 316–326.

Rice, J., Ottensmeier, C.H., Stevenson, F.K. (2008). DNA vaccines: precision tools for activating effective immunity against cancer, *Nat Rev Cancer*, **8**, 108–120.

Sardesai, N.Y., Weiner, D.B. (2011). Electroporation delivery of DNA vaccines: prospects for success, *Curr Opin Immunol*, **23**, 421–429.

Satkauskas, S., Ruzgys, P., Venslauskas, M.S. (2012). Towards the mechanisms for efficient gene transfer into cells and tissues by means of cell electroporation, *Expert Opin Biol Ther*, **12**, 275–286.

Stevenson, F.K., Ottensmeier, C.H., Rice, J. (2010). DNA vaccines against cancer come of age, *Curr Opin Immunol*, **22**, 264–270.

18

IMAGING OF GENE DELIVERY

Georges Vassaux[a,b,c,d], Peggy Richard-Fiardo[a,b,c], Béatrice Cambien[a,b,c] and Philippe Franken[a,b,c]

18.1 General Introduction

The past 15 years have seen the emergence of methodologies aimed at the visualization of gene transfer in live subjects. Indeed, monitoring the ectopic expression of a reporter gene in gene therapy was first suggested in 1995 (Tjuvajev *et al.*, 1995), whilst actual visualization of gene transfer in a live animal was reported for the first time in 1996 (Tjuvajev *et al.*, 1996). In this study, rat glioma cells were transduced *in vitro* with a recombinant retrovirus encoding the herpes simplex virus thymidine kinase gene (HSVtk) and injected to produce subcutaneous tumors. On the other flank of the animal, wild-type tumors were inoculated and transduced by intratumoral injection of the same retroviral vector encoding HSVtk. Administration of the radiotracer 2′-fluoro-2′-deoxy-1β-D-arabinofuranosyl-5-iodo-uracil (^{131}I-FIAU) was performed and its selective accumulation in the tumors was monitored by single-photon emission computed tomography (SPECT) (Tjuvajev *et al.*, 1996). This pioneering study demonstrated for the first time that a standard imaging modality, which is used routinely in nuclear medicine

[a] Faculté de médecine, Université de Nice Sophia-Antipolis, Nice, France
[b] Laboratoire TIRO, Institut de Biologie Environementale et Biotechnologie (iBEB), Direction des Sciences du Vivant (DSV), CEA, Nice, France
[c] Centre Antoine Lacassagne, Nice, France
[d] Email: georges.vassaux@unice.fr

departments worldwide, could detect the ectopic expression of a reporter gene in both *ex vivo*, genetically-engineered cells and *in vivo*, transduced cells. Subsequently, other imaging modalities such as optical imaging (bioluminescence, fluorescence), magnetic resonance imaging (MRI) and positron emission tomography (PET) have been used to detect the ectopic expression of a reporter gene in the live, anesthetized subject.

At the preclinical level, these techniques have been used to determine the biodistribution pattern of gene transfer allowed by different gene delivery vectors, and to assess the tissue-selectivity of promoter fragments driving the expression of reporter genes in order to design and screen new gene delivery systems and to guide the development of gene therapy strategies. In addition, the preclinical proofs-of-concepts have been translated to humans and gene therapy clinical trials with imaging branches have now been reported. This review describes the basic principles, as well as the main applications, of this technology.

18.2 Molecular Imaging of Gene Expression in the Live Subject: Basic Principles

Molecular imaging of gene expression requires three key components: (1) a reporter gene which will be expressed ectopically in the tissues susceptible to transfection/transduction by the gene therapy vector, (2) a tracer which will accumulate in the organ(s) in which the reporter gene is expressed, and (3) a scanner capable of detecting the presence of the tracer. The subject, usually anesthetized, is positioned in the scanner and data acquisition is performed. These data are then reconstructed into images that can be analyzed. For most applications, the comparison between the images obtained from a control subject and a subject administered the gene therapy vector will provide anatomical information on the site of gene transfer. In addition to these qualitative data, and using certain imaging modalities, it is also possible to obtain precise quantification of the extent of gene transfer.

Considering that this procedure can be repeated, longitudinal studies can be performed and can provide data on the kinetics of gene expression, monitored in each individual subject. In terms of animal experimentation, these studies can be performed on a single cohort, and so imaging reduces very significantly the number of animals involved.

As an alternative to imaging gene expression, another type of imaging relevant to gene therapy has been described. In certain cases, it is possible to label the gene delivery vector radioactively or with a fluorescent dye (Le *et al.*, 2006). These studies are very useful for determining the biodistribution of the vectors themselves, which does not necessarily correlate with gene transfer. This type of imaging, involving the fate of gene delivery vectors as biophysical objects, is beyond the scope of this review.

18.3 Imaging Modalities

Different imaging modalities have been described, each associated with its different reporter genes and specific tracers. For the purpose of this review, we will categorize them into "isotopic imaging" (i.e., the utilization of radiolabeled tracers and their dedicated scanners) and non-radioactive imaging modalities. This distinction is based on the possibility of extending isotopic imaging to humans, whilst the utilization of the non-radioactive methods has not been validated in humans.

18.3.1 Non-Radioactive Imaging

This group of methodologies includes optical imaging (bioluminescence and fluorescence) and MRI.

Bioluminescence imaging exploits charge-coupled device (CCD) cameras (Sweeney *et al.*, 1999), capable of detecting very weak, visible light. The main application of this technology has been the detection of luciferase reporter gene expression. This gene oxidizes its substrate, luciferin, emitting light, which is detected by the CCD camera. This technique is inexpensive and sensitive. The main limitation of this technology is that the detection is limited to structures located only a few centimeters deep within tissues. As a result, this methodology is particularly suited to experiments using small animals but is not generally applicable to larger animal models or to humans. Nevertheless, this technology has been successfully used to track gene expression and tumor metastasis, and to determine the biodistribution of various viral and non-viral gene delivery vectors. CCD cameras can also be adapted to detect fluorescence. In this case, the classically used reporter gene is the green fluorescent protein or its genetically engineered variants. *In vivo* fluorescent imaging can provide dynamic visualization of biological processes, however the same limitations as for bioluminescence imaging apply.

MRI is a powerful tool for clinical diagnostics which exploits very strong magnetic fields in association with radio-frequency fields to produce signals detectable by a scanner. MRI offers a very high spatial resolution (sub-milliliter range) as well as unlimited depth of sensitivity (Jacobs *et al.*, 1994). Very recently, a promising approach using ferritin as a reporter gene has been described. Ferritin is a ubiquitously expressed metaloprotein, assembled from 24 light and heavy subunits. This protein accumulates endogenous Fe(II) from the intracellular, labile iron pool of the organism and stores it as ferrihydrite, which accumulates at the site of expression. This methodology has been validated using adeno-associated virus (AAV) (Vande Velde *et al.*, 2011). However, in the light of the difficulties encountered using a lentivirus encoding ferritin (Vande Velde *et al.*, 2011), and considering that this modality detects live as well as dead cells, this approach appears to be limited in comparison with optical or nuclear techniques.

18.3.2 Isotopic Imaging

This group of methodologies uses radioactively labeled tracers to locate the anatomical site of transgene expression. In this context, two main techniques are used in gene therapy applications: SPECT and PET. These techniques differ with regard to the type of isotopes detected.

SPECT imaging uses arrays of detectors to identify photons emitted by some radioisotopes such as ^{99m}Tc, ^{123}I, ^{111}In, etc. Multiple, bi-dimensional images are collected from different angles and a computer is used to generate a tomographic reconstruction, yielding a three-dimensional dataset. These images can be manipulated to obtain, for example, slices containing the organ(s) of interest. Semi-quantitative analysis is possible. Some of the advantages of SPECT imaging are high resolution (sub-millimeter in preclinical scanners) and the ability to distinguish different radioisotopes with different radiation energies. As a result, SPECT offers the possibility of the simultaneous visualization of the biodistribution of two probes, related to two different biological phenomena. The main disadvantage of SPECT is its relatively low sensitivity.

PET scanners detect pairs of γ rays produced when positrons emitted from radioisotopes such as ^{18}F, ^{124}I, ^{11}C, etc. collide with local, neighboring electrons. PET is more sensitive, but shows a lower resolution, than SPECT.

A certain number of reporter genes, associated with specific radiotracers, have been described. These include: the dopamine receptor 2 (DR) gene, associated with its ^{99m}Tc- (Auricchio *et al.*, 2003) or ^{11}C- (Umegaki *et al.*, 2003) radiolabeled ligands for SPECT and PET imaging, respectively; the somatostatin receptor 2 (SSTR2), associated with its ^{111}In- and ^{90}Y-labeled ligands for γ-camera imaging (McCart *et al.*, 2004; Rogers *et al.*, 2002) or ^{99m}Tc-labeled ligand for SPECT (Hemminki *et al.*, 2002) or ^{94m}Tc- and ^{68}Ga-labeled ligands for PET imaging (Cotugno *et al.*, 2011; Rogers *et al.*, 2005); and the norepinephrine transporter with its ^{123}I- or ^{124}I-labeled epinephrine analogue for SPECT and PET imaging (Brader *et al.*, 2009). However, the two main reporter systems used in the field are the HSVtk and the Na/I symporter (NIS).

18.3.2.1 *The HSVtk reporter system*

Highly-specific substrates for the herpes virus thymidine kinase have been developed and these are commonly and routinely used as anti-herpetics. Compounds such as acyclovir, for example, have a much higher specificity for the HSVtk than for its mammalian enzyme counterpart. This type of substrate is, thus, monophosphorylated by the HSVtk, following which cellular kinases further phosphorylate the nucleotide to its tri-phosphorylated state, which is incorporated into nascent DNA chains. This process inhibits and terminates further chain elongation by DNA polymerases and eventually leads to cell death.

Because of these specificities, as well as their well-known safety profiles, these thymidine analogues have been radiolabeled and used as tracers to detect gene transfer, mainly in cancer gene therapy.

From the imaging point of view, ^{18}F-(PET) (Burton *et al.*, 2008; Mullerad *et al.*, 2006; Sato *et al.*, 2008) or ^{131}I-(SPECT) (Lan *et al.*, 2010) labeled thymidine analogues have been used to image transgene expression. In principle, the radiolabeled substrate can freely cross the plasma membrane and, once inside the cell, is phosphorylated by HSVtk. The resulting product is more polar and is trapped in the cell. The scanner detects the accumulation of this phosphorylated product over time, providing images of the anatomical site of transgene expression.

18.3.2.2 *The NIS reporter system*

Iodide concentration occurs mainly in thyroid tissue. The thyroid gland concentrates iodide by a factor of 20–40 times as compared with plasma, under physiological conditions (Hingorani *et al.*, 2010). Iodide is actively transported across the plasma membrane into the cytoplasm of thyroid follicular cells and is subsequently translocated passively from the cytoplasm into the follicular lumen. The cell/colloid interface within the follicular lumen is the main site of hormone biosynthesis and involves the coupling of iodide to tyrosine residues on thyroglobulin present within the follicular colloid.

NIS is the plasma membrane glycoprotein that mediates the active transport of iodide into cells (Baril *et al.*, 2010; Hingorani *et al.*, 2010). The symporter co-transports two sodium ions, along with one iodide ion. Energetically, the transmembrane sodium gradient serves as the driving force for iodide uptake. NIS functionality is therefore dependent on the electrochemical sodium gradient maintained by the oubaine-sensitive Na^+/K^+ adenosinetrphosphatase pump. The efflux of iodide from the apical membrane to the follicular lumen is driven by pendrin (the Pendred syndrome gene product) and possibly by other, unknown efflux proteins (apical anion transporters) (Hingorani *et al.*, 2010). Thiocyanate and perchlorate are competitive inhibitors of iodide accumulation in the thyroid due to their similarity both in size and charge to iodide ions. Perchlorate is the most potent inhibitor of iodide accumulation in a variety of *in vivo* and *in vitro* systems. Iodide organification within the follicular lumen is mediated by the enzyme, thyroperoxidase (TPO), involving the oxidation of iodide and its binding to tyrosine residues within the thyroglobulin backbone, followed by the oxidative coupling of iodotyrosines to generate thyroid hormones. The unique property of thyroid follicular cells to trap and concentrate iodide enables imaging and also the effective therapy of differentiated thyroid cancers using radioiodide (Hingorani *et al.*, 2010). In addition to the thyroid, NIS is mainly endogenously expressed in the stomach and the salivary glands.

In the field of oncology, NIS expression is crucial to the treatment of thyroid cancer. During the past 50 years, the utilization of radioiodide in the management of this pathology has shown great success. Thyroid cancer patients treated with therapeutic doses of ^{131}I have a very significantly lower mortality than those who are not given this treatment. This observation remains true for patients with metastatic disease: the treatment of these patients with ^{131}I leads to a ten-year survival of more than 80%. Unfortunately, thyroid cancer patients with no, or poor, radioiodide uptake have a much poorer prognosis (Hingorani *et al.*, 2010).

NIS expression is also suspected in different types of malignancies: Cholangiocarcinomas (Liu *et al.*, 2007) and breast cancers (Wapnir *et al.*, 2003, 2004) have been reported to express NIS, although for the latter a controversy exists (Peyrottes *et al.*, 2009). These observations suggest that NIS-mediated radionuclide therapy may also be applied to the treatment of non-thyroidal cancers.

In imaging, NIS used as a reporter gene is combined with tracer, rather than therapeutic, radioisotope doses. SPECT imaging can be performed using ^{123}I (Boland *et al.*, 2000) and PET imaging using ^{124}I (Groot-Wassink *et al.*, 2002). In addition to iodide, other anions (such as pertechnetate) can be concentrated in cells by NIS. Radioisotopes of these anions (such as $^{99m}TCO_4$, produced and used routinely in nuclear medicine departments worldwide) can then be used to visualize, by SPECT, the endogenous expression of NIS (thyroid, stomach) as well as its ectopic expression after gene transfer (Merron *et al.*, 2007). Hence, the main advantages of NIS imaging are its versatility and lack of requirement for extensive radiochemistry, as the radioisotope is the actual tracer. These points have important logistical and economic consequences, crucial for the dissemination of this technology. The main drawback of the approach is that, unlike other gene reporter/tracer systems, the isotope detected by the scanner is not trapped within the cell. NIS concentrates iodide in cells (up to 40 times the extracellular concentration) but, when iodide is removed from the extracellular milieu, a rapid efflux occurs. For an accurate measurement and comparison of gene transfer *in vivo*, under two different conditions, it is therefore essential to compare data obtained in conditions of equivalence of the extracellular radiotracer concentration. In theory, this can only be achieved by collecting samples of arterial blood at the same time as the imaging is performed. This procedure dramatically complicates the process, to the point that it may preclude any routine utilization of the methodology, particularly in small animals. However, kinetic experiments have shown that the concentration of radioisotope in the plasma is constant during the first hour after injection of radioactive iodide or pertechnetate *in vivo* (Zuckier *et al.*, 2004). It is, thus, possible to compare scans performed days apart (and, therefore, to compare the level of gene transfer or its duration), if the scans are performed within the first hour after radiotracer administration (Groot-Wassink *et al.*, 2004b).

18.4 Preclinical Applications

18.4.1 Visualization of Gene Transfer: Biodistribution Studies

The simplest application of molecular imaging to gene therapy is the determination of the pattern of gene expression allowed by a specific gene delivery vector. The route of administration and dose of vector injected can vary and the propensity of molecular imaging to allow longitudinal studies can be exploited to evaluate transgene expression kinetics using a reduced number of experimental animals, while allowing individual monitoring.

In this context, replication-incompetent, recombinant adenoviruses are the class of delivery vectors that have been most associated with molecular imaging. In addition, this type of gene delivery vector has been a key tool in the validation of most of the reporter/tracer systems. For example, differences in gene expression patterns upon intravenous versus intraperitoneal administration were determined using optical imaging (Johnson *et al.*, 2006). This imaging modality was also used to visualize adenovirus-mediated gene transfer upon intramuscular (Wu *et al.*, 2001) or cardiac (Wu *et al.*, 2002) injection. Adenovirus-mediated transfer was also key to the validation of imaging of gene transfer using the HSVtk reporter gene by SPECT (Tjuvajev *et al.*, 1996) and PET (Gambhir *et al.*, 1998), and the NIS reporter gene by PET (Groot-Wassink *et al.*, 2002, 2004b), SPECT (Dwyer *et al.*, 2005) and γ-camera (Boland *et al.*, 2000). Other systems that have also been validated using this delivery vectors include the dopamine D2 receptor and its radiolabeled ligands (Yaghoubi *et al.*, 2001), SSTR2 and its radiolabeled ligands (Chen *et al.*, 2010b; Cotugno *et al.*, 2011; McCart *et al.*, 2004; Rogers *et al.*, 2002, 2005; Singh *et al.*, 2011) and the norepinephrine transporter and radiolabeled epinephrine analogues (Buursma *et al.*, 2005).

Optical and/or PET and/or SPECT imaging have also been used to visualize *in vivo* gene transfer mediated by naked DNA electroporation (Nakamoto *et al.*,

2000), by recombinant lentivirus (De *et al.*, 2003), by AAV (Auricchio *et al.*, 2003; Cotugno *et al.*, 2011; Zincarelli *et al.*, 2008), by AAV–phage hybrid (Hajitou *et al.*, 2006) and by ultrasound/microbubble sonoporation (Watanabe *et al.*, 2010).

In addition to the above-mentioned, non-replicating gene-transfer agents, another class of replication-competent vectors has been developed: the so-called "oncolytic viruses", whose application is restricted to cancer gene therapy. These viruses are genetically designed to replicate selectively in cancer cells, while leaving normal tissues unaffected (Garcia-Aragoncillo *et al.*, 2010). In this context, molecular imaging of the propagation of these viruses is paramount for obvious safety reasons and also to monitor the speed of propagation and the length of time during which the virus is infective in the tumor. Information obtained through molecular imaging can then be used to improve the design of the next generation of agents.

Molecular imaging has, thus, been used in the context of oncolytic adenoviruses: The luciferase gene monitored using optical imaging (Hemminki *et al.*, 2002), the NIS gene using SPECT (Merron *et al.*, 2007, 2010; Peerlinck *et al.*, 2009) and the HSVtk reporter gene using PET (Freytag *et al.*, 2007). In addition, optical, SPECT and PET imaging have been used to visualize the propagation of oncolytic vaccinia virus (Brader *et al.*, 2009), oncolytic measles virus (Ong *et al.*, 2007) and vesicular stomatitis virus (Goel *et al.*, 2007). Analysis of the data obtained has shown that, overall, these viruses are indeed capable of replication and propagation in experimental tumors, whilst leaving normal tissues largely unaffected. Their rates of propagation are very variable and the peak of virus presence in the tumor varies from as little as two days for oncolytic adenoviruses (Peerlinck *et al.*, 2009) to as much as 9–10 days for oncolytic measles viruses (Dingli *et al.*, 2004).

18.4.2 Characterization of Promoter Selectivity and Strength

In order to drive the expression of therapeutic transgenes, promoter fragments are necessary. For many different applications, ubiquitous promoters, i.e., allowing a (strong) expression of the transgene in all the tissues transduced or transfected by the gene delivery vector, are sufficient. However, for special applications (toxic transgenes, induction of neovascularization or driving the replication of oncolytic viruses, for example), the utilization of tissue- or cell-selective promoters is required. This requirement has prompted the development of a specific field aiming at the characterization of promoter fragments tailored for *in vivo* gene therapy: *promoter pharmacology*.

As a general rule, the promoter fragment of a gene, chosen for its relevant expression pattern, is studied in the first instance *in vitro*. Selected promoter fragments are then incorporated in a relevant gene delivery vector and molecular imaging provides rapid information regarding the anatomical site of gene transfer and the level of gene expression. Because of the toxic nature of the therapeutic transgenes used, cancer gene therapy is the most "in-demand" field for cancer-selective promoters and various promoter fragments have already been characterized *in vivo*, using molecular imaging. These include promoter fragments from components of the telomerase complex (Groot-Wassink *et al.*, 2004a), the survivin gene (Ahn *et al.*, 2011), the carcinoembryonic antigen gene (Qiao *et al.*, 2002), the cyclo-oxygenaze 2 gene (Liang *et al.*, 2004), the mucin gene (Huyn *et al.*, 2009), the α-fetoprotein gene (Willhauck *et al.*, 2008) and the progression-elevated gene 3 (Bhang *et al.*, 2011). Other applications of this technology include the characterization of heart-specific (Chen *et al.*, 2010a) and liver-specific (Wilber *et al.*, 2005) promoters.

18.4.3 Molecular Imaging to Screen and Validate New, Efficient, Non-Viral Vectors for Gene Therapy

Non-viral gene therapy aims at the design and generation of "virus-like" particles, capable of packaging, protecting and delivering the genetic material carrying the therapeutic transgene to (target) cells. Non-viral formulations are potentially far superior to viral gene delivery vectors as they are chemically defined, potentially safer, and can be produced using the standards of the "classical" pharmacological industry. During the past 20 years, a very significant amount of resources has been invested in the development of non-viral gene delivery vectors for *in vivo* applications. Unfortunately, these efforts have not been rewarded and, on the whole and for most applications, viral vectors remain the most efficient.

One of the reasons for this lack of success may be that the process of developing these reagents has not been appropriately adapted. Classically, researchers in this field have used organic compounds capable of generating complexes with DNA. They have characterized the biophysical "objects" generated and have monitored the efficacy of their formulations by testing them *in vitro*, using cultured cell lines. As a result, many transfection reagents are now commercially available which are capable of transfecting the majority of cell types *in vitro*. Unfortunately, the vast majority of these reagents are unable to promote high, consistent levels of gene transfer *in vivo*, highlighting the uncoupling between *in vitro* and *in vivo* testing. A rational way forward may, therefore, be directly to test *in vivo* those compounds with a known biophysical structure and suspected activity. In this context, molecular imaging could be the technology of choice to obtain rapid results (and, thus, to be able to screen a significant number of new formulations), whilst keeping the number of experimental animals to a minimum through longitudinal studies capable of producing kinetic data on a single cohort of animals. Examples of success using this approach can be found in the scientific literature with the development of new, non-viral, gene-transfer vectors for cancer gene therapy which are capable of transfecting tumors upon systemic administration (Chisholm *et al.*, 2009; Klutz *et al.*, 2009). It is interesting that these formulations, which are active *in vivo*, are far less efficient *in vitro* than commercially available ones. Whether this strategy can be extrapolated to other types of tissues remains to be evaluated.

18.4.4 How Can Imaging Increase Therapeutic Efficacy?

In cancer gene therapy, the proof-of-principle of the selective replication of oncolytic viruses has been demonstrated both at the preclinical and clinical level. Evidence of infection, replication and propagation of the virus in experimental (Peerlinck *et al.*, 2009) and real-patient (Khuri *et al.*, 2000) tumors has been reported. However, the efficacy of these viruses remains limited and combining oncolytic viruses with other therapeutic strategies is now being heavily investigated. One of the approaches that has been proposed is to arm oncolytic viruses with therapeutic transgenes which could kill tumor cells that are refractory to the action of the virus. In this context, the inclusion of the NIS gene in the genome of oncolytic viruses has been proposed (Harrington *et al.*, 2008). By analogy to the treatment of thyroid cancers (see Section 18.3.2.2), administering therapeutic doses of ^{131}I to tumor-bearing subjects after the administration of oncolytic viruses encoding NIS has been envisaged. In this way, tumor cells infected with the virus would express NIS and would take up radioiodide, irradiating the surrounding cells (whether or not they are infected by the virus). However, mistimed irradiation could result in the loss in efficacy and even in virostatic effects. In this context, the development of molecular imaging strategies capable of monitoring the propagation of oncolytic viruses in tumors is essential in determining the optimal window of time for administration of the radioisotope. Typically, ^{131}I should be administered when both the spread of the virus and NIS expression are at their peak, and NIS imaging is the technique of choice in determining the correct timing.

We have recently demonstrated this principle using an oncolytic adenovirus (Peerlinck *et al.*, 2009). This virus (AdIP2) has been genetically designed to express NIS and to replicate selectively in cells in which the Wingless–Integration (Wnt) signaling pathway is constitutively activated. As this molecular defect is observed in most colorectal cancer cells (Segditsas *et al.*, 2006), the indication of this virus would be in liver metastasis of colorectal cancers. SPECT imaging of NIS expression has shown that the peak of NIS expression in the tumor is obtained 48 h after virus administration (Fig. 18.1). Based on this information, the administration of ^{131}I was performed 48 h after virus administration, resulting in a very significant reduction of tumor burden (Fig. 18.2). This therapeutic strategy was also successfully demonstrated for oncolytic measles (Dingli *et al.*, 2004; Dingli *et al.*, 2005; Hasegawa *et al.*, 2006; Li *et al.*, 2010) and vesicular stomatitis (Goel *et al.*, 2007) viruses.

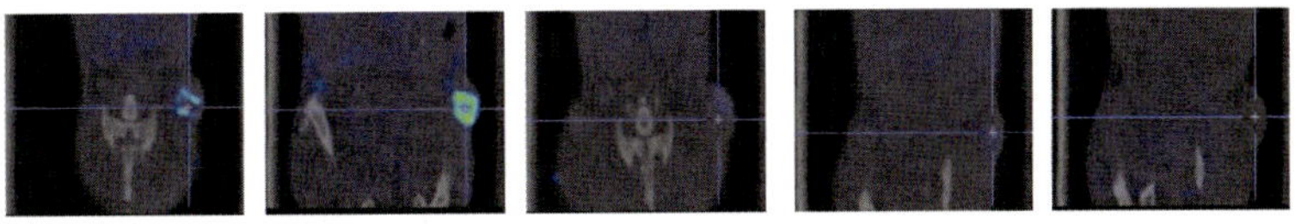

FIGURE 18.1 ■ Kinetics of spread of AdIP2 *in vivo*. Balb-c nude mice were seeded subcutaneously with HCT116 cells (2×10^6). When the tumors reached around 1 cm^2 in size, 10^9 plaque forming units of the Wnt-selective replicating adenovirus AdIP2 were injected. Mice (n = 10) were scanned using a dedicated small-animal SPECT/CT scanner and representative scans showing $^{99m}TcO_4^-$ accumulation in the tumor are shown. The data demonstrate that the peak of NIS expression is reached 48 h after virus administration. (For further details, see Peerlinck *et al.*, 2009.)

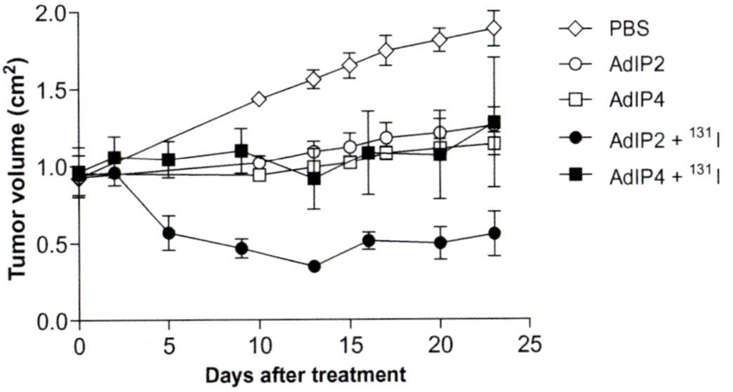

FIGURE 18.2 ■ Comparison of the viro-radiotherapy induced by the NIS-positive (AdIP2) and NIS-negative (AdIP4) oncolytic adenoviruses. Balb-c nude mice were seeded subcutaneously with HCT116 cells (2×10^6). When the tumors reached around 1 cm^2 in size, 10^9 plaque-forming units of AdIP2 or AdIP4 were injected. After 48 h a single dose of $^{131}I^-$ (1.5 mCi) was administered intraperitoneally. The data represent means ± standard error of the mean of the tumor measurement, with six animals per experimental group. Two-way analysis of variance statistical analysis was performed. These results are representative of two experiments. These data validate the concept of image-guided viro-radiotherapy in the context of an oncolytic adenovirus. (For further details, see Peerlinck *et al.*, 2009.)

18.5 Clinical Studies

To date, only three clinical gene therapy studies involving imaging have been published. These involved cancer patients administered either a non-viral gene delivery vector (Jacobs *et al.*, 2001), a replication-incompetent adenovirus (Penuelas *et al.*, 2005) or an oncolytic adenovirus (Barton *et al.*, 2008).

The first published study reporting the visualization of gene transfer in patients was published in 2001 (Jacobs *et al.*, 2001). In this phase I/II study, patients with recurrent glioblastoma were given a non-viral liposomal vector delivering the HSVtk gene administered intratumorally. PET imaging was performed using ^{124}I-FIAU as a radiotracer. Of the five patients involved in this study, only one showed specific ^{124}I-FIAU accumulation in the tumor, strongly indicating the expression of the transgene. Administration of the prodrug, ganciclovir, resulted in signs of necrosis (measured using fluorodeoxyglucose (FDG)-PET and ^{11}C-methionine (MET)-PET) in the area of specific ^{124}I-FIAU accumulation, suggesting an HSVtk-mediated response. In the four other patients, no signs of transgene expression were observed, suggesting the low efficacy of the gene delivery vector used in this study (Jacobs *et al.*, 2001).

In 2005, PET imaging of the intratumoral administration of an adenovirus encoding HSVtk was reported (Penuelas *et al.*, 2005). In this study, performed on seven patients with hepatocellular carcinoma, transgene expression in the tumor was dependent on the injected dose and was detectable in all patients who received more than 10^{12} viral particles. No transgene expression was detected in distant organs or in surrounding, cirrhotic tissues. When the PET-imaging procedure was repeated after treatment with the prodrug, valganciclovir, no expression was observed, suggesting that this treatment efficiently eliminated the transduced tumor cells. In one patient, re-administration of the adenovector was performed but no PET signal was detected in the injected tumor nodule. This study established the vector dose required to be able to detect transgene expression by PET, and demonstrated that hepatocarcinoma is a permissive tumor for adenoviral infection and that non-tumoral tissue is spared when the vector is administered by intratumoral injection.

The most-recent study was published in 2008 and involved nine patients with clinically-localized prostate cancers (Barton *et al.*, 2008). Intra-prostatic injection of an oncolytic adenovirus encoding NIS as a reporter gene was performed and NIS expression was monitored by SPECT, using $^{99m}TCO_4^-$ as a tracer. The whole procedure was concluded to be safe, and gene expression was detected in seven out of nine patients who had received injections of 10^{12} viral particles. In the case of the three patients who received 10^{11} viral particles, no signal was detectable. The volume of gene expression increased to a mean of 6.6 cm^3, representing an average of 18% of the volume of the prostate. Both the volume and the intensity of gene expression peaked one to two days after virus administration and remained detectable in the prostate up to seven days after administration. Whole-body SPECT imaging demonstrated that there were no signs of extra-prostatic viral dissemination (Barton *et al.*, 2008).

18.6 Conclusions

Molecular imaging of gene expression has developed from an academic concept into a real clinical possibility in less than 15 years. Preclinical and clinical studies have demonstrated that the methodologies are robust and relevant, and can provide unique information which can be used to monitor the safety of the approach and may help to improve therapeutic efficacy. With the development of improved gene delivery vectors, molecular imaging is likely to become a key tool in the implementation of human gene therapy.

References

Ahn, B.C., Ronald, J.A., Kim, Y.I., *et al.* (2011). Potent, tumor-specific gene expression in an orthotopic hepatoma rat model using a survivin-targeted, amplifiable adenoviral vector, *Gene Ther*, **18**, 606–612.

Auricchio, A., Acton, P.D., Hildinger, M., *et al.* (2003). *In vivo* quantitative noninvasive imaging of gene transfer by single-photon emission computerized tomography, *Hum Gene Ther*, **14**, 255–261.

Baril, P., Martin-Duque, P., Vassaux, G. (2010). Visualization of gene expression in the live subject using the Na/I symporter as a reporter gene: applications in biotherapy, *Br J Pharmacol*, **159**, 761–771.

Barton, K.N., Stricker, H., Brown, S.L., *et al.* (2008). Phase I study of noninvasive imaging of adenovirus-mediated gene expression in the human prostate, *Mol Ther*, **16**, 1761–1769.

Bhang, H.E., Gabrielson, K.L., Laterra, J., *et al.* (2011). Tumor-specific imaging through progression elevated gene-3 promoter-driven gene expression, *Nat Med*, **17**, 123–129.

Boland, A., Ricard, M., Opolon, P., *et al.* (2000). Adenovirus-mediated transfer of the thyroid sodium/iodide symporter gene into tumors for a targeted radiotherapy, *Cancer Res*, **60**, 3484–3492.

Brader, P., Kelly, K.J., Chen, N., *et al.* (2009). Imaging a genetically engineered oncolytic vaccinia virus (GLV-1h99) using a human norepinephrine transporter reporter gene, *Clin Cancer Res*, **15**, 3791–3801.

Burton, J.B., Johnson, M., Sato, M., *et al.* (2008). Adenovirus-mediated gene expression imaging to directly detect sentinel lymph node metastasis of prostate cancer, *Nat Med*, **14**, 882–888.

Buursma, A.R., Beerens, A.M., de Vries, E.F., *et al.* (2005). The human norepinephrine transporter in combination with ^{11}C-m-hydroxyephedrine as a reporter gene/reporter probe for PET of gene therapy, *J Nucl Med*, **46**, 2068–2075.

Chen, I.Y., Gheysens, O., Ray, S., *et al.* (2010a). Indirect imaging of cardiac-specific transgene expression using a bidirectional two-step transcriptional amplification strategy, *Gene Ther*, **17**, 827–838.

Chen, R., Parry, J.J., Akers, W.J., *et al.* (2010b). Multimodality imaging of gene transfer with a receptor-based reporter gene, *J Nucl Med*, **51**, 1456–1463.

Chisholm, E.J., Vassaux, G., Martin-Duque, P., *et al.* (2009). Cancer-specific transgene expression mediated by systemic injection of nanoparticles, *Cancer Res*, **69**, 2655–2662.

Cotugno, G., Aurilio, M., Annunziata, P., *et al.* (2011). Noninvasive repetitive imaging of somatostatin receptor 2 gene transfer with positron emission tomography, *Hum Gene Ther*, **22**, 189–196.

De, A. Lewis, X.Z., Gambhir, S.S. (2003). Noninvasive imaging of lentiviral-mediated reporter gene expression in living mice, *Mol Ther*, **7**, 681–691.

Dingli, D., Peng, K.W., Harvey, M.E., *et al.* (2004). Image-guided radiovirotherapy for multiple myeloma using a recombinant measles virus expressing the thyroidal sodium iodide symporter, *Blood*, **103**, 1641–1646.

Dingli, D., Peng, K.W., Harvey, M.E., *et al.* (2005). Interaction of measles virus vectors with Auger electron emitting radioisotopes, *Biochem Biophys Res Commun*, **337**, 22–29.

Dwyer, R.M., Schatz, S.M., Bergert, E.R., *et al.* (2005). A preclinical large animal model of adenovirus-mediated expression of the sodium-iodide symporter for radioiodide imaging and therapy of locally recurrent prostate cancer, *Mol Ther*, **12**, 835–841.

Freytag, S.O., Barton, K.N., Brown, S.L., *et al.* (2007). Replication-competent adenovirus-mediated suicide gene therapy with radiation in a preclinical model of pancreatic cancer, *Mol Ther*, **15**, 1600–1606.

Gambhir, S.S., Barrio, J.R., Wu, L., *et al.* (1998). Imaging of adenoviral-directed herpes simplex virus type 1 thymidine kinase reporter gene expression in mice with radiolabeled ganciclovir, *J Nucl Med*, **39**, 2003–2011.

Garcia-Aragoncillo, E., Hernandez-Alcoceba, R. (2010). Design of virotherapy for effective tumor treatment, *Curr Opin Mol Ther*, **12**, 403–411.

Goel, A., Carlson, S.K., Classic, K.L., *et al.* (2007). Radioiodide imaging and radiovirotherapy of multiple myeloma using VSV(Δ51)-NIS, an attenuated vesicular stomatitis virus encoding the sodium iodide symporter gene, *Blood*, **110**, 2342–2350.

Groot-Wassink, T., Aboagye, E.O., Glaser, M., *et al.* (2002). Adenovirus biodistribution and noninvasive imaging of gene expression *in vivo* by positron emission tomography using human sodium/iodide symporter as reporter gene, *Hum Gene Ther*, **13**, 1723–1735.

Groot-Wassink, T., Aboagye, E.O., Wang, Y., *et al.* (2004a). Noninvasive imaging of the transcriptional activities of human telomerase promoter fragments in mice, *Cancer Res*, **64**, 4906–4911.

Groot-Wassink, T., Aboagye, E.O., Wang, Y., *et al.* (2004b). Quantitative imaging of Na/I symporter transgene expression using positron emission tomography in the living animal, *Mol Ther*, **9**, 436–442.

Hajitou, A., Trepel, M., Lilley, C.E., *et al.* (2006). A hybrid vector for ligand-directed tumor targeting and molecular imaging, *Cell*, **125**, 385–398.

Harrington, K.J., Melcher, A., Vassaux, G., *et al.* (2008). Exploiting synergies between radiation and oncolytic viruses, *Curr Opin Mol Ther*, **10**, 362–370.

Hasegawa, K., Pham, L., O'Connor, M.K., *et al.* (2006). Dual therapy of ovarian cancer using measles viruses expressing carcinoembryonic antigen and sodium iodide symporter, *Clin Cancer Res*, **12**, 1868–1875.

Hemminki, A., Zinn, K.R., Liu, B., *et al.* (2002). *In vivo* molecular chemotherapy and noninvasive imaging with an infectivity-enhanced adenovirus, *J Natl Cancer Inst*, **94**, 741–749.

Hingorani, M., Spitzweg, C., Vassaux, G., *et al.* (2010). The biology of the sodium iodide symporter and its potential for targeted gene delivery, *Curr Cancer Drug Targets*, **10**, 242–267.

Huyn, S.T., Burton, J.B., Sato, M., *et al.* (2009). A potent, imaging adenoviral vector driven by the cancer-selective mucin-1 promoter that targets breast cancer metastasis, *Clin Cancer Res*, **15**, 3126–3134.

Jacobs, A., Voges, J., Reszka, R., *et al.* (2001). Positron-emission tomography of vector-mediated gene expression in gene therapy for gliomas, *Lancet*, **358**, 727–729.

Jacobs, R.E., Fraser, S.E. (1994). Magnetic resonance microscopy of embryonic cell lineages and movements, *Science*, **263**, 681–684.

Johnson, M., Huyn, S., Burton, J., *et al.* (2006). Differential biodistribution of adenoviral vector *in vivo* as monitored by bioluminescence imaging and quantitative polymerase chain reaction, *Hum Gene Ther*, **17**, 1262–1269.

Khuri, F.R., Nemunaitis, J., Ganly, I., *et al.* (2000). A controlled trial of intratumoral ONYX-015, a selectively-replicating adenovirus, in combination with cisplatin and 5-fluorouracil in patients with recurrent head and neck cancer, *Nat Med*, **6**, 879–885.

Klutz, K., Russ, V., Willhauck, M.J., *et al.* (2009). Targeted radioiodine therapy of neuroblastoma tumors following systemic nonviral delivery of the sodium iodide symporter gene, *Clin Cancer Res*, **15**, 6079–6086.

Lan, X., Liu, Y., He, Y., *et al.* (2010). Autoradiography study and SPECT imaging of reporter gene HSV1-tk expression in heart, *Nucl Med Biol*, **37**, 371–380.

Le, L.P., Le, H.N., Dmitriev, I.P., *et al.* (2006). Dynamic monitoring of oncolytic adenovirus *in vivo* by genetic capsid labeling, *J Natl Cancer Inst*, **98**, 203–214.

Li, H., Peng, K.W., Dingli, D., *et al.* (2010). Oncolytic measles viruses encoding interferon β and the thyroidal sodium iodide symporter gene for mesothelioma virotherapy, *Cancer Gene Ther*, **17**, 550–558.

Liang, Q., Yamamoto, M., Curiel, D.T., *et al.* (2004). Noninvasive imaging of transcriptionally restricted transgene expression following intratumoral injection of an adenovirus in which the COX-2 promoter drives a reporter gene, *Mol Imaging Biol*, **6**, 395–404.

Liu, B., Herve, J., Bioulac-Sage, P., *et al.* (2007). Sodium iodide symporter is expressed at the preneoplastic stages of liver carcinogenesis and in human cholangiocarcinoma, *Gastroenterology*, **132**, 1495–1503.

McCart, J.A., Mehta, N., Scollard, D., *et al.* (2004). Oncolytic vaccinia virus expressing the human somatostatin receptor SSTR2: molecular imaging after systemic delivery using ^{111}In-pentetreotide, *Mol Ther*, **10**, 553–561.

Merron, A., Baril, P., Martin-Duque, P., *et al.* (2010). Assessment of the Na/I symporter as a reporter gene to visualize oncolytic adenovirus propagation in peritoneal tumours, *Eur J Nucl Med Mol Imaging*, **37**, 1377–1385.

Merron, A., Peerlinck, I., Martin-Duque, P., *et al.* (2007). SPECT/CT imaging of oncolytic adenovirus propagation in tumours *in vivo* using the Na/I symporter as a reporter gene, *Gene Ther*, **14**, 1731–1738.

Mullerad, M., Eisenberg, D.P., Akhurst, T.J., *et al.* (2006). Use of positron emission tomography to target prostate cancer gene therapy by oncolytic herpes simplex virus, *Mol Imaging Biol*, **8**, 30–35.

Nakamoto, Y., Saga, T., Misaki, T., *et al.* (2000). Establishment and characterization of a breast cancer cell line expressing Na^+/I^- symporters for radioiodide concentrator gene therapy, *J Nucl Med*, **41**, 1898–1904.

Ong, H.T., Hasegawa, K., Dietz, A.B., *et al.* (2007). Evaluation of T cells as carriers for systemic measles virotherapy in the presence of antiviral antibodies, *Gene Ther*, **14**, 324–333.

Peerlinck, I., Merron, A., Baril, P., *et al.* (2009). Targeted radionuclide therapy using a Wnt-targeted replicating adenovirus encoding the Na/I symporter, *Clin Cancer Res*, **15**, 6595–6601.

Penuelas, I., Mazzolini, G., Boan, J.F., *et al.* (2005). Positron emission tomography imaging of adenoviral-mediated transgene expression in liver cancer patients, *Gastroenterology*, **128**, 1787–1795.

Peyrottes, I., Navarro, V., Ondo-Mendez, A., *et al.* (2009). Immunoanalysis indicates that the sodium iodide symporter is not overexpressed in intracellular compartments in thyroid and breast cancers, *Eur J Endocrinol*, **160**, 215–225.

Qiao, J., Doubrovin, M., Sauter, B.V., *et al.* (2002). Tumor-specific transcriptional targeting of suicide gene therapy, *Gene Ther*, **9**, 168–175.

Rogers, B.E., Parry, J.J., Andrews, R., *et al.* (2005). MicroPET imaging of gene transfer with a somatostatin receptor-based reporter gene and ^{94m}Tc-Demotate 1, *J Nucl Med*, **46**, 1889–1897.

Rogers, B.E., Zinn, K.R., Lin, C.Y., *et al.* (2002). Targeted radiotherapy with [^{90}Y]-SMT 487 in mice bearing human nonsmall cell lung tumor xenografts induced to express human somatostatin receptor subtype 2 with an adenoviral vector, *Cancer*, **94**, 1298–1305.

Sato, M., Figueiredo, M.L., Burton, J.B., *et al.* (2008). Configurations of a two-tiered amplified gene expression system in adenoviral vectors designed to improve the specificity of *in vivo* prostate cancer imaging, *Gene Ther*, **15**, 583–593.

Segditsas, S., Tomlinson, I. (2006). Colorectal cancer and genetic alterations in the Wnt pathway, *Oncogene*, **25**, 7531–7537.

Singh, S.P., Han, L., Murali, R., *et al.* (2011). SSTR2-based reporters for assessing gene transfer into non-small cell lung cancer: evaluation using an intrathoracic mouse model, *Hum Gene Ther*, **22**, 55–64.

Sweeney, T.J., Mailander, V., Tucker, A.A., *et al.* (1999). Visualizing the kinetics of tumor-cell clearance in living animals, *Proc Natl Acad Sci USA*, **96**, 12044–12049.

Tjuvajev, J.G., Finn, R., Watanabe, K., *et al.* (1996). Noninvasive imaging of herpes virus thymidine kinase gene transfer and expression: a potential method for monitoring clinical gene therapy, *Cancer Res*, **56**, 4087–4095.

Tjuvajev, J.G., Stockhammer, G., Desai, R., *et al.* (1995). Imaging the expression of transfected genes *in vivo*, *Cancer Res*, **55**, 6126–6132.

Umegaki, H., Ishiwata, K., Ogawa, O., *et al.* (2003). Longitudinal follow-up study of adenoviral vector-mediated gene transfer of dopamine D2 receptors in the striatum in young, middle-aged, and aged rats: a positron emission tomography study, *Neuroscience*, **121**, 479–486.

Vande Velde, G., Rangarajan, J.R., Toelen, J., *et al.* (2011). Evaluation of the specificity and sensitivity of ferritin as an MRI reporter gene in the mouse brain using lentiviral and adeno-associated viral vectors, *Gene Ther*, **18**, 594–605.

Wapnir, I.L., Goris, M., Yudd, A., *et al.* (2004). The Na^+/I^- symporter mediates iodide uptake in breast cancer metastases and can be selectively down-regulated in the thyroid, *Clin Cancer Res*, **10**, 4294–4302.

Wapnir, I.L., van de Rijn, M., Nowels, K., *et al.* (2003). Immunohistochemical profile of the sodium/iodide symporter in thyroid, breast, and other carcinomas using high density tissue microarrays and conventional sections, *J Clin Endocrinol Metab*, **88**, 1880–1888.

Watanabe, Y., Horie, S., Funaki, Y., *et al.* (2010). Delivery of Na/I symporter gene into skeletal muscle using nanobubbles and ultrasound: visualization of gene expression by PET, *J Nucl Med*, **51**, 951–958.

Wilber, A., Frandsen, J.L., Wangensteen, K.J., *et al.* (2005). Dynamic gene expression after systemic delivery of plasmid DNA as determined by *in vivo* bioluminescence imaging, *Hum Gene Ther*, **16**, 1325–1332.

Willhauck, M.J., Sharif Samani, B.R., Klutz, K., *et al.* (2008). α-Fetoprotein promoter-targeted sodium iodide symporter gene therapy of hepatocellular carcinoma, *Gene Ther*, **15**, 214–223.

Wu, J.C., Inubushi, M., Sundaresan, G., *et al.* (2002). Optical imaging of cardiac reporter gene expression in living rats, *Circulation*, **105**, 1631–1634.

Wu, J.C., Sundaresan, G., Iyer, M., *et al.* (2001). Noninvasive optical imaging of firefly luciferase reporter gene expression in skeletal muscles of living mice, *Mol Ther*, **4**, 297–306.

Yaghoubi, S.S., Wu, L., Liang, Q., *et al.* (2001). Direct correlation between positron emission tomographic images of two reporter genes delivered by two distinct adenoviral vectors, *Gene Ther*, **8**, 1072–1080.

Zincarelli, C., Soltys, S., Rengo, G., *et al.* (2008). Analysis of AAV serotypes 1–9 mediated gene expression and tropism in mice after systemic injection, *Mol Ther*, **16**, 1073–1080.

Zuckier, L.S., Dohan, O., Li, Y., *et al.* (2004). Kinetics of perrhenate uptake and comparative biodistribution of perrhenate, pertechnetate, and iodide by NaI symporter-expressing tissues *in vivo*, *J Nucl Med*, **45**, 500–507.

PART III
THERAPEUTIC APPLICATIONS

19

ONCOLYTIC ADENOVIRUSES FOR CANCER GENE THERAPY

Gunnel Hallden[a], *Yaohe Wang*[a], *Han-Hsi Wong*[a] *and Nick R. Lemoine*[a,b]

19.1 Cancer and Oncolytic Adenoviruses

Cancer is still a leading cause of death in the Western world despite the many efforts to develop new strategies for better treatments and earlier detection. A major problem in treating cancers is the development of resistance to current standard therapeutics, including cytotoxic drugs and radiation therapy. More efficacious therapies with different mechanisms of action to overcome treatment resistance are therefore urgently needed.

A relatively novel and promising therapeutic platform is virotherapy with oncolytic adenoviruses. Over the last two decades several engineered viral mutants have been evaluated in clinical trials targeting various tumor types and were demonstrated to be safe, with some efficacy. Oncolytic mutants were specifically engineered to infect, replicate in and lyse tumor cells, leaving normal tissue relatively unharmed (Fig. 19.1). This approach has been applied to numerous viral species, including adenovirus, measles, herpes and poxviruses, to name a few. However, most of the work has focused on adenoviral vectors, especially serotype 5 (Ad5), because of: the ease of genetically engineering its small, linear and well-characterized 36 kb genome (Figs. 19.2(a), 19.2(b)); its natural tropism for epithelial cells and the carcinomas derived from them; the lack of integration into the host cell genome; the clinical safety record with only flu-like side effects; and the ease

[a] Barts Cancer Institute, Barts & The London School of Medicine & Dentistry, Queen Mary University of London, London EC1M 6BQ, UK
[b] Email: director@qmcr.qmul.ac.uk

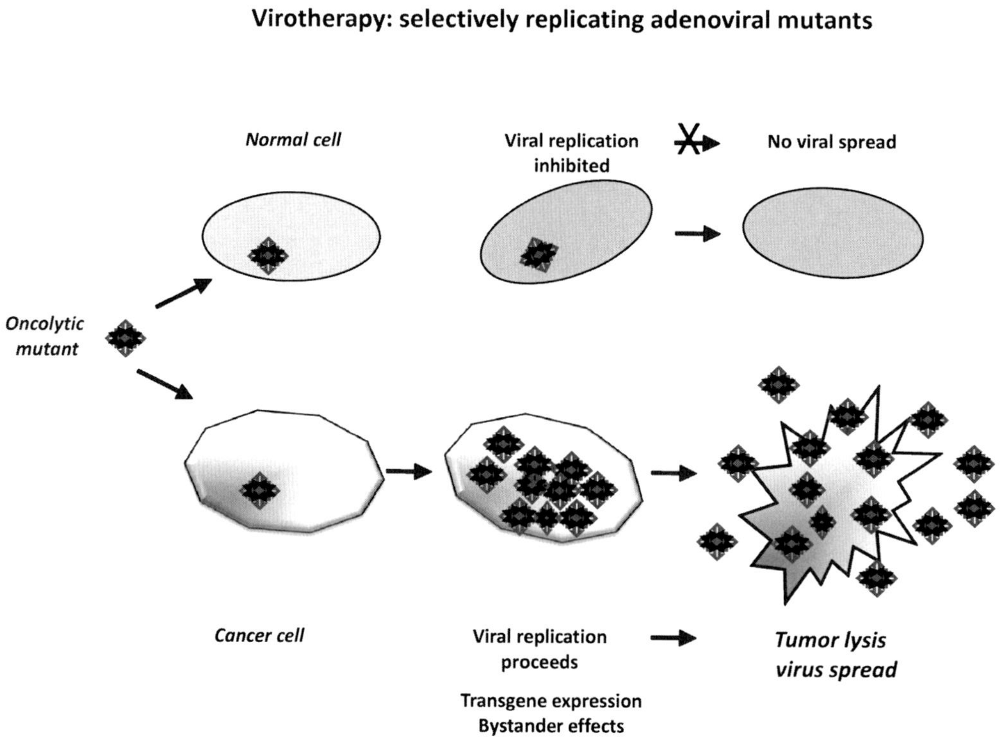

FIGURE 19.1 ■ Principles of oncolytic adenoviral targeting and lysis of cancer cells. Adenoviruses can be engineered to replicate selectively in cancer cells and to have limited toxicity in normal cells. Replication-selective oncolytic mutants can amplify the genome, express viral proteins, allow their assembly into virions and finally lyse the cancer cells and spread to surrounding cells. Replication and spread cannot proceed in normal cells.

of production under good manufacturing practice (GMP). In addition, no cross-resistance with current standard therapies occurs, and on the contrary, early viral genes appear to enhance the apoptosis-inducing effects of other cytotoxic therapies and to reverse resistance. In combination with standard anticancer therapeutics, synergistic cell killing can often be achieved specifically in cancer cells. Additionally, adenoviral mutants can be engineered to not only lyse cancer cells but also express therapeutic transgenes to promote elimination of tumors. Another advantage of adenovirus is the ability to infect both proliferating and non-proliferating tumor cells, an important consideration in many solid tumors with only sub-populations of cells that are actively dividing (e.g., prostate cancer). Here we will address the problems and possibilities for further improvements of oncolytic adenoviruses to target cancers and the most promising recent developments focusing on viruses as multimodal therapeutic agents.

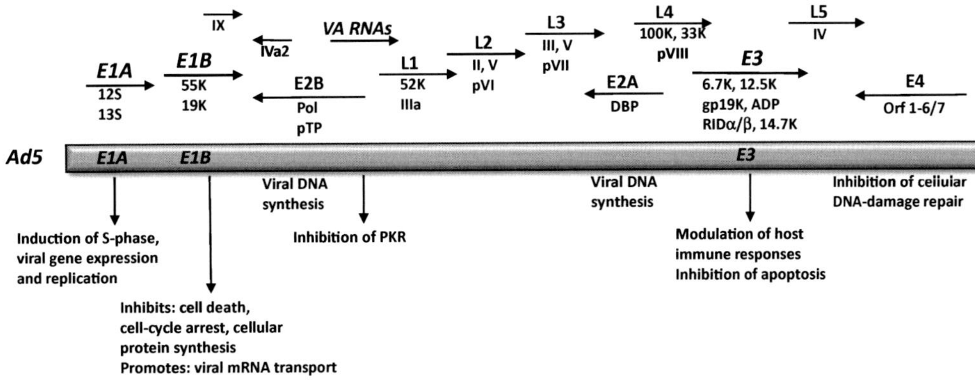

(a)

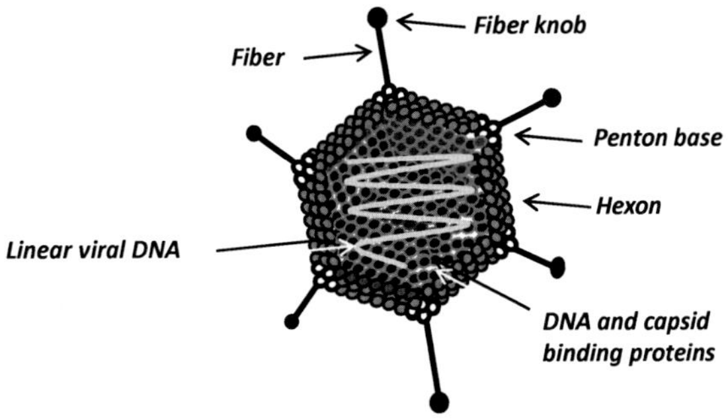

(b)

FIGURE 19.2 ■ Genome organization and structure of Adenovirus serotype 5 (Ad5). (a) Ad5 genome and functions. Graphic representation of the adenoviral 36 kDa linear genome. Viral genes are grouped into early (E1A, E1B, E2A, E2B, E3 and E4) and late (L1–5) genes, referring to the time of expression during infection. E1A is essential for viral genome amplification, protein synthesis and replication, and drives the expression of other early viral genes that are essential for viral propagation and the late genes (L1–5) from the major late promoter (not shown). The late genes code mainly for structural proteins such as hexon, penton and fiber. (b) Adenovirus structure illustrating the capsid organization and the location of viral receptor binding proteins. The fiber knob binds to cellular Coxsackie and adenovirus receptor (CAR)-receptors and the penton base binds to αvβ3- and αvβ5-integrins. Hexon is the main capsid protein that is essential for viral integrity and is stabilized by several additional coat proteins. The linear 36 kDa viral genome is attached to the capsid through binding to capsid-associated proteins and specialized DNA binding proteins.

19.2 Development of Oncolytic Adenoviruses

To date, several replication-selective adenoviral mutants have been demonstrated to kill cancer cells potently and specifically in culture and in *in vivo* tumor models.

Some of these mutants were also evaluated in clinical trials in thousands of cancer patients, targeting different tumor types. While all tested oncolytic mutants were found to be safe with limited side effects, clinical efficacy was modest when they were administered alone with improved responses in combination with chemotherapeutics or radiation therapy (reviewed by Parato *et al.*, 2005; Aghi and Martuz, 2005). Therefore, most current research is focused on generating more efficacious mutants whilst retaining robust safety profiles.

There are several approaches to engineering replication-selective, adenoviral mutants: (1) deletion of viral genes that are essential for the viral life-cycle to proceed in normal cells but are functionally complemented by the altered gene expression in cancer cells, (2) insertion of tumor-/tissue-specific promoters to control expression of early viral genes that drive replication, (3) modification of viral tropism to specifically target tumor antigens and infect cancer cells only, (4) insertion of microRNA (miRNA) target sequences to suppress expression in normal cells and (5) combining any or all of the approaches above, with or without expression of therapeutic genes e.g., small RNAs, anti-angiogenesis factors, cytokines or prodrug-converting enzymes. Common deletion and insertion sites in the Ad5 genome are illustrated in Fig. 19.3(a).

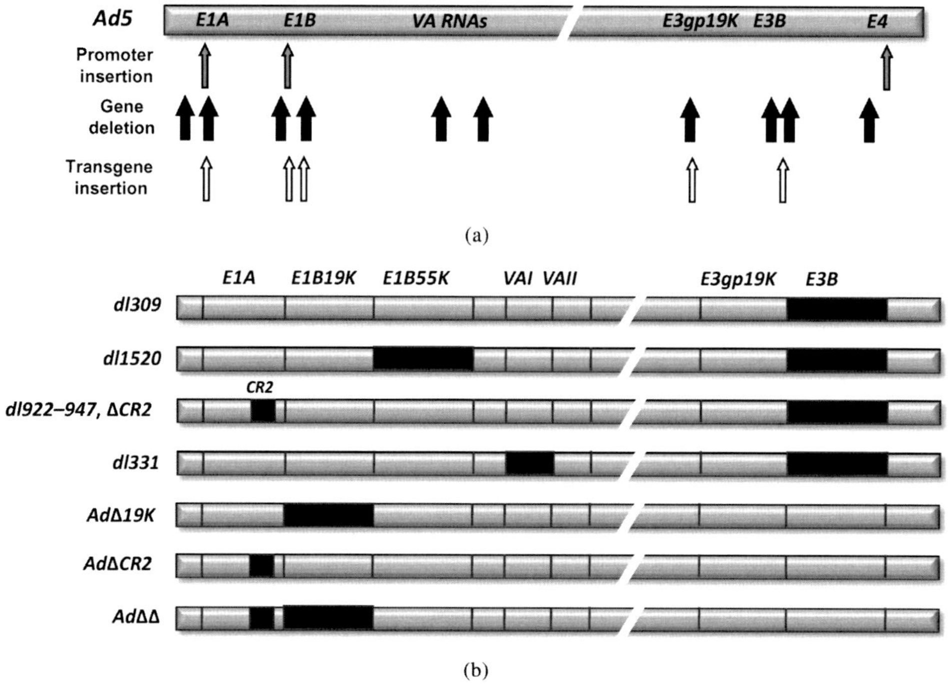

FIGURE 19.3 ■ Outline of the Ad5 genome illustrating common sites for gene insertions and deletions in oncolytic mutants. (a) Gene regions in the Ad5 genome that are frequently used for insertion of promoters and transgenes, and gene-deletions when constructing complementation-deleted replication-selective mutants. (b) Deletion-mutants described in the text illustrating the characteristic functional deletions incorporated in each virus. Black boxes indicate the deleted gene-regions.

19.2.1 Oncolytic Adenoviral Deletion Mutants

The most common strategy to generate oncolytic mutants is deletion of genes or gene-regions that are essential for viral replication in normal cells but not in cancer cells. Thus, viral propagation can only proceed in the presence of cellular pathways already deregulated during carcinogenesis, such as mutations of p53, retinoblastoma protein (pRb), p16 and other cell-cycle and apoptosis regulatory factors.

The most intensively investigated oncolytic adenoviruses to date, ONYX-015 and H101, are based on the *dl*1520 variant, and both are Ad5 with the E1B55K-gene deleted (Fig. 19.3(b)). E1B55K is essential for p53-binding and degradation and also for nuclear export of viral messenger RNA (mRNA) and host-cell protein synthesis shut-off, functions that are crucial for viral propagation in normal cells (Fig. 19.4). The majority of cancers present with both p53-inactivating mutations and alterations in mRNA export, rendering the *dl*1520 mutants tumor-selective. Most of the clinical trials with oncolytic mutants have been with ONYX-015 or H101 that proved to be safe in humans by all routes of administration. H101 is now licensed in China for treatment of head and neck cancers (Oncorine; Shanghai Sunway Biotech). However, to achieve significant antitumor efficacy with either *dl*1520 mutant combinations with cytotoxic drugs such as 5-fluorouracil (5-FU), cisplatin or gemcitabine were necessary. One reason for the poor efficacy of these mutants is that viral replication was not only attenuated in normal cells but also in the majority of tumor cells, limiting lysis and intratumoral spread. It is now known that E1B55K has additional important functions in supporting the viral life-cycle and it is the loss of these that contributes to the poor efficacy. In addition, *dl*1520 mutants have deletion of the E3B-genes that are essential for defending the infected host cell against innate immune responses (Figs. 19.2(a), 19.3(b) and 19.5). In hosts with intact immune systems, the absence of the E3B-genes results in premature elimination of infected cells by infiltrating macrophages, further preventing viral propagation and spread.

To improve on viral efficacy several new mutants have been engineered with the E3-genes retained for prolonged survival *in vivo*. Additionally, smaller specific gene-deletions in the E1A-region are now replacing the older prototype deletion of the E1B55K gene, which is retained to increase viral potency. Several of these mutants target the pRb/p16 cell-cycle pathway by deletion of the small Conserved Region 2 in the E1A gene (E1ACR2; 24 nt) that is essential for pRb-binding and inactivation in normal cells (Figs. 19.3(b) and 19.4). E1ACR2-binding to pRb results in S-phase entry so that viral replication can proceed. However, the cell cycle is almost always deregulated in cancer cells (e.g., Kirsten rat sarcoma oncogene (*K-ras*)-activation, *p53*mut, *p16*mut) and consequently replication of E1ACR2-deleted mutants can proceed in tumor but not in normal cells with intact cell-cycle control. While all tested E1ACR2-deleted oncolytic adenoviral mutants

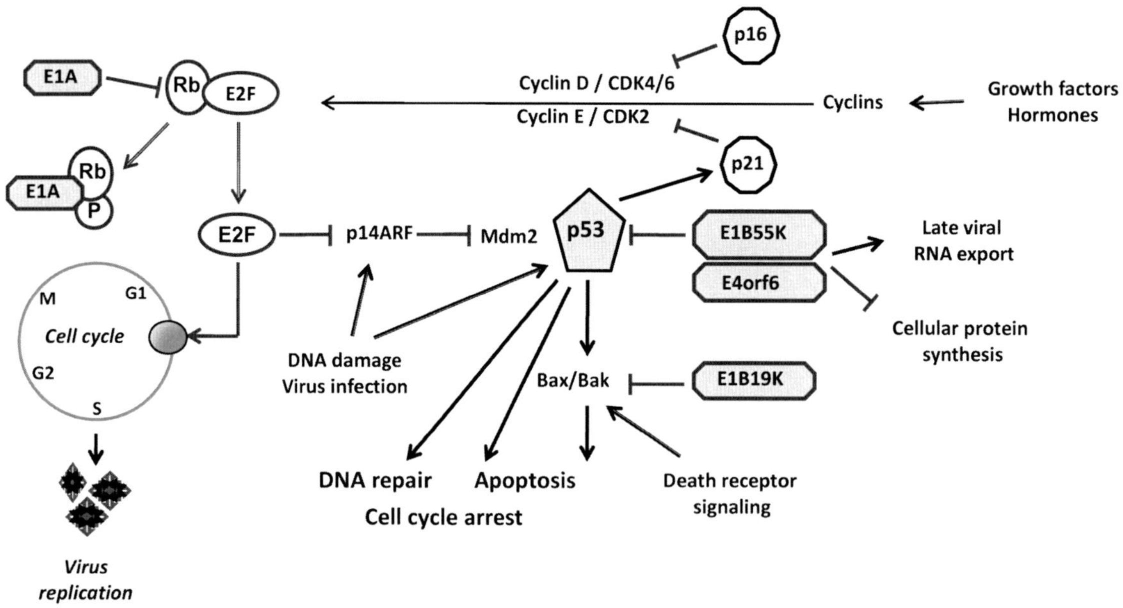

FIGURE 19.4 ■ Interaction of Ad5 proteins with cellular proteins. Function of early viral genes (E1A, E1B and E4) in directing the cellular machinery to favor viral propagation. The retinoblastoma protein (Rb) is normally hypophosphorylated and binds to E2F transcription factors to regulate the G1-to-S checkpoint of the cell cycle. In response to mitogenic stimuli such as growth factors and hormones, cyclins and cyclin-dependent kinases (CDKs) are activated in turn phosphorylating Rb, and E2F is released. E2F expression drives the cell into S-phase with protein synthesis, DNA amplification and cell-cycle progression. One of these E2F-dependent up-regulated proteins is p14 alternate reading frame (p14ARF), which inhibits mouse double minute 2 (Mdm2). Mdm2 binds to p53 and directs the complex towards degradation. In response to stress signals such as virus infection or DNA damage the p53 transcription factor is up-regulated and activated, resulting in the expression of proteins that induce apoptosis (B-cell leukemia 2-associated X protein (Bax)), cell-cycle arrest (p21-CDK-interacting protein 1/wild-type p53-activated fragment 1 (p21-CIP1/WAF1) via its inhibition of CDK2) or DNA repair. p16 CDK4 inhibitor A (p16INK4A) is a tumor suppressor that inactivates CDK4/6. The adenoviral E1A protein binds to Rb to release E2F, forcing the cell into S-phase so that viral DNA can be replicated. E1A also promotes the acetylation of Rb by binding to p300/cyclic AMP-responsive element-binding protein binding protein (CBP), resulting in Rb binding to Mdm2 to inhibit p53 activation. The cell cycle is frequently deregulated in cancer cells and consequently mutants that cannot bind Rb (E1A conserved region 2 (*E1ACR2*)-deleted e.g., *dl*922–947, AdΔ24) can selectively replicate in and eliminate proliferating cancer cells but not normal non-proliferating cells. The anti-apoptotic E1B19K protein binds to and inhibits Bax/B-cell leukemia 2-homologous antagonist/killer (Bak) preventing mitochondrial depolarization. The tumor selectivity of *E1B19K*-deleted mutants (AdΔ19K) is due to multiple defects in the apoptotic pathways, while survival of the virus in normal cells is limited because of rapid apoptosis-induction in response to tumor necrosis factor-α (TNF-α). The E1B55K protein associates with the adenovirus E4 open reading frame 6 (E4orf6) protein to form an E3 ubiquitin ligase complex that targets p53 for degradation. The complex also induces expression of cyclin E as well as simultaneous inhibition of cellular mRNA export to prevent host cell protein synthesis and promotion of late viral mRNA export. *E1B55K*-deleted mutants can replicate selectively in tumors because of non-functional p53 activation, cyclin E overexpression and E1B55K-independent late viral RNA export in cancer, but not normal, cells.

(e.g., *dl*922-947, AdΔ24; Fig. 19.3(b)) have higher efficacy compared with *dl*1520 mutants, higher levels of replication in normal cells and higher liver toxicity were also reported. Nevertheless, the first clinical evaluation of an E1ACR2-deleted mutant, Ad5-Δ24RGD was recently completed in a phase I trial for recurrent malignant ovarian adenocarcinomas (http://www.clinicaltrials.gov/NCT00562003).

Additional modifications of the E1ACR2-deleted mutants have been attempted to increase tumor selectivity. For example, a double-deleted mutant with both the E1ACR2-deletion and the E1B55K-gene deleted was developed but resulted in a virus that was as attenuated as the original *dl*1520 mutant. We recently developed another mutant with the E1ACR2-region and a second anti-apoptotic gene (E1B19K) deleted, but with retained E3-genes (AdΔΔ; Fig. 19.3(b)). E1B19K is a B-cell leukemia-2 (Bcl-2) homologue that is redundant in the majority of cancers with altered apoptosis pathways (Fig. 19.4). The AdΔΔ mutant replicates at much lower levels in normal cells and is as potent as wild-type virus in both prostate and pancreatic preclinical cancer models.

Adenovirus also produces the virus-associated (VA) RNAs. These are RNA polymerase III transcripts that, amongst other functions, are obligatory for efficient translation of viral and cellular mRNAs by blocking the double-stranded RNA-activated protein kinase (PKR), a natural host antiviral defense system (Fig. 19.2(a)). Lack of VAI expression may lead to viral gene expression defects, or low levels of gene expression later in infection. Deletion of VA-RNAI leads to a six-fold to ten-fold reduction in replication efficiency in certain cell lines. In non-transformed cells, adenovirus lacking VAI and VAII were more than 100-fold less potent than wild-type virus making VAI-deleted mutants appealing as candidates for oncolytic vectors. We have shown that a VAI-deleted Ad5 mutant (*dl*331; Fig. 19.3(b)) is able to selectively target Epstein–Barr virus (EBV)-associated tumors such as Burkitt lymphoma and nasopharyngeal carcinoma. EBV tumors express the RNAs EBV-encoded RNA 1 (EBER1) and EBER2, with EBER1 complementing *dl*331 and enabling the synthesis of viral proteins. Interestingly, antitumoral efficacy *in vitro* and *in vivo* was superior to wild-type Ad5 in these tumors. In addition, VAI-RNA-deleted adenoviruses may be potential therapeutic agents targeting tumors with activating *ras*-mutations because the pathways activated downstream can inhibit PKR phosphorylation and activation.

19.2.2 Tumor-Specific Promoter-Driven Mutants

Tumor-specific promoter-driven oncolytic adenoviral mutants have been engineered to target a variety of tissue and tumor types. The most intensively investigated promoter-constructs are the androgen-response elements (ARE) comprised of modified androgen receptor-activated promoter and enhancer sequences in various combinations, most often derived from the prostate-specific antigen (PSA), prostate-specific membrane antigen (PSMA) and probasin promoter/enhancer regions. Insertion of these AREs to drive viral E1A-expression and/or other early viral genes such as E1B or E4 generates prostate-selective oncolytic viruses with replication limited to late-stage androgen-insensitive prostate cancers. The most selective and efficacious of these mutants, CG7870 (Cell Genesys, USA), has a relatively large PSA promoter and upstream enhancer region inserted between the native E1A promoter and coding region in addition to the rat probasin promoter and the minimal PSA enhancer/promoter to drive E1B expression, and intact E3-immunomodulatory genes. CG7870 has been evaluated in several clinical trials with both intratumoral and intravenous delivery. Even though the treatment was not curative, some efficacy was reported from the trials with 27% of patients free from tumor progression six months later. Limited responses were also observed in several clinical trials targeting a variety of solid tumors with viruses containing other tumor-specific promoters driving viral replication such as the human telomerase reverse transcriptase (hTERT; Telomelysine), the osteocalcin (Ad-OC-E1a) and various human E2F promoters.

A major problem when exchanging the native E1A promoter with tissue-specific/tumor-specific promoters is that cellular promoters are not optimized for adenoviral gene expression and consequently viral propagation is always attenuated compared with wild-type virus. Another difficulty with adenovirus is its limited cloning capacity that only allows for a 5% increase in packaged viral DNA. This problem has been a major obstacle in constructing potent tumor-specific-promoter-driven mutants because many promoter constructs are rather large, including tissue-specific enhancer or insulator sequences to shield them from viral enhancers, which necessitates deletion of viral genes resulting in attenuation of viral propagation. One recently developed mutant with a relatively large and complex chimeric promoter/enhancer sequence is Ad[i/PPT-E1A], comprising the T-cell receptor γ-chain alternate reading frame protein (TARP) promoter, the PSMA and PSA enhancer sequences (PPT) and part of the H19 insulator. This mutant potently expressed viral genes both in the presence and absence of androgens. However, due to the large size of the insert essential viral genes had to be deleted such as the E1B55K and/or E3-genes. While the promoter construct in Ad[i/PPT-E1A] caused high levels of gene expression, viral propagation appeared to be attenuated and would probably not be sufficient for elimination of tumors in the clinic, similar to other mutants with cellular promoters.

19.2.3 Tumor-Specific Tropism-Modified Mutants

Infection of cells by most adenoviruses in subgroups A, C, D, E and F, including Ad5 (subgroup C), is facilitated by binding to the Coxsackie and adenovirus receptor (CAR) on the cell surface membrane via the knob portion of the viral fiber protein (Fig. 19.2(b)). Following the initial attachment, viral penton proteins bind to cellular integrins through arginine–glycine–aspartic acid (RGD)-sequences, an interaction which appears to be essential for internalization of virus. The expression levels of CAR have long been thought of as critical in determining the transduction efficiency of adenovirus, in particular for Ad5, *in vivo*. While CAR is ubiquitously expressed in epithelial cells, it is often down-regulated in many cancer types due to activation of the Ras-associated factor–mitogen-activated protein kinase (Raf–MAPK) pathway. To circumvent this problem, various attempts have been made to increase the delivery of Ad5 to target tumor tissues by modifying structural tropism determinants. Firstly, given that most subgroup B adenoviruses bind to CD46, a receptor often up-regulated in a number of tumor types, including breast, cervical, liver, lung, endometrial and hematological malignancies, several chimeric variants of oncolytic Ad5, such as Ad5/35, Ad5/3 or Ad5/11, have been developed to exploit the fiber tropism of subgroup B viruses, and they have all shown encouraging results. Other retargeting strategies have included the insertion of RGD motifs, a high-affinity, αvβ6-selective peptide, basic fibroblast growth factor (FGF2), epidermal growth factor (EGF) and transactivator (TAT) peptide from human immunodeficiency virus type 1 etc. into the region of adenovirus fiber knob in attempts to improve delivery to malignant tissue (reviewed by Waehler *et al.* 2007).

19.3 Enhancing the Potency of Oncolytic Adenoviral Mutants

To eliminate tumors in cancer patients, the efficacy of most oncolytic adenoviral mutants needs to be greatly improved without losing tumor selectivity. The following approaches are most likely to generate improved therapies and need further exploration and development: (1) generation of more potent viral mutants that will also enhance the efficacy of current standard therapeutics, (2) utilization of the virus–tumor microenvironment interactions by modifications of tropism, antigen targeting, stromal degradation and anti-angiogenesis targeting, (3) exploiting the host immune response by engineering mutants to stimulate antitumor immunity.

19.3.1 Arming Replication-Selective Oncolytic Mutants with Therapeutic Genes

The selective replication and spread of oncolytic adenoviruses in cancer cells can also result in potent, local and long-lasting expression of transgenes, and therefore greatly improve on the therapeutic efficacy of the mutant. Arming oncolytic adenoviruses with anticancer genes has been a major focus in cancer virotherapy. In order to retain viral functions, therapeutic genes are ideally expressed from the E1B or E3 regions either under direct control of the corresponding viral promoters or by insertion of cellular promoters (Fig. 19.3a). Numerous armed viral mutants have been constructed to express genes such as prodrug-converting enzymes, anti-angiogenesis genes, immunomodulatory cytokines, tumor suppressors or pro-apoptotic genes, small RNAs and imaging genes.

Tumor-selective mutants expressing prodrug-converting enzymes amplify both the viral DNA and the transgene to high levels in the infected cells and, consequently, conversion of the non-toxic prodrug to toxic drug occurs only in the tumor, thereby minimizing unwanted systemic effects. Cell killing is also frequently enhanced through synergistic interactions between virus and active drug that promote viral spread. On the other hand, drugs that damage cellular DNA might also attenuate viral DNA synthesis with detrimental effects on the viral life cycle and need to be carefully assessed. The three most commonly used prodrug-converting enzyme drug systems used with adenoviral mutants are herpes simplex virus thymidine kinase (HSVtk) with ganciclovir (GCV) or acyclovir (ACV); bacterial or yeast-derived cytosine deaminase sometimes combined with uracil phosphoribosyltransferase (UPRT) with 5-fluorocytosine (5-FC); and *Escherichia coli* nitroreductase (NTR) with CB1954. Several mutants expressing these enzymes have been evaluated in early phase trials targeting various cancers. Despite the fact that viral replication of these mutants is often attenuated due to compromised viral gene expression, promising outcomes were reported, probably because of the positive virus and drug interactions that can activate cell death pathways synergistically.

Anti-angiogenic agents also represent a promising strategy for anticancer therapies when expressed from oncolytic mutants. The development of angiogenesis inhibitors has become a broad and active area of cancer research. Despite great promise, results with peptide inhibitors or anti-angiogenic gene therapy delivered by plasmids or non-replicating viruses alone in clinical trials had proved disappointing. The belief that more sustained exposure to the agents locally in tumor masses might be more efficacious has led to the development of anti-angiogenic gene therapies for cancer using replication-selective oncolytic viruses. This approach prevents further growth by angiogenesis inhibition, while eradicating tumors by viral cell lysis. Several replication-selective oncolytic adenovirus mutants expressing anti-angiogenic components, such as vascular endothelial growth factor (VEGF) promoter-targeted transcriptional repressor zinc-finger protein (ZFP), soluble McDonough feline sarcoma-like tyrosine kinase 1 (Sflt-1), interleukin (IL)-8 specific short hairpin RNA (shRNA), endostatin, endostatin–angiostatin fusion gene and mutated Kringle 5 of human plasminogen have been developed and tested in preclinical tumor models and demonstrated superior antitumor efficacy over unarmed parental oncolytic mutants.

Administration of oncolytic adenoviruses to tumor-bearing animals with an intact immune system can, in addition to tumor cell lysis, also result in tumor-specific immunity by encouraging production of tumor-antigen-specific clusters of differentiation (CD)8$^+$ T-cells. Combining immunotherapy and virotherapy could therefore further enhance the antitumor efficacy. The efficacy of the immunotherapeutic treatment relates to two key factors: the quantity of T-cells produced by the host in reaction to tumor antigens and the length of time that tumor-specific T-cells persist. Around a dozen immunomodulatory genes have been characterized as potential targets to increase substantially the number and therapeutic function of immune T-cells *in vivo*. A plethora of immunostimulatory genes have been inserted into the genome of oncolytic adenoviruses with the aim of stimulating effective antitumoral immune responses. Recent examples include the heat shock protein, B7-1, chemokine (C-C motif) ligand 5 (CCL5), granulocyte macrophage colony-stimulating factor (GM-CSF), interferon (IFN)-β, IL-12, 4-1BB ligand (4-1BBL), IL-18, and IL-24.

Oncolytic adenoviruses could also be armed with tumor suppressor or pro-apoptotic genes that are frequently lost in cancer. One example is the p16 cyclin-dependent kinase (CDK) 4 inhibitor A (p16^{INK4A})-armed mutant, which demonstrated good growth inhibition of gastric tumor xenografts. The tumor suppressor gene p53 has also been inserted into replication-selective mutants, and these agents showed tumor selectivity with efficient p53 expression and better antitumor efficacy than their unarmed equivalents in preclinical tumor models. Nonetheless, targeting a single gene is unlikely to have a major effect on survival, given that in advanced cancers there is a redundancy of targets due to the large number of genetic alterations accumulated. Hence, rather than targeting genes individually, targeting a core set of signaling pathways and miRNAs that regulate several tumor-associated genes is likely to have a greater and more lasting effect on therapeutic outcomes. Viruses that enhance the apoptotic pathways have also been generated. For example, a chimeric Ad5/35 carrying the gene encoding the TNF-related apoptosis-inducing ligand (TRAIL) has been developed for treatment of leukemia, gastric cancer and pancreatic cancer, with good results *in vivo*.

A reciprocal approach is to ablate the function of oncogenes post-transcriptionally by arming the mutants with shRNA. Several oncolytic adenoviruses have been armed with shRNA, targeting Ras, hTERT, Ki-67, survivin and Apollon. All of these have been reported to show efficient antitumoral effects *in vitro* and *in vivo*.

19.3.2 Combining Replication-Selective Oncolytic Mutants with Chemotherapeutics or Other Agents

Another strategy to enhance the antitumor efficacy of oncolytic mutants is to administer virus concurrently with standard clinical therapies. This approach is promising because cell killing can be greatly improved in cancer cells at low doses of cytotoxic drugs or radiation that do not have significant toxicity in normal cells. In addition, cancer cells that are highly insensitive or resistant to treatment with current therapeutics can be sensitized to the treatments with replication-selective mutants enabling tumor elimination. Several oncolytic mutants have been demonstrated to interact synergistically in cancer cells and in tumor xenografts *in vivo*, both in animals with and without intact immune responses. Synergistic cell killing appears to occur readily in cancer cells but not in normal primary cells.

The exact mechanisms for the cancer-specific enhancement of cell killing with adenoviral mutants are not fully understood. It is clear that both the degree and the molecular mechanisms for the synergistic responses are dependent on the specific gene alterations in the tumor cells, especially those involving drug metabolism, virus susceptibility, apoptosis and cell-cycle regulation. In addition, host immune factors can play an important role. It is established that the viral E1A gene administered alone can potently induce apoptosis while the intact wild-type virus does not cause apoptotic death due to the simultaneous expression of the early anti-apoptotic proteins E1B19K, E1B55K and E314.7K (Fig. 19.2(a)). Many cytotoxic drugs such as cisplatin, mitoxantrone and docetaxel, to name a few, can potently induce E1A expression levels either through increased cellular uptake of virus by transcriptional effects on the viral attachment and internalization receptors or through stimulation of E1A transcription. In contrast, the majority of cytotoxic drugs often initially attenuate viral replication. At later stages viral replication might proceed since many drugs have relatively short half-lives under physiological conditions. While E1A expression has convincingly been demonstrated to cause synergistic cell killing in combination with numerous chemotherapeutics both in the presence and absence of viral replication, synergy has not been reported with non-replicating E1A-deleted mutants. The E1A protein binds to hundreds of cellular factors during the course of infection, but it is not known whether specific sets of these interactions are essential for synergy to occur. However, it has been demonstrated that the pRb- and the p300-binding regions of E1A can be deleted with retained ability to synergistically enhance drug-induced cancer cell killing. Even though E1A expression is essential for the synergistic effects and viral replication is often attenuated in the presence of chemotherapeutics and radiotherapy, replication will naturally also contribute to the enhanced cell killing.

One mutant that caused synergistic efficacy with both paclitaxel and docetaxel as well as with radiotherapy is the clinically evaluated prostate-specific-promoter-driven CG7870 mutant, which caused greatly enhanced cell killing in prostate cancer models. Synergistic, or more than additive, effects were demonstrated for several E1ACR2-deleted mutants such as Ad5Δ24 and *dl*922-947 in combination with TRAIL, RAD001 and temozolomide, irinotecan, 5-FC/5-FU, docetaxel, gemcitabine and mitoxantrone. The E1B55K-deleted *dl*1520 mutants have also been demonstrated to cause synergistically improved cell killing in several model systems in combination with, for example, cisplatin, 5-FC/5-FU, gemcitabine and docetaxel. In the many clinical trials with the ONYX-015 and H101 mutants the positive interactions with drugs were confirmed in patients, resulting in higher antitumor efficacy than expected from simple additive effects or from historical data. The modified *dl*1520 variant Ad5-CD/TKrep expressing two suicide genes was recently evaluated in prostate cancer patients with localized disease in combination with prodrugs or radiation therapy. Long-term benefits to patients were reported with the Ad5-CD/TKrep virus in combination with both drugs and radiation.

Most clinical trials to date have tested mutants with the immunomodulatory E3B-genes deleted, and it was concluded that part of the synergistic effects *in vivo* involved drug-induced effects on the host immune system. Cytotoxic drugs can modify the functions of immune cells such as macrophages, natural killer (NK)-cells, cytotoxic and regulatory T-cells. Both paclitaxel and cisplatin have been reported to promote T-cell and macrophage activities through cytokine induction. For example, paclitaxel was shown to stimulate activated macrophages to release TNF-α and chemokines, which in turn induced apoptosis in tumor cells. We previously demonstrated that the enhancement of viral potency was indeed greater with ΔE3B-deleted mutants than with viruses with intact E3-region, resulting in restored virus efficacy. However, absolute tumor growth inhibition never reached that of similar combinations wild-type virus or the *dl*922-947 mutant. These effects were not observed in immunodeficient animals, clearly indicating a role for host immune factors. However, the deleted E3B genes do not only suppress the host immune-defense but also play a role in preventing apoptosis through its E314.7K gene product (Fig. 19.5). In light of later findings, it is now clear that adenovirus can stimulate drug-induced apoptosis when one or several of its anti-apoptotic genes are deleted. We recently demonstrated that mutants with the viral anti-apoptotic E1B19K-gene deleted (AdΔ19K, AdΔΔ; Fig. 19.3(b)) could greatly improve antitumor efficacy in combination with gemcitabine, docetaxel and mitoxantrone through enhancement of drug-induced apoptosis. In wild-type viruses, the apoptotic functions of E1A expression is balanced by the timely expression of the anti-apoptotic genes while deletion of one of these could greatly promote drug-induced apoptosis. Even though deletion of anti-apoptotic genes can compromise virus production through prematurely induced cell death, preventing efficient replication, the synergistic effects with cytotoxic drugs frequently compensate and higher levels of efficacy can be reached.

We speculate that in addition to potential immunomodulatory effects viral genes and cytotoxic drugs activate cellular mechanisms and pathways that converge to enhance drug-induced cell killing. However, due to the plethora of cellular effects exerted by cytotoxic drugs and the many E1A-binding partners,

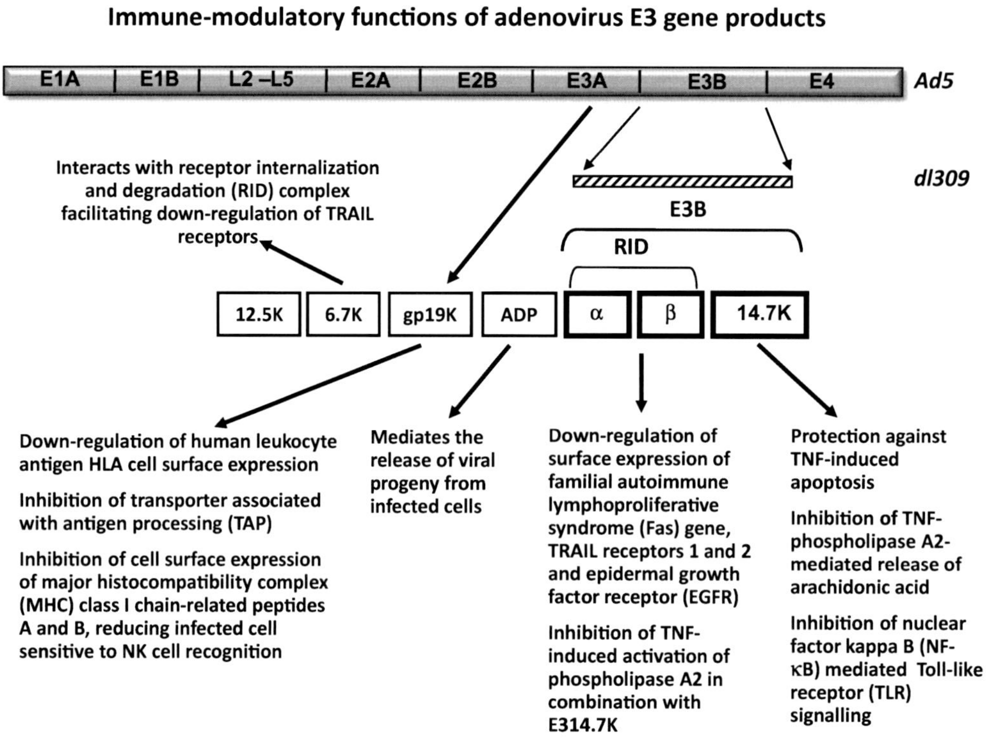
Immune-modulatory functions of adenovirus E3 gene products
E1A
E1B
L2 –L5
E2A
E2B
E3A
E3B
E4
Ad5
dl309
E3B
RID
Interacts with receptor internalization and degradation (RID) complex facilitating down-regulation of TRAIL receptors
12.5K
6.7K
gp19K
ADP
α
β
14.7K
Down-regulation of human leukocyte antigen HLA cell surface expression
Inhibition of transporter associated with antigen processing (TAP)
Inhibition of cell surface expression of major histocompatibility complex (MHC) class I chain-related peptides A and B, reducing infected cell sensitive to NK cell recognition
Mediates the release of viral progeny from infected cells
Down-regulation of surface expression of familial autoimmune lymphoproliferative syndrome (Fas) gene, TRAIL receptors 1 and 2 and epidermal growth factor receptor (EGFR)
Inhibition of TNF-induced activation of phospholipase A2 in combination with E314.7K
Protection against TNF-induced apoptosis
Inhibition of TNF-phospholipase A2-mediated release of arachidonic acid
Inhibition of nuclear factor kappa B (NF-κB) mediated Toll-like receptor (TLR) signalling

FIGURE 19.5 ■ Ad5 E3 genes and their function.

only limited insights into these interactions have been delineated to date. For example, the therapeutic toxicity of topoisomerase II inhibitors such as mitoxantrone, etoposide and doxorubicin occurs mainly through stabilization of cleavable complexes between topoisomerase II and DNA, rather than direct enzyme inhibition or DNA intercalation. The resulting DNA cross-links and breaks occur more efficiently in cells actively replicating DNA in S-phase and consequently, adenovirus is likely to promote drug-induced cell death by stimulating S-phase entry. The E1A protein has also been demonstrated to increase the expression of topoisomerase II, perhaps further stimulating DNA breaks in the presence of drugs. The therapeutic efficacy of docetaxel and paclitaxel is through stabilization of microtubules and binding to centromeres that cause cell damage in S-, G2- and M-phases. It is possible that these events promote cellular uptake and transport of virus, and/or that virus enhances the drug-induced mitotic aberrations resulting in increased cell death. It was reported that low concentrations of drugs such as taxanes that suppress microtubule dynamics greatly enhanced the microtubule-dependent adenovirus transport to the nucleus.

Cellular uptake of adenovirus has also been reported to promote its own nuclear trafficking through integrin-related activation of phosphoinositide 3-kinase (PI3K) and Ras-related C3 botulinum toxin substrate 1 (Rac1). These reports indicate that virus might enhance the taxane-induced microtubule-stabilization by acting on the same structural cellular networks. In addition to drug-specific mechanisms, most drugs induce alterations in cell-cycle progression, oxidative stress and DNA damage, ultimately leading to apoptotic death. In contrast, adenoviruses prevent apoptosis during early stages of infection and induce cell lysis once progeny virions are assembled. It is likely that the increased S-phase activity in response to virus can further promote DNA-damage during synthesis in response to drugs.

Recently, Rajecki *et al.* (2011) reported on very interesting findings when combining glucocorticoids, the chemotherapeutic cyclophosphamide and oncolytic adenoviruses in cancer patients with various tumor types. The study was performed under the Advanced Therapy Access Program (Finnish Medicine Agency) and was not a clinical trial. When administering various E1ACR2-deleted viral mutants including E2F- or Coxsackie-promoter driven, with or without GM-CSF expression with both serotype 5 and the chimeric Ad5/3, no serious adverse events were observed in any patient. Moreover, there was a tendency to higher efficacy when the mutants were administered without glucocorticoids compared with concurrent treatment, probably because of suppression of cytotoxic T-cells in response to the steroids. Glucocorticoids do not have antitumor efficacy but are frequently used in the clinic to alleviate cancer-related ailments. In addition, the glucocorticoids appeared to slow the production of neutralizing antibodies but also attenuate viral replication. It was concluded that the combinations of cyclophosphamide and glucocorticoids with a variety of oncolytic mutants were safe and did not significantly attenuate viral efficacy although clinical trials are required to confirm the efficacy. The same research team has previously developed several combination mutants such as Ad5-Δ24-GMCSF that was subsequently evaluated in 20 patients with advanced tumors that were resistant to all current therapies. While safety and some efficacy were demonstrated, the most promising aspect was the potent development of virus-mediated antitumor immunity. Similar findings were also reported with the chimeric Ad5/3-Δ24-GMCSF mutant in combination with cyclophosphamide.

Synergistic interactions have also been observed when combining oncolytic viruses of different species. A recent example is the administration of vesicular stomatitis virus and vaccinia virus in preclinical models where each mutant enhanced the replication of the other. In addition, this strategy circumvents the neutralization by antibodies generated after the first virus administration. It is possible that a similar strategy might be applicable to adenoviral mutants in combination with other biological agents.

19.4 Development of a New Generation of Oncolytic Mutants to Enhance Antitumor Efficacy

19.4.1 Inserting Cellular miRNAs as Transgenes and Viral Suppressors

An area of virotherapy which is not yet fully explored but is very exciting and promising is the incorporation of miRNA sequences in viral mutants either to limit unwanted toxicity or to inhibit oncogene expression. An RNA-binding sequence for the hepatocyte-selective miRNA mir-122 was incorporated in four copies in the E1A transcriptional cassette, resulting in up to 80-fold reduction of viral gene expression *in vivo* in murine hepatocytes while potency in tumor cells was unaffected. Similar modifications using miR-122 to control E1A-expression also greatly reduced viral gene expression in human hepatocytes while viral propagation could proceed in neuroendocrine tumors. These findings are encouraging since E1A expression was sufficiently repressed to prevent liver toxicity in mice and major cytotoxicity in human hepatocytes. However, it is not clear whether liver detargeting is critical in humans. Recently, Seymour and co-workers demonstrated that adenovirus binds with high affinity to human erythrocytes when delivered intravenously, minimizing the amount of free virus systemically (Carlisle *et al.*, 2009). Human adenoviral mutants do not bind to murine erythrocytes and consequently the murine liver is exposed to higher doses of virus, resulting in significant toxicity. Therefore, liver toxicity might not be a major concern in the clinic in contrast to murine *in vivo* studies. Indeed, no significant liver toxicity has so far been reported from clinical studies. However, Waddington *et al.* (2008) identified coagulation factor X as an adenoviral hexon binding-partner that is essential for mediating hepatocyte transduction, but not uptake, in Kupffer cells, suggesting that high-affinity uptake in the liver does occur, at least in model systems. The balance between adenoviral erythrocyte binding and factor X binding is not yet understood. Nevertheless, it is possible that inhibition of viral E1A-gene expression by insertion of tissue-specific miRNA target sequences could allow for administration of higher doses of oncolytic mutants to improve on efficacy without toxicity to normal tissue.

Another strategy to use miRNAs in virotherapy is by expressing tumor suppressor miRNAs. For example, when miR-145 (frequently down-regulated in prostate cancer) was inserted in an adenoviral mutant (non-replicating), increased levels of the apoptosis-inducing factor (AIF) and reduced cell growth resulted. Attenuated expression of the B-cell leukemia 2/adenovirus E1B 19K interacting protein 3 (BNIP3) through binding of miR-145 to the 3′ untranslated region (UTR) was suggested to cause the suppression of progression by relieving

BNIP3-dependent inhibition of AIF expression. We anticipate that regulatory miRNAs will be further explored as a powerful strategy to improve on virotherapy in the future.

19.4.2 Subgroup B Adenovirus Serotype 11 as an Alternative Oncolytic Mutant

In contrast to other species of adenoviruses which utilize CAR as the primary attachment receptor, most subgroup B adenoviruses (such as Ad11) use CD46, which is ubiquitously expressed in all human nucleated cells and is up-regulated in a number of tumor types. Several chimeric oncolytic Ad5 have been developed (with fibers derived from subgroup B adenoviruses, but the remainder of the particle from Ad5) to target CD46. Although the use of intact subgroup B adenoviruses as oncolytic agents is still underexplored it has great potential. They have different tropism and levels of infectivity in comparison with the chimeric viruses, and are more beneficial in terms of a reduced propensity for neutralization by antibodies that are mainly directed against the hexon protein. Ad11 can be classified into the prototype strain Ad11p and the less common Ad11a, with the former having better binding affinities than Ad11a in several human cell lines. It has several distinct advantages over Ad5 for development as an oncolytic virus. The prevalence of neutralizing antibodies within the human population is lower for Ad11 (10–31% versus 45–90% for Ad5), with no cross-reactivity between them. When injected intravenously into CD46-transgenic mice, there was an almost complete absence of liver transduction and toxicity. Evidence also suggests that Ad11 attaches to other receptor(s), originally named "receptor X" and recently identified as desmoglein 2. Ad11 is the only subgroup B adenovirus that uses both CD46 and receptor X, suggesting that it could potentially infect a wider range of tumor cells and overcome the problem of receptor down-regulation that confounds the use of Ad5. Finally, Ad11 can transduce dendritic cells with higher efficiency. This is beneficial in terms of cancer immunotherapy, whereby a stronger immune response could be elicited against an encoded tumor-specific antigen. Therefore, Ad11 should be further exploited as an alternative oncolytic adenovirus for cancer treatment.

19.4.3 Manipulation of the Host Immune Response

For oncolytic virus-based therapeutics, the host immune response is a double-edged sword. On the one hand, a vigorous host immune response to the virus can result in rapid viral clearance before the virus is able to exert a therapeutic effect. The efficacy of multiple injections of the same virus may be further limited by a neutralizing antibody response. On the other hand, the host immune response may be critical to the efficacy of virotherapy. This may be mediated via innate immune effectors, adaptive antiviral immune responses eliminating infected cells or adaptive antitumor immune responses. It appears that the innate immune response plays an important role in virus clearance, whereas T-cell-mediated responses are largely responsible for the antitumoral effect.

Activation of components of the host innate immune system has been shown to occur as early as ten minutes after infection with an adenoviral vector. This has the effect of rapidly removing the virus from the body before it has a chance to exert significant antitumor potency. The wild-type adenovirus contains the E3 region that encodes proteins that enable the virus to evade some of the effects of the host innate immune response (Fig. 19.5). The E3 region protein gp19 kDa inhibits major histocompatibility complex (MHC) class I expression on the cell surface, which allows infected cells to avoid cytotoxic T-lymphocyte-mediated (CTL) killing. The E3B proteins (10.4 and 14.5 kDa [receptor internalization and degradation (RID)-complex] and 14.7 kDa) inhibit apoptosis mediated by familial autoimmune lymphoproliferative syndrome (Fas) ligand (FasL), TRAIL and/or TNF. Given the fact that E3 is not necessary for adenovirus replication *in vitro* and E3 deletions provide more space to arm adenovirus vectors, the majority of oncolytic adenoviruses in clinical trials have deletions within the E3B region. However, *in vivo* data from our laboratory have demonstrated that the antitumor potency of the E3B-deleted mutant (*dl*309; Fig. 19.3(b)) is inferior to that of wild-type virus (Ad5). Tumors treated with *dl*309 showed a significant infiltration of macrophages compared with Ad5-treated tumors. A recent study demonstrated that specimens of glioma from a patient before and after treatment with ONYX-015 (E3B-deleted) showed increased macrophage infiltration after treatment, and depletion of peripheral macrophages and brain microglia increased tumor titers of the virus. These observations imply that blockage of macrophage function may be an effective approach to enhance the antitumor efficacy of oncolytic adenoviruses, especially for those with E3B deletions. It is of note that pattern recognition receptors located on cells such as macrophages, dendritic cells and NK-cells act to identify foreign material and stimulate the onset of an innate immune response. Activation of pattern recognition receptors stimulates an intracellular cascade of downstream signaling pathways. The main pathways involved are the nuclear factor kappa B (NFκB), alpha serine/threonine-protein kinase thyoma (Akt) and the IFN pathways. Selective inhibitors that block these pathways may have the potential to combine with oncolytic adenovirus for an effective cancer treatment.

Numerous experiments have been conducted to modify the adaptive immune response in favor of virus replication and tumor lysis. One method is the use of an immunosuppressive agent, such as cyclophosphamide, which has been shown to improve virus spread and antitumoral efficacy. Various data suggest that pre-existing antibodies decrease virus spread after intravenous delivery, but have a lesser effect on intratumoral injection. Although antibodies could prevent possible toxicity, they could also reduce efficacy. Possible ways to circumvent this include plasmapheresis to deplete antibodies and the use of other viral strains with a lower prevalence of antibodies in the human population, such as Ad11 as described above.

Most interestingly, it has been shown that administration of oncolytic viruses such as measles, vaccinia and adenovirus (our unpublished data) can not only infect and lyse tumor cells, but also result in tumor-specific immunity. Oncolytic virotherapy may therefore be considered as a method to achieve vaccination *in situ*, enabling the adaptive immune response to clear residual disease and provide long-term surveillance against relapse. Therefore, it seems likely that the combination of oncolytic adenoviruses and immunotherapy may render an effective induction of tumor-specific immunity. As described above, several oncolytic adenoviruses have been armed with different immunomodulatory genes, however, the antitumor efficacy of the new viruses has still been relatively disappointing. Given the distinct effects of different cytokines, an innovative and rationally designed combination of oncolytic adenovirus armed with different cytokines, as well as sequential use of other viruses, may have an enhanced impact on antitumor efficacy.

19.4.4 Targeting the Tumor Environment

The size of adenovirus (90 nm in diameter) is larger than that of other anticancer agents such as chemicals and antibodies. After intratumoral injection, effective virus spread could be impaired by the extracellular matrix, areas of fibrosis and necrosis, and surrounding normal cells in the tumor bed. Other factors such as injection volume, tumor size and type could also affect the effectiveness of intratumoral injection. Coadministration of the enzyme hyaluronidase with oncolytic mutants during intratumoral injection has been exploited to degrade the major constituent of the extracellular matrix, hyaluronan, resulting in enhanced virus spread *in vivo*. Induction of cancer cell death with an apoptosis-inducing agent prior to injection of oncolytic HSV produced channels for effective virus spread. This might also be worth testing in combination with oncolytic adenoviruses. Recently, tumor-associated stromal cells have been shown to play a role in either enhancing or reducing the efficacy of oncolytic adenoviruses, depending on the tumor type. Hypoxia, a common feature in tumor tissues, has been found to reduce the replicative and oncolytic potential of mutants despite the unaltered expression of surface receptors. In this regard it might be wiser to develop oncolytic viruses that are not attenuated in hypoxia, such as subgroup B Ad11, vaccinia virus or HSV.

Recently our group has shown that a well-documented tumor-associated gene, carcinoembryonic antigen-related cell adhesion molecule 6 (CEACAM6) which is overexpressed in a variety of human cancer types, antagonizes the Rous sarcoma oncogene (Src) signaling pathway, down-regulates cancer cell cytoskeleton proteins, and blocks adenovirus trafficking to the nucleus. Knockdown of CEACAM6 by small interfering RNA (siRNA) significantly enhanced the antitumoral potency of oncolytic mutants. The protein p21 cyclin-dependent kinase-interacting protein 1/wild-type p53-activated fragment 1 (p21CIP1/WAF) normally inhibits cyclin-dependent kinase 2 (CDK2) and blocks the progression of the cell cycle from G1- to S-phase. siRNA knockdown of p21CIP1/WAF has been reported to increase adenovirus replication and oncolysis. Therefore, targeting tumor-associated genes or intracellular pathways that affect the adenovirus life cycle is an approach that should be further exploited in the future.

19.4.5 Improvement of Systemic Delivery of Oncolytic Adenoviruses

Death from cancer is often due to metastatic disease. Treatment of multiple metastases or inaccessible tumors requires the systemic delivery of therapeutic agents that are able to stay in the circulation without depletion through breakdown or entering non-target cells, while selectively infecting tumor cells. The ability of oncolytic adenoviruses to infect tumors after systemic delivery is limited by host defenses such as tissue-resident macrophages, complement, antibodies, antiviral cytokines released from peripheral blood mononuclear cells and non-specific uptake by other tissues such as the liver and spleen. Adhesion to red blood cells could also lead to therapeutic inhibition.

After intravenous delivery, the liver, a part of the reticuloendothelial system, is the predominant site of Ad5 sequestration with significant hepatocyte transduction. Ad5 is known to cause liver toxicity in mice, and its use has raised some concerns after the death of Jesse Gelsinger in 1999 from Ad5-based gene therapy injected directly into the hepatic artery. However, it has to be noted that Ad5 has consistently been observed not to cause significant liver toxicity in cancer patients. A landmark study in mice by Waddington *et al.* (2008) showed that liver transduction is mediated by interaction of the adenoviral hexon protein with the blood coagulation factor X. Several techniques to reduce liver uptake have been performed. Kupffer cell depletion (by pre-dosing mice with non-replicating Ad5) and warfarin treatment (to inhibit vitamin K-dependent coagulation factors) significantly increased the antitumoral effect of systemically delivered oncolytic Ad5 in nude mice. Good results have also been demonstrated by coating Ad5 with high-molecular-weight polyethylene glycol or by genetic modification of the hexon protein to ablate blood factor binding for liver detargeting.

Beyond these approaches, a so-called "Trojan horse" strategy involving infection of cells *in vitro* and systemic administration of these cells back is another approach for systemic delivery of oncolytic viruses, including oncolytic adenovirus. Cells that have been tested include mesenchymal stem cells and tumor cells.

Whilst the cell carrier approach has yielded promising data *in vivo*, numerous issues must be considered before clinical application, including the best cell type to use, ease of infection, tumor-targeting capabilities, protection of virus from the host immune response, virus delivery and potential tumorigenicity.

19.5 Conclusions

Adenovirus administration to cancer patients is safe, with low liver toxicity. Indications of efficacy even after systemic delivery, albeit modest, have been achieved with both tumor-specific-promoter-driven and complementation-deletion replicating adenoviral mutants. In several clinical trials, significant efficacy resulted from combining mutants with cytotoxic drugs or radiation treatment. Synergistic interactions enhancing drug-induced apoptosis have also been

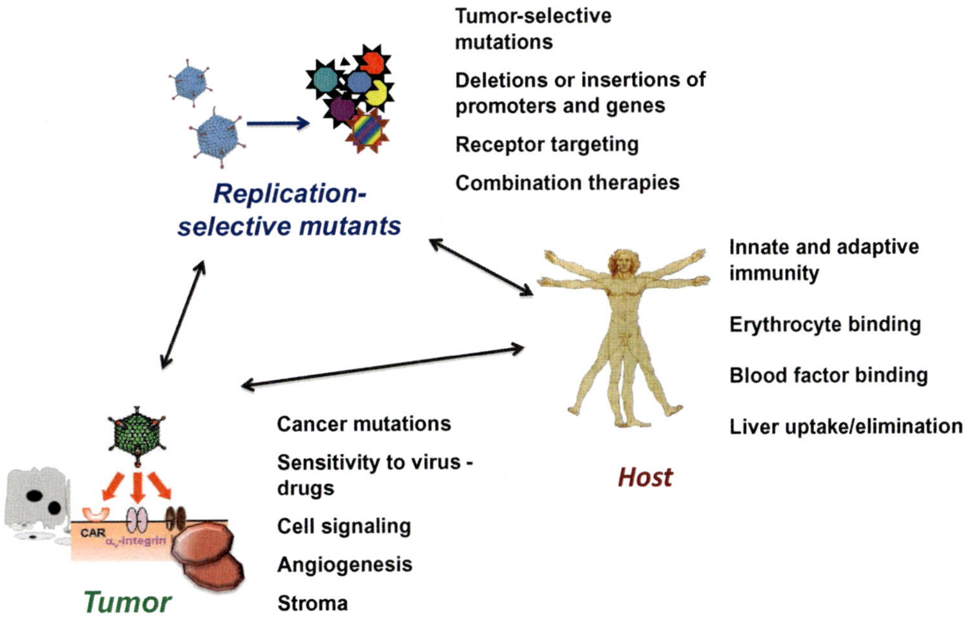

FIGURE 19.6 ■ Interactions between the oncolytic adenoviral mutants, the host physiological functions and the tumor and its microenvironment. At least three aspects need to be considered to improve the potency of replication-selective adenoviruses. Firstly, virus must be optimized through modifications of key viral genes such as E1ACR2 and E1B19K deletions, and retaining the E3-genes. Secondly, host functions that will prevent the mutant from reaching the tumor must be reduced such as the host immune response, erythrocyte binding and the rapid elimination by the liver. Thirdly, gene alterations in the tumor could be identified to better select appropriate virus mutations, including insertion of angiogenesis inhibitors and factors affecting the stroma to promote spread of the mutant within the tumor.

demonstrated in several preclinical studies. The most promising results were reported for mutants expressing toxic or immunostimulatory transgenes. We anticipate that future developments will involve multimodal mutants with several optimized gene deletions, tumor-selective promoters and therapeutic genes combined in a replicating adenovirus. In particular, the discovery of miRNAs with altered expression in tumor tissue has not yet been fully explored in combination with viral mutants. For example, expression of tumor-suppressor miRNAs as transgenes preventing oncogene translation or insertion of suppressor miRNA binding sites in the viral genome to eliminate unwanted expression in normal cells.

A greater challenge is the delivery of viral mutants to distant metastatic sites. Systemically administered adenovirus mainly binds to erythrocytes in humans, and is eventually eliminated by the liver without major liver toxicity. Several approaches have been explored to increase the quantities of free circulating virus so that the target tumors can be reached, such as modifications of the fiber and hexon proteins or by polymer coating of the viral particle in combination with tumor targeting. Further developments in this area are underway to ensure safe and efficient systemic delivery such as delivery of adenovirus-infected mesenchymal stem cells.

The different approaches to improve oncolytic adenoviruses, which depend on the interactions between the oncolytic adenoviral mutants, the host physiological functions and the tumor and its microenvironment, are schematized in Fig. 19.6. Another challenge is the high cost involved in GMP-manufacturing and testing for clinical evaluation of oncolytic viruses, which has significantly impeded the progress in optimization of mutants and development of delivery strategies. Although none of the viruses evaluated to date appeared sufficiently potent for widespread clinical application, great progress has been made recently in the development of adenovirus-based therapies and the remaining issues are likely to be solved in the near future.

For Further Reading

Original Articles

Carlisle, R.C., Di, Y., Cerny, A.M., *et al.* (2009). Human erythrocytes bind and inactivate type 5 adenovirus by presenting Coxsackie virus-adenovirus receptor and complement receptor 1, *Blood*, **113**, 1909–1918.

Heise, C., Hermiston, T., Johnson, L., *et al.* (2000). An adenovirus E1A mutant that demonstrates potent and selective systemic anti-tumoral efficacy, *Nat Med*, **6**, 1134–1139.

Hiley, C., Yuan, M., Lemoine, N.R., *et al.* (2010). Lister strain vaccinia virus, a potential therapeutic vector targeting hypoxic tumours, *Gene Therapy*, **17**, 281–287.

Leitner, S., Sweeney, K., Öberg, D., *et al.* (2009). Oncolytic adenoviral mutants with E1B19K gene deletions enhance gemcitabine-induced apoptosis in pancreatic carcinoma cells and anti-tumor efficacy *in vivo*, *Clin Cancer Res*, **15**, 1730–1740.

Öberg, D., Yanover, E., Sweeney, K., *et al.* (2010). Improved potency and selectivity of an oncolytic E1ACR2 and E1B19K deleted adenoviral mutant (AdΔΔ) in prostate and pancreatic cancers, *Clin Cancer Res*, **16**, 541–553.
O'Shea, C.C., Choi, S., McCormick, F., *et al.* (2005). Adenovirus overrides cellular checkpoints for protein translation, *Cell Cycle*, **4**, 883–888.
Rajecki, M., Raki, M., Escutenaire, S., *et al.* (2011). Safety of glucocorticoids in cancer patients treated with oncolytic adenoviruses, *Mol Pharm*, **8**, 93–103.
Small, E.J., Carducci, M.A., Burke, J.M., *et al.* (2006). A phase I trial of intravenous CG7870, a replication-selective, prostate-specific antigen-targeted oncolytic adenovirus, for the treatment of hormone-refractory, metastatic prostate cancer, *Mol Ther*, **14**, 107–117.
Waddington, S.N., McVey, J.H., Bhella, D., *et al.* (2008). Adenovirus serotype 5 hexon mediates liver gene transfer, *Cell*, **132**, 397–409.
Wang, Y., Gangeswaran, R., Zhao, X., *et al.* (2009). CEACAM6 attenuates adenovirus infection by antagonizing viral trafficking in cancer cells, *J Clin Invest*, **119**, 1604–1615.
Wang, Y., Hallden, G., Hill, R., *et al.* (2003). E3 gene manipulations affect oncolytic adenovirus activity in immunocompetent tumor models, *Nat Biotechnol*, **21**, 1328–1335.
Wang, Y., Xue, S.A., Hallden, G., *et al.* (2005). Virus-associated RNA I-deleted adenovirus, a potential oncolytic agent targeting EBV-associated tumors, *Cancer Res*, **65**, 1523–1531.

Reviews

Aghi, M., Martuza, R.L. (2005). Oncolytic viral therapies — the clinical experience, *Oncogene*, **52**, 7802–7816.
Ekblad, M., Hallden, G. (2010). Adenovirus-based therapy for prostate cancer, *Curr Opin Mol Ther*, **12**, 421–431.
Freytag, S.O., Stricker, H., Movsas, B., *et al.* (2007). Prostate cancer gene therapy clinical trials, *Mol Ther*, **15**, 1042–1052.
Parato, K.A., Senger, D., Forsyth, P.A., *et al.* (2005). Recent progress in the battle between oncolytic viruses and tumours, *Nat Rev Cancer*, **5**, 965–976.
Schenk, T.E., Horwitz, M.S. (2001). 'Adenoviridae' in Knipe, D.M., Howley, P.M. (eds.), *Fields Virology*, Philadelphia PA: Lippincott Williams and Wilkins, pp. 2265–2325.
Tysome, J., Lemoine, N.R., Wang, Y. (2009). Combining anti-angiogenic therapy and virotherapy, *Curr Opin Mol Ther*, **11**, 664–669.
Waehler, R., Russell, S.J., Curiel, D.T. (2007). Engineering targeted viral vectors for gene therapy, *Nat Rev Genet*, **8**, 573–587.
Wong, H.S. Lemoine, N.R., Wang, Y. (2010). Oncolytic viruses for cancer therapy, overcoming the obstacles, *Viruses*, **2**, 78–106.

20

PROGRESS IN DNA VACCINE APPROACHES FOR CANCER IMMUNOTHERAPY

Geoffrey D. Hannigan[a] and David B. Weiner[a,b]

20.1 Introduction

Despite numerous clinical and research advances, cancer remains one of the major health problems faced around the globe. In the USA alone, one out of every four deaths is due to cancer and it is estimated that last year over 1.5 million people were newly diagnosed within this spectrum of disease (Jemal *et al.*, 2010). In an attempt to eradicate this major health problem, three treatments continue to be our major approaches toward cancer therapy: Chemotherapy, radiation therapy and surgery. In addition to these approaches, new biological therapies are currently being explored, including monoclonal antibody targeting therapy, angiogenesis inhibitor therapy, gene therapy, targeted cancer therapy and others. Most recently, engagement of the effector immune response through targeted immune therapy has gained interest.

According to the National Cancer Institute there are, in 2011, nine Food and Drug Administration (FDA) approved immunotherapeutic agents for use in cancer treatment (Table 20.1). Cancer immunotherapy, which involves stimulating a patient immune response against cancerous tissues or cells, includes antibody treatments, cytokine treatments, autologous cell therapies and vaccines. Cancer vaccination approaches include dendritic cell vaccines, viral vector vaccines,

[a] Department of Pathology and Laboratory Medicine, Chair, Gene Therapy and Vaccine Program, CAMB, 505 SCL- 422 Curie Blvd, University of Pennsylvania, Philadelphia, PA 19104 USA
[b] Email: dbweiner@mail.med.upenn.edu

TABLE 20.1 Current FDA Approved Cancer Immunotherapies

Antibody	Brand Name	Approval Date	Type	Target	Approved Treatment(s)
Alemtuzumab	Campath	2001	Humanized	CD52	Chronic lymphocytic leukemia
Bevacizumab	Avastin	2004	Humanized	Vascular endothelial growth factor	Colorectal cancer
Cetuximab	Erbitux	2004	Chimeric	Epidermal growth factor receptor	Colorectal cancer
Gemtuzumab ozogamicin	Mylotarg	2000	Humanized	CD33	Acute myelogenous leukemia (with calicheamicin)
Ibritumomab tiuxetan	Zevalin	2002	Murine	CD20	Non-Hodgkin lymphoma (with yttrium-90 or indium-111)
Panitumumab	Vectibix	2006	Human	Epidermal growth factor receptor	Colorectal cancer
Rituximab	Rituxan, Mabthera	1997	Chimeric	CD20	Non-Hodgkin lymphoma
Trastuzumab	Herceptin	1998	Humanized	ErbB2	Breast cancer
T-cell immune therapy	Provenge	2011	Cell therapy-inducing vaccine	PAP antigen of prostate cancer	Prostate cancer

bacterial vector vaccines, peptide vaccines and DNA vaccines, which will be the topic of this review.

DNA vaccination was introduced to the public in 1992 when it was reported that a mouse antibody response could be generated as a side effect against human growth hormone (hGH) by inoculating murine tissue with an expression plasmid containing the hGH gene (Tang *et al.*, 1992). At the same time three major reports, describing the use of plasmid vaccines to target pathogens or tumors, were presented at the annual vaccine meeting at the Cold Spring Harbor Laboratory (Kutzler *et al.*, 2008). These reports described the ability of plasmid DNA inoculation to elicit humoral and cellular immune responses against cancer, influenza and human immunodeficiency virus (HIV)-1 proteins (Kutzler *et al.*, 2008). Together these findings motivated further animal investigations into the use of DNA vaccination which lead to several clinical trials.

The first human trial of DNA-based vaccination was initiated in 1994. The study reported the safety of human DNA vaccination by demonstrating that a DNA vaccine containing two HIV-1 genes was well tolerated in participants (MacGregor *et al.*, 1998). At the same time, an early targeted cancer study was also initiated. In this study, patients suffering from cutaneous T-cell lymphoma (CTCL) were DNA vaccinated using a plasmid which encoded the VB region from their own tumor. Two of the six patients exhibited a partial response to the therapy. Several clinical trials investigating DNA vaccination against many diseases including cancers, malaria, influenza, human papilloma virus (HPV) and hepatitis ensued, all of which produced weak immune responses without clear clinical benefit (Ferraro *et al.*, 2011b; Mincheff *et al.*, 2000). Despite the clinical disappointment of early DNA vaccine-induced immunity, the results did provide a proof of concept that DNA vaccines are capable of providing a safe immune response against delivered antigens. Working off of this concept, researchers have continued to advance DNA vaccine strategies by improving plasmid delivery methods (e.g., electroporation), utilizing adjuvants and addressing improvements in gene design (Ferraro *et al.*, 2011b; Kutzler *et al.*, 2008). These modifications have continued to increase interest in DNA vaccine research, as suggested by the increase in new publications in the field over the past 20 years (Fig. 20.1). Additionally, cancer DNA vaccines have accounted for 17.5% of human clinical trial DNA vaccine publications in the past five years (Fig. 20.2).

Because current DNA vaccines have advanced far beyond their original designs, modern DNA vaccines are often referred to as second-generation DNA vaccines. Second-generation DNA vaccination offers numerous conceptual advantages over many immunotherapeutic vaccine strategies. The first major advantage is the observed clinical tolerability, safety and DNA's inability to induce a host immune response against itself (Alam *et al.*, 2010). In contrast, some virus-based vaccines may provide specific safety challenges to demonstrate attenuation and may exhibit lower boosting through induction of an anti-vector host immune response (Johnson *et al.*, 2007). The second advantage of DNA vaccination is that, like virus-based vaccines but unlike peptide-based vaccines, the antigen protein is expressed inside the host cell which ensures post-translational modification and, as a result, customized humoral and cellular immune responses (Alam and McNeel, 2010). As a practical advantage, DNA vaccines are relatively inexpensive to produce and they are easy to construct, manipulate and store.

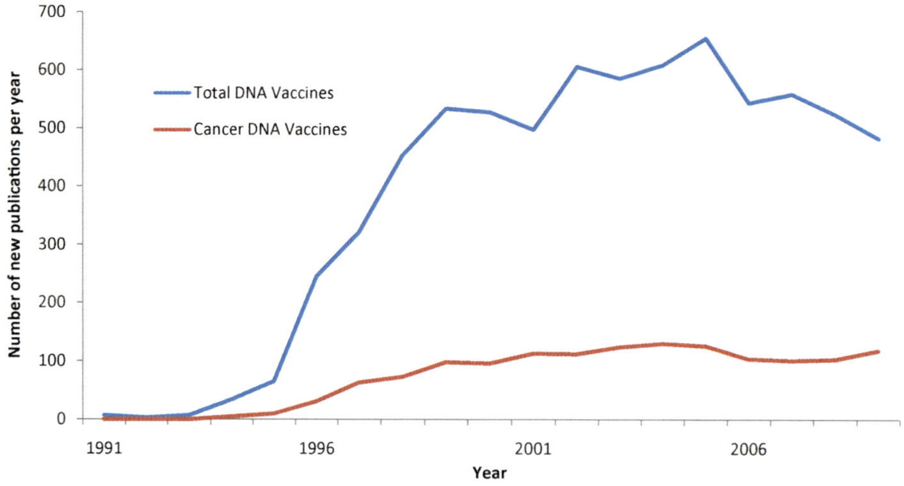

FIGURE 20.1 ■ Number of publications related to DNA vaccination since 1991. The first DNA vaccine-related paper was published in 1992 and the first DNA cancer vaccine paper was published in 1994. As of 2010, in total, 1,407 DNA vaccine papers had been published while 7,262 DNA cancer vaccine papers had been published. Publications determined through www.pubmed.gov.

Additionally, DNA is relatively straightforward to manipulate compared with many other vaccine platforms. Finally, the bacterial plasmid DNA that serves as the backbone for DNA vaccines has been shown to stimulate the adaptive immune system and act like a vaccine adjuvant (Krieg *et al.*, 1995). For these reasons, and provided it is possible to achieve improved immune potency, DNA vaccination remains an attractive immunotherapeutic method for cancer therapy.

Just as it is important to understand the advantages of DNA vaccination, it is equally important to understand the mechanism by which DNA vaccination focuses the immune response on cancer. DNA vaccination relies primarily on inducing a cytotoxic T-lymphocyte (CTL) response targeting cancerous cells. The mechanism of this induced response is outlined in Fig. 20.3. The process begins when the DNA vaccine, which contains the gene for the target protein, is delivered to the host and transfects immune and non-immune cells. Some of the expressed target proteins are degraded to provide fragments that are presented on major histocompatibility complex (MHC)1 surface molecules which are recognized by CTLs and result in CTL activation. Other expressed target proteins are secreted from the cell. Additionally, some transfected cells die and are engulfed by antigen-presenting cells (APCs) which go on to cross-present the antigen.

Some of the secreted target proteins are taken up by APCs, which present fragments of the target protein on MHC2 surface molecules. Other secreted proteins

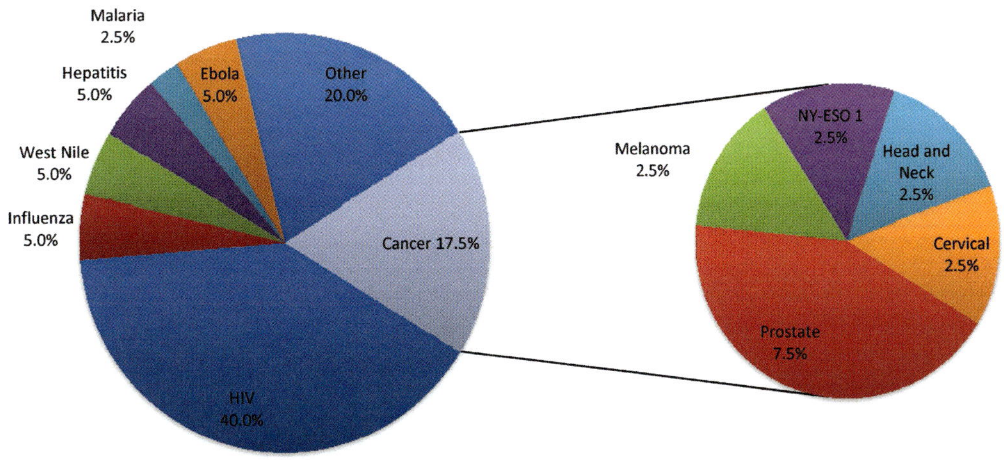

FIGURE 20.2 ■ The estimated distribution of human clinical trial DNA vaccine publications over the last five years. Of the 40 papers published, seven were related to cancer DNA vaccination. Three articles (7.5%) were related to prostate cancer and one (2.5%) was related to each of the other four categories. It is important to note that head and neck cancer consists primarily of varying types of head and neck squamous cell carcinomas and New York esophageal squamous cell carcinoma 1 (NY-ESO 1) is a cancer vaccine target related to melanoma, prostate cancer, lung cancer and others. Publications determined through www.pubmed.gov.

bind to B-cell receptors to aid in B-cell activation. Finally, if the directly transfected cell is an APC, it is able to present the peptide fragments on MHC1 and MHC2 molecules, thereby providing co-stimulation for both arms of the cellular immune response.

Presentation of target protein fragments on MHC2 molecules activates T helper 1 (TH1) cells which are consequently able to co-stimulate antigen-bound B-cells and thereby activate them. Once activated, the B-cell is able to expand and produce target-protein-specific antibodies. Overall, TH1 cells are able to be activated through MHC2 presentation by activated B-cells, APCs and perhaps other transfected cells. When activated, TH1 cells secrete interferon (IFN)-γ to inhibit TH2 cell proliferation (helper cells that are involved in humoral response) and secrete IL-2 to enhance CTL proliferation. Once CTLs have been activated and their proliferation enhanced, the CTLs can traffic to tumors and recognize the tumor cells that express the target tumor antigen by presentation on their MHC1 molecules. This recognition causes the CTL to destroy the cancer cell. It is the goal of the DNA vaccine approach to drive these complex immune responses to find and destroy tumor cells.

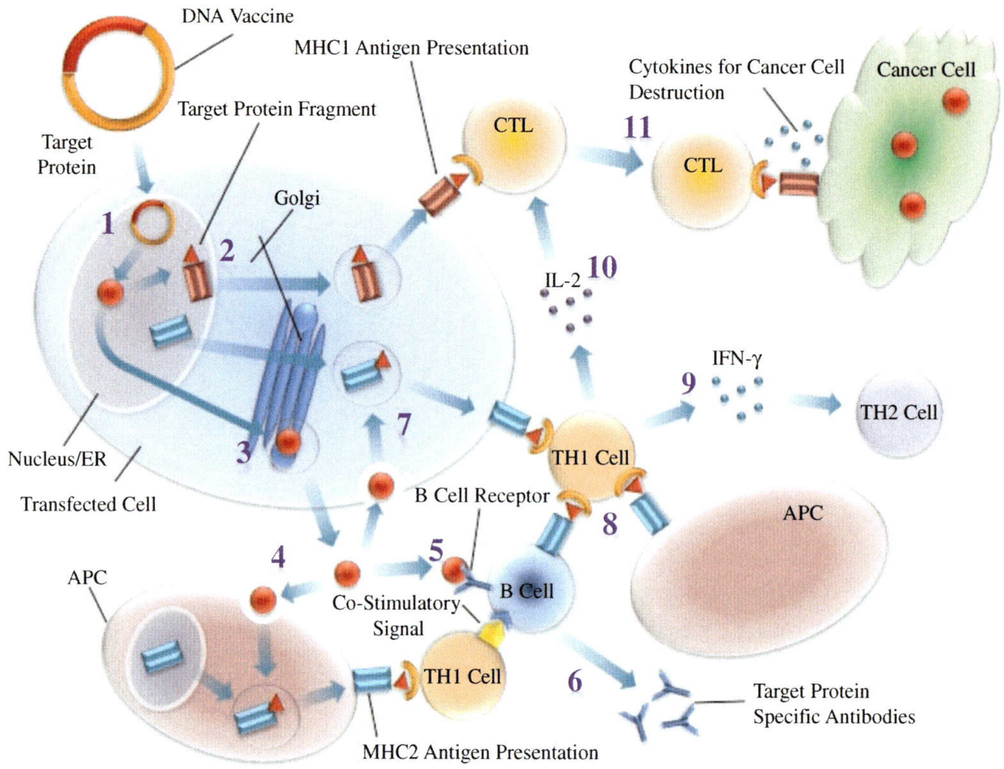

FIGURE 20.3 ■ Mechanistic overview of the general cellular immune response induced by cancer DNA vaccines. (1) The DNA vaccine is delivered to the host and transfects host immune and non-immune cells. (2) Some target proteins are degraded and presented on major histocompatibility complex (MHC)1 molecules. (3) Proteins that are not degraded can be secreted from the host cell. (4) APCs are capable of taking up the secreted proteins and presenting them on MHC2 molecules. (5) B-cells are activated by two signals. First the antigen binds B-cell receptors, then an MHC2 activated T helper type 1 (TH1) cell provides a co-stimulatory signal for the B-cell activation. (6) Activated B-cells produce antigen-specific antibodies. (7) If the original transfected cell is an APC, it is capable of taking up the target protein and presenting it on an MHC2 molecule. (8) TH1 cells can be activated by MHC2 presentation by APCs, activated B-cells and antigen-presenting transfected cells. (9) Activated TH1 cells secrete interferon (IFN)-γ which inhibits proliferation of TH2 cells. (10) Activated TH1 cells also secrete IL-2 which enhances proliferation of cytotoxic T-lymphocytes (CTL). (11) CTLs, which are activated by MHC1 antigen presentation, recognize the target antigen presented by cancer cell MHC1 molecules and destroy the cancer cell.

Beyond providing a basic understanding of DNA vaccines and their mechanism of action, this review provides a summary of the published developments of cancer DNA vaccines in both human and animal studies. We focus on four of the major types of non-viral-induced cancers, three of which have been the subject of recent human clinical trial publications (Fig. 20.2). These four cancer types are prostate cancer, melanoma, non-Hodgkin B-cell lymphoma and breast cancer. We would also like to note that due to length restrictions, some references were excluded from this review.

20.2 Prostate Cancer

Prostate cancer is a significant national health problem for American men. Last year prostate cancer accounted for the majority of new cancer cases (28%) in men and was the second most common cause of male death due to cancer (11%) (Jemal *et al.*, 2010). Because prostate cancer affects so many men in America and around the world, its treatment continues to grow as a research priority, as evidenced by the increase in National Institutes of Health (NIH) funding dollars and total funding percentage since 2007 (Fig. 20.4). Currently the treatment options for men with prostate cancer include active surveillance, surgical removal, radiation therapy, cytotoxic chemotherapy and androgen deprivation (Gupta *et al.*, 2011). In addition to these treatment options, immunotherapy is becoming an increasingly attractive and promising method.

There are four major factors that make prostate cancer an ideal target for immunotherapy. These factors are the presence of multiple organ- or cancer-specific antigens (meaning there is minimal risk of generating an immune response to

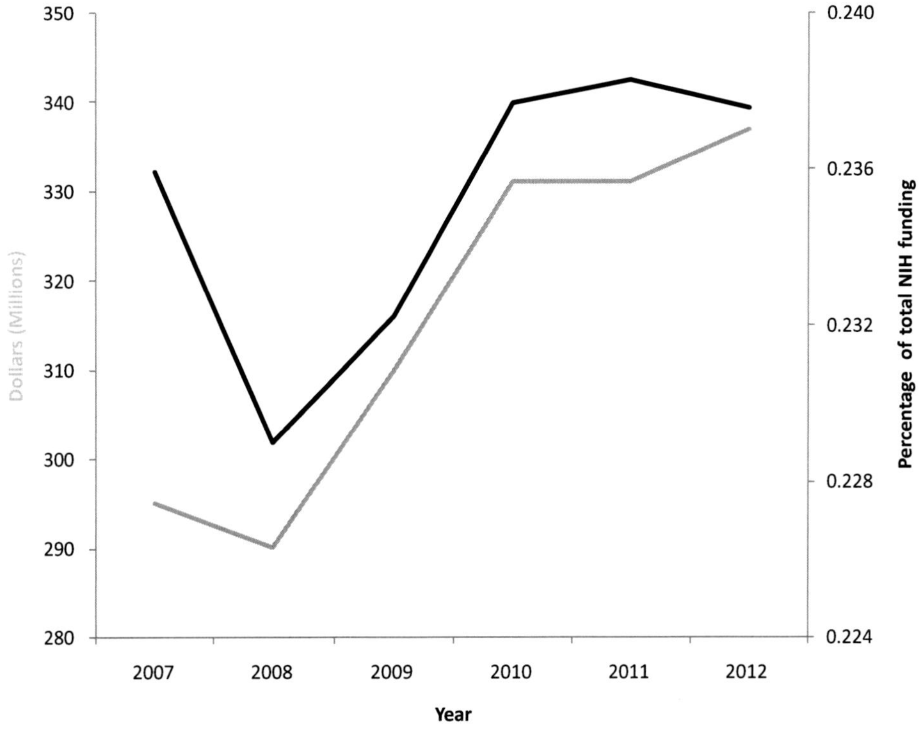

FIGURE 20.4 ■ USA NIH grant (millions of dollars) toward prostate cancer research (gray) and the corresponding percent dedicated from total grant dollars (black). Prostate cancer research dollars have increased from 295 to 337 million dollars. This equates to an increase in percentage from 0.236% to 0.238% of total NIH funding. Data taken from the NIH Estimates of Funding for Various Research, Condition and Disease Categorizations (RCDCs).

undesired tissues), the slow progression of the disease (leaving more time to generate an immune response), possible presence of nascent antibody immune response (leaving a potential for immune amplification) and the lack of necessity of the prostate organ (which reduces the consequences of an unintentional autoimmune response) (Gupta *et al.*, 2011). For these reasons immunotherapy, and especially cancer vaccination, has been actively pursued against this target.

One of the most significant current advancements in prostate cancer vaccination came this year with Sipuleucel-T (also referred to as Provenge), the first cancer immunotherapy to be approved by the FDA (Gupta *et al.*, 2011). Sipuleucel-T is an autologous immunotherapy that, by patient dendritic cell treatment *ex vivo*, increases prostatic acid phosphatase (PAP) targeting by the immune system (Gupta *et al.*, 2011). Although this therapy was shown to be effective in treating castration-refractory prostate cancer, it conversely showed no evidence of a measurable antitumor effect, objective response rate or serological response (Garcia, 2011). Furthermore it is for use in asymptomatic or minimally symptomatic men only (Garcia, 2011). Despite its drawbacks, Sipuleucel-T is a beneficial treatment that will additionally open doors for future immunotherapies because it is the first FDA-approved cancer vaccine (Garcia, 2011). One such future immunotherapy may in fact be a PAP-targeting DNA vaccine.

PAP belongs to a family of acid phosphatases and is predominantly expressed in normal and cancerous prostate tissue, making it an ideal target for DNA vaccination. In one of the early DNA vaccine studies, rats immunized with a xenogeneic human PAP (hPAP)-expressing plasmid developed a significant cluster of differentiation (CD)4$^+$ and CD8$^+$ T-cell hPAP-specific response, a result not seen in the vaccinia-based vaccine to which it was compared (Johnson *et al.*, 2007). Furthermore, multiple immunizations expressing the autologous rPAP gene resulted in a PAP-specific immune response without the need for a heterologous or xenogeneic approach (Johnson *et al.*, 2007).

These early reports in mouse studies led to a dose-escalation phase I clinical trial in which patients were administered a DNA vaccine expressing full length human PAP as well as granulocyte–macrophage colony stimulating factor (GM-CSF) as an adjuvant (McNeel *et al.*, 2009). The trial, as well as follow-up observations, reported that the DNA construct was safe and elicited an hPAP-specific T-cell response (Becker *et al.*, 2010; McNeel *et al.*, 2009). A follow-up study by the same investigators additionally indicated that multiple immunizations were needed to induce a PAP-specific IFN-γ secreting T-cell immune response, and the immune response could be further amplified by subsequent boosting (Becker *et al.*, 2010). Overall this series of reports suggest the PAP-targeted DNA vaccine is safe and elicits necessary T-cell responses, which together warrants further clinical evaluation (Bilusic *et al.*, 2011).

One of the primary targets of prostate cancer immunotherapy is the prostate-specific antigen (PSA). PSA is a prostate-tissue-specific serine protease that is recognized as both a biomarker for prostate cancer progression as well as an immunotherapeutic target. Currently an important PSA-targeting vaccine is the poxvirus-based vaccine PSA-TRICOM (PROSTVAC).

PSA-TRICOM uses recombinant poxviruses to stimulate an immune response against the PSA antigen. Using the poxvirus platform for this vaccine construct is advantageous because it is impressively immunogenic, it is capable of carrying large amounts of genetic material and it has been extensively characterized and studied. In addition to these advantages, the PSA-TRICOM immunogenicity is increased by including three co-stimulatory molecules that enhance T-cell activation (Kantoff *et al.*, 2010). Furthermore, the immune response is increased by priming with vaccinia pox virus and boosting with a fowl pox construct, a virus which will not induce a neutralizing antibody response (Bilusic *et al.*, 2011).

In a phase II clinical trial, PSA-TRICOM was shown to reduce the death rate of men with metastatic castration-resistant prostate cancer (mCRPC) by 44% and improved median overall survival by 8.5 months (Kantoff *et al.*, 2010). Furthermore, a National Cancer Institute phase II clinical trial reported that 13 out of 29 evaluable patients experienced greater than a two-fold increase in PSA-specific T-cell response, and five of the 13 experienced a greater than six-fold increased response (Gulley *et al.*, 2010). The beneficial effects on mCRPC men will further be evaluated in a phase III clinical trial expected to begin this year (Bilusic *et al.*, 2011). Although PSA-TRICOM is a promising advancement, it alone is only part of the solution and there remains clear room for additional immunotherapeutic approaches.

One of the initial studies to evaluate the efficacy of a PSA-targeted DNA vaccine was reported in 1998 when an expression vector, containing the human PSA gene, was injected into Balb/c mice (Kim *et al.*, 1998). Significant humoral and cellular responses were observed, which indicated PSA to be a promising DNA vaccine target (Kim *et al.*, 1998). This investigation team went on to evaluate the vaccine in rhesus macaques and also reported the effects of certain adjuvants on the vaccine's immunogenicity.

The team reported, in rhesus macaques, an increased PSA-specific antibody and TH1 type cytokine IFN-γ response (Kim *et al.*, 2001). In both mice and rhesus macaques, co-immunization with an interleukin (IL)-2 expression plasmid increased the PSA-specific antibody response while co-immunization with an IL-12 expression plasmid reduced the antibody responses (Kim *et al.*, 2001). Co-immunization of mice, but not rhesus macaques, with IL-2, IL-12 or IL-18 greatly increased T helper cell proliferation compared with the vaccine without adjuvants (Kim *et al.*, 2001).

In a later study, Roos and colleagues reported that co-injecting the adjuvants IL-2 and GM-CSF with an expression plasmid, containing full-length hPSA, induced protection against tumor growth in 80% of the C57Bl/6 (H-2b) mice treated (Roos *et al.*, 2005). They also reported that, *in vitro*, the presence of IL-2 and/or GM-CSF with the PSA plasmid in splenocytes had no effect on the generation of PSA-specific IFN-γ-secreting $CD8^+$ T-cells (Roos *et al.*, 2005). These findings led investigators to perform a phase I clinical trial which demonstrated that, when given with GM-CSF and IL-2 in doses of 900 μg, their vaccine was capable of eliciting cellular and humoral immune responses (Pavlenko *et al.*, 2004). In a follow-up study, they reported that the patients who received the highest vaccine and adjuvant dose (900 μg) had a higher cellular immune response, and those T-cells were able to recognize both naturally processed PSA proteins and PSA peptides (Miller *et al.*, 2005). This phase I clinical trial suggested that their PSA-targeting DNA vaccine construct was capable of inducing PSA-specific immune responses in certain prostate cancer patients and warranted further trials.

Another target for prostate cancer immunotherapy is the prostate-specific membrane antigen (PSMA). PSMA is a Type II membrane glycoprotein that is expressed in healthy prostate tissue but significantly overexpressed in cancerous prostate tissue. PSMA has also been reported as being strongly expressed in vascular endothelium of multiple carcinomas but not in non-cancerous vascular endothelium (Liu *et al.*, 1997). Because of its increased expression in cancer tissue, PSMA has become an attractive target for prostate cancer immunotherapy. In fact a recently reported dendritic-cell-based vaccine, which utilized dendritic cells transduced with adenoviruses expressing truncated PSMA and the T-cell co-stimulatory molecule 4-1BB ligand (4-1BBL), was shown to be effective in mice and shows promise for future development (Kuang *et al.*, 2010a). Other PSMA-directed dendritic cell vaccine approaches have also been explored.

A recent approach to DNA vaccination against PSMA has been a DNA fusion gene vaccine. In a recent study it was reported that expression of PSMA human leukocyte antigen (HLA)-A*0201-binding epitopes ($PSMA_{27}$, $PSMA_{663}$ and $PSMA_{711}$), hybridized to the first domain of fragment C of the tetanus toxin gene, elicited increased PSMA-specific T-cell responses in HHD mice (Vittes *et al.*, 2011). Mice immunized with the full-length PSMA gene construct failed to elicit a PSMA-specific T-cell response to the same epitopes (Vittes *et al.*, 2011). Furthermore, all of the vaccine (tetanus fusion)-induced T-cells were cytotoxic against peptide-expressing tumor cells and all except $PSMA_{711}$ were cytotoxic against endogenous PSMA-expressing tumor cells (Vittes *et al.*, 2011). This suggested that the presence of a tetanus toxin fragment significantly increased DNA vaccine efficacy and further studies are ongoing.

Another improvement to PSMA DNA vaccination is the development of the truncated PSMA gene splice variant. Initial studies showed that most antibodies to PSMA recognize epitopes in the 43–570 amino acid region, but are rapidly internalized upon binding to the cell, which naturally led to the question of whether the epitopes beyond this region may induce novel antibodies that are not rapidly internalized (Kuratsukuri *et al.*, 2002a; Kuratsukuri *et al.*, 2002b). To address this question, Kuratsukuri and colleagues compared a PSMA gene only containing the C-terminal 180 amino acids (PSMc) to a full-length gene in a DNA vaccine in Balb/c mice (Kuratsukuri *et al.*, 2002b). The PSMc vaccination successfully induced adaptive humoral immunity and was able to significantly inhibit human prostate cancer growth in athymic mice (Kuratsukuri *et al.*, 2002b). Recently, in a mouse study, the antitumor efficacy of the PSMc splice variant was enhanced by co-expression of 4-1BBL (Kuang *et al.*, 2010b). In a human phase I/II clinical trial, a PSMc-expressing DNA vaccine co-expressed with CD86 and co-immunized with GM-CSF adjuvants showed enhanced patient immune responses (Mincheff *et al.*, 2000). Boosting with a PSMc-containing adenoviral vaccine further enhanced patient immune response (Mincheff *et al.*, 2000). A follow-up study on these patients reported that the vaccine induced a significant humoral response against PSMA in most patients (Todorova *et al.*, 2005).

An additional advancement in PSMA immunogenic induction has been the xenogeneic DNA vaccination technique. In this method, full-length human PSMA, which shares 84% homology to mouse PSMA, was expressed in mouse tissue and caused expression of mouse PSMA (mPSMA) (as well as human PSMA (hPSMA)) antibodies while the vaccine expressing mPSMA did not induce an mPSMA antibody response due to an inherent tolerance to mPSMA (Gregor *et al.*, 2005). Because the data suggest that highly homologous antigens from different species may be effective in eliciting an immune response when given as a DNA vaccine, further studies of this approach will be ongoing (Gregor *et al.*, 2005).

In an interesting synergistic approach, Ferraro and colleagues designed a synergistic DNA vaccine that expressed both the PSMA and PSA antigen and was delivered by electroporation. This was the first report, to our knowledge, of a multiple-antigen DNA vaccine against PSMA and PSA (Ferraro *et al.*, 2011a). In this method, 10 μg of an optimized PSA vaccine plasmid and 20 μg of an optimized PSMA vaccine plasmid were intramuscularly injected (followed immediately by electroporation) into BALB/c mice (Ferraro *et al.*, 2011a). The vaccine was reported as inducing a PSA-specific antibody response as well as inducing a robust PSA- and PSMA-specific cellular response (Ferraro *et al.*, 2011a). This study will probably warrant further investigation of the preclinical efficacy of the PSA+PSMA DNA vaccine construct.

Another major target for prostate cancer DNA vaccines is the six transmembrane epithelial antigen of the prostate (STEAP). STEAP is a membrane protein not expressed by normal tissue, but highly expressed by prostate cancer cells, and is furthermore expressed by renal and bladder cancer cells. STEAP was recently included in a dendritic cell vaccine mouse study, but has also been evaluated in DNA vaccine constructs (Krupa *et al.*, 2011). A study evaluating the efficacy of a mouse STEAP (mSTEAP)-based DNA vaccine found that, although a DNA vaccine alone can significantly prolong mouse survival, the vaccine was most effective when boosted with a Venezuelan equine encephalitis (VEE)-vector-based vaccine (Garcia-Hernandez *et al.*, 2007). The antitumor effects were due largely to induction of an mSTEAP-specific $CD8^+$ T-cell response as well as production of IFN-γ, tumor necrosis factor (TNF)-α and IL-2 cytokines by CD4+ T-cells (Garcia-Hernandez *et al.*, 2007). Furthermore, high IL-12 levels in the tumor were associated with a significant increase in antitumor response, an observation that further suggests the efficacy of IL-12 as an adjuvant (Garcia-Hernandez *et al.*, 2007; Kutzler *et al.*, 2008). Finally, the vaccine was able to significantly slow the progression of murine tumors established 31 days prior to immunization (Garcia-Hernandez *et al.*, 2007).

An additional target antigen for DNA vaccines, as well as virus-based vaccines, is the prostate stem cell antigen (PSCA). PSCA is related to the lymphocyte antigen 6/thymus cell surface antigen-1 (Ly-6/Thy-1) gene family and, although it is expressed in many tissues, it is overexpressed in prostate cancer tissue. One of the first studies to investigate a DNA vaccine expressing PSCA reported that expressing the full human PSCA gene in mice induced an antigen specific antibody and $CD8^+$ T-cell response (Zhang *et al.*, 2007). Furthermore, the co-expression or co-immunization of the human heat shock protein 70 (HSP70) greatly increased the $CD8^+$ T-cell induction, but did not significantly affect the $CD4^+$ TH1/TH2 T-cell response, which suggests that HSP70 would be a promising adjuvant candidate for future studies (Zhang *et al.*, 2007). The optimal expression of the genes was achieved with PSCA coupled to the N-terminus of the HSP70 gene (Zhang *et al.*, 2007). Electroporation was also shown to increase the efficacy of PSCA-targeting DNA vaccines (Ahmad *et al.*, 2009).

Soon after these findings, another study reported a PSCA vaccine regimen that consisted of priming with a PSCA expression plasmid and boosting with PSCA expressing VEE virus vector (Garcia-Hernandez *et al.*, 2008). The vaccine regimen induced a significant immune response characterized by an increase in tumor-infiltrating $CD4^+$ and $CD8^+$ T-cells which correlated with improved survival and slower tumor growth (Garcia-Hernandez *et al.*, 2008). The vaccine induced a 90% survival rate among immunized mice and the long-term immunity did not induce a significant autoimmune reaction (Garcia-Hernandez *et al.*, 2008).

The overall current state of prostate cancer immunotherapy, and specifically DNA vaccination, remains an exciting and promising field of research. Due to the prostate-organ and cancer-antigen specificity and consequent lack of off-target immune responses, prostate cancer will likely remain an ideal target for immunotherapy. Enhanced prostate cancer DNA vaccines will likely increase gene expression and antitumor potency by incorporating multiple antigens and plasmid delivery enhancements such as electroporation. Furthermore the recent FDA approval of Provenge will likely open doors for more cancer immunotherapies and we will see exciting advancements and applications of prostate cancer DNA vaccines.

20.3 Melanoma

Melanoma, the cancer of melanocytes (the pigment-producing cells of the skin), also poses a major health problem in the USA. Last year, melanoma was the fifth most diagnosed (5%) cancer type in the USA in men and women and accounted for the most deaths among the skin cancer types (Jemal *et al.*, 2010). Additionally, over the past three decades the presence of melanoma has nearly tripled in the population of the USA and prognosis for those with advanced cancer continues to remain poor (Jemal *et al.*, 2010). Current treatments for melanoma include adjuvant therapy with IFN-α (the overall survival benefits of which remain controversial, possibly due to trial design shortcomings), biochemotherapy (which slightly improves response rate but does not have a clear affect on overall survival and does show increased toxicity), the Bedford Lab drug "dacarbazine" (which has demonstrated response rates of less than 15%), and immunotherapy including high-dose IL-2 (which is promising but requires further study) (Jilaveanu *et al.*, 2009). Despite the current and historical disappointment in melanoma treatment, there continue to be multiple exciting therapeutic approaches being explored, including the CTL antigen (CTLA)-4 monoclonal antibodies "ipilimumab" and "tremelimumab", an adjuvant therapy using pegylated interferon-α2b and various drugs used to inhibit signaling pathways by mutation (Eggermont and Robert, 2011). In addition to these new methods, work on canine melanoma treatment with xenogeneic human tyrosinase has led to a human-tyrosinase-based DNA vaccine that is planned to be issued a conditional license by the US Department of Agriculture for treatment of canine melanoma (McNeel *et al.*, 2009).

Tyrosinase (TYR) is one of the most commonly targeted antigens for melanoma immunotherapy. TYR, the key enzyme in melanin synthesis, is a prototypical differentiation antigen that is expressed by melanocytes and is recognized by autologous cytotoxic T-lymphocytes (Chen *et al.*, 1995). TYR is expressed homogenously by most melanoma species, making it an ideal target for immunotherapies including DNA vaccination (Chen *et al.*, 1995).

Early preclinical trials suggested that DNA expression of a xenogeneic TYR antigen induces significant autologous immune responses against melanoma tissue in animal models and, as stated above, is expected to result in a canine vaccine granted conditional license by the US Department of Agriculture (Goldberg *et al.*, 2005; Liao *et al.*, 2006). Inspired by the preclinical work, Wolchok and colleagues conducted the first clinical trial, tour knowledge, evaluating the safety and immune efficacy of a xenogeneic DNA vaccine targeting human TYR (Wolchok *et al.*, 2007). The regimen consisted of both mouse TYR and human TYR expression.

The vaccine regimen was reported as safe and induced human-TYR-specific $CD8^+$ T-cell responses in 7 of 18 patients (Wolchok *et al.*, 2007). Similarly, Weber and co-workers reported a clinical trial investigating the induced immune response and safety of a DNA vaccine expressing TYR and Melan-A/melanoma-associated antigen recognized by T-cells (MART)-1 epitopes (Weber *et al.*, 2008). Although the vaccine was well tolerated, the immunogenicity induced by the vaccine was inadequate and further supported the notion that adjuvants or some other methodology must be used to enhance DNA vaccine efficacy. Most recently, Manley and colleagues reported an effective oral staging system combined with a xenogeneic TYR DNA vaccine regimen for the treatment of canine digit melanoma, a process that could be translated to provide the vaccine potency needed for humans (Manley *et al.*, 2011).

Another major target antigen for melanoma DNA vaccination is glycoprotein gp100. Gp100 is a melanocyte differentiation protein that is present only in melanocytes and melanoma tissue, making it an attractive target antigen for melanoma immunotherapy. Despite promising preclinical trials, an early clinical trial reported that a DNA vaccine encoding the gp100 antigen was an ineffective strategy due to its inability to elicit an autologous-antigen-specific T-cell response (Rosenberg *et al.*, 2003). In a later clinical trial, researchers co-immunized patients with gp100 and an adjuvant GM-CSF expression vector delivered by particle-mediated epidermal delivery (PMED) (Cassaday *et al.*, 2007). The results suggested that, in addition to the vaccine being well tolerated, the GM-CSF caused greater infiltration of dendritic cells into the vaccination site, and the overall regimen induced a modest immune response at best (Cassaday *et al.*, 2007). This was a promising indication that gp100 targeting melanoma DNA vaccines could be efficacious in humans.

In another clinical trial, researchers attempted to enhance the gp100-specific immune response by expressing a xenogeneic gp100 protein (Yuan *et al.*, 2009). Researchers injected either a mouse gp100 plasmid followed by human gp100 plasmid, or vice versa into patients. The xenogeneic approach was found to be safe, minimally toxic and induced an increase in gp100 specific $CD8^+$ T-cells in 5 of 18 patients as well as increased the frequency of $CD8^+$ IFN-γ^+ T-cells in 1 of 18 patients (Yuan *et al.*, 2009).

Later, a pilot clinical trial investigated the enhancement of a xenogeneic mouse gp100 DNA with delivery by either PMED or intramuscular injection in the arm (Ginsberg *et al.*, 2010). The researchers reported a similar efficacy and safety for the two delivery methods, although a much greater dose was required for intramuscular delivery for a response equal to PMED (Ginsberg *et al.*, 2010). Another recent clinical trial suggested that anti-CTLA-4 antibodies, such as ipilimumab, may boost cancer vaccine efficacy if administered after initial DNA vaccine treatment (Yuan *et al.*, 2011). One patient, who was initially vaccinated with a gp100-targeting DNA vaccine, was immunized with ipilimumab and demonstrated a significantly increased antigen-specific response (Yuan *et al.*, 2011). This suggests that CTLA-4 blockade could act as an effective boost for gp100 DNA vaccines, although further research is needed to confirm these results.

In a preclinical trial investigating anti-angiogenesis adjuvants, Chan and colleagues reported that when mice received intratumoral injections of a DNA vaccine, including gp100 as well as tyrosinase related protein 2 (TRP2) and invariant chain-pan human leukocyte antigen-DR reactive epitope (Ii-PADRE), as well as anti-angiogenesis factors, the mice experienced a greater than 50% survival rate over a 90-day trial (Chan *et al.*, 2009). They also showed that intratumoral injection was more effective than intramuscular (Chan *et al.*, 2009). Another study related to the TRP2 antigen showed that a DNA vaccine construct targeting TRP2 and co-expressed with DNA-dependent activator of interferon regulatory factors (DAI), triggered an increase in IFN-γ producing $CD8^+$ T-cells (Lladser *et al.*, 2011). Also, when the mice that survived the initial post-vaccination challenge were challenged again three months later, a surprising 70% of the adjuvant group survived while 0% of the adjuvant-deprived mice survived (Lladser *et al.*, 2011). This suggests that DAI, as well as other intracellular pattern recognition receptors, show promise as cancer DNA vaccine adjuvants.

Additionally, last year a research team reported on the efficacy of the Mel-3 poly-epitope vaccine, which includes Melan-A, tyrosinase, melanoma-associated antigen (MAGE) and New York esophageal squamous cell carcinoma (NY-ESO) epitopes (Dangoor *et al.*, 2010). Researchers reported that, when given in high dose, and when the DNA vaccine was boosted with a viral Mel-3 vaccine, the vaccine induced a 91% T-cell responder rate and was well tolerated (Dangoor *et al.*, 2010). This significant clinical efficacy is expected to lead to further clinical trials evaluating this construct.

This year Weber and colleagues reported results of a phase I clinical trial evaluating a cancer vaccine regimen that targeted both a melanoma and a prostate-cancer-specific antigen, making it novel in its targeting of both cancer cells and solid tumor tissue (Weber *et al.*, 2011). These targeted antigens were PSMA (described above) and preferentially expressed antigen in melanoma (PRAME) which has not yet been explored as a therapeutic target in the clinic (Weber *et al.*, 2011). Because PRAME is overexpressed by a variety of cancer tissue types and has been shown to play a role in cancer progression, and PSMA is capable of being expressed by solid tumors of certain cancer types as well as overexpressed by prostate cancer cells, these targets were ideal choices for a cancer vaccine against multiple tumor types. The vaccine construct included a DNA priming with peptide boosts (Weber *et al.*, 2011). Overall the vaccine was reported as being safe and well tolerated, and resulted in immune responses in half of the immunized patients with advanced, progressive cancer (Weber *et al.*, 2011). The results warrant further clinical investigation of this construct and such investigation will include a comparison between expression of the antigens together and separately (Weber *et al.*, 2011). Although this approach is still in the early stages of development, it presents an intriguing potential for synergistic DNA vaccines.

Finally, a different approach to melanoma DNA immunotherapy, which expresses the cytokine IL-12, is being evaluated for its clinical potential. IL-12 is an ideal cytokine for anti-melanoma treatment because of the antitumor effects it induces through cellular immunity activation. Initial preclinical trials, which attempted to inject IL-12 cytokine directly into the animals, reported promising results. Unfortunately, early clinical trials suggested disappointing antitumor effects and significant toxicity (Colombo *et al.*, 2002). This caused a considerable loss of interest in the field.

A more promising approach of IL-12 DNA vaccination with electroporation has generated a somewhat renewed interest in the potential of IL-12. Preclinical trials in mouse melanoma models reported that intratumoral injection of IL-12 expression plasmid, immediately followed by electroporation, resulted in a 47% cure of tumor bearing mice (Lucas *et al.*, 2002). No toxic effects were observed and the results were supported by other preclinical mouse studies (Heller *et al.*, 2006; Lucas *et al.*, 2002, 2003). In 2008, the first human phase 1 trial was reported in which patients received escalating doses of IL-12 DNA vaccine and presented minimal toxicity (Daud *et al.*, 2008). Additionally, 10% of patients showed complete regression of all metastases and an additional 42% showed partial response (Daud *et al.*, 2008). Although this approach may be developed into a stand-alone therapy, it will most likely provide a much needed complementation to other melanoma DNA vaccine approaches.

Like many DNA vaccine approaches, melanoma-targeting DNA vaccines continue to make important and exciting strides toward effective melanoma immunotherapy. Future melanoma DNA vaccine potency will probably be enhanced by multi-antigenic approaches, cytokine and adjuvant enhancements (e.g., IL-12) and plasmid delivery enhancements (e.g., electroporation). Although the potential and promise of this field are exciting, we must not neglect the concern for autoimmune effects of the vaccine-induced immune response.

Autoimmune effects, such as vitiligo (loss of skin pigmentation due to melanocyte death) and systemic autoimmunity have already been associated with current common immunotherapies such as ipilimumab (Chang *et al.*, 2011; Hodi *et al.*, 2010; Yee *et al.*, 2000). However, vitiligo is often associated with tumor regression so we expect to see additional work focusing on improved potency as DNA vaccines enter this area.

20.4 Non-Hodgkin B-Cell Lymphoma

A significant cancer problem in the USA, as well as around the world, is non-Hodgkin lymphoma (NHL). NHL describes a wide variety of cancers that do not

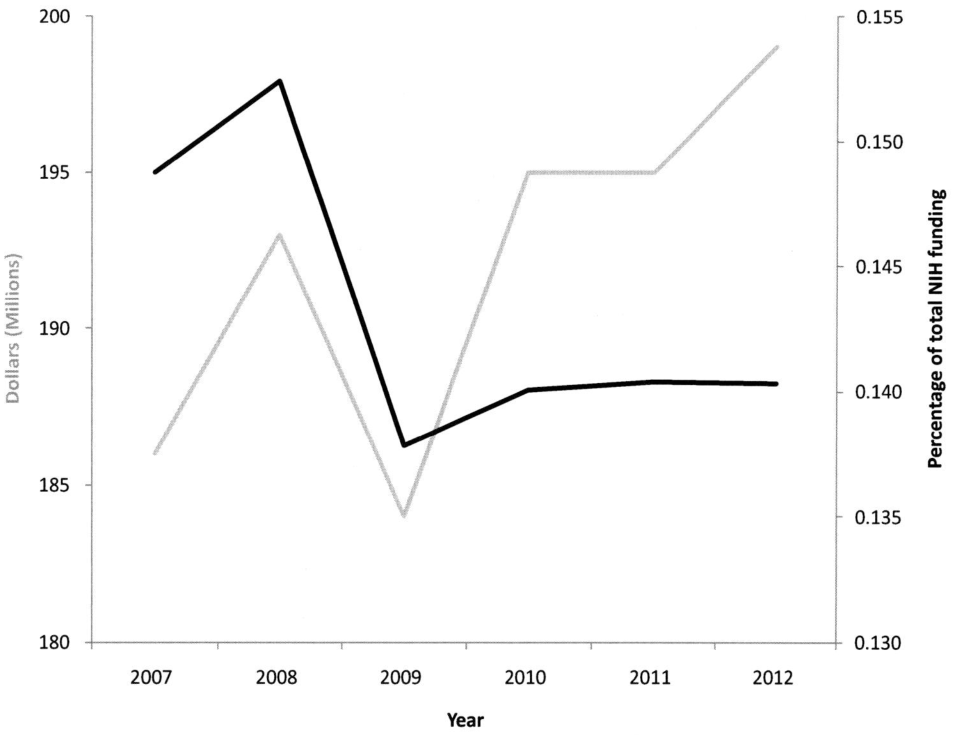

FIGURE 20.5 ■ USA NIH grant (millions of dollars) toward lymphoma-related research (gray) and the corresponding percent dedicated from total grant dollars (black). Lymphoma research dollars have increased from 186 to 199 million dollars. This equates to a decrease in percentage from 0.149% to 0.140% of total NIH funding. Data taken from the NIH Estimates of Funding for Various RCDCs.

include Hodgkin Lymphoma. Last year NHL accounted for 4% of new cancer cases and likewise accounted for 4% of cancer-related deaths in the USA (Jemal *et al.*, 2010). Since 2007, NIH grants toward lymphoma have increased from 186 million to 199 million dollars (Fig. 20.5). Interestingly, this increase in grant dollars equates to an overall decrease in NIH funding percent, perhaps due to a large increase in other funding categories while lymphoma research funding remained relatively stagnant. The conventional treatment for NHL is chemotherapy which, despite its efficacy in destroying the actively dividing cells, often fails to cure the cancer due to a rare group of "cancer stem cells" which are capable of renewing and sustaining cancer cell presence (Brody *et al.*, 2011). This, along with the increasing toxicity of chemotherapy treatments, has inspired researchers to explore immunotherapeutic approaches.

One of the most significant recent immunotherapeutic advances in treatment for this diverse lymphoma is the use of monoclonal antibodies (mAbs). mAb treatment was first described in human trials in 1982 in which anti-idiotype antibodies (idiotype is a B-cell-tumor-specific antigen) were used in the clinic to treat B-cell lymphoma (Miller *et al.*, 1982). These and other findings eventually led to the development of rituximab which, in 1997, became the first FDA-approved mAb for treatment of B-cell lymphoma, and is currently considered a standard therapy for NHL (Leget *et al.*, 1998). This treatment has provided non-toxic treatment for low-grade cancers and has also been used successfully in combination with chemotherapy (Czuczman *et al.*, 1999). Since the success of rituximab, there has been ongoing research to improve its efficacy, but this has been met with only limited success, leading some researchers to explore additional approaches (Brody *et al.*, 2011).

A promising alternative NHL immunotherapy is cancer vaccination. Although idiotype-protein-based vaccines have been the most pursued method to date (specifically the idiotype keyhole limpet hemocyanin (KLH) protein-based vaccine), DNA vaccination is currently being explored as a viable alternative (Briones, 2008). DNA vaccination is a particularly attractive alternative to idiotype protein vaccination, which requires *in vitro* protein expression, because of DNA's rapid production by polymerase chain reaction (PCR) (Briones, 2008). In NHL DNA vaccine preparation, the idiotype variable regions are PCR amplified from the patient's tumor sample and inserted into an expression plasmid (Briones, 2008).

An early murine study conducted by Syrengelas and colleagues in 1996 demonstrated that immunization with a DNA plasmid encoding the idiotype of a murine B-cell lymphoma induced a humoral-antigen-specific response and protected the mice from subsequent lymphoma challenge in a response similar to the idiotype KLH protein vaccine response (Syrengelas *et al.*, 1996). The team also reported that fusing GM-CSF to the idiotype protein improved efficacy and noted that a xenogeneic construct was required for immunogenicity (Syrengelas *et al.*, 1996). Further studies in murine lymphoma models went on to show that fusion of chemokines, including interferon inducible protein 10 (IP-10), monocyte chemotactic protein 3 (MCP-3) and fragment C of tetanus toxin to the idiotype gene in a DNA vaccine significantly increased the antigen-specific humoral response and offered protection to cancer challenge that either equaled or surpassed that of the KLH peptide vaccine (Biragyn *et al.*, 1999; King *et al.*, 1998). These promising studies warranted the initiation of clinical trials.

The first lymphoma DNA vaccine clinical trial was a phase I/II clinical trial reported in 2002. The study, which was conducted on 12 follicular B-cell lymphoma patients in remission after chemotherapy treatment, consisted of vaccinating the patients with a chimeric idiotype-encoding plasmid (containing their tumor-specific variable region and a xenogeneic murine constant region) which was also co-injected into patients with or without GM-CSF (Timmerman *et al.*, 2002b). Although the vaccination regimen resulted in an anti-murine immunoglobulin response, the anticancer response was unsatisfactory (Timmerman *et al.*, 2002b). Despite the unsatisfactory immune response, the suggested safety of the vaccine process, along with previous clinical activity of similar dendritic-cell- and protein-based vaccines, warranted further investigation with modifications to the DNA vaccine construction, delivery and adjuvant accompaniment (Timmerman *et al.*, 2002a, 2002b).

Working off of preclinical and previous clinical trial observations, Rice and co-workers conducted a clinical trial in the UK in 2008 which evaluated a DNA vaccine containing patient anti-idiotype fused with fragment C (FrC) of the tetanus toxin (Rice *et al.*, 2008). The vaccine was administered alone to 25 patients . It resulted in no safety concerns and induced an anti-FrC immune response in 72% of the patients. The authors of this trial evaluated anti-idiotype responses in 16 patients and found that 6 out of 16 (38%) were able to elicit anti-idiotype cellular and/or humoral responses (Rice *et al.*, 2008). Although relatively low clinical responses were observed, their improvement is expected to result from new delivery methods such as electroporation (Rice *et al.*, 2008). Due to its advantages over a protein-based anti-idiotype lymphoma vaccine, DNA vaccination will probably result in more studies in the future.

Due to its advantages over other treatment options, it is likely that NHL DNA vaccination will continue to be actively pursued. Due to the approaches promising overall efficacy, we expect future DNA vaccine constructs will continue to be based on the inclusion of patient-specific tumor antigens fused with adjuvant-like protein components. A possible concern for future DNA vaccine construction is progressive multifocal leukoencephalopathy (PML) which is already a concern in monoclonal antibody NHL treatment (e.g., Rituximab).

PML is a serious demyelinating brain infection caused by reactivation of latent John Cunningham viral infections. Some researchers have suggested a correlation between treatment with Rituximab and an increased probability of developing PML, although the actual risk remains undetermined (Clifford *et al.*, 2011; Keene *et al.*, 2011; Tuccori *et al.*, 2010). Although PML development remains rare, it still presents a possible clinical concern and NHL DNA vaccine constructs will need to be monitored for similar adverse effects.

20.5 Breast Cancer

Breast cancer poses a major health threat to American women. Last year breast cancer was the most diagnosed cancer type in American women (28%) and caused the second highest amount of cancer related deaths (15%) (Jemal *et al.*, 2010). Due to its prevalence, the NIH has continued to support breast cancer research by increasing its grant dollars from 729 to 778 million dollars (Fig. 20.6). This increase equates to a decrease in overall funding percentage in a similar way to NHL

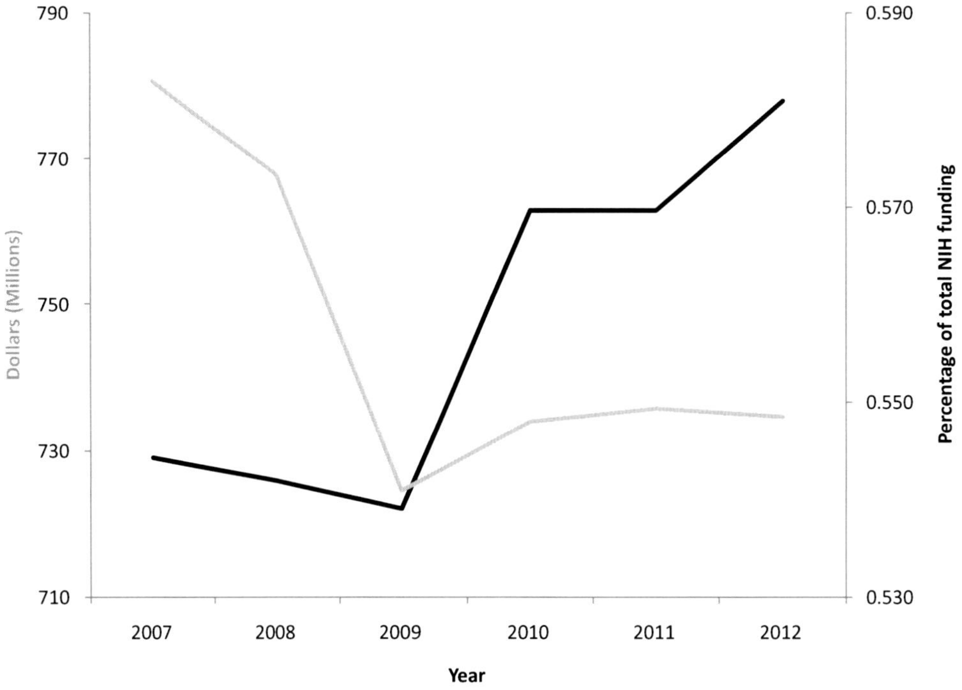

FIGURE 20.6 ■ USA NIH grant (millions of dollars) toward breast cancer research (gray) and the corresponding percentage dedicated from total NIH grant dollars (black). Breast cancer research dollars have increased from 729 to 778 million dollars. This equates to a decrease in percentage from 0.583% to 0.548% of total NIH funding. Data taken from the NIH Estimates of Funding for Various RCDCs.

research, and has probably occurred in a manner similar to that described above. Another speculation is that, due to the somewhat satisfactory current breast cancer treatment, more "pressing" issues are being funded by the NIH. Because breast cancer still remains a health threat in America, as well as around the world, research continues being conducted to improve diagnosis and treatment.

The current standard treatments for breast cancer include surgery, medications such as chemotherapy and antibody therapy and radiation treatment. Although breast cancer cure rates have increased, an estimated 20–30% of women experience a recurrence of their disease, an occurrence which has prompted an increased interest in immunotherapy (Soliman, 2010). Attempted methods of vaccination have included peptide-, virus- and dendritic-cell-based vaccines and DNA vaccines (Chen *et al.*, 2011; Met *et al.*, 2011; Miles *et al.*, 2011).

The major antigen target for breast cancer vaccination has been human epidermal growth factor receptor 2 (HER-2). HER-2 (expressed from the HER-2/neuregulin (neu) gene) is a protein receptor with tyrosine kinase activity and homology to the epidermal growth factor receptor and is present in many kinds of malignancies (Curigliano *et al.*, 2006). HER-2 is a satisfactory DNA vaccine target because it is highly expressed in cancerous tissue (notably breast cancer), it is localized on the cell surface, and it is involved in cancer progression/prognosis. HER-2 has been extensively studied as an immunotherapeutic target antigen and has been shown to be an effective target in multiple vaccine strategies; work which led to the investigation of its effects as a DNA vaccine (Curigliano *et al.*, 2006). HER-2 breast cancer vaccines have also been shown to be well received by patients in clinical trials (Disis *et al.*, 2009; Peoples *et al.*, 2005).

In one of the initial studies, Amici and colleagues reported that intramuscular injection of a plasmid expressing HER-2 was able to elicit significant protection against breast cancer tumors in mice (Amici *et al.*, 2000). IL-12 was also shown to be an effective adjuvant when co-expressed with the vaccine construct (Amici *et al.*, 2000). Further studies went on to show that HER-2 is an effective and promising DNA vaccine target antigen (Cappello *et al.*, 2003; Renard *et al.*, 2003). Since these initial reports, effort has been focused on improving the efficacy of the vaccine.

Recently, exciting advances have been made in improving HER-2 DNA vaccine efficacy. It has previously been reported that breast cancer treatment with the human monoclonal antibody trastuzumab is clinically effective although resistance can develop against the antibody over time (Nahta *et al.*, 2006; Vogel *et al.*, 2002). Inspired by this knowledge, Orlandi and colleagues investigated the adjuvant properties of trastuzumab by delivering a HER-2 DNA vaccine in combination with the trastuzumab-similar antibody 7.16.4 mAb (Orlandi *et al.*, 2011). They found that addition of the 7.16.4 mAb to the DNA vaccine regimen significantly reduced the sizes of established tumors compared with the DNA vaccine alone (Orlandi *et al.*, 2011). They also found that the combination elicited an antigen-specific $CD4^+$ and $CD8^+$ immune response, but the response was not significantly better than that produced to the DNA vaccine alone (Orlandi *et al.*, 2011). Overall the results suggested that the addition of trastuzumab is a promising method to enhance HER-2 DNA vaccine efficacy, although further studies will be required to confirm this claim (Orlandi *et al.*, 2011).

Another approach to improving the HER-2 DNA vaccine efficacy was investigated recently by Pakravan and colleagues. This approach attempted to use gp96, a member of the HSP90 family which has been a successful adjuvant in bacteria and virus DNA vaccines, as an adjuvant in the HER-2 DNA vaccine (Pakravan *et al.*, 2010b). To accomplish this, human gp96 was co-expressed with the rat HER-2 vaccine in mice (Pakravan *et al.*, 2010b). The results indicated that the adjuvant initially decreased tumor sizes, but, surprisingly, later increased tumor size to the point that the vaccine was more effective without the adjuvant (Pakravan *et al.*, 2010b). The authors suggest this could be due to an immune system failure at the final stages of the tumor growth, or because of antigen loss and tumor escape. Despite these disappointing results, it was reported that when the C-terminal domain of gp96 was fused to HER-2, CTL activity and IFN-γ secretion was increased and tumor size was significantly reduced (Pakravan *et al.*, 2010a).

This suggested that fusion of C-terminal gp96 to HER-2 could significantly enhance DNA vaccine clinical efficacy (Pakravan *et al.*, 2010a). It is also worth noting that addition of the *N*-terminal domain of gp96 to HER-2, either by co-administration or fusion, did not enhance efficacy of the vaccine.

Another promising target antigen for DNA vaccination against breast cancer is Mage-b. Mage-b has been found to be expressed on breast cancer tissue almost exclusively which made it an ideal vaccine target antigen (Park *et al.*, 2002). Inspired by this promise, a research group led by Sypniewska began investigating the efficacy of a Mage-b-targeting DNA vaccine.

In 2005, Sypniewska and co-workers reported that although immunizing mice with a DNA vaccine expressing Mage-b had only minimal effects on primary tumor size and progression it elicited a significant reduction in the number of metastases in the mice (Sypniewska *et al.*, 2005). This was a significant immune response because, while primary tumors are often capable of being treated by surgery, metastases are not surgically treatable and are the causes of most breast cancer deaths (Sypniewska *et al.*, 2005).

This research group went on to enhance the Mage-b DNA vaccine efficacy by co-expressing GM-CSF with the vaccine and intraperitoneally injecting it into mice, before the immunization, with thioglycolate broth (TGB) which is known to more efficiently recruit APCs to the immunization site (Gravekamp *et al.*, 2008). This immunization regimen reduced metastases number by 69% compared with the control vector+GM-CSF+TGB, and reduced tumor growth by 47% compared with the control vector+GM-CSF+TGB (Gravekamp *et al.*, 2008). Furthermore, the vaccine regimen significantly increased the $CD8^+$ CTL response in primary tumors and metastases, it increased the amount of Fas-ligand in primary tumors, and decreased the $CD4^+$ CTL response (Gravekamp *et al.*, 2008). Further analysis of this vaccine regimen is required in order to understand what immune mechanisms played a role in the response.

Finally, the efficacy of the Mage-b vaccine was also enhanced by delivery using *Listeria monocytogenes* as a vector. The research team showed that this regimen dramatically reduced the number of metastases by 88% compared with the vector control group, but had no significant effect on the primary tumors (Kim *et al.*, 2008). Overall, this research group has made significant and exciting advances in breast cancer DNA vaccination against Mage-b. Both DNA vaccination using naked plasmid DNA and vaccination delivered by *L. monocytogenes* show promise as vaccination attempts against breast cancer. Further research will be needed to determine how best to utilize these advances, each with its own set of advantages, in breast cancer treatment.

Because of its prevalence we can expect breast cancer treatment research to continue to generate interest, especially in immunotherapy. We expect that in the coming years, breast cancer DNA vaccine potency will be increased using multi-antigen-targeting constructs to address both primary tumor and metastasis cancer forms. Furthermore, DNA vaccine in combination with other common

TABLE 20.2 Summary of the Cancer DNA Vaccine Improvements Mentioned in this Review

Cancer Type	Modification	Reported Target Antigens	Immune Effects
Prostate cancer	IL-2	PSA	Increased antitumor immune response
	IL-18	PSA	Increased antitumor immune response
	Toxin fusion	PSMA	Overall increased DNA vaccine efficacy
	Splice variation	PSMA	Increased antitumor immune response
	4-1BBL	PSMc	Overall increased DNA vaccine efficacy
	CD-86	PSMc	Increase immune response
	hHSP70	PSCA	Increased $CD8^+$ T-cell response
Melanoma	Intracellular PRRs	TRP2	Increased $CD8^+$ T-cell response, increased mouse survival
	CTLA-4 boost	gp100	Increased induced-antigen-specific survival rate
	Anti-angiogenesis factors	gp100, TRP2	Increased cancer survival rate
Breast cancer	Trastuzumab	HER-2	Reduction in established tumor size
	gp96	HER-2	C-terminal fusion enhances immune response and reduces tumor size; N-terminal fusion and full-length expression are ineffective
	Thioglycolate Broth	MAGE-b	Increased induced antigen-specific immune response; reduction in metastases number; reduced tumor growth
Multiple cancers	Viral boosting	PSMc, STEAP PCSA, Mel-3 Poly epitope	Increased immune response
	Xenogenic antigen	PSA, PSMA, PAP PSCA, TYR, gp100 NHL Id mAb	Increased immune response
	IL-12	PSA, HER-2	Reduced antigen specific antibody response; Increased Th-cell proliferation not rhesus macaques); Increased antitumor efficacy
	GM-CSF	PSA, PSMc, PAP, gp100, MAGE-b	Increased tumor growth protection

therapeutic approaches, such as monoclonal antibody therapy and surgical procedures, will probably generate effective synergistic approaches.

For some cancer immunotherapies, there exists a possible concern of autoimmune-related adverse effects. The target of concern here is HER-2 as it is expressed on many tissues throughout the body, although it is overexpressed on breast cancer tissue. It has been suggested that CTLs against HER-2 preferentially recognize overexpressed tumor-related HER-2 and autoimmunity disorders do not occur (Bernhard *et al.*, 2002). Rather, because many patients have a pre-existing tolerance to HER-2 and researchers therefore attempt to cause immune-suppression, immune-suppression may have the potential to result in a problematic autoimmune response (Jacob *et al.*, 2009; Ladjemi *et al.*, 2010). These possibilities should be monitored in breast cancer DNA vaccine development as the field continues to gain more traction.

20.6 Concluding Remarks

Cancer immunotherapy with DNA vaccines has made great advances in recent years. By modifying the delivery, expression and adjuvant approaches to cancer DNA vaccines, researchers have made great strides toward successfully treating and preventing four of the most problematic cancers in the USA as well as around the world. The advances discussed in this review are outlined in Table 20.2.

It is likely that we will observe a combination of target antigens used in DNA vaccine treatment of the four main cancers. Newer strategies will employ enhanced delivery methods (i.e., electroporation), and may contain a combination or novel formulation of the adjuvants discussed above. It is likely that results of DNA vaccination studies will be used to expand or supplement current treatment strategies, such as chemotherapy, to offer more complete eradication of cancerous cells. By combining these techniques, the ability of newer and more potent DNA approaches to target cancers will become clearer. It is likely, based on the preliminary success in this field, that we can expect a great deal of exciting studies and growth over the next few years in DNA approaches for cancer immunotherapy.

References

Ahmad, S., Casey, G., Sweeney, P., *et al.* (2009). Prostate stem cell antigen DNA vaccination breaks tolerance to self-antigen and inhibits prostate cancer growth, *Mol Ther*, **17**, 1101–1108.

Alam, S., McNeel, D.G. (2010). DNA vaccines for the treatment of prostate cancer, *Expert Rev Vaccines*, **9**, 731–745.

Amici, A., Smorlesi, A., Noce, G., *et al.* (2000). DNA vaccination with full-length or truncated neu induces protective immunity against the development of spontaneous mammary tumors in HER-2/neu transgenic mice, *Gene Ther*, **7**, 703–706.

Becker, J.T., Olson, B.M., Johnson, L.E., *et al.* (2010). DNA vaccine encoding prostatic acid phosphatase (PAP) elicits long-term T-cell responses in patients with recurrent prostate cancer, *J Immunother*, **33**, 639–647.

Bernhard, H., Salazar, L., Schiffman, K., *et al.* (2002). Vaccination against the HER-2/neu oncogenic protein, *Endocr Relat Cancer*, **9**, 33–44.

Bilusic, M., Heery, C., Madan, R.A. (2011). Immunotherapy in prostate cancer: Emerging strategies against a formidable foe, *Vaccine*, **29**, 6485–6497.

Biragyn, A., Tani, K., Grimm, M.C., *et al.* (1999). Genetic fusion of chemokines to a self tumor antigen induces protective, T-cell dependent antitumor immunity, *Nat Biotechnol*, **17**, 253–258.

Briones, J. (2008). Therapeutic vaccines for non-Hodgkin B-cell lymphoma, *Clin Transl Oncol*, **10**, 543–551.

Brody, J., Kohrt, H., Marabelle, A., *et al.* (2011). Active and passive immunotherapy for lymphoma: proving principles and improving results, *J Clin Oncol*, **29**, 1864–1875.

Cappello, P., Triebel, F., Iezzi, M., *et al.* (2003). LAG-3 enables DNA vaccination to persistently prevent mammary carcinogenesis in HER-2/neu transgenic BALB/c mice, *Cancer Res*, **63**, 2518–2525.

Cassaday, R.D., Sondel, P.M., King, D.M., *et al.* (2007). A phase I study of immunization using particle-mediated epidermal delivery of genes for gp100 and GM-CSF into uninvolved skin of melanoma patients, *Clin Cancer Res*, **13**, 540–549.

Chan, R.C., Gutierrez, B., Ichim, T.E., *et al.* (2009). Enhancement of DNA cancer vaccine efficacy by combination with anti-angiogenesis in regression of established subcutaneous B16 melanoma, *Oncol Rep*, **22**, 1197–1203.

Chang, G.Y., Kohrt, H.E., Stuge, T.B., *et al.* (2011). Cytotoxic T lymphocyte responses against melanocytes and melanoma, *J Transl Med*, **9**, 122.

Chen, Y., Xie, Y., Chan, T., *et al.* (2011). Adjuvant effect of HER-2/neu-specific adenoviral vector stimulating $CD8^+$ T and natural killer cell responses on anti-HER-2/neu antibody therapy for well-established breast tumors in HER-2/neu transgenic mice, *Cancer Gene Ther*, **18**, 489–499.

Chen, Y.T., Stockert, E., Tsang, S., *et al.* (1995). Immunophenotyping of melanomas for tyrosinase: implications for vaccine development, *Proc Natl Acad Sci USA*, **92**, 8125–8129.

Clifford, D.B., Ances, B., Costello, C., *et al.* (2011). Rituximab-associated progressive multifocal leukoencephalopathy in rheumatoid arthritis, *Arch Neurol*, **68**, 1156–1164.

Colombo, M.P., Trinchieri, G. (2002). Interleukin-12 in anti-tumor immunity and immunotherapy, *Cytokine Growth Factor Rev*, **13**, 155–168.

Curigliano, G., Spitaleri, G., Pietri, E., *et al.* (2006). Breast cancer vaccines: A clinical reality or fairy tale?, *Ann Oncol*, **17**, 750–762.

Czuczman, M.S., Grillo-Lopez, A.J., White, C.A., *et al.* (1999). Treatment of patients with low-grade B-cell lymphoma with the combination of chimeric anti-CD20 monoclonal antibody and CHOP chemotherapy, *J Clin Oncol*, **17**, 268–276.

Dangoor, A., Lorigan, P., Keilholz, U., *et al.* (2010). Clinical and immunological responses in metastatic melanoma patients vaccinated with a high-dose poly-epitope vaccine, *Cancer Immunol Immunother*, **59**, 863–873.

Daud, A.I., DeConti, R.C., Andrews, S., *et al.* (2008). Phase I trial of interleukin-12 plasmid electroporation in patients with metastatic melanoma, *J Clin Oncol*, **26**, 5896–5903.

Disis, M.L., Wallace, D.R., Gooley, T.A., *et al.* (2009). Concurrent trastuzumab and HER2/neu-specific vaccination in patients with metastatic breast cancer, *J Clin Oncol*, **27**, 4685–4692.

Eggermont, A.M., Robert, C. (2011). New drugs in melanoma: It's a whole new world, *Eur J Cancer*, **47**, 2150–2157.

Ferraro, B., Cisper, N.J., Talbott, K.T., *et al.* (2011a). Co-delivery of PSA and PSMA DNA vaccines with electroporation induces potent immune responses, *Human Vaccines*, **7**, 120–127.

Ferraro, B., Morrow, M.P., Hutnick, N.A., *et al.* (2011b). Clinical applications of DNA vaccines: current progress, *Clin Infect Dis*, **53**, 296–302.

Garcia, J.A. (2011). Sipuleucel-T in patients with metastatic castration-resistant prostate cancer: an insight for oncologists, *Ther Adv Med Oncol*, **3**, 101–108.

Garcia-Hernandez, M. de, L., Gray, A., Hubby, B., *et al.* (2007). *In vivo* effects of vaccination with six-transmembrane epithelial antigen of the prostate: a candidate antigen for treating prostate cancer, *Cancer Res*, **67**, 1344–1351.

Garcia-Hernandez, M. de, L., Gray, A., Hubby, B., *et al.* (2008). Prostate stem cell antigen vaccination induces a long-term protective immune response against prostate cancer in the absence of autoimmunity, *Cancer Res*, **68**, 861–869.

Ginsberg, B.A., Gallardo, H.F., Rasalan, T.S., *et al.* (2010). Immunologic response to xenogeneic gp100 DNA in melanoma patients: comparison of particle-mediated epidermal delivery with intramuscular injection, *Clin Cancer Res*, **16**, 4057–4065.

Goldberg, S.M., Bartido, S.M., Gardner, J.P., *et al.* (2005). Comparison of two cancer vaccines targeting tyrosinase: plasmid DNA and recombinant alphavirus replicon particles, *Clin Cancer Res*, **11**, 8114–8121.

Gravekamp, C., Leal, B., Denny, A., *et al.* (2008). *In vivo* responses to vaccination with Mage-b, GM-CSF and thioglycollate in a highly metastatic mouse breast tumor model, 4T1, *Cancer Immunol Immunother*, **57**, 1067–1077.

Gregor, P.D., Wolchok, J.D., Turaga, V., *et al.* (2005). Induction of autoantibodies to syngeneic prostate-specific membrane antigen by xenogeneic vaccination, *Int J Cancer*, **116**, 415–421.

Gulley, J.L., Arlen, P.M., Madan, R.A., *et al.* (2010). Immunologic and prognostic factors associated with overall survival employing a poxviral-based PSA vaccine in metastatic castrate-resistant prostate cancer, *Cancer Immunol Immunother*, **59**, 663–674.

Gupta, S., Carballido, E., Fishman, M. (2011). Sipuleucel-T for therapy of asymptomatic or minimally symptomatic, castrate-refractory prostate cancer: an update and perspective among other treatments, *Onco Targets Ther*, **4**, 79–96.

Heller, L., Merkler, K., Westover, J., *et al.* (2006). Evaluation of toxicity following electrically mediated interleukin-12 gene delivery in a B16 mouse melanoma model, *Clin Cancer Res*, **12**, 3177–3183.

Hodi, F.S., O'Day, S.J., McDermott, D.F., *et al.* (2010). Improved survival with ipilimumab in patients with metastatic melanoma, *N Engl J Med*, **363**, 711–723.

Jacob, J.B., Kong, Y.C., Nalbantoglu, I., *et al.* (2009). Tumor regression following DNA vaccination and regulatory T cell depletion in neu transgenic mice leads to an increased risk for autoimmunity, *J Immunol*, **182**, 5873–5881.

Jemal, A., Siegel, R., Xu, J., *et al.* (2010). Cancer statistics, 2010, *CA Cancer J Clin*, **60**, 277–300.

Jilaveanu, L.B., Aziz, S.A., Kluger, H.M. (2009). Chemotherapy and biologic therapies for melanoma: do they work?, *Clin Dermatol*, **27**, 614–625.

Johnson, L.E., Frye, T.P., Chinnasamy, N., *et al.* (2007). Plasmid DNA vaccine encoding prostatic acid phosphatase is effective in eliciting autologous antigen-specific CD8+ T cells, *Cancer Immunol Immunother*, **56**, 885–895.

Kantoff, P.W., Schuetz, T.J., Blumenstein, B.A., *et al.* (2010). Overall survival analysis of a phase II randomized controlled trial of a Poxviral-based PSA-targeted immunotherapy in metastatic castration-resistant prostate cancer, *J Clin Oncol*, **28**, 1099–1105.

Keene, D.L., Legare, C., Taylor, E., *et al.* (2011). Monoclonal antibodies and progressive multifocal leukoencephalopathy, *Can J Neurol Sci*, **38**, 565–571.

Kim, J.J., Trivedi, N.N., Wilson, D.M., *et al.* (1998). Molecular and immunological analysis of genetic prostate specific antigen (PSA) vaccine, *Oncogene*, **17**, 3125–3135.

Kim, J.J., Yang, J.S., Dang, K., *et al.* (2001). Engineering enhancement of immune responses to DNA-based vaccines in a prostate cancer model in rhesus macaques through the use of cytokine gene adjuvants, *Clin Cancer Res*, **7**, 882s–889s.

Kim, S.H., Castro, F., Gonzalez, D., *et al.* (2008). Mage-b vaccine delivered by recombinant Listeria monocytogenes is highly effective against breast cancer metastases, *Br J Cancer*, **99**, 741–749.

King, C.A., Spellerberg, M.B., Zhu, D., *et al.* (1998). DNA vaccines with single-chain Fv fused to fragment C of tetanus toxin induce protective immunity against lymphoma and myeloma, *Nat Med*, **4**, 1281–1286.

Krieg, A.M., Yi, A.K., Matson, S., *et al.* (1995). CpG motifs in bacterial DNA trigger direct B-cell activation, *Nature*, **374**, 546–549.

Krupa, M., Canamero, M., Gomez, C.E., *et al.* (2011). Immunization with recombinant DNA and modified vaccinia virus Ankara (MVA) vectors delivering PSCA and STEAP1 antigens inhibits prostate cancer progression, *Vaccine*, **29**, 1504–1513.

Kuang, Y., Weng, X., Liu, X., *et al.* (2010a). Anti-tumor immune response induced by dendritic cells transduced with truncated PSMA IRES 4-1BBL recombinant adenoviruses, *Cancer Lett*, **293**, 254–262.

Kuang, Y., Zhu, H., Weng, X., *et al.* (2010b). Antitumor immune response induced by DNA vaccine encoding human prostate-specific membrane antigen and mouse 4-1BBL, *Urology*, **76**, 510 e511–e516.

Kuratsukuri, K., Sone, T., Wang, C.Y., *et al.* (2002a). Inhibition of prostate-specific membrane antigen (PSMA)-positive tumor growth by vaccination with either full-length or the C-terminal end of PSMA, *Int J Cancer*, **102**, 244–249.

Kuratsukuri, K., Wang, C.Y., Sone, T., *et al.* (2002b). Induction of antibodies against prostate-specific membrane antigen (PSMA) by vaccination with a PSMA DNA vector, *Eur Urol*, **42**, 67–73.

Kutzler, M.A., Weiner, D.B. (2008). DNA vaccines: Ready for prime time? *Nat Rev Genet*, **9**, 776–788.

Ladjemi, M.Z., Jacot, W., Chardes, T., *et al.* (2010). Anti-HER2 vaccines: new prospects for breast cancer therapy, *Cancer Immunol Immunother*, **59**, 1295–1312.

Leget, G.A., Czuczman, M.S. (1998). Use of rituximab, the new FDA-approved antibody, *Curr Opin Oncol*, **10**, 548–551.

Liao, J.C., Gregor, P., Wolchok, J.D., *et al.* (2006). Vaccination with human tyrosinase DNA induces antibody responses in dogs with advanced melanoma, *Cancer Immun*, **6**, 8.

Liu, H., Moy, P., Kim, S., *et al.* (1997). Monoclonal antibodies to the extracellular domain of prostate-specific membrane antigen also react with tumor vascular endothelium, *Cancer Res*, **57**, 3629–3634.

Lladser, A., Mougiakakos, D., Tufvesson, H., *et al.* (2011). DAI (DLM-1/ZBP1) as a genetic adjuvant for DNA vaccines that promotes effective antitumor CTL immunity, *Mol Ther*, **19**, 594–601.

Lucas, M.L., Heller, R. (2003). IL-12 gene therapy using an electrically mediated nonviral approach reduces metastatic growth of melanoma, *DNA Cell Biol*, **22**, 755–763.

Lucas, M.L., Heller, L., Coppola, D., *et al.* (2002). IL-12 plasmid delivery by *in vivo* electroporation for the successful treatment of established subcutaneous B16.F10 melanoma, *Mol Ther*, **5**, 668–675.

MacGregor, R.R., Boyer, J.D., Ugen, K.E., *et al.* (1998). First human trial of a DNA-based vaccine for treatment of human immunodeficiency virus type 1 infection: Safety and host response, *J Infect Dis*, **178**, 92–100.

Manley, C.A., Leibman, N.F., Wolchok, J.D., *et al.* (2011). Xenogeneic murine tyrosinase DNA vaccine for malignant melanoma of the digit of dogs, *J Vet Intern Med*, **25**, 94–99.

McNeel, D.G., Dunphy, E.J., Davies, J.G., *et al.* (2009). Safety and immunological efficacy of a DNA vaccine encoding prostatic acid phosphatase in patients with stage D0 prostate cancer, *J Clin Oncol*, **27**, 4047–4054.

Met, O., Balslev, E., Flyger, H., *et al.* (2011). High immunogenic potential of p53 mRNA-transfected dendritic cells in patients with primary breast cancer, *Breast Cancer Res Treat*, **125**, 395–406.

Miles, D., Roche, H., Martin, M., *et al.* (2011). Phase III multicenter clinical trial of the sialyl-TN (STn)-keyhole limpet hemocyanin (KLH) vaccine for metastatic breast cancer, *Oncologist*, **6**, 1092–1100.

Miller, A.M., Ozenci, V., Kiessling, R., *et al.* (2005). Immune monitoring in a phase 1 trial of a PSA DNA vaccine in patients with hormone-refractory prostate cancer, *J Immunother*, **28**, 389–395.

Miller, R.A., Maloney, D.G., Warnke, R., *et al.* (1982). Treatment of B-cell lymphoma with monoclonal anti-idiotype antibody, *N Engl J Med*, **306**, 517–522.

Mincheff, M., Tchakarov, S., Zoubak, S., *et al.* (2000). Naked DNA and adenoviral immunizations for immunotherapy of prostate cancer: A phase I/II clinical trial, *Eur Urol*, **38**, 208–217.

Nahta, R., Esteva, F.J. (2006). HER2 therapy: Molecular mechanisms of trastuzumab resistance, *Breast Cancer Res*, **8**, 215.

Orlandi, F., Guevara-Patino, J.A., Merghoub, T., *et al.* (2011). Combination of epitope-optimized DNA vaccination and passive infusion of monoclonal antibody against HER2/neu leads to breast tumor regression in mice, *Vaccine*, **29**, 3646–3654.

Pakravan, N., Soleimanjahi, H., Hassan, Z.M. (2010a). GP96 C-terminal improves Her2/neu DNA vaccine, *J Gene Med*, **12**, 345–353.

Pakravan, N., Langroudi, L., Hajimoradi, M., *et al.* (2010b). Co-administration of GP96 and Her2/neu DNA vaccine in a Her2 breast cancer model, *Cell Stress Chaperones*, **15**, 977–984.

Park, J.W., Kwon, T.K., Kim, I.H., *et al.* (2002). A new strategy for the diagnosis of MAGE-expressing cancers, *J Immunol Methods*, **266**, 79–86.

Pavlenko, M., Roos, A.K., Lundqvist, A., *et al.* (2004). A phase I trial of DNA vaccination with a plasmid expressing prostate-specific antigen in patients with hormone-refractory prostate cancer, *Br J Cancer*, **91**, 688–694.

Peoples, G.E., Gurney, J.M., Hueman, M.T., *et al.* (2005). Clinical trial results of a HER2/neu (E75) vaccine to prevent recurrence in high-risk breast cancer patients, *J Clin Oncol*, **23**, 7536–7545.

Renard, V., Sonderbye, L., Ebbehoj, K., *et al.* (2003). HER-2 DNA and protein vaccines containing potent Th cell epitopes induce distinct protective and therapeutic antitumor responses in HER-2 transgenic mice, *J Immunol*, **171**, 1588–1595.

Rice, J., Ottensmeier, C.H., Stevenson, F.K. (2008). DNA vaccines: precision tools for activating effective immunity against cancer, *Nat Rev Cancer*, **8**, 108–120.

Roos, A.K., Pavlenko, M., Charo, J., *et al.* (2005). Induction of PSA-specific CTLs and anti-tumor immunity by a genetic prostate cancer vaccine, *Prostate*, **62**, 217–223.

Rosenberg, S.A., Yang, J.C., Sherry, R.M., *et al.* (2003). Inability to immunize patients with metastatic melanoma using plasmid DNA encoding the gp100 melanoma-melanocyte antigen, *Hum Gene Ther*, **14**, 709–714.

Soliman, H. (2010). Developing an effective breast cancer vaccine, *Cancer Control*, **17**, 183–190.

Sypniewska, R.K., Hoflack, L., Tarango, M., *et al.* (2005). Prevention of metastases with a Mage-b DNA vaccine in a mouse breast tumor model: potential for breast cancer therapy, *Breast Cancer Res Treat*, **91**, 19–28.

Syrengelas, A.D., Chen, T.T., Levy, R. (1996). DNA immunization induces protective immunity against B-cell lymphoma, *Nat Med*, **2**, 1038–1041.

Tang, D.C., DeVit, M., Johnston, S.A. (1992). Genetic immunization is a simple method for eliciting an immune response, *Nature*, **356**, 152–154.

Timmerman, J.M., Czerwinski, D.K., Davis, T.A., *et al.* (2002a). Idiotype-pulsed dendritic cell vaccination for B-cell lymphoma: clinical and immune responses in 35 patients, *Blood*, **99**, 1517–1526.

Timmerman, J.M., Singh, G., Hermanson, G., *et al.* (2002b). Immunogenicity of a plasmid DNA vaccine encoding chimeric idiotype in patients with B-cell lymphoma, *Cancer Res*, **62**, 5845–5852.

Todorova, K., Ignatova, I., Tchakarov, S., *et al.* (2005). Humoral immune response in prostate cancer patients after immunization with gene-based vaccines that encode for a protein that is proteasomally degraded, *Cancer Immun*, **5**, 1.

Tuccori, M., Focosi, D., Blandizzi, C., *et al.* (2010). Inclusion of rituximab in treatment protocols for non-Hodgkin's lymphomas and risk for progressive multifocal leukoencephalopathy, *Oncologist*, **15**, 1214–1219.

Vittes, G.E., Harden, E.L., Ottensmeier, C.H., *et al.* (2011). DNA fusion gene vaccines induce cytotoxic T-cell attack on naturally processed peptides of human prostate-specific membrane antigen, *Eur J Immunol*, **41**, 2447–2456.

Vogel, C.L., Cobleigh, M.A., Tripathy, D., *et al.* (2002). Efficacy and safety of trastuzumab as a single agent in first-line treatment of HER2-overexpressing metastatic breast cancer, *J Clin Oncol*, **20**, 719–726.

Weber, J., Boswell, W., Smith, J., *et al.* (2008). Phase 1 trial of intranodal injection of a Melan-A/MART-1 DNA plasmid vaccine in patients with stage IV melanoma, *J Immunother*, **31**, 215–223.

Weber, J.S., Vogelzang, N.J., Ernstoff, M.S., *et al.* (2011). A Phase 1 study of a vaccine targeting preferentially expressed antigen in melanoma and prostate-specific membrane antigen in patients with advanced solid tumors, *J Immunother*, **34**, 556–567.

Wolchok, J.D., Yuan, J., Houghton, A.N., *et al.* (2007). Safety and immunogenicity of tyrosinase DNA vaccines in patients with melanoma, *Mol Ther*, **15**, 2044–2050.

Yee, C., Thompson, J.A., Roche, P., *et al.* (2000). Melanocyte destruction after antigen-specific immunotherapy of melanoma: direct evidence of T cell-mediated vitiligo, *J Exp Med*, **192**, 1637–1644.

Yuan, J., Ginsberg, B., Page, D., *et al.* (2011). CTLA-4 blockade increases antigen-specific $CD8^+$ T cells in prevaccinated patients with melanoma: three cases, *Cancer Immunol Immunother*, **60**, 1137–1146.

Yuan, J., Ku, G.Y., Gallardo, H.F., *et al.* (2009). Safety and immunogenicity of a human and mouse gp100 DNA vaccine in a phase I trial of patients with melanoma, *Cancer Immunol Immunother*, **9**, 5.

Zhang, X., Yu, C., Zhao, J., *et al.* (2007). Vaccination with a DNA vaccine based on human PSCA and HSP70 adjuvant enhances the antigen-specific $CD8^+$ T-cell response and inhibits the $PSCA^+$ tumors growth in mice, *J Gene Med*, **9**, 715–726.

21

ADOPTIVE IMMUNOTHERAPY AND CAR-T CELLS: A REVOLUTIONARY CELL/GENE THERAPY TO TREAT CANCER

Daniel Scherman [a]

21.1 Introduction

Cancer immunotherapy is based on treating tumors by an indirect way: The aim is not to directly target tumor cells, but to use or educate the immune system to eliminate the tumor cells by a selectively activated immune reaction involving, in particular, T-lymphocytes.

Spectacular success in the cancer immunotherapy field is already proved by its extended clinical practice through the use of "check-point inhibitors", which are monoclonal antibodies (mAb) targeting identified T-cells activating pathways. Adoptive immunotherapy, which uses cell and gene therapy tools, represents the next rupture step of this strategy.

An overview of the immunological synapse is necessary for the understanding of checkpoint inhibitors and adoptive immunotherapy. This presentation will be restricted to T-lymphocytes (CD4 and CD8) because these are actually the main cells in use for adoptive immunotherapy, but other components of the immune response, such as natural killer (NK) cells are already considered for future developments of adoptive immunotherapy.

[a] Laboratory of Chemical and Biological Technologies for Health, UTCBS, Pharmacy School, Paris Descartes University, CNRS, Inserm, 4 avenue de l'Observatoir, 75006, Paris, France
Email: daniel.scherman@parisdescartes.fr

21.2 The Immunological Synapse

The immunological synapse (or immune synapse) represents the interacting contact zone between an antigen-presenting cell (APC) or a target cell such as a tumor cell, and a T-lymphocyte. The interaction of several surface molecular components from a naïve CD4/CD8 T-lymphocyte with an APC elicits an activation state which leads to the differentiation, proliferation, and activation of either $CD4^+$ helper T-cells or killer $CD8^+$ cytotoxic T-cells.

In particular, when exposed to infected or dysfunctional somatic cells, cytotoxic $CD8^+$ T-cells release the cytotoxins, perforin, granzymes, and granulysin. Through the action of perforin, granzymes enter the cytoplasm of the target cell and their serine protease function triggers the caspase cascade involving a series of cysteine proteases that lead to apoptosis (programmed cell death).

An illustration of an immunological synapse between an APC and a T-cell is displayed in Fig. 21.1. The major histocompatibility (MHC) complex protein at the APC surface presents a peptide fragment (8–10 amino acid residues for MHC class I and 13–25 residues for MHC class II) derived from a foreign viral protein or tumor cell mutated protein. The TCR binds to the peptide/MHC complex, each TCR subtypes bearing specific recognition capacity for a given peptide/MHC complex. This interaction occurs between an APC and a "naïve" $CD4^+$ $CD8^+$ T-lymphocyte, which will then specialize into a $CD4^+$ helper T-cell if the MHC complex is of class II (i.e., presenting extracellular antigens), or into a cytotoxic $CD8^+$ T-cell if the interaction occurs through an MHC class I surface protein.

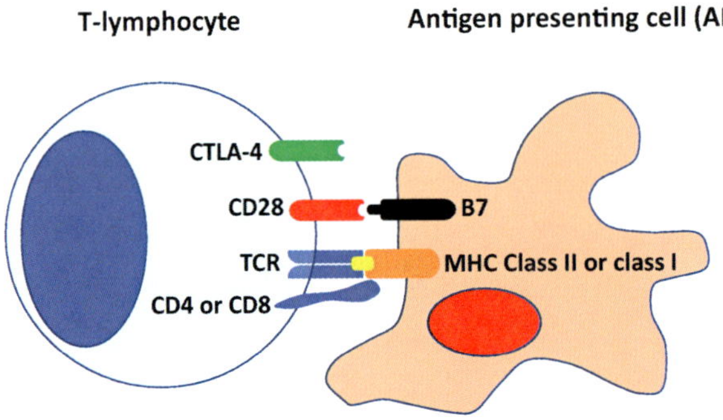

FIGURE 21.1 ■ Key molecular interactions involved in T-cell recognition and activation. Complete activation of effector T-cells (either $CD4^+$ or $CD8^+$) requires two signals transmitted by the APCs. In the first process, a specific peptide antigen (yellow) presented by the MHC protein, binds to by the T-cell receptor (TCR) complex (first signal). The second necessary signal is called "costimulatory signal". One of the main costimulatory pathways occurs through the binding of the B7 molecule present on the APC surface to the CD28 receptor expressed on T-cells (red). The B7 molecule is also called CD80 for the subtype B7-1 and CD86 for the subtype B7-2.

Both CD4 and CD8 molecules at the T-cell surface stabilize the immunological synapse interaction and are involved in the phosphorylation of TCR CD3ζ and of zap70 which insures intracellular transduction of the activation signal (see Section 21.3).

T-cell activation and proliferation is mediated by antigen-specific signals from the TCR-CD3 complex in combination with additional costimulatory signals provided by several co-receptors such as the extracellular protein CD28, the inducible co-stimulator ICOS, the *cytotoxic T-lymphocyte antigen-4* (*CTLA-4, or CD152-cluster of differentiation 152*), the *Programmed Death* receptor (PD-1), and other membrane proteins. While CD28 and ICOS provide positive signals that promote and sustain cytotoxic T-cell responses, on the contrary, CTLA-4 and PD-1 limit response and antagonize cytotoxic T-cell activation.

The interaction of APC surface protein B7 with the T-cell costimulatory CD28 is critical for T-cell activation and signal transduction. Remarkably, B7 can interact either with CD28, thus delivering a positive activating signal, or with CTLA-4, which has in contrary an inhibitory and anergizing effect in T-cells (Fig. 21.2). This duality of B7 recognition is essential in order for the immune system to be controlled and for the activation to be terminated: It is the balance between stimulatory and inhibitory co-signals which determines the ultimate nature of T-cell responses, thus controlling inflammation and autoimmunity.

Figure 21.2 helps us understand the therapeutic rationale for using mAb able to interfere with immunological synapse function. The mAb Ipilimimab and

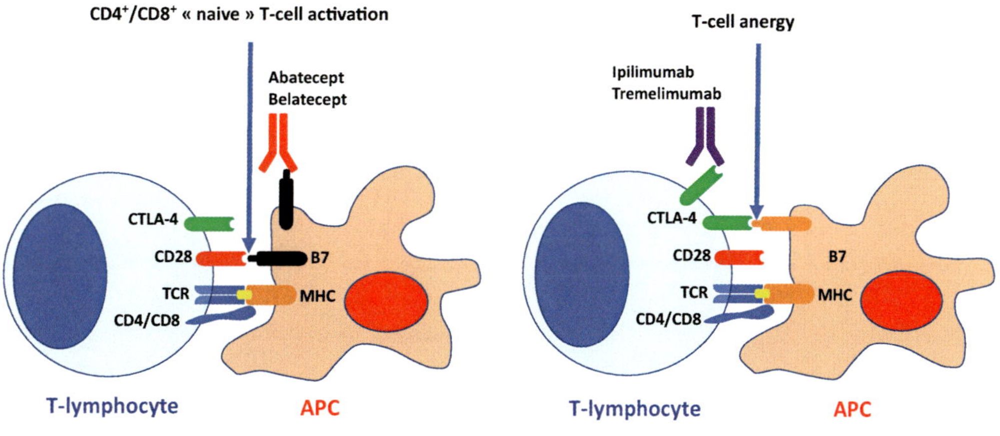

FIGURE 21.2 ■ Molecular interactions involved in T-cell activation or T-cell anergy. When APC-B7 is interacting with T-cell CD28, this induces T-cell activation. The mAb Abatecept and Belatecept bind to B7 and prevent this activation. They are clinically used to treat a variety of autoimmune diseases. In contrary, the binding of the APC-B7 to T-cell CTLA-4 induces T-cell anergy. This anergizing process is blocked by the mAb Ipilimumab and Tremelimumab, which have been proved to be active against several tumors such as melanoma.

Tremelibumab target CTLA-4 and inhibit B7 binding to CTLA-4 by steric effect: Hence, they potentialize T-cell activation and have been shown to be very efficient in the treatment of various cancers, in association with other anticancer drugs. On the other hand, Abatacept and Belatacept mAbs bind to B7 and thus anergize T-cells: They have demonstrated potency in autoimmune disorders, such as various arthritis, including rheumatoid arthritis and graft rejection.

Only the co-receptor CD28 and its ligand B7 are presented in Figs. 21.1 and 21.2 because they are the most relevant to the T-cells bearing a "chimeric antigen receptor" (CAR-T cells) technology described in this chapter, but other very important pathways, such as PD-1 and PD-1 ligand have been characterized as check-point inhibitors targets, leading to extensively used mAb for anticancer treatment.

21.3 TCR Structure and Activation Mechanism

The "TCR complex" schematized in Fig. 21.3 comprises eight polypeptides.

- The TCR receptor *per se* is a heterodimer constituted by the two chains TCR-α and TCR-β, which bind to the MHC-borne peptide epitope presented by APCs.
- The cluster of differentiation 3 co-receptor (CD3) helps to activate T-cells and consists of four distinct chains: CD3γ, CD3δ, and two CD3ε chains.

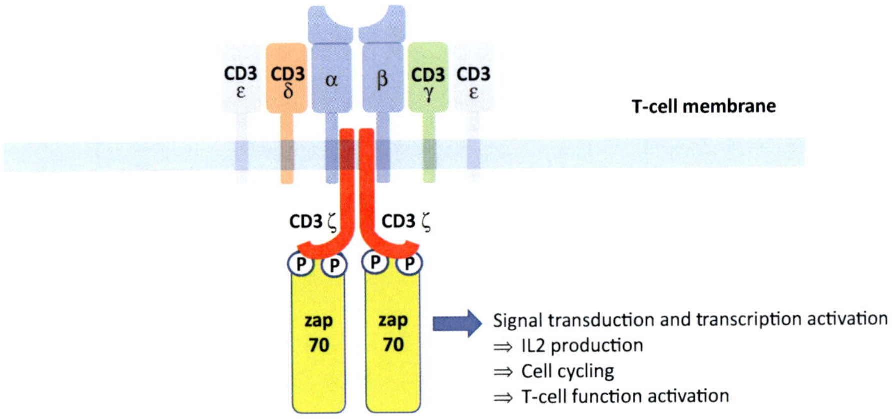

FIGURE 21.3 ■ Molecular assembly of the TCR complex and intracellular signal transduction. The two CD3ζ chains possess long intracellular domains bearing phosphorylation sites called ITAMs. When the TCR-α/β binds to the MHC-presented epitope, the ITAMS of the two CD3ζ are phosphorylated, which leads to zap 70 binding and intracellular signal transduction. All subsequent intracellular signalling events linked to the TCR complex are mediated through this CD3ζ activation.

- Two ζ-chains (zeta-chain) (also called CD3ζ by many authors) are responsible for intracellular signal transduction.

All the CD3 chains contain in their cytoplasmic tail at least one characteristic sequence motives for tyrosine phosphorylation called immunoreceptor tyrosine-based activation motifs (ITAMs) that play a major role in the initiation of signalling process. The CD3γ, CD3δ, and the two CD3ε chains harbor one ITAM motif whereas each CD3ζ chain has three ITAMS. Upon TCR interaction with a peptide/MHC molecule, the ITAMs are phosphorylated which leads to the binding of zap70 and the initiation an intracellular signalling pathway. This pathway will not be detailed here; it ultimately leads to IL-2 cytokine production and T-cell function activation. Supplementary stimulatory signals originating from costimulatory molecules such as CD28 are necessary for the complete activation of the immune response, involving not only T-cell functional activation, but also proliferation and enhancement of survival.

21.4 The General Principle of Adoptive Immunotherapy

In adoptive immunotherapy, tumor cells which express specific antigens are targeted and killed by a selected lymphocyte population. The difficulty arises from the selection of such a competent T-cell population, and this has promoted the development of several revolutionary technologies.

Primary attempts dating from the 1980s consisted of collecting the patient's tumor tissue, then isolating and amplifying from this tissue the T-cells naturally present and involved in the antitumor immune response, without any genetic modification. Such a population of amplified tumor infiltrating lymphocytes (TILs) showed promising therapeutic results in treating metastatic melanomas and several carcinomas such as renal, ovarian, colorectal, and pancreatic carcinomas. The problem is that efficient TILs are not found in all tumors, this being the cause of an important interindividual variability in the number and potency of the produced helper and cytotoxic T-cells.

The transplantation of non-autologous T-cells was also attempted in haematological malignancies, and allogeneic donor T-cells were sometimes shown to eradicate the disease via the graft-versus-leukaemia (GVL) effect. But this approach has a low success rate, and bears the important risk of graft-versus-host disease (GVHD), in which the grafted lymphocytes destroy cells of the patient, considered as non-self by the graft.

In order to increase the population of tumor-competent T-cells, the gene of a TCR bearing a specific affinity to an antigen expressed by the tumor cells was introduced into the patient's own T-lymphocytes. This autologous engineered T-cell population is restricted to a single TCR and is thus of potential higher efficacy and more general use than the TILs. This strategy was described to be active in myeloma and metastatic melanoma, but although it is still very actively pursued, it suffers from several drawbacks.

First, TCR-α/β displays a generally low affinity for the MHC molecules, and current efforts aim at isolating TCRs with optimal specificity and affinity.

Second, since the TCR-engineered T-lymphocytes need to bind to an MHC protein on targeted tumor cells to implement their cytotoxic functions, this brings the limitation linked to human leukocyte antigen (HLA) restriction. The introduced TCR must match the MHC HLA-specificity of the patient, which imposes to identify multiple TCRs for any given antigen in order to adapt to the different HLA haplotypes of the patients.

Third, tumor cells' subpopulations usually escape killing by down-regulating the expression at their surface of MHC molecules.

CAR-T cells were introduced to circumvent these shortcomings.

21.5 CAR-T Cells Adoptive Immunotherapy

CARs are fusion proteins engineered to bind to an antigen, for instance, a protein at the surface of a tumor cell, and to initiate T-cell activation, thus mimicking a TCR. The seminal pioneering work was that of Zelig Eshhar at the Weizmann Institute (Israel), who initially called these CARs "T-bodies". These first-generation CARs showed very promising efficiency in eradicating hematological tumors in animal models, but the subsequent clinical trials were disappointing.

A first-generation T body (CAR/T cell) is displayed in Fig. 21.4. It results from the fusion of several components, each having a specific functionality, which makes possible an infinite number of combinations that can be optimized for each targeted cell type.

- A signal peptide is necessary to direct the nascent protein to the endoplasmic reticulum and Golgi apparatus lumen, and then to the cell surface.
- The ectodomain contains mainly a single-chain variable fragment (ScFv) derived from an antibody which displays a high affinity to the targeted antigen. This ScFv is the key recognition moiety of the epitope/MHC complex. Other ligand to the targeted cell surface proteins can be used.
- A flexible spacer allows the scFv or other ligand to orientate in different directions in order to favor antigen binding. The spacer is an empirically defined

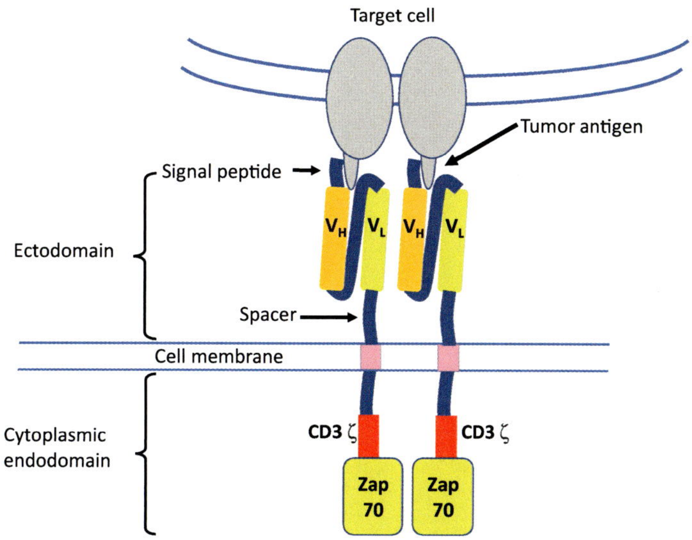

FIGURE 21.4 ■ First generation CAR or T-body. The cytoplasmic endodomain derived from the TCR CD3ζ moiety recruits zap70 and initiates the T-cell activation/proliferation process.

part of the engineered recombinant receptor, but it plays a major role in the targeted antigen binding efficacy. Several often-used fragments consist of the hinge region from IgG1, the CH_2CH_3 region of immunoglobulin, or portions of CD3.

- The transmembrane domain is a hydrophobic alpha helix usually derived from the original molecule of the signaling endodomain, such as CD3ζ. Alternatively, the transmembrane domain of CD28 or that of the tumor necrosis factor alpha (TNF-α) family has been used.
- The intracellular signaling endodomain protrudes into the cell and transmits the desired signal. It was initially speculated that the CD3ζ domain bearing the ITAMs and responsible for initiating the intracellular signal transduction would be sufficient to trigger T-cell activation upon binding to a target cell through ScFv recognition. It is assumed that the interaction between two CAR-T receptors associated to two targeted antigens leads to a pseudo-dimerization which recruits two adjacent zap70 molecules, thus mimicking in a perfect manner the TCR.

In spite of early promising results on preclinical models, disappointing clinical results led to the conclusion that signaling through such first-generation CAR receptors was not sufficient to trigger a productive immune response. This suggested that more molecular components of the immunological synapses are

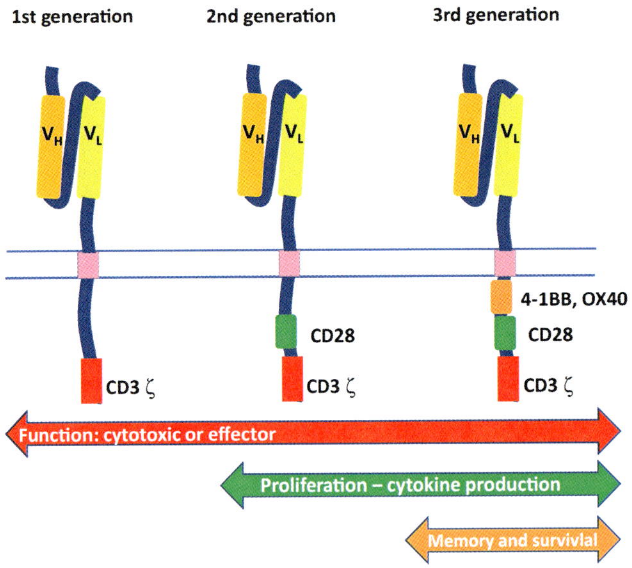

FIGURE 21.5 ■ Successive generations of CARs. The cytoplasmic endodomain derived from the TCR CD3ζ moiety recruits zap70 and initiates the T-cell activation process. Additional activation is conferred by the intracellular moiety of the CD28 costimulatory molecule which recruits PI3 kinase. Even more potent CAR-T cells can be produced by fusing another costimulatory signal, such as 4-1BB which recruits TRAF2.

necessary. Significant advances by several teams, in particular those of June and Sadelain, led to the "second generation" CAR-T cells which showed extraordinary success in B-lymphocytes malignancies such as acute or chronic lymphocytic leukemia.

The breakthrough came from the addition to the fusion intracellular endodomain of another activating moiety. Typical second-generation and third-generation CARs are schematized in Fig. 21.5.

The "second-generation" CAR genes add intracellular signalling domains from various costimulatory protein receptors (e.g., CD28, 4-1BB, ICOS). An improved antitumor activity was observed. The functional properties of CAR-T cells may depend on the co-stimulation signal used. For instance, CD28 promotes strong and fast proliferation of T-cell but limited persistence, while 4-1BB costimulatory fusion fragment induces a milder but more persistent CAR-T-cell response.

More recently introduced, the "third-generation" CARs combine multiple signaling domains, such as CD3ζ/CD28/4-1BB or CD3ζ/CD28/OX40, to further increase potency. Importantly, while CD3ζ recruits zap70 for intracellular signal transduction, the other added costimulatory factors recruit different intracellular relays, such as PI3 kinase for CD28 intracellular fragment and TRAF2 for the

4-1BB intracellular fragments. This results in additive or synergistic activation, and it is generally considered that first-generation CAR-T-cells induce the differentiated T-cell functions, such as cytotoxicity for CD8 T-cells, second-generation CAR-T cells possess stronger proliferation capacity, mainly due to IL2 production, and third-generation CAR-T cells possess survival and memory capacities. As already mentioned, burst effect or persistent CAR-T cells response might be modulated by selecting different costimulatory moieties.

21.6 CAR-T Cells Production and Administration

The success of the CAR-T cell strategy was built on many years of tediously acquired knowledge on T-cell biology and T-cell *in vitro* manipulations. A large number of steps had to be optimized one by one, comprising leukapheresis, apheresis washing, magnetic bead labelling, T-cell enrichment, cell growth formulation, cell stimulation, viral transduction, cell expansion, cell harvest, and release testing.

Briefly, blood cells are harvested from GM-CSF-stimulated patients by leukapheresis. The preparation is then enriched in T-cells, and purity is a critical factor, in particular, ensuring the elimination of any contamination by tumor cells. Various techniques have been used, initially density gradient separation, then centrifugation-based techniques separating according to cell size. More recent sophisticated techniques are still based on centrifugation, such as counterflow centrifugal elutriation.

The enriched T-cells are then transduced after a couple of days in culture in the presence of IL-2 and OKT3 antibody (antihuman CD3 murine mAb) by using a lentiviral vector carrying the CAR-T gene. An alternative to the use of lentiviral vectors to deliver the CAR-T gene consists of using a transposon plasmid carrying the CAR sequence associated to a plasmid encoding a transposase such as the "sleeping beauty" transposase. The transposase mediates CAR gene stable integration into the T-cell genome in a manner similar to that of a lentivirus, which ensures the stability of expression during the multiple divisions of the *in vitro* and *in vivo* CAR-T cells expansion.

CAR-T cells are then cultured and expanded *in vitro* for several days (in the presence of IL-2 cytokine using stimulation by with anti-CD3/CD28 antibodies coated on beads, and finally injected into the patient together with IL-2. Several variants of these techniques have been proposed, and automated devices that integrate all these preparation steps within a 7–9 days' process are now commercially available.

21.7 Clinical Results

A major advantage of CAR-T cells is that they circumvent HLA restriction, since no MHC molecule is involved in the recognition of the target cell, which involves a direct interaction of the ScFv with the targeted antigen. In addition, higher binding affinity is observed when compared to MHC binding. This and other factors explain the strikingly efficient results that have been obtained in the first clinical trials concerning acute and chronic lymphoblastic leukemia (ALL and CLL), in which the bone marrow produces too many B-lymphocytes, and lymphoma, also a blood cancer developed from B-lymphocytes and characterized by enlarged lymph nodes.

The targeted marker on B-lymphocytes is CD19, which is appealing since it is carried only by B-lymphocytes, whose temporary elimination is not a life-threatening. Moreover, temporary B-cells elimination might allow to avoid antibody-mediated response against CAR-T cells, which carry a non-self CAR gene.

Following preclinical work by the group of Michel Sadelain and others, in 2013, the group of Carl June published the result of a clinical trial using CD19 CAR-T cells for acute lymphoblastic leukemia in two patients. Complete remission was observed for the two patients, and in the two following years, five clinical studies reported from 70% to 100% of complete remissions in ALL patients who had no other therapeutic options.

These results generated a tremendous excitation, in particular because B-ALL is the most common malignancy in children, and that complete and prolonged remission was observed in patients with no other therapeutic option. This was the start of one of the fastest drug development process in history, and in 2018 two drugs based on antiCD19 CARs were approved by FDA for lymphoblastic leukemia and for refractory aggressive non-Hodgkin lymphoma. KYMRIAH® (tisagenlecleucel) is a CD19-directed genetically modified autologous T-cell immunotherapy indicated for the treatment of adult patients with relapsed or refractory (r/r) large B-cell lymphoma after two or more lines of systemic therapy including diffuse large B-cell lymphoma (DLBCL) not otherwise specified, high-grade B-cell lymphoma, and DLBCL arising from follicular lymphoma. Yescarta™ (Axicabtagene Ciloleucel) was the first CAR-T therapy approved by the FDA for the treatment of adult patients with relapsed or refractory large B-cell lymphoma after two or more lines of systemic therapy.

21.8 Toxicity Challenges and Safety Considerations

Although future prospects for the treatment of a large variety of tumors are exciting, several challenges have still to be overcome. First, toxicity due to the treatment has been observed, mainly concerning severe cytokine release syndrome (CRS), which can lead to death and neurotoxicity. This toxicity is further increased by the conditioning regimen necessary to deplete the naïve T-cell population of the patient before administering the CAR-T cells. The use of corticoids and anticytokine antibodies such as anti-IL-6 appears to alleviate most CRS manifestations, but their use is limited in order to avoid inhibiting CAR-T cells themselves, which is the case for glucocorticoids. This CRS toxicity was acutely observed when using the most aggressive third-generation CARs, and the difficulty to handle these tremendously potent CAR-T cells have precluded them from further clinical development so far.

Another limitation for the generalization of CAR-T cells is the identification of selective targets. Ideally, the target might be present only on the tumor tissue, since off-target effects might lead to severe unwanted damage to a life-essential tissue. For instance, the targeting of human epithelial growth factor receptor 2 HER2/neu (ErbB-2) has been investigated in an advanced colon cancer patient with liver metastases. The patient was treated with an anti-HER2/neu second-generation CAR containing a humanized Herceptin fragment and an optimized costimulatory signal. The patients suffered severe lung inflammation and oedema, linked to the fact that HER2 antigens are also present on lung epithelial cells, which led to CRS and to the patient's death. In another example, patients with metastatic renal carcinoma were treated with a CAR-T cell targeting the renal marker carbonic-anhydrase IX (CAIX). Hepatotoxicity was observed, due to the presence of CAIX on biliary epithelium. These two examples stress for necessary caution in selecting an antigen for CAR-T cell therapy.

Solutions have been proposed to alleviate such off-target effects, mainly based on combinatorial antigen recognition. The first strategy consists of making CARs dependent on the simultaneous presence of two antigens which might be selectively present only on tumor cells. On the contrary, a second strategy is based on the specific presence of a given target on normal, but not tumor cells, and consists of equipping the CAR-T cell with two CARs: An activating CAR recognizing the antigen present on both tumor and normal cells, and an additional inhibitory CAR (iCAR) containing a death signal such as PD-1 or CTLA4. The role of this iCAR is to attenuate or even suppress any CAR-T cells attack to normal cells (Fig. 21.6).

One way to improve the safety of a CAR-T cell treatment is to use a suicide gene approach, which would allow to destroy the CAR-T cells on demand. Ideally, this response should occur in a very short time, in order to control CRS or a GVH syndrome. The first approach is to introduce a gene whose products will kill the CAR-T cells upon administration of a chemical drug. Classical suicide genes are herpes-simplex thymidine kinase (HSV-TK) gene or the inducible caspase-9 (iC9) gene, the latter leading to a faster action than HSV-TK. Another elegant approach is to co-transduce T-cells with a surface marker such as truncated epidermal growth factor receptor (tEGFR). This truncated receptor is devoid of signal

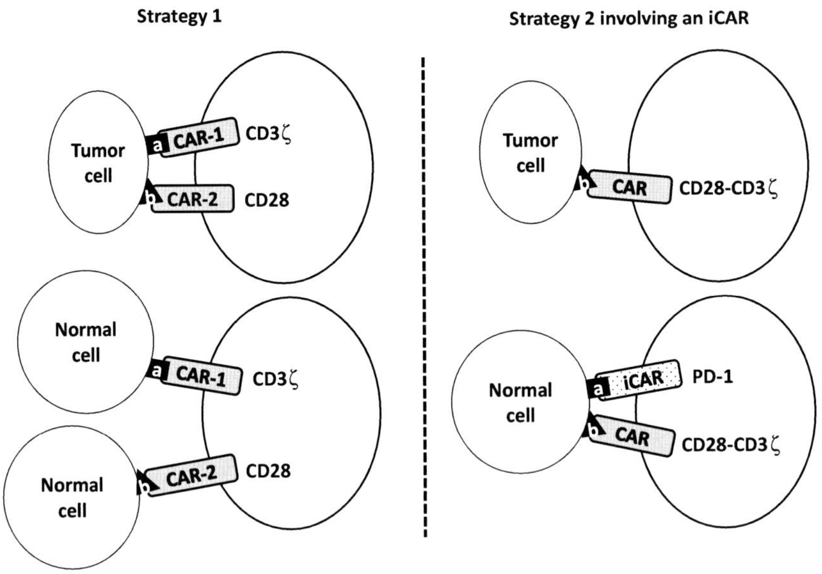

FIGURE 21.6 ■ Proposed strategies to improve CAR-T cell specificity and alleviate off-target effects. In the first strategy (left), the simultaneous presence of two antigens **a** and **b** is necessary to activate both CD33ζ and the costimulatory signal CD28. If antigens **a** or **b** are present alone on different normal cells, the CAR-T cell will not kill that cell. In the second strategy (right), the selective presence of antigen **a** on a normal cell will trigger the PD-1 mediated signal of the inhibitory iCAR, leading to the death of the CAR-T cell.

transduction domain but can be targeted by mAb already in the clinic, thus allowing antibody-dependent cell-mediated cytotoxicity processes against CAR-T cells when necessary.

21.9 Tumor Resistance and the Challenge of Solid Tumors

Another limitation for CAR-T cell treatment comes from the CD19 down-regulation in a subpopulation of tumor cells that will eventually escape the CAR-T cell treatment and cause relapse. For instance, several 2017–2018 reports indicated that up to 30% of patients who relapse after CD19 CAR-T cell therapy have CD19 negative disease. Since there are other markers specific of B-cells, such as CD20 and CD22, CARs have been designed against these markers, leading to successful CAR-T cells treatment in these relapsing patients. More recently, bispecific CD19–CD20 and even trispecific CAR-T cells have shown interesting results in CD19–CAR relapsed patients. One way to produce these divalent or trivalent CAR-T cells is to introduce a construct containing 2A viral sequences for simultaneous expression of equal concentration of each CAR at the cell membrane.

Finally, although CAR-T cells have been tremendously successful for treating several of the most severe haematological malignancies, multiple challenges must still be circumvented in order to generalize their use to solid tumors.

The first problem concerns the identification of suitable target antigen. Several clinical trials or advanced preclinical work involve targeting L1CAM for neuroblastoma, receptor tyrosine kinase-like orphan receptor 1 (ROR1) for triple-negative breast cancer, MUC16 (mucin 16) for primary peritoneal and Fallopian tube cancer, LeY for advanced solid tumors. Actively investigated targets, among many others, are carcinoembryonic antigen (CEA) for various carcinomas including colon cancer liver metastasis and mucin 1 (MUC1) and alpha-fetoprotein for liver cancer.

Although a ScFv moiety in the fusion protein is the most common targeting head in prototype CARs, other ligands have been envisioned. For instance, the proliferation inducing ligand APRIL has been fused either as a monomer or as a fusion trimer to target two proteins highly expressed on multiple myeloma tumor cells, BCMA and TACI (transmembrane activator and CAML interactor) which are members of the TNF receptor superfamily.

A second challenge specific to solid tumors is the poor CAR-T cells accessibility within the tumor tissue. This is caused not only by a mechanical barrier which hamper T-cell penetration into non-vascularized deep hypoxic tumor zones, but also to a general immunosuppressive environment linked to hypoxic conditions and release of anergizing cytokines. A proposed approach to overcome this central limitation is based on the co-treatment with small molecules or checkpoint inhibitors which interfere with immunosuppressive pathways. A complementary approach has been to load CAR-T cells with an immunostimulatory cytokine gene, such as IL-12, under the dependence of the CAR-T cell intracellular activation signal. The cytokine is then released at the immediate vicinity of the tumor. However, some toxicity has been observed in early clinical trials involving IL-12 CAR-T cells, and special caution should be given when using this potentially synergistic strategy.

21.10 Conclusion

In conclusion, CAR-T cells represent a revolutionary incremental advance in our fight against cancer, and further significant development of this technology should appear in the near future, including the use of other "CAR" cells, such as NK-cells, considering the tremendous effort and attention given to adoptive immunotherapy.

Progress should also come from the use of "off-the-shelf universal" heterologous T-cells, thus avoiding the tedious steps involved in obtaining autologous

CAR-T cells, and because the T-cells in patients having received heavy anticancer drug treatment are sometimes exhausted, leading to low potency CAR-T cells.

Major progress is expected from genome editing techniques which would allow the improvement and refining of CAR-T cells' function, for instance, by ablating their TCR in order to avoid GVH, by deleting their MHC molecules in order to allow heterologous "universal" CAR-T cells to survive in the host environment without being rejected.

Finally, the application of CAR-T cells should be extended in the future to facilitate the treatment of autoimmune, inflammatory, and infectious diseases.

For Further Reading

Eshhar, Z., Waks, T., Gross, G. (2014). The emergence of T-bodies/CAR T cells, *Cancer J*, **20**, 123–126.

Garfall, A.L., Stadtmauer, E.A., Hwang, W.T., *et al.* (2018). Anti-CD19 CAR T cells with high-dose melphalan and autologous stem cell transplantation for refractory multiple myeloma, *JCI Insight*, **3**, doi: 10.1172/jci.insight.120505, (Epub ahead of print).

June, C.H., O'Connor, R.S., Kawalekar, O.U., *et al.* (2018). CAR T cell immunotherapy for human cancer, *Science*, **359**, 1361–136.

Köhler, M., Greil, C., Hudecek, M., et al. (2018). Current developments in immunotherapy in the treatment of multiple myeloma, *Cancer*, **124**, 2075–2085.

Li, K., Lan, Y., Wang, J., *et al.* (2017). Chimeric antigen receptor-engineered T cells for liver cancers, progress and obstacles, *Tumour Biol*, 1–8.

Sadelain, M., Rivière, I., Riddell, S. (2017). Therapeutic T cell engineering, *Nature*, **545**, 423–431.

Wang, H., Kadlecek, T.A., Au-Yeung, B.B., http://cshperspectives.cshlp.org/content/2/5/a002279.full.

22

GENE THERAPY FOR SEVERE COMBINED IMMUNODEFICIENCIES (SCID)

***Salima Hacein-Bey-Abina*[a,b,g], *Alain Fischer*[c,d,e,h] *and Marina Cavazzana*[c,d,f,i]**

22.1 Introduction

Severe combined immunodeficiencies (SCIDs) comprise a group of inherited Mendelian disorders characterized by an early onset and a profound block in T-lymphocytes development. Given that adaptive immunity is abrogated, patients with SCID are prone to recurrent infections caused by viral, fungal, and bacterial pathogens such as cytomegalovirus (CMV), adenoviruses, and Streptococcus pneumoniae, and also by opportunistic organisms such as *Pneumocystis jiroveci* (Fischer *et al.*, 2015). Moreover, patients can develop severe, systemic, and often fatal disease when inadvertently given live vaccines for rotavirus, poliovirus, or Tuberculosis bacillus (the Bacillus Calmette–Guérin vaccine) (Marciano *et al.*, 2014; Shearer *et al.*, 2014; Fischer *et al.*, 2015). Without an appropriate treatment, the outcome of the disease is fatal before the age of one year. Typical SCIDs are defined by a complete absence of T-lymphocytes in the blood and lymphoid

[a] UTCBS, CNRS UMR 8258, INSERM U1022, Faculté de Pharmacie de Paris, Université Paris Descartes, Chimie ParisTech, Paris, France
[b] Clinical Immunology Laboratory, Groupe Hospitalier Universitaire Paris-Sud, Kremlin-Bicêtre Hospital, AP-HP, Le Kremlin Bicêtre, France
[c] Paris Descartes Sorbonne Paris Cité University, Imagine Institute, Paris, France
[d] INSERM UMR 1163, Paris, France
[e] Pediatric Hematology-Immunology and Rheumatology Unit, Necker-Enfants Malades Hospital, AP-HP, Paris, France
[f] Biotherapy Clinical Investigation Center, Groupe Hospitalier Universitaire Ouest, AP-HP, INSERM, Paris, France
[g] Email: salima.hacein-bey@aphp.fr; [h] Email: alain.fischer@aphp.fr; [i] Email: m.cavazzana@aphp.fr

organs, associated with other lymphocyte subsets defects depending on the gene type mutated. In T(–)B(+) SCIDs forms, even though B-lymphoid lineage may not be intrinsically affected, the absence of T-helper cells prevents B-cell from normal antibody production, leading to the "combined" deficits of cellular and humoral immunity. The underlying genetic defects and molecular mechanisms have been increasingly well characterized, for most of SCIDs, over the past 30 years (Notarangelo *et al.*, 2009; Fischer *et al.*, 2015). To date, mutations in 16 genes have been reported to account for nearly all cases of typical SCIDs.

The incidence of SCIDs, as measured in population-based newborn screening, is much higher than the estimates published before screening which were one in 75,000 live births (Stephan *et al.*, 1993; Buckley *et al.*, 1997). Indeed, universal newborn screening in the United States has allowed the reassessment of this value to one in 58,000 live births and has led to the improvement of survival outcome (Chinen *et al.*, 2014; Kwan *et al.*, 2014; Cicalese and Aiuti, 2015).

Allogeneic hematopoietic stem cell transplantation (HSCT) has been the gold standard therapy for the SCIDs as *in vivo* differentiation of normal transplanted HSC allows the replacement of dysfunctional lymphoid lineages (Slatter and Gennery, 2013). HSCT is most successful with a survival rate of 90% when a human leukocyte antigen (HLA)-genoidentical sibling donor is available (Gennery *et al.*, 2010). The rapid T-cell compartment restoration and the absence of graft-versus-host disease (GVHD) (the GVHD that results from the immune conflict between the donor's immune cells and the recipient) account for the excellent prognosis for HLA-identical HSCT in this setting. The use of matched, unrelated donors (MUDs) or haplo-identical donors is associated with a higher risk of morbidity and mortality because of GVHD and/or slow kinetics of T-cell development (Antoine *et al.*, 2003; Gennery *et al.*, 2010; Buckley, 2011; Pai *et al.*, 2014). For the significant number of patients without an appropriate donor, an attractive alternative approach has been therefore developed based on autologous transplantation of genetically modified hematopoietic stem and progenitor cells (HSPC).

Here, we review the results of gene therapy for the most frequent forms of SCIDs, X-linked SCID (SCID-X1), and adenosine deaminase (ADA) deficiency (ADA-SCID), which have served as models for implementing hematopoietic stem cell gene therapy.

22.2 Gene Therapy for SCID-X1

22.2.1 The Rationale for Gene Therapy of X-Linked Severe Combined Immunodeficiency (SCID-X1)

X-linked SCID is the most common SCID and is characterized by a complete lack of T- and natural killer (NK) cells, although B-cell development is normal

(Buckley, 2002). SCID-X1 is caused by defects in *IL2R*, the gene encoding the interleukin-2 receptor gamma chain (γc) (Noguchi *et al.*, 1993). This γc is shared by the cell surface receptors for six interleukins (IL-2, IL-4, IL-7, IL-9, IL-15, and IL-21) (Sugamura *et al.*, 1996). The pathophysiological features in SCID-X1 are explained by the absence of the survival and proliferation signals normally provided by IL-7 and IL-15 in lymphoid progenitors given that these two cytokines play a critical role in the development of T-cells and NK cells, respectively (Leonard and Warren, 1996). As with other SCIDs, allogeneic HSCT from an HLA-identical donor has a high survival rate, but transplantation from matched unrelated or mismatched related donors in patients severely affected by opportunistic agents is associated with lower survival rates (Antoine *et al.*, 2003; Gennery *et al.*, 2010; Buckley, 2011; Pai *et al.*, 2014). These limitations have accelerated (at least in part) the application of gene therapy to SCID-X1.

SCID-X1 was thought to be the most accurate model for assessing gene therapy because a selective growth advantage on the lymphoid precursors has been observed in many cases of spontaneous reversion of mutations in patients with primary immunodeficiencies (PIDs). This latter phenomenon (referred to as "natural gene therapy") has been reported in patients with ADA deficiency, SCID-X1, and Wiscott–Aldrich syndrome (WAS) (Stephan *et al.*, 1996; Bousso *et al.*, 2000; Speckmann *et al.*, 2008). This reversion in a single cell *in vivo* led to the development of functional T-cells and improvement of disease phenotype related to the emergence from a single T-cell precursor of a seemingly diversified T-cell repertoire (estimated to be circa 1% of normal values) (Bousso *et al.*, 2000). These findings have had a significant impact on the prospects for stem cell gene therapy of immunodeficiencies, and the presence of a selective advantage together with the long-life span of mature T-cells have been considered as important success factors because a few transduced T-cell precursors may give rise to a full and stable T-cell pool in treated patients. The other important SCID-X1 feature that prompted researchers to consider that no adverse effect related to the constitutive expression of the γc chain in all hematopoietic lineages would be observed (even in the absence of tight regulation of the transgene by its own promoter) is the fact that the γc chain is involved in di- or trimeric complex receptors, within which its expression and function is controlled by the other subunits.

The biological and clinical knowledge generated in HSCT has also been exploited in the development of gene therapy protocols; this explains why the first genetic diseases to have been "corrected" by gene transfer were SCID. The goal of *ex vivo* gene transfer of a functional copy of the defective gene into a patient's HSPCs with a viral vector is to correct progenitor cells and enable the subsequent, stable production of differentiated cells. That is why, beside the favorable disease characteristics, gene transfer technological advances had made it possible to perform clinical trials. Significant advances in retroviral vector technology as well as in gene transfer methods had been made in the early 1990s. These vectors were developed initially from murine oncoretroviruses (γ-retroviruses)

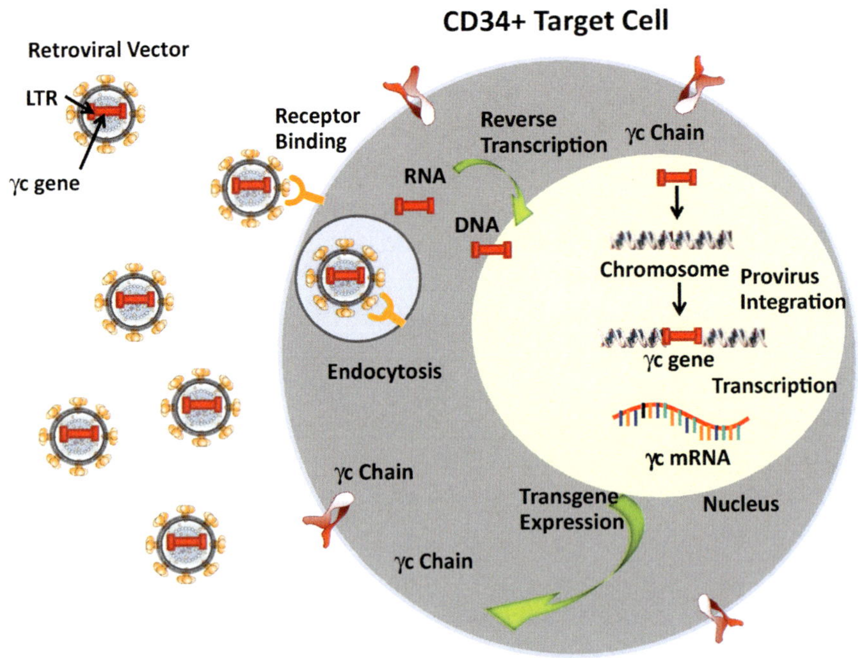

FIGURE 22.1 ■ Retroviral vector delivery of γc subunit to CD34$^+$ hematopoietic cells.

that are able to integrate into the genome of dividing cells (the integration pattern of these retroviruses was supposed to be random at time of initiation of clinical trials). In the first generation of vector constructs, based on the Moloney murine leukemia virus (MLV), transgene expression was driven by the viral promoter/enhancer sequences within the long terminal repeats (LTRs). Relatively high titers in vector production were achieved thanks to improvement in the construction of helper-free packaging cell lines as well as in retroviral vector design. The conditions for *ex vivo* stem cell transduction were optimized by using (i) combinations of cytokines that promotes CD34$^+$ cell population activation without losing its renewal capacity (Dao *et al.*, 1997) and (ii) a recombinant fragment of fibronectin which enhances vector-target cells' co-localization (Hanenberg *et al.*, 1996). (See Fig. 22.1).

22.2.2 SCID-X1 Clinical Trials Results

22.2.2.1 *Immune reconstitution*

Following extensive *in vitro* and *in vivo* preclinical studies (Cavazzana-Calvo *et al.*, 1996; Hacein-Bey *et al.*, 1996, 1998, 2001; Soudais *et al.*, 2000), the first clinical gene therapy protocol for SCID-X1 was conducted at Necker Hospital (Paris, France) between 1999 and 2002 (Cavazzana-Calvo *et al.*, 2000; Hacein-Bey-Abina *et al.*, 2002). Ten children under the age of one year with typical SCID-X1 (diagnosis was based on peripheral blood lymphocyte counts and confirmed by γc gene mutation analysis) but with no HLA-identical sibling were enrolled. The trial was based on the defective Moloney MLV derived vector (MFG (B2)-γc) within which the γc transgene was under the transcriptional control of the LTR. 30–150 ml of bone marrow was harvested under general anesthesia and the selected $CD34^+$ cells were activated by culture in X-vivo 10 medium (BioWhittaker, Walkersville, NJ) that contained 4% fetal calf serum (Stem Cell Technologies, Vancouver, Canada) in the presence of stem cell factor (300 ng/mL, Amgen, Thousand Oaks, CA), Flt-3 ligand (300 ng/mL, Immunex, Seattle, WA), polyethylene glycol-megakaryocyte growth and differentiation factor (100 ng/mL, Amgen), and IL-3 (60 ng/mL, Novartis). The cells were then transduced with viral supernatant on human recombinant fibronectin (50 ng/mL) (CH-296, Takara Shuzo, Shiga, Japan) coated sterile bags (PL-2417, Nexell-Therapeutics, Irvine, CA) in the presence of the same cytokines and protamine sulfate (4 ng/mL, Choay Sanofi, Gentilly, France). Viral supernatant was replaced every 24 h during the three-day transduction period. Cultured cells expanded by 5- to 8-fold and an average dose of 9×10^6 per kilogram CD34(+) γc(+) cells were infused back to the patients without preparative conditioning.

Among the 10 treated patients, one patient presented with persistent splenomegaly (caused by a disseminated Bacillus Calmette Guérin infection) that led to treatment failure. Transduced cells injected into this patient were trapped in the enlarged spleen, thereby preventing seeding of bone marrow and thymus (Hacein-Bey-Abina *et al.*, 2002, 2010). Another patient received a very low number of transduced progenitor cells that led to inefficient T-cell development (Ginn *et al.*, 2005). In the other eight evaluable patients, T-cell counts increased progressively to reach normal values for ages after 3–6 months and were still within the normal range at the last follow-up (Hacein-Bey-Abina *et al.*, 2010; Cavazzana *et al.*, 2016). In some patients, circulating $CD3^+$ T-cells were detected as early as one month after treatment. Distinct T-cell subsets were detected including regulatory T-cells, γδ T-cells, and NKT-cells. T-cell functional characteristics were also satisfactory as attested by normal proliferation *in vitro* in response to mitogens (PHA and anti-CD3) and vaccine antigens. In all cases, a Gaussian distribution of the CDR3 length expression profiles of T-cell receptor Vβ and Vα families was observed. Thymopoiesis is still ongoing more than 15 years after treatment (as evidenced by the continued detection of circulating recent thymic emigrants) in all patients except in one who received the lowest number of genetically modified progenitor cells. Indeed, a correlation between $CD34^+$ γc^+ cell dose infused to the patient and the robustness of T-cell reconstitution has been demonstrated in this trial. The persistence of naïve T-cells over the years strongly suggests that there is persistent production of T-cells from a pool of efficiently transduced T-cell progenitors. The restoration of functional immune responses conferred a sustained clinical benefit as evidenced by recovery from infections with a poor prognosis (such as VZV infections and disseminated BCG) during the first few months' post-treatment. Immunity recovery allowed children to live in normal environmental conditions (Hacein-Bey-Abina *et al.*, 2010; Cavazzana *et al.*, 2016).

As observed following non-myeloablative HSCT, NK cell counts first increased 2–4 months after gene therapy, then fell gradually after a year to very low levels. This phenomenon could be attributed to the weaker proliferative and/or survival capacity of NK cell progenitors (relative to that of T-cell progenitors), combined with a shorter life span of NK cell compared to T-cells. These NK cell were nevertheless shown to express γc and exert normal cytotoxic activity. No clinical consequences of low NK-cell counts have so far been observed (Hacein-Bey-Abina *et al.*, 2010; Cavazzana *et al.*, 2016).

Engraftment of B-cell was low as expected by the absence of a conditioning regimen required for effective HSPC engraftment. A few transduced B- and myeloid cells (<1%) were detected in the blood one year after gene therapy but then disappeared after 1–2 years (Hacein-Bey-Abina *et al.*, 2002, 2010). Despite the lack of persistent γc^{+} B-cell, humoral immunity was found to be sufficient to withdraw half of the patients from immunoglobulin (Igs) replacement therapy. Partial Igs production occurs likely through γc (IL21)-independent pathways. However, the development of full humoral responses and the long-term maintenance of normal levels of memory B-cells (associated with stable production of high-affinity antibodies against specific antigens) can only be achieved by stable engraftment of genetically modified stem cells. This would require access to the hematopoietic niches by conditioning the patient prior to transplantation (Hacein-Bey-Abina *et al.*, 2002, 2010; Cavazzana *et al.*, 2016) as envisaged now for further gene therapy of SCID-X1.

A similar trial was performed on 10 patients at the Great Ormond Street Hospital (London, the United Kingdom) (Gaspar *et al.*, 2004, 2011). The main difference with the French trial consisted of the use of a slightly different vector with either an amphotropic envelope (Hacein-Bey-Abina *et al.*, 2010) or the gibbon ape leukemia virus envelope (Gaspar *et al.*, 2011). The vector modification did not seem to induce any difference. When assessing both trials together, 17 of the 20 treated SCID-X1 patients displayed sustained correction of the T-cell immunity that was able to restore the patients' health status and enabled them to lead a normal life with long-lasting beneficial effects (today median follow-up: 16 years) (Hacein-Bey-Abina *et al.*, 2010; Gaspar *et al.*, 2011; Cavazzana *et al.*, 2016). The finding that virtually all T-lymphocytes were stably transduced while B- and myeloid cells were not, demonstrates that a selective advantage was conferred to T-lymphoid progenitors by γc expression. In addition, the sustained detection of naive T-cells over time indicates that transduced T-cell progenitors with self-renewal capacity persists in the thymus allowing long-term ongoing thymopoiesis (Martins *et al.*, 2012; Peaudecerf *et al.*, 2012).

Given the efficacy of gene therapy in patients with a typical SCID-X1, a similar protocol was applied to treat five older patients (aged 10–20 years) who displayed either atypical SCID-X1 (caused by hypomorphic mutations) or had partially failed HSCT. The treatment failed to correct the T-cell immune deficiency or show any significant clinical benefit for these patients, probably as a consequence of the loss of thymic function (Thrasher *et al.*, 2005; Chinen *et al.*, 2007).

22.2.2.2 Insertional genotoxicity in SCID-X1 trials

Unfortunately, six of the 20 treated patients (five in the Paris trial and one in the London trial) developed T-cell leukemia 2–15 years (2–5.5 years post-gene therapy in five of them) after gene therapy (Hacein-Bey-Abina *et al.*, 2003, 2008; Howe *et al.*, 2008; Six *et al.*, 2017). Chemotherapy eliminated five of the six leukemic clones, leading to sustained complete remission and restoration of a well-differentiated polyclonal γc^{+} T-cell population (Hacein-Bey-Abina *et al.*, 2010; Gaspar *et al.*, 2011; Cavazzana *et al.*, 2016). One patient died 26 months after leukemia occurrence, with refractory leukemia despite multiple chemotherapy cycles, two successive bone marrow transplantations from a MUD and treatment with a monoclonal anti-TCRδ1 antibody produced specifically to target the patient T-cell clone (carrying the Vγ9δ1 TCR) (Hacein-Bey-Abina *et al.*, 2003, 2008, 2010).

The occurrence of these severe adverse events prompted the halt of these trials, and a significant investment by the teams involved in these clinical trials and those specializing in retroviruses biology made it possible to understand the mechanisms underlying these side effects. Genetic analysis showed that the malignant cells had one or two provirus integrations within or near tumor-promoting genes causing their transcriptional transactivation (Hacein-Bey-Abina *et al.*, 2008; Howe *et al.*, 2008). Indeed, competent LTRs being part of the retroviral vector, their strong enhancer sequences mediated transactivation of the targeted proto-oncogenes. Among these genes, *LMO-2* (LIM domain only-2) had been targeted in five out the six cases (Hacein-Bey-Abina *et al.*, 2003, 2008; Fischer *et al.*, 2004; Howe *et al.*, 2008; Six et *al.*, 2017). The role of *LMO2* in oncogenesis had been well established from analysis of primary T-cell leukemia (Royer-Pokora *et al.*, 1991) and murine models. In one patient, two insertions were found, one in *LMO2* locus and the other in *BMI1*. The *BMI1* product is known to regulate normal and leukemic stem cell proliferation, including through repression of the *CDKN2* locus (Jacobs *et al.*, 1999; Lessard and Sauvageau, 2003). Cyclin D2 over-expression was observed in one patient and had been reported in cases of primary T-ALL as a result of TCRB or TCRA/D-related translocations (Clappier *et al.*, 2006). A variety of genetic alterations were detected in all cases in addition to vector integration, such as the *SIL-TAL1* fusion and *NOTCH1*-activating mutations (Weng *et al.*, 2004). Biallelic deletion of the *CDKN2* locus was also found in blast cells of two analyzed patients. The inactivation of this tumor suppressor gene is the most prevalent genomic abnormality in common T-ALL (Cayuela *et al.*, 1996). All these abnormalities and oncogenic rearrangements found in patients' blast cells fit with a multistep oncogenesis model of T-ALL (Armstrong and Look, 2005; Mullighan *et al.*, 2007), in which oncogenes were first activated by vector insertional mutagenesis, followed by accumulation of secondary genome rearrangements, including point mutations as well as gene deletions and amplifications.

Since these events, large-scale analysis of retroviral integration site (RIS) patterns has been performed on peripheral blood samples provided by patients included in all the various retrovirus-based gene therapy trials. These studies clearly showed that γ-retroviral vectors integrate semi-randomly throughout the genome into transcriptionally active promoters and regulatory regions of highly expressed genes involved in proliferation and cell survival, with a symmetrical accumulation around the transcription start site in the coding region (Deichmann *et al.*, 2007; Wang *et al.*, 2010). Moreover, the integrations characteristically accumulate around enhancer and promoter regions.

22.2.2.3 *The next generation of SCID-X1 gene therapy trials*

These trials have taught us that retroviral vector-based γc gene transfer into HSPC enables the long-term correction of the patients' immune system defect. The trials have also highlighted the significant toxicity associated with the LTR's strong enhancer effect present in the MLV-based retroviral vector, which was responsible for the six cases of leukemia that occurred in the SCID-X1 trials (Hacein-Bey-Abina *et al.*, 2003, 2008, 2010; Howe *et al.*, 2008; Six et *al.*, 2017) but also in other gene therapy trials for WAS and chronic granulomatous disease (CGD) (Stein *et al.*, 2010; Díez *et al.*, 2011). This drawback led to the development of self-inactivating (SIN) retroviral and lentiviral vectors (LVs). Indeed, in order to improve safety while maintaining the efficacy achieved with the γ-retroviral vector in the first SCID-X1 trials, the MLV vector was deleted from its LTR U3 enhancer (Thornhill *et al.*, 2008; Zychlinski *et al.*, 2008). In the new designed SIN-retroviral vector, the expression of the γc complementary DNA was driven by the human EF1α (elongation factor 1α) short promoter. This internal, weak promoter capable of inducing transcription of the therapeutic gene alone has been shown to be less genotoxic in *in vitro* assays of the clonogenicity of myeloid precursors (Modlich *et al.*, 2009; Cattoglio *et al.*, 2010; Kustikova *et al.*, 2010). An international multicenter (Paris, London, Cincinnati, Boston, and Los Angeles) phase I/II trial based on this SIN-γc vector has been initiated in 2010. Nine patients without an HLA-identical donor and with severe therapy-resistant infections have been enrolled. Among the nine treated patients, one had died from an overwhelming adenoviral infection four months after treatment (prior to full T-cell reconstitution). Of the remaining eight patients, seven showed a significant improvement in clinical condition, which was associated with an immune reconstitution of the T-cell compartment comparable to that obtained in the previous trials based on the parental MLV vector. With up to seven years of follow-up, no adverse events have been observed. This is corroborated by the very significant decrease in the findings of integration sites located within oncogenes in the T-cell population of these patients. These data provide a strong basis suggesting a higher safety profile of this SIN-γc vector (Deichmann *et al.*, 2007; Hacein-Bey-Abina *et al.*, 2014).

SIN-LVs have also been developed as their integration features bring additional safeguard. Indeed, the LVs have been shown to be randomly distributed across actively transcribed genes without targeting regulatory elements of the genome (Biffi *et al.*, 2011; Moiani *et al.*, 2012). Moreover, LVs are capable of infecting non-dividing cells (Naldini *et al.*, 1996) thereby increasing the rate of stem cells transduction and thus enabling to reduce the number of *ex vivo* transduction cycles. Limited $CD34^+$ cell culture preserves the stemness of the pluripotent HSPCs, which is crucial for the success of long-term transplantation. A SIN-LV has been developed by Sorrentino and his team (Zhou *et al.*, 2010) and was used in a clinical trial for older SCID-X1 who had failed previous haplo-identical hematopoietic stem cell transplant in infancy (De Ravin *et al.*, 2016). Five patients were enrolled and received low-dose busulfan conditioning. In the first two treated patients, with 2–3 years follow-up, a significant stable gene marking in multiple hematopoietic lineages coincided with reconstitution of NK- and B-cells, a limited but significant increase in T-cell counts (likely because of poor thymic functions), sustained restoration of humoral function, specific vaccine responses, and marked clinical improvement, in contrast to restricted T-cell gene marking in prior MLV-based retroviral vector gene therapy trials without conditioning. Similar gene marking levels have been achieved in the three other patients included in this SIN-LV trial, albeit with only 6–9 months of follow-up. Lentiviral gene therapy with reduced-intensity conditioning appears safe and can restore somehow humoral immune function in patients who previously received a haplo-identical transplant (De Ravin *et al.*, 2016).

22.3 Gene Therapy for ADA–SCID

ADA deficiency is an autosomal recessive SCID with an estimated frequency of 1 Hz in 250,000 live births (Gaspar *et al.*, 2009). It leads to an accumulation of adenosine and deoxyadenosine which is converted inside cells into deoxy-ADP and deoxy-ATP. Accumulation of toxic metabolites in the intracellular as well as extracellular compartments induces the premature apoptosis of lymphocyte precursors, causing a SCID phenotype. ADA–SCID represents the second most frequent form of SCIDs accounting for approximately 15–20% of all cases of SCID worldwide (Gaspar *et al.*, 2009). Allogeneic HSCT from matched sibling donors constitutes the gold standard in ADA–SCID therapy (around 88% success), but survival is significantly reduced following transplants from matched unrelated (66%) or haplo-identical donors (43%) (Hassan *et al.*, 2012; Pai *et al.*, 2014).

ADA deficiency has also been treated since the early 1980s with enzyme replacement therapy (ERT) in the form of weekly or bi-weekly intramuscular injection of polyethylene glycol-conjugated ADA (PEG–ADA). This treatment has been shown to be life-saving for most patients; however, long-term follow-up suggests incomplete immune recovery with progressive decline over time. In addition, the prohibitive prize of this treatment makes it difficult to provide universally (Gaspar *et al.*, 2009).

Historically, ADA deficiency was the first disease for which attempts of gene therapy were performed. The first two trials, started in the early 1990s, were based on γ-retroviral vectors-mediated gene-corrected autologous T-cells or HSPCs (Blaese *et al.*, 1995; Bordignon *et al.*, 1995). Despite the fact that this first therapy was not sufficient to correct the immunodeficiency, this innovative approach demonstrated the procedure's feasibility and the safe *in vivo* persistence of transduced T-cells for more than 12 years (Blaese *et al.*, 1995; Bordignon *et al.*, 1995; Hoogerbrugge *et al.*, 1996; Kohn *et al.*, 1998; Muul *et al.*, 2003; Biasco *et al.*, 2015). Thanks to improvement in transduction procedure and in gene therapy study designs, the Italian group from SR-Tiget (Milan) performed later on a successful trial that led to correction of the immunodeficiency (Aiuti *et al.*, 2002, 2009; Cicalese *et al.*, 2016). The major differences between this trial and the previous ADA–SCID trials (performed ten years earlier) consisted of (i) improvement in vector design and transduction conditions (as for SCID-X1, see Section 22.2.1), (ii) the addition of a mild, preconditioning, chemotherapy regimen with busulfan (4 mg/kg IV), in order to make space for the transduced progenitors and achieve a higher rate of engraftment, and (iii) the withdrawal of ERT before gene therapy to leverage the growth-selective advantage of transduced cells (Aiuti *et al.*, 2002, 2009; Cicalese *et al.*, 2016). The updated results of the 18 patients treated by the SR–Tiget group have been recently published. All the patients are alive with a median of 6.9 years follow-up (range, 2.3–13.4 years) and most of them (15 out of 18) remained off ERT thanks to restoration of ADA activity. Moreover, no event indicative of leukemic transformation was reported (Cicalese *et al.*, 2016; Ferrua and Aiuti, 2017).

Based on the positive efficacy and safety data collected from these 18 patients, *ex vivo* HSPC gene therapy for the treatment of ADA–SCID, consisting of a single infusion of an "autologous CD34$^+$ cells transduced with a γ-retroviral vector encoding for the human ADA cDNA sequence" received marketing approval in Europe in May 2016 under the commercial name of Strimvelis™ (GlaxoSmithKline, GSK). This advanced therapy medicinal product (ATMP) is the first to be approved by the European Medicines Agency (EMA). This was achieved thanks to a joint effort among GSK, the Italian Telethon Foundation, and San Raffaele Scientific Institute, at SR–Tiget, where the medicinal product was originally developed. Strimvelis™ is available for ADA–SCID patients without a suitable HLA-matched related donor and the first patient was treated in March 2017 at the San Raffaele Hospital in Milan (Ferrua and Aiuti, 2017; Stirnadel-Farrant *et al.*, 2018). However, it should be stressed that the product is sold at a high prize, making the overall accessibility of the procedure questionable.

Other trials using autologous $CD34^+$ cells transduced with a γ-retroviral vector encoding the ADA were performed in the United Kingdom (Great Ormond Street Hospital (GOSH) in London) and the United States (National Institutes of Health and Children's Hospital of Los Angeles). Patients received also a mild conditioning regimen (Melphalan in UK and Busulfan in USA). More than 50 ADA–SCID patients have been treated to date in different centers and are all reported to be alive more than 13 years after treatment. In most of them, improvement of cellular and humoral responses was observed as well as sustained systemic metabolic detoxification without the need of PEG–ADA (Aiuti *et al.*, 2002, 2009; Gaspar *et al.*, 2006; Candotti *et al.*, 2012; Carbonaro-Sarracino et *al.*, 2014; Cicalese *et al.*, 2016, 2018; Ferrua and Aiuti, 2017; Shaw *et al.*, 2017). Importantly, to date, none of the patients treated in these trials developed leukemia despite the fact that the very same first generation γ-retroviral backbones were used and a similar pattern of RIS was detected (including common IS inside or near proto-oncogenes such as LMO2, MECOM, BCL2, and CCND2 locus) (Aiuti *et al.*, 2007, 2009; Cassani *et al.*, 2009; Gaspar *et al.*, 2011; Candotti *et al.*, 2012; Cooper *et al.*, 2017). Facts such as common IS involving proto-oncogenes were not able to induce clonal dominance and selective expansion of malignant cell clones (in contrast to SCID-X1, WAS and CGD diseases) indicate that the ADA-deficiency by itself reduces the risk of leukemogenesis through a disease-related toxic environment (Mukherjee and Thrasher, 2013; Cicalese and Aiuti, 2015; Ferrua and Aiuti, 2017). However, safety monitoring will be continued for a long term in all patients. Analysis of common IS also revealed that many of them were shared between lymphoid and myeloid lineages providing evidence of multilineage progenitor engraftment. This is confirmed by the significant proportions of B, NK, and myeloid cells found to stably express the transgene (Biasco *et al.*, 2011; Candotti *et al.*, 2012). The results achieved in these studies confirmed the positive outcome and the key role of non-myeloablative conditioning in favoring multilineage engraftment of gene-corrected HSPC (Ferrua and Aiuti, 2017).

Even though, no adverse event was registered in ADA trials based on γ-retroviral vectors, the gene therapy field moved toward the use of SIN-LVs in order to improve efficacy. Two-phase I/II clinical trials based on an LV were initiated in 2012 in the United Kingdom and the United States (Gaspar *et al.*, 2014; Kohn and Gaspar, 2017). This vector contains a codon-optimized cDNA of *ADA* gene under the transcriptional control of the EF1a (short form) promoter (Carbonaro *et al.*, 2014). The interim results of the first five patients treated after conditioning with low-dose busulfan (5 mg/kg IV) indicated that, at a mean follow-up of one year, a significant immunological and metabolic recovery was achieved with improved T-cell counts and normalization in mitogen responses (Gaspar *et al.*, 2014). Results have now been confirmed in more than 30 patients, all free of ERT.

22.4 Other Gene Therapy Applications for PID

The clinical success achieved in SCID-X1 and ADA deficiency set the stage for a remarkable resurgence in support for gene therapy. These seminal clinical trials established the feasibility of gene therapy in humans and helped overcome a variety of regulatory and political barriers. Other SCID diseases have also been considered for development of gene therapy strategy. Among them, the T-B-NK$^+$ SCID was an obvious target given that it accounts for 30% of all types of SCID. Encouraging preclinical results have been reported for Rag-2 and Artemis deficiencies and, to a lesser extent, Rag-1 deficiency (Lagresle-Peyrou *et al.*, 2006; Mostoslavsky *et al.*, 2006; Benjelloun *et al.*, 2008; Pike-Overzet *et al.*, 2011; van Til *et al.*, 2012). The extension of gene therapy to other forms of PIDs was initiated several years ago with very good results in some applications like in the WAS (Aiuti *et al.*, 2013; Hacein-Bey-Abina *et al.*, 2015). Preclinical set-up is also underway for the following deficiencies: Perforin, Munc 13-4, CD40L, or FOXP3, as well as for HLH.

22.5 Moving Toward Genome Engineering

The ultimate goal in gene therapy would be to achieve DNA recombination in specific, safe sites and thus correct *in situ* mutations without the more or less random introduction of an additional copy of the gene into the host genome, thus placing the corrected gene in its physiological environment.

Breakthroughs in biotechnologies, such as genome editing, have emerged in recent years. Among them, CRISPR/Cas9 (a bacterial clustered regularly interspaced short palindromic repeats (CRISPR) RNA together with a CRISPR-associated (Cas) protein) system is currently widely used for genome engineering of eukaryotic cells in order to correct gene mutations or to engineer T-cells for cancer immunotherapy. The preclinical studies published recently by Naldini and his team demonstrated the functional validation of the editing of human CD34$^+$ hematopoietic progenitor cells for *IL2RG (γc)* gene correction in a SCID-X1 mouse model by using advanced generation zinc-finger nucleases (ZFNs) or CRISPR/Cas9 nucleases. They reported that ~35% and ~20% homology directed repair (HDR) was achieved in the whole CD34$^+$ treated population and in SCID-X1 mice upon transplantation, respectively. Their work also showed the functionality of the IL2RG-edited lymphoid progeny in mice (Schiroli *et al.*, 2017). Nevertheless, a bottleneck of this approach consists of the fact that gene editing requires the homologous recombination (HR) apparatus that is only available in dividing cells while the vast majority of HSC are not cycling. As such, the approach may be efficient for SCID diseases that do not require many transduced progenitor cells to achieve correction of the T-cell immunodeficiency. Transposition to other inherited hematopoietic conditions will be much harder to implement.

Despite the great promise of the gene editing technology, an efficient delivery technology will be needed for wide and successful applications to human patients, while the off-target toxicity will need accurate evaluation.

22.6 Conclusion

Gene therapy started with the simple idea that replacing a defective gene with a functional copy can cure a disease. The road was long and chaotic to finally translate this concept into reality. The SCID-X1 disease has represented a good model to the proof of concept of gene transfer efficiency in hematopoietic system to treat monogenic diseases. The clinical achievements, together with (i) improvement in vector designs and large-scale manufacturing of LVs and (ii) the development of gene editing technologies, contribute to the revitalization of the gene therapy field. The future clinical development of these cell-based therapies will be ensured by establishment of scalable technological platforms that (i) enable different levels of pharmaceutical-grade processes production and (ii) combine with the growing capacity for targeting gene integration. These new gene/cell therapy products might fulfill the promise of efficient and safe therapies for currently incurable diseases that will be accessible to the largest number of patients.

References

Aiuti, A., Biasco, L., Scaramuzza, S., Ferrua, F., Cicalese, M.P., Baricordi, C., *et al.* (2013). Lentivirus-based gene therapy of hematopoietic stem cells in Wiskott-Aldrich syndrome, *Science*, **341**, 1233151.

Aiuti, A., Cassani, B., Andolfi, G., Mirolo, M., Biasco, L., Recchia, A., *et al.* (2007). Multilineage hematopoietic reconstitution without clonal selection in ADA-SCID patients treated with stem cell gene therapy, *J Clin Invest*, **117**, 2233–2240.

Aiuti, A., Cattaneo, F., Galimberti, S., Benninghoff, U., Cassani, B., Callegaro, L., *et al.* (2009). Gene therapy for immunodeficiency due to adenosine deaminase deficiency, *N Engl J Med*, **360**, 447–458.

Aiuti, A., Slavin, S., Aker, M., Ficara, F., Deola, S., Mortellaro, A., *et al.* (2002). Correction of ADA-SCID by stem cell gene therapy combined with nonmyeloablative conditioning, *Science*, **296**, 2410–2413.

Antoine, C., Müller, S., Cant, A., Cavazzana-Calvo, M., Veys, P., Vossen, J., *et al.* (2003). Long-term survival and transplantation of haemopoietic stem cells for immunodeficiencies: Report of the European experience 1968–99, *Lancet*, **361**, 553–560.

Armstrong, S.A., Look, A.T. (2005). Molecular genetics of acute lymphoblastic leukemia, *J Clin Oncol*, **23**, 6306–6315.

Benjelloun, F., Garrigue, A., Demerens-de Chappedelaine, C., Soulas-Sprauel, P., Malassis-Séris, M., Stockholm, D., *et al.* (2008). Stable and functional lymphoid reconstitution in artemis-deficient mice following lentiviral artemis gene transfer into hematopoietic stem cells, *Mol Ther*, **16**, 1490–1499.

Biasco, L., Ambrosi, A., Pellin, D., Bartholomae, C., Brigida, I., Roncarolo, M.G., *et al.* (2011). Integration profile of retroviral vector in gene therapy treated patients is cell-specific according to gene expression and chromatin conformation of target cell, *EMBO Mol Med*, **3**, 89–101.

Biasco, L., Scala, S., Ricci, L.B., Dionisio, F., Baricordi, C., Calabria, A., *et al.* (2015). In vivo tracking of T cells in humans unveils decade-long survival and activity of genetically modified T memory stem cells, *Sci Transl Med*, **7**, 273ra13–273ra13.

Biffi, A., Bartolomae, C.C., Cesana, D., Cartier, N., Aubourg, P., Ranzani, M., *et al.* (2011). Lentiviral vector common integration sites in preclinical models and a clinical trial reflect a benign integration bias and not oncogenic selection, *Blood*, **117**, 5332–5339.

Blaese, R.M., Culver, K.W., Miller, A.D., Carter, C.S., Fleisher, T., Clerici, M., *et al.* (1995). T lymphocyte-directed gene therapy for ADA-SCID: Initial trial results after 4 years, *Science*, **270**, 475–480.

Bordignon, C., Notarangelo, L.D., Nobili, N., Ferrari, G., Casorati, G., Panina, P., *et al.* (1995). Gene therapy in peripheral blood lymphocytes and bone marrow for ADA–immunodeficient patients, *Science*, **270**, 470–475.

Bousso, P., Wahn, V., Douagi, I., Horneff, G., Pannetier, C., Deist, F.L., *et al.* (2000). Diversity, functionality, and stability of the T cell repertoire derived in vivo from a single human T cell precursor, *Proc Natl Acad Sci*, **97**, 274–278.

Buckley, R.H. (2002). Primary immunodeficiency diseases: Dissectors of the immune system, *Immunol Rev*, **185**, 206–219.

Buckley, R.H. (2011). Transplantation of hematopoietic stem cells in human severe combined immunodeficiency: Long-term outcomes, *Immunol Res*, **49**, 25–43.

Buckley, R.H., Schiff, R.I., Schiff, S.E., Markert, M.L., Williams, L.W., Harville, T.O., *et al.* (1997). Human severe combined immunodeficiency: Genetic, phenotypic, and functional diversity in one hundred eight infants, *J Pediatr*, **130**, 378–387.

Candotti, F., Shaw, K.L., Muul, L., Carbonaro, D., Sokolic, R., Choi, C., *et al.* (2012). Gene therapy for adenosine deaminase–deficient severe combined immune deficiency: Clinical comparison of retroviral vectors and treatment plans, *Blood*, **120**, 3635–3646.

Carbonaro, D.A., Zhang, L., Jin, X., Montiel-Equihua, C., Geiger, S., Carmo, M., *et al.* (2014). Preclinical demonstration of lentiviral vector-mediated correction of immunological and metabolic abnormalities in models of adenosine deaminase deficiency, *Mol Ther*, **22**, 607–622.

Carbonaro-Sarracino, D., Shaw, K., Sokolic, R., *et al.* (2014). Clinical gene therapy trials for adenosine deaminase-deficient severe combined immune deficiency (ADA-SCID), *J Clin Immunol*, **34**(2), S313.

Cassani, B., Montini, E., Maruggi, G., Ambrosi, A., Mirolo, M., Selleri, S., *et al.* (2009). Integration of retroviral vectors induces minor changes in the transcriptional activity of T cells from ADA-SCID patients treated with gene therapy, *Blood*, **114**, 3546–3556.

Cattoglio, C., Pellin, D., Rizzi, E., Maruggi, G., Corti, G., Miselli, F., *et al.* (2010). High-definition mapping of retroviral integration sites identifies active regulatory elements in human multipotent hematopoietic progenitors, *Blood*, **116**, 5507–5517.

Cavazzana, M., Six, E., Lagresle-Peyrou, C., André-Schmutz, I., Hacein-Bey-Abina, S. (2016). Gene therapy for X-linked severe combined immunodeficiency: Where do we stand? *Hum Gene Ther*, **27**, 108–116.

Cavazzana-Calvo, M., Hacein-Bey, S., Basile, G. de S., Coene, C.D., Selz, F., Deist, F.L., *et al.* (1996). Role of interleukin-2 (IL-2), IL-7, and IL-15 in natural killer cell differentiation from cord blood hematopoietic progenitor cells and from gamma c transduced severe combined immunodeficiency X1 bone marrow cells, *Blood*, **88**, 3901–3909.

Cavazzana-Calvo, M., Hacein-Bey, S., Basile, G. de S., Gross, F., Yvon, E., Nusbaum, P., *et al.* (2000). Gene therapy of human severe combined immunodeficiency (SCID)-X1 disease, *Science*, **288**, 669–672.

Cayuela, J.M., Madani, A., Sanhes, L., Stern, M.H., Sigaux, F. (1996). Multiple tumor-suppressor gene 1 inactivation is the most frequent genetic alteration in T-cell acute lymphoblastic leukemia, *Blood*, **87**, 2180–2186.

Chinen, J., Davis, J., De Ravin, S.S., Hay, B.N., Hsu, A.P., Linton, G.F., *et al.* (2007). Gene therapy improves immune function in preadolescents with X-linked severe combined immunodeficiency, *Blood*, **110**, 67–73.

Chinen, J., Notarangelo, L.D., Shearer, W.T. (2014). Advances in basic and clinical immunology 2013, *J Allergy Clin Immunol*, **133**, 967–976.

Cicalese, M.P., Aiuti, A. (2015). Clinical applications of gene therapy for primary immunodeficiencies, *Hum Gene Ther*, **26**, 210–219.

Cicalese, M.P., Ferrua, F., Castagnaro, L., Pajno, R., Barzaghi, F., Giannelli, S., *et al.* (2016). Update on the safety and efficacy of retroviral gene therapy for immunodeficiency due to adenosine deaminase deficiency, *Blood*, **128**, 45–54.

Cicalese, M.P., Ferrua, F., Castagnaro, L., Rolfe, K., De Boever, E., Reinhardt, R.R., *et al.* (2018). Gene therapy for adenosine deaminase deficiency: A comprehensive evaluation of short- and medium-term safety, *Mol Ther*, **26**, 917–931.

Clappier, E., Cuccuini, W., Cayuela, J.-M., Vecchione, D., Baruchel, A., Dombret, H., *et al.* (2006). Cyclin D2 dysregulation by chromosomal translocations to TCR loci in T-cell acute lymphoblastic leukemias, *Leukemia*, **20**, 82–86.

Cooper, A.R., Lill, G.R., Shaw, K., Carbonaro-Sarracino, D.A., Davila, A., Sokolic, R., *et al.* (2017). Cytoreductive conditioning intensity predicts clonal diversity in ADA-SCID retroviral gene therapy patients, *Blood*, **129**, 2624–2635.

Dao, M.A., Hannum, C.H., Kohn, D.B., Nolta, J.A. (1997). FLT3 ligand preserves the ability of human CD34+ progenitors to sustain long-term hematopoiesis in immune-deficient mice after ex vivo retroviral-mediated transduction, *Blood*, **89**, 446–456.

De Ravin, S.S., Wu, X., Moir, S., Kardava, L., Anaya-O'Brien, S., Kwatemaa, N., *et al.* (2016). Lentiviral hematopoietic stem cell gene therapy for X-linked severe combined immunodeficiency, *Science Transl Med*, **8**, 335ra57.

Deichmann, A., Hacein-Bey-Abina, S., Schmidt, M., Garrigue, A., Brugman, M.H., Hu, J., *et al.* (2007). Vector integration is nonrandom and clustered and influences the fate of lymphopoiesis in SCID-X1 gene therapy, *J Clin Invest*, **117**, 2225–2232.

Díez, I.A., Zychlinski, D., Coci, E.G., Galla, M., Modlich, U., Dewey, R.A., *et al.* (2011). Development of novel efficient SIN vectors with improved safety features for Wiskott–Aldrich syndrome stem cell based gene therapy, *Mol Pharm*, **8**, 1525–1537.

Ferrua, F., Aiuti, A. (2017). Twenty-five years of gene therapy for ADA-SCID: From bubble babies to an approved drug, *Hum Gene Ther*, **28**, 972–981.

Fischer, A., Abina, S.H.-B., Thrasher, A.J., Von Kalle, C., Cavazzana (2004). LMO2 and gene therapy for severe combined immunodeficiency, *New Engl J Med*, **350**, 2526–2527.

Fischer, A., Notarangelo, L.D., Neven, B., Cavazzana, M., Puck, J.M. (2015). Severe combined immunodeficiencies and related disorders, *Nat Rev Dis Primers*, 15061.

Gaspar, H.B., Aiuti, A., Porta, F., Candotti, F., Hershfield, M.S., Notarangelo, L.D. (2009). How I treat ADA deficiency, *Blood*, **114**, 3524–3532.

Gaspar, H.B., Bjorkegren, E., Parsley, K., Gilmour, K.C., King, D., Sinclair, J., *et al.* (2006). Successful reconstitution of immunity in ADA-SCID by stem cell gene therapy following cessation of PEG-ADA and use of mild preconditioning, *Mol Ther*, **14**, 505–513.

Gaspar, B., Buckland, k., Rivat, C., *et al.* (2014). Immunological and metabolic correction after Lentiviral vector mediated haematopoietic stem cell gene therapy for ADA deficiency, *J Clin Immunol*, **34**(2), S167.

Gaspar, H.B., Cooray, S., Gilmour, K.C., Parsley, K.L., Adams, S., Howe, S.J., *et al.* (2011). Long-term persistence of a polyclonal T Cell repertoire after gene therapy for X-Linked severe combined immunodeficiency, *Sci Transl Med*, **3**, 97ra79–97ra79.

Gaspar, H.B., Parsley, K.L., Howe, S., King, D., Gilmour, K.C., Sinclair, J., *et al.* (2004). Gene therapy of X-linked severe combined immunodeficiency by use of a pseudotyped gammaretroviral vector, *Lancet*, **364**, 2181–2187.

Gennery, A.R., Slatter, M.A., Grandin, L., Taupin, P., Cant, A.J., Veys, P., *et al.* (2010). Transplantation of hematopoietic stem cells and long-term survival for primary immunodeficiencies in Europe: Entering a new century, do we do better? *J Allergy Clin Immunol*, **126**, 602–610.e11.

Ginn, S.L., Curtin, J.A., Kramer, B., Smyth, C.M., Wong, M., Kakakios, A., *et al.* (2005). Treatment of an infant with X-linked severe combined immunodeficiency (SCID-X1) by gene therapy in Australia, **182**, 6.

Hacein-Bey, H., Cavazzana-Calvo, M., Deist, F.L., Dautry-Varsat, A., Hivroz, C., Riviere, I., *et al.* (1996). Gamma-c gene transfer into SCID X1 patients' B-cell lines restores normal high-affinity interleukin-2 receptor expression and function, *Blood*, **87**, 3108–3116.

Hacein-Bey, S., Basile, G.D.S., Lemerle, J., Fischer, A., Cavazzana-Calvo, M. (1998). γc Gene transfer in the presence of stem cell factor, FLT-3L, interleukin-7 (IL-7), IL-1, and IL-15 cytokines restores T-cell differentiation from γc(−) X-linked severe combined immunodeficiency hematopoietic progenitor cells in murine fetal thymic organ cultures, *Blood*, **92**, 4090–4097.

Hacein-Bey, S., Gross, F., Nusbaum, P., Hue, C., Hamel, Y., Fischer, A., *et al.* (2001). Optimization of retroviral gene transfer protocol to maintain the lymphoid potential of progenitor cells, *Hum Gene Ther*, **12**, 291–301.

Hacein-Bey-Abina, S., Garrigue, A., Wang, G.P., Soulier, J., Lim, A., Morillon, E., *et al.* (2008). Insertional oncogenesis in 4 patients after retrovirus-mediated gene therapy of SCID-X1, *J Clin Invest*, **118**, 3132–3142.

Hacein-Bey-Abina, S., Gaspar, H.B., Blondeau, J., Caccavelli, L., Charrier, S., Buckland, K., *et al.* (2015). Outcome following gene therapy in patients with severe Wiskott-Aldrich syndrome, *J Am Med Assoc*, **313**, 1550–1563.

Hacein-Bey-Abina, S., Hauer, J., Lim, A., Picard, C., Wang, G.P., Berry, C.C., *et al.* (2010). Efficacy of gene therapy for X-linked severe combined immunodeficiency, *New Engl J Med*, **363**, 355–364.

Hacein-Bey-Abina, S., Kalle, C.V., Schmidt, M., McCormack, M.P., Wulffraat, N., Leboulch, P., *et al.* (2003). LMO2-associated clonal T Cell proliferation in two patients after gene therapy for SCID-X1, *Science*, **302**, 415–419.

Hacein-Bey-Abina, S., Le Deist, F., Carlier, F., Bouneaud, C., Hue, C., De Villartay, J.-P., *et al.* (2002). Sustained correction of X-linked severe combined immunodeficiency by ex vivo gene therapy, *New Engl J Med*, **346**, 1185–1193.

Hacein-Bey-Abina, S., Pai, S.-Y., Gaspar, H.B., Armant, M., Berry, C.C., Blanche, S., *et al.* (2014). A modified γ-retrovirus vector for X-linked severe combined immunodeficiency, *New Engl J Med*, **371**, 1407–1417.

Hanenberg, H., Xiao, X.L., Dilloo, D., Hashino, K., Kato, I., Williams, D.A. (1996). Colocalization of retrovirus and target cells on specific fibronectin fragments increases genetic transduction of mammalian cells, *Nat Med*, **2**, 876–882.

Hassan, A., Booth, C., Brightwell, A., Allwood, Z., Veys, P., Rao, K., *et al.* (2012). Outcome of hematopoietic stem cell transplantation for adenosine deaminase–deficient severe combined immunodeficiency, *Blood*, **120**, 3615–3624.

Hoogerbrugge, P.M., van Beusechem, V.W., Fischer, A., Debree, M., Deist, F. le, Perignon, J.L., *et al.* (1996). Bone marrow gene transfer in three patients with adenosine deaminase deficiency, *Gene Ther*, **3**, 179–183.

Howe, S.J., Mansour, M.R., Schwarzwaelder, K., Bartholomae, C., Hubank, M., Kempski, H., *et al.* (2008). Insertional mutagenesis combined with acquired somatic mutations causes leukemogenesis following gene therapy of SCID-X1 patients, *J Clin Invest*, **118**, 3143–3150.

Jacobs, J.J.L., Scheijen, B., Voncken, J.-W., Kieboom, K., Berns, A., van Lohuizen, M. (1999). Bmi-1 collaborates with c-Myc in tumorigenesis by inhibiting c-Myc-induced apoptosis via INK4a/ARF, *Genes Dev*, **13**, 2678–2690.

Kohn, D.B., Gaspar, H.B. (2017). How we manage adenosine deaminase-deficient severe combined immune deficiency (ADA SCID), *J Clin Immunol*, **37**, 351–356.

Kohn, D.B., Hershfield, M.S., Carbonaro, D., Shigeoka, A., Brooks, J., Smogorzewska, E.M., *et al.* (1998). T lymphocytes with a normal ADA gene accumulate after transplantation of transduced autologous umbilical cord blood CD34+ cells in ADA-deficient SCID neonates, *Nat Med*, **4**, 775–780.

Kustikova, O., Brugman, M., Baum, C. (2010). The genomic risk of somatic gene therapy, *Semin Cancer Biol*, **20**, 269–278.

Kwan, A., Abraham, R.S., Currier, R., Brower, A., Andruszewski, K., Abbott, J.K., *et al.* (2014). Newborn screening for severe combined immunodeficiency in 11 screening programs in the United States, *J Am Med Assoc*, **312**, 729–738.

Lagresle-Peyrou, C., Yates, F., Malassis-Séris, M., Hue, C., Morillon, E., Garrigue, A., *et al.* (2006). Long-term immune reconstitution in RAG-1-deficient mice treated by retroviral gene therapy: A balance between efficiency and toxicity, *Blood*, **107**, 63–72.

Leonard, M., Warren, J. (1996). The molecular basis of X-linked severe combined immunodeficiency: Defective cytokine receptor signaling, *Annu Rev Med*, **47**, 229–239.

Lessard, J., Sauvageau, G. (2003). *Bmi-1* determines the proliferative capacity of normal and leukaemic stem cells, *Nature*, **423**, 255–260.

Marciano, B.E., Huang, C.-Y., Joshi, G., Rezaei, N., Carvalho, B.C., Allwood, Z., *et al.* (2014). BCG vaccination in SCID patients: Complications, risks and vaccination policies, *J Allergy Clin Immunol*, **133**, 1134–1141.

Martins, V.C., Ruggiero, E., Schlenner, S.M., Madan, V., Schmidt, M., Fink, P.J., *et al.* (2012). Thymus-autonomous T cell development in the absence of progenitor import, *J Exp Med*, **209**, 1409–1417.

Modlich, U., Navarro, S., Zychlinski, D., Maetzig, T., Knoess, S., Brugman, M.H., *et al.* (2009). Insertional transformation of hematopoietic cells by self-inactivating lentiviral and gammaretroviral vectors, *Mol Ther*, **17**, 1919–1928.

Moiani, A., Paleari, Y., Sartori, D., Mezzadra, R., Miccio, A., Cattoglio, C., *et al.* (2012). Lentiviral vector integration in the human genome induces alternative splicing and generates aberrant transcripts, *J Clin Invest*, **122**, 1653–1666.

Mostoslavsky, G., Fabian, A.J., Rooney, S., Alt, F.W., Mulligan, R.C. (2006). Complete correction of murine Artemis immunodeficiency by lentiviral vector-mediated gene transfer, *Proc Natl Acad Sci*, **103**, 16406–16411.

Mukherjee, S., Thrasher, A.J. (2013). Gene therapy for PIDs: Progress, pitfalls and prospects, *Gene*, **525**, 174–181.

Mullighan, C.G., Goorha, S., Radtke, I., Miller, C.B., Coustan-Smith, E., Dalton, J.D., *et al.* (2007). Genome-wide analysis of genetic alterations in acute lymphoblastic leukaemia, *Nature*, **446**, 758–764.

Muul, L.M., Tuschong, L.M., Soenen, S.L., Jagadeesh, G.J., Ramsey, W.J., Long, Z., *et al.* (2003). Persistence and expression of the adenosine deaminase gene for 12 years and immune reaction to gene transfer components: Long-term results of the first clinical gene therapy trial, *Blood*, **101**, 2563–2569.

Naldini, L., Blömer, U., Gallay, P., Ory, D., Mulligan, R., Gage, F.H., *et al.* (1996). In vivo gene delivery and stable transduction of nondividing cells by a lentiviral vector, *Science*, **272**, 263–267.

Noguchi, M., Yi, H., Rosenblatt, H.M., Filipovich, A.H., Adelstein, S., Modi, W.S., *et al.* (1993). Interleukin-2 receptor γ chain mutation results in X-linked severe combined immunodeficiency in humans, *Cell*, **73**, 147–157.

Notarangelo, L.D., Fischer, A., Geha, R.S., Casanova, J.-L., Chapel, H., Conley, M.E., *et al.* (2009). Primary immunodeficiencies: 2009 update, *J Allergy Clin Immunol*, **124**, 1161–1178.

Pai, S.-Y., Logan, B.R., Griffith, L.M., Buckley, R.H., Parrott, R.E., Dvorak, C.C., *et al.* (2014). Transplantation outcomes for severe combined immunodeficiency, 2000–2009, *New Engl j Med*, **371**, 434–446.

Peaudecerf, L., Lemos, S., Galgano, A., Krenn, G., Vasseur, F., Di Santo, J.P., *et al.* (2012). Thymocytes may persist and differentiate without any input from bone marrow progenitors, *J Exp Med*, **209**, 1401–1408.

Pike-Overzet, K., Rodijk, M., Ng, Y.-Y., Baert, M.R.M., Lagresle-Peyrou, C., Schambach, A., *et al.* (2011). Correction of murine Rag1 deficiency by self-inactivating lentiviral vector-mediated gene transfer, *Leukemia*, **25**, 1471–1483.

Royer-Pokora, B., Loos, U., Ludwig, W.D. (1991). TTG-2, A new gene encoding a cysteine-rich protein with the LIM motif, is overexpressed in acute T-cell leukaemia with the t(11;14)(p13;q11), *Oncogene*, **6**, 1887–1893.

Schiroli, G., Ferrari, S., Conway, A., Jacob, A., Capo, V., Albano, L., *et al.* (2017). Preclinical modeling highlights the therapeutic potential of hematopoietic stem cell gene editing for correction of SCID-X1, *Sci Transl Med*, **9**, eaan0820.

Shaw, K.L., Garabedian, E., Mishra, S., Barman, P., Davila, A., Carbonaro, D., *et al.* (2017). Clinical efficacy of gene-modified stem cells in adenosine deaminase–deficient immunodeficiency, *J Clin Invest*, **127**, 1689–1699.

Shearer, W.T., Dunn, E., Notarangelo, L.D., Dvorak, C.C., Puck, J.M., Logan, B.R., *et al.* (2014). Establishing diagnostic criteria for SCID, leaky SCID, and OMENN syndrome: The primary immune deficiency treatment consortium experience, *J Allergy Clin Immunol*, **133**, 1092–1098.

Six, E., Gandemer, V., Magnani, A., *et al.* (2017). LMO2 associated clonal T cell proliferation 15 years after gamma-Retrovirus mediated gene therapy for SCIDX1, *Mol Ther*, **25**(5S1), 347.

Slatter, M.A., Gennery, A.R. (2013). Advances in hematopoietic stem cell transplantation for primary immunodeficiency, *Expert Rev Clin Immunol*, **9**, 991–999.

Soudais, C., Shiho, T., Sharara, L.I., Guy-Grand, D., Taniguchi, T., Fischer, A., *et al.* (2000). Stable and functional lymphoid reconstitution of common cytokine receptor γ chain deficient mice by retroviral-mediated gene transfer, *Blood*, **95**, 3071–3077.

Speckmann, C., Pannicke, U., Wiech, E., Schwarz, K., Fisch, P., Friedrich, W., *et al.* (2008). Clinical and immunologic consequences of a somatic reversion in a patient with X-linked severe combined immunodeficiency, *Blood*, **112**, 4090–4097.

Stein, S., Ott, M.G., Schultze-Strasser, S., Jauch, A., Burwinkel, B., Kinner, A., *et al.* (2010). Genomic instability and myelodysplasia with monosomy 7 consequent to *EVI1* activation after gene therapy for chronic granulomatous disease, *Nat Med*, **16**, 198–204.

Stephan, J.L., Vlekova, V., Le Deist, F., Blanche, S., Donadieu, J., De Saint-Basile, G., *et al.* (1993). Severe combined immunodeficiency: A retrospective single-center study of clinical presentation and outcome in 117 patients, *J Pediatr*, **123**, 564–572.

Stephan, V., Wahn, V., Le Deist, F., Dirksen, U., Bröker, B., Müller-Fleckenstein, I., *et al.* (1996). Atypical X-linked severe combined immunodeficiency due to possible spontaneous reversion of the genetic defect in T cells, *New Engl J Med*, **335**, 1563–1567.

Stirnadel-Farrant, H., Kudari, M., Garman, N., Imrie, J., Chopra, B., Giannelli, S., *et al.* (2018). Gene therapy in rare diseases: The benefits and challenges of developing a patient-centric registry for Strimvelis in ADA-SCID, *Orphanet J Rare Dis*, **13**.

Sugamura, K., Asao, H., Kondo, M., Tanaka, N., Ishii, N., Ohbo, K., *et al.* (1996). The interleukin-2 receptor γ chain: Its role in the multiple cytokine receptor complexes and T cell development in XSCID, *Annu Rev Immunol*, **14**, 179–205.

Thornhill, S.I., Schambach, A., Howe, S.J., Ulaganathan, M., Grassman, E., Williams, D., *et al.* (2008). Self-inactivating gammaretroviral vectors for gene therapy of X-linked severe combined immunodeficiency, *Mol Ther*, **16**, 590–598.

Thrasher, A.J., Hacein-Bey-Abina, S., Gaspar, H.B., Blanche, S., Davies, E.G., Parsley, K., *et al.* (2005). Failure of SCID-X1 gene therapy in older patients, *Blood*, **105**, 4255–4257.

van Til, N.P., Boer, H. de, Mashamba, N., Wabik, A., Huston, M., Visser, T.P., *et al.* (2012). Correction of murine rag2 severe combined immunodeficiency by lentiviral gene therapy using a codon-optimized RAG2 therapeutic transgene, *Mol Ther*, **20**, 1968–1980.

Wang, G.P., Berry, C.C., Malani, N., Leboulch, P., Fischer, A., Hacein-Bey-Abina, S., *et al.* (2010). Dynamics of gene-modified progenitor cells analyzed by tracking retroviral integration sites in a human SCID-X1 gene therapy trial, *Blood*, **115**, 4356–4366.

Weng, A.P., Ferrando, A.A., Lee, W., Morris, J.P., Silverman, L.B., Sanchez-Irizarry, C., *et al.* (2004). Activating mutations of NOTCH1 in human T cell acute lymphoblastic leukemia, *Science*, **306**, 269–271.

Zhou, S., Mody, D., DeRavin, S.S., Hauer, J., Lu, T., Ma, Z., *et al.* (2010). A self-inactivating lentiviral vector for SCID-X1 gene therapy that does not activate LMO2 expression in human T cells, *Blood*, **116**, 900–908.

Zychlinski, D., Schambach, A., Modlich, U., Maetzig, T., Meyer, J., Grassman, E., *et al.* (2008). Physiological promoters reduce the genotoxic risk of integrating gene vectors, *Mol Ther*, **16**, 718–725.

23

GENE THERAPY OF THE β-HEMOGLOBINOPATHIES

Emmanuel Payen[a,b,h], Charlotte Colomb[a,b], Olivier Negre[a,b,c], Marina Cavazzana-Calvo[d,e], Salima Hacein-Bey-Abina[d,e], Yves Beuzard[a,b,c] and Philippe Leboulch[a,b,f,g]

23.1 The β-Hemoglobinopathies

After spreading from their native areas in Africa, the Mediterranean and Asia, inherited hemoglobin diseases are now common worldwide. Almost three newborns per one thousand births are severely affected and a total of approximately 350,000 diseased children are born each year (Modell and Darlison, 2008). β-hemoglobin disorders fall into two large groups of hemoglobin (Hb) mutations: structural variants, in which amino acid changes produce abnormal Hb such as Hb S, E, D or C and β-thalassemias, in which β-globin chain production is low or non-existent. Patients who inherit two different mutations may be severely affected whereas heterozygote carriers are generally asymptomatic. A disease-related β-globin gene variant is present in approximately 5% of the world's population but major geographical differences exist due to the selection of heterozygous individuals who are substantially protected against severe malaria. In some populations where malaria is endemic, the prevalence of Hb defects can be as high as 40% (Angastiniotis and Modell, 1998). Most affected patients in developing

[a]CEA, Institute of Emerging Diseases and Innovative Therapies (iMETI), Fontenay aux Roses, France
[b]Inserm U962 and University Paris XI, CEA-iMETI, Fontenay aux Roses, France
[c]Bluebird Bio-France, CEA-iMETI, Fontenay aux Roses, France
[d]Department of Biotherapy, Hôpital Necker, Paris, France
[e]Inserm U768 and University Paris V, Paris, France
[f]Harvard Medical School and Genetics Division, Department of Medicine, Brigham & Women's Hospital, Boston, MA, USA
[g]Email: pleboulch@rics.bwh.harvard.edu
[h]Email: emmanuel.payen@cea.fr

countries die before the age of five years, whereas most of the affected children born in high-income countries survive but live with a chronic and severe disorder (Platt *et al.*, 1994; Modell *et al.*, 2008; Telfer *et al.*, 2009). Indeed, even under intense and specific care, severe complications are frequent (Roseff, 2009; Borgna-Pignatti, 2010; Taylor *et al.*, 2010) and only 50–65% of β-thalassemic patients were alive past the age of 35 years in a recent survey from the UK and Italy (Modell *et al.*, 2008; Borgna-Pignatti, 2010).

Allogeneic hematopoietic stem cell (HSC) transplantation offers a curative treatment for patients with severe Hb disorders who have a healthy, human leukocyte antigen (HLA)-matched, sibling donor (Isgro *et al.*, 2010). However, only approximately 25% of patients have such a suitable donor. Furthermore, complications such as graft rejection and acute or chronic graft-versus-host diseases are major issues (Caocci *et al.*, 2010; Luznik *et al.*, 2010). For subjects lacking a matched donor, many of these drawbacks could be avoided by gene therapy of HSCs. Because direct genetic correction is not yet feasible in HSCs, gene addition by vector-based transfer and chromosomal integration of a normal globin gene remains the approach of choice. However, efficient modification of HSCs and high expression of globin genes in erythroid cells have been major technical challenges that required resolution before any human trials could be planned. Furthermore, findings on the risk of insertional mutagenesis in hematopoietic cells (Hacein-Bey-Abina *et al.*, 2008) have highlighted the need for careful consideration of the risk:benefit ratio of such strategies.

23.1.1 The β-Thalassemias

β-thalassemias have a very wide range of clinical severity, from the severe transfusion-dependent thalassemia major (β-TM) to the highly variable non-transfusion-dependent thalassemia intermedia (β-TI). They can be caused by a number of mutations at the β-globin locus (Giardine *et al.*, 2007), resulting in either no β-globin production (β^0) or reduced levels of synthesis (β^+) that lead to imbalanced α:non-α-globin chain ratios. Excess α-chain damages the cell membrane, leading to apoptosis (Mathias *et al.*, 2000), hemolysis (Vigi *et al.*, 1969) and anemia. As the transition from fetal (γ-globin) to adult (mainly β-globin) chain synthesis takes place during the perinatal period (Weatherall, 1974), patients with β-TM begin to have life-threatening anemia a few years after birth, concurrent to the physiological switch from γ- to β-globin expression. Regular transfusions and a strict iron chelation program are warranted to assure patient survival. Without treatment, β^0 patients succumb within two years while clinical manifestations secondary to decreased oxygen delivery to tissues, massive erythropoiesis and iron overload are present in all other severe cases (Fucharoen and Winichagoon, 2000). The major genetic modulators contributing to reduced severity of the disease are inheritance of mild α-thalassemia (Winichagoon *et al.*, 1985; Camaschella *et al.*, 1995, 1997) or the ability to produce fetal Hb after birth (Cappellini *et al.*, 1981; Winichagoon *et al.*, 1993). Both of these conditions reduce the magnitude of α-chain excess in erythroid cells. Therefore, molecules designed to activate γ-globin gene expression (Perrine, 2008; El-Beshlawy *et al.*, 2009) are of major interest for this disease. Unfortunately, the increased Hb levels produced by such classes of therapeutics are limited and insufficient for a number of patients (Perrine *et al.*, 2010).

23.1.2 Sickle Cell Disease

The first case of a patient with atypical elongated red blood cells (RBCs) was reported as early as 1910 by James Herrick (Herrick, 1910), and these findings led to the description of sickle cell anemia in 1922 (Mason, 1922). The molecular origin of the disease was demonstrated by Pauling in 1949 (Pauling *et al.*, 1949) and was identified ten years later at the amino-terminus of the β-globin chain (Ingram, 1957; Hunt and Ingram, 1959). It was subsequently attributed to a nucleotide change from GAG to GTG (Marotta *et al.*, 1977). This valine for glutamic acid substitution at codon 6 is responsible for HbS ($\alpha_2\beta_2^S$) polymerization, which is the primary molecular event leading to RBC defects, sickling, hemolysis, increased blood viscosity, vaso-occlusion, painful crises, strokes and multi-organ damage (Rees *et al.*, 2010). Current treatments for sickle cell disease include chronic blood transfusion (Wahl and Quirolo, 2009), inhibition of Hb polymerization by agents that induce fetal Hb expression (Charache *et al.*, 1995) and management of painful vaso-occlusive crises (Mousa *et al.*, 2010). Whereas it is well known that the thermodynamics of HbS polymerization in RBCs can be modulated by 2,3-diphosphoglycerate, oxygen concentration, pH and temperature (Eaton and Hofrichter, 1990), these factors are not easily controllable and, except for theoretical discussions, are not preferential targets for gene therapy (Garel *et al.*, 1994). On the other hand, as the clinical severity of sickle cell disease is modulated by variations in the proportion of HbA ($\alpha_2\beta_2$), HbF ($\alpha_2\gamma_2$) or HbA2 ($\alpha_2\delta_2$), gene transfer into HSCs of genes encoding globin chains capable of inhibiting HbS formation in erythroid cells is the leading strategy. Since HbF is more effective than HbA at decreasing HbS polymerization (Goldberg *et al.*, 1977; Poillon *et al.*, 1993), many efforts have been made to overexpress either γ-globin or modified forms of β-globin chains which have enhanced anti-polymerizing activity (Takekoshi *et al.*, 1995).

23.2 Erythroid-Specific and High-Level Expression of Globin Genes

Expression of the globin genes is erythroid-specific, highly efficient and developmentally regulated. Two switches characterize Hb production in humans, one from embryonic to fetal and one from fetal to postnatal stages (Baron, 1996). These expression pattern and regulatory processes are thought primarily to be due to changes in the control of transcription and in RNA processing. The goal of β-like-globin gene therapy is to mimic the physiological expression of the normal adult β-globin gene after gene addition. Thus, expression should ideally occur in the erythroid cell lineage only, within the correct time-frame (from pro-erythroblast to mature erythroblast), and at the appropriate level (the highest for human genes). Hence, effective gene therapy is likely to require vectors that support high levels of globin synthesis, knowing that erythroid cells each contain approximately 300 million Hb tetramers and 15,000 copies of globin messenger RNA (mRNA) at their peak content (Clissold *et al.*, 1977). Much work has been necessary to identify the minimal regulatory sequences required to achieve therapeutic levels of transgenic expression and yet keep the sequences short enough to be introduced either into γ-retroviral vectors (γ-RV) exemplified by murine leukemia virus (MLV)-based retroviral vectors or, more recently, into lentiviral vectors.

23.2.1 Globin Gene Regulating Elements

Among the five human β-like globin genes (HBE, HBG2, HBG1, HBD and HBB), HBB (encoding the adult human β-globin β^A) is the only one to be expressed at a high level after birth. Most of the *cis*-regulatory elements that control tissue specificity and developmental regulation are conserved among species and are located in the promoter region upstream and adjacent to the gene, downstream of the polyadenylation signal and in the second intron (Chada *et al.*, 1985; Magram *et al.*, 1985; Townes *et al.*, 1985; Kollias *et al.*, 1986). These sequences activate β-globin expression in a controlled manner through specific interactions with γ- or β-globin gene promoter elements (Choi and Engel, 1986; Behringer *et al.*, 1987; Kollias *et al.*, 1987; Trudel and Constantini, 1987). However, using such elements to control β-globin gene transcription in transgenic mice resulted in erythroid specificity but at low levels of expression (ranging from 0.1 to 3%) compared with the transcription level of the endogenous mouse β-globin gene. Similarly, the transplantation of hematopoietic cells modified by vectors containing identical regulating elements into mice gave rise to low levels of human β-globin gene expression in erythroid cells (Dzierzak *et al.*, 1988; Karlsson *et al.*, 1988; Bender *et al.*, 1989).

The discovery of active chromatin domains referred to as DNAse I "super" hypersensitive sites (HS) (Tuan *et al.*, 1985; Forrester *et al.*, 1986, 1987) shed light on the organization and regulation of the β-like globin genes and yielded clues for designing more effective expression systems. Most of these HS are located far upstream of the globin locus (6–20 kb 5′ of the ε-globin gene, HBE). They are specific to erythroid cells and are independent of the developmental stage of erythropoiesis. If the human β-globin gene (including its promoter, introns and

3′ regulatory sequences) is inserted immediately downstream of the 5′ HS in transgenic mice, these opening chromatin domains confer physiological levels of human β-globin expression in erythroid cells (Grosveld *et al.*, 1987; Talbot *et al.*, 1989), but do not, by themselves, direct the normal developmental expression pattern of β-globin (Behringer *et al.*, 1990). The large regulatory region of the β-globin gene locus that confers these activating properties and contains the four HS located at −6.1 kb (HS1), −10.9 kb (HS2), −14.9 kb (HS3) and −18 kb (HS4) with respect to the ε-globin gene transcription initiation start site has been called the locus activation region (LAR) (Forrester *et al.*, 1987), the dominant control region (DCR) (Grosveld *et al.*, 1987) and was finally renamed the locus control region (LCR).

Subsequent efforts have been aimed at dissecting the molecular mechanisms of LCR function and at delineating the core elements sufficient for efficient β-globin expression in gene transfer constructs. Individual HS regions (one to two kilo bases each) were shown to have between 10 and 50% of the activity of the full LCR *in vitro* and in transgenic mice (Ryan *et al.*, 1989; Collis *et al.*, 1990; Fraser *et al.*, 1990). Interestingly, the four HS of the LCR, isolated and combined to create a 6.5 kb "mini-LCR", retain full functional activity in transgenic mice regardless of their orientation relative to the β-globin gene (Talbot *et al.*, 1989). An even smaller combination of the four HS, reduced to form a 2.5 kb "micro-LCR", retained the ability to direct high-level expression *in vitro* (Forrester *et al.*, 1989), albeit with a high degree of variegation. HS2, HS3 or HS4 were introduced individually into MLV vectors and directed relatively high-level β-globin gene expression in erythroid cell lines, but with a high degree of clone-to-clone variation (Novak *et al.*, 1990). Similar vectors also transduced mouse HSCs with low efficiency (Plavec *et al.*, 1993). Indeed, oncoretroviral vectors containing LCR sequences together with the β-globin regulatory elements were difficult to produce at high titers and were very unstable (Novak *et al.*, 1990; Gelinas *et al.*, 1992), both of which are obstacles for the use of these vectors in gene therapy (Mulligan, 1993). Reducing the sizes of the chromatin regulatory domains was unsuccessful, as the core elements had a much lower enhancer effect than the full LCR (Philipsen *et al.*, 1990; Talbot *et al.*, 1990; Caterina *et al.*, 1991; Chang *et al.*, 1992), and produced very low levels of β-globin expression in oncoretroviral vectors (Chang *et al.*, 1992). Identifying and removing the DNA sequences responsible for vector instability and low titers (e.g., repeated sequences, cryptic splice signals, polyA sites), either by site-directed mutagenesis (Leboulch *et al.*, 1994) or by empirical segment permutation (Sadelain *et al.*, 1995), improved the transduction efficiency of murine HSCs. However, condensing the HS2, HS3 and HS4 domains (to a total of less than one kb) and/or reducing the size of the 3′ β-globin enhancer may be responsible for the high clonal variation in β-globin gene expression observed *in vitro* (Sadelain *et al.*, 1995) and *in vivo* (Raftopoulos *et al.*, 1997). Vectors with these modifications remained sub-optimal at transducing mouse HSCs and had limited overall expression capacity in erythroid cells after long-term transplantation in mice (Raftopoulos *et al.*, 1997; Rivella and Sadelain, 1998). Fortunately, as will be discussed below, using larger LCR domains and lentiviral vectors has been highly beneficial for increasing the level of β-globin gene expression *in vivo* (May *et al.*, 2000; Pawliuk *et al.*, 2001).

23.2.2 Other Combinations and Erythroid-Specific Regulating Elements

Several combinations of β-like globin genes and erythroid-specific promoter/enhancer elements have been investigated for their effects on retroviral titer and on *cis*-linked β-like globin gene expression. The γ^A-globin gene HBG1 (including a long 5′ promoter region and a short 3′ flanking sequence) lacking its second intron was well expressed in cell lines (Rixon *et al.*, 1990). This may allow retroviral vectors to be produced at a higher titer than with the β-globin gene. Introduction of the HS40 α-globin enhancer element into the 3′ long terminal repeat (LTR) of an MLV vector carrying the γ^A-globin gene (with introns), a short promoter and a long 3′ flanking sequence (Ren *et al.*, 1996) conferred high-level expression to the γ^A-gene and low clone-to-clone variability *in vitro*. In mouse transplant experiments, however, γ^A-globin was not detected in adult RBCs (Lung *et al.*, 2000). A nano-LCR construct (including HS2, HS3 and HS4 of the β-globin LCR in a total of 1.1 kb) introduced upstream of the β-globin promoter and the γ^A-globin gene, gave rise to high vector instability and low level of expression (Emery *et al.*, 1998) even after specific sequences responsible for these drawbacks were identified and removed. The best combination of titer, stability and expression for the γ^A-globin gene *in vitro* was obtained using the β-globin promoter, a partial deletion of intron 2, a short and mutated 3′ flanking region and the HS40 α-enhancer (Emery *et al.*, 1999, 2002; Li *et al.*, 1999). However, low-level expression due to cell-to-cell variability in γ^A-globin gene expression levels was observed in primary mouse erythroid cells (Emery *et al.*, 1999) and in RBCs after mouse transplant experiments (Emery *et al.*, 2002).

These observations emphasized the sensitivity of MLV-derived globin expression vectors to the effects of variegation and silencing (Raftopoulos *et al.*, 1997; Rivella and Sadelain, 1998; Pannell *et al.*, 2000). Other erythroid specific elements have been tested to control gene expression. The ankyrin-1 promoter, which may be intrinsically shielded from position effects (Sabatino *et al.*, 2000b), was introduced in retroviral vectors to activate γ^A-globin expression (Sabatino *et al.*, 2000a). However, expression levels were found to be similar to those obtained with optimized retroviral vectors using globin promoters. Many other combinations of promoters (α-spectrin, ankyrin-1, β-globin or ζ-globin), enhancers (α-globin locus HS40, β-globin locus HS2, GATA autoregulatory sequence or the erythroid 5-aminolevulinate synthase (eALAS) intron 8 enhancer) and RNA stabilizing elements were shown to confer high-level erythroid-specific expression *in vitro* and are reasonable candidates to direct globin gene expression (Moreau-Gaudry *et al.*, 2001). However, these combinations have not been fully investigated *in vivo* and no data are available on their ability to limit position-effect variegation.

23.2.3 Position-Effect Variegation and Chromatin Insulators

To address the issue of position-dependent epigenetic silencing mediated by the chromatin surrounding the integrated vector and/or the vector itself (Neff *et al.*, 1997; Pannell *et al.*, 2000), the effects of a distinct class of genomic elements referred to as chromatin insulators (Eissenberg and Elgin, 1991; Corces, 1995) have been investigated. Chromatin insulator DNA elements were discovered on the basis of their ability to confine regulatory sequences and to protect genes against extragenic illegitimate regulating influences that would impair the regulatory constraints of the gene. They function by blocking promiscuous interactions between the enhancer of a gene and the promoter of another gene (Geyer and Corces, 1992; Kellum and Schedl, 1992) and/or by preventing the incursion of heterochromatin signals into open domains (Sun *et al.*, 1999). First discovered in *Drosophila* as structures defining the boundaries of discrete chromatin domains (Udvardy *et al.*, 1985; Holdridge and Dorsett, 1991), there were subsequently identified near a number of vertebrate genes (West *et al.*, 2002). The prototype insulator is the chicken β-globin cHS4 insulator domain (Chung *et al.*, 1993), which can act both as an enhancer blocker (Chung *et al.*, 1993, 1997) and as a barrier against the propagation of silencing modifications (Pikaart *et al.*, 1998; Taboit-Dameron *et al.*, 1999). Thus, insulators might provide position-independent expression capabilities to transgenes and may also buffer the action of external enhancers on the internal promoters intended to control transgene expression. In addition, insulator elements may protect endogenous surrounding genes from potentially adverse effects of the enhancers used to drive transgene expression. Indeed, a 1.2 kb fragment containing the cHS4 element that was introduced into the 3′ LTR of a MLV retroviral vector protected the vector from chromosomal position effects *in vitro* and in transplant experiments (Emery *et al.*, 2000; Rivella *et al.*, 2000). It also increased the likelihood that a γ-globin transgene was expressed in RBCs (Emery *et al.*, 2002). Nevertheless, the level of expression per cell was low and highly variable when the cHS4 element was used in combination with the HS40 enhancer and the β-globin promoter (Emery *et al.*, 2002). These studies eventually led to the conclusion that the limited amount of DNA sequences that can be stably transferred via oncoretroviral systems, such as MLV-based vectors, severely limits their effectiveness in globin gene expression and their use for gene therapy of the hemoglobinopathies, in spite of the presence of chromatin insulators.

23.3 The Usefulness of Lentiviral Vectors

The major issues surrounding the use of murine-retrovirus-derived vectors for transgenic globin gene expression include: (1) their limited cargo capacity for transferred genes and regulatory sequences, (2) their instability in provirus transfer and low titers and (3) their poor efficiency in transducing HSCs. In addition, murine leukemia viruses, and their derived vectors, can enter the nucleus only when cells divide and the nuclear membrane breaks down, and this is an impediment to the *ex vivo* transduction of HSCs since most are quiescent at any given time. Although lentiviruses also belong to the category of retroviruses, they have several specific properties that offer advantages over MLV-based vectors for gene therapy. In particular, human immunodeficiency virus (HIV) and other lentiviruses can infect cell-cycle-arrested cells (Lewis *et al.*, 1992) by interacting with the nuclear import machinery and thereby mediating the active transport of the viral preintegration complex through the nucleopore (Bukrinsky *et al.*, 1993; Gallay *et al.*, 1995a, 1995b). Replication-defective HIV-derived vectors, produced by transient transfection procedures, were shown to be capable of transducing cells arrested at the G_1–S boundary of the cell cycle and to a lesser extent at the G_0 stage (Naldini *et al.*, 1996a, 1996b). Lentiviral vectors pseudotyped with vesicular stomatitis virus glycoprotein G (VSV-G) can transduce a broad range of tissues and cell types including brain, liver, muscle and HSCs (Kafri *et al.*, 1997; Miyoshi *et al.*, 1999; Park *et al.*, 2000). Lentiviral vector biosafety has been considerably enhanced by: (1) replacing the lentiviral envelope by that of another class of virus (e.g., VSV-G) upon pseudotyping, (2) eliminating all or most of the viral coding sequences and minimizing viral *cis*-elements within the transfer vector, (3) splitting the genetic elements needed for production of the viral core protein and supplying them from several packaging and envelope helper constructs and (4) removing viral promoter/enhancer elements from the transfer vector. These modifications have virtually abolished the risk of producing replication-competent lentiviruses (Delenda *et al.*, 2002; Sinn *et al.*, 2005; Westerman *et al.*, 2007) even during large-scale manufacturing for clinical trials (Cavazzana-Calvo *et al.*, 2010). Another major advantage of lentiviral vectors over γ-retroviruses such as MLV-based vectors is their capacity to accept the insertion of large and complex DNA sequences (Kumar *et al.*, 2001). These vectors are able to package full-length unspliced RNA due to the presence of a strong RNA export element (rev responsive element (RRE)) that binds the viral reverse transcriptase (REV) protein (Cullen, 1998).

For more than ten years, the 20 kb globin LCR has been extensively studied, restricted and manipulated in order to fit within the size limit of murine oncoretroviral vectors, yet the result was only a rather imperfect compromise between β-globin gene expression/variegation and vector stability (Rivella *et al.*, 1998). Side-by-side comparisons of lentiviral vectors containing either minimal LCR genomic regions

(1 kb), such as those that could be introduced in MLV-based vectors, versus vectors containing longer genomic fractions of different HS (3 kb), confirmed indeed that the vector insert size limitation was a major issue for transgenic β-globin expression. Longer LCR elements improved the likelihood of β-globin expression, resulted in much higher expression levels *in vivo* and appeared to be more resistant to silencing (May *et al.*, 2000). Thus, soon after their initial launch, lentiviral vectors emerged as an important advance toward gene therapy of the β-hemoglobinopathies.

23.4 Goals for the Gene Therapy of β-Thalassemia and Sickle Cell Disease

The key objectives of gene therapy approaches to ameliorate and even cure β-thalassemia and sickle cell disease (i.e., efficacy of gene expression and proportion of modified cells) are similar but not strictly identical. A confounding factor to consider for the gene therapy of these diseases is that the pre-therapeutic levels of endogenous β-globin have a major effect on the capability of the vector to express its encoded protein: it is easier to obtain large amounts of therapeutic β-globin per RBC in β-thalassemia, where competition for transcription and translation is limited due to decreased or absent levels of endogenous β-globin expression, than in sickle cell disease, in which the β^S-globin chains are expressed at physiological levels.

At the individual cellular level, the first requirement is to achieve levels of the expressed therapeutic proteins that are high enough to interfere with the disease processes. In β-thalassemia, the aim is to achieve cellular levels of β-globin expression that are sufficient to bind much of the unpaired α-chains, and this level varies whether there is complete absence of endogenous β-globin expression (β^0-thalassemia or Cooley anemia) or not (e.g., β^+-thalassemia, β^E/β^0-thalassemia) and, quite obviously, with the degree of severity of the thalassemic syndrome (transfusion-dependent thalassemia major versus thalassemia intermedia). In sickle cell disease, the goal is to express enough of a β-like globin to inhibit intracellular HbS polymerization by mere dilution of the β^S-globin chains or, preferentially, by disrupting the polymerization process and increasing the solubility of the Hb species.

The increased solubility induced by HbF ($\alpha_2\gamma_2$) and HbA2 ($\alpha_2\delta_2$) is higher than that induced by HbA ($\alpha_2\beta_2^A$) (Nagel *et al.*, 1980) because the hybrid heterotetramers FS ($\alpha_2\gamma\beta^S$) and A2S ($\alpha_2\delta\beta^S$) are excluded from the polymer whereas AS ($\alpha_2\beta^A\beta^S$) is not (Poillon *et al.*, 1993). The hydrophobic acceptor pocket, containing the amino acids Phe and Leu, located at β85 and β88 respectively, is essential in promoting the contacts between the tetramers of HbS, through its interaction with the sickle cell mutation β6Val (Adachi *et al.*, 1994b). Another essential amino acid located near the hydrophobic acceptor pocket is the β87 residue (Adachi *et al.*, 1994a), which is a threonine in β^A and β^S and a glutamine in δ- and γ-globin chains. This residue, even if not involved directly in the β6Val interaction, plays an important role in the exclusion of the hybrid heterotetramer FS and A2S in the formation of Hb polymers (Adachi *et al.*, 1996a, 1996b; Reddy *et al.*, 1997). Indeed, purification of modified Hb and *in vitro* polymerization studies showed that $\beta^{A\text{-}E22A;T87Q}$ is as efficient as the γ-chain at inhibiting HbS polymerization (McCune *et al.*, 1994). The δ-like E22A modification does not seem to be essential, as the γ-chain of HbF has an Asp instead of a Glu, and HbF, like HbA2, inhibits polymerization of deoxy-HbS and may be slightly better (Adachi *et al.*, 1996b). Indeed, the $\beta^{A\text{-}T87Q}$-globin chain has been shown to be as efficient as the γ-chain at inhibiting HbS polymerization (Pawliuk *et al.*, 2001).

With regard to the proportion of cells that need to be corrected, differences are also evident between sickle cell disease and the β-thalassemias, as follows. In β-thalassemia patients, therapeutic benefit has been observed with a limited degree of mixed chimerism after allogeneic hematopoietic transplantation from HLA-identical sibling donors. Hence, donor-derived hematopoiesis levels of 20–30% elevated the production of Hb to a level sufficient to avoid RBC transfusions (Andreani *et al.*, 2000). This was consistent with the preferential survival of normal erythroid cells against the high apoptotic rate of erythroid precursors and RBC hemolysis in β-thalassemia (Centis *et al.*, 2000). Also, in murine models of β-thalassemia, a 10–20% proportion of normal donor cells resulted in significant improvement of the phenotype (Persons *et al.*, 2001).

In sickle cell anemia, frequent blood transfusions reduce the incidence of painful crises and the risk of stroke (Styles and Vichinsky 1994; Pegelow *et al.*, 1995) but do not prevent all complications, suggesting that a higher degree of chimerism with corrected cells is required for a global therapeutic effect on this disease. In patients with 50% chimerism after allogenic marrow transplantation, most of the abnormal peripheral blood cells are replaced by donor cells (Wu *et al.*, 2007), and this is consistent with the poor survival of abnormal RBCs (Kean *et al.*, 2003) and with the ineffective erythropoiesis observed in patients and in animal models (Blouin *et al.*, 1999; Wu *et al.*, 2005).

23.5 Correction of Mouse Models of β-Thalassemia and Sickle Cell Disease

23.5.1 β-Thalassemia

In mice, the switches from primitive to definitive erythropoiesis and from embryonic (βh1 and εy2) to adult (β^{major} and β^{minor}) β-globin gene expression occur between 14 and 15 days of gestation (Craig and Russell, 1964; Barker, 1968), and there is no fetal-like Hb. Accordingly, absence of adult globin genes in mice leads to fetal lethality, which creates difficulties in using a thalassemia major (β^0/β^0) mouse model for gene therapy and transplantation studies. Mice lacking either of the two β^{major} alleles ($Hbb^{th1/th1}$) or which are heterozygotes for the deletion of both β^{major} and β^{minor} ($Hbb^{th3/+}$), recapitulate the cellular and hematological defects observed in human β-thalassemia intermedia (β^0/β^+) (Rouyer Fessard *et al.*, 1990; Yang *et al.*, 1995) and have been utilized in preclinical studies. Two slightly different lentiviral vectors, containing the human β-globin gene with introns, the human β-globin promoter and about 3 kb of the LCR were used to correct the disease (May *et al.*, 2000; Imren *et al.*, 2002). The second vector (May *et al.*, 2000, 2002; Imren *et al.*, 2002) had been first optimized in the context of mouse models of sickle cell disease (Pawliuk *et al.*, 2001), as explained in Section 23.5.2. Transplantation of genetically modified mouse HSC from β-thalassemic donors into myeloablated recipients resulted in pan-cellular and long-term expression of human β-globin. The levels of β-globin gene expression in these studies were high enough to significantly decrease the free α-chain concentration, to improve the RBC morphology, to reduce extramedullary hematopoiesis, to correct ineffective erythropoiesis and anemia and to prevent iron overload. These results constituted a major advance in the field and could be attributed to features of the lentiviral vectors that allowed transgene delivery to be much improved over that obtained with oncoretroviral vectors, including greater transduction efficiency of mouse HSCs, higher levels of β-globin production, and decreased gene silencing.

A lentiviral vector was also used to transfer the γ-globin gene (exons and introns, including partially deleted intron 2), whose expression was controlled by the β-globin promoter, the γ-globin 3′ untranslated region and a smaller portion of the LCR (1.7 kb). However, this vector produced lower expression levels than the β-globin lentiviral vectors (Persons *et al.*, 2003). In comparison, a similar γ-globin vector with an extended LCR (3.2 kb) was shown to be more efficient (Hanawa *et al.*, 2004). In the complete absence of endogenous mouse β-globin gene expression, which is the most severe context of mouse β-thalassemia and is obtained by engrafting adult normal recipient with $Hbb^{th3/th3}$ fetal liver cells, the expression level of β-globin per vector copy in transduced RBC was shown be approximately half that of the hemizygous endogenous Hb production (Rivella *et al.*, 2003), although the caveat of such quantitative assertions is that only RBC containing the highest levels of therapeutic globin expression appear in the circulation as a consequence of *in vivo* competitive selection.

23.5.2 Sickle Cell Disease

Many mouse models of sickle cell anemia have been produced. As no model perfectly reflects the human disease, several of them have been used for preclinical studies. A common issue that had to be tackled in generating these mouse models is that the murine endogenous β-like globins are strongly anti-sickling and must be either superseded by "super-sickling" transgenic constructs or ablated after "knocking-out" the genes that encode them. The transgenic SAD mouse model (which expresses the mouse globin chains and the human α-chain together with a supersickling β^{SAD}-globin, but appear healthy) (Trudel *et al.*, 1991), reproduces the systemic microvascular occlusions, ineffective erythropoiesis, hemolysis, renal complications and organ damage observed in sickle cell patients, and has a slightly shortened lifespan (Trudel *et al.*, 1994; Blouin *et al.*, 1999). This mouse model is very useful to induce vaso-occlusive complications under hypoxia and for testing and comparing therapeutic approaches (Siciliano *et al.*, 2011). However, the presence of the mouse α-chain, which has anti-sickling activity, delays the clinical manifestations and hampers the making of quantitative correlations between a given therapeutic product, its expressed levels, and the various parameters of disease correction.

Two transgenic knockout mice (the Berkeley and Townes models, see below) were generated to express exclusively human sickle Hb in the absence of endogenous murine β-like globins. These animals harbor features similar to the SAD mice (Paszty *et al.*, 1997; Ryan *et al.*, 1997). They are anemic, have high levels of reticulocytosis and contain many irreversibly sickled cells (two characteristics missing in SAD mice), and develop significant pathologies shortly after birth. Unfortunately, the α:non-α chain ratio is higher than normal, and the mean corpuscular Hb (MCH) level is low, indicating the presence of an associated β-thalassemia-like phenotype. The low MCH level may not only protect against polymer formation but may also favor transgenic β-globin expression in hypochromic cells in gene therapy models. Furthermore, dyserythropoiesis and the low RBC survival associated with the thalassemia phenotype may contribute in and of itself to cell selection and phenotype correction upon gene therapy and bone marrow transplantation. Thus, discerning the beneficial effects of gene therapy strategies in this model may be difficult.

A third transgenic knockout mouse model that expresses human globin only with a balanced α:non-α chain ratio has, however, persisting expression of γ-chain in adulthood, and this anomaly may also complicate the evaluation of therapeutic protocols and anti-sickling agents (Nagel *et al.*, 2001). The more recent mouse model, created by knockin of adult mouse α and β genes and replacement with human α and $\gamma^{+}\beta^{S}$ genes respectively (Wu *et al.*, 2006), should be a more ideal model to test the efficacy of anti-sickling agents without the confounding effects of hypochromia correction and imbalanced-globin chain synthesis.

As the β-globin gene (including its promoter, exons, intron and 3′ untranslated region) seemed to be expressed more efficiently than the γ-globin constructs in adult cells, and since the oxygen affinity of $HbA\beta^{A\text{-}T87Q}$ is very similar to that of HbA, a lentiviral vector encoding the mutated $\beta^{A\text{-}T87Q}$-globin chain was produced and tested in two mouse models of sickle cell anemia, Berkeley and SAD (Pawliuk *et al.*, 2001). HSC transduced with the lentiviral vector were transplanted into lethally irradiated recipients. The human $\beta^{A\text{-}T87Q}$ protein level was higher in RBC from Berkeley donor cells than in RBC coming from SAD cells. This is most likely due to the associated β-thalassemic phenotype in Berkeley mice which favors efficient translation of the transgenic human $\beta^{A\text{-}T87Q}$-globin mRNA. The transgenic β-globin protein levels were 75% and 40% of the level of sickle Hb in recipients of Berkeley and SAD cells, respectively. Importantly, inhibition of RBC sickling at low oxygen pressure was more efficient in the modified Berkeley cells than in unmodified cells. Furthermore, kinetic studies of polymerization as a function of Hb concentration showed that transduced cells (SAD and Berkeley) behave similarly to asymptomatic heterozygote human AS RBC whereas non-transduced mouse RBC resembled homozygote SS cells. RBC densities, hematological parameters and urine concentration defects were also corrected. A vector encoding a triple anti-sickling mutant ($\beta^{A\text{-}G16D;E22A;T87Q}$) was used to correct the Townes model and gave similar results (Levasseur *et al.*, 2003, 2004). Another strategy proposed is to knockdown concurrently endogenous β^{S}-globin chain expression by co-expression of a specifically targeted short hairpin RNA (shRNA) (Dykxhoorn *et al.*, 2006; Samakoglu *et al.*, 2006).

23.6. Limited Expression Level Per Cell and Position Effect Variegation

Analysis of β-globin expression efficiency was performed in normal mice after transduction and transplantation of $\beta^{A\text{-}T87Q}$ modified hematopoietic stem cells. Surprisingly, even with three vector copies per modified cell, the level of human β-globin mRNA was 71% of the level of mouse β-globin mRNA and the protein level was only 16% of the mouse β-globin chain level (Pawliuk *et al.*, 2001). β-globin expression levels were similar in the RBC of transplanted SAD mice. In mice containing a human β-globin transgene, low β-globin expression levels were also observed (Alami *et al.*, 1999). The reasons why human transgenic β-globin synthesis is lower than that of the mouse endogenous β-globin are not clear but these results suggest that it may be difficult to express the therapeutic protein in erythroid cells of patients with sickle cell anemia who do not have an associated β-thalassemic phenotype, primarily because of competition for limiting endogenous factors between RNA species to be expressed. Indeed, in Berkeley mice deficient in endogenous β-globin synthesis, the level of transgenic protein per cell was much higher than that in RBC of SAD or C57BL/6 mice (Pawliuk *et al.*, 2001). In β-thalassemic erythroid cells, β-globin transcript levels can be lower than normal and β-globin synthesis either low or absent.

Consequently, transgenic β-globin synthesis competes favorably for transcription and translation. In cells where endogenous rates of globin transcription and translation are normal (as they are in cells from normal individuals or patients with sickle cell anemia), production by the cellular β-globin gene may outcompete that of the transgene. Indeed, subtle differences in expression timing and/or translation efficiency between endogenous and transgenic β-globin synthesis may reduce the efficiency of exogenous Hb tetramer assembly.

In β-thalassemic and sickle cell mouse models, at least two to three copies per cell of a vector harboring large LCR sequences were necessary to obtain substantial phenotype correction (May *et al.*, 2000; Pawliuk *et al.*, 2001; Imren *et al.*, 2002; Persons *et al.*, 2003; Hanawa *et al.*, 2004). Analysis of individual erythroid cell clones showed wide variations in the amount of transgenic β- or γ-globin production. These results suggest that pancellular expression in those models was the result of balanced expression from polyclonal stem cell reconstitution with multiple chromosomal integration sites rather than from position-independent expression (Imren *et al.*, 2002). Thus, for all lentiviral vectors, even those which contain extensive LCR sequences and produce higher expression levels than the oncoretroviral vectors, position effect variegation was still an issue, and chromatin insulators were therefore evaluated in this context.

The 1.2 kb cHS4 chromatin insulator domain of the chicken β-globin locus contains a 250 bp core element which, when placed as tandem copies on either side of a reporter gene, was shown to retain both insulator activities in chicken cells: enhancer blocking and protection against position effects (Bell *et al.*, 1999; Recillas-Targa *et al.*, 2002). A lentiviral vector encoding β-globin and flanked on both sides by the 1.2 kb cHS4 fragment efficiently expressed transgenic β-globin in human thalassemic erythroid cells *in vitro* (Puthenveetil *et al.*, 2004) and had a two-fold higher probability than a vector containing the non-insulated transgene of being expressed *in vitro* and in mouse RBC (Arumugam *et al.*, 2007). When insulated with one copy of a 400 bp insulator region (including the 250 bp core element), the vector was also more efficient at expressing β-globin *in vitro* than the non-insulated vector (Aker *et al.*, 2007). However, one copy of the 250 bp core element was reported not to have an insulating effect in one study (Aker *et al.*, 2007) while another said otherwise (Hanawa *et al.*, 2009).

23.7 Genotoxic Potential of Lentiviral Vectors Expressing Globin Genes

γ-retrovirus-based vector integration into chromatin and subsequent activation of flanking oncogenes by viral enhancers have been involved in the lymphoproliferative and myelodysplastic syndromes observed in patients undergoing gene therapy for severe combined immune deficiency (Hacein-Bey-Abina *et al.*, 2003, 2008; Howe *et al.*, 2008), Wiskott–Aldrich syndrome (Avedillo Diez, *et al.*, 2011) and chronic granulomaous disease (Ott *et al.*, 2006). Similar observations have been made in animal models (Li, 2002; Modlich 2008). As a fallout of these adverse events, the mechanisms of gene activation by specific viral and cellular elements as well as the effects of vector design and insertional preference of γ-retroviral and lentiviral vectors have been investigated to assess the individual risks of insertional mutagenesis. The most important conclusion to date is that all vectors used in gene therapy should have their viral enhancers deleted from LTRs: the so-called self-inactivating (SIN) vectors. Secondly, when possible, cellular gene promoters are preferred over the use of strong and ubiquitous viral promoters for controlling transgene expression (Zychlinski *et al.*, 2008).

Importantly, the genotoxic potential of lentiviral vectors is believed to be lower than that of γ-retroviral vectors on the basis of several studies, a phenomenon that may be, in part, explained by the preferential targeting of gene promoter regions by γ-retrovectors as compared to lentiviral vectors in hematopoietic cells (Montini *et al.*, 2006, 2009). Furthermore, in lentiviral β-globin expression vectors, β-globin transgene expression is controlled by an erythroid-specific promoter. Nevertheless, concerns remain because LCR-containing vectors have been found near protooncogenes in human hematopoietic cells (Imren *et al.*, 2004) and to disturb the expression of genes in mouse β-thalassemic erythroid cells, which are up to 600 kb away from the integration site (Hargrove *et al.*, 2008). The LCR has also been shown to be in an active chromatin configuration prior to erythroid specification in hematopoietic cells (Jimenez *et al.*, 1992). Careful experiments aimed at precisely quantifying the risk of insertional mutagenesis by the LCR were performed in mouse hematopoietic progenitors. It was confirmed that SIN-lentiviral vectors are significantly safer than γ-retroviral vectors, and it was clearly shown that erythroid-specific LCR elements confer considerably less genotoxicity than strong viral enhancers (Arumugam *et al.*, 2009a). Surrounding the vector with cHS4 elements slightly protected the cells (by two-fold) from *in vitro* immortalization, with the 250 bp core element and the full-length 1.2 kb fragment having equivalent activities (Arumugam *et al.*, 2009b). However, the efficacy of this chromatin insulator as an element conferring enhancer blocking activity is controversial, as it seems to be promoter-dependent and does not simply depend on it being inserted between an activating enhancer and an active promoter (Desprat and Bouhassira, 2009). Furthermore, depending on the site of chromosome integration, an insulator may decrease the expression level of the gene it is designed to protect against position-effect variegation and worsen the risk of insertional mutagenesis (Desprat and Bouhassira, 2009). Therefore, the potential benefit of chromatin insulators is as yet unresolved. On the one hand, these sequences may offer higher probabilities of transgene expression and the prevention of insertional mutagenesis, while on the other they may result in lower transgene expression levels and lower vector titers. Thus, the risk:benefit ratio of currently known chromatin insulators has yet to be defined and warrants further study.

23.8 Transfusion Independence in a β-Thalassemic Patient Treated by Gene Therapy

A "Phase I/II, open label, safety and efficacy study on the administration of autologous CD34⁺ cells transduced *ex vivo* with a lentiviral vector encoding $\beta^{A\text{-}T87Q}$-globin (LentiGlobin™) in patients with β-hemoglobinopathies" (Fig. 23.1) received regulatory approval in 2005 (Bank *et al.*, 2005).

The trial is to include five patients with transfusion-dependent β-thalassemia major and five patients with sickle cell disease. The first treated patient who did not receive back-up cells (the first trial patient did) received a transplant of transduced cells on June 7, 2007 (Cavazzana-Calvo, 2010). He was 18 years old at the time of treatment, suffered from severe β^E/β^0-thalassemia and required monthly RBC transfusions since early childhood to stay alive. He was splenectomized at six years old, did not have a positive response to hydroxyurea and did not have a possible sibling donor. Bone marrow cluster of differentiation (CD)34⁺ cells were harvested, prestimulated and transduced with the LentiGlobin™ vector. The transduction efficiency, evaluated on myeloid progenitors one week after transduction, was 0.6 vector copies per cell. Transduced CD34⁺ cells (3.9×10^6 cells/kg)

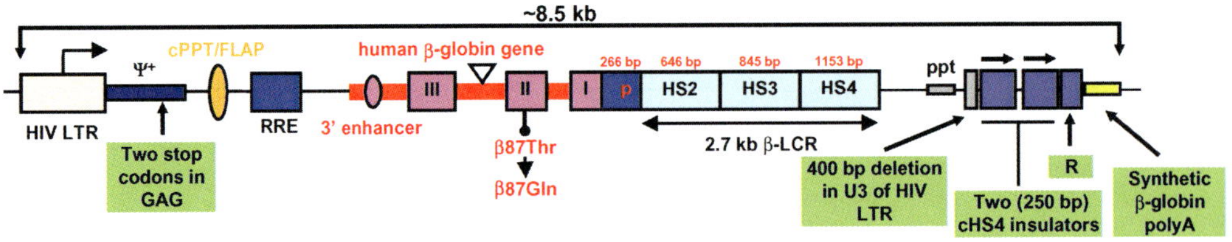

FIGURE 23.1 ■ (a) Diagram of the human β-globin ($\beta^{A\text{-}T87Q}$) lentiviral vector (LentiGlobin™, LG). The 3′ β-globin enhancer, the 372bp IVS2 deletion, the $\beta^{A\text{-}T87Q}$ mutation (ACA[Thr] to CAG[Gln]) and DNase I hypersensitive sites (HS) 2, HS3 and HS4 of the human β-globin locus control region (LCR) are indicated. Safety modifications including the two stop codons in the Ψ+ signal, the 400 bp deletion in the U3 of the right HIV LTR, the rabbit β-globin polyA signal and the 2 × 250 bp cHS4 chromatin insulators are indicated. HIV LTR, human immunodeficiency type-1 virus long terminal repeat; Ψ+, packaging signal; cPPT/flap, central polypurine tract; RRE, REV-responsive element; bp, human β-globin promoter; ppt, polypurine tract; R, repeat element of the RNA HIV genome.

were injected after conditioning by intravenous Busulfex®. The percentage of modified peripheral blood cells in this patient increased slowly and stabilized to approximately 10% at 30 months after transplantation. In the bone marrow three years post-transplant, the hematopoietic compartment with the highest proportion of modified cells was the erythroid lineage (about 30%). This result was expected as genetically modified cells are likely to have a selective advantage over unmodified cells because they are less prone to dyserythropoiesis (premature death of abnormal cells). Remarkably, one year after gene therapy and bone marrow transplant, the patient became transfusion-independent and the Hb level stabilized to approximately 9 g/dl (Fig. 23.2). The number of peripheral blood erythoblasts decreased and RBC survival improved. The MCH level is within the normal range and the therapeutic $\beta^{A\text{-}T87Q}$-globin output on a per-gene basis is between 70% and 100% of the normal value. The patient is now phlebotomized monthly to reduce iron overload, and plasma ferritin levels have decreased significantly (unpublished data). In spite of these repeated phlebotomies, the patient has not required nor received any blood transfusion for three years, this being four years after transplantation and gene therapy (Cavazzana-Calvo 2010).

Chromosomal integration sites (IS) analysis of the vector in peripheral blood and in purified hematopoietic cell compartments showed the relative dominance of an IS in the high mobility group AT-hook 2 (*HMGA2*) gene (Cavazzana-Calvo 2010). This specific IS was detected in myeloid cells but not in B- or in T-cells. The corresponding myeloid cell clone was first detected four months post-bone marrow transplant (BMT), stabilized since month 15, and now represents approximately 3% of the circulating nucleated cells (Fig. 23.3).

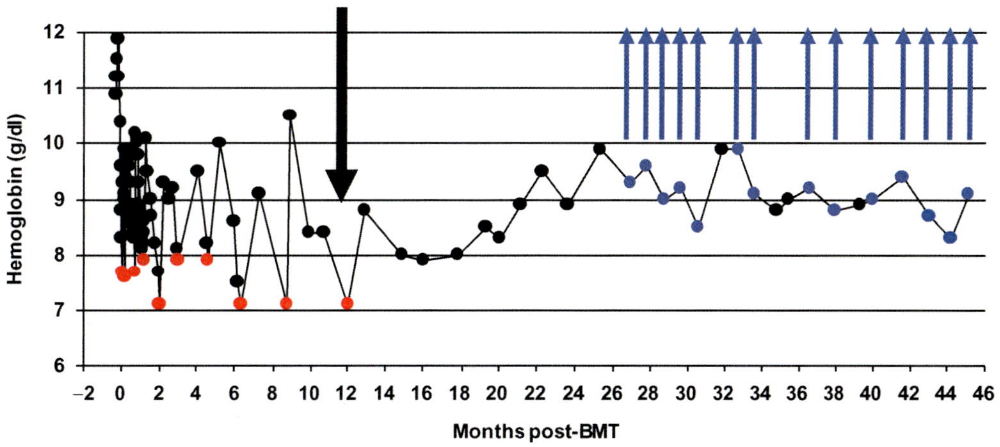

FIGURE 23.2 ■ Total Hb concentrations in whole blood. Red dots are transfusion time points. Black vertical arrow, the last time the patient was transfused (12 months post-bone marrow transplant (BMT). Blue arrows, phlebotomies (200 ml each) to remove excess iron.

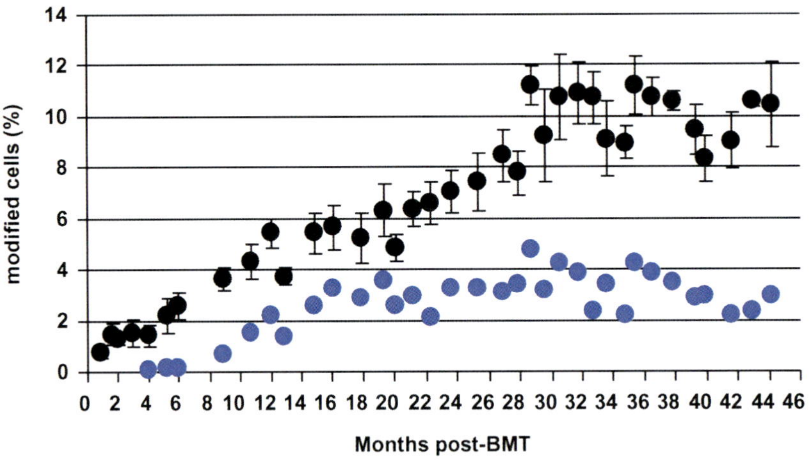

FIGURE 23.3 ■ LentiGlobin™ modified cells. Percentage (mean ± standard deviation) of modified peripheral blood cells at all sites (black dots) and at the HMGA2 site (blue dots).

The vector is inserted within the third intron of the gene, in direct orientation. We found that HMGA2 expression is primarily induced in erythroid cells upon LCR activation originating from the integrated vector. In addition, a cryptic splice site, present in the vector, leads to the production of an mRNA containing only exons 1, 2 and 3 of the five exons containing the *HMGA2* gene upon polyadenylation of this aberrant transcript at a polyA site located within the integrated LentiGlobin™ vector. Hence, removal of target sites for lethal-7 (let-7) microRNA (miRNA) in exon 5 further increases the level of the truncated form of *HMGA2* mRNA in erythroid cells. As translocation events in the *HMGA2* gene and overexpression of a truncated form of HMGA2 have been involved in neoplasia (Cleynen and Van de Ven, 2008), though mostly benign, careful and regular follow up of this patient is being pursued. So far, no breach of hematopoietic homeostasis has been observed. Other reassuring observations include the facts that (1) *HMGA2* clonal dominance has been observed for 17 years in a paroxysmal nocturnal hemoglobinuria patient with clonal yet benign hematopoietic expansion (T. Kinoshita, personal communication), (2) several *HMGA2* insertion sites were identified in X-linked severe combined immunodeficiency (X-SCID) patients treated by gene therapy and were not associated with leukemia (Wang *et al.*, 2010) and (3) truncated *HMGA2* genes are associated with benign tumor phenotypes (Cleynen and Van de Ven, 2008). The observation that a similar proportion of cells bear the integrated LentiGlobin™ vector within the *HMGA2* gene in all myeloid compartments, including long-term culture initiating cells (LTC-IC), whereas HMGA2 expression at the RNA and protein levels is only detected in the erythroid lineage, led us to conclude that myeloid-biased hematopoiesis may have been triggered in this patient or may simply be physiological and revealed by vector marking (Cavazzana-Calvo *et al.*, 2011).

This first-in-man gene therapy of the β-hemoglobinopathies has hence shown promise with the first evidence of long-term clinical benefit (complete transfusion independence for over three years) and has increased our knowledge of the dynamic of human hematopoiesis, its genetic control and the occurrence of endogenous gene activation upon lentiviral vector integration.

23.9 Lessons for Hematopoietic Gene Therapy

Gene therapy with γ-retroviral vectors and deep sequencing analysis of IS retrieved *in vivo* after hematopoietic transplantation with transduced cells have revealed that clonal skewing is the rule rather than the exception (Boztug *et al.*, 2010; Stein *et al.*, 2010; Wang *et al.*, 2010). In some patients with insertional mutagenesis and further accumulation of additional genetic lesions, leukemogenesis can develop (Hacein-Bey-Abina *et al.*, 2008; Howe *et al.*, 2008; Stein *et al.*, 2010). The probability of leukemic development seemingly depends on several factors that include the differentiation stage of the transduced cell, pre-existing genetic lesions, the level of expression of the activated gene, the oncogenic potential of the transgene itself and the immune state of the patient. Experiments conducted with lentiviral vectors *in vitro* and in mice showed that lentiviral vectors are safer than γ-retroviral vectors in that regard. The globin lentiviral vectors in particular did not induce any adverse event in a mouse model of β-thalassemia nor did it increase the proportion of IS near oncogenes (Ronen *et al.*, 2011). However, as reviewed above, myeloid-biased cell expansion and partial clonal dominance were observed in one β-thalassemia patient treated by gene therapy (Cavazzana-Calvo *et al.*, 2010). In a recent clinical trial for adrenoleukodystrophy involving a lentiviral vector (Cartier *et al.*, 2009), polyclonal hematopoietic reconstitution was observed, with no apparent clonal skewing, although integration within the *HMGA2* gene was also identified at a low frequency (unpublished data). The vector contained a ubiquitous and strong viral promoter and a high number of unique IS were detected. In the β-thalassemic patient treated by gene therapy, the total number of unique IS was lower and departure from polyclonality was observed (Cavazzana-Calvo *et al.*, 2010).

How could we reconcile the apparent paradox in the observation of (1) oligoclonality and clonal skewing associated with a lower number of transduced cells (and consequently a lower probability of insertional mutagenesis/oncogenesis) versus (2) polyclonality, no clonal skewing associated with higher numbers of insertion sites? One hypothesis is that there was no effect of the dysregulation of the *HMGA2* gene and the myeloid-biased expansion observed in the β-thalassemic patient. A second hypothesis is that a substantial portion of the genetically modified cells present in patients after hematopoietic gene therapy with vectors currently utilized result from the selection of progenitor/stem cells with minor and benign genetic dysregulations, which may confer an advantage in homing, proliferation and/or differentiation. Cell culture, pre-existing disease state, hematopoietic reconstitution and cell differentiation processes may alter event distribution. The molecular mechanisms involved could be either epigenetic or mutagenic, as both mechanisms may result in protooncogene activation and increased cell fitness. Apparent polyclonality may thus be the results of polyclonal co-selection. Risk:benefit ratios of a given vector in a given disease will ultimately be provided by larger series of gene therapy patients.

References

Adachi, K., Konitzer, P., Surrey, S. (1994a). Role of γ87 Gln in the inhibition of hemoglobin S polymerization by hemoglobin F, *J Biol Chem*, **269**, 9562–9567.

Adachi, K., Reddy L.R., Surrey S. (1994b). Role of hydrophobicity of phenylalanine β85 and leucine β88 in the acceptor pocket for valine β6 during hemoglobin S polymerization, *J Biol Chem*, **269**, 31563–31566.

Adachi, K., Pang, J., Konitzer, P., *et al.* (1996a). Polymerization of recombinant hemoglobin F γE6V and hemoglobin F γE6V, γQ87T alone, and in mixtures with hemoglobin S, *Blood*, **87**, 1617–1624.

Adachi, K., Pang, J., Reddy, L.R., *et al.* (1996b). Polymerization of three hemoglobin A_2 variants containing Val^6 and inhibition of hemoglobin S polymerization by hemoglobin A_2, *J Biol Chem*, **271**, 24557–24563.

Aker, M., Tubb, J., Groth, A.C., *et al.* (2007). Extended core sequences from the cHS4 insulator are necessary for protecting retroviral vectors from silencing position effects, *Hum Gene Ther*, **18**, 333–343.

Alami, R., Gilman, J.G., Feng, Y.Q., *et al.* (1999). Anti-β^s-ribozyme reduces β^s mRNA levels in transgenic mice: Potential application to the gene therapy of sickle cell anemia, *Blood Cells Mol Dis*, **25**, 110–119.

Andreani, M., Nesci, S., Lucarelli, G., *et al.* (2000). Long-term survival of ex-thalassemic patients with persistent mixed chimerism after bone marrow transplantation, *Bone Marrow Transplant*, **25**, 401–404.

Angastiniotis, M., Modell, B. (1998). Global epidemiology of hemoglobin disorders, *Ann N Y Acad Sci*, **850**, 251–269.

Arumugam, P.I., Scholes, J., Perelman, N., *et al.* (2007). Improved human β-globin expression from self-inactivating lentiviral vectors carrying the chicken hypersensitive site-4 (cHS4) insulator element, *Mol Ther*, **15**, 1863–1871.

Arumugam, P.I., Higashimoto, T., Urbinati, F., *et al.* (2009a). Genotoxic potential of lineage-specific lentivirus vectors carrying the β-globin locus control region, *Mol Ther*, **17**, 1929–1937.

Arumugam, P.I., Urbinati, F., Velu, C.S., *et al.* (2009b). The 3′ region of the chicken hypersensitive site-4 insulator has properties similar to its core and is required for full insulator activity, *PLoS One*, **4**, e6995.

Avedillo Diez, I., Zychlinski, D., Coci, E.G., *et al.* (2011). Development of novel efficient SIN vectors with improved safety features for Wiskott-Aldrich syndrome stem cell based gene therapy, *Molecular pharmaceutics*, **8**, 1525–1537.

Bank, A., Dorazio, R., Leboulch, P. (2005). A phase I/II clinical trial of β-globin gene therapy for β-thalassemia, *Ann N Y Acad Sci*, **1054**, 308–316.

Barker, J.E. (1968). Development of the mouse hematopoietic system. I. Types of hemoglobin produced in embryonic yolk sac and liver, *Dev Biol*, **18**, 14–29.

Baron, M.H. (1996). Developmental regulation of the vertebrate globin multigene family, *Gene Expr*, **6**, 129–137.

Behringer, R.R., Hammer, R.E., Brinster, R.L., *et al.* (1987). Two 3′ sequences direct adult erythroid-specific expression of human β-globin genes in transgenic mice, *Proc Natl Acad Sci USA*, **84**, 7056–7060.

Behringer, R.R., Ryan, T.M., Palmiter, R.D., *et al.* (1990). Human γ- to β-globin gene switching in transgenic mice, *Genes Dev*, **4**, 380–389.

Bell, A.C., West A.G., Felsenfeld, G. (1999). The protein CTCF is required for the enhancer blocking activity of vertebrate insulators, *Cell*, **98**, 387–396.

Bender, M.A., Gelinas R.E., Miller, A.D. (1989). A majority of mice show long-term expression of a human β-globin gene after retrovirus transfer into hematopoietic stem cells, *Mol Cell Biol*, **9**, 1426–1434.

Blouin, M.J., De Paepe, M.E., Trudel, M. (1999). Altered hematopoiesis in murine sickle cell disease, *Blood*, **94**, 1451–1459.

Borgna-Pignatti, C. (2010). The life of patients with thalassemia major, *Haematologica*, **95**, 345–348.

Boztug, K., Schmidt, M., Schwarzer, A., *et al.* (2010). Stem-cell gene therapy for the Wiskott–Aldrich syndrome, *N Engl J Med*, **363**, 1918–1927.

Bukrinsky, M.I., Haggerty, S., Dempsey, M.P., *et al.* (1993). A nuclear localization signal within HIV-1 matrix protein that governs infection of non-dividing cells, *Nature*, **365**, 666–669.

Camaschella, C., Kattamis, A.C., Petroni, D., *et al.* (1997). Different hematological phenotypes caused by the interaction of triplicated α-globin genes heterozygous β-thalassemia, *Am J Hematol*, **55**, 83–88.

Camaschella, C., Mazza, U., Roetto, A., *et al.* (1995). Genetic interactions in thalassemia intermedia: analysis of β-mutations, α-genotype, γ-promoters, and β-LCR hypersensitive sites 2 and 4 in Italian patients, *Am J Hematol*, **48**, 82–87.

Caocci, G., Efficace, F., Ciotti, F., *et al.* (2010). Prospective assessment of health related quality of life in pediatric beta-thalassemia patients following hematopoietic stem cell transplantation, *Biol Blood Marrow Transplant*, **17**, 861–866.

Cappellini, M.D., Fiorelli, G., Bernini, L.F. (1981). Interaction between homozygous β^0 thalassaemia and the Swiss type of hereditary persistence of fetal haemoglobin, *Br J Haematol*, **48**, 561–572.

Cartier, N., Hacein-Bey-Abina, S., Bartholomae, C.C., *et al.* (2009). Hematopoietic stem cell gene therapy with a lentiviral vector in X-linked adrenoleukodystrophy, *Science*, **326**, 818–823.

Caterina, J.J., Ryan, T.M., Pawlik, K.M., *et al.* (1991). Human β-globin locus control region: analysis of the 5′ DNase I hypersensitive site HS 2 in transgenic mice, *Proc Natl Acad Sci USA*, **88**, 1626–1630.

Cavazzana-Calvo, M., Fischer, A., Bushman, F.D., *et al.* (2011). Is normal hematopoiesis maintained solely by long-term multipotent stem cells? *Blood*, **117**, 4420–4424.
Cavazzana-Calvo, M., Payen, E., Negre, O. *et al.* (2010). Transfusion independence and HMGA2 activation after gene therapy of human β-thalassaemia, *Nature*, **467**, 318–322.
Centis, F., Tabellini, L., Lucarelli, G., *et al.* (2000). The importance of erythroid expansion in determining the extent of apoptosis in erythroid precursors in patients with β-thalassemia major, *Blood*, **96**, 3624–3629.
Chada, K., Magram, J., Raphael, K., *et al.* (1985). Specific expression of a foreign β-globin gene in erythroid cells of transgenic mice, *Nature*, **314**, 377–380.
Chang, J.C., Liu, D., Kan, Y.W. (1992). A 36-base-pair core sequence of locus control region enhances retrovirally transferred human β-globin gene expression, *Proc Natl Acad Sci USA*, **89**, 3107–3110.
Charache, S., Terrin, M.L., Moore, R.D., *et al.* (1995). Design of the multicenter study of hydroxyurea in sickle cell anemia. Investigators of the Multicenter Study of Hydroxyurea, *Control Clin Trials*, **16**, 432–446.
Choi, O.R., Engel, J.D. (1986). A 3′ enhancer is required for temporal and tissue-specific transcriptional activation of the chicken adult β-globin gene, *Nature*, **323**, 731–734.
Chung, J.H., Bell, A.C., Felsenfeld, G. (1997). Characterization of the chicken β-globin insulator, *Proc Natl Acad Sci USA*, **94**, 575–580.
Chung, J.H., Whiteley, M., Felsenfeld, G. (1993). A 5′ element of the chicken β-globin domain serves as an insulator in human erythroid cells and protects against position effect in Drosophila, *Cell*, **74**, 505–514.
Cleynen, I., Van de Ven, W.J. (2008). The HMGA proteins: a myriad of functions, *Int J Oncol*, **32**, 289–305.
Clissold, P.M., Arnstein, H.R., Chesterton, C.J. (1977). Quantitation of globin mRNA levels during erythroid development in the rabbit and discovery of a new β-related species in immature erythroblasts, *Cell*, **11**, 353–361.
Collis, P., Antoniou, M., Grosveld, F. (1990). Definition of the minimal requirements within the human β-globin gene and the dominant control region for high level expression, *EMBO J*, **9**, 233–240.
Corces, V.G. (1995). Chromatin insulators. Keeping enhancers under control, *Nature*, **376**, 462–463.
Craig, M.L., Russell, E.S. (1964). A developmental change in hemoglobins correlated with an embryonic red cell population in the mouse, *Dev Biol*, **10**, 191–201.
Cullen, B.R. (1998). Retroviruses as model systems for the study of nuclear RNA export pathways, *Virology*, **249**, 203–210.
Delenda, C., Audit, M., Danos, O. (2002). Biosafety issues in lentivector production. *Curr Top Microbiol Immunol*, **261**, 123–141.
Desprat, R., Bouhassira, E.E. (2009). Gene specificity of suppression of transgene-mediated insertional transcriptional activation by the chicken HS4 insulator, *PLoS One*, **4**, e5956.
Dykxhoorn, D.M., Schlehuber, L.D., London, I., *et al.* (2006). Determinants of specific RNA interference-mediated silencing of human β-globin alleles differing by a single nucleotide polymorphism, *Proc Natl Acad Sci USA*, **103**, 5953–5958.
Dzierzak, E.A., Papayannopoulou, T.R., Mulligan, C. (1988). Lineage-specific expression of a human β-globin gene in murine bone marrow transplant recipients reconstituted with retrovirus-transduced stem cells, *Nature*, **331**, 35–41.
Eaton, W.A., Hofrichter, J. (1990). Sickle cell hemoglobin polymerization, *Adv Protein Chem*, **40**, 63–279.

Eissenberg, J.C., Elgin, S.C. (1991). Boundary functions in the control of gene expression, *Trends Genet*, **7**, 335–340.

El-Beshlawy, A., Hamdy, M., El Ghamarawy, M. (2009). Fetal globin induction in β-thalassemia, *Hemoglobin*, **33**, 197–203.

Emery, D.W., Chen, H., Li, Q., Stamatoyannopoulos, G. (1998). Development of a condensed locus control region cassette and testing in retrovirus vectors for $^{A}\gamma$-globin, *Blood Cells Mol Dis*, **24**, 322–339.

Emery, D.W., Morrish, F., Li, Q., *et al.* (1999). Analysis of gamma-globin expression cassettes in retrovirus vectors, *Hum Gene Ther*, **10**, 877–888.

Emery, D.W., Yannaki, E., Tubb, J., *et al.* (2000). A chromatin insulator protects retrovirus vectors from chromosomal position effects, *Proc Natl Acad Sci USA*, **97**, 9150–9155.

Emery, D.W., Yannaki, E., Tubb, J., *et al.* (2002). Development of virus vectors for gene therapy of β chain hemoglobinopathies: flanking with a chromatin insulator reduces γ-globin gene silencing *in vivo*, *Blood*, **100**, 2012–2019.

Forrester, W.C., Novak, U., Gelinas, R., *et al.* (1989). Molecular analysis of the human β-globin locus activation region, *Proc Natl Acad Sci USA*, **86**, 5439–5443.

Forrester, W.C., Takegawa, S., Papayannopoulou, T., *et al.* (1987). Evidence for a locus activation region: the formation of developmentally stable hypersensitive sites in globin-expressing hybrids, *Nucleic Acids Res*, **15**, 10159–10177.

Forrester, W.C., Thompson, C., Elder, J.T., *et al.* (1986). A developmentally stable chromatin structure in the human β-globin gene cluster, *Proc Natl Acad Sci USA*, **83**, 1359–1363.

Fraser, P., Hurst, J., Collis, P., *et al.* (1990). DNaseI hypersensitive sites 1, 2 and 3 of the human β-globin dominant control region direct position-independent expression, *Nucleic Acids Res*, **18**, 3503–3508.

Fucharoen, S., Winichagoon, P. (2000). Clinical and hematologic aspects of hemoglobin E β-thalassemia, *Curr Opin Hematol*, **7**, 106–112.

Gallay, P., Swingler, S., Aiken, C., *et al.* (1995a). HIV-1 infection of nondividing cells: C-terminal tyrosine phosphorylation of the viral matrix protein is a key regulator, *Cell*, **80**, 379–388.

Gallay, P., Swingler, S., Song, J., *et al.* (1995b). HIV nuclear import is governed by the phosphotyrosine-mediated binding of matrix to the core domain of integrase, *Cell*, **83**, 569–576.

Garel, M.C., Arous, N., Calvin, M.C., *et al.* (1994). A recombinant bisphosphoglycerate mutase variant with acid phosphatase homology degrades 2,3-diphosphoglycerate, *Proc Natl Acad Sci USA*, **91**, 3593–3597.

Gelinas, R., Frazier, A., Harris, E. (1992). A normal level of beta-globin expression in erythroid cells after retroviral cells transfer. *Bone Marrow Transplant*, **9 Suppl 1**, 154–157.

Geyer, P.K., Corces, V.G. (1992). DNA position-specific repression of transcription by a Drosophila zinc finger protein, *Genes Dev*, **6**, 1865–1873.

Giardine, B., van Baal, S., Kaimakis, P., *et al.* (2007). HbVar database of human hemoglobin variants and thalassemia mutations: 2007 update, *Hum Mutat*, **28**, 206.

Goldberg, M.A., Husson, M.A., Bunn, H.F. (1977). Participation of hemoglobins A and F in polymerization of sickle hemoglobin, *J Biol Chem*, **252**, 3414–3421.

Grosveld, F., van Assendelft, G.B., Greaves, D.R., *et al.* (1987). Position-independent, high-level expression of the human β-globin gene in transgenic mice, *Cell*, **51**, 975–985.

Hacein-Bey-Abina, S., Garrigue, A., Wang, G.P., *et al.* (2008). Insertional oncogenesis in 4 patients after retrovirus-mediated gene therapy of SCID-X1, *J Clin Invest*, **118**, 3132–3142.

Hacein-Bey-Abina, S., von Kalle, C., Schmidt, M., *et al.* (2003). A serious adverse event after successful gene therapy for X-linked severe combined immunodeficiency, *N Engl J Med*, **348**, 255–256.

Hanawa, H., Hargrove, P.W., Kepes, S., *et al.* (2004). Extended β-globin locus control region elements promote consistent therapeutic expression of a γ-globin lentiviral vector in murine β-thalassemia, *Blood*, **104**, 2281–2290.

Hanawa, H., Yamamoto, M., Zhao, H., *et al.* (2009). Optimized lentiviral vector design improves titer and transgene expression of vectors containing the chicken β-globin locus HS4 insulator element, *Mol Ther*, **17**, 667–674.

Hargrove, P.W., Kepes, S., Hanawa, H., *et al.* (2008). Globin lentiviral vector insertions can perturb the expression of endogenous genes in β-thalassemic hematopoietic cells, *Mol Ther*, **16**, 525–533.

Herrick, J. (1910). Peculiar elongated and sickle-shaped red blood corpuscles in a case of severe anemia, *Arch Intern Med*, **6**, 517–521.

Holdridge, C., Dorsett, D. (1991). Repression of hsp70 heat shock gene transcription by the suppressor of hairy-wing protein of *Drosophila melanogaster*, *Mol Cell Biol*, **11**, 1894–1900.

Howe, S.J., Mansour, M.R., Schwarzwaelder, K., *et al.* (2008). Insertional mutagenesis combined with acquired somatic mutations causes leukemogenesis following gene therapy of SCID-X1 patients, *J Clin Invest*, **118**, 3143–3150.

Hunt, J.A., Ingram V.M. (1959). A terminal peptide sequence of human haemoglobin?, *Nature*, **184 Suppl 9**, 640–641.

Imren, S., Fabry, M.E., Westerman, K.A., *et al.* (2004). High-level β-globin expression and preferred intragenic integration after lentiviral transduction of human cord blood stem cells, *J Clin Invest*, **114**, 953–962.

Imren, S., Payen, E., Westerman, K.A., *et al.* (2002). Permanent and panerythroid correction of murine β thalassemia by multiple lentiviral integration in hematopoietic stem cells, *Proc Natl Acad Sci USA*, **99**, 14380–14385.

Ingram, V.M. (1957). Gene mutations in human haemoglobin: the chemical difference between normal and sickle cell haemoglobin, *Nature*, **180**, 326–328.

Isgro, A., Gaziev, J., Sodani, P., *et al.* (2010). Progress in hematopoietic stem cell transplantation as allogeneic cellular gene therapy in thalassemia, *Ann N Y Acad Sci*, **1202**, 149–154.

Jimenez, G., Griffiths, S.D., Ford, A.M., *et al.* (1992). Activation of the β-globin locus control region precedes commitment to the erythroid lineage, *Proc Natl Acad Sci USA*, **89**, 10618–10622.

Kafri, T., Blomer, U., Peterson, D.A., *et al.* (1997). Sustained expression of genes delivered directly into liver and muscle by lentiviral vectors, *Nat Genet*, **17**, 314–317.

Karlsson, S., Bodine, D.M., Perry, L., *et al.* (1988). Expression of the human β-globin gene following retroviral-mediated transfer into multipotential hematopoietic progenitors of mice, *Proc Natl Acad Sci USA*, **85**, 6062–6066.

Kean, L.S., Manci, E.A., Perry, J., *et al.* (2003). Chimerism and cure: hematologic and pathologic correction of murine sickle cell disease, *Blood*, **102**, 4582–4593.

Kellum, R., Schedl P. (1992). A group of scs elements function as domain boundaries in an enhancer-blocking assay, *Mol Cell Biol*, **12**, 2424–2431.

Kollias, G., Hurst, J., deBoer, E., *et al.* (1987). The human β-globin gene contains a downstream developmental specific enhancer, *Nucleic Acids Res*, **15**, 5739–5747.

Kollias, G., Wrighton, N., Hurst, J., *et al.* (1986). Regulated expression of human $^{A}\gamma$-, β-, and hybrid γβ-globin genes in transgenic mice: manipulation of the developmental expression patterns, *Cell*, **46**, 89–94.

Kumar, M., Keller, B., Makalou, N., *et al.* (2001). Systematic determination of the packaging limit of lentiviral vectors, *Hum Gene Ther*, **12**, 1893–1905.
Leboulch, P., Huang, G.M., Humphries, R.K. (1994). Mutagenesis of retroviral vectors transducing human β-globin gene and β-globin locus control region derivatives results in stable transmission of an active transcriptional structure, *EMBO J*, **13**, 3065–3076.
Levasseur, D.N., Ryan, T.M., Pawlik, K.M., *et al.* (2003). Correction of a mouse model of sickle cell disease: Lentiviral/antisickling β-globin gene transduction of unmobilized, purified hematopoietic stem cells, *Blood*, **102**, 4312–4319.
Levasseur, D.N., Ryan, T.M., Reilly, M.P., *et al.* (2004). A recombinant human hemoglobin with anti-sickling properties greater than fetal hemoglobin, *J Biol Chem*, **279**, 27518–27524.
Lewis, P., Hensel, M., Emerman, M. (1992). Human immunodeficiency virus infection of cells arrested in the cell cycle, *Embo J*, **11**, 3053–3058.
Li, Q., Emery, D.W., Fernandez, M., *et al.* (1999). Development of viral vectors for gene therapy of β-chain hemoglobinopathies: optimization of a γ-globin gene expression cassette, *Blood*, **93**, 2208–2216.
Li, Z., Dullmann, J., Schiedlmeier, B., Schmidt, M., *et al.* (2002). Murine leukemia induced by retroviral gene marking, *Science*, **296**, 497.
Lung, H.Y., Meeus, I.S., Weinberg, S., *et al.* (2000). *In vivo* silencing of the human γ-globin gene in murine erythroid cells following retroviral transduction, *Blood Cells Mol Dis*, **26**, 613–619.
Luznik, L., Jones, R.J., Fuchs, E.J. (2010). High-dose cyclophosphamide for graft-versus-host disease prevention, *Curr Opin Hematol*, **17**, 493–499.
Magram, J., Chada, K., Costantini, F. (1985). Developmental regulation of a cloned adult β-globin gene in transgenic mice, *Nature*, **315**, 338–340.
Marotta, C.A., Wilson, J.T., Forget, B.G., *et al.* (1977). Human β-globin messenger RNA. III. Nucleotide sequences derived from complementary DNA, *J Biol Chem*, **252**, 5040–5053.
Mason, V.R. (1922). Sickle cell anemia, *JAMA*, **79**, 1318–1320.
Mathias, L.A., Fisher, TC., Zeng, L., *et al.* (2000). Ineffective erythropoiesis in β-thalassemia major is due to apoptosis at the polychromatophilic normoblast stage, *Exp Hematol*, **28**, 1343–1353.
May, C., Rivella, S., Callegari, J., *et al.* (2000). Therapeutic haemoglobin synthesis in β-thalassaemic mice expressing lentivirus-encoded human β-globin, *Nature*, **406**, 82–86.
May, C., Rivella, S., Chadburn, A., *et al.* (2002). Successful treatment of murine β-thalassemia intermedia by transfer of the human β-globin gene, *Blood*, **99**, 1902–1908.
McCune, S.L., Reilly, M.P., Chomo, M.J., *et al.* (1994). Recombinant human hemoglobins designed for gene therapy of sickle cell disease, *Proc Natl Acad Sci USA*, **91**, 9852–9856.
Miyoshi, H., Smith, K.A., Mosier, D.E., *et al.* (1999). Transduction of human $CD34^+$ cells that mediate long-term engraftment of NOD/SCID mice by HIV vectors, *Science*, **283**, 682–686.
Modell, B., Darlison, M. (2008). Global epidemiology of haemoglobin disorders and derived service indicators, *Bull World Health Organ*, **86**, 480–487.
Modell, B., Khan, M., Darlison, M., *et al.* (2008). Improved survival of thalassaemia major in the UK and relation to T2* cardiovascular magnetic resonance, *J Cardiovasc Magn Reson*, **10**, 42.
Modlich, U., Schambach, A., Brugman, M.H., *et al.* (2008). Leukemia induction after a single retroviral vector insertion in Evi1 or Prdm16, *Leukemia*, **15**, 19–28.

Montini, E., Cesana, D., Schmidt, M., *et al.* (2006). Hematopoietic stem cell gene transfer in a tumor-prone mouse model uncovers low genotoxicity of lentiviral vector integration, *Nat Biotechnol*, **24**, 687–696.

Montini, E., Cesana, D., Schmidt, M., *et al.* (2009). The genotoxic potential of retroviral vectors is strongly modulated by vector design and integration site selection in a mouse model of HSC gene therapy, *J Clin Invest*, **119**, 964–975.

Moreau-Gaudry, F., Xia, P., Jiang, G., *et al.* (2001). High-level erythroid-specific gene expression in primary human and murine hematopoietic cells with self-inactivating lentiviral vectors, *Blood*, **98**, 2664–2672.

Mousa, S.A., Momen, A., Al Sayegh, F., *et al.* (2010). Management of painful vaso-occlusive crisis of sickle-cell anemia: consensus opinion, *Clin Appl Thromb Hemost*, **16**, 365–376.

Mulligan, R.C. (1993). The basic science of gene therapy, *Science*, **260**, 926–932.

Nagel, R.L., Fabry, M.E. (2001). The panoply of animal models for sickle cell anaemia, *Br J Haematol*, **112**, 19–25.

Nagel, R.L., Johnson, J., Bookchin, R.M., *et al.* (1980). β-Chain contact sites in the haemoglobin S polymer, *Nature*, **283**, 832–834.

Naldini, L., Blomer, U., Gage, F.H., *et al.* (1996a). Efficient transfer, integration, and sustained long-term expression of the transgene in adult rat brains injected with a lentiviral vector, *Proc Natl Acad Sci USA*, **93**, 11382–11388.

Naldini, L., Blomer, U., Gallay, P., *et al.* (1996b). *In vivo* gene delivery and stable transduction of nondividing cells by a lentiviral vector, *Science*, **272**, 263–267.

Neff, T., Shotkoski, F., Stamatoyannopoulos, G. (1997). Stem cell gene therapy, position effects and chromatin insulators, *Stem Cells*, **15 Suppl 1**, 265–271.

Novak, U., Harris, E.A., Forrester, W., *et al.* (1990). High-level β-globin expression after retroviral transfer of locus activation region-containing human β-globin gene derivatives into murine erythroleukemia cells, *Proc Natl Acad Sci USA*, **87**, 3386–3390.

Ott, M.G., Schmidt, M., Schwarzwaelder, K., *et al.* (2006). Correction of X-linked chronic granulomatous disease by gene therapy, augmented by insertional activation of MDS1-EVI1, PRDM16 or SETBP1, *Nat Med*, **12**, 401–409.

Pannell, D., Osborne, C.S., Yao, S., *et al.* (2000). Retrovirus vector silencing is *de novo* methylase independent and marked by a repressive histone code, *EMBO J*, **19**, 5884–5894.

Park, F., Ohashi, K., Chiu, W., *et al.* (2000). Efficient lentiviral transduction of liver requires cell cycling *in vivo*, *Nat Genet*, **24**, 49–52.

Paszty, C., Brion, C.M., Manci, E., *et al.* (1997). Transgenic knockout mice with exclusively human sickle hemoglobin and sickle cell disease, *Science*, **278**, 876–878.

Pauling, L., Itano, H.A., Singer, S.J., *et al.* (1949). Sickle cell anemia a molecular disease, *Science*, **110**, 543–548.

Pawliuk, R., Westerman, K.A., Fabry, M.E., *et al.* (2001). Correction of sickle cell disease in transgenic mouse models by gene therapy, *Science*, **294**, 2368–2371.

Pegelow, C.H., Adams, R., McKie, J.V. *et al.* (1995). Risk of recurrent stroke in patients with sickle cell disease treated with erythrocyte transfusions, *J Pediatr*, **126**, 896–899.

Perrine, S.P. (2008). Fetal globin stimulant therapies in the beta-hemoglobinopathies: principles and current potential, *Pediatr Ann*, **37**, 339–346.

Perrine, S.P., Castaneda, S.A., Chui, D.H., *et al.* (2010). Fetal globin gene inducers: novel agents and new potential, *Ann N Y Acad Sci*, **1202**, 158–164.

Persons, D.A., Allay, E.R., Sabatino, D.E., *et al.* (2001). Functional requirements for phenotypic correction of murine β-thalassemia: implications for human gene therapy, *Blood*, **97**, 3275–3282.

Persons, D.A., Allay, E.R., Sawai, N., *et al.* (2003). Successful treatment of murine β-thalassemia using *in vivo* selection of genetically modified, drug-resistant hematopoietic stem cells, *Blood*, **102**, 506–513.

Philipsen, S., Talbot, D., Fraser P, *et al.* (1990). The β-globin dominant control region: hypersensitive site 2, *EMBO J*, **9**, 2159–2167.

Pikaart, M.J., Recillas-Targa F., Felsenfeld, G. (1998). Loss of transcriptional activity of a transgene is accompanied by DNA methylation and histone deacetylation and is prevented by insulators, *Genes Dev*, **12**, 2852–2862.

Platt, O.S., Brambilla, D.J., Rosse, W.F., *et al.* (1994). Mortality in sickle cell disease. Life expectancy and risk factors for early death, *N Engl J Med*, **330**, 1639–1644.

Plavec, I., Papayannopoulou, T., Maury, C., *et al.* (1993). A human β-globin gene fused to the human β-globin locus control region is expressed at high levels in erythroid cells of mice engrafted with retrovirus-transduced hematopoietic stem cells, *Blood*, **81**, 1384–1392.

Poillon, W.N., Kim, B.C., Rodgers, G.P., *et al.* (1993). Sparing effect of hemoglobin F and hemoglobin A2 on the polymerization of hemoglobin S at physiologic ligand saturations, *Proc Natl Acad Sci USA*, **90**, 5039–5043.

Puthenveetil, G., Scholes, J., Carbonell, D., *et al.* (2004). Successful correction of the human β-thalassemia major phenotype using a lentiviral vector, *Blood*, **104**, 3445–3453.

Raftopoulos, H., Ward, M., Leboulch, P., *et al.* (1997). Long-term transfer and expression of the human β-globin gene in a mouse transplant model, *Blood*, **90**, 3414–3422.

Recillas-Targa, F., Pikaart, M.J., Burgess-Beusse, B., *et al.* (2002). Position-effect protection and enhancer blocking by the chicken β-globin insulator are separable activities, *Proc Natl Acad Sci USA*, **99**, 6883–6888.

Reddy, L.R., Reddy, K.S., Surrey, S., *et al.* (1997). Role of β87 Thr in the β6 Val acceptor site during deoxy Hb S polymerization, *Biochemistry*, **36**, 15992–15998.

Rees, D.C., Williams, T.N., Gladwin M.T. (2010). Sickle-cell disease, *Lancet*, **376**, 2018–2031.

Ren, S., Wong, B.Y., Li, J., *et al.* (1996). Production of genetically stable high-titer retroviral vectors that carry a human γ-globin gene under the control of the α-globin locus control region, *Blood*, **87**, 2518–2524.

Rivella, S., Callegari, J.A., May, C., *et al.* (2000). The cHS4 insulator increases the probability of retroviral expression at random chromosomal integration sites, *J Virol*, **74**, 4679–4687.

Rivella, S., May, C., Chadburn, A., *et al.* (2003). A novel murine model of Cooley anemia and its rescue by lentiviral-mediated human β-globin gene transfer, *Blood*, **101**, 2932–2939.

Rivella, S., Sadelain, M. (1998). Genetic treatment of severe hemoglobinopathies: the combat against transgene variegation and transgene silencing, *Semin Hematol*, **35**, 112–125.

Rixon, M.W., Harris E.A., Gelinas, R.E. (1990). Expression of the human γ-globin gene after retroviral transfer to transformed erythroid cells, *Biochemistry*, **29**, 4393–4400.

Ronen, K., Negre, O., Roth, S., *et al.* (2011). Distribution of lentiviral vector integration sites in mice following therapeutic gene transfer to treat β-thalassemia, *Mol Ther*, **19**, 1273–1286.

Roseff, S.D. (2009). Sickle cell disease: a review, *Immunohematology*, **25**, 67–74.

Rouyer Fessard, P., Leroy Viard, K., Domenget, C., *et al.* (1990). Mouse β thalassemia, a model for the membrane defects of erythrocytes in the human disease, *J Biol Chem*, **265**, 20247–20251.

Ryan, T.M., Behringer, R.R., Martin, N.C., *et al.* (1989). A single erythroid-specific DNase I super-hypersensitive site activates high levels of human β-globin gene expression in transgenic mice, *Genes Dev*, **3**, 314–323.

Ryan, T.M., Ciavatta, D.J., Townes, T.M. (1997). Knockout-transgenic mouse model of sickle cell disease, *Science*, **278**, 873–876.

Sabatino, D.E., Seidel, N.E., Aviles-Mendoza G.J., *et al.* (2000a). Long-term expression of γ-globin mRNA in mouse erythrocytes from retrovirus vectors containing the human γ-globin gene fused to the ankyrin-1 promoter, *Proc Natl Acad Sci USA*, **97**, 13294–13299.

Sabatino, D.E., Wong, C., Cline, A.P., *et al.* (2000b). A minimal ankyrin promoter linked to a human γ-globin gene demonstrates erythroid specific copy number dependent expression with minimal position or enhancer dependence in transgenic mice, *J Biol Chem*, **275**, 28549–28554.

Sadelain, M., Wang, C.H., Antoniou, M., *et al.* (1995). Generation of a high-titer retroviral vector capable of expressing high levels of the human β-globin gene, *Proc Natl Acad Sci USA*, **92**, 6728–6732.

Samakoglu, S., Lisowski, L., Budak-Alpdogan, T., *et al.* (2006). A genetic strategy to treat sickle cell anemia by coregulating globin transgene expression and RNA interference, *Nat Biotechnol*, **24**, 89–94.

Siciliano, A., Malpeli, G., Platt, O.S., *et al.* (2011). Abnormal modulation of cell protective systems in response to ischemic/reperfusion injury is important in the development of mouse sickle cell hepatopathy, *Haematologica*, **96**, 24–32.

Sinn, P.L., Sauter, S.L., McCray, P.B., Jr. (2005). Gene therapy progress and prospects: development of improved lentiviral and retroviral vectors — design, biosafety, and production, *Gene Ther*, **12**, 1089–1098.

Stein, S., Ott, M.G., Schultze-Strasser, S., *et al.* (2010). Genomic instability and myelodysplasia with monosomy 7 consequent to EVI1 activation after gene therapy for chronic granulomatous disease, *Nat Med*, **16**, 198–204.

Styles, L.A., Vichinsky, E. (1994). Effects of a long-term transfusion regimen on sickle cell-related illnesses, *J Pediatr*, **125**, 909–911.

Sun, F.L., Elgin, S.C. (1999). Putting boundaries on silence, *Cell*, **99**, 459–462.

Taboit-Dameron, F., Malassagne, B., Viglietta, C., *et al.* (1999). Association of the 5′HS4 sequence of the chicken β-globin locus control region with human EF1 α gene promoter induces ubiquitous and high expression of human CD55 and CD59 cDNAs in transgenic rabbits, *Transgenic Res*, **8**, 223–235.

Takekoshi, K.J., Oh, Y.H., Westerman, K.W., *et al.* (1995). Retroviral transfer of a human β-globin/δ-globin hybrid gene linked to β locus control region hypersensitive site 2 aimed at the gene therapy of sickle cell disease, *Proc Natl Acad Sci USA*, **92**, 3014–3018.

Talbot, D., Collis, P., Antoniou, M., *et al.* (1989). A dominant control region from the human β-globin locus conferring integration site-independent gene expression, *Nature*, **338**, 352–355.

Talbot, D., Philipsen, S., Fraser, P., *et al.* (1990). Detailed analysis of the site 3 region of the human β-globin dominant control region, *EMBO J*, **9**, 2169 –177.

Taylor, L.E., Stotts, N.A., Humphreys, J., *et al.* (2010). A review of the literature on the multiple dimensions of chronic pain in adults with sickle cell disease, *J Pain Symptom Manage*, **40**, 416–435.

Telfer, P.T., Warburton, F., Christou, S., *et al.* (2009). Improved survival in thalassemia major patients on switching from desferrioxamine to combined chelation therapy with desferrioxamine and deferiprone, *Haematologica*, **94**, 1777–1778.

Townes, T.M., Lingrel, J.B., Chen, H.Y., *et al.* (1985). Erythroid-specific expression of human β-globin genes in transgenic mice, *EMBO J*, **4**, 1715–1723.

Trudel, M., Costantini, F. (1987). A 3′ enhancer contributes to the stage-specific expression of the human β-globin gene, *Genes Dev*, **1**, 954–961.

Trudel, M., De Paepe, M.E., Chretien, N., *et al.* (1994). Sickle cell disease of transgenic SAD mice, *Blood*, **84**, 3189–3197.

Trudel, M., Saadane, N., Garel, M.C., *et al.* (1991). Towards a transgenic mouse model of sickle cell disease: hemoglobin SAD, *Embo J*, **10**, 3157–3165.

Tuan, D., Solomon, W., Li, Q., *et al.* (1985). The "β-like-globin" gene domain in human erythroid cells, *Proc Natl Acad Sci USA*, **82**, 6384–6388.

Udvardy, A., Maine, E., Schedl, P. (1985). The 87A7 chromomere. Identification of novel chromatin structures flanking the heat shock locus that may define the boundaries of higher order domains, *J Mol Biol*, **185**, 341–358.

Vigi, V., Volpato, S., Gaburro, D., *et al.* (1969). The correlation between red-cell survival and excess of α-globin synthesis in β-thalassemia, *Br J Haematol*, **16**, 25–30.

Wahl, S., Quirolo, K.C. (2009). Current issues in blood transfusion for sickle cell disease, *Curr Opin Pediatr*, **21**, 15–21.

Wang, G.P., Berry, C.C., Malani, N., *et al.* (2010). Dynamics of gene-modified progenitor cells analyzed by tracking retroviral integration sites in a human SCID-X1 gene therapy trial, *Blood*, **115**, 4356–4366.

Weatherall, D.J. (1974). The genetic control of protein synthesis: The haemoglobin model, *J Clin Pathol Suppl* (*R Coll Pathol*), **8**, 1–11.

West, AG., Gaszner, M., Felsenfeld, G. (2002). Insulators: many functions, many mechanisms, *Genes Dev*, **16**, 271–288.

Westerman, K.A., Ao, Z., Cohen, E.A., *et al.* (2007). Design of a trans protease lentiviral packaging system that produces high titer virus, *Retrovirology*, **4**, 96.

Winichagoon, P., Fucharoen, S., Weatherall D., *et al.* (1985). Concomitant inheritance of α-thalassemia in β^0- thalassemia/Hb e disease, *Am J Hematol*, **20**, 217–222.

Winichagoon, P., Thonglairoam, V., Fucharoen, S., *et al.* (1993). Severity differences in β-thalassaemia/haemoglobin E syndromes: implication of genetic factors, *Br J Haematol*, **83**, 633–639.

Wu, C.J., Gladwin, M., Tisdale, J., *et al.* (2007). Mixed haematopoietic chimerism for sickle cell disease prevents intravascular haemolysis, *Br J Haematol*, **139**, 504–507.

Wu, C.J., Krishnamurti, L. Kutok, J.L., *et al.* (2005). Evidence for ineffective erythropoiesis in severe sickle cell disease, *Blood*, **106**, 3639–3645.

Wu, L.C., Sun, C.W., Ryan, T.M., *et al.* (2006). Correction of sickle cell disease by homologous recombination in embryonic stem cells, *Blood*, **108**, 1183–1188.

Yang, B., Kirby, S., Lewis, J., *et al.* (1995). A mouse model for β^0-thalassemia, *Proc Natl Acad Sci USA*, **92**, 11608–11612.

Zychlinski, D., Schambach, A., Modlich, U., *et al.* (2008). Physiological promoters reduce the genotoxic risk of integrating gene vectors, *Mol Ther*, **16**, 718–725.

24

GENE THERAPY FOR HEMOPHILIA A AND B

***Nisha Nair*[a]*, Marinee Chuah*[a,b,c] *and Thierry VandenDriessche*[a,b,d]**

24.1 Hemophilia A and B

Hemophilia A and B are congenital bleeding disorders caused by a deficiency of functional clotting factor, FVIII and FIX, respectively. Hemophilia results in an X-linked bleeding diathesis (Fig. 24.1), caused by a mutation in the corresponding clotting factor genes. It affects an estimated 400,000 individuals worldwide (according to the World Federation of Hemophilia). Hemophilia A affects nearly 80–85% of the patients, whereas the remaining 15% are afflicted by hemophilia B. As FVIII and FIX are essential cofactors in the blood coagulation cascade, patients suffer from recurrent bleeding and chronic damage to soft tissues, joints and muscles. This progresses towards chronic synovitis, crippling arthropathy and physical disability. The bleeding could also be fatal in the case of intracranial hemorrhage.

Current therapy for hemophilia is based on protein substitution therapy (PST), where plasma-derived or recombinant clotting factor concentrates are administered parenterally. Although PST considerably improved the patient's quality of life and life expectancy, it has its own limitations. The treatment is non-curative, and therefore the patient is always at risk of bleeding episodes and chronic joint damage. Another constraint is the short half-life of the clotting factors, which demands repeated infusion of relatively large doses. Consequently some patients can produce neutralizing antibodies (inhibitors) against the injected FVIII or FIX

[a] Department of Gene Therapy & Regenerative Medicine, Free University of Brussels (VUB), Building D — 3rd floor, Laarbeeklaan 103, B-1090 Brussels
[b] Center for Molecular & Vascular Biology, University of Leuven, Belgium
[c] Email: marinee.chuah@vub.ac.be
[d] Email: thierry.vandendriessche@vub.ac.be

protein. These inhibitors can render further therapy ineffective. These drawbacks of conventional PST make gene therapy an interesting alternative for hemophilia treatment (Mátrai *et al.*, 2010a).

24.2 Hemophilia as a Target for Gene Therapy

As a monogenetic hereditary disorder, hemophilia has been considered as a particularly significant target for gene therapy. Most aspects of the disease, including its genetics (Fig. 24.1), are well understood.

The introduction of a functional FVIII or FIX gene copy into the target cells via gene therapy may provide a cure and eliminate the need for repeated clotting factor infusions. Also there is no need to normalize circulating clotting factor levels to obtain a therapeutic effect, as a slight increase in plasma clotting factor levels (above 1% threshold) suffices to reduce the risk of mortality and morbidity. Indeed, patients with moderate hemophilia have 1–5% of the normal clotting factor in their blood and tend to have bleeding episodes predominantly only after an injury. Patients with mild hemophilia have 5–50% of normal clotting factor levels and they bleed only in case of severe injury. Consequently a modest increase in clotting factor levels by gene therapy can profoundly improve the clinical symptoms. Moreover, therapeutic efficacy of gene therapy can relatively easily be

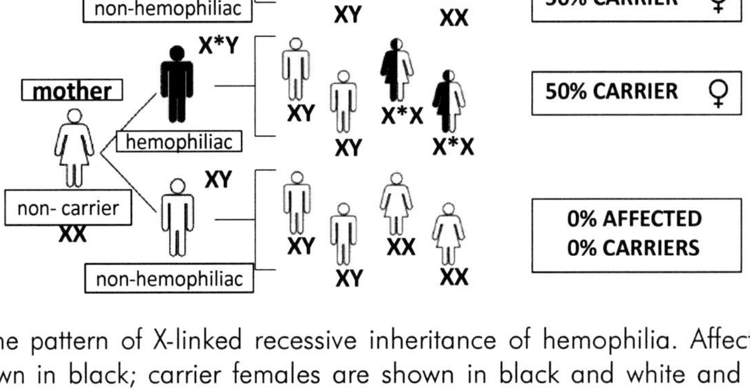

FIGURE 24.1 ■ The pattern of X-linked recessive inheritance of hemophilia. Affected males or females are shown in black; carrier females are shown in black and white and non-affected males or females are shown in white.

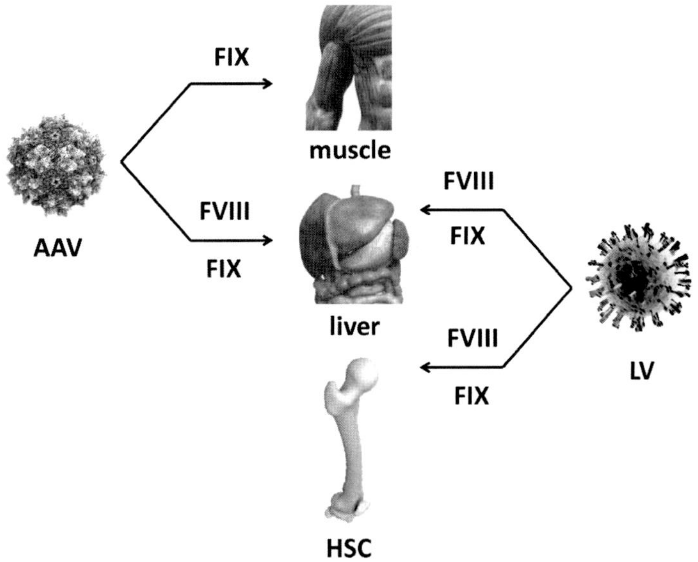

FIGURE 24.2 ■ Gene therapy for hemophilia A and B using adeno-associated virus (AAV) and lentivirus-based vectors (LV). AAV has been used for muscle-directed gene delivery of FIX or liver-directed FVIII and FIX gene delivery. FVIII delivery into the circulation from the muscle is inefficient. LVs have been explored for liver-directed and hematopoietic stem cell (HSC)-based gene delivery of FVIII and FIX. AAV vectors can be derived from different serotypes that enhance transduction efficiencies in the respective target tissues (i.e., AAV8 for liver, AAV1 or AAV6 for muscle). Though AAV genomes are composed of single-stranded DNA they can be converted into transcription-competent self-complementary DNA (scAAV) to enhance transduction. All these approaches resulted in prolonged therapeutic FVIII or FIX levels in preclinical mouse models and some were also validated in hemophilic dog (AAV–FIX, AAV–FVIII, LV–FIX) or primate models (AAV–FIX). Only AAV-FIX gene therapy has so far resulted in prolonged therapeutic FIX expression in severe hemophilia B patients.

ascertained based on robust clinical end points, like circulating clotting factor levels and bleeding frequency. Also there are several well-validated small and large hemophilic animal models available for preclinical testing.

Several target cells and vectors that have been used for hemophilia gene therapy. The most promising approaches are summarized in Fig. 24.2 and discussed in more detail below.

24.3 Target Cells

24.3.1 Hepatocytes

The liver is the main physiological site of FVIII and FIX synthesis and hence hepatocytes are well-suited target cells for hemophilia gene therapy. From this site, FVIII and FIX protein can easily enter into the circulation. The hepatic niche may favor the induction of immune tolerance towards the transgene product. The induction of immune tolerance may also, at least in part, depend on the induction of regulatory T-cells. This antigen-specific immune tolerance to the transgene product could be a key factor in establishing long-term therapeutic clotting factor levels (Mingozzi *et al.*, 2003). Another advantage of targeting liver cells is that all post-translational modifications required to faithfully replicate the normal FVIII or FIX processing pathways are implemented efficiently and precisely. A point of concern regarding liver-directed gene transfer would be to avoid unintentional transduction of antigen-presenting cells (APCs) present in the liver microenvironment, since ectopic FVIII or FIX expression in APCs increases the risk of inducing an immune response against the transduced cells and the resultant transgene product.

24.3.2 Muscle Cells

Muscle does not normally express FVIII or FIX. It is an easily accessible site with a robust secretory capacity. Muscle cells are also equipped with the cellular machineries for post-translational modifications to generate functional transgene product. Nevertheless, proteolytic cleavage and glycosylation are slightly impaired compared with hepatocytes. Unlike liver-directed gene transfer, muscle-directed approaches appear to have a higher risk of inducing immune response against the transgene product.

24.3.3 Endothelial Cells

It has been shown that endothelial cells naturally express FVIII. One of the sources of endothelial cells is the liver sinusoidal endothelial cells (LSECs). These cells are considered to be an important endogenous source of FVIII. Alternative sources of endothelial cells include blood outgrowth endothelial cells (BOECs). These are circulating endothelial cells which could easily be isolated from peripheral blood. Following genetic engineering, they can secrete functionally active FVIII and FIX proteins (Lin *et al.*, 2002; Matsui *et al.*, 2007). These genetically modified BOECs could engraft *in vivo* and have the potential to express the transgene long term.

24.3.4 Hematopoietic Stem Cells

Hematopoietic stem cells (HSCs) are an attractive source of target cells for hemophilia gene therapy. HSCs can self-renew and differentiate into all blood lineages during hematopoiesis. Another advantage is their accessibility for genetic modification and ease of transplant. Furthermore there is a possibility of inducing immune hyporesponsiveness to the transgene product. However, the use of HSCs for gene transfer is limited by the toxic effects of the conditioning regimens needed to achieve efficient engraftment of HSCs. HSC-derived erythrocytes, megakaryocytes and their platelet progeny could then serve as a delivery platform to secrete the clotting factors in the circulation following HSC-targeted gene transfer.

24.4 Gene Delivery Systems

Gene therapy requires an efficient and innocuous gene transfer system for clinical applications. The gene delivery system should give rise to long-term therapeutic FVIII or FIX expression without any toxic side effects. Several such vectors have been developed for gene therapy of hemophilia.

24.4.1 Non-Viral Vectors

Non-viral vectors typically rely on a plasmid-based gene delivery system, where only the naked DNA is delivered without any viral protein. Consequently the non-viral approach may be less immunogenic and potentially safer than viral vectors. The DNA is delivered to the target cells via various physical and chemical methods. The non-viral gene transfer method is simple, but the efficiency is generally low compared with most viral-vector-mediated gene transfer approaches. Moreover, non-viral transfection is typically short lived and gives transient expression of the transgene, unless selection is applied on *ex vivo*-transfected cells. A phase I clinical trial for hemophilia A has been conducted with stably transfected autologous fibroblasts electroporated with FVIII-expressing plasmids. Transfected cells were selected, expanded and subsequently implanted into the patient. This resulted in a modest and transient improvement in FVIII activity, without any adverse events (Roth *et al.*, 2001).

Nevertheless efficient *in vivo* gene delivery of non-viral vectors remains a bottleneck. Typically, for hepatic gene delivery, plasmids are administered by hydrodynamic injection. In this case, a hydrodynamic pressure is generated by rapid injection of a large volume of DNA solution into the circulation, in order to deliver the gene of interest in the liver. Efforts are being made to adapt hydrodynamic injection towards a clinically relevant modality by reducing the volume of injection along with maintaining localized hydrodynamic pressure for gene transfer. Alternative approaches based on targetable nanoparticles are being explored to achieve target-specific delivery of FVIII or FIX into LSECs and hepatocytes.

Expression could be prolonged or enhanced by incorporating episomal persistence promoting elements (matrix attachment regions, (MARs)) or by removing bacterial backbone sequences which interfere with long term expression (i.e., minicircle DNA). Finally, to increase the stability of FVIII or FIX expression after non-viral transfection, transposons could be used that result in stable genomic transgene integration. We and others have shown that transposons could be used to obtain stable clotting factor expression following *ex vivo* or *in vivo* gene therapy (Yant *et al.*, 2000; Mátés *et al.*, 2009; Chuah *et al.*, 2009, VandenDriessche *et al.*, 2009).

24.4.2 Retroviral and Lentiviral Vectors

Retroviral vectors are derived from Moloney murine leukemia virus (MoMLV). They can transduce a wide variety of target cells and integrate into the host genome, provided that the target cells are actively dividing. The first proof-of-concept that hemophilia A mice could be cured by gene therapy had been established with γ-retroviral vectors after hepatic gene delivery in neonatal mice (VandenDriessche *et al.*, 1999). Similarly, efficient γ-retroviral transduction could be obtained in the liver of neonatal hemophilia A and B dogs. A phase I clinical trial was conducted in adults with severe hemophilia A, involving systemic administration of γ-retroviral vectors encoding FVIII. The procedure was well tolerated by the patients, but only a minimal increase in FVIII activity could be detected. Another phase I clinical trial for hemophilia B was conducted with autologous skin fibroblasts, transduced *ex vivo* with FIX-encoding γ-retroviral vectors. The treated subjects had slightly increased FIX plasma levels with no significant side effects. Other approaches based on retroviral vectors rely on the use of other target cells, such as HSC or mesenchymal stem cells, resulting in circulating clotting factors following transplantation (reviewed in Mátrai *et al.*, 2010a, 2010b).

To overcome the need for active cell division to achieve efficient transduction, lentiviral vectors were developed. These vectors can transduce quiescent non-cycling hepatocytes *in vivo*, leading to relatively efficient gene transfer into the adult liver after local or even after systemic gene delivery. However, lentiviral vectors can also efficiently transduce APCs (VandenDriessche *et al.*, 2002). Inadvertent transgene expression in APCs can increase the risk of immune responses against the clotting factors, inhibiting long-term expression. In particular, antibodies to FIX could be detected when ubiquitously expressed promoters were used, whereas the utilization of a hepatocyte-specific promoter reduced this immunological risk, consistent with long-term FIX expression (Follenzi *et al.*, 2004). By including an additional layer of post-transcriptional control, mediated by the use of endogenous microRNA, non-specific expression in APCs could be further reduced, resulting in prolonged transgene expression (Brown *et al.*, 2007).

Alternatively, lentiviral vectors could be used for *ex vivo* gene transfer into HSCs. Long-term therapeutic FVIII expression could be demonstrated in hemophilia A mice after transplantation of transduced HSCs encoding a hybrid human/porcine FVIII transgene (Spencer *et al.*, 2011). Remarkably, this hybrid transgene could significantly increase FVIII expression levels compared with the (human) FVIII transgene. Using lentiviral transduction of HSCs, FVIII and FIX expression can be specifically directed to platelets, resulting in phenotypic correction of the bleeding diathesis in hemophilia A or B mice (Shi *et al.*, 2007; Montgomery and Shi, 2012). Also, phenotypic correction of hemophilia A mice could be achieved in the presence of high-titer inhibitory antibodies after lentiviral platelet-directed (human) FVIII gene therapy. These strategies could be especially beneficial to treat hemophilia patients with inhibitors.

The most important safety concern related to all integrating vectors, such as γ-retroviral and lentiviral vectors, is that of insertional mutagenesis and oncogene activation (or tumor-suppressor gene inactivation) resulting from genomic integration (Mátrai *et al.*, 2010b). The vector design or the presence of transcriptionally active long terminal repeat (LTR) can influence this genotoxic risk. Removal of approximately 400 bp of the LTR region to abolish its transcriptional activity (i.e., self-inactivating vector or SIN design) coupled with the use of a moderately active promoter in an internal position may lower the risk of insertional oncogenesis. By contrast, the potential oncogenic risk of lentiviral vectors may be reduced compared with γ-retroviral vectors because they integrate more distantly from transcriptional start sites. An alternative approach to minimize the risk of insertional mutagenesis is the use of integration-defective lentiviral vectors (IDLVs), which harbor an inactivating mutation in the integrase. Liver-directed gene therapy using IDLVs resulted in long-term Factor IX (FIX) transgene expression and sustained induction of immune tolerance to the transgene protein (Mátrai *et al.*, 2011).

24.4.3 Adenoviral Vectors

Adenoviral vectors have been intensively studied for hemophilia gene therapy. The latest version high-capacity adenoviral vectors are devoid of any viral genes resulting in reduced adaptive immune responses and improved stability of transgene expression. There is an added advantage of the minimized risk of oncogene activation, as they are essentially non-integrating. One disadvantage of adenoviral vectors is their robust interaction with APCs. This interaction triggers innate immune responses which is a safety concern. High-capacity adenoviral vectors encoding Factor VIII (FVIII) or FIX from a liver-specific promoter gave rise to robust supra-physiological clotting factor expression levels with limited toxicity in hemophilic mice (Ehrhardt and Kay, 2002; Chuah *et al.*, 2003; Brown *et al.*, 2004). These results were later confirmed in large-animal models, in particular hemophilic dogs.

A phase I clinical trial was conducted, involving systemic administration of high-capacity adenoviral vectors encoding FVIII from a liver-specific promoter to a severe hemophilia A patient. Although very low FVIII levels (approximately 1%) may have been obtained, the trial was stopped upon detection of a transient inflammatory response with hematological and liver abnormalities. Therefore, the interaction of (high-capacity) adenoviral vectors with the innate immune system needs to be minimized to enable further clinical use. For instance, localized hepatic delivery results in efficient, long-term transgene expression in non-human primates, while minimizing such side effects.

24.4.4 Adeno-Associated Viral (AAV) Vectors

AAV is a naturally occurring replication-defective non-pathogenic virus with a single-stranded DNA genome. AAV vectors have a favorable safety profile and are capable of achieving persistent transgene expression. Long-term expression is predominantly mediated by episomally retained AAV genomes. More than 90% of the stably transduced vector genomes were extrachromosomal, mostly organized as high-molecular-weight concatemers. Therefore, the risk of insertional oncogenesis is minimal, especially in the context of hemophilia gene therapy where no selective expansion of transduced cells is expected to occur. The major limitation of these AAV vectors is the limited packaging capacity of the vector particles (i.e., approximately 4.7 kb), constraining the size of the transgene expression cassette to obtain functional vectors. Several immunologically distinct AAV serotypes have been isolated from human and non-human primates, although most vectors for hemophilia gene therapy were initially derived from the most prevalent AAV serotype 2. The first clinical success of AAV-based gene therapy for congenital blindness underscores the potential of this gene transfer technology.

The muscle was one of the first targets for AAV-mediated gene therapy for hemophilia B (High *et al.*, 2011). On the basis of encouraging preclinical studies in hemophilic mice and dogs, yielding sustained FIX expression and phenotypic correction, severe hemophilia B patients received injections at multiple intramuscular sites with AAV vectors encoding FIX from a ubiquitous promoter. FIX expression could be detected at the site of injection in all patients, even ten years post-treatment. However, systemic FIX levels were mostly below the therapeutic range. No antibodies to the FIX transgene product were observed. Hence, this study clearly demonstrates that long-term muscle-directed gene transfer is feasible if the vector dose per site is limited and patients are carefully selected, enrolling only those with a missense mutation. However, systemic expression levels were low, as the use of higher vector doses is limited owing to the prohibitively high number of injections required. To overcome the limitations of multiple intramuscular injections, an intravascular delivery system of AAV vectors encoding the FIX transgene to skeletal muscle was developed (Arruda *et al.*, 2010). This procedure was performed under transient immunosuppression and resulted in widespread transduction of muscle in hemophilia B dogs and sustained dose-dependent therapeutic levels of canine FIX transgene up to ten-fold higher than those obtained by intramuscular delivery. Correction of bleeding time correlated clinically with a significant reduction of spontaneous bleeding episodes. None of the dogs receiving the AAV vector under transient immunosuppression developed inhibitory antibodies to canine FIX, whereas a transient inhibitor was detected following vector delivery with no immunosuppression. Collectively, these results demonstrate the feasibility of this approach for the treatment of hemophilia B and highlight the importance of immunosuppression to prevent immune response to the FIX transgene product.

AAV-mediated hepatic gene transfer is an attractive alternative for gene therapy of hemophilia. Preclinical studies with the AAV vectors in murine and canine models of hemophilia have demonstrated persistent therapeutic expression, leading to partial or complete correction of the bleeding phenotype. Particularly, hepatic transduction conveniently induces immune tolerance to FIX that required induction of regulatory T-cells (Tregs). Long-term correction of the hemophilia phenotype without inhibitor development was achieved in inhibitor-prone null mutation hemophilia B dogs treated with liver-directed AAV2–FIX gene therapy (Mount *et al.*, 2002). In order to further reduce the vector dose, more potent FIX expression cassettes have been developed. This could be accomplished by using stronger promoter/enhancer elements, codon-optimized FIX or self-complementary, double-stranded AAV vectors (scAAV) that overcome one of the limiting steps in AAV transduction (i.e., single-stranded to double-stranded AAV conversion). Alternative AAV serotypes could be used (e.g., AAV8) that result in increased transduction into hepatocytes and improve intra-nuclear vector import (Gao *et al.*, 2002). Furthermore, recent studies indicate that mutations of the surface-exposed tyrosine residues allow the vector particles to evade phosphorylation and subsequent ubiquitination and, thus, prevent proteasome-mediated degradation, which resulted in a ten-fold increase in hepatic expression of FIX in mice (Zhong *et al.*, 2008). As far as FVIII is concerned, the packaging constraints of AAV initially hampered the production of AAV vectors for hemophilia A gene therapy, considering the large size of the FVIII transgene. However, this limitation could eventually be overcome by using small regulatory elements to drive expression of a B-domain-deleted form of FVIII (Jiang *et al.*, 2006). AAV-vector-mediated gene transfer resulted in sustained FVIII expression in mice and dogs, though high vector doses were required.

These liver-directed preclinical studies paved the way toward the use of AAV vectors for clinical gene therapy in patients suffering from severe hemophilia B. Hepatic delivery of AAV-FIX vectors resulted in therapeutic FIX levels (maximum 12% of normal levels) in subjects receiving AAV-FIX by hepatic artery catheterization (Kay *et al.*, 2000). However, the transduced hepatocytes were able to present AAV capsid-derived antigens in association with major histocompatibility complex (MHC) class I- to T-cells (Manno *et al.*, 2006). Although antigen presentation was modest, it was sufficient to flag the transduced hepatocytes for T-cell-mediated destruction.

An important milestone has now been reached in a recent gene therapy clinical trial for hemophilia B (Nathwan *et al.*, 2011; VandenDriessche and Chuah, 2012). In this trial, patients suffering from severe hemophilia B (<1% FIX) received injections by peripheral vein administration of AAV8 encoding a codon-optimized FIX. This AAV8 serotype can efficiently transduce hepatocytes, does not interact as efficiently with APCs as AAV2 and has limited cross-reactivity with pre-existing anti-AAV2 antibodies. Subjects received low, intermediate or high scAAV8–FIX vector doses, with two participants in each cohort. All subjects expressed FIX above the 1% threshold for several months after gene therapy. In particular, sustained FIX levels varied between 2% and 11% of normal levels in all the treated subjects. Though this is the first study that yields sustained FIX expression levels after gene therapy, immune-mediated clearance of AAV-transduced hepatocytes remains a concern, warranting the use of transient immune suppression.

24.5 Perspectives

In recent years substantial progress has been made in the field of gene therapy. Various preclinical studies using different vectors and target cells demonstrate unequivocally that gene therapy can result in a sustained therapeutic effect in animal models of hemophilia A or B. Lentiviral and AAV vectors are the most promising vectors to date for hemophilia gene therapy, based on the levels and duration of expression and the immune consideration. Hepatocytes, muscle cells or HSCs are particularly attractive targets.

Convincing evidence continues to emerge from clinical trials that gene therapy is effective in patients suffering from hemophilia B, resulting in sustained therapeutic FIX levels. However, there are still a number of hurdles that need to be overcome. Although sustained therapeutic FIX levels were obtained in this trial, these levels are not sufficient to achieve a *bona fide* cure of hemophilia and to prevent bleeding in the face of trauma or injury. Consequently, improved vector designs are warranted to achieve full hemostatic correction in patients with severe hemophilia B. Moreover, the proportion of functional infectious vector particles versus empty particles could be further improved. Last but not least, the AAV capsids themselves could be engineered further to minimize the risk of inducing dose-limiting T-cell responses and liver toxicity. It is tempting to speculate that the same approach could ultimately also be used for gene therapy of the more common hemophilia A form but more efficient vector designs are needed to help accomplish this goal.

Acknowledgments

Some of the work presented in this review was funded with grants from EU FP6 (INTHER), EUFP7 (PERSIST), Bayer Hemophilia Special Project Awards, FWO, VUB GOA (EPIGEN) and AFM. We thank our team members and collaborators for their valuable contributions. Thierry VandenDriessche and Marinee Chuah are joint corresponding authors.

References

Arruda, V.R., Stedman, H.H., Haurigot, V., *et al.* (2010). Peripheral transvenular delivery of adeno-associated viral vectors to skeletal muscle as a novel therapy for hemophilia B, *Blood*, **115**, 4678–4688.

Brown, B.D., Shi, C.X., Powell, S., *et al.* (2004). Helper-dependent adenoviral vectors mediate therapeutic factor VIII expression for several months with minimal accompanying toxicity in a canine model of severe hemophilia A, *Blood*, **103**, 804–810.

Brown, B.D., Cantore, A., Annoni, A., *et al.* (2007). A microRNA-regulated lentiviral vector mediates stable correction of hemophilia B mice, *Blood*, **110**, 4144–4152.

Chuah, M.K., Schiedner, G., Thorrez, L., *et al.* (2003). Therapeutic factor VIII levels and negligible toxicity in mouse and dog models of hemophilia A following gene therapy with high-capacity adenoviral vectors, *Blood*, **101**, 1734–1743.

Ehrhardt, A., Kay, M.A. (2002). A new adenoviral helper-dependent vector results in long-term therapeutic levels of human coagulation factor IX at low doses *in vivo*, *Blood*, **99**, 3923–3930.

Follenzi, A., Battaglia, M., Lombardo, A., *et al.* (2004). Targeting lentiviral vector expression to hepatocytes limits transgene-specific immune response and establishes long-term expression of human antihemophilic factor IX in mice, *Blood*, **103**, 3700–3709.

Gao, G.P., Alvira, M.R., Wang, L., *et al.* (2002). Novel adeno-associated viruses from rhesus monkeys as vectors for human gene therapy, *Proc Natl Acad Sci USA*, **99**, 11854–11859.

High, K.A. (2011). Gene therapy for haemophilia: a long and winding road, *J Thromb Haemost*, **9 Suppl 1**, 2–11.

Jiang, H., Lillicrap, D., Patarroyo-White, S., *et al.* (2006). Multiyear therapeutic benefit of AAV serotypes 2, 6, and 8 delivering factor VIII to hemophilia A mice and dogs, *Blood*, **108**, 107–115.

Kay, M.A., Manno, C.S., Ragni, M.V., *et al.* (2000). Evidence for gene transfer and expression of factor IX in haemophilia B patients treated with an AAV vector, *Nat Genet*, **24**, 257– 261.

Lin, Y., Chang, L., Solovey, A., *et al.* (2002). Use of blood outgrowth endothelial cells for gene therapy for hemophilia A, *Blood*, **99**, 457–462.

Manno, C.S., Pierce, G.F., Arruda, V.R., *et al.* (2006). Successful transduction of liver in hemophilia by AAV-Factor IX and limitations imposed by the host immune response, *Nat Med*, **12**, 342–347.

Mátés, L., Chuah, M.K., Belay, E., *et al.* (2009). Molecular evolution of a novel hyperactive Sleeping Beauty transposase enables robust stable gene transfer in vertebrates, *Nat Genet*, **41**, 753–761.

Mátrai, J., Chuah, M.K., VandenDriessche, T. (2010a). Preclinical and clinical progress in hemophilia gene therapy, *Curr Opin Hematol*, **17**, 387–392.

Mátrai, J., Chuah, M.K., VandenDriessche, T. (2010b). Recent advances in lentiviral vector development and applications, *Mol Ther*, **18**, 477–490.

Mátrai, J., Cantore, A., Bartholomae, C.C., *et al.* (2011). Hepatocyte-targeted expression by integrase-defective lentiviral vectors induces antigen-specific tolerance in mice with low genotoxic risk, *Hepatology*, **53**, 1696–1707.

Matsui, H., Shibata, M., Brown, B. (2007). *Ex vivo* gene therapy for hemophilia A that enhances safe delivery and sustained *in vivo* factor VIII expression from lentivirally engineered endothelial progenitors, *Stem Cells*, **25**, 2660–2669.

Mingozzi, F., Liu, Y.L., Dobrzynski, E., *et al.* (2003). Induction of immune tolerance to coagulation factor IX antigen by *in vivo* hepatic gene transfer, *J Clin Invest*, **111**, 1347–1356.

Montgomery, R.R., Shi, Q. (2012). Platelet and endothelial expression of clotting factors for the treatment of hemophilia, *Thromb Res*, **129 Suppl 2**, S46–S48.

Mount, J.D., Herzog, R.W., Tillson, D.M. (2002). Sustained phenotypic correction of hemophilia B dogs with a factor IX null mutation by liver-directed gene therapy, *Blood*, **99**, 2670–2676.

Roth, D.A., Tawa, N.E., Jr., O'Brien, J.M., *et al.* (2001). Factor VIII Transkaryotic Therapy Study Group. Nonviral transfer of the gene encoding coagulation factor VIII in patients with severe hemophilia A, *N Engl J Med*, **344**, 1735–1742.

Shi, Q., Wilcox, D.A., Fahs, S.A., *et al.* (2007). Lentivirus-mediated platelet-derived factor VIII gene therapy in murine haemophilia A, *J Thromb Haemost*, **5**, 352–361.

Spencer, H.T., Denning, G., Gautney, R.E., *et al.* (2011). Lentiviral vector platform for production of bioengineered recombinant coagulation factor VIII, *Mol Ther*, **19,** 302–309.

VandenDriessche, T., Vanslembrouck, V., Goovaerts, I., *et al.* (1999). Long-term expression of human coagulation factor VIII and correction of hemophilia A after *in vivo* retroviral gene transfer in factor VIII-deficient mice, *Proc Natl Acad Sci USA*, **96**, 10379–10384.

VandenDriessche ,T., Ivics, Z., Izsvák, Z., *et al.* (2009). Emerging potential of transposons for gene therapy and generation of induced pluripotent stem cells, *Blood*, **114**, 1461–1468.

VandenDriessche, T., Chuah, M.K. (2012). Clinical progress in gene therapy: sustained partial correction of the bleeding disorder in patients suffering from severe hemophilia B, *Hum Gene Ther*, **23**, 4–6.

VandenDriessche, T., Thorrez, L., Naldini, L. (2002). Lentiviral vectors containing the human immunodeficiency virus type-1 central polypurine tract can efficiently transduce nondividing hepatocytes and antigen-presenting cells in vivo, *Blood*, **100**, 813–822.

Yant, S.R., Meuse, L., Chiu, W., *et al.* (2000). Somatic integration and long-term transgene expression in normal and haemophilic mice using a DNA transposon system, *Nat Genet*, **25**, 35–41.

Zhong, L., Li, B., Mah, C.S., *et al.* (2008). Next generation of adeno-associated virus 2 vectors: point mutations in tyrosines lead to high-efficiency transduction at lower doses, *Proc Natl Acad Sci USA*, **105**, 7827–7832.

25

EXPERIMENTAL AND CLINICAL OCULAR GENE THERAPY

Alexis-Pierre Bemelmans[a,b,c,d,f] *and José-Alain Sahel*[b,c,d,e,g]

25.1 Introduction

In the early 1990s, new classes of viral vectors for gene transfer capable of efficiently targeting post-mitotic cells appeared at the forefront of biomedical research. The research community was immediately interested in the transduction of neurons, that is the post-mitotic cells. Around the world several teams selected the retina as the prime target among different parts of the central nervous system (CNS). Responsible for capturing and conveying to the brain images of the surrounding world, the retina indeed exhibits several characteristics that make it an ideal target for gene therapy. First, surgical approaches are relatively easy for both small laboratory animals and humans. Second, the retina is the location of many diseases, whether hereditary or acquired, for which treatments are as yet inadequate or non-existent. Finally, there are many techniques for functional exploratory examination of the retina, making it possible to assess the efficiency of gene transfer and/or therapeutic strategy. From the successful princeps studies demonstrating gene transfer to neurons (Fink *et al.*, 1992; Le Gal La Salle *et al.*, 1993), different types of vectors have been developed that can effectively and specifically target different cell types of the eye and retina. These vectors

[a]Molecular Imaging Research Center, URA2210 CEA/CNRS, Commissariat à l'Energie Atomique, Fontenay aux Roses F 92265, France
[b]Institut de la Vision, INSERM, U968, Paris F 75012, France
[c]UPMC Université Paris 06, UMR S 968, Institut de la Vision, Paris F 75012, France
[d]CNRS, UMR 7210, Paris F 75012, France
[e]Centre Hospitalier National d'Ophtalmologie des Quinze Vingts, INSERM DHOS CIC 503, Paris F 75012, France
[f]Email: alexis.bemelmans@cea.fr
[g]Email: j.sahel@quinze-vingts.fr

have provided proof of concept of gene therapy efficacy in numerous models of retinal neurodegenerative diseases. Based on these results, several clinical trials targeting inherited retinal degeneration, but also acquired pathologies, have been initiated, and for some have proven an undeniable therapeutic benefit following gene transfer treatment. In this chapter, we review the different approaches to transfer genes for treatment of inherited or acquired disease, with particular focus on diseases that target the retina, i.e., the part of the eye that belongs to the CNS.

25.2 The Eye, an Ideal Target for *in vivo* Gene Transfer

The globe of the eye is composed of different transparent media whose optical properties determine the image of the surrounding objects that is captured by the retina. The retina is the only part of the CNS for which one can visualize neurons without invasive means, or even without anesthetizing the animal or patient.

Compared with most other targets of gene therapy, the eye appears as a relatively small and confined organ. Forming a compartment, the retina allows for localized gene transfer using limited amounts of vector, thus greatly reducing its spread outside the target tissue, compared to other targets of gene transfer. Both the anatomical isolation from the rest of the CNS and the presence of a blood–retinal barrier further limit the risks of vector dissemination, an important concern for biosafety. Inside the eye, one can also distinguish several subdivisions that can be individually targeted, thanks to the high precision of ocular surgery (Fig. 25.1).

Within the cornea, the endothelium can be easily transduced by a simple injection into the anterior chamber of the eye. In contrast, the corneal stroma is more difficult to reach, although recent surgical tools, such as femtosecond laser methods, can be applied for specific delivery of vectors. To reach the posterior chamber of the eye, intravitreal injections can easily be performed to target the anatomical structures adjacent to the vitreous humor, such as the ciliary body or retina. Regarding this latter structure, its highly organized strata allow one to accurately predict the target cells depending on the spreading characteristics of the vector used. It has thus been shown in animals that certain vectors are capable of targeting all layers of the retina upon intravitreal injection, while others lead to a restricted targeting pattern of the cells in the inner retinal surface. It is therefore possible to administer these vectors in the subretinal space to target the outer retinal layers such as the retinal pigmented epithelium (RPE) and photoreceptor (PR). The subretinal space, located between the RPE and PR, is a virtual space created during the injection procedure that resolves spontaneously within days post injection.

It is thus possible, using different surgical approaches made available by modern techniques of retinal and eye surgery, to target specifically and efficiently a large

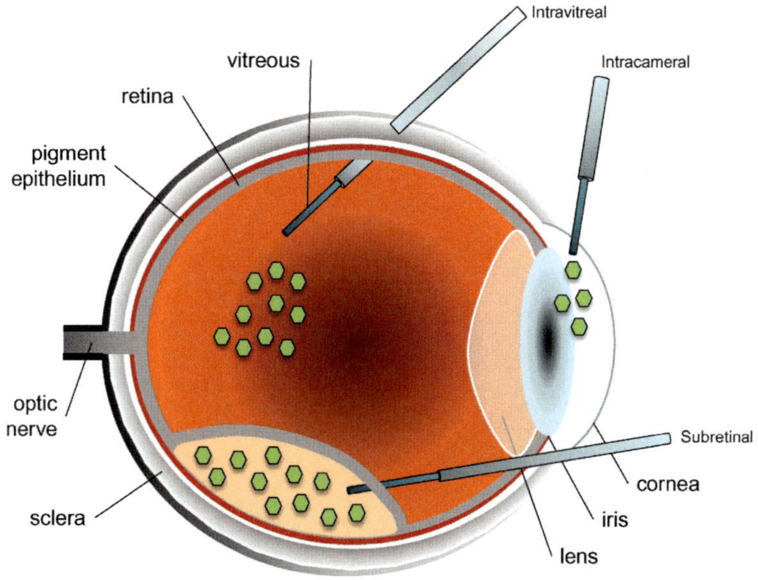

FIGURE 25.1 ■ Three different routes of vector administration in the eye. Intracameral targets the anterior chamber, while intravitreal and subretinal target the posterior chamber and the retina. The gene transfer vector is symbolized by green hexagons.

number of cell types of the eye, and in particular of the retina, whether the cell types are highly represented or, on the contrary, whether they form discrete populations.

25.3 Vectors for Retinal Gene Transfer

25.3.1 Non-Viral Vectors

In theory, non-viral vectors have many advantages, among which are: A large cloning capacity, fully mastered manufacturing processes and reduced or absent immune and inflammatory responses. In practice, however, various non-viral vectors tested to date in ocular gene therapy resulted in inefficient gene transfer. For this reason, we will not describe in this chapter the detail of the current state of research on non-viral vectors as applied to the eye. A recent review of the subject is available in Charbel Issa and MacLaren (2012).

25.3.2 Viral Vectors

Viral vector means a virus from which a substantial part of the genomic sequence has been deleted making it safe for laboratory and clinical use and allowing the introduction of a gene of interest. Typically, viral vectors are non-replicative and therefore their production requires a system that provides *in trans* all the elements necessary for their construction. Below we list the main vectors currently employed in the field of ocular gene therapy.

25.3.2.1 *Adenovirus-derived vectors*

The adenoviral vectors were among the first to be tested for transduction of neurons by direct gene transfer in the brain and retina (Bennett *et al.*, 1994). These vectors have a relatively large cloning capacity ranging from 8 kilobases (kb) for the first generation to several tens of kilo bases for the latest generation of vectors, called *gutless*. Subretinal injections of adenoviral vector lead to efficient transduction of the RPE but only limited transduction of the PR. These vectors, however, were shown to be relatively inflammatory and immunogenic, leading to a rapid clearance of the transduced cells by the host immune system (Hoffman *et al.*, 1997). In addition, the latest generation of adenoviral vectors requires a sophisticated production system that is difficult to implement. Accordingly, adenoviral vectors are now rarely used in the field of ocular gene therapy.

25.3.2.2 *Lentivirus-derived vectors*

The prototypical lentiviral vector is derived from human immunodeficiency virus 1 (HIV-1). These vectors are almost completely devoid of viral coding sequences, the only sequences remaining in the vector genome being those required *in cis* for the production of viral particles and transduction of target cells. This is an important advantage in terms of biosafety, and it also provides a cloning capacity for the transgene in the range of 8–9 kb, which is sufficient for most of the gene therapy applications currently under development. In contrast to vectors derived from conventional oncoretroviruses, those derived from lentiviruses, also called complex retroviruses, are capable of transducing post-mitotic cells, as their pre-integration complex is able to cross the nuclear membrane (Naldini *et al.*, 1996; Zennou *et al.*, 2000). Early studies describing lentiviral vectors have shown effective targeting of neurons of the CNS following an intracerebral injection. Following these initial tests, lentiviral vectors were applied in the retina, with very positive effects: it turned out that their neuronal tropism differed from that seen in the brain. In rodents, subretinal injection of lentiviral vector indeed leads to a very effective targeting of the RPE, resulting in virtually 100% transduction of this layer in the region that is the target of retinal detachment (Auricchio *et al.*, 2001). Targeting of retinal neurons, however, is much more limited. In this regard, the results reported to date show discrete transduction of PRs, and very limited transduction for other neuronal layers (Bemelmans *et al.*, 2005; Kostic *et al.*, 2003). Classically, lentiviral vectors are pseudotyped by the envelope glycoprotein of the vesicular stomatitis virus (VSV-G), which leads to more resistant virions but importantly increases the infectivity of the vector and the variety of cell types that can be transduced. There are other envelope glycoproteins that allow pseudotyping of lentiviral vectors, although those that have been tested so far do not increase the targeting of retinal neurons (Bemelmans *et al.*, 2005).

When administered via the intravitreal route, studies in rodents show that lentiviral vectors are ineffective for targeting neurons of the inner retina. At best they are capable of transducing a few cells near the injection site, and only as a result of the trauma induced by the needle or injection catheter within the retinal layers. It is therefore likely that retinal neurons possess the surface receptors needed for entry of the vector into the cell, but that this step is inhibited by physical barriers existing at the inner and/or outer retinal sides.

In summary, lentiviral vectors are suitable for conditions requiring targeting of the RPE. One should acknowledge, however, the notable exception represented by lentiviral vectors derived from the equine infectious anemia virus (EIAV). The few studies reporting its use for retinal gene transfer indeed show effective targeting of the PR layer after subretinal injection (Kong *et al.*, 2008). Thus, on the basis of these results, three clinical trials targeting PRs are underway, two targeting monogenic diseases, the third targeting an acquired pathology, the wet form of age-related macular degeneration (wet-AMD; see Section 25.4).

25.3.2.3 *Adeno-associated-virus-derived vectors*

Vectors derived from adeno-associated virus, or AAV, are the ones that currently attract the most attention in the field of ocular gene therapy. The genome of this vector is completely devoid of viral coding sequences and conserves only the inverted terminal repeats (ITRs) from the original virus, i.e., slightly less than 300 bp of genetic material. The viral particle is composed of an icosahedral protein capsid that contains a single-stranded genomic DNA, and has a theoretical cloning capacity of 4.7 kb. Despite this cloning capacity, which may appear relatively limited, the AAV vector has taken a major place among the vectors used in ocular gene therapy. The reasons are listed below.

First, there are a variety of different serotypes of AAV each of which has distinct tropic properties. For example, after a subretinal injection, serotype 4 specifically targets RPE cells, while serotypes 7 and 8 are more specific for PRs, and serotype 5 can target both layers with equivalent efficacy (Allocca *et al.*, 2007; Weber *et al.*, 2003). The use of polymerase chain reaction (PCR) with degenerate primers has furthermore allowed isolation of several tens of distinct AAV serotypes from primates and other mammals (Cearley *et al.*, 2008). It is likely that this diversity will quickly lead to the identification of serotypes specific enough to target the different retinal cell types.

Second, AAVs have a very good biosafety profile. The wild-type virus is non-pathogenic, and its immunogenicity is low. Serotype 2 AAV is known to integrate into a specific site on chromosome 19 in humans, but the vector in turn is mainly non-integrative, meaning that the vast majority of the vector's genomes persist in the episomal form in the nucleus of the target cell, thus dramatically reducing the risk of insertional mutagenesis.

Third, AAVs provide long-term expression in the target cell. As long as the target cell survives and does not divide, the vector will persist inside the cell, and the transgene, if placed under the transcriptional control of an appropriate promoter, will be expressed throughout this period. Thus, in cases where gene therapy targets retinal neurons that do not renew themselves, we may assume that transgene expression will persist indefinitely in the treated patients, which has already been shown in animal models. While the long-lasting transgene expression is an advantage in most cases, other instances may require the use of special systems that regulate transgene expression, thus giving the option to discontinue the treatment at a given time point.

Fourth, production of the AAV vector is relatively easy, which has enabled many laboratories to easily access this technology, apply it in a variety of experimental paradigms, and allow its development for gene therapy in general and for its applications to the eye in particular. Thus, the therapeutic efficacy of AAV vectors has been demonstrated in many experimental models of retinal disease (see Section 25.4 below).

Finally, AAVs have been — and still are — the subject of much research and development, with improvements that have significantly increased their transduction potential. Three notable developments should be mentioned: (1) double-stranded, or self-complementary, vectors, which have a cloning capacity reduced by 50%, but have a greater infectivity, bypassing the step of complementary strand synthesis during infection of the target cell (McCarty *et al.*, 2001); (2) point mutations of tyrosine residues exposed on the surface of the capsid, which help prevent ubiquitination of viral particles in the cell (Petrs-Silva *et al.*, 2009); (3) the method of "capsid shuffling" that, by mixing the sequences of several serotypes, provides hybrid vectors with entirely new properties (Koerber *et al.*, 2009).

25.4 Inherited Retinal Degeneration: From Animal Models to Clinical Trials

The most advanced programs of ocular gene therapy concern inherited retinal degeneration, and especially those of the recessive type. In these cases of recessive retinal disease, the strategy is relatively straightforward: The approach is to provide a copy of the normal gene in order to stop or slow down the neurodegenerative process. This implies that treatment must be administered early in the course of the disease. Gene therapy strategies to be implemented for dominant or acquired diseases are less obvious. However, there are many tracks that are currently being explored using animal models of these diseases.

25.4.1 Initial Proof of Concept in Animal Models

The feasibility of gene therapy has been demonstrated in more than two dozen animal models of recessive inherited retinal degeneration (Smith *et al.*, 2012). Early studies in this area pertained to the *Rd1* and *Rds* mouse strains, two models of retinitis pigmentosa (RP) due to a mutation in the phosphodiesterase 6b (*Pde6b*) and peripherin 2 (*Prph2*) genes, respectively. These princeps studies showed the feasibility of gene supplementation in the case of autosomal recessive RP. The therapeutic benefits observed in animals were, however, limited or transient, probably due to the choice of models, with a narrow therapeutic window due to the speed of PR degeneration and to the AAV vector used, which did not allow transgene expression at sufficiently high levels.

The emblematic proof of concept of gene supplementation feasibility in models of inherited retinal degeneration came from a study on the Briard dog model of RPE65 deficiency. This enzyme is expressed in the RPE and is responsible for a key step in the recycling of 11-*cis*-retinal, the chromophore allowing the capture of photons by the PR visual pigments. In humans, mutations in the *RPE65* gene cause a profound visual loss, present at birth, and called Leber congenital amaurosis (LCA). Using the Briard dog model, which bears a spontaneous mutation in the *Rpe65* gene, J. Bennett's team (University of Pennsylvania) has demonstrated in a seminal work published in 2001 that the subretinal injection of an AAV encoding a non-mutated version of the gene allows restoration of vision in these animals (Acland *et al.*, 2001). This work was subsequently confirmed by several independent teams, using both the dog model and mouse models deficient in RPE65 (Bemelmans *et al.*, 2006; Le Meur *et al.*, 2007; Narfstrom *et al.*, 2003; Rakoczy *et al.*, 2003). Moreover, results showing the long-term efficiency of this strategy were quickly reported (Acland *et al.*, 2005).

25.4.2 Clinical Trials of *RPE65* Gene Supplementation

LCA due to RPE65 deficiency has promptly emerged as an ideal target for gene therapy because of (1) the therapeutic efficacy demonstrated in animal models, (2) the relatively slow degeneration observed in patients suggesting a large therapeutic window and (3) the efficiency of AAV vector to transduce the target cells, i.e., the RPE. On this basis, three independent clinical trials were started in the USA and in Europe. Despite some differences in the vector construction, the dose administered and the design of the experiment, the three clinical trials were based on the administration of an AAV vector encoding the human *RPE65* cDNA in the subretinal space of young adult patients, i.e., at an advanced stage of the pathology (Bainbridge *et al.*, 2008; Hauswirth *et al.*, 2008; Maguire *et al.*, 2008).

The results of these clinical trials have demonstrated remarkable biosafety of the gene therapy procedure. In addition, improved light sensitivity of the retina, and improved visual function of patients, spectacular in some cases, were observed reproducibly. However, at a functional level, the improvements achieved in humans did not reach those observed in animal models, particularly considering the electroretinogram, which was largely restored in dogs and mice, whereas the improvements seen in humans were very small or non-existent. It is possible that discrepancies between clinical trials and studies on animals are due to an overly limited expression of the transgene in humans. It is therefore likely that improved therapeutic efficacy of *RPE65* gene supplementation will come from vector improvements, which should allow the construct to reach a greater proportion of the retina and/or obtain a higher expression in the target cells.

As a safety measure, all patients received the treatment in one eye only, the one with the lowest vision. Consequently, very shortly after evidence of safety and efficacy of the treatment was demonstrated, the question was raised as to whether treatment in the second eye should take place for these patients. As the risk of an immune response against the vector may lead to the production of neutralizing antibodies, it effectively appeared unreasonable to treat these patients with a second injection, although the subretinal space can be considered as an immune-privileged site. To investigate the possibility of treating the second eye, Bennett's team studied in detail the immune response following subretinal re-administration of the AAV vector in the Briard dog model of LCA, as well as in macaques, a non-human primate that possesses an immune system very similar to that of humans. This study showed that in both cases re-administration of the vector was safe and enabled efficient gene transfer, even if antibodies against the vector existed in these animals before the treatment began (Amado *et al.*, 2010). This latter study involves two important results. First, it is possible to treat the second eye of patients with LCA who have already been treated by gene therapy for their most affected eye. Second, patients with pre-existing antibodies against AAV could be eligible for gene therapy programs using this vector, influencing the current state of gene therapy even beyond the scope of LCA.

25.4.3 Other Inherited Retinal Degenerations

The success of gene therapy for RPE65 deficiency has led numerous teams to explore similar strategies for many other recessive or loss-of-function inherited retinal degenerative diseases. The proofs of concept were obtained for a broad variety of these diseases on the basis of the availability of (1) a vector to target the diseased cells and (2) an animal model sufficiently representative of the pathology.

One of the first models to benefit from this proof of concept was the Royal College of Surgeons (RCS) rat, a popular model of inherited retinal degeneration in which a mutation causes dysfunction of monocytes and tissues of epithelial and reproductive origin tyrosine kinase (MERTK) in the RPE, subsequently leading to PR degeneration. It has indeed been demonstrated that adenoviral, lentiviral and AAV vectors are able to slow degeneration by introducing a copy of the normal gene into RPE cells (Smith *et al.*, 2003; Tschernutter *et al.*, 2005; Vollrath *et al.*, 2001). However, due to the rapid evolution of the disease in this particular model reducing the therapeutic window, it has not yet been possible to completely halt the neurodegenerative process by gene transfer.

In the 2000s, new AAV serotypes allowing for more effective targeting of PRs appeared (Allocca *et al.*, 2007), which helped to achieve therapeutic levels of gene supplementation in different genetic models involving genes expressed in these cells (reviewed in Smith *et al.* (2012)). This has been demonstrated in particular for mutations in the genes guanylate cyclase 2D (*GUCY2D*) and aryl hydrocarbon receptor-interacting protein-like 1 (*AIPL1*), responsible for LCA, in phosphodiesterase β-subunit (β-*PDE*), responsible for RP, and in cyclic nucleotide gated channel α 3 (*CNGA3*) and cyclic nucleotide gated channel β 3 (*CNGB3*), responsible for congenital stationary night blindness.

As mentioned in Section 25.3.2.2, EIAV-derived lentiviral vectors have also shown a remarkable ability to transduce PRs. The greater cloning capacity of these vectors compared with AAV has led researchers to consider them for supplementation strategies for which the target genes or cDNAs overwhelm the AAV genome size. A first proof of concept came from the supplementation of the ATP-binding cassette sub-family A member 4 (*Abca4*) cDNA (6.8 kb) in the *Abca4*$^{-/-}$ mouse strain, a model of Stargardt disease, a macular dystrophy caused by mutations in the retina-specific ABC transporter (*ABCR*) gene (Kong *et al.*, 2008). Based on this study, two clinical trials sponsored by Oxford Biomedica are under way for Stargardt disease caused by ABCR mutations and Usher syndrome caused by myosin VIIA (*MYOVIIA*) mutations, a syndromic form of RP.

Lastly, it is important to mention mitochondrial diseases causing retinal degeneration such as Leber hereditary optic neuropathy (LHON), in which mutations in the genes reduced nicotinamide adenine dinucleotide dehydrogenase 1 (*ND1*), *ND4* or *ND6*, encoding subunits of the mitochondrial respiratory chain, lead to a selective degeneration of retinal ganglion cells (RGC). Clinically, the disease is characterized by a sudden onset of degeneration in adults that leads to loss of vision within weeks or months. Both eyes are still affected sequentially, degeneration in the second eye usually appearing only several months after onset of the disease. Consequently, the therapeutic window is wide enough to treat the second eye. In rodents, it is possible to reproduce the disease by expressing a mutated copy of the *ND4* gene, and to correct this condition by overexpressing the normal version of the gene (Ellouze *et al.*, 2008). It therefore appears that the disease process depends on the ratio between mutated and normal proteins and it is possible to counteract degeneration by increasing the expression level of the latter. Based on these results, clinical trials are currently being prepared for LHON. The aim will be to overexpress the *ND4* gene by intravitreal injection of an AAV serotype 2, a vector that can effectively target RGC in rodents and non-human primates, although in the latter case, transduction is limited to the central retina (Hellstrom *et al.*, 2009; Yin *et al.*, 2011).

25.5 Future Directions: Generic Gene Therapy of the Retina

Studies in different animal models, as well as clinical trials currently underway, have demonstrated the feasibility of retinal gene therapy both in terms of safety for the patient and therapeutic efficacy. However, for ocular gene therapy to enter routine clinical use, it is necessary to develop therapeutic products targeting much larger populations of patients. One could imagine a single gene transfer vector to treat all or a large proportion of RP, whatever the gene involved. The prevalence of RP is indeed around 1 in 4,000 in Europe and the USA. Conversely, one can consider gene therapy products targeting much more common diseases and therefore representing major issues of public health, such as AMD or glaucoma.

25.5.1 Strategies for Dominant Mutations

In the case of dominant inherited retinal degeneration, i.e., resulting in a gain-of-function or a dominant-negative effect, it is not enough to express the normal protein, and one must also suppress expression of the mutant protein to obtain a therapeutic effect. Different strategies of this type are currently under investigation using animal models of rhodopsin mutations, the major cause of autosomal dominant RP in humans, representing 25–30% of all cases. The proposed gene therapy strategies are to remove the endogenous gene expression using molecular tools and express a normal gene copy artificially rendered insensitive to the mechanism that inhibits endogenous gene expression. The first among these molecular tools is RNA interference, which has demonstrated its capacity to inhibit expression of the endogenous mutated rhodopsin, while allowing expression at the therapeutic level of a transgenic rhodopsin modified thanks to degeneration of the genetic code (Kiang *et al.*, 2005). This strategy has improved symptoms in a mouse model expressing a mutated version of the human rhodopsin gene (Chadderton *et al.*, 2009). Other technologies of the same type based on the use of ribozymes for the suppression of endogenous rhodopsin expression have also been investigated (LaVail *et al.*, 2000; Lewin *et al.*, 1998). These strategies can theoretically handle all of the targeted gene mutations, a great advantage considering the genetic heterogeneity of these diseases. For instance, these are more than 150 different mutations described for rhodopsin that lead to autosomal dominant RP.

Another avenue being considered is to act directly at the genomic level by repairing the mutations using various tools including oligonucleotides, zinc-finger proteins, or meganucleases. These strategies have a significant handicap, however, compared with those mentioned above: They will need the development of a specific tool for each mutation, which does not allow one to develop a unique therapeutic tool capable of handling all mutations of the same gene.

25.5.2 Neuroprotection

A large number of trophic factors, e.g., ciliary neurotrophic factor (CNTF), glial-cell-line-derived neurotrophic factor (GDNF) and rod-derived cone viability factor (RdCVF), have demonstrated their ability to protect retinal neurons. Gene therapy represents *a priori* an ideal solution for delivery of these factors into the eye. This administration must be local to avoid off-target side-effects of these powerful factors, as well as providing constant and sustained expression for maximum therapeutic benefit. It has been shown in different mouse models that gene transfer of CNTF is neuroprotective for PR but also for the RGC (Liang *et al.*, 2001; van Adel *et al.*, 2003). Similarly, gene transfer of GDNF has been proven to slow PR degeneration in a transgenic rat model expressing a mutated version of rhodopsin (McGee Sanftner *et al.*, 2001). Discovered more recently, RdCVF is a neurotrophic factor that was identified by its ability to promote survival of cone PRs (Leveillard *et al.*, 2004). It has been shown that intraocular injection of recombinant RdCVF promotes the survival of cone PR in the P23H rat (Yang *et al.*, 2009).

The long-term effects of overexpressing a trophic factor in the retina are not currently known and will certainly vary from case to case. This complicates the design of the toxicology studies to be carried out before considering clinical application of these factors. Clinical application may require the use of a system to provide for fine regulation of transgene expression. Despite these drawbacks, the use of trophic factors remains a very promising track, because it is applicable to many diseases, even those for which the pathophysiology is not completely understood. This is the case in glaucoma, for example, for which gene transfer of brain-derived neurotrophic factor (BDNF) or CNTF could provide a therapeutic solution (Liu *et al.*, 2009).

25.5.3 Ocular Neovascularization

Appearance of pathological blood vessels, or neovascularization, is an essential symptom in the onset of blindness for eye disease representing a crucial public health issue such as wet-AMD or diabetic retinopathy. Vascular endothelial growth factor (VEGF) has been identified as a major player in the onset of neovascularization. Consequently, a new class of pharmacological agents, designed to block VEGF action, has been put into use, dramatically improving the management of wet-AMD patients. The molecule with the best therapeutic effects and having met the greatest success to date is Lucentis®, a monoclonal antibody capable of neutralizing VEGF (Patel *et al.*, 2011). To be active it must be administered by intravitreal injection at regular intervals every few weeks, which ultimately poses the risk of infection and inflammation for the patients. Moreover, it would be beneficial to administer regular and prolonged release of the molecule by gene therapy rather than the bolus regimen provided by monthly injections. Different teams have thus invested in generating vectors for molecules capable of blocking VEGF action. These include an AAV vector encoding a soluble form of a VEGF receptor developed by the firm Genzyme, and a lentiviral vector developed by Oxford Biomedica, encoding the cDNA of endostatin and angiostatin, two factors leading to regression of pathological neovascularization. These factors are currently being investigated in phase I clinical trials (Campochiaro, 2012; Maclachlan *et al.*, 2011).

Similar to the administration of trophic factors by gene transfer, the advantage of anti-neovascularization gene therapy is the ability to treat the target cells without necessarily targeting them directly with the vector, but instead through the bystander effects provided by these secreted molecules. It is thus possible to transform cells such as Müller glia or cells of the ciliary body in "biological factories" that will secrete the therapeutic protein in the ocular environment. Moreover, a single therapeutic tool could be beneficial for the treatment of several diseases involving the occurrence of neovascularizations. This is currently the case for Lucentis®, which is also being tested for corneal neovascularization (Stevenson *et al.*, 2012). A single vector would thus be beneficial for several diseases and a considerably larger number of patients.

25.5.4 Optogenetics

Optogenetics is an emerging area of neuroscience based on targeted expression in neurons of bacterial ion channels whose opening is triggered by light and leads to depolarization of the neuronal cell. There are also proton pumps activated by light that hyperpolarize cells. These tools have led to significant progress in our understanding of the neural circuits of the CNS.

In the field of vision research, it quickly became obvious that these new tools could help restore vision in blind people or at a very late stage of retinal pathology. It has been shown in rodents that targeted expression by gene transfer of channelrhodopsin, a cationic channel cloned from *Chlamydomonas reinhardtii*, can transform RGCs or bipolar cells into PR-like cells, capable of responding to light and of conveying visual information to the brain (Bi *et al.*, 2006; Lagali *et al.*, 2008). More recently, B. Roska's team (Friedrisch Miescher Institute) showed that it is possible to reactivate the PRs rendered insensitive to light as a result of neurodegenerative processes, by targeted expression of halorhodopsin, a proton pump cloned from *Natromonas pharaonis* (Busskamp *et al.*, 2010). This latter approach has the advantage of reactivating the entire retinal circuitry, with the potential to restore "close to normal" vision. Although these tools are designed to restore circuit functionality, they depend on survival of the PR cell bodies throughout the degenerative process.

Many conditions must therefore be met for optogenetic tools to restore visual function to blind or near-blind patients. These include ensuring the safety of vectors designed to provide long-term expression of bacterial proteins that could elicit an immune response leading to significant destruction of transduced cells. Furthermore, the success of this technology will also be based on the development of light-emitting diode (LED)-goggles to project on the retina an image of the surrounding world with the optimal brightness and wavelength required to stimulate the newly created PRs (Cepko, 2010).

25.6 Conclusion

The initial ocular gene therapy studies have focused on genetic diseases affecting a small number of patients but holding the advantage of having relevant animal models with which the feasibility of gene transfer has been established relatively easily. This work has led to clinical trials demonstrating both safety and efficacy of gene therapy for the eye, opening the way to implement new strategies for dominant or acquired diseases, requiring more complex approaches such as RNA interference or transfer of genes encoding neurotrophic factors or optogenetic tools, described above.

Finally, our growing understanding of the pathophysiological processes leading to blindness in very common diseases such as glaucoma or the dry form of AMD suggests that new therapeutic targets, amenable to gene transfer, will be identified in the coming years.

Acknowledgments

This work is supported by the laboratory of excellence LIFESENSES, and by a Wynn-Gund Translational Research Award from National Neurovision Research Institute/Foundation Fighting Blindness.

The authors are grateful to Katie Matho for her critical reading of the manuscript.

References

Acland, G.M., Aguirre, G.D., Bennett, J., *et al.* (2005). Long-term restoration of rod and cone vision by single dose rAAV-mediated gene transfer to the retina in a canine model of childhood blindness, *Mol Ther*, **12**, 1072–1082.

Acland, G.M., Aguirre, G.D., Ray, J., *et al.* (2001). Gene therapy restores vision in a canine model of childhood blindness, *Nat Genet*, **28**, 92–95.

Allocca, M., Mussolino, C., Garcia-Hoyos, M., *et al.* (2007). Novel adeno-associated virus serotypes efficiently transduce murine photoreceptors, *J Virol*, **81**, 11372–11380.

Amado, D., Mingozzi, F., Hui, D., *et al.* (2010). Safety and efficacy of subretinal readministration of a viral vector in large animals to treat congenital blindness, *Sci Transl Med*, **2**, 21ra16.

Auricchio, A., Kobinger, G., Anand, V., *et al.* (2001). Exchange of surface proteins impacts on viral vector cellular specificity and transduction characteristics: the retina as a model, *Hum Mol Genet*, **10**, 3075–3081.

Bainbridge, J.W., Smith, A.J., Barker, S.S., *et al.* (2008). Effect of gene therapy on visual function in Leber's congenital amaurosis, *N Engl J Med*, **358**, 2231–2239.

Bemelmans, A.P., Bonnel, S., Houhou, L., *et al.* (2005). Retinal cell type expression specificity of HIV-1-derived gene transfer vectors upon subretinal injection in the adult rat: influence of pseudotyping and promoter, *J Gene Med*, **7**, 1367–1374.

Bemelmans, A.P., Kostic, C., Crippa, S.V., *et al.* (2006). Lentiviral gene transfer of RPE65 rescues survival and function of cones in a mouse model of Leber congenital amaurosis, *PLoS Med*, **3**, e347.

Bennett, J., Wilson, J., Sun, D., *et al.* (1994). Adenovirus vector-mediated *in vivo* gene transfer into adult murine retina, *Invest Ophthalmol Vis Sci*, **35**, 2535–2542.

Bi, A., Cui, J., Ma, Y.P., *et al.* (2006). Ectopic expression of a microbial-type rhodopsin restores visual responses in mice with photoreceptor degeneration, *Neuron*, **50**, 23–33.

Busskamp, V., Duebel, J., Balya, D., *et al.* (2010). Genetic reactivation of cone photoreceptors restores visual responses in retinitis pigmentosa, *Science*, **329**, 413–417.

Campochiaro, P.A. (2012). Gene transfer for ocular neovascularization and macular edema, *Gene Ther*, **19**, 121–126.

Cearley, C.N., Vandenberghe, L.H., Parente, M.K., *et al.* (2008). Expanded repertoire of AAV vector serotypes mediate unique patterns of transduction in mouse brain, *Mol Ther*, **16**, 1710–1718.

Cepko, C. (2010). Neuroscience. Seeing the light of day, *Science*, **329**, 403–404.

Chadderton, N., Millington-Ward, S., Palfi, A., *et al.* (2009). Improved retinal function in a mouse model of dominant retinitis pigmentosa following AAV-delivered gene therapy, *Mol Ther*, **17**, 593–599.

Charbel Issa, P., MacLaren, R.E. (2012). Non-viral retinal gene therapy: a review, *Clin Experiment Ophthalmol*, **40**, 39–47.

Ellouze, S., Augustin, S., Bouaita, A., *et al.* (2008). Optimized allotopic expression of the human mitochondrial ND4 prevents blindness in a rat model of mitochondrial dysfunction, *Am J Hum Genet*, **83**, 373–387.

Fink, D.J., Sternberg, L.R., Weber, P.C., *et al.* (1992). *In vivo* expression of β-galactosidase in hippocampal neurons by HSV-mediated gene transfer, *Hum Gene Ther*, **3**, 11–19.

Hauswirth, W.W., Aleman, T.S., Kaushal, S., *et al.* (2008). Treatment of leber congenital amaurosis due to RPE65 mutations by ocular subretinal injection of adeno-associated virus gene vector: short-term results of a phase I trial, *Hum Gene Ther*, **19**, 979–990.

Hellstrom, M., Ruitenberg, M.J., Pollett, M.A., *et al.* (2009). Cellular tropism and transduction properties of seven adeno-associated viral vector serotypes in adult retina after intravitreal injection, *Gene Ther*, **16**, 521–532.

Hoffman, L.M., Maguire, A.M., Bennett, J. (1997). Cell-mediated immune response and stability of intraocular transgene expression after adenovirus-mediated delivery, *Invest Ophthalmol Vis Sci*, **38**, 2224–2233.

Kiang, A.S., Palfi, A., Ader, M., *et al.* (2005). Toward a gene therapy for dominant disease: validation of an RNA interference-based mutation-independent approach, *Mol Ther*, **12**, 555–561.

Koerber, J.T., Klimczak, R., Jang, J.H., *et al.* (2009). Molecular evolution of adeno-associated virus for enhanced glial gene delivery, *Mol Ther*, **17**, 2088–2095.

Kong, J., Kim, S.R., Binley, K., *et al.* (2008). Correction of the disease phenotype in the mouse model of Stargardt disease by lentiviral gene therapy, *Gene Ther*, **15**, 1311–1320.

Kostic, C., Chiodini, F., Salmon, P., *et al.* (2003). Activity analysis of housekeeping promoters using self-inactivating lentiviral vector delivery into the mouse retina, *Gene Ther*, **10**, 818–821.

Lagali, P.S., Balya, D., Awatramani, G.B., *et al.* (2008). Light-activated channels targeted to ON bipolar cells restore visual function in retinal degeneration, *Nat Neurosci*, **11**, 667–675.

LaVail, M.M., Yasumura, D., Matthes, M.T., *et al.* (2000). Ribozyme rescue of photoreceptor cells in P23H transgenic rats: long-term survival and late-stage therapy, *Proc Natl Acad Sci USA*, **97**, 11488–11493.

Le Gal La Salle, G., Robert, J.J., Berrard, S., *et al.* (1993). An adenovirus vector for gene transfer into neurons and glia in the brain, *Science*, **259**, 988–990.

Le Meur, G., Stieger, K., Smith, A.J., *et al.* (2007). Restoration of vision in RPE65-deficient Briard dogs using an AAV serotype 4 vector that specifically targets the retinal pigmented epithelium, *Gene Ther*, **14**, 292–303.

Leveillard, T., Mohand-Said, S., Lorentz, O., *et al.* (2004). Identification and characterization of rod-derived cone viability factor, *Nat Genet*, **36**, 755–759.

Lewin, A.S., Drenser, K.A., Hauswirth, W.W., *et al.* (1998). Ribozyme rescue of photoreceptor cells in a transgenic rat model of autosomal dominant retinitis pigmentosa, *Nat Med*, **4**, 967–971.

Liang, F.Q., Dejneka, N.S., Cohen, D.R., *et al.* (2001). AAV-mediated delivery of ciliary neurotrophic factor prolongs photoreceptor survival in the rhodopsin knockout mouse, *Mol Ther*, **3**, 241–248.

Liu, X., Rasmussen, C.A., Gabelt, B.T., *et al.* (2009). Gene therapy targeting glaucoma: where are we?, *Surv Ophthalmol*, **54**, 472–486.

Maclachlan, T.K., Lukason, M., Collins, M., *et al.* (2011). Preclinical safety evaluation of AAV2-sFLT01 — a gene therapy for age-related macular degeneration, *Mol Ther*, **19**, 326–334.

Maguire, A.M., Simonelli, F., Pierce, E.A., *et al.* (2008). Safety and efficacy of gene transfer for Leber's congenital amaurosis, *N Engl J Med*, **358**, 2240–2248.

McCarty, D.M., Monahan, P.E., Samulski, R.J. (2001). Self-complementary recombinant adeno-associated virus (scAAV) vectors promote efficient transduction independently of DNA synthesis, *Gene Ther*, **8**, 1248–1254.

McGee Sanftner, L.H., Abel, H., Hauswirth, W.W., *et al.* (2001). Glial cell line derived neurotrophic factor delays photoreceptor degeneration in a transgenic rat model of retinitis pigmentosa, *Mol Ther*, **4**, 622–629.

Naldini, L., Blomer, U., Gallay, P., *et al.* (1996). *In vivo* gene delivery and stable transduction of nondividing cells by a lentiviral vector, *Science*, **272**, 263–267.

Narfstrom, K., Katz, M.L., Bragadottir, R., *et al.* (2003). Functional and structural recovery of the retina after gene therapy in the RPE65 null mutation dog, *Invest Ophthalmol Vis Sci*, **44**, 1663–1672.

Patel, R.D., Momi, R.S., Hariprasad, S.M. (2011). Review of ranibizumab trials for neovascular age-related macular degeneration, *Semin Ophthalmol*, **26**, 372–379.

Petrs-Silva, H., Dinculescu, A., Li, Q., *et al.* (2009). High-efficiency transduction of the mouse retina by tyrosine-mutant AAV serotype vectors, *Mol Ther*, **17**, 463–471.

Rakoczy, P.E., Lai, C.M., Yu, M.J., *et al.* (2003). Assessment of rAAV-mediated gene therapy in the Rpe65$^{-/-}$ mouse, *Adv Exp Med Biol*, **533**, 431–438.

Smith, A.J., Bainbridge, J.W., Ali, R.R. (2012). Gene supplementation therapy for recessive forms of inherited retinal dystrophies, *Gene Ther*, **19**, 154–161.

Smith, A.J., Schlichtenbrede, F.C., Tschernutter, M., *et al.* (2003). AAV-Mediated gene transfer slows photoreceptor loss in the RCS rat model of retinitis pigmentosa, *Mol Ther*, **8**, 188–195.

Stevenson, W., Cheng, S.F., Dastjerdi, M.H., *et al.* (2012). Corneal neovascularization and the utility of topical VEGF inhibition: ranibizumab (lucentis) vs bevacizumab (avastin), *Ocul Surf*, **10**, 67–83.

Tschernutter, M., Schlichtenbrede, F.C., Howe, S., *et al.* (2005). Long-term preservation of retinal function in the RCS rat model of retinitis pigmentosa following lentivirus-mediated gene therapy, *Gene Ther*, **12**, 694–701.

van Adel, B.A., Kostic, C., Deglon, N., *et al.* (2003). Delivery of ciliary neurotrophic factor via lentiviral-mediated transfer protects axotomized retinal ganglion cells for an extended period of time, *Hum Gene Ther*, **14**, 103–115.

Vollrath, D., Feng, W., Duncan, J.L., *et al.* (2001). Correction of the retinal dystrophy phenotype of the RCS rat by viral gene transfer of *Mertk*, *Proc Natl Acad Sci USA*, **98**, 12584–12589.

Weber, M., Rabinowitz, J., Provost, N., *et al.* (2003). Recombinant adeno-associated virus serotype 4 mediates unique and exclusive long-term transduction of retinal pigmented epithelium in rat, dog, and nonhuman primate after subretinal delivery, *Mol Ther*, **7**, 774–781.

Yang, Y., Mohand-Said, S., Danan, A., *et al.* (2009). Functional cone rescue by RdCVF protein in a dominant model of retinitis pigmentosa, *Mol Ther*, **17**, 787–795.

Yin, L., Greenberg, K., Hunter, J.J., *et al.* (2011). Intravitreal injection of AAV2 transduces macaque inner retina, *Invest Ophthalmol Vis Sci*, **52**, 2775–2783.
Zennou, V., Petit, C., Guetard, D., *et al.* (2000). HIV-1 genome nuclear import is mediated by a central DNA flap, *Cell*, **101**, 173–185.

For Further Reading

Original Articles

Proof of concept in animal models

Acland, G.M., Aguirre, G.D., Ray, J., *et al.* (2001). Gene therapy restores vision in a canine model of childhood blindness, *Nat Genet*, **28**, 92–95.
Chadderton, N., Millington-Ward, S., Palfi, A., *et al.* (2009). Improved retinal function in a mouse model of dominant retinitis pigmentosa following AAV-delivered gene therapy, *Mol Ther*, **17**, 593–599.
Ellouze, S., Augustin, S., Bouaita, A., *et al.* (2008). Optimized allotopic expression of the human mitochondrial ND4 prevents blindness in a rat model of mitochondrial dysfunction, *Am J Hum Genet*, **83**, 373–387.

Clinical trials

Bainbridge, J.W., Smith, A.J., Barker, S.S., *et al.* (2008). Effect of gene therapy on visual function in Leber's congenital amaurosis, *N Engl J Med*, **358**, 2231–2239.
Maclachlan, T.K., Lukason, M., Collins, M., *et al.* (2011). Preclinical safety evaluation of AAV2-sFLT01 — a gene therapy for age-related macular degeneration, *Mol Ther*, **19**, 326–334.
Maguire, A.M., Simonelli, F., Pierce, E.A., *et al.* (2008). Safety and efficacy of gene transfer for Leber's congenital amaurosis, *N Engl J Med*, **358**, 2240–2248.

Optogenetics

Bi, A., Ma, Y.P., Olshevskaya, E., *et al.* (2006). Ectopic expression of a microbial-type rhodopsin restores visual responses in mice with photoreceptor degeneration, *Neuron*, **50**, 23–33.
Busskamp, V., Picaud, S., Sahel, J.A., *et al.* (2010). Genetic reactivation of cone photoreceptors restores visual responses in retinitis pigmentosa, *Science*, **329**, 413–417.
Lagali, P.S., Balya, D., Awatramani G.B., *et al.* (2008). Light-activated channels targeted to ON bipolar cells restore visual function in retinal degeneration, *Nat Neurosci*, **11**, 667–675.

Reviews

Busskamp, V., Picaud, S., Sahel, J.A., *et al.* (2012). Optogenetic therapy for retinitis pigmentosa, *Gene Ther*, **19**, 169–175.

Charbel Issa, P., MacLaren, R.E. (2012). Non-viral retinal gene therapy: a review, *Clin Experiment Ophthalmol*, **40**, 39–47.

Mohan, R.R., Tovey, J.C., Sharma, A., *et al.* (2012). Gene therapy in the cornea: 2005–present, *Prog Retin Eye Res*, **31**, 43–64.

Smith, A.J., Bainbridge, J.W., Ali, R.R. (2012). Gene supplementation therapy for recessive forms of inherited retinal dystrophies, *Gene Ther*, **19**, 154–161.

Wu, Z., Asokan, A., Samulski, R.J. (2006). Adeno-associated virus serotypes: vector toolkit for human gene therapy, *Mol Ther*, **14**, 316–327.

26

GENE THERAPY OF NEUROLOGICAL DISEASES

Lisa M. Stanek[a], Lamya S. Shihabuddin[a] and Seng H. Cheng[a,b]

26.1 Considerations for Treating Diseases of the Central Nervous System

Gene therapy is emerging as a potentially promising approach for the treatment of neurodegenerative disorders previously considered refractory to prevailing conventional therapies. Advances have been made in the design and construction of expression cassettes and cellular and viral transgene carriers, as well as the characterization of the target cells for neuronal gene therapy. With the advent and refinement of this technology platform there is now the realistic prospect of delivering therapeutic genes to the brain for neuroprotection, restoration of neuronal function, or replacement of deficient proteins as treatments for central nervous system (CNS) diseases. Since the first clinical gene therapy trial in 1989, more than 1,300 clinical trials have been carried out worldwide. Of these, approximately 100 trials were directed against either neurological diseases or brain tumors.

The complexities of the brain pose unique challenges for designing therapeutic interventions for neurological diseases. In addition to high risk and limited access to the brain, the combination of compartmentalized functional domains and complex circuitry means that even the delivery mode itself can be disruptive and potentially dangerous. There are, however, some practical advantages for intervention in the nervous system. These include a reduced immune response to vectors and transgene products in the brain parenchyma as compared with other peripheral sites of injection, focal targets that require limited gene delivery

[a]Genzyme, A Sanofi company, 49 New York Avenue, Framingham, MA 01701-9322, USA
[b]Email: seng.cheng@genzyme.com

within the brain and the reduced ability of vectors injected in the brain to access germ line tissues. Gene therapy for neurological disease is not yet standard clinical practice and its limitations are still under experimental investigation. Critical issues for clinical success, such as the mode of delivery, stability and regulation of transgene expression and potential toxicity are still under investigation. Experimental studies in animal models continue to advance our understanding of these issues. While CNS gene therapy presents a number of unique challenges and clinical applications remain in their infancy, collaborative efforts between clinicians and basic researchers will lead the way to the realization of this goal.

26.1.1 Design of Expression Cassettes for Optimal Expression in the CNS

The development of viral gene transfer vectors that facilitate safe and sustained expression of therapeutic transgenes in specific target cell populations is continually advancing. Gene therapy for the nervous system is particularly challenging due to the post-mitotic nature of neuronal cells and the restricted accessibility of the brain itself. The prototypical vectors derived from oncoretroviruses, such as the Moloney murine leukemia virus (MoMLV), are commonly used for gene transfer into host cell genomes; however, because these viruses only efficiently transduce dividing cells, they are deemed not as useful for transducing the non-dividing cells of the nervous system. The advent of recombinant viral vectors such as those derived from adenovirus, adeno-associated virus (AAV), herpes simplex virus (HSV) and lentivirus addresses this limitation, permitting their usage in the adult CNS. Each of these more commonly used viral vectors exhibits distinct biological properties and offer advantages and disadvantages for gene delivery to the CNS (Table 26.1).

For gene therapy to be considered a practical approach for treating neurological disorders, several obstacles that are related to the unique attributes of the CNS need to be overcome. One of the obstacles is the great diversity of cell types in the CNS, many of which perform critical physiological functions and are highly sensitive to change. This underscores the importance of restricting expression of the therapeutic gene products to particular cell types, thus ensuring therapeutic effects in the desired cells while limiting side effects caused by gene expression in non-target cells. The approach of restricting gene expression to specific cell populations can be attained through the use of a cell-specific promoter. In addition to offering cell-type-specific gene expression, the promoters, because of their cellular authentic sequences, may also be less likely to activate host cell defense machinery. Thus, they may be less sensitive to cytokine-induced promoter

TABLE 26.1 Characteristics of the Most Common Vectors Used in CNS Gene Therapy (see Chapters 9–12 for more details)

Vector	Vector Type	Insert Size (kb)	Expression	Advantages	Disadvantages
Adeno-associated virus (AAV)	Recombinant	4.5	Stable	Low immunogenicity and toxicity. Transduces non-dividing cells. Small size facilitates CNS distribution.	Limited transgene capacity.
Adenovirus	Recombinant	36	Transient	Transduces non-dividing cells. Wide host-cell range.	Immunogenic, short-term expression. Does not integrate.
Retrovirus	MoMLV-based (No virus genes)	8	Stable	Integration into genome of dividing cells. High transduction efficiency.	Random integration. Limited transgene capacity. Variable expression.
Lentivirus	HIV-based	8	Stable	Transduces non-dividing cells. High transduction efficiency. Stable long-term expression.	Potential safety risk and reversion to wild-type virus. Low titers.
Herpes Simplex Virus (HSV)	Recombinant	>30	Stable	Transduces non-dividing cells. Broad host range. Large capacity.	Immunogenic. Safety concerns. Short-term expression.

inactivation than viral promoters. As such, an improved stability of gene expression can be expected.

26.1.2 Promoter Choice

The selection of the promoter often depends upon a compromise between the level of expression required, the cell type targeted and the size of the promoter that can be accommodated into the viral packaging cassette. With advancements in viral packaging and purification technologies it is now possible to generate very high viral vector titers, making the compromise less about achieving optimal transduction levels and more about targeting specific cell types in the nervous system. The most frequently used promoter to effect transgene expression is the 0.7 kb human cytomegalovirus (CMV) promoter, a traditional choice in heterologous expression systems due to its strong transcriptional activity in a variety of different cell types (Baskar *et al.*, 1996). Nearly all of the original studies examining the potential of recombinant AAV (rAAV) as a vector for gene transfer into the brain employed the CMV promoter. However, for CNS applications, the CMV promoter is often not ideal. Transgene expression from CMV-based transcriptional cassettes has been shown to vary across different brain regions, and expression levels decline over time, probably due to silencing of the virus-derived promoter by methylation. Modifications of the CMV promoter have been engineered in an effort both to enhance and to stabilize gene expression in the CNS. Higher expression levels and persistence of expression can be achieved by stabilizing the CMV promoter through the addition of an intron to improve the efficiency of RNA processing.

Researchers continue to identify and develop promoters that facilitate optimal AAV-mediated expression in the brain. Examples of efficient promoters for rAAV-based expression in the CNS include: A chimeric promoter combining the CMV enhancer element with the chicken β-actin (CBA) promoter, neuron-specific enolase (NSE) promoter, platelet-derived growth factor β-chain (PDGF) promoter, human β-glucuronidase (GUSB) promoter and human synapsin I (hSYN) promoter (Fitzsimons *et al.*, 2002). These promoters purportedly confer sustained transgene expression primarily in neuronal cell types. Transduction of glia may be desirable for gene therapy of demyelinating diseases such as multiple sclerosis and Canavan disease. Several glial-specific promoters such as those of glial fibrillary acidic protein (GFAP), mouse myelin basic protein (MBP) and mouse F4/80 have all been used to achieve selective rAAV-mediated expression in specific glial cell types. The most commonly used promoters and their defining characteristics are highlighted in Table 26.2.

TABLE 26.2 Characteristics of Promoters Used in rAAV Expression Cassettes

Promoter	Size (kb)	Cell Type-specific Expression	Level of rAAV-mediated Expression	Stability of rAAV-mediated Expression
Human CMV	0.7	Predominantly neurons (5% Glia)	Lower than NSE, PDGF, CBA, RSV	Expression decreases over time
Rous sarcoma virus (RSV)	0.4	Neurons	Higher than CMV	Expression detected 5 weeks post-injection
NSE	2.2	Predominantly neurons	Higher than CMV (300-fold) and GFAP (20-fold), lower than CBA (3-fold)	Expression stable >25 months
PDGF	1.5	Neurons	Higher than CMV	Expression stable >12 weeks
GUSB	0.4	Predominantly neurons (some glia)	Higher than CMV, equivalent to NSE	Expression stable >24 months
hSYN	0.48	Neurons	Higher than CMV, equal to CBA	Expression detected 5 weeks post-injection
CBA	1.7	Neurons and glia	Three-fold higher than NSE	Expression stable >18 months
GFAP	2.2	Predominantly glia	Higher than CMV (13-fold), lower than NSE (20-fold) in rat brain	Expression stable >3 months
MBP	1.9	Oligodendrocytes	Approximately 1,000 oligodendrocytes transduced per mouse brain	Expression detected 3 months post-injection
F4/80	0.67	Microglia	Microglial expression surrounding injection site	Expression detected 3 weeks post-injection

26.1.3 Regulatory Elements

Optimizing transgene expression at the post-transcriptional level can enhance the potency of gene delivery systems. Post-transcriptional enhancement frequently involves the addition of introns to the 3′ end of the RNA of interest to stimulate increased gene expression (Choi *et al.*, 1991). Although the exact mechanism of this stimulation is not known, intronic sequences, or the process of splicing itself, may promote 3′ processing and/or facilitate cytoplasmic transport. Cell-type-specific promoters are often combined with enhancer elements, such as the woodchuck hepatitis virus post-transcriptional regulatory element (WPRE) or short intronic sequences, for improved transport of the transgene messenger RNA (mRNA) from the nucleus to the cytoplasm. Although the addition of WPRE adds an additional 600 bp to the expression cassette, this is outweighed by the marked enhancement in expression levels that can be realized. In the context of rAAV vectors, WPRE has been shown to increase transgene expression several fold following injection into the brain parechyma (Paterna and Bueler, 2002). The specific mechanism of this enhancement, while not well understood, is known to be posttranscriptional in nature and may involve the facilitation of RNA processing and/or export. Frequently the delivery of two genes from the same vector construct is desired. This can be achieved using another common regulatory element called an internal ribosome entry site (IRES). An IRES initiates translation in a cap-independent manner thereby allowing synthesis of two proteins from a single bicistronic mRNA. Several IRESs have been used successfully in rAAV vectors including those from the encephalomyocarditis virus, poliovirus and hepatitis C virus.

26.1.4 Serotype Selection

Over the past decade, rAAV vectors have rapidly advanced to the forefront as the gene transfer vector of choice for multiple disease indications (see Chapter 11). In addition to their excellent safety profile, broad cellular tropism and ability to establish long-term transgene expression, the exponential progress of AAV-based vectors has been catalyzed by the isolation of several naturally occurring AAV serotypes and over 100 AAV variants from different animal species. There are currently ten serotypes of AAV described with different tropisms or ability to transduce specific cell types and organs. Although numerous AAV serotypes have been identified with different tropisms, studies have shown that these vectors share some common properties including genome size, organization and inverted terminal repeats (Bartlett *et al.*, 2000). These serotypes are ideally suited for development into clinical gene therapy vectors due to their diverse tissue tropisms and potential to evade pre-existing neutralizing antibodies to the viral capsids. Utilization of alternative AAV serotypes can lower the vector load, due to their higher transduction efficiency, and potentially help evade pre-existing neutralizing antibodies generated as a result of immune response to natural infection or prior treatment with rAAV-based vectors.

Researchers have found significant differences in tissue distribution and pharmacokinetics (onset, duration of expression and elimination) of the various AAV serotypes. In all of these studies, the tissue distribution of AAV serotypes has been shown to vary depending on the route of delivery. From a clinical and therapeutic point of view, it is desirable to know the tissue distribution and pharmacokinetics of the various AAV serotypes in order to exploit this natural variation in tropism for targeting different brain regions. There is active research underway to examine this variability among the serotypes and their ability to transduce various cells and tissues using different routes of administration. Further work is also ongoing to understand the factors that contribute to the differences in transduction efficiencies with the different serotypes of AAV.

26.2 CNS Delivery

The route by which the gene transfer vectors can be efficiently delivered to the CNS remains a significant challenge to gene therapy for treatment of neurological disease. Although various peripheral organ systems can be efficiently transduced following systemic delivery of viral vectors, robust transduction of the CNS using this method remains elusive. Challenges to delivery into the brain include (1) the limited and risk-prone access through the skull, (2) the presence of compartmentalized and discrete functional domains, (3) complex circuitry, (4) sensitivity to volumetric changes and (5) the presence of highly specialized barriers.

The barriers surrounding the CNS regulate brain homeostasis and the transport of endogenous and exogenous factors by controlling their selective and specific uptake, efflux and metabolism in the brain. There are three barriers that limit drug transport to the brain parenchyma. These are the blood–brain barrier (BBB), localized in the capillaries in the brain (Fig. 26.1); the blood–cerebrospinal-fluid barrier (BCSFB), which is presented by the choroid plexus epithelium in the ventricles; and the ependyma, which is an epithelial layer of cells covering the brain tissue in the ventricles and limits the transport of factors from the CSF to the brain tissue. Unfortunately, most candidate therapeutics for brain disorders are unable to cross these barriers from the systemic circulation. Consequently, specialized delivery and targeting strategies must be developed to enhance the transport and distribution of gene therapy agents into the brain.

Three main routes of delivery have been extensively examined in animal models, and in some cases clinical trials, to address both focal and global delivery of viral vectors. These include craniotomy-based methods involving direct stereotactic injection into the brain, intrathecal delivery and, more recently,

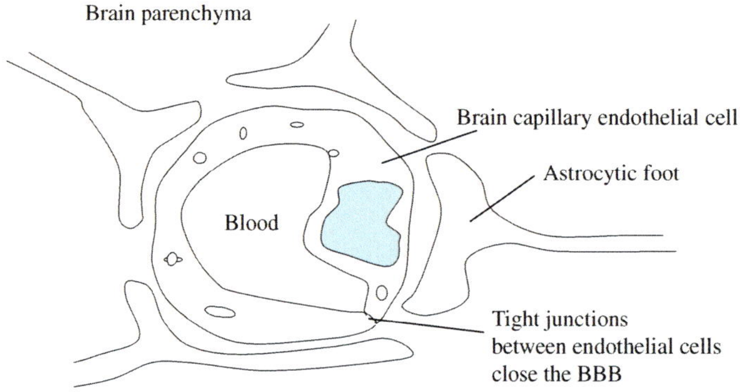

FIGURE 26.1 ■ The blood–brain barrier (BBB) results from the existence of tight junctions between the brain capillary endothelial cells. Only hydrophobic compounds, or hydrophilic molecules such as glucose or large neutral amino acids recognized and transported by specific carriers, or transcytosed proteins such as transferrin, can cross the BBB.

intravascular or systemic delivery. Additionally, there are also several strategies that have been implemented to maximize or facilitate gene transfer into the CNS through the use of osmotic agents as well as endogenous transport systems within the brain.

26.2.1 Craniotomy-Based Methods

26.2.1.1 *Intraparenchymal delivery*

Direct surgically guided injection of gene transfer vectors into the brain has been the delivery route of choice in all gene therapy trials for CNS diseases conducted to date. This method not only bypasses the BBB altogether, but it limits the potential for systemic toxicity. Additionally, the quantity of virus required for focal delivery to the target therapeutic area is minimized. Many of the delivery techniques currently used in gene therapy applications are derived from neurosurgical operations designed to lesion brain areas or implant devices such as deep-brain-stimulating electrodes. Stereotactic methods permit precise three-dimensional targeting of brain structures, and have been aided by advanced imaging techniques and computer-assisted reconstruction and navigation. Current neurosurgical methods depend largely on computed tomography (CT) and magnetic resonance imaging (MRI) monitoring; however, new advances in functional MRI and electrophysiological monitoring have further improved the accuracy of these surgeries.

Although intraparenchymal injection is the most direct approach to bypass the BBB, this technique is often limited by the small volumes (and therefore the dose) that can be injected into focal brain areas. Diffusion of viral vectors is minimal in brain tissue, thus limiting transduction to cells that are within millimeters of the injection site (Davidson and Breakefield, 2003). Some vectors, however, such as vectors derived from HSV (Chapter 12), can traffic by retrograde or anterograde transport within neuronal processes and traverse greater distances. Additionally, neuronal processes can also serve to spread transgene products, particularly those that are secreted. For example, retinal neurons infected in the eye with rAAV vectors project back through the optic tract and release corrective lysosomal enzyme to the brain proper. Vector spread in the CNS may also be increased by distributing the vectors in columns, increasing the number of injection sites or by using fluid convection (bulk flow), which can be achieved by maintaining constant pressure gradients during intraparenchymal infusion (Bobo *et al.*, 1994). The latter method can yield a more homogeneous and significantly increased distribution of viral vectors (Bankiewicz *et al.*, 2000). Despite the potential for viral vector spread beyond the injection site, the possibility of distribution throughout the entire CNS using direct intraparenchymal delivery methods remains unlikely, and thus this method of delivery may only be sufficient for targeting diseases with focal and discrete pathogenesis.

26.2.1.2 *Intracerebroventricular delivery*

Intracerebroventricular (ICV) injection of gene transfer agents is another emerging method for potentially effecting broad CNS delivery of therapeutics. Studies in animal models suggest that ICV delivery may be a promising approach to treating neurodegenerative diseases that affect broad regions of the CNS. Injection of viral vectors into the ventricles leads to their spread via the CSF and typically results in periventricular delivery (Passini and Wolfe, 2001). This delivery strategy is thus well suited for delivery of secretory proteins throughout the CNS. Recently it was demonstrated that ICV delivery of AAV vectors encoding insulin-like growth factor (IGF)-1 and vascular endothelial growth factor (VEGF)-165 results in efficient transduction of ependymal cells in adult mice and widespread delivery of these trophic factors throughout the CNS (Dodge *et al.*, 2010).

Although ICV injection bypasses the BBB, the BCSFB presents a limitation to penetration and diffusion of therapeutic agents from the ventricles into the brain parenchyma. This barrier segregates the brain from the CSF space and is functionally and anatomically distinct from the BBB. The immune system within the ventricular lining is more robust than that in the normal brain parenchyma and consists of specialized glial cells that contribute to a unique and heterogeneous epithelial layer providing a physical barrier against pathogen infiltration. Recombinant retrovirus, known for its immunogenicity, has produced proliferation of reactive astrocytes, loss of ependymal cells and lymphocytes infiltration near the ventricular wall in the adult mouse brain following ICV delivery. Less immunogenic viruses such as AAV, have demonstrated the ability to achieve extensive ependymal transduction and mediate long-term gene expression without any apparent toxicity. The use of ICV injection as a route to effect gene transfer will need to be carefully evaluated in both small- and large-animal models prior to clinical use. Presently, there is significant research underway to identify viral vectors and serotypes with acceptable safety profiles that can facilitate broad CNS biodistribution following ICV delivery.

26.2.2 Intrathecal Delivery

Intrathecal gene delivery into the spinal cord may produce widespread distribution of viral vectors and allow diffusion of secreted transgenic proteins throughout the CSF. This route of delivery is particularly attractive because access to the CSF space is minimally invasive and frequently performed on human patients. Administration of viral vectors directly into the CSF circulation bypasses the BBB and allows for transduction of neuroepithelial cells and the CNS delivery of transgene products through the ventricular circulation. It has been shown that administration of recombinant adenoviral vectors into the CSF of non-human primates leads to efficient transduction of the meninges covering the brain and spinal cord (Driesse *et al.*, 1999), making this delivery strategy particularly useful for gene therapy applications for secreted proteins.

Intrathecally administered viral vectors have successfully led to the production of amounts of bioactive proteins such as interleukin (IL)-2, IL-4, nerve growth factor (NGF) and B-cell leukemia/lymphoma-associated protein 2 (BCL-2). In rodent models, intrathecal administration of poliovirus-based vectors into the CSF has also been shown to result in significant transgene expression in both the brain and spinal cord. Just as the case for ICV delivery, intrathecal delivery methods must also contend with the brain-CSF barrier, which will limit penetration and diffusion of therapeutic agents from the intrathecal space into the parenchyma. The host immune response to intrathecally administered viral vectors must also be carefully examined prior to use in treating chronic diseases where prolonged expression or repeated injections of the therapeutic vectors are required. Encouragingly, it has been shown that intrathecal injection of a helper-dependent adenoviral vector into the CSF of naïve rhesus monkeys results in the transduction of neuroepithelial cells and long-term transgene expression in the absence of inflammatory or immune responses (Butti *et al.*, 2008). Delivery into an immune-privileged area, such as the CSF space, where vector-induced inflammatory or immunogenic responses may be less pronounced, represents a safe and feasible method to deliver therapeutic vectors to the CNS for prolonged periods in human subjects.

26.2.3 Systemic Delivery

Systemic or intravenous administration of gene transfer vectors represents the least invasive means of delivering transgenes throughout the CNS, but this strategy is severely hindered by the presence of the BBB. This anatomical and physiological barrier is formed by tight junctions between the endothelial cells of the CNS and restricts the transport of molecules, including conventional gene therapy vectors, from the bloodstream to the CNS. Systemic delivery may only be possible by either temporarily disrupting the BBB, typically achieved using osmotic agents, or accessing endogenous transport systems within the BBB that serve to permit selective transport under normal conditions. Some vectors may be better suited than others for this delivery approach. Recombinant AAV9 vectors that package a double-stranded genome, called "self-complementary" (sc) AAV vectors provide substantial advantages compared with their conventional single-stranded (ss) counterparts for brain transduction (McCarty, 2008). Recently, it was shown that a double-stranded vector (scAAV9) could produce efficient spinal cord transduction after intravenous delivery in both neonate and adult rodents and cats. The unprecedented potential of this non-invasive gene delivery strategy opens up promising new perspectives for the treatment of CNS disorders. It should be cautioned, however, that much work still needs to be done to assess the long-term safety of these approaches. The presence of pre-existing neutralizing antibodies to AAV in the systemic circulation will present a significant hurdle to the development of this delivery strategy.

26.2.4 Bypassing the Blood–Brain Barrier

An alternative to direct injection of viral vectors into the brain to correct global neurodegenerative disease is to take advantage of the vasculature of the host. A less invasive approach to facilitate global delivery of viral vectors to the CNS involves transport through the vascular endothelium using transport/carrier systems. Yet another approach involves transiently disrupting the tight junctions of the vascular endothelia using osmotic and chemical disruption agents, allowing for direct vector access to the underlying parenchyma.

26.2.4.1 *Transport/carrier systems*

The BBB is a dynamic organ that combines restricted diffusion to the brain of endogenous and exogenous compounds with specialized transport mechanisms for essential nutrients. Endogenous BBB transport systems include carrier-mediated transporters (CMT) such as glucose and amino acid carriers, receptor-mediated transcytosis (RMT) for insulin or transferrin and active efflux transporters such as *p*-glycoprotein.

Small water-soluble nutrients and vitamins can rapidly traverse the BBB via CMT. The CMT systems mediate the blood-to-brain transport of nutrients such as glucose, amino acids and purine bases, and utilize the GLUT1 glucose transporter, the LAT1 large neutral amino acid transporter, the MCT1 monocarboxylic acid transporter, the CNT2 concentrative nucleoside transporter, in addition to many other small molecule transporters (Pardridge, 2001). The CMT systems can serve as portals of entry for small-molecule drugs that have molecular structures similar to endogenous nutrients (such as L-3,4-dihydroxyphenylalanine (L-DOPA)). In addition to the CMT systems, certain large peptides or plasma proteins are selectively transported across the BBB via RMT systems. These transporters include the insulin receptor, the transferrin receptor (TfR) and the leptin receptor (Pardridge, 2006). Through the use of genetic engineering to develop recombinant fusion proteins, scientists have capitalized on these receptor systems to bypass the BBB. This approach, known as the molecular Trojan horse (MTH) approach involves fusing a therapeutic peptide or protein drug to a MTH, which is a second peptide or peptidomimetic monoclonal antibody that binds a specific receptor on the BBB. The Trojan horse enables RMT of the fusion protein across the BBB, thereby facilitating delivery of the protein drug into the brain to exert the desired pharmacological effect.

The MTH approach has traditionally been used in conjunction with non-viral gene delivery systems. CMT systems have also been co-opted for gene therapy applications using viral vectors. For example, it has been shown that viral vectors can traverse the BBB by targeting to the TfR (Xia *et al.*, 2000). The TfR is present on brain vascular endothelium and *in vitro*, *in vivo* and *ex vivo* studies have shown that antibody or transferrin conjugates with specificity for the TfR allowed for delivery of entities to brain capillary endothelial cells. Adenoviruses with motifs targeting the TfR have also been shown to allow for transport through the brain microcapillary endothelium. Modification of the virus for targeting to the TfR can be accomplished through the use of bifunctional antibodies or by genetically modifying the virus to display a specific TfR binding motif. Experiments also showed that adenovirus capsids modified to contain a carboxyl-terminal poly-Lysine tract allowed for improved gene transfer via facilitated binding to cell surface heparan sulfate proteoglycans. Several short motifs were used to support human TfR-targeted transduction using recombinant adenovirus vectors (Xia *et al.*, 2000). It will be important to evaluate whether these epitopes can also be used to improve gene transfer of other encapsidated viruses, such as AAV. Targeting recombinant viral vectors to a receptor expressed on brain microvessel endothelial (BME) cells could be a step toward evaluating whether the vascular system can be used to facilitate global distribution of genes to the CNS, thereby eliminating the need for multiple and highly invasive parenchymal injections of viral gene delivery vectors.

26.2.4.2 *Osmotic disruption*

The BBB plays an important role in preventing microorganisms and large molecules from entering the CNS. As such, it represents a major obstacle to developing new therapies for CNS diseases since it impedes effective delivery of therapeutic agents, including viral vectors, into the CNS. Many studies have been performed to interrupt the BBB, with the ultimate goal of enabling peripherally delivered therapeutic substances to cross the BBB.

Mannitol, a well-characterized osmotic agent, has long been administered by intravascular infusion in routine medical practice for various purposes, the most important of which has been the temporary opening of the BBB. Intravenous infusion of highly concentrated mannitol (25%) is used to reduce the intracranial pressure in patients with traumatic brain diseases, by temporarily increasing vascular osmotic pressure. Many human clinical studies have been performed with intra-arterial infusion of mannitol to open the BBB and enhance CNS delivery of chemotherapeutic reagents, with no obvious damage to the BBB. Mannitol has been used in preclinical studies to achieve the entrance of a wide range of substances into the brain, including enzymes, antibodies and viral vectors. Recent studies using systemic delivery of sscAAV vectors have incorporated mannitol infusions to successfully achieve broad distribution of AAV-mediated transgene expression in the rodent CNS (Valori *et al.*, 2010). However, as the BBB serves to protect the brain, breach of this barrier can have toxic consequences through fluid influx (leading to edema), changes in electrolyte balance and access to blood-borne pathogens. Further optimization of this procedure, combining systemic scAAV gene delivery with mannitol infusion, could facilitate the development of treatments for global CNS diseases, especially diseases involving both the somatic system and the CNS, such as lysosomal storage disorders (LSD).

26.3 Therapeutic Strategies and Examples of Gene Therapy of Neurological Diseases

It is evident that gene therapy is emerging as a potentially potent technology platform for the clinical treatment of a number of neurological disorders previously considered refractory to conventional therapeutic interventions. Therapeutics for these indications include the underlying causative genes and neurotrophic factors that affect the survival and function of neurons, as well as aberrant secondary metabolic and neurotransmitter functions. Recent advances in the field have led to the emergence of two primary approaches. The first involves either replacing the missing or defective gene product or removing the disease-causative gene product. The second approach involves providing factors that prevent neurodegeneration or increase function. Both of these methods have been used with some success in both preclinical and clinical studies designed to test the efficacy of gene transfer strategies in the amelioration of CNS disorders.

26.3.1 Replacement of Missing/Defective Gene Product

26.3.1.1 *Lysosomal storage diseases*

LSD are inherited diseases marked by deficiencies of lysosomal enzymes. This loss results in storage of undegraded substrates in the lysosomes and consequent cellular dysfunction and death. There are more than 40 forms of LSD, with a combined incidence estimated at 1 in 7,000 live births. The majority of LSD are caused by the deficient activity of a single lysosomal enzyme. As lysosomal enzymes are ubiquitously expressed, a deficiency in a single enzyme can affect multiple organ systems and result in a broad spectrum of clinical manifestations. LSD are generally classified by their accumulated substrate and they include sphingolipidoses, glycoproteinoses, mucolipidoses, mucopolysaccharidoses and others.

With a few exceptions, these disorders lead to a severe neurodegenerative phenotype, which complicates potential treatment strategies. Gene therapy represents a promising approach for treating LSD, as it has the potential to provide a permanent source of the deficient enzyme. Some features of LSD make them attractive candidates for gene therapy. Lysosomal enzymes are intracellular proteins localized within membrane-bound vesicles; however a small proportion of the mature enzymes are secreted from the cell. The secreted enzymes can be endocytosed and targeted to the lysosome via the cation-independent mannose-6-phosphate receptor, present on most cell types (Kornfeld, 1992). Genetic modification of a depot organ (such as liver, lung, or muscle) may allow for production and secretion of therapeutic levels of the deficient enzyme throughout the body (Cheng and Smith, 2003). This secretion and endocytosis mechanism, referred to as cross-correction, forms the basis for most therapeutic approaches that have been developed for LSD to date, such as bone marrow transplantation (BMT) and enzyme replacement therapy (ERT). BMT and ERT are relatively effective at treating the systemic disease; however, the efficacy of ERT in treating CNS disease is still under investigation. The development of novel strategies to deliver genes to neurons throughout both the peripheral and CNS would be necessary to achieve the continuous expression of the proper therapeutic enzyme and correction of the metabolic defect. The possibility of combining disparate approaches such as BMT and CNS-directed gene therapy might increase treatment efficacy in LSD with CNS involvement.

Both small- and large-animal models of LSD have been generated that recapitulate characteristics of the human disease. The results of CNS-directed gene therapy in these models have been encouraging. Viral vectors including adenovirus, AAV, lentivirus and herpes virus have all been delivered directly to the CNS of both small-(rodents) and large-(canine and feline) animal models of LSD with some success. Injection of viral vectors into the brain parenchyma or into the ventricular system reduced lysosomal storage in the CNS in several animal models of LSD. Functional improvements have also been observed in mouse models following virus-mediated transduction of the CNS. Mice with mucopolysaccharidosis (MPS) IIIB showed significant improvements in behavioral tests of anxiety following CNS-directed AAV-mediated gene therapy (Cressant *et al.*, 2004). Intracranial injection of AAV2 encoding human acid sphingomyelinase (ASM) in the mouse model of Neimann–Pick A disease showed a large reduction of storage product with improved rotarod scores (Passini *et al.*, 2007). Large-animal models of LSD have been used to determine the scalability of gene therapy to a more appropriately sized organism. Dog models of MPS I have shown widespread reduction in lysosomal storage following two injections into each cerebral hemisphere of an AAV vector expressing L-iduronidase (IDUA) (Ciron *et al.*, 2006). Although it may take more injections of the viral vector to effectively treat the CNS disease in human subjects, these data suggest that this direct delivery approach is clinically feasible.

Research exploring the potential of gene therapy for LSD has been performed largely in preclinical animal models; however human experimentation has commenced to evaluate viral-vector-mediated gene therapy of Canavan disease and late-infantile neuronal ceroid lipofuscinosis (LINCL). Canavan disease, a childhood leukodystrophy, is a monogeneic, autosomal recessive disease in which the gene coding for the enzyme aspartoacylase (ASPA) is defective. The lack of functional enzyme leads to an increase in the substrate molecule, *N*-acetyl-aspartate (NAA) within the CNS, which impairs normal myelination and results in spongiform degeneration of the brain. A clinical trial designed to deliver a rAAV vector encoding ASPA directly to affected regions of the brain in patients with Canavan disease has been initiated (Janson *et al.*, 2002). This gene transfer study represents, to our knowledge, the first clinical use of AAV in the human brain and the first instance of viral gene transfer for a neurodegenerative disease. A similar study in LINCL has also been initiated using a rAAV vector encoding tripeptidyl peptidase I (Souweidane *et al.*, 2010). Although there were indications of adverse events associated with the procedure, subjective analysis of the treated subjects suggested a reduced rate of neurological decline. Data generated by these studies are informative toward the design of future human clinical trials employing viral-vector-mediated gene augmentation strategies in the CNS.

26.3.1.2 *Spinal muscular atrophy*

Spinal muscular atrophy (SMA), the most common autosomal recessive neurodegenerative disease affecting children, results in impaired motor neuron function (Burghes and Beattie, 2009). SMA is caused by loss or mutation of the telomeric copy of the survival motor neuron gene (*SMN1*) and retention of the *SMN2* gene. The lack of SMN protein results in motor neuronal degeneration in the ventral (anterior) horn of the spinal cord, leading to weakness of the proximal muscles responsible for crawling, walking, neck control and swallowing and the involuntary muscles that control breathing and coughing. While the precise roles of SMN and the relative contributions of cell body and axon functions that drive SMA disease is under investigation, it is reasonable to predict that the reconstitution of SMN levels in motor neurons should be corrective. Preclinical studies with small-molecule drugs that increase SMN levels have shown modest therapeutic benefit in animal models (Butchbach *et al.*, 2010). While some small-molecule drugs were shown to be well tolerated in SMA patients, clinical trials have not resulted in a robust efficacy, underscoring the need to generate new therapeutic strategies for this disease. Gene therapy is a potentially powerful approach for SMA because the genetic defect is known and the therapeutic gene to be delivered is clearly defined (SMN1). The most direct approach for gene therapy of SMA is to utilize a viral vector that encodes the full-length SMN protein to provide a continuous source of SMN.

A double transgenic knockout mouse model of SMA has been generated by replacing the mouse *SMN* gene; which is functionally equivalent to human *SMN1*, with two copies of the *SMN2* gene, resulting in mice with severe SMA. Using this aggressive mouse model, a number of studies have shown extension in lifespan and improved motor function following viral vector-mediated delivery of SMN. A one time postnatal delivery of SMN1, using either recombinant lentivirus or AAV vectors, resulted in significantly increased survival of this animal model (Valori *et al.*, 2010). Survival was further increased when injections were performed with a scAAV vector, a rAAV vector with double-stranded DNA genome that results in earlier onset of gene expression compared with traditional ssAAV vectors (Valori *et al.*, 2010).

Recently, it was demonstrated that a scAAV9 vector can transduce approximately 60% of motor neurons following temporal facial vein injections in neonatal mice and lead to significant functional and survival benefits in SMA mice (Foust *et al.*, 2010). To help move this treatment toward clinical trials, researchers also tested the ability of the scAAV9 vector to reach the spinal cord in a non-human primate. Extensive scAAV9-mediated motor neuron transduction was seen after injection into a newborn cynomolgus macaque, demonstrating that scAAV9 traverses the immature BBB in a non-human primate and emphasizes the clinical potential of scAAV9-mediated gene therapy for SMA (Foust *et al.*, 2010).

Although these studies strongly support the development of AAV-based technologies as a potential therapy for SMA, several barriers still remain. Determinations of safety and toxicity of these vectors as well as the timing of therapeutic intervention for a meaningful therapeutic benefit to patients need to be performed.

26.3.2 Prevention of Degeneration or Replacement of Function

For those diseases for which the underlying basis is still unknown but nevertheless exhibit a neurodegenerative phenotype, neuroprotective strategies could be considered. The identification of genes involved in neuronal differential, growth and survival has allowed for the development of potential approaches to retard and possibly reverse some neurodegenerative diseases. Gene delivery of neuropractive factors is currently under active investigation for Parkinson's disease (PD), Alzheimer's disease (AD) and amyotrophic lateral sclerosis (ALS). These strategies typically utilize genes for neurotrophic factors such as NGF, glial-derived neurotrophic factor (GDNF), brain-derived neurtrophic factor (BDNF) and others. Several neurotrophic factors have shown promise as therapeutic agents in animal models of these diseases.

26.3.2.1 *Huntington's disease*

Huntington's disease (HD) is an autosomal dominant neurodegenerative disorder resulting from a polyglutamine repeat expansion (CAG codon, Q) in exon 1 of the interesting transcript 15 (IT15) gene that confers a toxic gain of function on the protein huntingtin (Htt) (Gusella *et al.*, 1986). HD is a progressive disorder that typically develops in midlife, with the age of onset inversely correlated to CAG expansion length. Hallmark features of the disease include cognitive and behavioral disturbance, involuntary movements (chorea), neuronal inclusions, cortical thinning and severe striatal degeneration. Discovery of the gene that causes HD in 1993 prompted a surge in translational research, generation of a number of HD mouse models and creation of a genetic test to detect the mutant gene. The use of predictive genetic testing to identify pre-symptomatic carriers provides a unique opportunity in HD for early intervention, years prior to disease onset and neurodegeneration. The absence of effective therapies for HD however, have made at-risk individuals reluctant to take this test.

The use of gene therapy to reduce the expression of the mutant *huntingtin* gene product could slow down or prevent the onset of HD. Davidson and colleagues demonstrated that RNA interference (RNAi) might provide a powerful approach to shut down the disease-causing gene in HD and prevent accumulation of the toxic Htt protein (Harper *et al.*, 2005). RNAi is a form of post-transcriptional gene silencing mediated by short double-stranded RNA. Researchers have capitalized on this natural biological process to reduce the expression of target mRNAs using exogenously applied small interfering RNAs (siRNAs), short hairpin RNAs (shRNAs), or artificial microRNAs (miRNAs) which can be delivered to cells using viral vectors. By reducing striatal expression of mutant Htt using AAV1-delivered shRNA, amelioration of both motor and neuropathological abnormalities in a HD mouse model was achieved (Harper *et al.*, 2005). Therapies to directly reduce mutant gene expression through similar RNAi-based strategies may be generally applicable to the treatment of other dominant neurodegenerative disorders. The same laboratory previously demonstrated that RNAi can improve neuropathology and behavioral deficits in a mouse model of spinocerebellar ataxia type 1 (SCA1), a dominant neurodegenerative disorder that affects a population of neurons distinct from those degenerating in HD.

However, despite promising results in animal models, several issues still need to be addressed before the RNAi technology can be translated to the clinic. A major concern with this approach for HD is the possibility for toxicity incurred by the silencing of the endogenous wild-type *HTT* allele in affected patients. It is not yet known whether coincident reduction of the normal allele will have deleterious effects. The possible necessity for allele-specific silencing would add considerable complexity to this approach. All RNAi approaches tested to date have targeted both mutant and wild-type *HTT* alleles. Researchers are currently working on identifying RNA sequences that could selectively reduce the pathogenic allele (Gagnon *et al.*, 2010). Another consideration is the long-term effects of RNAi in the brain. Since this approach uses the cells' own molecular machinery that governs naturally occurring RNAi to accomplish targeted gene silencing, it is still unclear if this will be well tolerated with chronic treatment. Strategies that safely regulate siRNA expression if needed are still being addressed.

In addition to targeting the mutant huntingtin protein, gene therapy directed at a number of downstream targets might prove effective in delaying disease progression in HD. Positive outcomes have been seen in preclinical studies with drugs that increase transcription of neuroprotective genes (histone deacetylases), prevent apoptosis (caspase inhibitors), enhance energy metabolism (coenzyme Q, remacemide and creatine) and inhibit the formation of polyglutamine aggregates (trehalose, Congo red and cystamine). Gene therapy strategies focused on neuroprotection have also shown promise in HD mouse models. Viral-vector-mediated delivery of neurotrophic factors such as BDNF, neurotrophin (NT) 4/5 and ciliary neurotrophic factor (CNTF) have shown some benefit in animal models of HD. Use of gene therapy to modulate these downstream target genes and neurotrophic factors, in addition to reducing the disease-causing mutant *HTT* allele, provides a potential therapeutic strategy for the treatment of HD.

26.3.2.2 *Parkinson's disease*

PD is a slowly progressive neurodegenerative disease with no known single etiology. The predominant pathological feature is loss of dopaminergic neurons in the substantia nigra that project to the striatum and consequent decrease in the neurotransmitter dopamine (Figs. 26.2 and 26.3). Patients exhibit a range of clinical symptoms, mostly affecting motor function, including resting tremor, rigidity, akinesia, bradykinesia and postural instability.

The well-established link between the selective degeneration of dopaminergic neurons and the motor neurological deficits in PD provided defined targets for several gene therapy approaches using rAAV or lentivirus vectors. Some approaches are intended to increase the level of endogenous dopamine via direct delivery of genes involved in neurotransmitter synthesis (e.g., aromatic amino acid decarboxylase (AADC), tyrosine hydroxylase (TH) and guanosine triphosphate cyclohydrolase (GCH1) (Azzouz *et al.*, 2002) or enhance the function of the prodrug levodopa (L-DOPA) via delivery of amino acid decarboxylase that converts L-DOPA to dopamine (Bankiewicz *et al.*, 2006) (Fig. 26.4).

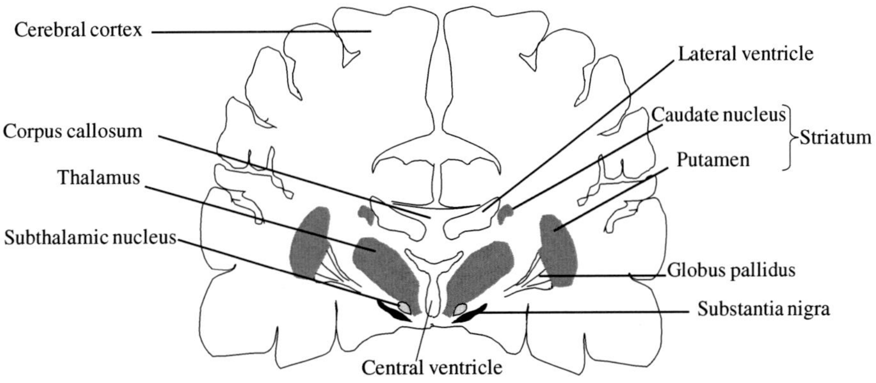

FIGURE 26.2 ■ Frontal view of human brain, displaying the major brain structures. The population of dopaminergic neurons originating from the substantia nigra is massively depleted in PD. This concerns particularly the dopaminergic projections to the striatum (caudate and putamen), which control movement. This dopamine depletion results in the movement disorders typical of PD.

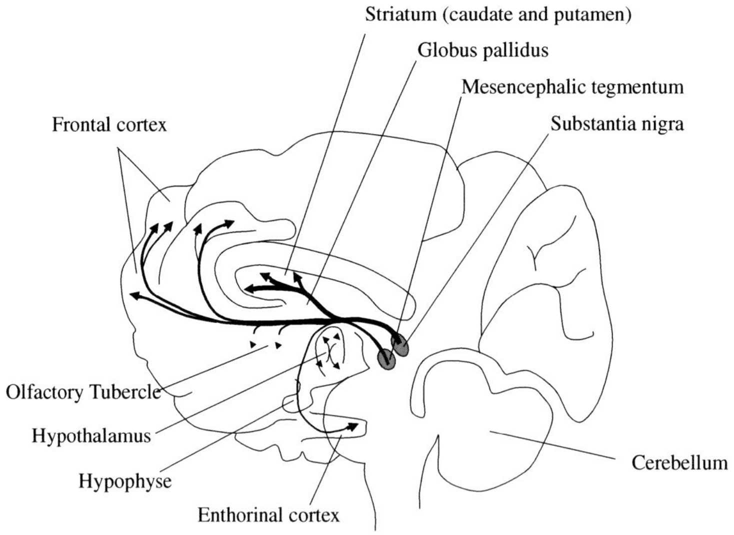

FIGURE 26.3 ■ In addition to their striatal projections, the dopaminergic neurons of the substantia nigra also project to the associative frontal cortex, the enthorinal cortex, which is involved in memory, and the olfactory tubercle which is a multi-sensory processing center in the olfactory cortex that plays a role in reward behaviors. The humor and memory impairments observed in PD are linked to the dopamine depletion of these brain structures.

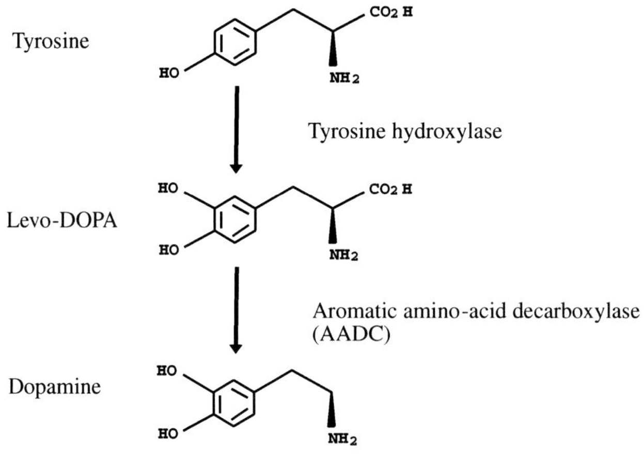

FIGURE 26.4 ■ The dopamine biosynthesis pathway.

Others gene therapy strategies in PD are intended to normalize basal ganglia circuitry by reducing the PD-related overactivity of specific brain structures by delivering glutamic acid decarboxylase (GAD), which decreases the amount of excitatory glutamic amino acid, to the subthalamic nucleus. Lastly, gene-therapy-mediated delivery of neurotrophic-factors, such as GDNF or neuturin (NTN), aimed at protecting the degenerating nigrostriatal pathway has also been widely investigated. These therapies seek to increase dopaminergic function and are potentially curative by slowing disease progression. On the basis of positive preclinical data, four different gene therapy approaches are currently in phase I or phase II clinical testing (Bjorklund and Kordower, 2010).

As PD is characterized by degeneration of dopaminergic neurons, one approach is to reconstitute the defective enzymes in the Parkinsonian striatum. Striatal delivery of AAV (Kirik *et al.*, 2002) or lentiviral vectors (Azzouz *et al.*, 2002), expressing the genes required for the synthesis of dopamine from tyrosine: Tyrosine hydroxylase, an oxidase which converts tyrosine to DOPA, GCH1 required for the synthesis of tetrahydrobiopterin, an essential AADC cofactor and AADC that converts DOPA to dopamine, resulted in motor improvement in rat and primate models of PD (Fig. 26.4). These studies served as the basis for an ongoing phase I/II clinical trial utilizing a multicistronic lentivirus vector to transfer the three genes: TH, AADC and GCH1 into striatal neurons of PD patients (clinical trial NCT00627588).

Gene delivery of AADC has also been pursued to increase the efficiency of converting peripheral L-DOPA to dopamine in the striatum. Preclinical studies demonstrated that delivery of human AADC via a rAAV vector (AAV-hAADC2) restored the ability of the striatum to convert L-DOPA to dopamine and corrected PD-like behavior in rat and primate models of disease. Two phase I open-label clinical trials, showing safety and preliminary amelioration of symptoms in 15 patients with advanced PD, were recently published employing this strategy (Muramatsu *et al.*, 2010). The data warrant further evaluation in a randomized, controlled phase II trial.

Motor abnormalities of PD are caused by alteration in basal ganglia network activity, including disinhibition of the subthalamic nucleus (STN). Preliminary efficacy studies in rats and primates, showed that AAV2-mediated expression of GAD, that is the enzyme which catalyzes the synthesis of the neurotransmitter gamma aminobutyric acid (GABA) in the subthalamic nucleus, resulted in neuroprotection of nigral dopamine neurons and rescued the Parkinsonism behavior phenotype. This approached also advanced to phase I and II clinical trials (clinical trial identifiers: NCT00643890 and NCT00195143). The open-label phase I trial established safety, tolerability and preliminary efficacy (Kaplitt *et al.*, 2007). A phase II randomized, double-blind, placebo-controlled trial to confirm the efficacy results is ongoing.

The well-established link between selective degeneration of nigrostriatal dopaminergic neurons and the neurological motor deficits in PD patients provides defined targets for neurotrophic-factor-based gene therapy. Neuroprotection has been achieved in rodent and primate animal models of PD by delivery of GDNF, a protein thought to support dopaminergic neuron survival and function, using AAV or lentiviral vectors (Bjorklund and Kordower, 2010). Another

GDNF member that supports dopaminergic neurons, NTN, has also been shown to effectively protect dopaminergic neurons in rodent and non-human primate models of PD. Safety, dosing and tolerability of intraputaminal delivery of AAV-NTN (Cere-120) were demonstrated in humans in a phase I clinical trial (Marks *et al.*, 2008). However, phase II clinical testing did not demonstrate any significant differences in the primary endpoint (Unified Parkinson's Disease Rating Scale (UPDRS)-motor off score) at 12 months between treated patients and controls. Further phase II trials with AAV-NTN are ongoing. These trials enrolled patients with advanced PD and presumably significant neuron loss; however, neuroprotective therapies may be more effective in patients at an earlier stage of disease.

26.3.2.3 *Alzheimer's disease*

AD is a neurodegenerative disorder associated with progressive functional decline, dementia and neuronal loss. Characteristic pathological manifestations include extracellular accumulation and aggregation of the amyloid beta (Aβ) peptide into plaques and intracellular accumulation of hyperphosphorylated tau, forming neurofibrillary tangles. The multifactorial causes of AD offer a variety of possible targets for gene therapy including the deployment of the neurotrophic factors, NGF and BDNF to promote neuronal survival. Another approach is to use Aβ-degrading enzymes, such as neprilysin, endothelin-converting enzyme and insulin-degrading enzyme, in order to halt Aβ-mediated toxicity.

One of the most critical elements for decline in memory and cognition in AD appears to be loss of cholinergic neurons, especially in the septohippocampal and basal cortical pathways. The administration of NGF prevents the death of basal cholinergic neurons and improves learning and memory in lesioned and aged rats and primates. Following these promising results, recombinant NGF was administered intraventricularly to AD patients; however, the trial was terminated as a result of the nociceptive response and other adverse effects triggered by uncontrolled distribution of NGF throughout the brain (Lim *et al.*, 2010). Towards further clinical use of NGF for AD, preclinical studies demonstrated that controlled and sustained release of NGF can confer a beneficial effect with less associated toxicity. A phase I clinical trial of AAV-NGF in AD patients demonstrated that AAV-NGF was safe and well tolerated, and cognitive testing suggested a reduced rate of cognitive decline (Bakay *et al.*, 2007). A phase II multi-center, sham-surgery-controlled trial with AAV-NGF is currently in progress.

New research also highlights the neuroprotective effects of BDNF in several AD animal models (Nagahara *et al.*, 2009). Rodent and primate models of AD administered lentiviral vectors expressing BDNF in the enthorinal cortex showed a reversal in synaptic loss and ameliorated learning and memory deficits without changing the levels of insoluble Aβ. It is known that BDNF levels are decreased in

AD and are inversely correlated to Aβ loads in animal models. It has recently been suggested, however, that increasing BDNF levels is unlikely to reverse Aβ pathology (Nagahara *et al.*, 2009). In this regard, short-term treatment with BDNF-based therapies may provide some therapeutic benefit for AD; however more research must be done to evaluate the extent of improvement.

There is a body of evidence from preclinical and clinical studies to support a major pathophysiological role of Aβ in AD (Hardy and Selkoe, 2002). Other factors, including hyperphosphorylated tau, apolipoprotein E (APOE)-associated lipid metabolism and inflammation (Ballatore *et al.*, 2007), are likely to contribute to AD pathology. In the majority of AD cases, accumulation of Aβ is caused by an imbalance between Aβ generation and clearance. Several proteases have been implicated in Aβ degradation. Among these proteases, neprilysin has been identified as the major Aβ-degrading enzyme, while the contributions of insulin degrading enzyme (IDE) and endothelin-converting enzyme (ECE) appear to be minor. Gene therapy involving Aβ-degrading enzymes represents a possible therapeutic strategy. Several studies in animals have indeed shown that gene delivery of neprilysin leads to efficient degradation of Aβ in both neprilysin knockout mice and amyloid β precursor protein (APP) transgenic mice, with demonstrable improvement in behavior in the APP mice (Nilsson *et al.*, 2010). Further studies in primates are warranted to evaluate the potential of neprilysin as a potential gene therapy target for AD.

26.3.2.4 *Amyotrophic lateral sclerosis (ALS)*

ALS is a devastating neurodegenerative disease that results from the progressive loss of motor neurons in the spinal cord, brainstem and cerebral cortex. This loss of motor neurons leads to gradual muscle weakness and atrophy, with death often occurring 2–5 years after diagnosis as a result of respiratory failure. To date, no effective therapy is available. Many hypotheses have been formulated to explain the observed selective degeneration of motor neurons, including protein misfolding and aggregation of mutant proteins, defective axonal transport, excitotoxicity, mitochondrial dysfunction, apoptosis and inadequate trophic support. Neurotrophic factors are considered a promising therapeutic approach for ALS since they can support neuronal maintenance and survival under conditions of stress without targeting a specific pathogenic mechanism. However, the delivery of neurotrophic factors through conventional delivery routes has been largely ineffective and associated with undesirable side effects (Boillee *et al.*, 2006). The weakness of these approaches was largely due to inadequate delivery to the appropriate cell populations, which has led to the call for the development of more efficient viral-mediated gene delivery techniques.

In animal models, efficacy has been achieved following either intraparenchymal or ICV delivery of AAV or lentivirus vectors in order to genetically modify target cells to secrete trophic factors within the CNS (Dodge *et al.*, 2008). Other methods harnessed the ability of certain viral vectors to be retrogradely transported from distal axonal projections following intramuscular delivery. Use of those strategies to deliver IGF-1 and/or VEGF slowed disease progression in the $SOD1^{G93A}$ mouse model of ALS, even when initiated after disease onset (Nanou and Azzouz, 2009). Not surprisingly, success was dependent on the trophic factors used. In addition, lentiviral and AAV vectors expressing functional RNAi have been used to knock-down levels of mutant superoxide dismutase (SOD1), following intraspinal or intramuscular injection into the $SOD1^{G93A}$ mouse model of familial ALS, to result in beneficial therapeutic outcome (Towne and Aebischer, 2009). Although these studies are promising, more research is needed to translate the use of viral-mediated gene transfer into a therapeutic strategy in ALS patients.

26.4 Critical Issues

Gene therapy as a therapeutic approach for CNS indications is currently the focus of intense investigation by institutions and companies around the world. This emerging technology holds great promise by offering radical new ways of treating debilitating and incurable diseases of the CNS. Gene therapy may hold the potential to treat, cure or ultimately prevent disease by changing the pattern of gene expression within the brain and/or spinal cord. Despite its promise, effectiveness in clinical trials has been limited, mainly due to physiological barriers unique to the CNS. Although experimental studies have demonstrated "proof of concept" with regards to the feasibility of this approach, a number of issues need to be addressed before it can become standard clinical practice.

26.4.1 Safety Concerns

Much of the data generated on the use of gene therapy for CNS disorders comes from studies of animal models of the respective diseases. While data from many of these studies support the use of this technology platform for CNS disorders, very little is known about potential side effects from short- or long-term gene delivery using these expression systems in humans. Several human gene therapy trials have now been conducted with a focus on safety. The main safety issues identified by these studies are vector/delivery-mediated toxicity, potential for inadvertent genetic modification of the germ line and generation of new epidemiological viral agents. It is clear that the most immediate risk in a human brain gene therapy trial typically relates to the surgical procedure. These risks include hemorrhage, stroke, anesthetic complications, infection, paralysis, coma or death. Although these neurosurgical risks are serious, they are not necessarily unique to gene therapy trials. Neurosurgical procedures are now routinely performed on human patients to treat a number of neurological conditions and the use of neuroradiological methods, such as computer-assisted imaging CT, MRI and positron emission tomography (PET), have greatly improved the precision and safety of these procedures.

Another potential risk may arise from the vector itself engendering an adverse host immune reaction against the vector and/or transgene product. This possibility was highlighted by the death of a patient after hepatic–arterial infusion of an adenovirus vector during a gene therapy trial for ornithine transcarbamylase (OTC) deficiency (Somia and Verma, 2000). This death was attributed to a massive cytokine response to the adenovirus vector, resulting in disseminated intravascular coagulation (Schnell *et al.*, 2001). This event clearly highlights the potential for immune-mediated toxicity from adenoviral vectors when delivered systemically and illustrates the importance of careful vector selection in combination with delivery technique. Immune-mediated vector toxicity has also been noted in experimental animals after direct injection of recombinant adenovirus and HSV-1 vectors into the brain. In contrast to the immunogenic adenovirus and HSV vectors, AAV vectors have shown an impressive lack of pathogenicity in both animal models and humans and AAV-based gene transfer vectors have gained increasing popularity as a vector of choice for human gene therapy trials. Clearly, several factors have a major influence on the immune response to viral vectors. These factors include route of administration, promoter choice, transgene sequence, immune status of human subjects and vector selection. Having been clearly defined, these factors can now be manipulated to reduce the risk of immune reactions in humans.

Other safety considerations with gene therapy include the possibility that introducing viral proteins into the CNS may activate replication of endogenous pathogenic viruses, or that replication-competent recombinant virus may be generated with uncharted pathogenic potential due to altered tropism or the incorporation of new genes. Analysis of human brains at autopsy suggests that almost half bear latent HSV-1 and potential activation and replication of this virus in the brain could cause fatal encephalitis. Experimental studies, however, have found that HSV vectors do not activate latent HSV-1 and thus have minimized these concerns. Another safety concern is that of insertional mutagenesis by integrating vectors such as lentivirus. Safety considerations when using these types of vectors must take into account the rare but distinct possibility that a replicating vector may be able to replicate in the relatively immune-privileged brain, causing direct toxicity by viral replication or gene expression. The risk of generating a "new breed" of virus with altered tropism that has the potential to spread among individuals would be a major epidemiological concern. These concerns have been addressed through the generation of "self-inactivating lentiviruses" which have significant modification to the wild-type genome and thus improved safety profiles (Iwakuma *et al.*, 1999).

Another safety consideration for insertional viral vectors would be the oncogenic potential of vectors, through deletion/mutation of cellular tumor suppressor genes or activation of oncogenes in the host cell genome through insertion of viral promoters. This risk, however, is lessened in the CNS since most cells in the brain are no longer dividing and DNA replication is an important component of mutagenesis and the recombination events leading to transformation. Furthermore, many cells in the nervous system are terminally differentiated such that oncogene activation would not trigger cell division. There are some populations of cells in the adult brain, such as astrocytes, that do have the potential to divide. Oncogenic potential remain a greater concern for gene delivery during the neonatal/childhood period when astrocytes are rapidly proliferating. Due to the risks of lentiviral vector recombination and insertional oncogenesis, other viral vectors have been gaining an important place in gene therapy approaches, especially non-integrating adenovirus and AAV.

Finally, regulation of both the level and location of gene expression represents a critical safety requirement for gene therapy. Expression in non-target cells or overexpression of gene products may prove pathogenic. For example, unregulated delivery of trophic growth factors for neurodegenerative diseases may produce unwanted consequences. High-level NGF expression in the brain can lead to ingrowth of peripheral sensory neurites into the brain parenchyma, leading to pain and disruption of local circuitry. These concerns are currently being addressed through the use of regulatable and cell-type-specific promoters.

26.4.2 Control of Gene Expression

Regulated and sustained transgene expression are important considerations for gene delivery to the brain. Due to the complicated and potentially risky neurosurgical procedure, the number of invasive interventions needs to be limited. As a result, transgenes should optimally confer sustained therapeutic levels of expression. Given that most studies of experimental animals do not extend beyond months to one year, it is unclear whether the promoter constructs currently available can achieve long-term stable transgene expression in these animal models. However, there is a report of therapeutic benefit for 8 years following a single administration of an AAV vector into a primate model of PD (Hadaczek *et al.*, 2010). Loss of expression over time could occur due to promoter inactivation or loss of episomal vector genomes and cytotoxicity due to low levels of viral gene expression. The strength of the promoter and its ability to avoid inactivation are key to achieving prolonged transgene expression. For many disorders, low levels of transgene product may be sufficient to restore normal function. The ability to control the level and longevity of gene expression can be accomplished using cell-type-specific promoters as well as inducible/repressible promoters. Cell-type-specific promoters can be employed to restrict transgene expression to specific cell populations and minimize inappropriate gene expression in non-target regions. With inducible and repressible promoters, the gene is transcribed when the promoter is either induced or not repressed. In the repressible tetracycline system, administration of this antibiotic prevents transgene expression. Currently these inducible systems have limited utility in the CNS due to the limited permeability of tetracycline across the BBB. Other inducible expression systems such as the inducible ecdysone promoter system (RheoSwitch) or the light-activated channel rhodopsin are currently in the preclinical testing stages and represent a means with which to achieve both real-time and persistent control over gene expression.

26.4.3 Risk:Benefit Ratios

The ethical requirement of a favorable risk:benefit ratio in clinical trials involve minimizing risks, enhancing potential benefits and ensuring that the risks to the subjects are justified by the potential benefits to the subjects and/or society. The major concerns during these trials fall into three categories: (1) risk of adverse events to the patient in the near or distant future, which may be worse than the condition being treated, (2) potential public health risks associated with the spread of novel viral agents and (3) the possibility of genetic contamination of the germ line (for systemic delivery). The last two concerns represent serious public health risks that regulatory agencies, such as the Food and Drug Administration (FDA), and advisory bodies, such as the National Institutes of Health Office of Biotechnology Activities (NIH OBA), will need to evaluate in order to envisage potential problems and minimize their likelihood. This can be achieved by carefully evaluating the preclinical studies and providing rigorous monitoring of clinical trials. With regard to patient risk, the neurosurgical procedure to administer the gene therapy agent is often a more immediate risk than even the most lethal of neurodegenerative disorders. Adverse surgical events, as well as inadvertent toxic insults to the brain can cause irreparable brain damage, compromise cognitive functions and induce physical disabilities. Therefore, the risk of the procedure must be evaluated in light of the severity of the disease and the age of the patient. During a phase I trial, it is optimistic to anticipate benefit from any gene therapy procedure for neurological disease and the hope that a particular procedure might benefit current or future patients is a balancing factor.

26.5 Conclusions

The success of gene therapy for treating CNS diseases depends upon knowledge gained from preclinical studies, evaluation of patient safety based on phase I clinical trials, identifying enhanced procedures for early diagnosis of degenerative diseases, developing non-invasive delivery methods and finally addressing ethical issues related to gene therapy for neurodegenerative disorders. As knowledge of vector biology is gained, clinical investigations will need to modify endpoint measures used to evaluate safety in humans. For gene therapy to ultimately be a standard practice of care, patient safety must remain the foremost concern. Today, over 100 clinical trials are ongoing or completed for Parkinson's disease and Alzheimer's disease alone (www.clinicaltrials.gov). The overall number of clinical trials worldwide using AAV, adenoviral, HSV or lentiviral vectors for human disease has gone from the first five proposed trials in 1994 to a cumulative 465 proposed, open or completed trials in 2009 (Lim *et al.*, 2010). The major hurdles to gene therapy, which include safety and efficacy, continue to be roadblocks to its widespread application as a standard medical treatment for neurodegenerative disease. Improvement of efficacy can be mediated in part by the development of more efficient vectors. AAV-based vectors have managed to bypass the main gene therapy hurdles, such as long-term and stable transgene expression in many tissues, safety, broad range of target diseases and lack of immunogenicity and pathogenicity; and are increasingly becoming the vector of choice for a wide range of gene therapy approaches.

Once a disease is diagnosed and a gene therapy agent chosen, the vector must be delivered to the appropriate brain region. Preclinical studies often use invasive procedures to demonstrate efficacy, but such procedures can be impractical to implement in humans. Current human clinical trials on viral gene delivery to the brain still rely on invasive neurosurgical procedures. Developing minimally invasive procedures for gene delivery while still ensuring sufficient transduction of target areas will help ensure the advancement of this therapeutic strategy. Vectors capable of delivering genes to the CNS via injection into the CSF or blood would prove most valuable. Research on vector engineering, both viral and non-viral, will lead to more specific transgene targeting and expression. Ideally, a vector should be capable of crossing the BBB, use a cellular receptor for cell-specific entry and deliver transgenic DNA that contains a drug-regulated, cell-restricted promoter to express a therapeutic gene. Finally, irrespective of the scientific issues, there are ethical issues regarding clinical trials that must also be addressed. Informed consent for neurological disease involving impaired cognitive function can be complicated at best. Despite these major hurdles, scientists are working to develop novel solutions to successfully advance gene therapy for CNS diseases into the clinic.

References

Azzouz, M., Martin-Rendon, E., Barber, R.D., *et al.* (2002). Multicistronic lentiviral vector-mediated striatal gene transfer of aromatic L-amino acid decarboxylase, tyrosine hydroxylase, and GTP cyclohydrolase I induces sustained transgene expression, dopamine production, and functional improvement in a rat model of Parkinson's disease, *J Neurosci*, **22**, 10302–10312.

Bakay, R., Arvanitakis, Z., Tuszynski, M., *et al.* (2007). Analyses of a phase 1 clinical trial of adeno-associated virus-nerve growth factor (CERE-110) gene therapy in Alzheimer's disease: 866, *Neurosurgery*, **61**, 216.

Ballatore, C., Lee, V.M., Trojanowski, J.Q. (2007). Tau-mediated neurodegeneration in Alzheimer's disease and related disorders, *Nat Rev Neurosci*, **8**, 663–672.

Bankiewicz, K.S., Eberling, J.L., Kohutnicka, M., *et al.* (2000). Convection-enhanced delivery of AAV vector in Parkinsonian monkeys; *in vivo* detection of gene expression and restoration of dopaminergic function using pro-drug approach, *Exp Neurol*, **164**, 2–14.

Bankiewicz, K.S., Forsayeth, J., Eberling, J.L., *et al.* (2006). Long-term clinical improvement in MPTP-lesioned primates after gene therapy with AAV-hAADC, *Mol Ther*, **14**, 564–570.

Bartlett, J.S., Wilcher, R. Samulski, R.J., (2000). Infectious entry pathway of adeno-associated virus and adeno-associated virus vectors, *J Virol*, **74**, 2777–2785.

Baskar, J.F., Smith, P.P., Ciment, G.S., *et al.* (1996). Developmental analysis of the cytomegalovirus enhancer in transgenic animals, *J Virol*, **70**, 3215–3226.

Bjorklund, T., Kordower, J.H. (2010). Gene therapy for Parkinson's disease, *Mov Disord*, **25 Suppl 1**, S161–S173.

Bobo, R.H., Laske, D.W., Akbasak, A., *et al.* (1994). Convection-enhanced delivery of macromolecules in the brain, *Proc Natl Acad Sci USA*, **91**, 2076–2080.

Boillee, S., Vande Velde, C., Cleveland, D.W. (2006). ALS: A disease of motor neurons and their nonneuronal neighbors, *Neuron*, **52**, 39–59.

Burghes, A.H., Beattie, C.E. (2009). Spinal muscular atrophy: Why do low levels of survival motor neuron protein make motor neurons sick?, *Nat Rev Neurosci*, **10**, 597–609.

Butchbach, M.E., Singh, J., Thorsteinsdottir, M., *et al.* (2010). Effects of 2,4-diaminoquinazoline derivatives on SMN expression and phenotype in a mouse model for spinal muscular atrophy, *Hum Mol Genet*, **19**, 454–467.

Butti, E., Bergami, A., Recchia, A., *et al.* (2008). Absence of an intrathecal immune reaction to a helper-dependent adenoviral vector delivered into the cerebrospinal fluid of non-human primates, *Gene Ther*, **15**, 233–238.

Cheng, S.H., Smith, A.E. (2003). Gene therapy progress and prospects: Gene therapy of lysosomal storage disorders, *Gene Ther*, **10**, 1275–1281.

Choi, T., Huang, M., Gorman, C., *et al.* (1991). A generic intron increases gene expression in transgenic mice, *Mol Cell Biol*, **11**, 3070–3074.

Ciron C., Desmaris, N., Colle, M.A, *et al.* (2006). Gene therapy of the brain in the dog model of Hurler's syndrome, *Ann Neurol*, **60**, 204–213.

Cressant, A., Desmaris, N., Verot L., *et al.* (2004). Improved behavior and neuropathology in the mouse model of Sanfilippo type IIIB disease after adeno-associated virus-mediated gene transfer in the striatum, *J Neurosci*, **24**, 10229–10239.

Davidson, B.L., Breakefield, X.O. (2003). Viral vectors for gene delivery to the nervous system, *Nat Rev Neurosci*, **4**, 353–364.

Dodge, J.C., Haidet, A.M., Yang, W., *et al.* (2008). Delivery of AAV-IGF-1 to the CNS extends survival in ALS mice through modification of aberrant glial cell activity, *Mol Ther*, **16**, 1056–1064.

Dodge, J.C., Treleaven, C.M., Fidler, J.A., *et al.* (2010). AAV4-mediated expression of IGF-1 and VEGF within cellular components of the ventricular system improves survival outcome in familial ALS mice, *Mol Ther*, **18**, 2075–2084.

Driesse, M.J., Kros, J.M., Avezaat, C.J., *et al.* (1999). Distribution of recombinant adenovirus in the cerebrospinal fluid of nonhuman primates, *Hum Gene Ther*, **10**, 2347–2354.

Fitzsimons, H.L., Bland, R.J., During, M.J. (2002). Promoters and regulatory elements that improve adeno-associated virus transgene expression in the brain, *Methods*, **28**, 227–236.

Foust, K.D., Wang, X., McGovern, V.L., *et al.* (2010). Rescue of the spinal muscular atrophy phenotype in a mouse model by early postnatal delivery of SMN, *Nat Biotechnol*, **28**, 271–274.

Gagnon, K.T., Pendergraff, H.M., Deleavey, G.F., *et al.* (2010). Allele-selective inhibition of mutant huntingtin expression with antisense oligonucleotides targeting the expanded CAG repeat, *Biochemistry*, **49**, 10166–10178.

Gusella, J.F., Gilliam, T.C., Tanzi, R.E., et *al.* (1986). Molecular genetics of Huntington's disease, *Cold Spring Harb Symp Quant Biol*, **51 Pt 1**, 359–364.

Hadaczek, P., Eberling, J.L., Pivirotto, P., *et al.* (2010). Eight years of clinical improvement in MPTP-lesioned primates after gene therapy with AAV2-hAADC, *Mol Ther*, **18**, 1458–1461.

Hardy, J., Selkoe, D.J. (2002). The amyloid hypothesis of Alzheimer's disease: Progress and problems on the road to therapeutics, *Science*, **297**, 353–356.

Harper, S.Q., Staber, P.D., He, X., *et al.* (2005). RNA interference improves motor and neuropathological abnormalities in a Huntington's disease mouse model, *Proc Natl Acad Sci USA*, **102**, 5820–5825.

Iwakuma, T., Cui, Y., Chang, L.J. (1999). Self-inactivating lentiviral vectors with U3 and U5 modifications, *Virology*, **261**, 120–132.

Janson, C., McPhee, S., Bilaniuk, L., *et al.* (2002). Clinical protocol. Gene therapy of Canavan disease: AAV-2 vector for neurosurgical delivery of aspartoacylase gene (ASPA) to the human brain, *Hum Gene Ther*, **13**, 1391–1412.

Kaplitt, M.G., Feigin, A., Tang, C., *et al.* (2007). Safety and tolerability of gene therapy with an adeno-associated virus (AAV) borne GAD gene for Parkinson's disease: An open label, phase I trial, *Lancet*, **369**, 2097–2105.

Kirik, D., Georgievska, B., Burger, C., *et al.* (2002). Reversal of motor impairments in parkinsonian rats by continuous intrastriatal delivery of L-dopa using rAAV-mediated gene transfer, *Proc Natl Acad Sci USA*, **99**, 4708–4713.

Kornfeld, S. (1992). Structure and function of the mannose 6-phosphate/insulinlike growth factor II receptors, *Annu Rev Biochem*, **61**, 307–330.

Lim, S.T., Airavaara, M., Harvey, B.K. (2010). Viral vectors for neurotrophic factor delivery: A gene therapy approach for neurodegenerative diseases of the CNS, *Pharmacol Res*, **61**, 14–26.

Marks, W.J., Jr., Ostrem, J.L., Verhagen, L., *et al.* (2008). Safety and tolerability of intraputaminal delivery of CERE-120 (adeno-associated virus serotype 2-neurturin) to patients with idiopathic Parkinson's disease: An open-label, phase I trial, *Lancet Neurol*, **7**, 400–408.

McCarty, D.M. (2008). Self-complementary AAV vectors; advances and applications, *Mol Ther*, **16**, 1648–1656.

Muramatsu, S., Fujimoto, K., Kato, S., *et al.* (2010). A phase I study of aromatic L-amino acid decarboxylase gene therapy for Parkinson's disease, *Mol Ther*, **18**, 1731–1735.

Nagahara, A.H., Merrill, D.A., Coppola, G., *et al.* (2009). Neuroprotective effects of brain-derived neurotrophic factor in rodent and primate models of Alzheimer's disease, *Nat Med*, **15**, 331–337.

Nanou, A., Azzouz, M. (2009). Gene therapy for neurodegenerative diseases based on lentiviral vectors, *Prog Brain Res*, **175**, 187–200.

Nilsson, P., Iwata, N., Muramatsu, S., *et al.* (2010). Gene therapy in Alzheimer's disease — potential for disease modification, *J Cell Mol Med*, **14**, 741–757.

Pardridge, W.M. (2006). Molecular Trojan horses for blood-brain barrier drug delivery, *Curr Opin Pharmacol*, **6**, 494–500.

Passini, M.A., Bu, J., Fidler, J.A., *et al.* (2007). Combination brain and systemic injections of AAV provide maximal functional and survival benefits in the Niemann-Pick mouse, *Proc Natl Acad Sci USA*, **104**, 9505–9510.

Passini, M.A., Wolfe, J.H. (2001). Widespread gene delivery and structure-specific patterns of expression in the brain after intraventricular injections of neonatal mice with an adeno-associated virus vector, *J Virol*, **75**, 12382–12392.

Paterna, J.C., Bueler, H. (2002). Recombinant adeno-associated virus vector design and gene expression in the mammalian brain, *Methods*, **28**, 208–218.

Schnell, M.A., Zhang, Y., Tazelaar, J., *et al.* (2001). Activation of innate immunity in nonhuman primates following intraportal administration of adenoviral vectors, *Mol Ther*, **3**, 708–722.

Somia, N., Verma, I.M. (2000). Gene therapy: Trials and tribulations, *Nat Rev Genet*, **1**, 91–99.

Souweidane, M.M., Fraser, J.F., Arkin, L.M., *et al.* (2010). Gene therapy for late infantile neuronal ceroid lipofuscinosis: Neurosurgical considerations. *J Neurosurg Pediatr*, **6**, 115–122.

Towne, C., Aebischerm P. (2009). Lentiviral and adeno-associated vector-based therapy for motor neuron disease through RNAi, *Methods Mol Biol*, **555**, 87–108.

Valori, C.F., Ning, K., Wyles, M., *et al.* (2010). Systemic delivery of scAAV9 expressing SMN prolongs survival in a model of spinal muscular atrophy, *Sci Transl Med*, **2**, 35ra42.

Xia, H., Anderson, B., Mao, Q., *et al.* (2000). Recombinant human adenovirus: Targeting to the human transferrin receptor improves gene transfer to brain microcapillary endothelium, *J Virol*, **74**, 11359–11366.

For Further Reading

Lim, S.T., Airavaara, M., Harvey, B.K. (2010). Viral vectors for neurotrophic factor delivery: A gene therapy approach for neurodegenerative diseases of the CNS, *Pharmacol Res*, **61**, 14–26.

Lundberg, C., Bjorklund, T., Carlsson, T., *et al.* (2008). Applications of lentiviral vectors for biology and gene therapy of neurological disorders, *Curr Gene Ther*, **8**, 461–473.

Manfredsson, F.P., Mandel, R.J. (2010). Development of gene therapy for neurological disorders, *Discov Med*, **9**, 204–211.

Maxwell, M.M. (2009). RNAi applications in therapy development for neurodegenerative disease, *Curr Pharm Des*, **15**, 3977–3991.

Terzi, D., Zachariou, V. (2008). Adeno-associated virus-mediated gene delivery approaches for the treatment of CNS disorders, *Biotechnol J*, **3**, 1555–1563.

27

GENETIC THERAPY OF MUSCLE DISEASES: DUCHENNE MUSCULAR DYSTROPHY

Takis Athanasopoulos[a], Susan Jarmin[a], Helen Foster[a], Keith Foster[a], Jagjeet Kang[a], Taeyoung Koo[a], Alberto Malerba[a], Linda Popplewell[a], Daniel Scherman[b] and George Dickson[a,c]

27.1 Introduction

Muscular disorders represent the ultimate challenge for innovative therapies, in particular gene therapy, first because of the very large volume of tissue to be treated, and second, because the most frequent genetic diseases occur in muscle. This high frequency originates from the function in muscle of the largest genes and proteins of the body, such as dystrophin or titin, which display a high mutation frequency because of their size. As a correlate, a strong challenge results from the difficulty of correcting large gene deficiency by gene therapy, because of the limited encapsidation capacity of viral gene delivery vectors. However, innovative solutions are in progress to treat genetic muscular disorders. They use either a gene replacement approach or a genetic pharmacology approach, such as the exon skipping strategy which has been introduced in Chapter 6, and which will be presented in greater detail in Section 27.3.

[a]School of Biological Sciences, Royal Holloway, University of London, Egham, Surrey, TW20 0EX, UK
[b]Laboratory of Chemical and Genetic Pharmacology and of Biomedical Imaging, Paris Descartes Pharmacy University, CNRS, Inserm, Chimie ParisTech, 4, avenue de l'Observatoire Paris Cedex 06, France
[c]Email: g.dickson@rhul.ac.uk

27.2 Attractiveness of Muscle Tissue for Gene Therapy Approaches

27.2.1 Muscle Cytoarchitecture Favors Gene Delivery

In contrast to protein therapy, where a parenterally administered protein drug targeting to extracellular receptors has ready access to its target, therapeutic genes need to reach the nucleus for transcription. This represents a major delivery hurdle because the therapeutic gene needs not only to cross the cell plasma membrane, but also needs to traffic across the cytosol and get access to the nucleus. In post-mitotic cells, the nuclear envelope is a further barrier, and the genetic therapeutic material has to be delivered through the nuclear pore (Fig. 27.1).

As schematized in Fig. 27.2(a), a specific aspect of differentiated mature muscle fibers is the location of the nucleus in the immediate vicinity of the plasma membrane. This facilitates the intracellular trafficking of both viral and non-viral vectors to the nucleus, and thus the delivery of therapeutic DNA (Bureau *et al.*, 2004).

A second characteristic of mature fibers comes from the fact that they result from the fusion of a large number of individual "satellite stem cells", and are thus multinucleated. Hence, the transfection or transduction of a minority of nuclei can still result in genetic correction, which raises the prospects for success of gene therapy in muscle tissue. Also, because their sizes are much larger than those of other cells of the body, muscle fibers are particularly efficiently transfected by techniques using mechanical (pressure-based hydrodynamic) and electrical (electrotransfer) forces, as presented in detail in Chapters 16 and 17 (Bigey *et al.*, 2002).

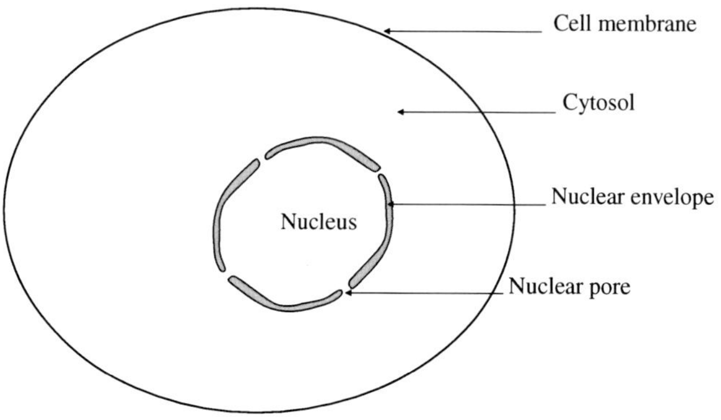

FIGURE 27.1 ■ Different barriers to gene delivery to the nucleus: (1) cell plasma membrane; (2) cytosol; (3) nuclear envelope.

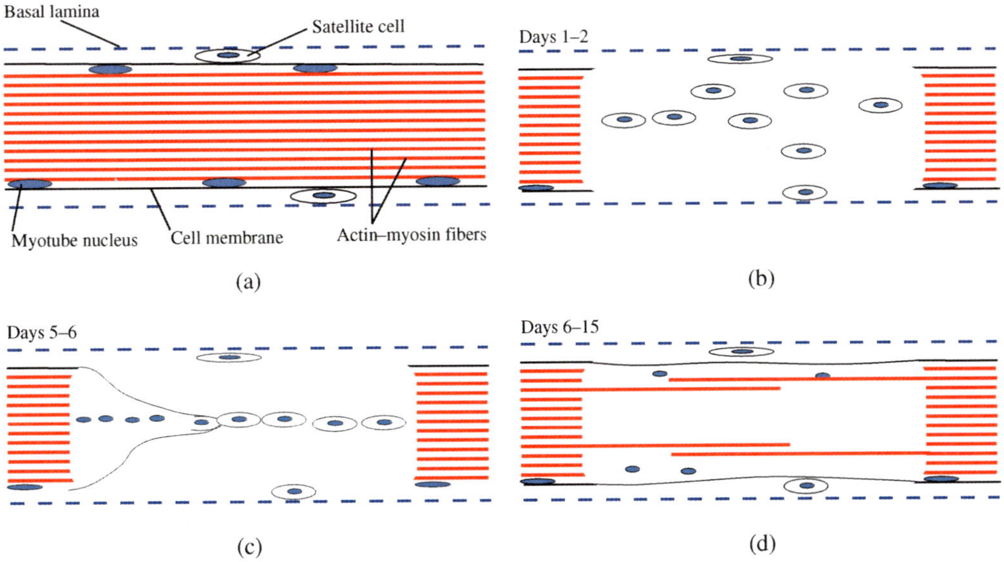

FIGURE 27.2 ■ (a) Satellite cells are localized between the muscle fiber plasma membrane and the basal lamina. (b) When a fiber is disrupted either through a mechanical injury caused by electrotransfer and/or hydrodynamic delivery of plasmid DNA, or through immune response against the viral vector or the exogenous transgene. This induces the proliferation of satellites cells. (c) Between days 5 and 6, satellite cells fuse. (d) Between days 6 and 15 a fiber is regenerated *in situ* within the basal lamina, in the absence of scar (inspired by Morgan and Partridge, 2003).

Another cytoarchitectural feature of muscle tissue is the presence of an extracellular basal lamina which surrounds each fiber (Fig. 27.2(a)). This basal lamina is not affected and maintains the basic fibrillary architecture of the muscle tissue when a fiber is disrupted, either through a mechanical injury caused by electrotransfer (Leroy-Willig *et al.*, 2005) and/or by hydrodynamic delivery of plasmid DNA, or by a cytotoxic immune response against a viral vector or an exogenous transgene. As shown in the regeneration sequence illustrated in Fig. 27.2, the muscle stem cells (satellite cells) are localized within the basal lamina boundary, and their expansion (Fig. 27.2(b)) and fusion (Fig. 27.2(c)) regenerates muscle fibers without permanent tissue disruption and scars (Fig. 27.2(d)) (Vilquin *et al.*, 2001).

27.2.2 Muscle as "Protein Factory" Target Tissue

Muscle tissue is well irrigated, and sustained high levels of intramuscular protein are observed after gene delivery, suggesting the use of muscle for "genetic" vaccination, as described in Chapter 20. Active immunization resulting from the intramuscular expression of an immunogenic transgene is the most frequent application, and is actually the object of several clinical trials for various anticancer and antiviral vaccines. In addition, passive immunization through the constant or regulated secretion by muscle of pathogen-neutralizing monoclonal antibodies has been described. Finally, the biotechnological use of intramuscular immunization of animals for polyclonal antibody production and/or for monoclonal derivation has also been exemplified (Trollet *et al.*, 2009).

After intramuscular delivery of a gene encoding a secreted protein, both intramuscular and blood-circulating proteins are detected at high levels. This has promoted the idea of using muscle as an endogenous "protein factory" for the treatment of various diseases such as haemophilia or anaemia. The therapy of haemophilia B by muscle secretion of factor IX after AAV-mediated gene delivery is described in Chapter 24. The potency of delivery of the erythropoietin (EPO) gene to muscle for the treatment of anaemia is exemplified in the results of the experiment shown in Fig. 27.3. A phenotypic correction of β-thalassemic mice following EPO-encoding plasmid electrotransfer has also been described (Payen *et al.*, 2001). Other examples of proteins secreted by transfected muscle include various cytokines and cytokine-neutralizing soluble receptors (Mallat *et al.*, 1999; Bettan *et al.*, 2000; Saidenberg-Kermanac'h *et al.*, 2003, Bloquel *et al.*, 2004a, 2004b, 2006; Potteaux *et al.*, 2006).

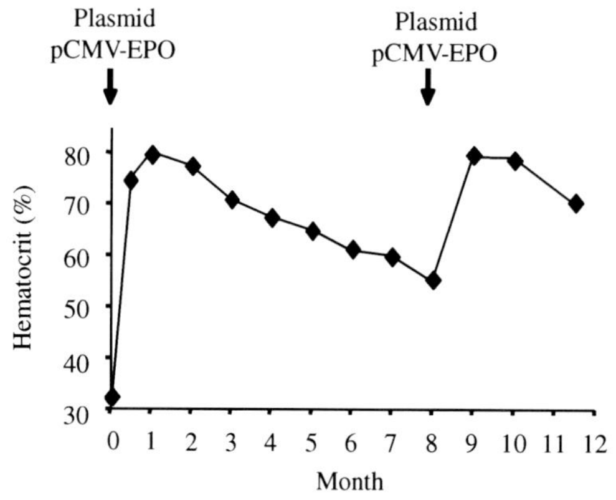

FIGURE 27.3 ■ Intramuscular electrotransfer of an Epo-encoding plasmid in mouse increased the plasmatic Epo level approximately 10–100-fold, as compared with naked DNA alone. As little as 1 μg of plasmid was sufficient to induce an increase from 47% to 80% in mouse haematocrit, which was stable for more than two months (Kreiss *et al.*, 1999). As shown in the figure, the re-administration of the plasmid after several months induced the same effect.

27.3 Genetic Pharmacology of the Mutated Dystrophin Pre-mRNA by Exon Skipping for Duchenne Muscular Dystrophy Treatment

27.3.1 Concepts and Chemistries

Duchenne muscular dystrophy (DMD) is a severe genetic disease, characterized by progressive, and ultimately lethal, muscle-wasting caused by the lack of functional dystrophin protein in skeletal muscles. The expression of this fundamental protein is prevented by frame-disrupting deletions or duplications or, less commonly, non-sense or missense mutations in the *DMD* gene. Mutations that maintain the reading frame of the gene allow expression of truncated, but semi-functional, dystrophin protein; such deletions are characteristic of the less severe muscular disease, Becker muscular dystrophy (BMD).

As yet, there is no gene therapy available within the clinic for the treatment of DMD. However, exon skipping is the most advanced of the gene therapies currently being developed, and holds great promise as a treatment for certain DMD deletions (Trollet *et al.*, 2009). As described in Chapter 6, the theory behind exon skipping is the use of antisense oligonucleotides (AOs) to sterically mask specific RNA sequence motifs on the pre-messenger RNA (pre-mRNA), so that assembly of the spliceosome on the target exon is prevented and the target exon is subsequently spliced out of the mature gene transcript. Where the deletions allow, AOs could thus be used to transform a DMD phenotype into a BMD phenotype by skipping certain exons to restore the reading frame of the *DMD* gene. Indeed, AOs have been shown to restore truncated dystrophin expressed *in vitro* in DMD patient cells (Arechavala-Gomeza *et al.*, 2007), and in animal models of the disease *in vivo* (Graham *et al.*, 2004). Interestingly, asymptomatic intragenic *DMD* deletions exist, suggesting that exon skipping may have the potential to revert a DMD phenotype to a normal phenotype.

Loss of dystrophin compromises the integrity of the muscle cell membrane and results in muscle fibers that are highly prone to contraction-induced injury. Consequently there are progressive rounds of degeneration and regeneration of the muscle leading to the replacement of muscle fibers with non-contractile fibrotic tissue and fatty infiltrates. Recent reports have suggested that a minimum of 30% of normal dystrophin levels need to be present uniformly in all myofibers to prevent muscular dystrophy in humans (Neri *et al.*, 2007), which is supported by data from transgenic *mdx* mice. Efficient systemic gene transfer of dystrophin is crucial if restoration of muscle function is to be achieved.

The chemistry of AO backbone used has a profound effect on efficacy, toxicity and half-life. The available backbone chemistries confer high biostability and target specificity and include phosphorodiamidate morpholino oligomers (PMOs), peptide nucleic acids (PNAs), locked nucleic acids (LNAs) and 2′-O-methyl phosphorothioate (2OMePS), which also have the advantage of recruiting RNAse H for RNA target degradation. LNA AOs have been shown to have potential toxicity issues, while PNAs are expensive and have poor water solubility. While both 2OMePS and PMO chemistries have excellent safety profiles, PMOs appear to produce more consistent and sustained exon skipping in the *mdx* mouse model of DMD (Malerba *et al.*, 2011b) in human muscle explants (McClorey *et al.*, 2006a), and dystrophic canine muscle cells *in vitro* (McClorey *et al.*, 2006b). PMOs also have the advantage of being easily conjugated to cell-penetrating peptides (CPPs) which improve delivery to mammalian cells, in particular to the heart, and enhance antisense activity (Wu *et al.*, 2009).

27.3.2 Preclinical Dystrophin Exon Skipping Studies in the *mdx* Mouse

The *mdx* mouse, the most commonly used animal model for DMD, carries a nonsense point mutation (C→T) in exon 23. Only rare revertant dystrophin-positive fibers are produced in its skeletal and cardiac muscles due to skipping of one or more exons correctly re-framing the transcript. The mature truncated mRNA lacking exon 23 in the *mdx* mouse does not produce a functional protein, making this mouse the ideal animal model for preclinical studies of the exon skipping strategy for DMD. Additionally, the sarcolemmal integrity seen within the *mdx* mouse is an added bonus for AO exon skipping studies. This therapeutic approach has been achieved by using different AO chemistries, among which the most promising are the phosphorothioate-linked 2OMePS and the PMO. The latter is particularly resistant to endonucleases and possesses a high affinity to the sequence target which makes it particularly suitable for *in vivo* applications. The sequence of the antisense oligonucleotide has been optimized over the years and it is very effective at inducing skipping (up to 95%). The choice of an effective dosing regimen for PMO administration is also a pivotal parameter to reduce the amount of AO necessary for systemic delivery.

It has recently been demonstrated in *mdx* mice that even a low dosage such as four weekly injections of 5 mg/kg PMO induces a significant increase in dystrophin expression (Malerba *et al.*, 2009). However, due to the mechanism of AO action, repeated chronic administration of AO is necessary to guarantee a continuous production of dystrophin. A recent report showed that systemic delivery of low clinically applicable doses of PMO for up to one year is safe in *mdx* mice and ameliorates the pathology of skeletal muscles (Malerba *et al.*, 2011b). Furthermore, the rescued dystrophin expression partially recovers limb strength and results in motor activity and movement behavior of *mdx* mice that is indistinguishable from that of normal wild-type C57BL10/mice (Fig. 27.4).

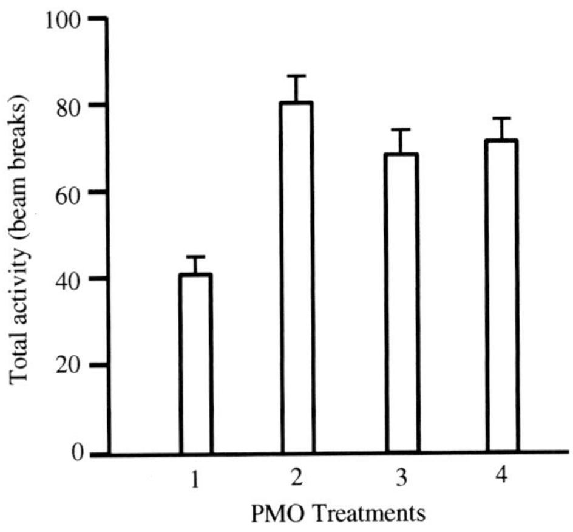

FIGURE 27.4. ■ Physical activity is normalized in *mdx* mice following chronic PMO administration. *Mdx* mice were treated for a period of 50 weeks over which five cycles of repeated low-dose (LD: 5 mg/kg/week) and high-dose (HD: 50 mg/kg/week) PMO were administered intravenously. Mice were analyzed after 50 weeks of treatment with open field behavioral activity cages. The graph shows the detailed measurement related to the parameter "total activity" for (1) MDX, (2) LD-MDX, (3) HD-MDX and (4) C57 non-dystrophic controls. Experimental details in Malerba *et al.* (2011a).

However, naked AOs are not able to enter the relatively undamaged cardiomyocytes and indeed no dystrophin expression is observed in cardiac muscles after systemic delivery of AOs (Malerba *et al.*, 2011b). Analysis of the cardiac muscles of *mdx* mice treated for one year with PMO to induce exon 23 skipping in skeletal muscle showed signs of increased fibrotic histopathology and cardiomyocyte permeability in cardiac muscles of PMO-treated mice (Malerba *et al.*, 2011a). Therefore, in the presence of skeletal muscle dystrophin restoration and the absence of cardiac muscle dystrophin restoration, the exercise-induced chronotrophic effects on the heart have the potential to accelerate the existing cardiac dysfunction and pathology noted in DMD patients. While the efficiency of some PMOs to induce exon skipping in human DMD RNAs or the functionality of derived human truncated dystrophin protein may be relatively low, these data suggest that reduced dystrophin expression may be sufficient to allow improvement of the day-to-day activities in treated DMD patients.

27.3.3 Preclinical Dystrophin Exon Skipping Studies in DMD Muscle Cultures

DMD is caused by a diverse array of mutations. According to the Leiden DMD database, deletions account for 72% of all mutations; those out-of-frame deletions for which exon skipping would restore the reading frame are summarized in Table 27.1. As exon skipping is mutation-specific, personalized AO therapy is required. The continued development and analysis of AOs for the targeting of other *DMD* exons is therefore vital.

There have been many studies published describing the in-depth analyses of AO arrays, designed using a number of tools for the targeted skipping of certain *DMD* exons (Aartsma-Rus *et al.*, 2009b; Popplewell *et al.*, 2009). Taken collectively, bioactive AOs are suggested to target certain serine/arginine (SR)-rich protein-binding motifs, bind to their target more strongly, either as a result of being longer or by being able to access their target site more easily, have their target sites within the exon, rather than intronically and nearer to the splice acceptor site. However, an AO possessing all of these properties will not necessarily be bioactive; thus the empirical analysis of designed AOs for targeted exon skipping is still essential, and provides the foundation for any AO clinical trial. The methodological details of such analysis have recently been described in Popplewell *et al.* (2011). Briefly, arrays of designed AOs are transfected using lipofection (i.e., cationic-lipid-mediated delivery) or nucleofection (i.e., electrotransfer) into either normal human skeletal muscle cells or DMD patient muscle cells, carrying appropriate deletions. Harvested RNA is subjected to nested reverse transcription polymerase chain reaction (RT-PCR) analysis for the establishment of specific exon skipping at the genetic level in both cell types; detection of dystrophin protein on Western blot or immunocytochemistry in DMD patient cells confirms restoration of dystrophin protein expression as a result of exon skipping. To be cost-effective and safe, the choice of AO for each targeted exon would be expected to produce skipping at low doses and to have persistence of action. Therefore, time-course and dose-response studies are routinely performed to ensure complete target optimization. Such detailed empirical analyses have been performed for a range of *DMD* exons, but in particular for exon 51 (Arechavala-Gomeza *et al.*, 2007; Aartsma-Rus *et al.*, 2009b), exon 53 (Popplewell *et al.*, 2010) and exon 45 (Aartsma-Rus *et al.*, 2003); skipping of these exons would have the potential to treat 15%, 9% and 13% of DMD patients respectively (see Table 27.1).

TABLE 27.1 DMD Deletions Suitable for Therapeutic Exon Skipping (Adapted from van Deutekom *et al.*, 2001).

Exon Skipping	Therapeutic for DMD Exonic Deletions	Frequency (%)
2	3–7	2
8	3–7, 4–7, 5–7, 6–7	4
43	44, 44–47	5
44	35–43, 45, 45–54	8
45	18–44, 44, 46–47, 46–48, 46–49, 46–51, 46–53	13
46	45	7
50	51, 51–55	5
51	50, 45–50, 48–50, 49–50, 52, 52–63	15
52	51, 53, 53–55	3
53	45–52, 48–52, 49–52, 50–52, 52	9

27.3.4 Clinical Trials of Dystrophin Exon Skipping in DMD Patients

On the basis of *in vitro* and *in vivo* preclinical studies in DMD patient cells and in animal models, a number of patient trials, phase I and more recently phase II, have been undertaken. In the first of these, four DMD patients carrying appropriate deletions received a single intramuscular injection of high dose of an AO with a 2'OMePS backbone (PRO051), which targets exon 51. Each patient showed myofiber expression of specific exon 51 skipping, dystrophin protein, which was detectable at 3–12% of normal levels four weeks after injection. No clinically adverse events were detected (van Deutekom *et al.*, 2007).

In the second trial, the AON AVI-4658, which has PMO backbone and targets a slightly different intraexonic sequence (+68+95) of exon 51, has been injected intramuscularly in a dose-escalating trial into nine DMD boys. At the higher doses, this PMO AO produced good levels of local dystrophin protein production in treated muscles; the intensity of dystrophin staining was up to 42% of that seen in healthy muscle. The treatment had no adverse effects (Kinali *et al.*, 2009). The clinical evaluation has been extended to 12-week systemic delivery of both exon 51 AOs and results have very recently been reported. Both chemistries showed no adverse effects and dose-dependent restoration of dystrophin production was clearly seen; functionality of this expressed dystrophin protein was established by the detection of other dystrophin-associated proteins at the sarcolemma (for AVI-4658), and by a modest, but not statistically significant, improvement in the patient six-minute walk test after 12 weeks of extended treatment (for PRO051) (Goemans *et al.*, 2011; Cirak *et al.*, 2011). However, such studies highlight some other interesting findings: the best responders were patients with a deletion of exon 49–50, suggesting that sequence context of deletions may influence exon skipping frequencies or that the resultant truncated dystrophin proteins have differential stabilities. On the basis of results seen in the *mdx* model using various dosing regimens over extended periods (Malerba *et al.*, 2009, Malerba *et al.*, 2011a, Malerba *et al.*, 2011b; Wu *et al.*, 2011), further phase II and phase III clinical studies are planned.

The drug company Sarepta has performed preclinical studies with AVI-5038 in collaboration with the charity Charley's Fund. AVI-5038 is a PMO conjugated to a cell-penetrating peptide (PPMO), designed to target skipping of exon 50 of the dystrophin gene. PPMOs have been used in the *mdx* animal model and shown to have improved deliverability to both skeletal and cardiac muscle, thus improving efficacy. Repeated weekly intravenous bolus injection over four weeks at a low dose of this PPMO was shown to be well tolerated; however, higher doses administered weekly for 12 weeks showed significant toxicological effects, particularly in relation to the kidney. As yet this problem has not been resolved, and an unconjugated version of the same PMO (AVI-4038) is currently being developed for clinical trials. There are a number of alternative peptide conjugates (Yin *et al.*, 2009; Kang *et al.*, 2011) that show promise as enhancers of deliverability and are undergoing rapid preclinical development; however, toxicological and immunogenic profiles of these new conjugates are yet to be established. The next planned UK phase I trial by the MDEX consortium will involve conjugation of a PMO developed for the targeted skipping of exon 53 (Popplewell *et al.*, 2010), and is supported by the Wellcome Trust. Prosensa–GlaxoSmithKline are currently performing a phase I trial using a 2OMePS AO for the targeted skipping of exon 45.

27.3.5 Destructive Exon Skipping to Inhibit Myostatin

Recent work in the field of skeletal muscle structure and function strongly suggests that this tissue controls its own mass through a regulatory mechanism that involves an endogenous negative muscle mass regulator called myostatin or growth differentiation factor 8 (GDF 8). As the natural mutations in the myostatin gene have led to increases in muscle mass in cattle, mice and dogs, as well as humans, various approaches have been explored in order to develop a strategy that would help in recovering loss of muscle mass and function in various muscle wasting conditions. Also, myostatin-null mice have significantly lower fat accumulation and increased insulin sensitivity, thereby elevating the likelihood that inhibiting myostatin signaling could be a potential therapy for Type II diabetes as well as obesity. Therefore, myostatin inhibition could have a very favorable influence on public health.

In the case of myostatin, destructive exon skipping is the aim rather than reading frame restoration as is the case in *DMD* skipping. Using safe and controlled exon skipping approach, the 374 nucleotide exon 2 of myostatin mRNA was skipped by AOs, resulting in an out-of-frame fusion of exons 1 and 3 in a murine skeletal muscle cell line, as well as in animal models. A significant increase in treated muscles was observed following systemic as well as intramuscular injections of different chemistries of AOs targeted at myostatin exon 2 (see Fig. 27.5).

A number of conjugates have been reported to enhance the deliverability of the AOs and the ones used for myostatin so far (Kang *et al.*, 2011) are vivo-morpholino (Malerba *et al.*, 2009) and B-PMO (Yin *et al.*, 2009). These studies indicate that (1) antisense-mediated destructive exon skipping can be induced in the myostatin RNA, (2) antisense AO treatment reduces myostatin bioactivity and enhances muscle mass *in vivo* and (3) AO-induced myostatin exon skipping is a potential therapeutic strategy to counter muscular dystrophy, muscular atrophy, cachexia and sarcopenia (Kang *et al.*, 2011). Future work in this field needs to be done to assess the electrophysiological effects of the antisense-mediated myostatin exon skipping in animals. A combined study to induce dual exon skipping has

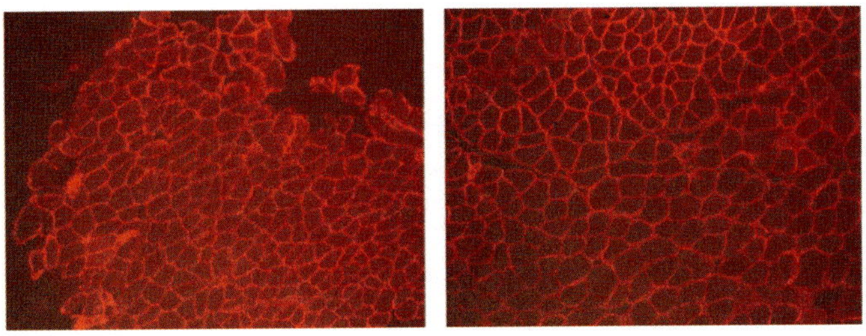

Control Soleus Muscle PMO-treated Soleus Muscle

FIGURE 27.5 ■ Systemic injection of PMO conjugated to octa-guanidine dendrimer (Vivo-PMO) results in a significant increase in muscle mass and myofiber size as a result of targeted myostatin exon skipping. Mice were treated with 6 mg of Vivo-PMO-MSTN per kg by five weekly intravenous injections and muscles were harvested ten days later. (Representative dystrophin immunohistology indicating increased myofiber cross-sectional area (CSA) *in vivo*-PMO-treated compared with control soleus muscle cryosections. Experimental details in Kang *et al.* (2011).

been reported and holds promise for a better strategy for muscular degeneration (Kemaladewi *et al.*, 2011; Malerba *et al.*, 2012).

27.3.6 Perspectives on Exon Skipping for Muscle Disease

AO-induced exon skipping has the potential, theoretically, to treat up to about 60% of all patients with DMD, converting their disease phenotype to the less-severe BMD phenotype (see Table 27.1). The progress of clinical trials from phase I to phase II for AOs targeting exon 51 of the *DMD* gene and the results reported from these trials are extremely positive; repeated systemic injection of PMO or 2OMe AO was well tolerated and with no adverse effects. Higher doses of both chemistries were required to see exon skipping and consequent restoration of dystrophin expression. The scope of the *DMD* exon skipping trials is currently being extended to include targeting of exons 45 and 53. However, it should be noted that only 8%, 4%, 13% or 9% of DMD patient mutations should be convertible into a BMD phenotype by single AO exon 45, 50, 51 or 53 skipping, respectively. Personalized molecular medicine for each skippable *DMD* deletion is necessary and this would require the optimization and clinical trial workup of many specific AONs. It has been suggested that multi-exon skipping, using cocktails of AONs or chemically linked AONs, around deletion hotspots (e.g., exons 45–55) may have the potential to treat approximately 65% of DMD patients. A multi-exon strategy has been shown to work in *mdx* mice, but this has not yet been achieved in DMD patient cells.

There are further obstacles to overcome for AO-induced exon skipping to be a viable gene therapy for DMD. The cost implications may end up being siginificant for many patients; since AOs are rapidly cleared from the circulation, regular administration of high doses of AO would be required for therapeutic effect. Secondly, although deliverability, particularly to the heart, is enhanced with the use of conjugated PMOs, their potential toxicological and immunogenic problems need to be addressed. Lastly, the need for personalized medicine will require the completion of many expensive, lengthy clinical trials of many AONs. Even so, the potential of combined AO-induced DMD and myostatin exon skipping being the first gene therapy for DMD available in the clinic is highly likely, and this certainly looks achievable in the very near future.

27.4 Microdystrophin and Myostatin Gene Therapy for Duchenne Muscular Dystrophy Using Adeno-associated Virus Vectors, Gene Therapy: Administration of the Dystrophin Gene and/or of a Protein Inhibiting Myostatin

27.4.1 Concepts and Vectorology

Adeno-associated virus (AAV)-mediated gene transfer is widely used for skeletal-muscle-directed gene therapy due to the high tropism of AAV vectors for muscle fibers. Transgenes delivered by AAV vectors to the muscle of animal models, including non-human primates have been shown to be stably expressed over several years, demonstrating the potential of AAV vectors to provide "life-long" gene expression. AAV vectors are being investigated for their potential to induce muscle "remodeling" in dystrophic muscle by the delivery of transgenes to inhibit the myostatin pathway and to also restore dystrophin expression in dystrophic muscle and thereby improve muscle function and stabilize or halt disease progression (Athanasopoulos *et al.*, 2004; Foster *et al.*, 2008). However, there is a limitation in the packaging capacity of foreign DNA into AAV vectors. The AAV capsid is able to package a genome of up to 5 kb and therefore it is not possible to transfer the 11 kb full-length dystrophin cDNA into AAV vectors. To overcome this limitation, truncated but functional mini- and micro-dystrophin cDNAs have been engineered by several research groups and have been tested in both murine and canine models of DMD.

In addition to functionality of the microdystrophin, further considerations need to be made in the treatment of DMD with AAV; these include the ability to efficiently deliver the transgene systemically to all muscles including the heart and diaphragm. Scale up of these processes is critical so that AAV microdystrophin vectors can be assessed in large-animal models for efficiency, functionality and immunogenicity before inclusion in clinical trials. Systemic delivery of AAV microdystrophin vectors has previously been demonstrated in mouse models following injection of very high titers of AAV expressing dystrophin under control of a constitutive viral promoter. However, AAV vectors are currently being developed that can achieve efficient gene transfer and expression at lower viral titers with a muscle-restrictive promoter driving microdystrophin expression that will be more applicable for use in a clinical setting.

27.4.2 Adeno-Associated Viral Vectors Expressing Microdystrophin in the *mdx* Mouse

A number of groups have conducted extensive studies in the *mdx* mouse following AAV microdystrophin gene transfer. When expressed at a sufficient level, many of these microdystrophin proteins have been demonstrated to improve, but not completely normalize, a range of markers of the dystrophic phenotype. Restoration of the dystrophin-associated protein complex (DAPC), stabilization of muscle degeneration and improvements in muscle function have been demonstrated following delivery of microdystrophin at different stages of disease progression (Athanasopoulos *et al.*, 2004). However, the lack of large portions of the dystrophin gene, including some functional domains, means that microdystrophin proteins are less able to restore specific force and protect dystrophic muscle from contraction-induced injury. Recently, the Dickson laboratory has made progress in this area through DNA sequence optimization and the design of modified microdystrophin genes (Fig. 27.6).

Sequence optimization led to increased levels of dystrophin mRNA and protein and resulted in normalization of specific force but not resistance to eccentric contractions. Further modification of this microdystrophin to include additional functional domains lead to additional improvements in function when administered at sub-optimal doses (Foster *et al.*, 2008; Koo *et al.*, 2011a) (Fig. 27.7).

Another hurdle in the path towards the generalized correction of DMD by AAV microdystrophin gene therapy is that high titers of AAV are required for efficient systemic gene transfer. Such high titers may be toxic and may cause immune responses against the viral proteins. In particular, there is a potential for humoral and cellular immune responses to AAV vectors and transgenes following injection into the *mdx* muscles (Yuasa *et al.*, 2002). To overcome these issues, transient immune suppression to enable re-infusion or use of muscle-specific promoters to avoid expression in antigen-presenting cells have been successfully used. We have recently demonstrated that optimization of the codon usage of a eukaryotic gene induces significant increases in levels of mRNA and dystrophin expression in *mdx* mice after intramuscular and systemic AAV gene transfer (Foster *et al.*, 2008). The

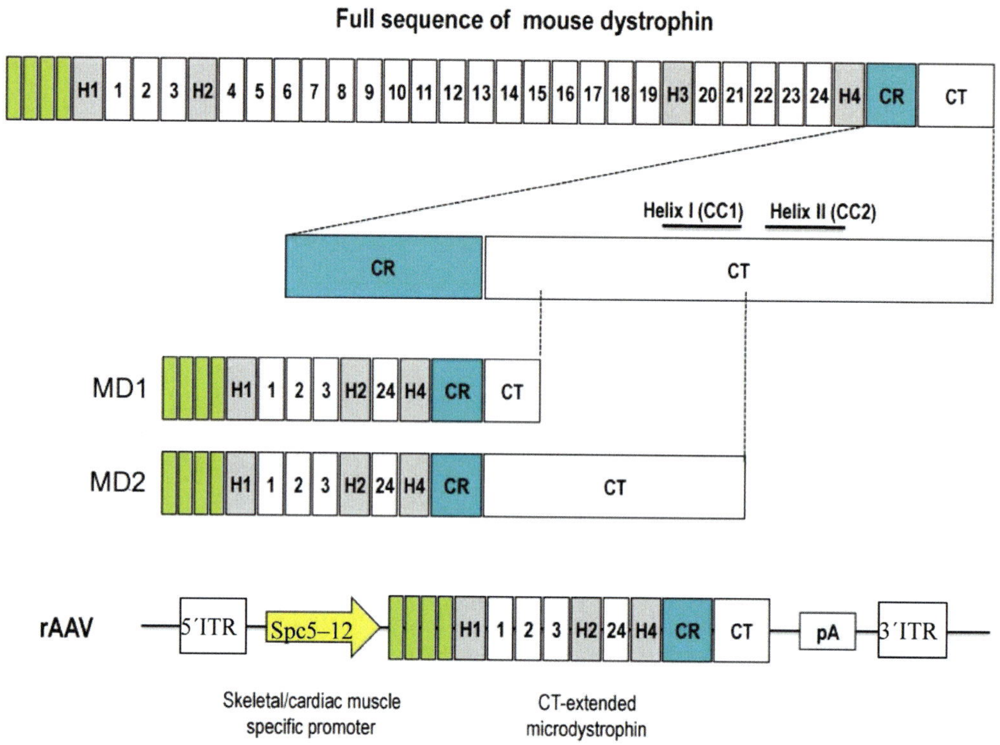

FIGURE 27.6 ■ Generation of mouse-specific microdystrophin cDNAs with CT domain extension. The full-length dystrophin protein is defined by four structural domains: N-terminus (NT), rod, cysteine-rich (CR) and carboxyl-terminal (CT) domains. The MD1 cDNA incorporates deletions of the Rod domain repeats 4–23 and the CT domain. The MD2 cDNAs incorporate additionally coiled coil helix 1 of the CT domain of dystrophin. At the CT end, each microdystrophin contains three amino acids of exon 79 followed by three stop codons.

higher expression of an mRNA sequence optimized version of dystrophin may decrease the amount of AAV vectors to be used in systemic administration thus reducing the risk of toxicity of the vector. The inclusion of specific domains identified as compulsory for the rescue of the functionality of the dystrophin may allow a reduction of the amount of AAV vectors to be administered for a successful therapy. We demonstrated that the delivery of AAV2/9-microdystrophin incorporating helix 1 of the coiled-coil motif in the CT domain of dystrophin increased the recruitment of some members of the DAPC, including α1-syntrophin and α-dystrobrevin, at the sarcolemma of skeletal muscle fibers of *mdx* mice and efficiently protected *mdx* muscles from muscle damage (Koo *et al.*, 2011a). An alternate strategy to detarget AAV transduction from non-target tissue can also lower the viral load required for efficient gene transfer to muscle (Asokan *et al.*, 2010).

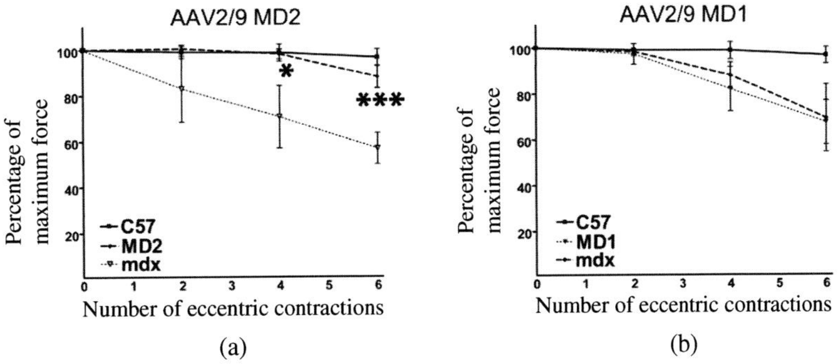

FIGURE 27.7 ■ Intramuscular injection of AAV2/9-MD2 leads to muscle protection from lengthening contraction in tibealis anterior (TA) muscles of *mdx* mice. TA muscles of two-month-old *mdx* mice were treated with injections of either 2×10^{10} vector genomes of AAV2/9-MD1, -MD2. Two months after injection, TA muscles were assessed for force deficit following a series of six eccentric contractions. Muscles treated with 2×10^{10} vector genomes of AAV2/9-MD2 (a) showed a significant improvement in their resistance to contraction-induced injury, while no change was observed in muscle injected with 2×10^{10} vector genomes of AAV2/9-MD1 (b) compared with saline-injected *mdx* muscles. (Mean ± standard error of the mean. n = 4–6, one-way ANOVA test, $^*p < 0.05$).

27.4.3 Adeno-Associated Viral Vectors Expressing Microdystrophin in a Dystrophic Dog Model

The golden retriever muscular dystrophy (GRMD) dog was the first characterized dog model of DMD. It has been reported that *GRMD* dogs eventually die due to cardiomyopathy. *GRMD* dogs have been identified as having complete dystrophin deficiency with higher genotypic/phenotypic similarity to human DMD disease than that of *mdx* mouse model. The *GRMD* model has been bred onto a beagle background (canine X-linked muscular dystrophy in Japan (*CXMDj*)), a smaller dog model which exhibits a similar phenotype to the *GRMD* dog.

To apply AAV-mediated gene therapy to DMD patients it is first desirable to test any therapy in large-animal models such as the dystrophic dog prior to human clinical trials. It is important to examine both the functionality of the microdystrophin and the immune responses against transgene or AAV vectors in a large-animal model such as the *CXMDj* or *GRMD* dogs, which are more clinically relevant models for DMD compared with the *mdx* mouse.

A major hurdle facing development of AAV-mediated gene delivery in muscle towards clinical applications is the immune response against the AAV capsid protein or transgene product. Profiling human sera demonstrates worldwide neutralizing antibodies to most AAV serotypes (Calcedo *et al.*, 2009). Strong immune responses to capsid have been exhibited following intramuscular injection of AAV2 or AAV6 vectors carrying various transgenes in dogs, in contrast to successful gene delivery in mouse models (Wang *et al.*, 2007a, 2007b). Another study demonstrated that a strong cellular and humoral immune response against transgene in the *CXMDj* dog model was activated following delivery of AAV2 (AAV2/2) or 8 (AAV2/8) encoding β-galactosidase (Ohshima *et al.*, 2009). These immune responses have not been encountered in small-animal models, but have been observed in juvenile and neonatal muscular dystrophy dogs and random-bred wild-type dogs (Wang *et al.*, 2007a, 2007b; Koo *et al.*, 2011b).

These new challenges call on efforts to improve targeted gene delivery, tissue-specific therapeutic gene expression, immune modulation and new vector development and large-scale production technologies. Immune response can be suppressed with a combination of anti-thymocyte globulin (ATG), cyclosporine (CSP) and mycophenolate mofetil (MMF) and allows long-term expression of a canine microdystrophin in the skeletal muscle of *CXMDj* dog (Wang *et al.*, 2007b).

It is important to evaluate the serotypes of AAV which produce efficient gene transfer at a low dose of virus in large animals prior to clinical trials. AAV2/8 is one of the promising vector serotypes for preclinical trials in canine models. Several research groups have compared the efficiency of gene transfer between AAV serotype 2 and 8 and found that AAV2/8 vectors showed efficient gene transfer into canine skeletal muscle more widely than AAV2 vectors (Oshima *et al.*, 2009).

Recently we have demonstrated that delivery of AAV2/8 expressing a canine-specific and mRNA sequence-optimized microdystrophin gene controlled by a muscle-specific promoter results in high and stable levels of microdystrophin expression in the *CXMDj* dog model of DMD for at least eight weeks. Following intramuscular injection of AAV2/8 microdystrophin large areas of dystrophin-positive fibers were observed (Fig. 27.8); this was associated with stabilization of the DAPC, a reduction in central nucleation (an important marker of muscle regeneration) and stabilization of myofiber permeability (Fig. 27.9). Importantly, the injections were carried out in the absence of any immunosuppressive regime and, unlike other studies in canine models of DMD, no evidence of immune response was observed (Koo *et al.*, 2011b). The design of these vectors marks an important step forward to clinical trials.

27.4.4 Clinical Trials of AAV Microdystrophin in DMD Patients

To date, one AAV microdystrophin clinical trial has been conducted in DMD patients (Mendell *et al.*, 2010). Patients received either 2×10^{10} vg/kg or 1×10^{11} vg/kg of AAV2.5 cytomegalovirus (CMV) microdystrophin via intramuscular injection into a biceps muscle. AAV2.5 is a variant of AAV2 which contains five amino

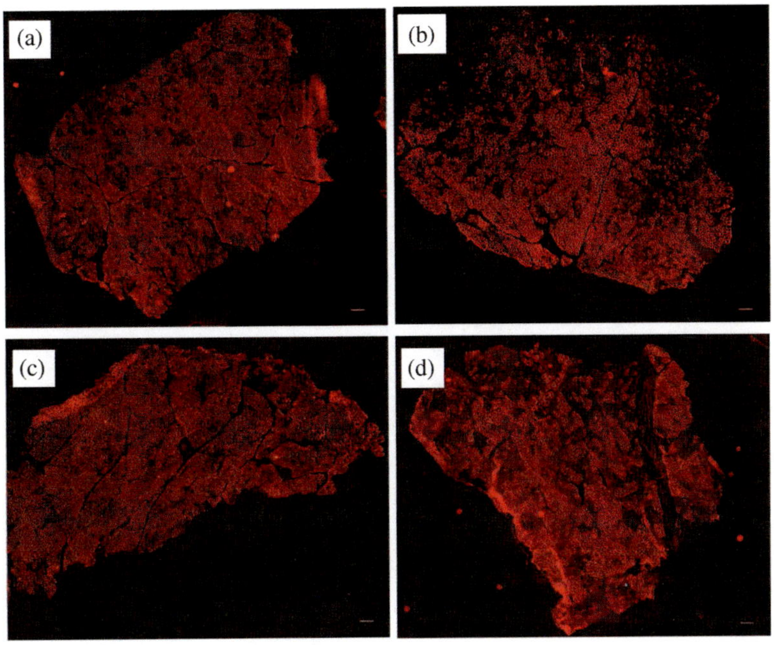

FIGURE 27.8 ■ Widespread expression of an mRNA sequence-optimized canine microdystrophin after intramuscular injection of AAV2/8-cMD1 in the *CXMDj* muscles. The (a, b) tibialis cranialis (TC), (c) extensor carpi radialis (ECR) and (d) gastrocnemius (GC) muscles of the CXMDj dog were injected intramuscularly with either 1×10^{13} (a, c), 1×10^{12} (b) or 2×10^{13} vector genome (d) of AAV2/8-cMD1. At four weeks (c) or eight weeks (a, b and d) after injection, dystrophin expression was evaluated by immunohistochemistry using NCL-dysB antibody. The dystrophin signal was visualized with Alexafluor 568-conjugated anti-mouse IgG antibody. Scale bar = 300 μm.

acids from AAV1 (one insertion and four substitutions) in the VP1 domain. The AAV2.5 vector expressed a microdystrophin gene under control of the constitutive CMV promoter. Microdystrophin expression was only detected in two out of six patients treated and at very low levels (one and three fibers respectively) although vector DNA was detected in all muscle biopsies. A possible explanation is that transduction efficiency of AAV2.5 vectors in human muscle may be low. However, unexpectedly, dystrophin-specific T-cells were detected in four patients. Further analysis revealed that two patients exhibited pre-existing dystrophin-specific T-cells and that gene transfer may have stimulated a memory T-cell response in these patients. In addition, microdystrophin-specific T-cells were also detected within a least one patient. The detection of both cluster of differentiation $CD4^+$ and $CD8^+$ dystrophin specific T-cells within these patients provides the most likely explanation for the very low levels of dystrophin expression that were observed (no dystrophin specific T-cells were detected in the highest expressing patient). This unexpected observation, in particular the presence of

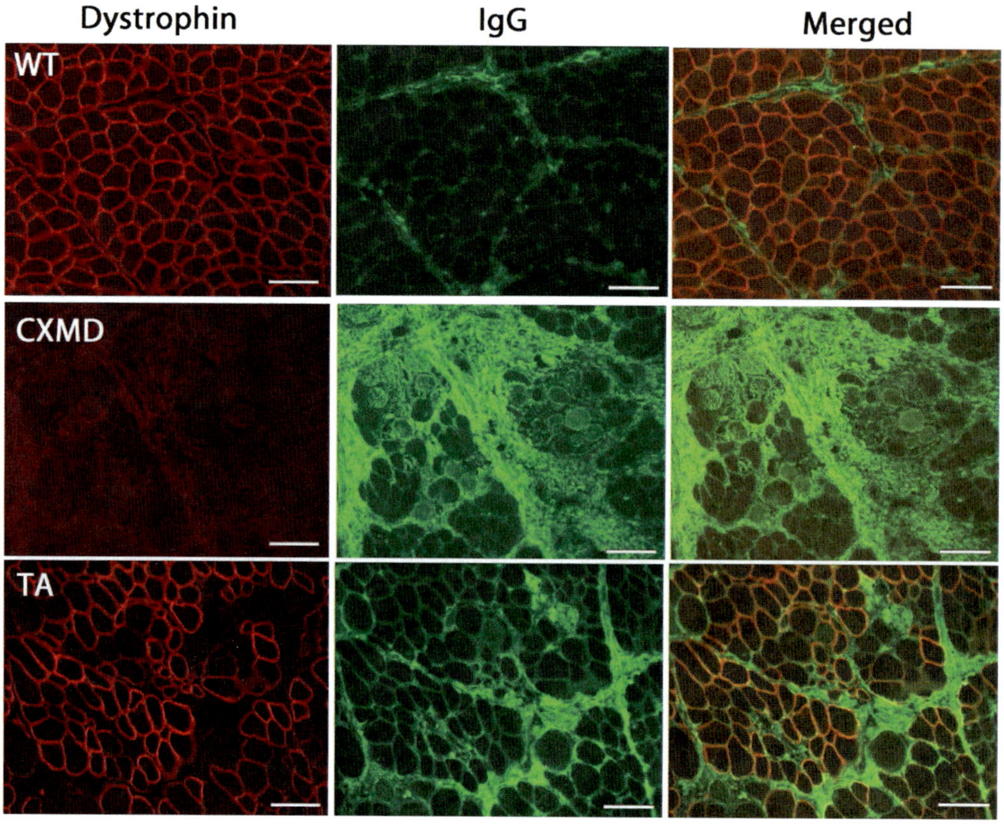

FIGURE 27.9 ■ Examination of muscle membrane integrity of muscle of *CXMDj* dog following intramuscular injection of AAV2/8-cMD1. The tibialis cranialis (TC) muscle of *CXMDj* dog was treated with injections of 1×10^{13} vg AAV2/8-cMD1. At eight weeks post injection, tissues were harvested, cryosectioned and subjected to immunohistochemistry of serial-sections to examine membrane integrity. Serial-sections were stained using NCL-dysB antibody against microdystrophin or an Alexa 488-anti-canine immunoglobulin G (IgG) antibody. The dystrophin signal was visualized with an Alexa 568-conjugated anti-mouse IgG. *Right panels* represent the merged figures between microdystrophin and anti-canine IgG. Age-matched TC muscles of wild-type dog and *CXMDj* dog were assessed in parallel as positive and negative controls, respectively. Magnification bar = 50 μm.

pre-existing dystrophin-specific T-cells, is in contrast to the proposed induction of tolerance to dystrophin by the presence of spontaneous revertant fibers (Ferrer *et al.*, 2000). However, is unclear it why pre-existing antibodies are detected in some patients and not others. It is speculated that this may be related to disease severity, levels of inflammation within the muscle and timing of the revertant fiber "event". Although this AAV-based clinical trial resulted in very few dystrophin-positive fibers, it has brought to the fore some important considerations for AAV microdystrophin and other dystrophin replacement strategies that the "cellular immune status" of patients with respect to both dystrophin and AAV must be carefully monitored both prior to and post treatment with any vector.

27.4.5 AAV-Based Inhibition of Myostatin in Models of Muscular Dystrophy

Myostatin, a member of the transforming growth factor-beta (TGF-β) superfamily of signaling molecules, is a potent inhibitor of muscle growth, expressed mainly by skeletal muscle. Inactivating mutations of myostatin increases muscle mass in species ranging from mice to men (McPherron *et al.*, 1997; Schuelke *et al.*, 2004). The mature region binds to the activin receptor Type II (ActRIIB), a serine/threonine kinase transmembrane receptor on target cells, leading to an intracellular signaling cascade via the SMAD2/3/4 complex, activation and translocation to the nucleus and regulated transcription of myogenic genes (Joulia-Ekaza *et al.*, 2007). Myostatin has also been shown to abolish signaling via SMAD3-independent pathways to counter the positive insulin-like growth factor (IGF)-1/phosphoinositide 3-kinases (PI3K)/Ak strain thyoma (AKT) mitogenic pathway.

Many strategies have been developed to decrease myostatin expression or its biological activity and are currently being evaluated in various muscle disorder models. These include targeting myostatin-binding proteins, including myostatin propeptide, follistatin, ActRIIB, growth and differentiation factor-associated serum protein-1 (GASP-1), and anti-myostatin antibodies as well as RNA interference system-based knockdown and strategies that knock down or inhibit the protein's function. Many of these strategies have yielded promising results and increased muscle mass in experimental animals. Within our own lab, an exon skipping regime to manipulate pre-mRNA levels has been evaluated *in vivo* (Kang *et al.*, 2011). Preliminary work in the *mdx* mouse demonstrates the feasibility of dual myostatin inhibition and dystrophin restoration by exon skipping strategies (Kemadelawi *et al.*, 2001). Myostatin inhibition in *GRMD* dogs leads to muscle hypertrophy and a reduction in muscle fibrosis (Bish *et al.*, 2011).

27.4.6 Perspectives on Gene Replacement Therapy for Neuromuscular Diseases

The design of any future AAV microdystrophin clinical trial must carefully consider not only the functionality of the microdystrophin gene, the AAV serotype used and the use of tissue-restrictive promoters but also the "dystrophin" immune status of the patients. Much progress has recently been made in many of these areas of AAV gene transfer for the treatment of DMD.

Design of the microdystrophin gene is of the upmost importance, with a number of research groups around the world having tested many microdystrophin gene configurations (Athanasopoulos *et al.*, 2004). However, the Dickson Laboratory has recently demonstrated a number of improvements to the microdystrophin gene configuration. mRNA sequence optimization of the microdystrophin gene, such as the inclusion of improved Kozak sequences, codon optimization and altering GC content had significant results in terms of gene expression and in turn improvement in muscle function (Foster *et al.*, 2008). In addition, further modifications have been made to the microdystrophin gene to include additional DAPC binding domains, which were demonstrated to further improve functionality (Koo *et al.*, 2011a). It is possible that the inclusion of other functional domains in the microdystrophin gene may further improve function. Such improvements in microdystrophin gene function and levels of gene expression may also allow a lower effective viral dose to be administered.

In addition to functionality of the microdystrophin, the choice of AAV serotype and route of vector administration are critically important. Many studies have compared the tropism and efficiency of gene transfer of different AAV serotypes in murine and canine muscle, with AAV2/8 and AAV2/9 appearing to have very high tropism in muscle of both species (Foster *et al.*, 2008; Koo *et al.*, 2011b; Bish *et al.*, 2008). However, when translating studies to non-human primates, tropisms identified in other animal models may not be consistent (Gao *et al.*, 2011) and may be an important factor in the success of clinical trials. Recent data also suggests that route of administration plays an important role in development immune responses to vector and transgene (Toromanoff *et al.*, 2010). Therefore, testing of AAV microdystrophin vectors by regional limb perfusion in a dystrophic dog model will be of upmost importance in developing clinically relevant vectors.

Although progress has been made in many areas such as microdystrophin design, use of tissue-restrictive promoters to limit immune responses and the development of systemic administration strategies, there are areas where progress and further considerations need to be made. From a safety perspective additional improvements need to be made to the efficiency of systemic gene transfer to allow viral doses to be lowered further. If effective gene transfer cannot be achieved via a single treatment or gene expression is not as long-lived as expected there may also be a need to re-administer AAV vectors. The inclusion of extra levels of gene regulation such as the incorporation of microRNA target sequences into vectors to prevent off-target expression may also be required. Finally, it is increasingly apparent that immune responses to AAV capsid proteins and, unexpectedly, the dystrophin protein, need to be considered and managed through the development of immune-suppressive regimes to allow effective clinical trials to be developed.

Whilst dystrophin restoration is the cornerstone for any DMD therapy, attention is being focused upon the environmental milieu of the muscle to counter the reduced muscle regeneration potential, the loss of muscle and the increase in fibro-fatty lesions. Dual administration strategies to address all the pathological changes within dystrophic muscle will be a natural extension to current phase II/III DMD clinical trials.

References

Aartsma-Rus, A., Janson, A.A., Kaman, W.E., *et al.* (2003). Therapeutic antisense-induced exon skipping in cultured muscle cells from six different DMD patients, *Hum Mol Genet*, **12**, 907–914.

Aartsma-Rus, A., Fokkema, I., Verschuuren, J., *et al.* (2009a). Theoretic applicability of antisense-mediated exon skipping for Duchenne muscular dystrophy mutations, *Hum Mutat*, **30**, 292–299.

Aartsma-Rus, A., Van Vliet, L., Hirschi, M., *et al.* (2009b). Guidelines for antisense oligonucleotide design and insight into splice-modulating mechanisms, *Mol Ther*, **17**, 548–553.

Arechavala-Gomeza, V., Graham, I.R., Popplewell, L.J., *et al.* (2007). Comparative analysis of antisense oligonucleotide sequences for targeted skipping of exon 51 during dystrophin pre-mRNA splicing in human muscle, *Hum Gene Ther*, **18**, 798–810.

Asokan, A., Conway, J.C., Phillips, J.L., *et al.* (2010). Reengineering a receptor footprint of adeno-associated virus enables selective and systemic gene transfer to muscle, *Nat Biotechnol*, **28**, 79–82.

Athanasopoulos, T., Graham, I.R., Foster, H., *et al.* (2004). Recombinant adeno-associated viral (rAAV) vectors as therapeutic tools for Duchenne muscular dystrophy (DMD), *Gene Ther*, **11 Suppl 1**, S109–S121.

Bettan, M., Emmanuel, F., Darteil, R., *et al.* (2000). High level protein secretion into blood circulation after electric-pulse mediated gene transfer into skeletal muscle, *Mol Ther*, **2**, 204–210.

Bigey, P., Bureau, M.F., Scherman, D. (2002). *In vivo* plasmid DNA electrotransfer, *Curr. Opin. Biotechnol*, **13**, 443–447.

Bish, L.T., Morine, K., Sleeper, M.M., *et al.* (2008). Adeno-associated virus (AAV) serotype 9 provides global cardiac gene transfer superior to AAV1, AAV6, AAV7, and AAV8 in the mouse and rat, *Hum Gene Ther*, **19**, 1359–1368.

Bish, L.T., Sleeper, M.M., Forbes, S.C., *et al.* (2011). Long-term systemic myostatin inhibition via liver-targeted gene transfer in golden retriever muscular dystrophy, *Hum Gene Ther*, **22**, 1499–1509.

Bloquel, C., Bejjani, R., Bigey, P. (2006). Plasmid electrotransfer of eye ciliary muscle: principles and therapeutic efficacy using hTNF-α soluble receptor in uveitis, *FASEB*, **20**, 389–391.

Bloquel, C., Bessis, N., Boissier, M.C., *et al.* (2004a). Gene therapy of collagen-induced arthritis by electrotransfer of human tumor necrosis factor-α soluble receptor I variants, *Hum Gene Ther*, **15**, 189–201.

Bloquel, C., Fabre, E., Bureau, M.F., *et al.* (2004b). Plasmid DNA electrotransfer for intracellular and secreted proteins expression: new methodological developments and applications, *J Gene Med*, **6 Suppl 1**, S11–S23.

Bureau, M.F., Naimi, S., Torero Ibad, R., *et al.* (2004). Intramuscular plasmid DNA electrotransfer: biodistribution and degradation, *Biochim Biophys Acta*, **1676**, 138–148.

Calcedo, R., Vandenberghe, L.H., Gao, G., *et al.* (2009). Worldwide epidemiology of neutralizing antibodies to adeno-associated viruses, *J Infect Dis*, **199**, 381–390.

Cirak, S., Arechavala-Gomeza, V., Guglieri, M., *et al.* (2011). Exon skipping and dystrophin restoration in patients with Duchenne muscular dystrophy after systemic phosphorodiamidate morpholino oligomer treatment: an open-label, phase 2, dose-escalation study, *Lancet*, **378**, 595–605.

Ferrer, A., Wells, K.E., Wells, D.J. (2000). Immune responses to dystropin: implications for gene therapy of Duchenne muscular dystrophy, *Gene Ther*, **7**, 1439–1446.

Foster, H., Sharp, P.S., Athanasopoulos, T., *et al.* (2008). Codon and mRNA sequence optimization of microdystrophin transgenes improves expression and physiological outcome in dystrophic mdx mice following AAV2/8 gene transfer, *Mol Ther*, **16**, 1825–1832.

Gao, G., Bish, L.T., Sleeper, M.M., *et al.* (2011). Transendocardial delivery of AAV6 results in highly efficient and global cardiac gene transfer in rhesus macaques, *Hum Gene Ther*, **22**, 979–984.

Goemans, N.M., Tulinius, M., van den Akker, J.T., *et al.* (2011). Systemic administration of PRO051 in Duchenne's muscular dystrophy, *N Engl J Med*, **364**, 1513–1522.

Graham, I., Hill, V.J., Manoharan, M., *et al.* (2004). Towards a therapeutic inhibition of dystrophin exon 23 splicing in mdx mouse muscle induced by antisense oligoribonucleotides (splicomers): target sequence optimisation using oligonucleotide arrays, *J Gene Med*, **6**, 1149–1158.

Joulia-Ekaza, D., Cabello, G. (2007). The myostatin gene: physiology and pharmacological relevance, *Curr Opin Pharmacol*, **7**, 310–315.

Kang, J.K., Malerba, A., Popplewell, L., *et al.* (2011). Antisense-induced myostatin exon skipping leads to muscle hypertrophy in mice following octa-guanidine morpholino oligomer treatment, *Mol Ther*, **19**, 159–164.

Kemaladewi, D.U., Hoogaars, W.M.H., van Heiningen, S.H., *et al.* (2011). Dual exon skipping in myostatin and dystrophin for Duchenne muscular dystrophy, *BMC Med Genomics*, **4**, 36.

Kinali, M., Arechavala-Gomeza, V., Feng, L., *et al.* (2009). Local restoration of dystrophin expression with the morpholino oligomer AVI-4658 in Duchenne muscular dystrophy: a single-blind, placebo-controlled, dose-escalation, proof-of-concept study, *Lancet Neurol*, **8**, 918–928.

Koo, T., Malerba, A., Athanasopoulos, T., *et al.* (2011a). Delivery of AAV2/9-microdystrophin genes incorporating helix 1 of the coiled-coil motif in the C-terminal domain of dystrophin improves muscle pathology and restores the level of α1-syntrophin and α-dystrobrevin in skeletal muscles of mdx mice, *Hum Gene Ther*, **22**, 1379–1388.

Koo, T., Takashi, O., Athanasopoulos, T., *et al.* (2011b). Long-term functional adeno-associated virus microdystrophin expression in the dystrophic CXMDj dog, *J Gene Med*, **13**, 497–506.

Kreiss, P., Bettan, M., Crouzet, J., *et al.* (1999). Erythropoietin secretion and physiological effect in mouse after intramuscular plasmid DNA transfer, *J Gene Med*, **1**, 245–250.

Leroy-Willig, A., Bureau, M.F., Scherman, D., *et al.* (2005). *In vivo* NMR imaging evaluation of efficiency and toxicity of gene electrotransfer in rat muscle, *Gene Ther*, **12**, 1434–1443.

Malerba, A., Boldrin, L., Dickson, G. (2011a). Long-term systemic administration of unconjugated morpholino oligomers for therapeutic expression of dystrophin by exon skipping in skeletal muscle: implications for cardiac muscle integrity, *Nucleic Acid Ther*, **21**, 293–298.

Malerba, A., Sharp, P.S., Graham, I.R., *et al.* (2011b). Chronic systemic therapy with low-dose morpholino oligomers ameliorates the pathology and normalizes locomotor behavior in mdx mice, *Mol Ther*, **19**, 345–354.

Malerba, A., Thorogood, F.C., Dickson, G., *et al.* (2009). Dosing regimen has a significant impact on the efficiency of morpholino oligomer-induced exon skipping in mdx mice, *Hum Gene Ther*, **20**, 955–965.

Mallat, Z., Besnard, S., Duriez, M., *et al.* (1999). Protective rôle of interleukin-10 in atherosclerosis, *Circulation Research*, **85**, 1–8.

McClorey, G., Fall, A.M., Moulton, H.M., *et al.* (2006a). Induced dystrophin exon skipping in human muscle explants, *Neuromuscul Disor*, **16**, 583–590.

McClorey, G., Moulton, H.M., Iversen, P.L., *et al.* (2006b). Antisense oligonucleotide-induced exon skipping restores dystrophin expression *in vitro* in a canine model of DMD, *Gene Ther*, **13**, 1373–1381.

McPherron, A.C., Lawler, A.M., Lee, S.J. (1997). Regulation of skeletal muscle mass in mice by a new TGF-β superfamily member, *Nature*, **387**, 83–90.

Mendell, J.R., Campbell, K. Rodino-Klapac, L., *et al.* (2010). Dystrophin immunity in Duchenne's muscular dystrophy, *N Engl J Med*, **363**, 1429–1437.

Morgan, J.E., Partridge, T.A. (2003). Muscle satellite cells, *Int J Biochem Cell Biol*, **35**, 1151–1156.

Neri, M., Torelli, S., Brown S., *et al.* (2007). Dystrophin levels as low as 30% are sufficient to avoid muscular dystrophy in the human, *Neuromuscul Disord*, **17**, 913–918.

Ohshima, S., Shin, J-H., Yuasa, K., *et al.* (2009). Transduction efficiency and immune response associated with the administration of AAV8 vector into dog skeletal muscle, *Mol Ther*, **17**, 73–80.

Popplewell, L.J., Adkin, C, Arechavala-Gomeza, V., *et al.* (2010). Comparative analysis of antisense oligonucleotide sequences targeting exon 53 of the human DMD gene: Implications for future clinical trials, *Neuromuscul Disord*, **20**, 102–110.

Popplewell, L.J., Graham, I.R., Malerba, A., *et al.* (2011). Bioinformatic and functional optimization of antisense phosphorodiamidate morpholino oligomers (PMOs) for therapeutic modulation of RNA splicing in muscle, *Methods Mol Biol*, **709**, 153–178.

Popplewell, L.J., Trollet, C., Dickson, G., *et al.* (2009). Design of phosphorodiamidate morpholino oligomers (PMOs) for the induction of exon skipping of the human DMD gene, *Mol Ther*, **17**, 554–561.

Potteaux, S., Deleuze, V., Merval, R., *et al.* (2006). In vivo electrotransfer of interleukin-10 cDNA prevents endothelial upregulation of activated NF-κB and adhesion molecules following an atherogenic diet, *Eur Cytokine Netw*, **17**, 13–18.

Saidenberg-Kermanac'h, N., Bessis, N., Deleuze, V., *et al.* (2003). Efficacy of interleukin-10 gene electrotransfer into skeletal muscle in mice with collagen-induced arthritis, *J Gene Med*, **5**, 164–171.

Schuelke, M.,Wagner, K.R., Stolz, L.E., *et al.* (2004). Myostatin mutation associated with gross muscle hypertrophy in a child, *N Engl J Med*, **350**, 2682–2688.

Toromanoff, A., Adjali, O., Larcher, T., *et al.* (2010). Lack of immunotoxicity after regional intravenous (RI) delivery of rAAV to nonhuman primate skeletal muscle, *Mol Ther*, **18**, 151–160.

Trollet, C., Athanasopoulos, T., Popplewell, L.J., *et al.* (2009). Gene therapy for muscular dystrophy: current progress and future prospects, *Expert Opin Biol Ther*, **9**, 849–866.

van Deutekom, J.C., Bremmer-Bout, M., Janson, A.A., *et al.* (2001). Antisense-induced exon skipping restores dystrophin expression in DMD patient derived muscle cells, *Hum Mol Genet*, **10**, 1547–1554.

van Deutekom, J.C., Janson, A.A., Ginjaar, I.B., *et al.* (2007). Local antisense dystrophin restoration with antisense oligonucleotide PRO051, *N Engl J Med*, **357**, 2677–2687.

Vilquin, J.T., Kennel, P.F., Paturneau-Jouas, M., *et al.* (2001). Electrotransfer of naked DNA in the skeletal muscles of animal models of muscular dystrophies, *Gene Ther*, **8**, 1097–1107.

Wang, Z., Allen, J.M., Riddell, S.R., *et al.* (2007a). Immunity to adeno-associated virus-mediated gene transfer in a random-bred canine model of Duchenne muscular dystrophy, *Hum Gene Ther*, **18**, 18–26.

Wang, Z., Kuhr C.S., Allen J.M., *et al.* (2007b). Sustained AAV-mediated dystrophin expression in a canine model of Duchenne muscular dystrophy with a brief course of immunosuppression, *Mol Ther*, **15**, 1160–1166.

Wu, B., Li, Y., Morcos, P.A., *et al.* (2009). Octa-guanidine morpholino restores dystrophin expression in cardiac and skeletal muscles and ameliorates pathology in dystrophic mdx mice, *Mol Ther*, **17**, 864–871.

Wu, B., Xiao, B., Cloer, C., *et al.* (2011). One-year treatment of morpholino antisense oligomer improves skeletal and cardiac muscle functions in dystrophic mdx mice, *Mol Ther*, **19**, 576–583.

Yin, H., Moulton, H.M., Betts, C. (2009). A fusion peptide directs enhanced systemic dystrophin exon skipping and functional restoration in dystrophin-deficient mdx mice, *Hum Mol Genet*, **18**, 4405–4414.

Yuasa, K., Sakamoto M., Miyagoe-Suzuki Y., *et al.* (2002). Adeno-associated virus vector-mediated gene transfer into dystrophin-deficient skeletal muscles evokes enhanced immune response against the transgene product, *Gene Ther*, **9**, 1576–1588.

28

NEW GENETIC APPROACHES TO TREATING CYSTIC FIBROSIS

Stephen L. Hart[a,c]**, Amy Walker**[a,d] **and Patrick T. Harrison**[b,e]

Cystic fibrosis (CF) is an autosomal recessive disease, which affects around 1 in 2,500 births with a carrier frequency of 1 in 25 Hz, making it the most common inherited disease in the European population (Leitch and Rodgers, 2013). CF is categorized by high sodium levels in sweat, pancreatic insufficiency, biliary and gastrointestinal disease and respiratory disease (Ratjen *et al.*, 2015). The clinical manifestations of CF result from mutations in the gene encoding the CF transmembrane regulator (CFTR) (Riordan *et al.*, 1989; Kerem *et al.*, 1989; Rommens *et al.*, 1989). CFTR is a cyclic AMP (cAMP) regulated ion channel important for transepithelial anion transport, particularly chloride (Anderson *et al.*, 1991) and bicarbonate ions (Poulsen *et al.*, 1994). CFTR channels are found in the secretory epithelia of many organs including the lung, pancreas, digestive, and reproductive tracts (Kerem *et al.*, 1989), and play a particularly important role in the lung where they contribute to airway surface liquid (ASL) homeostasis (Haq *et al.*, 2016). Patients harboring mutations in both alleles of the CFTR gene have impaired conductance of chloride and bicarbonate ions across the airway epithelium resulting in abnormally viscous secretions and impaired mucociliary clearance in the airway of the lungs (Ratjen *et al.*, 2015). Thickened mucus provides a microenvironmental niche for bacteria such as *Pseudomonas aeruginosa*, which permanently colonize the lung resulting in chronic inflammation leading, ultimately, to respiratory failure, the primary cause of morbidity and mortality (Ratjen *et al.*, 2015).

[a] Department of Genetics and Genomic Medicine, UCL Great Ormond Street Institute of Child Health, UCL, London, UK
[b] Department of Physiology, BioSciences Institute, University College Cork, Cork, Ireland
[c] Email: s.hart@ucl.ac.uk
[d] Email: amy.walker.16@ucl.ac.uk
[e] Email: p.harrison@ucc.ie

The discovery of the CFTR gene in 1989 by positional cloning (Kerem *et al.*, 1989) revolutionized the understanding of CF and led to fundamental insights into the pathophysiology of the disease, the classification of mutation classes, and their correlation with clinical manifestations and severity (Cutting *et al.*, 2015), while opening up new diagnostic and therapeutic opportunities (Ratjen *et al.*, 2015; Hart and Harrison, 2017; Griesenbach and Alton, 2015). Through its revelation, the gene itself became a potential therapeutic target and has led to the development of personalized medicines for CF such as the small molecule correctors of CFTR (Ramsey *et al.*, 2011; Wainwright *et al.*, 2015) and various approaches to gene therapy(Hart and Harrison, 2017; Griesenbach and Alton, 2015). These rapidly evolving technologies offer exciting potential for new CF therapies but it is important to consider the challenges that must be overcome before these novel strategies can be applied in the clinic.

28.1 The CF Gene

CFTR is located on chromosome 7q31 (Riordan *et al.*, 1989) and is 250 kB in length, comprising 27 exons encoding an mRNA of 6.5 kb and a protein of 1,480 amino acids (Online Mendelian Inheritance in Man; www.omim.org). Over 2,000 mutant variants are listed on the Cystic Fibrosis Mutation Database, CFTR2.org (Castellani, C.

TABLE 28.1 CFTR Mutation Classes and Examples (Adapted from De Boeck and Amaral, 2016)

Class	CFTR Defect	Example
I	Premature termination codon leading to reduced mRNA and protein	p.Gly542X; p.Trp1282X
II	Folding defect so the protein does not reach the apical membrane	p.Phe508del p.Asn1303Lys p.Ala561Glu
III	Gating defect preventing transport of anions	p.Gly551Asp p.Ser549Arg p.Gly1349Asp
IV	Conductance defect preventing transport of anions	p.ARG117His p.Arg334Trp p.Ala455Glu
V	Low levels of CFTR protein caused by splicing defects	c.3272-26A→G c.3849+10kb C→T
VI	Unstable protein	c.120del123 p.Phe580del
VII	No mRNA or protein	Dele2,3(21 kb), 1717-1G→A

and C. team, 2013), while more than 300 of these variants definitively cause CF (cftr2.org). Around 88% of CF patients on the Cystic Fibrosis Foundation Patient Registry have at least one copy of p.Phe508del which is caused by the deletion of three nucleotides, c.1521_1523delCTT, resulting in the loss of a single codon for the amino acid phenylalanine (Phe) (Foundation, 2017). Mutations in *CFTR* have been grouped into six distinct functional classes according to the method by which they disrupt the synthesis, trafficking, and function and stability of CFTR, although it was recently suggested this should be seven classes (De Boeck and Amaral, 2016) with the possibility of further subgroups (da Cunha *et al.*, 2016) (Table 28.1). While these classifications help in understanding of the CFTR defects at a cellular level, it is important to note that some mutations, such as p.Phe508del, the most common mutation of *CFTR*, are more complex and fall into several categories.

28.2 Current Treatment of CF

Patients with CF today are living much longer, healthier lives than 20 years ago, largely due to improved treatment plans and a better understanding of the disease pathology. Over half of CF patients are now adults and the life expectancy of a CF patient is over 40 years (Leitch and Rodgers, 2013). Treatment of the disease involves a multidisciplinary approach in specialist CF care centers where doctors, nurses, physiotherapists dietitians, and social workers, among others, work as a team to tackle this multifaceted disease (King *et al.*, 2018). The primary objectives of CF treatment include maintenance of lung function in the normal range, and prevention and control of lung infections while managing the associated complications. Virtually all CF patients develop life-long, repeated respiratory viral and bacterial infections with common pathogens including *Staphylococcus aureus* and *Pseudomonas aeruginosa*. Antibiotics are crucial in the treatment of both acute exacerbations and chronic infections in CF lung disease (King *et al.*, 2018).

Despite improvements, these treatments are extremely burdensome and so there is a demand for new improved therapies. CF modulator drugs that rescue defective CFTR proteins in most mutation classes are already having a clinical impact with the future possibility of more and better drugs (Fajac and De Boeck, 2017). In addition, developments in nucleic acid technologies, such as mRNA and gene editing, and delivery methods, both viral and non-viral, have reignited interest in the potential of genetic therapies.

28.2.1 CF Modulator Therapies

CFTR modulators include drugs that act as "potentiators", which modulate the gating of the chloride channel, and "correctors" which restore the folding and stability of CFTR (Hynds, R.E., *et al.*, 2016). Potentiators, such as Ivacaftor (also known by the trade name Kalydeco®), are effective for Class III CFTR mutations, such as p.Gly511Asp which display defective gating properties. Ivacaftor is widely approved for the treatment of CF individuals carrying at least one copy of a CFTR class III mutation enabling treatment of 5–6% of CF individuals, as well as those carrying p.Arg117His, a class IV mutation, enabling treatment of a further 2% of CF individuals (https://www.medicines.org.uk/emc/product/3040/smpc).

Correctors, such as Lumacaftor, enhance folding and trafficking of the CFTR protein to the plasma membrane but were ineffective for treatment of the common p.Phe508delta mutation because of multiple defects in the protein. In addition to misfolding and poor trafficking to the plasma membrane, this mutated protein displays aberrant glycosylation and decreased stability leading to rapid protein turnover in addition to defective gating (Farinha and Amaral, 2005; Jensen *et al.*, 1995; Lukacs *et al.*, 1993; Sharma *et al.*, 2001) making folding correctors ineffective for this mutation. Combinations of Ivacaftor and Lumacaftor are more effective in treating patients homozygous for the p.Phe508del mutation (Wainwright *et al.*, 2015) and marketed as Orkambi®. Another class of modulators, called "amplifiers", enhance translation of CFTR mRNA (Molinski *et al.*, 2017) where it was shown that drugs such as PTI-428 (Proteostasis Therapeutics; www.proteostasis.com) may be effective in combination with Orkambi® for treatment of patients with one copy of the p.Phe508delta mutation. A further class of CFTR modulators, termed "amplifiers", stabilize CFTR mRNA, increasing the amount of protein and are in Phase 2 clinical trials in combination with Orkambi®.

While modulator drugs are useful for rescuing defective CFTR proteins, Class I CFTR mutations do not produce a full-length protein due to the creation of premature termination codons (PTC), while those in class V produce lower levels of CFTR mRNA and consequently less protein. Read-through drugs such as Ataluren, which was identified by chemical library screening, insert an amino acid at internal PTC sites. Ataluren was evaluated in clinical trials for CF Class I mutations, although recent Phase 3 trials failed to show efficacy (Aslam *et al.*, 2017).

A further target for drug development includes alternative ion channels where targets include alternative chloride channels TMEM16A (Zhang *et al.*, 2014) and SLC26A9 (Anagnostopoulou *et al.*, 2012), as well as the epithelial sodium channel (ENaC) to reduce sodium uptake (Haq *et al.*, 2016; Mall and Galietta, 2015). These alternative ion channel strategies are all designed to correct the epithelial ion transport imbalance and restore hydration of the airway, reduce mucus thickening, and enhance mucociliary clearance. PTC therapeutics and alternative channel therapies have been reviewed in more detail recently (Fajac and De Boeck, 2017; De Boeck and Davies, 2017).

28.2.2 Genetic Approaches to CF Therapy

CF is a monogenic, recessively inherited pulmonary disorder and has long been regarded as a target for gene replacement therapy in the lung. Genetic therapies tackle the basic molecular defect of the disease with the potential for long-term therapy for all patients regardless of mutation class. It has been shown that as little as 5% expression of normal levels of CFTR corrects the ion transport defect *in vitro* (Johnson *et al.*, 1992), although expression at 25% of normal was required for correction of mucus transport in a human epithelial model (Zhang *et al.*, 2009). In another study, it was shown in air–liquid interface (ALI) epithelial cultures with epithelia containing 20% of wild type cells mixed with CF cells generated approximately 70% of the transepithelial chloride ion current of a fully wild-type epithelium (Farmen *et al.*, 2005). Thus, gene therapy efficiency in the 5–25% range may be sufficient to treat CF but the lack of suitable *in vivo* models has made it difficult to refine this target.

Gene therapy approaches have been evaluated over the last 25 years by viral or non-viral vectors, yet no clinically effective therapy has emerged till date (Griesenbach and Alton, 2015). Gene therapy with liposomal complexes with plasmid DNA has been investigated extensively in trials for CF, most recently in 2015 in year-long multidose trial (Alton *et al.*, 2015). The most significant outcome of this study was a small improvement in forced expiratory volume (FEV-1) compared to the control group, while the safety of repeated delivery was validated. Transgene expression was not detectable, although the CFTR expression vector contained an optimized hybrid promoter to maximize expression (Alton *et al.*, 2014). Gene therapy with viral vectors has been attempted with Adenovirus (Boucher *et al.*, 1994; Welsh *et al.*, 1994) and AAV2 (Moss *et al.*, 2004) but transgene expression was low or undetectable with either viral vector, even after repeated dosing in the case of AAV (Moss *et al.*, 2007).

It has become clear from clinical trials for CF gene therapy that the lung is far more resistant to gene transfer than initially anticipated, resulting in low levels of gene transfer. However, in recent years, improved understanding of the biology of the lung and its barriers, such as the periciliary brush (Button *et al.*, 2012), mucus composition (Duncan *et al.*, 2016), and epithelial biology is enabling the rational development of more efficient viral and non-viral vectors. Furthermore, new nucleic acid technologies including minicircle DNA, siRNA and mRNA, as well as CRISPR/Cas9 gene editing technologies, have also emerged so that higher and more sustained levels of correction and CFTR expression may be anticipated. These advances have led to renewed interest in genetic therapies for CF.

28.2.3 Viral Vectors for CF Gene Therapy

Retroviral vectors, such as the murine leukemia virus (MoMuLV), have proven effective in gene therapy for immunodeficiency diseases such as X-linked severe combined immunodeficiency disease (X-SCID) (Cavazzana *et al.*, 2016) and adenosine deaminase deficiency-SCID (ADA-SCID) (Ferrua and Aiuti, 2017). Lentivirus vectors (LV) are of particular interest for *in vivo* transduction of airway epithelial cells as, unlike MoMuLV, they integrate into the genome of non-dividing cells, such as airway epithelial cells and provide expression for the lifetime of the epithelial cells. Viral integration, however, also presents risks of insertional oncogenesis if they should integrate in the location of an oncogene, such as *LMO2*, as observed in clinical trials of X-SCID, that caused leukaemia (Hacein-Bey-Abina *et al.*, 2003). The latest generation of LV have self-inactivating (SIN) long terminal repeats that reduce their ability to activate oncogenes in the genome and so should be much safer (Miyoshi *et al.*, 1998; Montini *et al.*, 2009). LV are under development, including a VSV-G pseudotyped LV, containing a LacZ reporter, which is administered after dosing with a mild detergent, lysophosphatidylcholine, which has confirmed the transducibility of CF-relevant airway cell types in a preliminary study in marmosets, a non-human primate model (Farrow *et al.*, 2013). An aerosolized feline immunodeficiency virus (FIV) lentivirus pseudotyped with a GP64 envelope protein nebulized to CF new-born pigs was reported to provide a significant increase in cAMP-activated Cl^- current near to wild-type levels (Cooney *et al.*, 2016). Finally, a simian immunodeficiency virus (SIV) derived LV, pseudotyped with Sendai virus fusion protein (F) and Hemagglutinin/Neuraminidase (HN) envelope proteins was reported to demonstrate stable transduction of murine airway epithelium that persisted for a year (Griesenbach *et al.*, 2016) and are in preclinical development(Alton *et al.*, 2017).

AAV vectors, in contrast to wild type AAV, have a low chromosomal integration frequency but achieve long-term, episomal transgene expression through the formation of stable concatameric structures of the vector genome (Yang *et al.*, 1999). AAV has become the most widely used vector for *in vivo* gene delivery, including the first EU-approved gene therapy product, Glybera, for lipoprotein lipase deficiency, a metabolic disease. AAV has also been used in clinical trials for inherited retinal dystrophy caused by *RPE65* mutations, achieving improvements in vision (Ku and Pennesi, 2015) and for haemophilia restoring Factor IX to levels sufficient to reduce bleeding episodes (Nienhuis *et al.*, 2017). The immunogenicity of AAV and pre-existing immunity to AAV serotype used in gene therapy

(Halbert *et al.*, 2006) have so far limited its efficacy in repeated administration in the lung (Masat *et al.*, 2013). Current AAV vector development for CF is focused on improving lung epithelial transduction efficiency by use of alternative naturally occurring capsid variants (McClain *et al.*, 2016) as well as by viral capsid selection and promoter development (Kurosaki *et al.*, 2017). Recently, a new human parvovirus vector, human Bocavirus-Type-1 derived vector, was described (Yan *et al.*, 2017), that displays high tropism for the apical membrane of human airway epithelial cells. Bocavirus has a larger capsid than AAV, enabling it to package a full-length CFTR along with a full length, strong CMV-β-actin promoter.

28.3 Nanoparticles and Barriers to Transfection in the Lung

Nanoparticles, particularly liposomes (Caplen *et al.*, 1995; McLachlan *et al.*, 1996; Gill *et al.*, 1997; Porteous *et al.*, 1997; Knowles *et al.*, 1998; Hyde *et al.*, 2000; Noone *et al.*, 2000), but also compacted, PEGylated polylysine nanoparticles (Ziady *et al.*, 2003), have a long history of clinical development in CF with the most recent involving the GL67 lipid formulation (Alton *et al.*, 2015). Overall, the conclusion from these studies is that improved nanoparticle formulations for more efficient nucleic acid delivery are still required.

Challenges to nanoparticle delivery include nebulization and deposition in the correct region of the lung followed by mucus penetration through to the apical membrane of the airway epithelium, then endocytic uptake. The favored method of pulmonary delivery of gene therapy formulations is by inhalation as this is minimally invasive and allows pain-free access. Several studies have demonstrated the stability of nanoparticles to nebulization (Manunta *et al.*, 2011, 2013; Davies *et al.*, 2014). Nanoparticle formulations labeled with 99m-technetium were used for scintigraphic imaging of aerosol biodistribution in the lung of normal pigs as a similarly sized model of the human lung (Manunta *et al.*, 2013). Under optimal nebulization conditions, nanoparticle deposition was demonstrated in the central and lower airways, where CFTR is most abundantly expressed, and where transgene expression might thus be most effective.

Once deposited in the airways by nebulization, nanoparticles must then penetrate through the ASL which comprises mucus and the periciliary liquid layer (PCL), the watery layer that bathes the cilia (Boucher, 2004). Mucus, a gel-like layer containing mucin glycoproteins, entraps inhaled particles, including virus, bacteria, pollutants, and potentially nanoparticles, which are then removed rapidly in a proximal to distal direction by ciliary beating. CF mucus may present more of a physical barrier as it is thicker and stickier than non-CF mucus. Impaired CFTR-mediated bicarbonate secretion lowers the ASL pH and leads to retention of calcium ions by the mucins, maintaining them in a compacted form rather than the open-mesh network of normal mucus (Birket *et al.*, 2014; Tang *et al.*, 2016). Mucus viscosity is further contributed by chromosomal DNA released by dead inflammatory cells such as neutrophils (Shah *et al.*, 1996). Approaches to overcome the mucus barrier include treating patients in early childhood before the onset of mucus thickening, the use of mucolytics such as N-acetyl cysteine to mobilize the mucus (Suk *et al.*, 2011), magnetofection to pull paramagnetic nanoparticles through the mucus layer by a magnetic field (Castellani *et al.*, 2016), while attention to the chemical properties of the nanoparticles themselves may enable better penetration of mucus. For example, a dense coating of polyethylene glycol on the surface of the DNA nanoparticles that reduce charge-mediated interactions with mucus components, was shown to improving penetration of mucus (Suk *et al.*, 2014). Once through the mucus, the PCL, the periciliary brush presents a further barrier which is almost impenetrable to particles larger than 40 nm (Button *et al.*, 2012), thus, the production of smaller particles would likely be beneficial to transfection efficiency.

There are many diverse cell types in the epithelium including ciliated epithelial cells, goblet and club cells, as well as macrophages and neutrophils. Targeted transfection of the epithelial cells that express CFTR, is required to maximize efficacy of treatment. CFTR expression appears to be most evident in submucosal glands (Engelhardt *et al.*, 1992) but the narrow opening to the gland largely excludes gene therapy vectors (Pilewski *et al.*, 1995). Thus, gene therapy strategies are usually designed to transfect the surface, ciliated epithelium. *In vivo* studies demonstrated that liposomal transfection of the nasal epithelium in CF mice, which lacks submucosal glands, was sufficient to correct the electrophysiological defect (MacVinish *et al.*, 1997a, 1997b; Ziady *et al.*, 2002), while correction of human CF epithelial cells *in vitro* by retroviral transduction restored chloride transport across the polarized epithelium *in vitro*.

A further challenge is that epithelial cells are terminally differentiated with a half-life of approximately 6 months in mice (Rawlins and Hogan, 2008) although epithelial turnover may be even shorter in the CF lung. Transfecting terminally-differentiated cells such as the airway epithelium, presents two problems, the first of which is that plasmid-based, CFTR gene transfer is inefficient due to the highly selective nature of uptake through pores in the nuclear envelope limiting nuclear uptake of plasmid (Liu *et al.*, 2003; Meng *et al.*, 2004). This barrier is likely to be one of the main reasons why clinical gene therapy studies with liposomal nanoparticles have demonstrated such low levels of transgene expression. The second issue is that transfected cells have a limited life-span of a few weeks and so, periodically, the therapy will have to be readministered. Previous clinical and preclinical trials have shown that nanoparticle formulations are safe, well tolerated, and non-immunogenic, although transient inflammatory responses have been observed (Griesenbach *et al.*, 2015). Thus, in the respect that this will be a life-long therapy, nanoparticle formulations offer considerable advantages over viral vectors which in most cases are immunogenic or face challenges of pre-existing immunity.

In the next section, we consider the new genetic molecules and therapeutic strategies that can be delivered with nanoparticles, how the nanoparticles are being developed to break through lung barriers to transfection so that this combination of technologies might overcome the previous deficiencies in non-viral genetic therapies for CF.

28.4 Minicircle DNA

Gene therapy approaches using plasmid DNA face a number of cellular barriers from endocytic internalization, to endocytic escape to the cytoplasm, and finally, uptake into the nucleus. In addition, the bacterial DNA content of plasmid DNA can cause inflammation on delivery to the lung by interactions of hypomethylated CpG motifs with toll-like receptor 9 (TLR9). Minicircle DNA is devoid of almost all bacterial DNA, including the antibiotic bacterial selection gene (Darquet *et al.*, 1999; Bigger *et al.*, 2001; Chen *et al.*, 2005) and offers several benefits over plasmids; firstly, a CFTR minicircle will be much smaller than a plasmid, potentially enhancing potency and nuclear uptake efficiency combined with reduced inflammatory potential. Transfection studies demonstrated enhanced expression with minicircle DNA compared to plasmid DNA *in vitro* and in murine lung, which was more persistent *in vivo* with reduced inflammation (Munye *et al.*, 2016).

28.4.1 *In vitro* Transcribed mRNA

In vitro transcribed messenger RNA (mRNA) as a template for CFTR protein could be an effective therapeutic for patients with any CF-causing mutation (Johler *et al.*, 2015; Bangel-Ruland *et al.*, 2013). The structure of mRNA from 5′ to 3′ comprises the 5′ cap, 5′ untranslated region (UTR), coding region, 3′ UTR and polyA tail, and the structure of each can be readily optimized. Chemical modifications, particularly chemically modified bases such as pseudouridine have resulted in mRNA with improved stability, translational efficiency, and reduced immunogenicity (Anderson *et al.*, 2011). Alternatively, sequence engineering with natural nucleosides to maximize GC content was also reported to reduce inflammatory effect *in vivo* (Thess *et al.*, 2015). Equally important have been the developments in the UTRs of the mRNA template, including the 5′-capping procedure using anti-reverse capping reagents (ARCA), capping analogues that reduce the rate of decapping (Strenkowska *et al.*, 2016), the structure of the 5′ and 3′ UTRs and poly A region, (reviewed in Sahin *et al.*, 2014). The coding region itself may be optimized for higher levels of expression by codon optimization in the template DNA; this ensures use of codons for which there is a higher concentration of transfer RNA (tRNA) with the relevant anticodon although, interestingly, this approach does not always lead to higher expression (Mauro and Chappell, 2014).

CFTR mRNA therapies would have several important advantages over plasmid-based CF gene therapies (Johler *et al.*, 2015; Bangel-Ruland *et al.*, 2013; Antony *et al.*, 2015; Robinson *et al.*, 2018). Translation of mRNA in the cytoplasm means that, unlike plasmids, the nuclear envelope is not a barrier to transfection and so expression levels are potentially much higher. From a safety perspective, mRNA cannot integrate into the genome and so the risk of insertional mutagenesis is nullified in contrast to plasmid DNA, where there is a low but significant risk of insertional oncogenesis or germ line transmission. Production of mRNA is relatively simple and inexpensive without the need for bacterial fermentation processes required for plasmid production, apart from providing the plasmid template for *in vitro* transcription, although HPLC purification methods are important to achieve highest levels of purity and activity (Kariko *et al.*, 2011).

Protein production from the mRNA template is short-lived, with a half-life of several hours only (Pardi *et al.*, 2015), necessitating administration at more frequent intervals than plasmid-based expression vectors although this may not be a limiting factor depending on protein stability. For example, delivery of surfactant protein B (SpB) mRNA to murine lungs of a mouse model of SpB deficiency led to restoration of 71% of normal levels of SpB from twice weekly administration to the airways, which was sufficient to achieve complete survival of the mice for the duration of the experiments (Kormann *et al.*, 2011). Wild-type CFTR protein, once in the apical epithelial membrane, has a half-life of at least 24 h (Lukacs *et al.*, 1993) with others reporting more than 48 h (Heda *et al.*, 2001), although this is greatly reduced with CFTR protein containing the p.Phe508del mutation. Thus, it is likely that a single dose of CFTR mRNA would lead to CFTR channel activity of at least several days implying an approximately, weekly, repeated dosing requirement to maintain membrane CFTR protein levels and ion transport activity.

While naked mRNA delivery by intradermal injection for vaccination purposes has been reported (Probst *et al.*, 2007; Zeyer *et al.*, 2016), most applications of mRNA, particularly for lung delivery by nebulization (Johler *et al.*, 2015), will demand a nanoparticle formulation since even chemically modified mRNA remain susceptible to nuclease degradation and would be unlikely to penetrate the mucociliary barriers to epithelial transfection discussed before. Nanoparticle-mediated delivery *in vivo* has been described for pulmonary delivery of CFTR mRNA *in vitro* and *in vivo* (Robinson *et al.*, 2018) as well as other examples of pulmonary mRNA delivery (Johler *et al.*, 2015; Antony *et al.*, 2015; Mahiny *et al.*, 2015), suggesting the promise of this approach.

28.5 Gene Editing

Gene editing involves methods of introducing targeted, double-strand breaks (DSBs) by targeted nucleases into the mutated CFTR gene of CF epithelial cells, then providing a template DNA to enable the endogenous, homology-directed DNA repair (HDR) pathway to correct the genetic defect. In the absence of template DNA, non-homologous end joining (NHEJ) pathways may introduce indel mutations into the open reading frame creating frameshifts, often with PTC, thus inactivating the gene. Gene editing therapies were first developed with zinc-finger nucleases (ZFN) and then transcription activator-like effector nucleases (TALENs), both of which involve targeting by protein domains to both strands of DNA fused to dimeric *FokI* nuclease (Gaj *et al.*, 2013). Although both ZFN and TALENs are effective for gene editing, the ease of use and low cost of CRISPR/Cas9 gene editing systems has facilitated the more widely used gene editing formulation (Gaj *et al.*, 2013).

The term clustered regularly interspaced palindromic repeats (CRISPR) and their CRISPR-associated (Cas) nucleases was first used in 2002 (Jansen *et al.*, 2002) to define interspaced sequence repeats that had been reported initially in the 1980s (Nakata *et al.*, 1982). In 2012, the potential of harnessing the CRISPR–Cas9 system to facilitate targeted genome editing was realized by the Charpentier and Doudna labs (Jinek *et al.*, 2012), shortly after the use of the technology for engineering human cells was demonstrated (Mali *et al.*, 2013). The type II CRISPR system comprises a Cas9 nuclease and non-coding RNA elements which program the specificity for the site of nucleic acid cleavage (Ran *et al.*, 2013). Cas9 first recognizes the protospacer adjacent motif (PAM) sequence comprising any 5′-NGG-3′ sequence, where "N" is any nucleotide, and "GG" denotes two guanines, then unwinds the DNA to allow binding of the guide RNA (gRNA) to the target site where, provided there is a match, the Cas9 nuclease induces a dsb in the gene, 3 bp upstream of the PAM sequence (Sternberg *et al.*, 2014).

Initial demonstrations of applications of gene editing for CF exploited the HDR pathway to correct the p.Phe508del mutation, with both ZFNs (Lee *et al.*, 2012; Crane *et al.*, 2015) and TALENs (Camarasa and Galvez, 2016). ZFNs were also used to insert an entire CFTR cDNA into the chemokine (C-C motif) receptor 5 (CCR5) locus of inducible pluripotential stem cells (IPSCs) derived from human lung fibroblasts, leading to efficient expression from the CCR5 promoter, in a demonstration of the "safe harbor" strategy to insertion of CFTR cDNA (Ramalingam *et al.*, 2013). ZFNs were also used to insert a "superexon" in the CFTR locus, encoding exons 11 to 27 which restored CFTR function in a CF cell line (Bednarski *et al.*, 2016). More recently, CRISPR/Cas9 was used to correct *CFTR* in CF intestinal progenitor cells which were then used to make organoids for functional analysis by the organoid swelling assay, involving fluid uptake by forskolin-activated CFTR in the organoid membrane (Schwank *et al.*, 2013). In these studies, mutation-specific correction of the p.Phe508del was performed but there are more than 300 disease-causing mutations in CFTR (Cutting, 2015) and so it is unrealistic to envisage a specific therapy for each mutation. A further challenge to implementing gene editing technologies by HDR includes efficiency of repair where maximal efficiency levels of approximately 3% were achieved without selection. Approaches to extend the application to wider CF populations and to increase efficiency are in development.

An approach that could apply to larger numbers of CF patients by gene editing involves use of a template DNA to knock-in a block of exons, a so-called "superexon" (Bednarski *et al.*, 2016). The superexon donor can correct all downstream mutations from the insertion site at the endogenous locus and was demonstrated in the CFBE41o- cell line, which is homozygous for the p.Phe508del mutation. A therapeutic super-exon encompassing exons 11 to 27 of CFTR was introduced into the 5′ end of exon 27, after making DSBs with a ZFN integration strategy (Bednarski *et al.*, 2016). This resulted in expression of corrected CFTR mRNA from the endogenous CFTR promoter. CFTR functional correction by Ussing chamber analysis was performed to corroborate the ability of repaired CFTR channels to restore ion conductance. This approach to gene editing of CFTR has the potential to treat all patients either by delivery of a superexon or the whole cDNA, although efficiencies by HDR strategies may be limiting.

Although these studies present encouraging proof-of-concept data, their translational potential is limited because of the low efficiency of correction and their reliance on HDR mechanisms. HDR pathways are only functional in mitotic cells yet, as discussed above, it is most likely that cells targeted by aerosol-mediated systems will be deposited on the non-dividing, terminally-differentiated, surface epithelium. Even if the basal epithelium was more accessible, it is also unlikely to be replicating, thus, strategies that exploit non-HDR pathways are gaining more attention for *in vivo* gene editing. The NHEJ repair pathway is far more efficient than HDR but is also highly error prone, thus it is not suitable for precise repair but can be used to disrupt genomic target sites. The NHEJ pathway was exploited in CF to disrupt an aberrant intronic splice site in a CFTR mini gene with the 3849+10 kb C>T mutation, a rare type VI mutation, using two intronic flanking gRNAs to excise the cryptic splice site, restoring normal splicing of the CFTR gene (Anderson *et al.*, 2011). This pathway has been extended to gene correction strategies by homology independent targeted integration (HITI) method, which was shown to work *in vitro* and *in vivo* in non-mitotic cells of the eye and brain, with ten-fold higher efficiency levels than HDR repair pathways (Suzuki *et al.*, 2016). Although use of HITI has not been reported in CF lung applications, this approach would seem to be compatible with targeted integration of CFTR superexons or cDNA.

Off-target effects are caused by DSBs occurring at alternative PAM sites, or more randomly, followed by NHEJ repair introducing indels which could disrupt gene expression with unexpected consequences. These indels may occur at genomic sites 5′ of potential PAMs bearing sufficiently close homology to the gRNA. Whole genome sequencing has been used to assess the frequency of off-target effects and it was shown that off-target indels are quite rare in mice (Iyer *et al.*, 2015). Most of the current research is focused on more precise gene editing formulations, such as nickases that require two gRNAs and base editing approaches that do not require DSBs (Ran *et al.*, 2013; Komor *et al.*, 2016), and more effective ways to detect off-target events and their consequences such as GUIDEseq and others (Zhu *et al.*, 2017).

Approaches to gene editing that do not require DSBs are also under investigation such as the use of triplex-peptide nucleic acids (PNAs) delivered by nanoparticles. Transfection of human CFBE cells, treated *in vitro* serially three times with triplex-PNA nanoparticles, achieved targeted gene correction of the p.Phe508del with a frequency of almost 10% by deep sequencing while *in vivo* correction of the mouse nasal epithelium was 5% and lung epithelium 1% (McNeer *et al.*, 2015). In the situation where higher levels of transfection and indel formation are achieved e.g., 80% indel formation, p53-associated toxicity was observed in iPSCs (Robinson *et al.*, 2018). Thus, it was proposed where engineered iPS cell therapies are in development, their p53 status should be considered.

28.6 Delivery of the CRISPR/Cas9 System

Therapeutic gene editing by CRISPR is the area currently of greatest interest for CF owing to its ease of use and low cost. Two main strategies for CF therapy are under investigation, *in vivo* editing by viral or non-viral delivery, or *ex vivo* gene editing for cell-mediated therapy. Gene editing of the basal cell progenitor cell population *in vivo* would provide a long-term therapy but targeting of these cells will be challenging as they reside deep within the lung tissue. Thus, gene editing *in vivo* would likely be targeted to the differentiated epithelium thus requiring repeated delivery as cells are replaced. For cell therapy, basal cell might be corrected *ex vivo* but the challenge would then be to engraft them.

CRISPR gene editing formulations comprise two components that must be delivered in to the same cell to achieve DSBs, including Cas9 nuclease and gRNA. Cas9 nuclease is targeted by the gRNA to make DSBs at specific genomic sites, 3-nucleotides upstream of a 3-nucleotide, PAM, typically NGG. For delivery, Cas9 nuclease may be encoded in a viral or plasmid vector along with a gRNA. Cas9 derived from *Streptococcus pyogenes* (*Sp* Cas9; ~4.2 kb), however, is too large for insertion into AAV vectors with regulatory elements. AAV important vectors for *in vivo* gene editing and so after screening other bacterial species, Cas9 derived from *Staphylococcus aureus* (*Sa* Cas9), which is approximately 1 kb shorter than *Sp* Cas9, was identified as an effective alternative Cas9 for gene editing in AAV delivery constructs under a full length eukaryotic RNA polymerase II promoter (Ran *et al.*, 2015). Cas9 nuclease may also be delivered as a transcribed mRNA or even as the *in vitro* translated protein in a preassembled ribonucleoprotein (RNP) complex with synthetic, single gRNA 9 (sgRNA) molecules.

The gRNA comprises two RNA components, a transactivating CRISPR RNA (tracrRNA) and a CRISPR RNA (crRNA), that can be delivered either as a hybrid or as a sgRNA construct with a short hairpin connecting the two strands (Jiang and Doudna, 2017). DNA constructs encoding the gRNA can be packaged into viral vectors including AAV, or plasmid DNA, expressed from RNA polymerase III promoters, such as U6. Chemically modified synthetic sgRNAs combined with either Cas9 mRNA or in RNPs offer improved specificity and activity (Hendel *et al.*, 2015). RNP complexes can be delivered by non-viral nanoparticles or liposomes, along with donor DNA (Lee *et al.*, 2017; Zuris *et al.*, 2015). RNP delivery has the advantage that it eliminates the need to first transcribe RNA or translate to protein to elicit editing, which can enhance efficiency (Lee *et al.*, 2017; Zuris *et al.*, 2015). The transient nature of RNP–Cas9 enzyme activity also reduces the potential for off-target events as high levels of stable Cas9 expression, achieved from integrating retroviral vectors, are associated with increased off-target events (Ortinski *et al.*, 2017).

If gene editing by HDR or HITI is required, then a donor, or template, DNA is also required in the same cells as those transfected with CRISPR/Cas9. The DNA repair template may include plasmid or minicircle DNA, double-stranded linear DNA or single-stranded DNA molecules. Short oligonucleotide templates will likely enter the nucleus more easily than plasmids and thus provide highest efficiencies (Richardson *et al.*, 2016) and specificities, while for larger templates, minicircle DNA template is more efficient than plasmid DNA (Suzuki *et al.*, 2016). Viral vectors such as AAV may also be used to provide the template more efficiently (Mahiny *et al.*, 2015), although this approach, like viral gene replacement therapy, may be impacted by viral immunogenicity if repeat dosing is required.

28.7 Regenerative Lung Cell Therapy

Two types of cells are under investigation as CF therapeutics for modification by gene editing, followed by repopulation of the lung epithelium by cell engraftment, including basal epithelial cells and IPSCs. Basal cells are the progenitors of most of the other airway epithelial cell types and can be obtained from the airways, including the nasal epithelium, by brushing. Basal cells can then be cultured in media containing Rho-kinase inhibitors on fibroblast feeder layers that allow the cells to be expanded without losing their differentiation potential on ALI cultures (Hynds *et al.*, 2016; Miyashita *et al.*, 2013). After gene editing, the corrected cells would be selected and expanded prior to engraftment (LaRanger *et al.*, 2018). Selection may be performed by incorporation of antibiotic resistance markers, such as puromycin, into the donor template which enables enrichment of the corrected cells followed by clonal expansion.

IPS cells offer the opportunity to use a wider range of donor cell types from CF donors that can be differentiated to epithelial cell types suitable for engraftment after gene editing of CFTR. Gene editing of CFTR in IPSCs has so far been performed with ZFN (Crane *et al.*, 2015) and CRISPR/Cas9 *in vitro* (Firth *et al.*, 2015). For both cell therapy strategies, preconditioning of the lung to create a niche for engraftment of engineered basal cells will be required with agents such as polidocanol (Gui *et al.*, 2015), by irradiation (Rosen *et al.*, 2015) or sulfur dioxide treatment (Zuo *et al.*, 2015). Most of these strategies are currently problematic clinically in CF patients due to their severity in stripping the surface epithelium but more acceptable strategies may be possible.

Regenerative medicine approaches in the CF intestine are also under investigation. Organoids are 3D cellular clusters that are capable of self-renewal and organization, and retain functionality of the tissue of origin (Fatehullah *et al.*, 2016). In this study, a CRISPR/Cas9 based gene editing approach was used to correct the CFTR locus in cultured intestinal stem cells from CF patients, restoring CFTR functionality in clonally-expanded organoids (Schwank *et al.*, 2013). It was also shown in another study that GFP^+ colon organoids could be successfully reintroduced into a superficially damaged mouse colon, where the transplanted cells adhered and readily integrated to the damaged area (Yui *et al.*, 2012). Long-term (6 month) engraftment was observed with transplantation of organoids derived from a single colon stem cell after *in vitro* expansion. Taken together, this work provides a potential *ex vivo* cell therapy strategy for intestinal defects in CF patients.

28.8 Conclusions

Genetic approaches to therapy of CF have frustrated scientific and medical practitioners in this field for 25 years. However, recent scientific developments in understanding of the genetics of CF, lung epithelial biology, and nucleic acid therapeutics, particularly mRNA and gene editing, as well as developments in viral vector technologies, have offered causes for renewed optimism and corresponding surge of activity in genetic therapies for CF. The success of effective small molecule drugs such as Ivacaftor also encourages optimism that a genetic strategy which restores CFTR activity could be similarly as effective therapeutically as that drug. It could be argued that the advent of small molecule drugs means that genetic approaches are no longer needed for CF, that the future therapeutic ecosystem for CF will be populated entirely by small molecule drugs. However, there are many mutations for which there are no effective drugs available affecting approximately 5% of all patients and so therapies are needed for these patients. Moreover, gene therapy still offers the intriguing possibility of a single therapy suitable for all patients with administration only required at extended intervals of time — a prospect that continues to motivate.

References

Alton, E., *et al.* (2015). Repeated nebulisation of non-viral CFTR gene therapy in patients with cystic fibrosis: A randomised, double-blind, placebo-controlled, phase 2b trial, *Lancet Respir Med*, **3**(9), 684–691.

Alton, E.W., *et al.* (2014). Toxicology study assessing efficacy and safety of repeated administration of lipid/DNA complexes to mouse lung, *Gene Ther*, **21**(1), 89–95.

Alton, E.W., *et al.* (2017). Preparation for a first-in-man lentivirus trial in patients with cystic fibrosis, *Thorax*, **72**(2), 137–147.

Anagnostopoulou, P., *et al.* (2012). SLC26A9-mediated chloride secretion prevents mucus obstruction in airway inflammation, *J Clin Invest*, **122**(10), 3629–34.

Anderson, M.P., *et al.* (1991). Demonstration that CFTR is a chloride channel by alteration of its anion selectivity, *Science*, **253**(5016), 202–5.

Anderson, B.R., *et al.* (2011). Nucleoside modifications in RNA limit activation of 2′-5′-oligoadenylate synthetase and increase resistance to cleavage by RNase L, *Nucleic Acids Res*, **39**(21), 9329–38.

Antony, J.S., *et al.* (2015). Modified mRNA as a new therapeutic option for pediatric respiratory diseases and hemoglobinopathies, *Mol Cell Pediatr*, **2**(1), 11.

Aslam, A.A., *et al.* (2017). Ataluren and similar compounds (specific therapies for premature termination codon class I mutations) for cystic fibrosis, *Cochrane Database Syst Rev*, **1**, CD012040.

Bangel-Ruland, N., *et al.* (2013). Cystic fibrosis transmembrane conductance regulator-mRNA delivery: A novel alternative for cystic fibrosis gene therapy, *J Gene Med*, **15**(11-12), 414–26.

Bednarski, C., *et al.* (2016). Targeted integration of a super-exon into the CFTR locus leads to functional correction of a cystic fibrosis cell line model, *PLoS One*, **11**(8), e0161072.

Bigger, B.W., *et al.* (2001). An araC-controlled bacterial cre expression system to produce DNA minicircle vectors for nuclear and mitochondrial gene therapy, *J Biol Chem*, **276**(25), 23018–27.

Birket, S.E., *et al.* (2014). A functional anatomic defect of the cystic fibrosis airway, *Am J Respir Crit Care Med*, **190**(4), 421–32.

Boucher, R.C. (2004). New concepts of the pathogenesis of cystic fibrosis lung disease, *Eur Respir J*, **23**(1), 146–58.

Boucher, R.C., *et al.* (1994). Gene therapy for cystic fibrosis using E1-deleted adenovirus: A phase I trial in the nasal cavity, *Hum Gene Ther*, **5**(5), 615–39.

Button, B., *et al.* (2012). A periciliary brush promotes the lung health by separating the mucus layer from airway epithelia, *Science*, **337**(6097), 937–41.

Camarasa, M.V., Galvez, V.M. (2016). Robust method for TALEN-edited correction of pF508del in patient-specific induced pluripotent stem cells, *Stem Cell Res Ther*, **7**, 26.

Caplen, N.J., *et al.* (1995). Liposome-mediated CFTR gene transfer to the nasal epithelium of patients with cystic fibrosis, *Nat Med*, **1**(1), 39–46.

Castellani, C., Team, C. (2013). CFTR2: How will it help care? *Paediatr Respir Rev*, **14**(1), 2–5.

Castellani, S., *et al.* (2016). Magnetofection enhances lentiviral-mediated transduction of airway epithelial cells through extracellular and cellular barriers, *Genes*, **7**(11).

Cavazzana, M., *et al.* (2016). Gene therapy for X-linked severe combined immunodeficiency: Where do we stand? *Hum Gene Ther*, **27**(2), 108–16.

Chen, Z.Y., He, C.Y. Kay, M.A. (2005). Improved production and purification of minicircle DNA vector free of plasmid bacterial sequences and capable of persistent transgene expression in vivo, *Hum Gene Ther*, **16**(1), 126–31.

Cooney, A.L., *et al.* (2016). Lentiviral-mediated phenotypic correction of cystic fibrosis pigs, *JCI Insight*, **1**(14), e88730.

Crane, A.M., *et al.* (2015). Targeted correction and restored function of the CFTR gene in cystic fibrosis induced pluripotent stem cells, *Stem Cell Rep*, **4**(4), 569–77.

Cutting, G.R. (2015). Cystic fibrosis genetics: From molecular understanding to clinical application, *Nat Rev Genet*, **16**(1), 45–56.

da Cunha, M.F., *et al.* (2016). Analysis of nasal potential in murine cystic fibrosis models, *Int J Biochem Cell Biol*, **80**, 87–97.

Darquet, A.M., *et al.* (1999). Minicircle: An improved DNA molecule for in vitro and in vivo gene transfer, *Gene Ther*, **6**(2), 209–18.

Davies, L.A., *et al.* (2014). Aerosol delivery of DNA/liposomes to the lung for cystic fibrosis gene therapy, *Hum Gene Ther Clin Dev*, **25**(2), 97–107.

De Boeck, K., Amaral, M.D. (2016). Progress in therapies for cystic fibrosis, *Lancet Respir Med*, **4**(8), 662–74.

De Boeck, K., Davies, J.C. (2017). Where are we with transformational therapies for patients with cystic fibrosis? *Curr Opin Pharmacol*, **34**, 70–75.

Duncan, G.A., *et al.* (2016). The mucus barrier to inhaled gene therapy, *Mol Ther*, **24**(12), 2043–2053.

Engelhardt, J.F., *et al.* (1992). Submucosal glands are the predominant site of CFTR expression in the human bronchus, *Nat Genet*, **2**, 240–248.

Fajac, I., De Boeck, K. (2017). New horizons for cystic fibrosis treatment, *Pharmacol Ther*, **170**, 205–211.

Farinha, C.M., Amaral, M.D. (2005). Most F508del-CFTR is targeted to degradation at an early folding checkpoint and independently of calnexin, *Mol Cell Biol*, **25**(12), 5242–52.

Farmen, S.L., *et al.* (2005). Gene transfer of CFTR to airway epithelia: Low levels of expression are sufficient to correct Cl- transport and overexpression can generate basolateral CFTR, *Am J Physiol Lung Cell Mol Physiol*, **289**(6), L1123–30.

Farrow, N., *et al.* (2013). Airway gene transfer in a non-human primate: Lentiviral gene expression in marmoset lungs, *Sci Rep*, **3**, 1287.

Fatehullah, A., Tan, S.H., Barker, N. (2016). Organoids as an in vitro model of human development and disease, *Nat Cell Biol*, **18**(3), 246–54.

Ferrua, F., Aiuti, A. (2017). Twenty-five years of gene therapy for ADA-SCID: From bubble babies to an approved drug, *Hum Gene Ther*, **28**(11), 972–981.

Firth, A.L., *et al.* (2015). Functional gene correction for cystic fibrosis in lung epithelial cells generated from patient iPSCs, *Cell Rep*, 12(9):1385–90.

Foundation, C.F. (2017). The Clinical and Functional TRanslation of CFTR (CFTR2), https://www.cftr2.org.

Gaj, T., Gersbach, C.A., Barbas, C.F. (2013). 3rd, ZFN, TALEN, and CRISPR/Cas-based methods for genome engineering, *Trends Biotechnol.*

Gill, D.R., *et al.* (1997). A placebo-controlled study of liposome-mediated gene transfer to the nasal epithelium of patients with cystic fibrosis, *Gene Ther*, **4**(3), 199–209.

Griesenbach, U., Davies, J.C., Alton, E. (2016). Cystic fibrosis gene therapy: A mutation-independent treatment, *Curr Opin Pulm Med*, **22**(6), 602–9.

Griesenbach, U., Alton, E.W. (2015). Recent advances in understanding and managing cystic fibrosis transmembrane conductance regulator dysfunction, *F1000Prime Rep*, **7**, 64.

Griesenbach, U., Pytel, K.M. Alton, E.W. (2015). Cystic fibrosis gene therapy in the UK and elsewhere, *Hum Gene Ther*, **26**(5), 266–75.

Gui, L., *et al.* (2015). Efficient intratracheal delivery of airway epithelial cells in mice and pigs, *Am J Physiol Lung Cell Mol Physiol*, **308**(2), L221–8.

Hacein-Bey-Abina, S., *et al.* (2003). LMO2-associated clonal T cell proliferation in two patients after gene therapy for SCID-X1, *Science*, **302**(5644), 415–9.

Halbert, C.L., *et al.* (2006). Prevalence of neutralizing antibodies against adeno-associated virus (AAV) types 2, 5, and 6 in cystic fibrosis and normal populations: Implications for gene therapy using AAV vectors, *Hum Gene Ther*, **17**(4), 440–7.

Haq, I.J., *et al.* (2016). Airway surface liquid homeostasis in cystic fibrosis: Pathophysiology and therapeutic targets, *Thorax*, **71**(3), 284–7.

Hart, S.L., Harrison, P.T. (2017). Genetic therapies for cystic fibrosis lung disease, *Curr Opin Pharmacol*, **34**, 119–124.

Heda, G.D., Tanwani, M., Marino, C.R. (2001). The Delta F508 mutation shortens the biochemical half-life of plasma membrane CFTR in polarized epithelial cells, *Am J Physiol Cell Physiol*, **280**(1), C166–74.

Hendel, A., *et al.* (2015). Chemically modified guide RNAs enhance CRISPR-Cas genome editing in human primary cells, *Nat Biotechnol*, 33, 985–989.

Hyde, S.C., *et al.* (2000). Repeat administration of DNA/liposomes to the nasal epithelium of patients with cystic fibrosis, *Gene Ther*, **7**(13), 1156–65.

Hynds, R.E., *et al.* (2016). Expansion of human airway basal stem cells and their differentiation as 3D tracheospheres, *Methods Mol Biol*, doi: 10.1007/7651_2016_5.

Iyer, V., *et al.* (2015). Off-target mutations are rare in Cas9-modified mice, *Nat Methods*, **12**(6), 479.

Jansen, R., *et al.* (2002). Identification of genes that are associated with DNA repeats in prokaryotes, *Mol Microbiol*, **43**(6), 1565–75.

Jensen, T.J., *et al.* (1995). Multiple proteolytic systems, including the proteasome, contribute to CFTR processing, *Cell*, **83**(1), 129–35.

Jiang, F., Doudna, J.A. (2017). CRISPR-Cas9 structures and mechanisms, *Annu Rev Biophys*, **46**, 505–529.

Jinek, M., *et al.* (2012). A programmable dual-RNA-guided DNA endonuclease in adaptive bacterial immunity, *Science*, **337**(6096), 816–21.

Johler, S.M., *et al.* (2015). Nebulisation of IVT mRNA complexes for intrapulmonary administration, *PLoS One*, **10**(9), e0137504.

Johnson, L.G., *et al.* (1992). Efficiency of gene transfer for restoration of normal airway epithelial function in cystic fibrosis, *Nat Genet*, **2**(1), 21–5.

Kariko, K., *et al.* (2011). Generating the optimal mRNA for therapy: HPLC purification eliminates immune activation and improves translation of nucleoside-modified, protein-encoding mRNA, *Nucleic Acids Res*, **39**(21), e142.

Kerem, B.-S., *et al.* (1989). Identification of the cystic fibrosis gene: Genetic analysis, *Science*, **245**, 1073–1080.

King, C.S., *et al.* (2019). Critical care of the adult patient with cystic fibrosis, *Chest*, **155**(1), 202–214.

Knowles, M.R., *et al.* (1998). A double-blind, placebo controlled, dose ranging study to evaluate the safety and biological efficacy of the lipid-DNA complex GR213487B in the nasal epithelium of adult patients with cystic fibrosis, *Hum Gene Ther*, **9**(2), 249–69.

Komor, A.C., *et al.* (2016). Programmable editing of a target base in genomic DNA without double-stranded DNA cleavage, *Nature*, **533**(7603), 420–4.

Kormann, M.S., *et al.* (2011). Expression of therapeutic proteins after delivery of chemically modified mRNA in mice, *Nat Biotechnol*, **29**(2), 154–7.

Ku, C.A., Pennesi, M.E. (2015). Retinal gene therapy: Current progress and future prospects, *Expert Rev Ophthalmol*, **10**(3), 281–299.

Kurosaki, F., *et al.* (2017). Optimization of adeno-associated virus vector-mediated gene transfer to the respiratory tract, *Gene Ther*, **24**, 290–297.

LaRanger, R., *et al.* (2018). Reconstituting mouse lungs with conditionally reprogrammed human bronchial epithelial cells, *Tissue Eng Part A*, **24**(7-8), 559–568.

Lee, C.M., *et al.* (2012). Correction of the DeltaF508 mutation in the cystic fibrosis transmembrane conductance regulator gene by zinc-finger nuclease homology-directed repair, *Biores Open Access*, **1**(3), 99–108.

Lee, K., *et al.* (2017). Nanoparticle delivery of Cas9 ribonucleoprotein and donor DNA in vivo induces homology-directed DNA repair, *Nat Biomed Eng*, **1**, 889–901.

Leitch, A.E., Rodgers, H.C. (2013). Cystic fibrosis, *J R Coll Physicians Edinb*, **43**(2), 144–50.

Liu, G., *et al.* (2003). Nanoparticles of compacted DNA transfect postmitotic cells, *J Biol Chem*, **278**(35), 32578–86.

Lukacs, G.L., *et al.* (1993). The delta F508 mutation decreases the stability of cystic fibrosis transmembrane conductance regulator in the plasma membrane. Determination of functional half-lives on transfected cells, *J Biol Chem*, **268**(29), 21592–8.

MacVinish, L.J., *et al.* (1997a). Chloride secretion in the trachea of null cystic fibrosis mice: The effects of transfection with pTrial10-CFTR2, *J Physiol*, **499**(3), 677–87.

MacVinish, L.J., *et al.* (1997b). Normalization of ion transport in murine cystic fibrosis nasal epithelium using gene transfer, *Am J Physiol*, **273**(2), C734–40.

Mahiny, A.J., *et al.* (2015). *In vivo* genome editing using nuclease-encoding mRNA corrects SP-B deficiency, *Nat Biotechnol*, **33**(6), 584–6.

Mali, P., *et al.* (2013). RNA-guided human genome engineering via Cas9, *Science*, **339**(6121), 823–6.

Mall, M.A., Galietta, L.J. (2015). Targeting ion channels in cystic fibrosis, *J Cyst Fibros*, **14**(5), 561–70.

Manunta, M.D., *et al.* (2011). Nebulisation of receptor-targeted nanocomplexes for gene delivery to the airway epithelium, *PloS one*, **6**(10), e26768.

Manunta, M.D., *et al.* (2013). Airway deposition of nebulized gene delivery nanocomplexes monitored by radioimaging agents, *Am J Respir Cell Mol Biol*, **49**(3), 471–80.

Masat, E., Pavani, G. Mingozzi, F. (2013). Humoral immunity to AAV vectors in gene therapy: Challenges and potential solutions, *Discov Med*, **15**(85), 379–89.

Mauro, V.P., Chappell, S.A. (2014). A critical analysis of codon optimization in human therapeutics, *Trends Mol Med*, **20**(11), 604–13.

McClain, L.E., *et al.* (2016). Vector serotype screening for use in ovine perinatal lung gene therapy, *J Pediatr Surg*, **51**(6), 879–84.

McLachlan, G., *et al.* (1996). Laboratory and clinical studies in support of cystic fibrosis gene therapy using pCMV-CFTR-DOTAP, *Gene Ther*, **3**(12), 1113–23.

McNeer, N.A., *et al.* (2015). Nanoparticles that deliver triplex-forming peptide nucleic acid molecules correct F508del CFTR in airway epithelium, **6**, 6952.

Meng, Q., *et al.* (2004). Efficient transfection of non-proliferating human airway epithelial cells with a synthetic vector formulated with EGTA, *J Gene Med*, **6**, 210–221.

Miyashita, H., *et al.* (2013). Long-term maintenance of limbal epithelial progenitor cells using rho kinase inhibitor and keratinocyte growth factor, *Stem Cells Transl Med*, **2**(10), 758–65.

Miyoshi, H., *et al.* (1998). Development of a self-inactivating lentivirus vector, *J Virol*, **72**(10), 8150–7.

Molinski, S.V., *et al.* (2017). Orkambi® and amplifier co-therapy improves function from a rare CFTR mutation in gene-edited cells and patient tissue, *EMBO Mol Med*, **9**(9), 1224–43.

Montini, E., *et al.* (1998). The genotoxic potential of retroviral vectors is strongly modulated by vector design and integration site selection in a mouse model of HSC gene therapy, *J Clin Invest*, **119**(4), 964–75.

Moss, R.B., *et al.* (2004). Repeated adeno-associated virus serotype 2 aerosol-mediated cystic fibrosis transmembrane regulator gene transfer to the lungs of patients with cystic fibrosis: A multicenter, double-blind, placebo-controlled trial, *Chest*, **125**(2), 509–21.

Moss, R.B., *et al.* (2007). Repeated aerosolized AAV-CFTR for treatment of cystic fibrosis: A randomized placebo-controlled phase 2B trial, *Hum Gene Ther*, **18**(8), 726–32.

Munye, M.M., *et al.* (2016). Minicircle DNA provides enhanced and prolonged transgene expression following airway gene transfer, *Sci Rep*, **6**, 23125.

Nakata, A., Shinagawa, H. Amemura, M. (1982). Cloning of alkaline phosphatase isozyme gene (iap) of Escherichia coli, *Gene*, **19**(3), 313–9.

Nienhuis, A.W., Nathwani, A.C., Davidoff, A.M. (2017). Gene therapy for hemophilia, *Mol Ther*, **25**, 1163–7.

Noone, P.G., *et al.* (2000). Safety and biological efficacy of a lipid-CFTR complex for gene transfer in the nasal epithelium of adult patients with cystic fibrosis, *Mol Ther*, **1**(1), 105–14.

Ortinski, P.I., *et al.* (2017). Integrase-deficient lentiviral vector as an all-in-one platform for highly efficient CRISPR/Cas9-mediated gene editing, *Mol Ther Methods Clin Dev*, **5**, 153–164.

Pardi, N., *et al.* (2015). Expression kinetics of nucleoside-modified mRNA delivered in lipid nanoparticles to mice by various routes, *J Control Release*, **217**, 345–51.

Pilewski, J.M., *et al.* (1995). Adenovirus-mediated gene transfer to human bronchial submucosal glands using xenografts, *Am J Physiol*, **268**(4), L657–65.

Porteous, D.J., *et al.* (1997). Evidence for safety and efficacy of DOTAP cationic liposome mediated CFTR gene transfer to the nasal epithelium of patients with cystic fibrosis, *Gene Ther*, **4**(3), 210–8.

Poulsen, J.H., *et al.* (1994). Bicarbonate conductance and pH regulatory capability of cystic fibrosis transmembrane conductance regulator, *Proc Natl Acad Sci*, **91**, 5340–5344.

Probst, J., *et al.* (2007). Spontaneous cellular uptake of exogenous messenger RNA in vivo is nucleic acid-specific, saturable and ion dependent, *Gene Ther*, **14**(15), 1175–80.

Ramalingam, S., *et al.* (2013). Generation and genetic engineering of human induced pluripotent stem cells using designed zinc finger nucleases, *Stem Cells Dev*, **22**(4), 595–610.

Ramsey, B.W., *et al.* (2011). A CFTR potentiator in patients with cystic fibrosis and the G551D mutation, *N Engl J Med*, **365**(18), 1663–72.

Ran, F.A., *et al.* (2013). Double nicking by RNA-guided CRISPR Cas9 for enhanced genome editing specificity, *Cell*, **154**(6), 1380–9.

Ran, F.A., *et al.* (2013). Genome engineering using the CRISPR-Cas9 system, *Nat Protoc*, **8**(11), 2281–308.

Ran, F.A., *et al.* (2015). In vivo genome editing using Staphylococcus aureus Cas9, *Nature*, **520**(7546), 186–91.

Ratjen, F., *et al.* (2015). Cystic fibrosis, *Nat Rev Dis Primers*, **1**, 15010.

Ratjen, F., *et al.* (2015). Tiotropium Respimat in cystic fibrosis: Phase 3 and Pooled phase 2/3 randomized trials, *J Cyst Fibros*, **14**(5), 608–14.

Rawlins, E.L., Hogan, B.L. (2008). Ciliated epithelial cell lifespan in the mouse trachea and lung, *Am J Physiol Lung Cell Mol Physiol*, **295**(1), L231–4.

Richardson, C.D., *et al.* (2016). Enhancing homology-directed genome editing by catalytically active and inactive CRISPR-Cas9 using asymmetric donor DNA, *Nat Biotechnol*, **34**(3), 339–44.

Riordan, J.R., *et al.* (1989). Identification of the cystic fibrosis gene: Cloning and characterization of complementary DNA, *Science*, **245**, 1066–1073.

Robinson, E., *et al.* (2018). Lipid nanoparticle-delivered chemically modified mRNA restores chloride secretion in cystic fibrosis, *Mol Ther*, **26**(8), 2034–2046.

Rommens, J.M., *et al.* (1989). Identification of the cystic fibrosis gene: Chromosome walking and jumping, *Science*, **245**, 1059–1065.

Rosen, C., *et al.* (2015). Preconditioning allows engraftment of mouse and human embryonic lung cells, enabling lung repair in mice, *Nat Med*, **21**(8), 869–79.

Sahin, U., Kariko, K. Tureci, O. (2014). mRNA-based therapeutics--developing a new class of drugs, *Nat Rev Drug Discov*, **13**(10), 759–80.

Schwank, G., *et al.* (2013). Functional repair of CFTR by CRISPR/Cas9 in intestinal stem cell organoids of cystic fibrosis patients, *Cell Stem Cell*, **13**(6), 653–8.

Shah, P.L., *et al.* (1996). In vivo effects of recombinant human DNase I on sputum in patients with cystic fibrosis, *Thorax*, **51**(2), 119–25.

Sharma, M., *et al.* (2001). Conformational and temperature-sensitive stability defects of the delta F508 cystic fibrosis transmembrane conductance regulator in post-endoplasmic reticulum compartments, *J Biol Chem*, **276**(12), 8942–50.

Sternberg, S.H., *et al.* (2014). DNA interrogation by the CRISPR RNA-guided endonuclease Cas9, *Nature*, **507**(7490), 62–7.

Strenkowska, M., *et al.* (2016). Cap analogs modified with 1,2-dithiodiphosphate moiety protect mRNA from decapping and enhance its translational potential, *Nucleic Acids Res*, **44**(20), 9578–9590.

Suk, J.S., *et al.* (2011). N-acetylcysteine enhances cystic fibrosis sputum penetration and airway gene transfer by highly compacted DNA nanoparticles, *Mol Ther*, **19**(11), 1981–9.

Suk, J.S., *et al.* (2014). Lung gene therapy with highly compacted DNA nanoparticles that overcome the mucus barrier, *J Control Release*, **178**, 8–17.

Suzuki, K., *et al.* (2016). *In vivo* genome editing via CRISPR/Cas9 mediated homology-independent targeted integration, *Nature*, **540**(7631), 144–149.

Tang, X.X., *et al.* (2016). Acidic pH increases airway surface liquid viscosity in cystic fibrosis, *J Clin Invest*, **126**(3), 879–91.

Thess, A., *et al.* (2015). Sequence-engineered mRNA without chemical nucleoside modifications enables an effective protein therapy in large animals, *Mol Ther*, **23**(9), 1456–64.
Wainwright, C.E., *et al.* (2015). Lumacaftor-ivacaftor in patients with cystic fibrosis homozygous for phe508del CFTR, *N Engl J Med*, **373**, 1783–4.
Welsh, M.J., *et al.* (1994). Cystic fibrosis gene therapy using an adenovirus vector: *In vivo* safety and efficacy in nasal epithelium, *Hum Gene Ther*, **5**(2), 209–19.
Yan, Z., *et al.* (2017). Human bocavirus type-1 capsid facilitates the transduction of ferret airways by adeno-associated virus genomes, *Hum Gene Ther*, **28**(8), 612–625.
Yang, J., *et al.* (1999). Concatamerization of adeno-associated virus circular genomes occurs through intermolecular recombination, *J Virol*, **73**(11), 9468–77.
Yui, S., *et al.* (2012). Functional engraftment of colon epithelium expanded in vitro from a single adult Lgr5(+) stem cell, *Nat Med*, **18**(4), 618–23.
Zeyer, F., *et al.* (2016). mRNA-Mediated gene supplementation of toll-like receptors as treatment strategy for asthma *in vivo*, *PLoS One*, **11**(4), e0154001.
Zhang, L., *et al.* (2009). CFTR delivery to 25% of surface epithelial cells restores normal rates of mucus transport to human cystic fibrosis airway epithelium, *PLoS Biol*, **7**(7), e1000155.
Zhang, S., *et al.* (2014). Sinupret activates CFTR and TMEM16A-dependent transepithelial chloride transport and improves indicators of mucociliary clearance, *PLoS One*, **9**(8), e104090.
Zhu, L.J., *et al.* (2017). GUIDEseq: A bioconductor package to analyze GUIDE-Seq datasets for CRISPR-Cas nucleases, *BMC Genomics*, **18**(1), 379.
Ziady, A.-G., *et al.* (2002). Functional evidence of CFTR gene transfer in nasal epithelium of cystic fibrosis mice in vivo following lumenal application of DNA complexes targeted to the serpin-enzyme complex receptor, *Mol Ther*, **5**(4), 413–419.
Ziady, A.G., *et al.* (2003). Transfection of airway epithelium by stable PEGylated poly-L-lysine DNA nanoparticles in vivo, *Mol Ther*, **8**(6), 936–47.
Zuo, W., *et al.* (2015). p63(+)Krt5(+) Distal airway stem cells are essential for lung regeneration, *Nature*, **517**(7536), 616–20.
Zuris, J.A., *et al.* (2015). Cationic lipid-mediated delivery of proteins enables efficient protein-based genome editing *in vitro* and *in vivo*, *Nat Biotechnol*, **33**(1), 73–80.

29

INDUCED PLURIPOTENT STEM CELLS AND GENE TARGETING FOR REGENERATIVE MEDICINE

Jizhong Zou[a,b,c] *and Linzhao Cheng*[a]

29.1 Pluripotent Stem Cells (PSCs)

A stem cell is defined as a type of cell that can self-renew through division under certain conditions and is also capable of turning into a variety of specialized cell types under specific conditions, a process referred to as differentiation. Stem cells exist at various stages during animal development, including embryonic and adult stages. Based on their developmental potential from the most to the least, stem cells can be categorized as totipotent, pluripotent, multipotent, unipotent or nullipotent stem cells. From a fertilized egg to a 16-cell-stage embryo, cells in early mammalian embryogenesis are totipotent and can develop into both embryonic and extra-embryonic lineages. Further development into blastocyst separates inner cell mass that can give rise to embryonic stem cells (ESCs) *in vitro* from extra-embryonic trophectoderm that forms placenta *in vivo*. ESCs are the earliest PSCs that can be derived from mammalian embryogenesis. Following the normal developmental path, other PSCs, such as epiblast stem cells, embryonic germ cells and germline stem cells can be derived by *in vitro* culture of epiblast cells, primodial germ cells, and spermatogonial stem cells, respectively.

[a]Stem Cell Program, Institute for Cell Engineering, Johns Hopkins University School of Medicine, 733 North Broadway, BRB 780, Baltimore, MD 21205, USA
[b]Center for Refenerative Medicine National Institutes of Health, Bethesda, MD 20892, USA
[c]Email: jzou2@ jhmi.edu

Not only can PSCs be directly isolated from a normal developing animal, but they have also been created from non-pluripotent somatic cells by "reprogramming" techniques that turn specialized somatic cells back into a pluripotent state. Somatic cell nuclear transfer (SCNT), also known as somatic cloning, is the early reprogramming technique that made many headlines; it was the method used for the creation of "Dolly the sheep" by UK scientists, led by Dr Ian Wilmut, in 1996. Although demonstrated in many species, including monkeys, so far using SCNT to derive cloned ESC lines from primates remains technically challenging. In 2005, scientists at Harvard University fused adult skin cells with human ESCs to create "hybrid" cells showing ESC-like characteristics. However, these "hybrid" cells are tetraploid, i.e., containing double the amount of DNA of a normal human cell, and therefore are not suitable for clinical use. In 2006, a milestone of stem cell research was achieved by Japanese scientist Dr Shinya Yamanaka and his colleagues. They induced mouse somatic cells into a new kind of PSCs by overexpressing four exogenous transcription factors. He called these cells iPSCs. The generation of iPSCs opened a door to a promising future of personalized regenerative medicine.

Based on the definition, all PSCs share the characteristic that they can form many, if not all, of the somatic cell types and germline cells of developing embryos.

29.1.1 ESCs

The first derivation of mouse ESCs in 1981 started a new era of stem cell research. It set up standards for other PSCs discovered afterwards. For example, the techniques used by Dr James Thomson for the derivation of first human ESCs in 1998 followed the same concept as in the derivation of mouse ESCs, although certain technical details were different. Many pluripotency genes, such as master transcription factors octamer-binding protein 4 (*Oct4*), sex determining region Y-box 2 (*Sox2*) and *Nanog*, govern the core regulatory network in all kinds of PSCs, and were therefore chosen to create iPSCs. After PSCs were generated, quality controls were done using assays established for ESCs, such as *in vitro* embryoid body (EB) formation and *in vivo* teratoma formation to confirm that PSCs are capable of differentiating into three germ layers. When studying human PSCs, we learn from previous experiences with mouse ESCs, which were better understood in their self-renewal and differentiation process correlated to *in vivo* development. One should also keep in mind that human ESCs have some substantial differences from the conventional mouse ESCs that are derived from blastocysts. For example, cell surface marker stage-specific embryonic antigen (SSEA)-1 is specific for mouse ESCs but not human ESCs, whereas SSEA-4 is specific for human ESCs but not mouse ESCs.

Animal serum containing media plus leukemia inhibitory factor (LIF) is sufficient for the maintenance of undifferentiated mouse ESCs, but causes differentiation when culturing human ESCs long term.

Conventional ESC derivation involves explanting most cells from the inner cell mass of a blastocyst; a process that usually destroys the developing embryo and therefore causes ethical concerns. Improved methods include using a single blastomere before the blastocyst stage. Each blastomere is considered totipotent and removing one of eight to ten cells will not prevent the rest from developing into a normal blastocyst. Such practice has been used in pre-implantation genetic diagnosis (PGD) during *in vitro* fertilization (IVF) to detect suspected genetic abnormalities before the embryo is fully developed in the uterus. PGD-ESCs have been established as important models to study human diseases. Another method to circumvent destroying embryos utilizes a process called parthenogenesis, which stimulates an unfertilized egg (oocyte) into diploid ESC-like cells. Parthenogenesis is not yet well understood, and the success rate remains extremely low for many species.

29.1.2 iPSCs

Although PSCs were generated by several methods since the first mouse ESCs in 1981, an intriguing question was not answered for more than 20 years: Can somatic cells be reprogrammed to ESC-like cells by defined factors? The generation of iPSCs in 2006 provided a definitely positive answer with a very simple formula: Only four transcriptional factors, Oct4, Sox2, Kruppel-like factor 4 (Klf4) and cellular-myelocytomatosis viral oncogene homolog (c-Myc), were needed to reprogram mouse fibroblastic cells into ESC-like iPSCs. These four factors were selected by a systematic screening of 24 candidates that play critical roles in establishing and maintaining ESC identity. The same recipe was sufficient for other species, too. After 15 months, Dr Shinya Yamanaka's group published a paper using the same four factors to generate human iPSCs (Takahashi *et al.*, 2007). Published online the same day was another paper by Dr James Thomson's group who used human OCT4, SOX2, NANOG and lineage 28 (LIN28) to generate human iPSCs (Yu *et al.*, 2007).

While many other factors were discovered to contribute to iPSC generation, they also provoked studies of stem cell biology, mechanisms of pluripotency maintenance and differentiation. So far, iPSCs have been generated from cells from all three germ layers. Depending on somatic cells' endogenous gene expression and epigenetic proximity to pluripotency, various combinations of reprogramming factors, sometimes only Oct4, are needed to turn differentiated cells into iPSCs. Now iPSCs can be derived from "non-permissive" mouse strains or from other animals from which it was not possible to establish ESCs before, such as rats and pigs. New animal models will not only aid veterinary studies and agricultural development, but also provide better disease models that are closer to human physiological conditions.

Another huge advantage of iPSCs over ESCs is reflected in ethical and medical issues. There are only a limited number of human ESC lines available. Most of them, if not all, have been cultured under various conditions involving mouse components. Genetic and epigenetic abnormalities arise frequently after extended culture. Clonal variations could result in varied differentiation potential. While ESCs were intended to be used for regenerative medicine, differentiated cells from the limited number of human ESC lines were hardly immunocompatible with the patients with unique human leukocyte antigens. Only patient-specific human iPSCs can provide the ideal solution for personalized cell therapy. iPSCs derived directly from the patient can be used not only to provide an unlimited resource of stem cells for differentiation into either committed progenitors that can be used for transplantation or terminally differentiated cells for transfusion, but also to establish disease models *in vitro*. Since 2008, dozens of disease-specific iPSC lines have been generated including those for Parkinson's disease, diabetes and, in our lab, for sickle cell disease (SCD).

29.2 Reprogramming

The term "reprogramming", a process to reverse or change nuclear epigenetic status and cell fate, was coined before the generation of iPSCs, as SCNT and cell fusion were also techniques to convert somatic cells into PSCs. However, it gained the most popularity after iPSC derivation. Here we will focus on the techniques used in iPSC generation, which evolved from integrating to non-integrating, genetic to non-genetic methods.

29.2.1 Integrating DNA Vectors

29.2.1.1 *Retrovirus/Lentivirus*

Viruses were the first choice for delivering reprogramming factors due to their high efficiency of infection of animal cells. Moloney murine leukemia virus-based retroviral vectors were first used in the generation of mouse iPSCs since they can transduce mouse fibroblasts with >80% efficiency. They have the advantage of self-silencing in PSCs, which limits residual exogenous reprogramming genes' expression that may cause oncogenic transformation or affect differentiation; but they are much less efficient in transducing non-dividing cells unless pseudotyped with the vesicular stomatitis virus G (VSV-G) envelope protein. Constitutive lentiviral vectors were used in generating human iPSCs with high transduction efficiency, but lack silencing in pluripotent cells. Thus, inducible lentivectors were designed to control the reprogramming genes' expression level and duration by turning them on during reprogramming and off after iPSCs were obtained. Both retrovectors and lentivectors can easily integrate into the genome with multiple copies, and therefore bear a great risk of insertional mutagenesis. To solve this issue, the causes recombination–locus of X over P1 (Cre–loxP) system has been added to the single lentivector that co-expresses all the reprogramming factors in order to achieve excision of transgene integration, leaving one copy of the engineered viral long terminal repeat (LTR) containing the 34-bp loxP sequence in the genome. However, the exact insertion of viral vectors (and therefore the remaining loxP-containing LTR DNA), which often insert into transcriptionally active genes, is uncontrolled.

29.2.1.2 *Transposon*

Vector integration during iPSC generation appears to be helpful because it ensures robust and enduring reprogramming of genes' expression. The piggyBac (mobile DNA elements initially observed in Baculoviruses) transposon was an attractive option in 2009 to replace loxP flanked (floxed) lentivectors because they were able to generate mouse iPSCs with integration first and mediate footprint-free excision afterwards, creating "transgene-free" iPSCs. Although the first step of integration and reprogramming efficiency is high enough, the excision efficiency is extremely low and has not been demonstrated in high-quality human iPSCs where the reprogramming vectors were silenced after reprogramming.

29.2.2 Non-Integrating Vectors

29.2.2.1 *Episomal plasmid*

Non-integrating and non-viral vectors provide an alternative solution for achieving safe iPSC generation. Derived from Epstein–Barr virus (EBV), origin of replication P/Epstein–Barr nuclear antigen 1 (oriP/EBNA1)-based plasmids can maintain stable DNA replication gene expression in human cells as episomes. The EBV-derived episomal vectors appeared to stop replicating and were gradually lost in successfully reprogrammed, proliferating iPSCs. This is probably due to the epigenetic modification of ENBA1 gene expression and/or oriP functions resulting in DNA replication defects. Therefore, several groups successfully used the episomal vector to generate vector-free and transgene-free human iPSCs. Recently, EBV vectors were reported by two groups to reprogram human blood cells with an efficiency several hundred-fold higher than that of fibroblastic cells, and, therefore, to be a better choice than conventional plasmid-mediated reprogramming, which has extremely low efficiency even after repeated transfection in mouse embryonic fibroblast cells.

29.2.2.2 Adenovirus and RNA virus

Adenovirus is a double-stranded DNA virus that allows transient and high-level expression of transgenes, and was used to generate the first integration-free mouse iPSCs. However, like conventional plasmid-mediated reprogramming, the efficiency is also extremely low.

All of the viral and non-viral vectors mentioned above are DNA vectors that still have higher likelihood of integration than RNA viruses. Recently, a RNA virus called Sendai virus has been shown to reprogram unfractionated human blood T-cells to iPSC with high efficiency.

29.2.3 mRNA, Proteins or Small Molecules

Further efforts to eliminate DNA integration came from mRNA-mediated reprogramming. To extend the presence of mRNA encoding reprogramming factors, modifications were made to prevent degradation. Although the improved approach showed high efficiency for generation of human iPSCs, in one study multiple rounds (17–21 times) of transfection of 4–5 mRNAs were needed.

Non-nucleic-acid reprogramming was achieved in 2009 in mouse and human iPSCs by protein transduction via a poly-arginine tag. Reprogramming efficiency was extremely low even after repeated transduction, probably due to the low protein transduction efficiency. Another non-genetic reprogramming tool is expected to be chemical, such as the use of small molecules, as they have been shown to replace one or most of the reprogramming transcription factors. Some small molecules that are known for modulating chromatin remodeling, such as butyrate, can enhance reprogramming even in the presence of optimal genetic factors. It is possible that a small molecule combination may be sufficient to induce mouse and/or human iPSCs. However, any novel reprogramming tool has to be efficient and suitable for multiple cell types to gain popularity and for a meaningful future in clinical application. In addition, it is not clear whether non-DNA-based methods are free of aberrant induction of epigenetic and genetic changes during reprogramming.

Methodological improvements in reprogramming also stimulate the mechanistic studies of stem cell biology. New transcription factors, chemicals, or microRNA (miRNA) were found to be associated with establishing and maintaining ESC/iPSC pluripotency, gene expression profile and epigenetic states.

29.3 Genetic Modification of iPSCs

When iPSCs are used to study a gene's function in early human embryogenesis, model disease development or as candidates for gene therapy, genetic modification plays an important role and various genetic tools can be used.

29.3.1 Transgene Expression

iPSCs provide the cellular environment to study a gene's cell-type-specific expression since they can be differentiated into any somatic cell type. A reporter transgene introduced into the genome can be used to trace the differentiated lineage, or aid the isolation of desired cell types. For example, a ubiquitously expressed green fluorescent protein or luciferase reporter can be used for *in vivo* imaging of iPSC-derived cells in animal transplantation experiments. A neuronal-specific enhancer/promoter-driven expression of the reporter gene can be used to detect integration of such neurons in the animal's nervous system. Hematopoietic cells can be sorted if they express cell-type-specific transgenes.

For therapeutic purposes, transgenes can be used to restore functions in diseased cells. In 2009, the first "disease-corrected" Fanconi Anemia (FA) iPSCs were established by reprogramming patient's fibroblasts after gene correction with lentivector encoding *FANCA* or *FANCD2* expression. The genetic instability and apoptosis phenotypes of FA cells were reverted and normal hematopoietic progenitors were differentiated from corrected FA-iPSCs. The issues of virus-mediated transgene expression, silencing and insertional mutagenesis are still present in this approach and need to be solved before gene-corrected iPSCs and their derived cells can be used for therapeutic applications. One way to ensure faithful transgene expression is to use large non-viral vectors, such as bacterial or human artificial chromosomes (BACs or HACs), that contain (normally long) endogenous regulatory elements. In late 2009 an HAC with a complete genomic *dystrophin* locus showed persistent restoration of *dystrophin* expression in mouse and human Duchenne muscular dystrophy (DMD) iPSCs. The major challenge of using BAC or HAC is the construction and delivery of these megabase-size vectors, as well as ruling out all the potential random integrations.

Recently, promising discoveries were found in lentivector-mediated gene transfer in both iPSCs and hematopoietic stem cells (HSCs). In one study in 2010, 10% of lentivector insertion sites in β-thalassemia iPSCs were found to qualify for "safe harbor" genome loci, which are defined as >50 megabases (Mb) away from known coding sequence, miRNA and ultraconserved regions. Erythroid-specific β-globin transgenes delivered by lentivector and inserted in a "safe harbor" have shown therapeutic levels of β-globin expression in differentiated erythroid progenitor cells.

29.3.2 Gene Targeting

Random integrations of transgene into the genome, even if they happen to be in the "safe harbor" after labor-intensive screening, still pose risks of gene expression perturbation by viral sequences such as lentivirus LTR. A targeted genetic modification approach called gene targeting, mediated by the cell's native DNA repair process of homologous recombination (HR), is a better solution. The discovery of HR combined with mouse ESC technology, best used to generate mouse models for human diseases and study gene function, won the 2007 Nobel Prize in Physiology or Medicine for M.R. Capecchi, M.J. Evans and O. Smithies.

29.3.2.1 *Targeting vectors*

Conventional gene-targeting vectors are based on plasmids containing two homologous arms to align with the endogenous target sequence. A selection marker, such as a drug-resistance or fluorescence gene, is often included to select rare gene-targeting events. Although plasmid targeting vectors are now widely used to target mouse ESCs with relatively high efficiency, this approach is still difficult for human ESCs/iPSCs. Since the first report in human ESCs in 2003 there have been only a handful of reports of plasmid-mediated gene targeting in human ESCs, BAC and adenoviral vectors have also been used for gene targeting in human ESCs. Both of them can accommodate a large cargo size (>20 kiobases (kb)), which is supposed to increase the length of homologous arms therefore enhancing targeting efficiency. However, the difficulties in vector construction and less than 10^{-6} efficiency hampered their applications in other human PSCs.

Recombinant adeno-associated virus (rAAV), a single-stranded DNA virus, is one of few methods that have been reported to target human iPSCs. Upon infection of a target cell, the rAAV genome enters the nucleus with an episomal conformation as a linear monomer, or sometimes a circular double-stranded DNA or concatemer. Using double-strand breaks (DSBs) to mediate gene transfer as in the cases of other HR vectors, rAAVs have been shown to achieve gene targeting with up to 10^{-4}~10^{-5} efficiency in human ESCs and iPSCs, several magnitudes higher than that of the plasmid approach. However, it has to be noted that rAAV also has ten times higher chance of random integration in the host genome via non-homologous end joining (NHEJ), and its genome can only pack up to 4.3 kb DNA sequence. So far, three genes (hypoxanthine phosphoribosyltransferase 1 (*HPRT1*), high mobility group AT-hook 1 (*HMGA1*) and *NANOG*), all of them actively transcribed genes in human PSCs, have been targeted in human iPSCs using rAAV vectors. The targeting efficiency for silent genes still needs to be determined.

29.3.2.2 *Zinc-finger nuclease (ZFN)*

Since gene targeting using DSB induced HR, the techniques that can induce DSBs are likely to enhance gene targeting efficiency. ZFNs are engineered sequence-specific nucleases consisting of a customized array of zinc fingers engineered to bind to a specific DNA sequence and a non-specific DNA endonuclease (*Flavobacterium okeanokoites* I (FokI)) domain. Each zinc-finger domain recognizes 3–4 bp of DNA and a three to six-finger ZFN recognizes a 9–24 bp DNA sequence. When two ZFNs bind their cognate target sequence in the proper orientation, the nuclease domains are able to dimerize, become active and create a sequence-specific DSB. When a donor DNA that is homologous to the target on both sides of the DSB is provided, the genomic site can be modified by HR-mediated DNA repair, allowing the incorporation of exogenous sequences inserted between two homology arms.

ZFNs have been shown to stimulate gene targeting efficiency over 100-fold in cell types from many organisms including plants, fruitflies, zebrafish, rats and humans. In 2009, our group at Johns Hopkins University and Dr Rudolf Jaenisch's group at MIT demonstrated that ZFNs can also enhance gene targeting in human iPSCs over 1000-fold, resulting in nearly 100% targeted iPS clones after simple selection. Not only actively expressed genes, but also silent genes were successfully targeted with designed high-quality ZFNs. Gene addition and gene replacement, single-allele and bi-allelic modification were both feasible in ZFN-mediated gene targeting. With gene correction efficiency enhanced up to 0.1~1%, the specificity of ZFNs needs further investigation if ZFNs are to be used in clinical gene therapy. Based on limited reports, highly active ZFNs can occasionally induce NHEJ on the other allele not modified by HR, and rarely cause off-target mutations. These positive findings are based on several highly specific ZFNs and examination of top potential off-target loci. Whole-genome sequencing, which is now available at a reasonable price, will aid thorough investigation of genomic perturbation.

While we may see new technologies replacing ZFNs for safer gene targeting, targeted genomic modification is the future direction for replacing random integration in gene therapy. The major advantage of gene targeting is that a gene's endogenous regulatory elements will be preserved. Therefore, knockin reporters will faithfully represent spatio-temporal expression patterns of endogenous genes. Disease-causing mutations can be precisely repaired on-site without any insertional mutagenesis. Even when combined with the "safe-habor" approach, targeted gene addition reduces the uncertainty of which loci will be integrated at each time.

29.4 Application of iPS Cells in Regenerative Medicine

Adult stem cells, such as HSCs, have been used in regenerative medicine for many years. Compared with ESCs/iPSCs, adult stem cells are mostly mulitpotent and can only proliferate within a limited time period, and therefore cannot provide a renewable resource for cell research and therapy. Nevertheless, knowledge of their function, differentiation and genetic modification techniques will help us apply PSCs to regenerative medicine. Since iPSCs can now be derived from many somatic cell types of patients, they are not limited by ethical concerns or availability of immunocompatible donors.

29.4.1 Disease Modeling

In vitro cultured iPSCs are excellent candidates for disease modeling because they have human physiological characteristics that animal models do not have. For example, mouse cardiomyocytes beat much faster than their human counterparts. Furthermore, genetically engineered mouse models hardly have all the genetic background that human patients have, e.g., Down's syndrome trisomy mice failed to recapitulate neurodegeneration and cranial abnormalities. Even immortalized human cell lines cannot reveal the onset of the disease phenotype. Human PGD-ESCs could be a good human disease model as an alternative to iPSCs, except that not many disease-specific PGD-ESCs have been established, not to mention the difficulty of generating ESCs for each patient.

Since 2008, many disease-specific iPSCs have been generated, including those for SCD, Parkinson's disease, Huntington's disease, Down's syndrome and Type I diabetes. While more than two dozen diseases have been the subject of iPSC generation, only some genetic-disease-specific iPSCs have mimicked disease-specific phenotypes. For example, spinal muscular atrophy (SMA)-iPSCs can differentiate into neural lineages, which maintain a lack of survival of motor neuron 1 (SMN1) expression and the disease phenotype of selective motor neuron death. Myeloproliferative disease (MPD)-iPSCs showed abnormal erythroid expansion, as was observed in the patients' cells. Familial dysautonomia (FD) patient-specific iPSC-derived neural crest precursors expressed particularly low levels of normal inhibitor of kappa light polypeptide gene enhancer in B-cells, kinase complex-associated protein (*IKBKAP*) transcript, demonstrating tissue-specific mis-splicing. FD pathogenesis was also revealed by marked defects in neurogenic differentiation and migration behavior. Lentigines, electrocardiogram conduction abnormalities, ocular hypertelorism, pulmonic stenosis, abnormal genitalia, retardation of growth, and sensorineural deafness (LEOPARD) disease's hypertrophic cardiomyopathy syndrome phenotype was mimicked in that *in vitro*-derived cardiomyocytes from LEOPARD-iPSCs are larger and have a higher degree of sarcomeric organization compared with those from healthy iPSCs. These diseases are all caused by genetic mutations that make well-defined products; therefore their derived cells will carry the same mutations and are more likely to result in disease phenotypes. More complex neurodegenerative diseases, like Parkinson's disease, involve multiple factors and may need certain environmental factors or aging processes to be able to show desired phenotypes.

29.4.2 Drug Screening

When disease/patient-specific iPSCs can be differentiated into certain cell types where the disease phenotype is observed, novel drugs can be screened under *in vitro* culture conditions before being tested on animals or humans. Although it may not provide accurate dosage and administration information suitable for real patient treatment, this practice gives fast and economical assessment of drug efficacy. To date, several cases of iPSC usage in drug screening have been reported. In the FD-iPSCs disease model, the authors tested two candidate compounds that were known to reverse *IKBKAP* defective transcripts and found they also worked effectively in the FD-iPSCs-derived neural crest cells. In the Rett syndrome iPSC model (RTT-iPSCs), a selected drug that impairs ribosomal proofreading overcame the non-sense methyl CpG binding protein 2 (MeCP2) mutations in the majority of RTT-iPSC-derived neurons and rescued glutamatergic synapses in RTT neurons. The authors also found this potential broad-spectrum drug has a narrow window of dosage suitable for treatment. A recent study of type 2 long-QT syndrome iPSCs (LQTS-iPSCs) went even further to show three different kinds of drugs, a $Ca2^{+}$-channel blocker nifedipine, K_{ATP}-channel opener pinacidil and late Na^{+}-channel blocker ranolazine, can be used to ameliorate arrythymia in LQTS-iPSCs-derived cardiomyoctyes, thus demonstrating the potential of customized treatment using patient-specific iPSCs.

With more and more disease/patient-specific iPSCs being developed, establishment of disease-specific phenotypes and robust assays to measure effects of pharmacological agents are the prerequisite of using iPSCs for high-throughput screening.

29.4.3 Cell Therapy

Ultimately, iPSCs are an unlimited resource for obtaining somatic cells for treating disease and repairing damaged tissues in patients. However, undifferentiated iPSCs or incompletely reprogrammed cells can form tumors when injected into immunodeficient mice. Therefore, the quality and safety of human iPSCs become important issues when considering cell therapy.

Even at the beginning of iPSC technology development, Dr Shinya Yamanaka noticed that c-Myc, although enhancing reprogramming efficiency, resulted in a high frequency of tumor formation in chimeric mice. Nowadays, c-Myc and other oncogenes can be eliminated or replaced by potentially safer factors such as lung carcinoma myc related oncogene (l-Myc). Non-integrating and non-genetic reprogramming methods will further help to achieve safer iPSCs.

The quality of human iPSCs, in terms of their ESC-like characteristics and differentiation potential, also needs thorough investigation. Using ESCs as the standard, iPSCs were initially shown to have ESC-like signatures by simple assays such as pluripotency marker expression, EB and teratoma formation, and DNA and histone methylation. With advanced technologies, such as whole-genome gene expression and epigenetic profiling, more subtle differences are being discovered and studied. Epigenetic memory, which is the unique epigenetic signature shared between original somatic cells and reprogrammed iPSCs but not ESCs, was suggested to be responsible for the cell-type-of-origin-specific differentiation potential, at least in early iPSC passages. iPSC-specific CpG or non-CpG methylations detected by whole-genome sequencing are also found in late-passage human iPSCs. Although the number of cell lines used in these reports may not be large enough and some other reports suggested the discrepancy between iPSCs and ESCs was due to lab- or culture condition-specific variations, we should not assume each iPSC clone will behave identically during differentiation. If epigenetic memory is indeed a critical factor in differentiation, one should try to use the desired somatic cell type for reprograming into iPSCs at the first place. On the other hand, even if there are some abnormal genetic or epigenetic signatures, an iPSC still could differentiate well into one or a few desired somatic cell types.

Differentiated progenitor cells from iPSCs are ideal candidates for transplantation since they can repopulate multiple cell types in the entire lineage. Large quantities of iPSC lines can be established efficiently using HSCs from cord blood banks or skin biopsies from a sufficient number of donors to cover all the human leukocyte antigen (HLA) types to match those the patients needed. The lack of autologous stem cells in gene therapy can also be resolved by derivation of patient-specific iPSCs. If a defective gene needs to be repaired before differentiation and transplantation, the genetic modification options discussed earlier can be applied to iPSCs due to their unlimited *ex vivo* self-renewal ability. In fact, a proof-of-principle autologous treatment of SCD was elegantly demonstrated by Dr Rudolf Jeanisch's group using a humanized SCD mouse model where the diseased mouse fibroblasts were used to generate iPSCs whose disease-causing mutation was then repaired by HR before iPSCs were differentiated and transplanted into the original SCD mouse.

Technological and mechanistic studies of iPSCs are currently one of the most exciting and fastest-moving fields of biomedical research. With safer patient-specific iPSC derivation, high-efficiency precise genetic modification and directed iPSC differentiation, the bright future of human iPSCs in regenerative medicine will soon be realized.

For Further Reading

Original Articles

Chou, B.K., Mali, P., Huang, X., *et al.* (2011). Efficient human iPS cell derivation by a non-integrating plasmid from blood cells with unique epigenetic and gene expression signatures, *Cell Research*, **21**, 518–529.

Hanna, J., Wernig, M., Markoulaki, S., *et al.* (2007). Treatment of sickle cell anemia mouse model with iPS cells generated from autologous skin, *Science*, **318**, 1920–1923.

Hockemeyer, D., Soldner, F., Beard, C., *et al.* (2009). Efficient targeting of expressed and silent genes in human ESCs and iPSCs using zinc-finger nucleases, *Nature Biotechnology*, **27**, 851–857.

Raya, A., Rodriguez-Piza, I., Guenechea, G., *et al.* (2009). Disease-corrected haematopoietic progenitors from Fanconi anaemia induced pluripotent stem cells, *Nature*, **460**, 53–59.

Takahashi, K., Tanabe, K., Ohnuki M., *et al.* (2007). Induction of pluripotent stem cells from adult human fibroblasts by defined factors, *Cell*, **131**, 861–872.

Takahashi, K., Yamanaka, S. (2006). Induction of pluripotent stem cells from mouse embryonic and adult fibroblast cultures by defined factors, *Cell*, **126**, 663–676.

Yu, J., Vodyanik, M.A., Smuga-Otto K., *et al.* (2007). Induced pluripotent stem cell lines derived from human somatic cells, *Science*, **318**, 1917–1920.

Zou, J., Maeder, M.L., Mali P., *et al.* (2009). Gene targeting of a disease-related gene in human induced pluripotent stem and embryonic stem cells, *Cell Stem Cell*, **5**, 97–110.

Zou, J., Sweeney, C.L., Chou B.K., *et al.* (2011). Oxidase deficient neutrophils from X-linked chronic granulomatous disease iPS cells: Functional correction by zinc finger nuclease mediated safe harbor targeting, *Blood*, **117**, 5561–5572.

Reviews

Colman, A., Dreesen, O. (2009). Pluripotent stem cells and disease modeling, *Cell Stem Cell*, **5**, 244–247.

Giudice, A., Trounson, A. (2008). Genetic modification of human embryonic stem cells for derivation of target cells, *Cell Stem Cell*, **2**, 422–433.

Hanna, J., Saha, K., Jaenisch, R. (2010). Pluripotency and cellular reprogramming: Facts, hypotheses, unresolved issues, *Cell*, **143**, 508–525.

Saha, K., Jaenisch, R. (2009). Technical challenges in using human induced pluripotent stem cells to model disease, *Cell Stem Cell*, **5**, 584–595.

Ye, Z., Cheng, L. (2010). Potential of human iPS cells derived from blood and other postnatal cell types, *Regen Med*, **4**, 521–530.

PART IV
GENE VECTOR PRODUCTION

30

PRODUCTION AND PURIFICATION OF VIRAL VECTORS AND SAFETY CONSIDERATIONS RELATED TO THEIR USE

Otto-Wilhelm Merten[a,d], Matthias Schweizer[b], Parminder Chahal[c] and Amine Kamen[c]

30.1 General Overview

30.1.1 Introduction

The last decade has been characterized by a general improvement in vector manufacturing technology of high importance for the production of clinical-grade viral vectors for gene therapy treatment. Up to 2010, more than half of all clinical trials have made use of adenoviral, adeno-associated viral (AAV), lentiviral (LV) or retroviral vectors (http://www.wiley.com/legacy/wileychi/genmed/clinical/), with a clear trend of increased use of AAV and LVs.

In direct relation to the chapters presenting the different viral vectors this chapter briefly describes their general production and purification principles, followed by specific sections on production and purification of adenoviral, AAV and LVs. Because of the general trend to replace retroviral by LVs, retroviral vectors will not be treated here. This chapter will conclude with some basic biosafety considerations specific to the respective vectors.

[a] Généthon, 1, rue de l'Internationale, BP60, F-91002 Evry-Cedex, France
[b] Paul-Ehrlich-Institut, Division of Medical Biotechnology, Paul-Ehrlich Str. 51–59, D-63225 Langen, Germany
[c] Biotechnology Research Institute, National Research Council of Canada, Animal Cell Technology, 6100 Royalmount Avenue, Montreal, Quebec, H4P 2R2, Canada
[d] Email: omerten@genethon.fr

30.1.2 Viral Vector Production

Depending on the amount of the viral vector to be produced as well as on the cell system used, different culture systems are available for the production of viral vectors. In general, adenoviral, AAV and LVs are produced using a discontinuous mode after induction of the production. This induction is performed either by transfection methods, where the cells are transfected with one or several plasmids providing the helper functions as well as the vector construct, or by infection of the cells with one or several virus(es) providing the functions required for the production of the viral vector.

Depending on the growth characteristics of the producer cells, cells can either be cultured adherently or in suspension: in principle, all cell lines, except those derived from the blood system, tumor cells and some insect-derived cell lines, grow adherently. Since tumor cells have an increased tendency to grow in suspension they can be adapted to suspension growth (Griffiths, 1985). Suspension culture is characterized by a homogeneous concentration of cells, nutrients, metabolites and products, thereby facilitating scale-up and enabling accurate monitoring and control of the culture.

The difference between anchorage-dependent and suspension cells is the method of subcultivation, which can be as simple as dilution for suspension cells whereas surface adherent cells have to be detached from the surface in order to plate them on new plates. Different ways to detach surface adherent cells for subcultivation have been reviewed by Merten (2010).

From the point of view of scale-up, these differences become very important. For suspension cultures, the size of the cell culture system can easily be scaled up (from a spinner or shake flask via a lab scale stirred-tank reactor (STR) system to an industrial scale STR of up to 20,000–30,000 l (Merten, 2006)); however, for adherently growing cells this scale-up is relatively difficult when they cannot be adapted to suspension growth. In this case, industry has opted for microcarrier-based systems (Butler, 1988) providing large surfaces at large scale (e.g., for the production of viral vaccines (Montagnon, 1989)). Vector manufacturers have evaluated fixed-bed reactor systems or multiple process systems, based on, for instance, the use of roller bottles or Cell Factories (see Section 30.2.1.1).

Cell cultures can be performed using different culture modes:

(1) Batch: This is the simplest culture mode. Traditionally at a small scale, cultures are performed in a batch mode: A culture is initiated with a given starting cell number/concentration. Cell growth follows a classical profile with lag phase (onset of cell growth), exponential phase (exponential growth), stationary phase (reduction and stopping of cell growth due to accumulation of eventually toxic by-products and/or limitation of essential nutrients) and death phase (cell death via apoptosis or necrosis).

(2) Fed batch: An improved process mode is the fed batch. Fed-batch cultures are started as batch cultures and at a given culture period (e.g., at a certain cell density or the reduction of the concentration of an essential nutrient below a critical value) fresh medium is added in pulses or continuously. The culture duration is limited because feeding is limited by the maximum volume of the culture vessel. The advantages are progressive addition of nutrients and reduced accumulation of toxic by-products due to medium optimization. The duration and the cellular activities are increased with respect to batch culture; in addition, the fed-batch mode is much simpler than the perfusion mode, and is the preferred mode for large-scale production at an industrial level (Werner *et al.*, 1992).

(3) Continuous and perfusion culture: This culture mode is of potential interest for the large-scale production of biologicals. In principle, when the medium of a culture is continuously changed, a continuous culture is obtained. However, in the case of suspension cells, it is characterized by a continuous exchange of the biomass, leading to a constant dilution of the cells and eventually to the selection of faster-growing clones with a reduction or loss of productivity (in particular, for not completely stable cell lines). The main optimization of this culture system lies in the augmentation of the viable biomass leading to a perfusion culture. This mode is characterized by a continuous medium change after having obtained a certain predefined cell density. In order to keep the viable cell mass, the medium exchange is performed with cell-retention systems, allowing medium change without cell loss (Woodside *et al.*, 1998), leading to cell concentrations of 10^7–10^8 cells per ml. Advantages are high productivities and a generally short residence time of the product, of particular interest when the product is not stable at the production conditions.

Cell culture systems which can be used for the production of different viral vectors are briefly presented and discussed below.

30.1.2.1 *Currently used cell culture systems*

(1) *Small-scale/laboratory-scale cell culture systems*

Adherent as well as suspension cells are cultured in static cell culture systems including Petri dishes and T-flasks. In the case of T-flasks, the surface available ranges from 25 cm^2 to 225 cm^2. The general applications are routine subculture as well as small-scale productions of viral vectors, using batch, fed-batch or repeated batch culture modes.

In the event that larger quantities have to be produced (for research or developmental purposes), anchorage-dependent cells can be cultured in Cell Factories (or equivalent) (Fig. 30.1(a)) or roller bottles (Fig. 30.1(b)) providing culture surfaces ranging from 6,300 cm^2 to 25,100 cm^2 in the case of 10- or 40-stack Cell Factories, respectively, and 490 cm^2, 850 cm^2 and 1,750 cm^2 in the case of the roller system. Whereas the Cell Factories are static culture systems, the roller bottles are placed on a roller machine for turning the rollers with a maximum speed of 1.5–2 rotations/minute. These culture systems have been used for the transient production of preclinical and clinical batches of LV and AAV vectors.

When really large numbers of rollers have to be treated, then two solutions can be envisaged: (a) the use of automatic Cellmate processing systems as proposed by The Automation Partnership, a system which, for instance, has been used by the biopharma industry for the production of viral vaccines for human and veterinary use (http://www.automationpartnership.com), and (b) the use of the RollerCell system from Cellon (http://www.cellon.lu) which is a self-contained roller bottle processing system. This system is designed to automate all stages of roller bottle-based tissue culture — from cell inoculation, incubation, medium change and trypsinization through to final harvest. A single unit can process the equivalent of 200 standard roller bottles simultaneously. A standard pack (= 20 bottles) provides a culture surface of 36,000 cm^2 or 85,000 cm^2 when using flat-wall or expanded-wall-surface (= pleated roller bottles), respectively.

Cell Factories show their limitations with respect to the possible variations in the culture volume (600–1,000 mL for the ten-stack system), whereas the culture volume can be varied from about 125 mL to 500 mL for a 850 cm^2 roller bottle.

Two typical suspension culture systems are shake (Fig. 30.1(c)) and spinner flasks (Fig. 30.1(d)), allowing the easy modification of the culture volumes in function of production needs but limited by the oxygen transfer capacity of the respective system.

(2) *Large-scale cell culture systems*

Only those culture systems which are really used for vector production are briefly presented here: these are the STR, the fixed-bed reactor and the WAVE reactor. It should be mentioned here that disposable versions have been developed in recent years for all systems.

Stirred-tank reactor (STR) system

STR systems can be used for the cultivation of suspension and adherent cells when grown on microcarriers. In general, these reactors are equipped with an agitator and often with baffles. The largest available scale for industrial scale cultivation of animal cells is 20,000–30,000 L (Merten, 2006). An overview of engineering and scaling issues was published by Varley and Birch (1999). In the case that batch or continuous cultures without cell retention are performed, cell density is limited to $1–5 \times 10^6$ cells per ml. For increasing cell density, and thus reactor productivity, the reactor has to be equipped with a retention device in order to run the cultures in

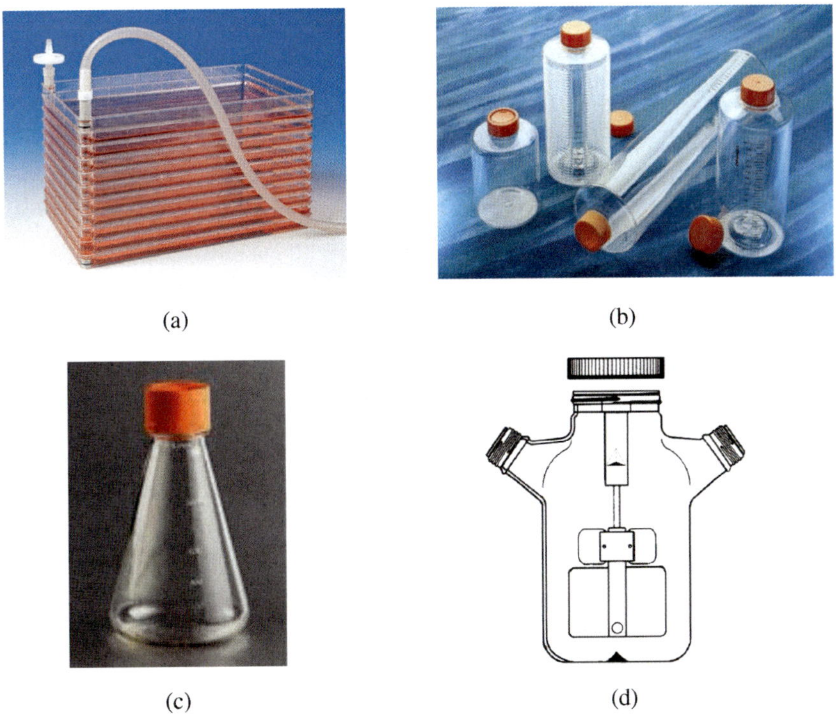

FIGURE 30.1 ■ Small-scale cell culture systems. (a) Cell Factory (10-stack version) (Corning). (b) Roller bottles: Four different sizes are shown: 490 cm^2, 850 cm^2, 1,750 cm^2, and 1,700 cm^2 Extended Surface Polystyrene Roller Bottle (Corning). (c) Shake flask (Corning). (d) Spinner flask (Bellco Glass).

perfusion mode (Woodside *et al.*, 1998). Using perfusion culture, the cell densities can be increased to 5×10^6–50×10^6 cells per ml (Merten, 1993).

Fixed-bed reactor system

Various biologicals have been produced in low- and high-density suspension processes: Monoclonal antibodies, recombinant proteins, viruses and viral vectors.

In order to increase the reactor cell density, the use of fixed- or packed-bed reactors is of interest because very high cell densities (0.5–2 $\times$ 10^8 cells per ml carrier) can be obtained. In principle, cells attach to solid or porous carriers or attach and are entrapped in the case of porous sponge-like materials, such as FibraCell (Kadouri and Zipori 1989). Thus, cells that do not readily attach can be cultivated in packed-bed reactors. The attached and/or entrapped cells grow on/in the carrier matrix and the culture medium, conditioned for optimal pH and pO_2, is circulated from a conditioning vessel to the fixed bed and back to the conditioning vessel (Fig. 30.2(a) — the CellCube system) or both systems (cell compartment and conditioning vessel) are integrated, as for the basket/fixed-bed reactor provided by New Brunswick Scientific (Eppendorf) (Figs. 30.2(b), and 30.2(c)) or ATMI. In general, these culture systems can be used in perfusion mode.

Such reactors have been used for the production of viral vaccines, retroviral vectors, adenoviral vectors, AAV vectors, recombinant proteins and monoclonal antibodies.

WAVE reactor system

The WAVE reactor system (Fig. 30.2(d)) is a scalable bioreactor system for the cultivation of animal, insect and plant cells using wave agitation induced by a

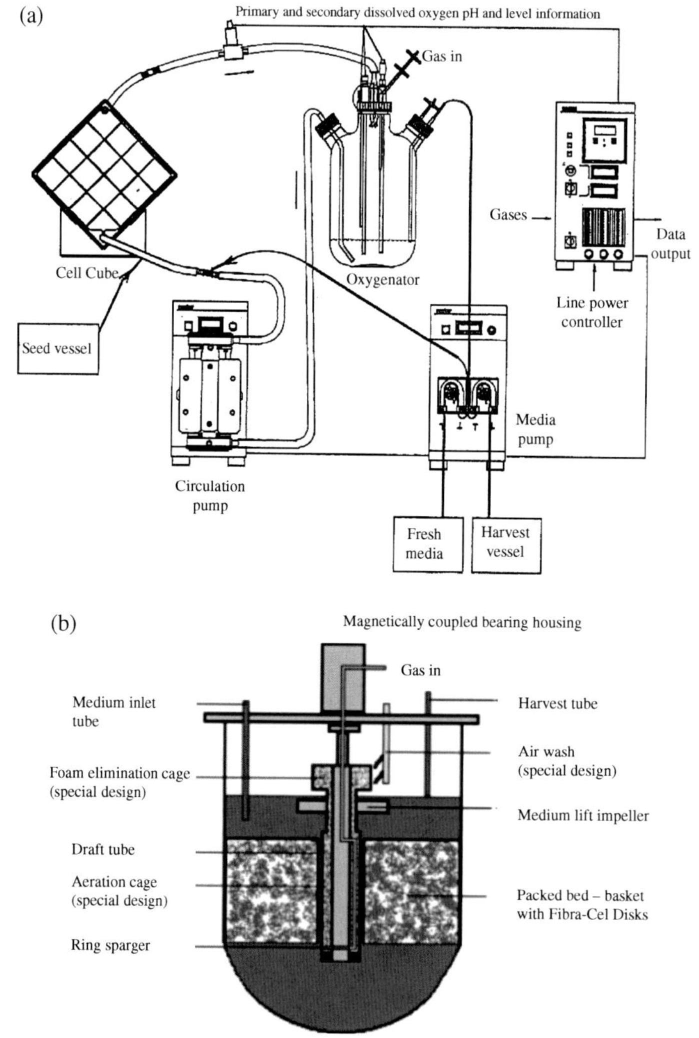

FIGURE 30.2 ■ Large-scale cell culture systems. (a) CellCube, lab version (module 25) (Corning). (b), (c) Packed-bed perfusion system with harvest and level tube (New Brunswick Scientific, reproduced with permission). (b) Principle of a packed-bed reactor. (c) Complete set-up of a packed-bed reactor. (d) WAVE reactor design (Singh, 1999).

(c)

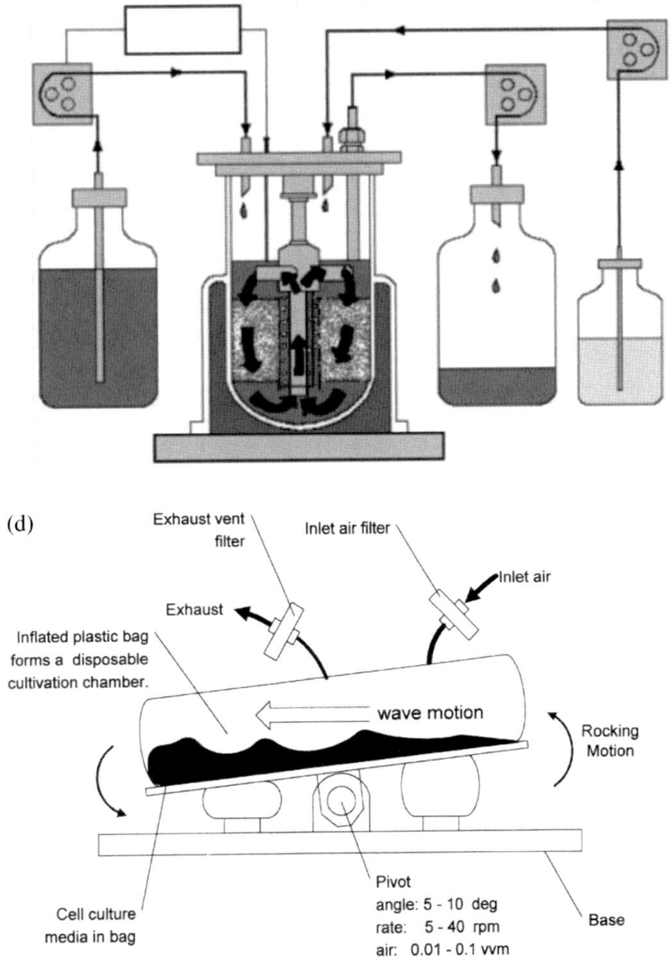

FIGURE 30.2 ■ *(Continued)*

rocking motion (Singh, 1999). This agitation system provides good nutrient distribution, off-bottom suspension and rather good oxygen transfer without damaging fluid shear or gas bubbles. In addition, it is easier to handle than an STR. The largest developed scale is the 500 L scale. The bioreactor is disposable, and therefore requires no cleaning or sterilization. It can be placed in an incubator requiring minimal instrumentation, thus reducing the investment costs. The main disadvantage is the limited mass (O_2) transfer capacity thus limiting the scalability to a maximum of 500 L at low cell densities.

The WAVE reactor system is well adapted for the production of biomass for starting production reactors or virus seed stocks for production purposes. Various cell types have been cultured in the WAVE reactor including: (1) recombinant NS0 cells in suspension; (2) human 293 cells in suspension for adenovirus production; (3) Sf9 insect cell/baculovirus system; and (4) human 293 cells on microcarriers (Singh, 1999).

30.1.3 Viral Vector Purification

Purification of viral vectors for delivery of therapeutic genes has relied on laboratory protocols, using essentially ultracentrifugation techniques developed to generate material in sufficient quantities to establish proofs of principles. However, to be used in clinical protocols, viral vector preparations need to comply with stringent standards that are rigorously scrutinized by regulatory agencies. Also, to achieve large and multi-center clinical trials, scalable upstream and downstream processes need to be implemented. Consequently, the downstream processing steps are designed to achieve high recovery yields of viral vectors with defined critical quality attributes that might affect the safety and efficacy of the final product. Many current viral vector downstream processes have been largely adapted from methods originally developed for purification of recombinant proteins and, to some extent, viral vaccines. However, these approaches, although conceptually satisfactory, are not without posing serious challenges. The physicochemical properties of viral vectors and especially their surface characteristics in the case of enveloped viral vectors are very different from those of simple proteins (Table 30.1). Also, it is essential to maintain viral delivery and gene expression functions as intact as possible throughout the sequence of purification steps. Although there might be a generic approach, for each vector type and serotype a strategic design and step-by-step optimization of the purification process is crucial to maximize the yield and quality of the final product.

Over the past 15 years, in order to support gene therapy clinical trials, important progress has been made in the area of upstream processing of viral vectors. In parallel, considerable efforts have been dedicated to develop purification strategies that meet the standards of current good manufacturing practices (GMPs).

TABLE 30.1. Properties of AAV, Adenoviral and LVs

	AAV	Adenoviral	LV
Size	~20 nm	~80 nm	~100 nm
Envelope	No	No	Yes
Stability	High	High	Low
Net charge at neutral pH	Positive	Negative	Negative
Buoyant density	1.39 in CsCl	1.34 in CsCl	1.16 in Sucrose

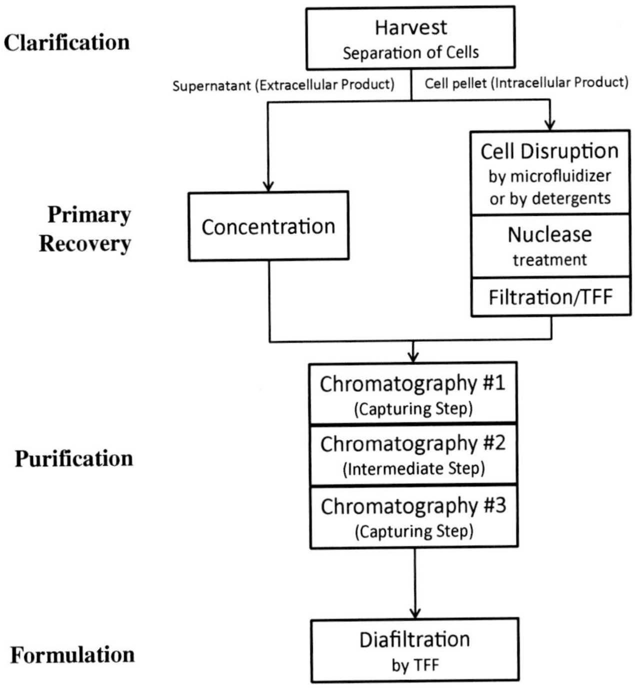

FIGURE 30.3 ■ General scheme for virus purification for large-scale operations.

The sequential downstream processing of viral vectors includes essentially the following steps (Fig. 30.3):

(1) Harvest of the viral particles from the cell culture. The non-enveloped viral particles, such as adenovirus and AAV, are produced intracellularly, which will require a cell-disruption step to release the viral particles and processing of the cell lysate. Enveloped viral vectors (i.e., LVs) are released into the culture supernatant so that only the supernatant is processed in this case.
(2) Clarification of the crude viral harvest from either the cell lysate or the supernatant to remove host cells and cell debris is essentially the first step of downstream processing. This step is achieved at large scale by either centrifugation or microfiltration.
(3) Concentration of the clarified viral stock to reduce the processing volume, especially in the case of low production yield and diluted viral material. This step is completed by ultracentrifugation, precipitation or ultrafiltration.
(4) Purification steps to separate the viral vector from the most abundant contaminants. Depending on the nature of the viral vector, the type of feed and the contaminants involved, more than one purification step is normally

required and would generally involve at least one ion exchange or affinity chromatography protocol.
(5) Polishing step to achieve the specified viral preparation purity — can be achieved by another ion exchange, hydrophobic interactions, size exclusion chromatography (SEC) or ultrafiltration step. An ultracentrifugation step may be added prior to the polishing step to remove empty capsids.
(6) A final concentration and buffer exchange step might be required to achieve a specified final product concentration and formulation.

This sequence of steps will involve different separation and purification techniques to maximize recovery yield and specific bioactivity while minimizing impurities and manufacturing costs. For each viral vector a specific purification strategy needs to be defined and validated to achieve a set of critical quality attributes of the product and meet safety and efficacy specifications for the intended use.

Here we review the current methods used for downstream processing of viral gene therapy products focusing on the three viral vectors that have been the most extensively used in clinical trials: Adenoviral vectors, AAV and LVs.

30.2 Adenoviral Vector Production and Purification

The first- and second-generation adenoviral vectors are propagated by infection of *trans*-complementing cell lines such as E1 complementing HEK 293. Whereas, third-generation or helper-dependent (adeno) virus (HDV) vector propagation relies on co-infection of complementing cell lines with HDV and a helper virus, involving a primary rescue of the HDV after transfection of HDV-DNA. Most of the production steps are common to all generations of adenoviral vectors; however, some steps are very specific to the last adenoviral vector generation. This section covers standard adenovirus production protocols and provides indications on HDV production.

30.2.1 Adenovirus Production

Any process development for an efficient, robust and scalable adenovirus production aims at maximizing the cell-specific vector yield, which ideally should attain 30,000 infectious virus particles (IVP) per cell as achieved by replicating wild-type adenovirus on a permissive cell line. The final production scale depends essentially on the vector quantity required per batch. For adenovirus production for gene therapy purposes, a scale associated with production of up to 10^{15} virus particles (VP) per batch may be up to 100 L.

The production of adenoviral vectors involves the following steps: Growing the cells to a desired cell density for infection; infecting with adenovirus stock at predetermined optimal multiplicity of infection (MOI); and, finally, harvesting the virus at a predetermined optimal time post infection.

There are a number of excellent serum-free formulations available commercially that have been optimized for adenovirus production using HEK293 or PER.C6 cell lines. In general, for potential clinical applications, it is recommended to use media that are free of animal-derived components and are preferably chemically defined.

For optimal cell growth and vector production, culture pH, pO_2 and culture and vector production temperatures have to be optimized. In addition, feeding at the time of infection or during vector production or medium change at the time of infection (this in particular for compensating for increased consumption rates during the first 24 h post infection (hpi)) may also be investigated for optimal vector production.

Virus assembly and DNA packaging occurs between 20 and 48 hpi. Viability (typically 80–40%) is usually used as a criterion to harvest the culture. Typically, the harvest is done at approximately 48 hpi, but is highly dependent on the MOI used, adenovirus serotype and the transgene expressed.

Commercial-scale cultivation of mammalian cells has been accomplished using different technologies: roller bottles, CellCube, hollow fibers, microcarriers and suspension-culture bioreactors (batch, fed-batch or perfusion mode). However, for products needed in large amounts, suspension culture is the most effective method over processes using adherent cells.

30.2.1.1 *Helper-dependent adenovirus production*

HDV vector propagation cannot rely only on complementing cell lines, but needs helper virus. The challenge is to remove the adenovirus helper from the HDV final preparation. The helper to HDV ratio should be better than $1{:}10^4$. It is beneficial for the downstream processing if during HDV production the adenovirus accumulation can be minimized. Standard HDV production is a two-step process: Rescue and amplification.

The replication of HDV at the rescue step is limited by the transfection efficiency and thus gives low yields relative to the infection processes. Therefore, multiple rounds of HDV amplifications are required to meet the demand for the HDV quantity required for large-scale operation. Normally, crude lysate from the preceding harvest is used to infect the cells for the next stage and this process is repeated until sufficient HDV is obtained.

The purpose of the rescue step is to recover HDV from HDV DNA plasmid. The complementing cells, which also express recombinase (e.g., enhanced flippase (FLPe)) are transfected with the linearized HDV genome, followed by an infection with an adenoviral vector, which has a packaging signal that is flanked by two recombination sites (flippase recognition target (FRT)). During vector production only the vector genome is packed into the viral capsids while the packaging signal of the helper vector is removed and thus cannot be packaged. When a cytopathic effect is visible at around 40 to 48 hpi, the viral lysate containing the HDV is recovered.

Amplification steps are carried on thereafter to build a viral HDV stock. Because the HDV titer is normally low at the end of the rescue step (10^2 to 10^5 infectious units (IU) of HDV per ml), further amplification of HDV is required. Typical amplification protocols consist of exhaustive passages of viral lysate on an increasing number of cells. A minimization and thus optimization of the amplification steps can be achieved by operating with a controlled MOI protocol for the HDV and the helper adenovirus.

30.2.2 Adenovirus Purification

Recombinant adenoviral vectors are produced intracellularly and released into the supernatant through cell lysis. Most of the reports on adenovirus vector processing in the current literature are about serotype 5. In general, these vectors are stable and they have traditionally been purified at a laboratory scale, after freeze–thaw cycles of the cell lysate, by two or three rounds of caesium chloride (CsCl) density-gradient ultracentrifugation to achieve clearance of host-cell nucleic acids, host-cell proteins, unassembled adenovirus proteins, unpackaged viral DNA and eventually products related to transgene expression. An additional benefit of the density gradient ultracentifugation is the separation of empty capsids from functional recombinant adenovirus providing there is careful collection of the specific bands. Thereafter, the viral product is dialyzed or desalted into the formulation solution to achieve preparations with ~10^{12} VP per mL for early-stage clinical trials. Recently, iodixanol medium has been proposed to address the problem of toxicity associated with CsCl. Also, this method has been used to separate helper-dependent adenovirus from helper adenovirus (Dormond *et al.*, 2010).

Except for vaccine manufacturing facilities, ultracentrifugation for large scale vector preparations is not a viable choice.

Scalable chromatographic purification techniques exploit size, surface charge and eventually specific receptor binding properties. In the case of adenovirus anion exchange, chromatography is the primary step used in many of the industrial processes that have been described (Altaras *et al.*, 2005). For cell lysis, repeated freeze/thaw is not scaleable and consequently other lysis methods, including non-ionic detergent addition (Tween-80 or Triton X-100), microfluidization and shear stress induced by high agitation rate in the cell culture vessel or microfiltration devices, have been applied for large-scale operations. Treatment with nucleases is the most common DNA clearance method employed for viral vector preparation. For adenovirus purification, Benzonase (EM Science) is often selected since it digests both DNA and RNA and remains relatively stable in a variety of buffers. Clarification has been generally achieved by centrifugation or microfiltration, whereas the most common method for feed volume reduction and concentration of adenovirus with or without buffer exchanging is ultrafiltration.

For example, ion exchange has been combined with SEC to achieve 99% purity and a typical overall recovery between 30% and 80% (Fig. 30.4). More specifically cell lysate was prepared by osmotic shock, treated with DNAse, centrifuged, conditioned and filtered before being applied to the ion exchange column. The column eluate was concentrated by ultrafiltration before the SEC polishing step (Kamen and Henry, 2004).

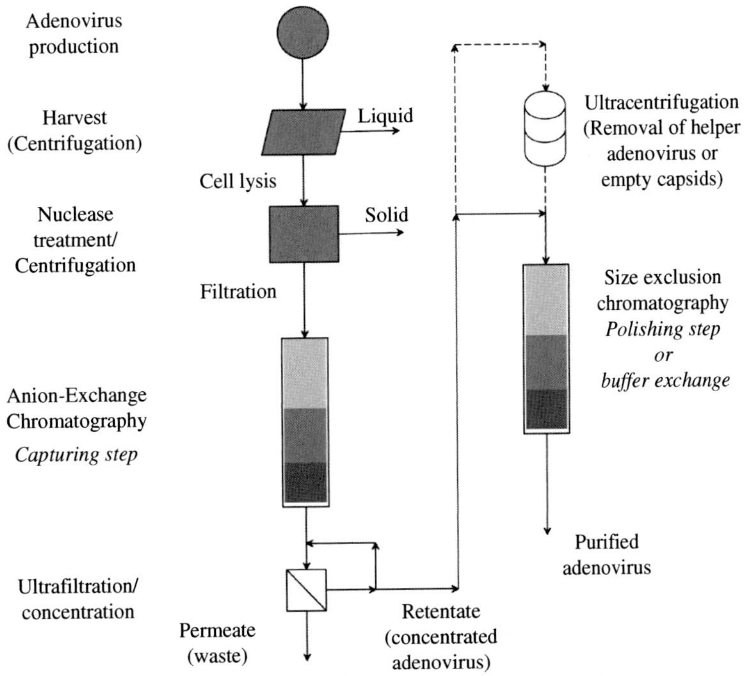

FIGURE 30.4. ■ Adenovirus purification flow sheet. Harvested culture is centrifuged and collected cells are lysed by osmotic shock or microfluidization. After Benzonase treatment and centrifugation, the supernatant is filtered and loaded on to the anion exchange column. The eluted virus is concentrated by tangential flow filtration (TFF) (ultrafiltration membrane) and is loaded to a size-exclusion chromatography column. If this is for the purification of HDV, then adenoviral helper is removed by ultracentrifugation prior to the polishing and buffer exchange step using SEC.

30.3 Adeno-Associated Virus Vector Production and Purification

30.3.1 Laboratory-Scale AAV Production

Basics of the design of a laboratory-scale production of AAV using transfection of HEK293 cells are presented in Chapter 11. It is particularly well suited for small-scale production for research and developmental purposes due to its high flexibility allowing the rapid modification of the serotype (cap) and transgene. In addition, specific productivities of 10^2–10^4 vector genomes (vg) per cell have been reported. The main drawback of the production of large vector lots is its limited scalability, although Cell Factory-based, roller bottle-based, CellCube-based and suspension cell-based transfection processes have been established. In addition, there is a relatively high incidence of recombination events between the plasmids used, leading to the generation of replication-positive recombinant AAV particles (rep+ rAAVs) and "replication competent" = pseudo-wild-type AAV (rcAAVs). For more details, see Galibert and Merten (2011).

30.3.2 Scalable AAV Vector Production

To overcome the limited scalability of the above-mentioned method, three different (scalable) AAV vector production systems have been developed, based on the use of (1) stable HeLa or A549 clones, (2) the recombinant herpes simplex expression system, and (3) the SF9/baculovirus system:

(1) Stable HeLa- or A549-cell clones contain both the AAV replication and capsid *rep-cap* genes as well as the AAV-vector sequence. The cells are infected with adenovirus for inducing AAV production. Specific vector production rates beyond 10^4 vg per cell have been reported. In the case of HeLa-cell-based producer cells, a suspension process of a scale of 250 L has been set up (Thorne *et al.*, 2009).
(2) A highly efficient recombinant herpes simplex virus (rHSV)-based rAAV complementation system has been reported (Kang *et al.*, 2009) which uses two rHSV vectors, one harboring the inverted terminal repeat (ITR)-flanked gene of interest (= rAAV-vector) and the second one bearing the *rep* and *cap* genes. The infection of BHK cells grown in suspension with both rHSV vectors leads to a specific production of up to 1.2×10^5 vg per cell. Scale-up to a 10 L scale (disposable WAVE reactor) has been reported with average volumetric productivities of 2.4×10^{14} DNase-resistant viral particles (DRP)/l (Thomas *et al.*, 2009).
(3) The infection of Sf9 insect cells growing in suspension with two baculoviruses, one harboring the ITR-flanked gene of interest (= rAAV-vector) and the second

one bearing the *rep* and *cap* genes, both inserted into the transposon Tn7 site of the bacmids, led to the production of AAV (Smith *et al.*, 2009) with specific productivities ranging from 10^4 to 10^5 vg per cell. The volumetric production yields are in the range of 1.1×10^{14} vg/l (AAV1) when using 10 L STR, thus achieving productivities similar to the HSV-1-based production system and larger production scales of up to 200 L have been established (results presented at the ESGCT Meeting, Versailles/F, October 25–29, 2012).

30.3.3 Adeno-Associated Virus Vector Purification

There are many distinct serotypes of AAV to target different tissues. This adds complexity to AAV purification. Surface proteins of different AAV serotypes are different; therefore the affinity chromatography method is specific to a particular serotype. In theory, ion exchange chromatography (IEC) and hydrophobic interaction chromatography (HIC) can work for all serotypes but might have different binding and elution characteristics for each serotype. SEC may be utilized in general for all of the serotypes as a polishing step because AAV can remain in the exclusion range and comes out first and the smaller molecules as contaminants would elute later. Ultracentrifugation is a method that can be used for all serotypes because the virus can be separated on the basis of density which is similar for all serotypes of AAV, however, it is not scalable and is of limited use for clinical vector production. Each of these methods has its advantages and disadvantages, for small- or large-scale applications.

30.3.3.1 *Primary recovery*

Since AAV is produced intracellularly in mammalian or insect cells it has to be released either by classical freeze–thaw cycles (at a small scale) or by mechanical (microfluidizer or equivalent) or chemical action (e.g., Triton X-100 at 0.1%). Both latter methods are scalable. Nuclease is usually added to digest the host cellular, nuclear and/or plasmid material during the extraction of AAV to avoid the formation of nucleic material–AAV complexes and to reduce the viscosity of the lysate.

Helper (adenovirus, herpesvirus, baculovirus) or contaminating viruses may be inactivated by raising the temperature to 56° C for 15 min.

30.3.3.2 *Ultracentrifugation*

The method that has been extensively used to purify viruses is based on CsCl isopycnic density gradient ultracentrifugation or on Iodixanol (non-toxic) density gradient ultracentrifugation. Ultracentrifugation allows easy separation of rAAV from helper viruses and has the supplementary advantage of being able to separate full from empty AAV particles. The ultracentrifugation method is serotype independent.

Although preclinical and early clinical studies have used CsCl-purified AAV vectors, recently, scalable chromatographic techniques have been used. Still, the advantage of ultracentrifugation may be realized as a final step when the process volume is significantly reduced to separate empty capsids from functional viral particles to improve the ratio of total particles to particles with complete viral genome, and thus the quality of the AAV vector preparation.

30.3.3.3 *Chromatography purification*

Chromatography is a well-established scalable process and has been adapted to separate contaminant proteins present in the virus solution based on physico-chemical properties such as affinity to a ligand, net charge, hydrophobicity and size. Since AAV has different serotypes, the surface characteristics are different for each type. It should be mentioned here that often more than one chromatographic step is needed for obtaining sufficient vector purity. The different methods employed for AAV purification are briefly presented.

(1) Affinity chromatography: This purification principle can make use of either the binding domains specific for a given serotype or antibodies or antibody fragments specific for one or several serotypes.

Affinity purification via specific binding domains: AAV2, 3 and 6 bind to cell surface via heparin sulfate proteoglycan receptor. Therefore, heparin-affinity chromatography has been successfully applied to purify these serotypes. In addition, AAV serotypes 4 and 5 show binding to 2,3-linked sialic acid, therefore, mucin-affinity chromatography is employed to purify these serotypes. The binding domains of other serotypes have not yet been identified, therefore, affinity chromatography is limited to only certain serotypes.

Affinity purification via antibodies: AVB Sepharose Highperformance medium is based on grafted antibody fragments against AAV from Llamas produced in yeast and coupled to N-hydroxysuccinimide-activated Sepharose Highperformance. AVB chromatography is used for the purification of AAV serotypes 1, 2, 3, 5, 6 and 8. Serotype-specific antibodies have also been used for purification.

Affinity chromatography is easily scalable, but the cost of the resin is high. These methods are serotype-specific and therefore their use is limited to the purification of selected serotypes.

(2) IEC: IEC is a simple and cost-effective technique to purify different serotypes of AAV. The method takes advantage of minor differences in net surface

charge properties of viral particles and proteins present in the solution at a given pH.

A general protocol for IEC using anion or cation resin involves loading at low-salt buffer conditions (normally less than 150 mM NaCl) and, after washing the unbound proteins, a gradient in step or continuous mode is generated towards higher salt concentration (normally 500 mM NaCl). The eluted fractions are identified for AAV and pooled for further purification. One IEC step is not sufficient to purify AAV serotypes. The eluted AAV is further processed by another IEC, HIC, or SEC.

A slightly higher negative charge due to the presence of DNA in full AAV capsids compared with empty capsids made it possible to utilize IEC to separate empty capsids from the AAV preparation. Thus for different AAV serotypes, the charge difference could be used to (partially or completely) separate empty from full particles.

Different workers have used different resins to demonstrate AAV purification by IEC. In the literature, it can be found that quarternary amine (Q)- (most often used) and diethylaminoethyl cellulose (DEAE)-based anion exchangers and sulphopropyl (SP)-, carboxymethyl (CM)-, and sulphonyl (S)-sepharose and cellufine (also referred to as affinity resin)-based cation exchangers have been used to purify not only AAV2, but also serotypes 1, 4, 5 and 6.

(3) SEC: Since AAV is a large particle (20 nm, 3.6×10^6 Da) compared with the contaminants present in the cell lysate, SEC can be utilized for its purification. It is mainly used as a polishing step where process volume has been significantly reduced from the previous steps. This method not only removes low-molecular-weight contaminants present in the solution, but the gentle process simultaneously gives a high yield and allows a buffer exchange to the final formulation. The size exclusion for globular proteins on Superdex 200 is 1.3×10^6 Da, therefore AAV, regardless of its serotype, elutes in the void volume. The proteins, which are generally of low molecular weight, elute at higher elution volumes. The capacity of the load is optimized based on the separation of AAV and contaminants. It is possible to load as high as 30% column volume for SEC when used as a polishing step in AAV purification scheme.

30.3.3.4 *Other considerations for the purification steps*

Recombinant AAV tends to form aggregates in the cell lysate during primary recovery from cell paste. This results in reduced recovery in density gradient ultracentrifugation. Addition of 150 mM NaCl or 37.5 mM $MgSO_4$ increases the ionic strength sufficiently to avoid binding to cell debris or other cellular proteins released after cell lysis.

Before running any type of chromatography system, the solution needs to be filtered through a 0.45/0.2 μm filter to avoid clogging of the column. Tangential flow filtration (TFF) may be employed before any chromatography steps. It can be used efficiently for buffer exchange between different steps and concentration to reduce operation volumes. Normally, >500 kDa molecular weight cut-off (MWCO) (or microfiltration membranes) can be used to remove cellular debris, whereas 100 kDa MWCO can be used to retain AAV to concentrate and remove low-molecular-weight contaminants.

Due to the various possible combinations of different separation principles, many different purification schemes of AAV vectors may be in use, such as cell lysis–clarification–AVB chromatography–TFF–sterile filtration (when using affinity chromatography), or cell lysis–clarification–IEC–multi-modal chromatography–IEC–TFF–sterile filtration (when using a scheme without affinity chromatography).

30.4 Lentiviral Vector Production and Purification

30.4.1 Production of LVs

Today, only transfection methods using two to five plasmids are routinely used for the small- and large-scale production of LVs because no good stable cell lines are available. Culture and production parameters can be optimized as, for instance, detailed in a review by Ansorge *et al.* (2009).

Large-scale production for preclinical and clinical use is essentially based on the use of multiple units of two or ten stack Cell Factories, which can be harvested once or several times after transfection. Due to feasibility considerations, large-scale production runs are often harvested two or three times, keeping the overall vector productivity, whereas for research-grade productions, up to five harvests are performed. The generally obtained vector titers in the unprocessed supernatants range from 1 to 5×10^7 infectious particles (ip) per ml, and are thus in the range published in previous reports using similar transient production processes with 293 T-cells. For more information the reader is referred to Schweizer and Merten (2010).

30.4.2 Purification of LVs

In contrast to Moloney murine leukemia virus (MLV)-based retroviral vectors, LVs are generally purified for further use in order to get rid of contaminating DNA (derived from the plasmids used but also from the lysed producer cells) and foreign serum- and producer-cell-derived proteins. In addition, several groups could show that purification not only ensures vector safety but significantly improves efficacy.

The glycoprotein of the vesicular stomatitis virus (VSV-G) is mostly used because it allows transduction of a wide range of target cells, and since VSV-G pseudotyped LVs are relatively insensitive to shear stress, traditional small-scale purification methods are based on ultracentrifugation of vector supernatant, allowing over 100-fold concentration. Although high vector concentrations can be achieved, long process times can lead to partial vector inactivation and these purification methods cannot be scaled up. Therefore, for large-scale processes, chromatography- and membrane-based methods have been developed.

For industrial applications, downstream processing protocols (= purification protocols) have been scaled up and in most of the cases, process steps traditionally used in the biotech industry have been developed and put together in order to obtain a high-performance downstream processing protocol: clarification, IEC, concentration/diafiltration using TFF, Benzonase for the degradation of DNA, a polishing step (often SEC) and final sterile filtration.

VSV-G pseudotyped LVs are purified using anion-exchange chromatography, whereby large-scale purification protocols make use of classical low-pressure column chromatography or membrane-based IEC. The vectors are eluted by a linear or step salt gradient. Scale-up is straightforward leading to pure vectors.

In order to concentrate the bulk product (supernatant or an intermediate product) TFF systems are employed, allowing also diafiltration for buffer exchange. At a small scale or at later stages of a large-scale purification protocol, centrifuge-based disposable devices are often used, whereas hollow fiber or flat membrane cartridges are used for larger scales. Ideally, the MWCO is 500–750 kDa (due to the size of the vector particle) allowing rapid concentration and buffer exchange with reduced vector inactivation.

Finally, as a polishing step, SEC can be employed, which leads to an efficient removal of all contaminants smaller than the pore size of the chromatographic matrix (e.g., when using a 500 kDa size exclusion) and thus to a further and efficient purification. Practically all large-scale purification protocols have adopted SEC as a final polishing step.

A supplementary process step is a Benzonase treatment during the downstream processing for removing cellular and plasmid-derived DNA contaminants. This step increases the overall biosafety feature of the process because the reduction of the size of cellular or plasmid DNA leads to a reduced risk of transfer of functional gene sequences to the target cells. Such a DNA degradation step has been implemented for all large-scale downstream processes.

According to published results, the performances are as follows: The concentration factor ranges between 10- and 50-fold and the overall process yields range from 16% to 40% for the protocols making use of an IEC step. The final vector concentrations range from 10^8 to 10^9 ip per mL.

For more detailed information on the overall purification of retroviral vectors in general as well as isolated purification steps, the reader is referred to the reviews by Segura *et al.* (2006) and Rodrigues *et al.* (2007), and, with respect to details on large-scale purification of lentiviral vectors, to Schweizer and Merten (2010).

30.5 Biosafety of Viral Vector

The use of viral vectors for treatment of human patients bears a number of vector-specific health risks. A major safety issue is due to the derivation of several viral vectors from human pathogens such as adenovirus, herpes virus, poxvirus and measles virus or even from Ebola virus or lentiviruses such as human immunodeficiency virus (HIV)-1 or -2. Accordingly, the occurrence of replication-competent parental virus in vector preparations has to be strictly avoided to prevent virus-specific pathogenicity. With the exception of oncolytic viruses, all vectors are designed to be replication-incompetent by deletion of viral genes required for virus replication. Ideally, the vector particle should exclusively contain the transgene which is transferred into the target cell and expressed there.

In order to improve the biological safety of vector production systems, different approaches have been used. In order to avoid eventual recombination events, vector constructs make use of the split genome approach for separating different viral genes on separate plasmids (e.g., plasmids used for transient LV production) or at different loci in the producer cell line (e.g., stable MLV producer cell lines). With respect to the development of complementing cell lines, sequence overlaps are as minimal as possible. For example, the development of the adenoviral producer cells from HEK293 over 911 to PerC6 is characterized by a considerable reduction of overlapping sequences with the adenoviral vector to be produced. Moreover, most of the viral sequences are removed, leaving only the really essential ones, such as the ITR for the AAV or the long terminal repeat (LTR) and ψ-region for the LV vector system.

The possibility of recombination events between vector sequences and endogenous or exogenous viruses, e.g., after superinfection of transduced cells, has also to be taken into account in addition to eventual recombination events between vector components.

There are several safety issues related to the administration of viral vectors:

(1) The direct administration of large (huge) vector quantities *in vivo* can trigger a severe immune response leading to organ failure and possibly death (as had been observed in the case adenoviral vectors infused into liver).
(2) The transduction of tissue, such as liver, can lead to the presentation of vector-derived antigens on the transduced cells (as seen for the transduction of liver by AAV for the treatment of hemophilia) leading to a cellular immune

response eliminating the transduced cells. In the case of pre-existing immunity against the vector, gene therapy might only be efficient when applying an immune-suppressing regimen. The expression of the transgene might also lead to an immunoreaction against this transgene.

(3) The expression of the transgene in unintended tissues might reveal toxicity. Such events can be avoided by the use of tissue-specific promoters, allowing gene expression exclusively in the intended target cell population, or the integration of microRNAs (miRNAs) into the vector construct for suppression of transgene expression in certain tissues.

(4) In the case of integrating vectors (retroviral vectors) integration may lead to destruction of cellular genes or alterations of expression of genes adjacent to the integration site, eventually leading to oncogene activation or inactivation of tumor-suppressor genes. Vector (MLV)-mediated insertional mutagenesis was observed in some cases in the X-linked severe combined immunodeficiency (SCID-X1) trial (in France and the UK) leading to leukemia. In all these cases, overexpression of a protooncogene adjacent to the integrated vector was observed. The replacement of MLV vectors by lentiviral vector showing different integration characteristics will improve the biosafety of the use of integrating vectors. In this context, the replacement of classical vector constructs (retroviral vectors in general) by self-inactivating (SIN) vectors will improve the safety profile of retroviral vectors because no functional retroviral LTR-promoter is integrated into the host cell DNA and cannot enhance the transcription of adjacent cellular genes, whereas the transgene is expressed by an internal heterologous promoter. Another strategy to prevent modulation of the expression of cellular genes flanking the integrated vector is the introduction of so-called insulator sequences into the transfer vector.

(5) The maximal vector copy number is a critical safety parameter, because a higher number is directly associated with a higher risk of vector integration into a "sensitive" gene (e.g., the promoter region of an oncogene or a gene associated with proliferation). Thus, based on statistical calculations, the vector copy number should typically be in the range of one to two copies per target cell.

(6) Finally a considerable improvement of the safety profile of integrating vectors will be site-specific integration, either by replacing the defective gene or by integration into safe sites within the host cell genome.

30.6. Perspectives

Although today we have relatively efficient manufacturing methods for viral vectors, which are characterized by an improved safety profile, more optimal manufacturing technologies are needed in view of advanced clinical trials and finally the marketing of viral vector-based medicines.

References

Dormond, E., Chahal, P., Bernier, A., *et al.* (2010). An efficient process for the purification of helper-dependent adenoviral vector and removal of helper virus by iodixanol ultracentrifugation, *J Virol Methods*, **165**, 83–89.

Kadouri, A., Zipori, D. (1989). 'Production of anti-leukemic factor from stroma cells in a stationary bed reactor on a new cell support', in Spier, R.E., Griffiths, B.J. (eds.) *Advances in Animal Cell Biology and Technology for Bioprocesses*, Sevenoaks: Butterworths, pp. 327–332.

Kamen, A., Henry, O. (2004). Development and optimization of an adenovirus production process, *J Gene Med*, **6 Suppl 1**, S184–S192.

Kang, W., Wang, L., Harrell, H., *et al.* (2009). An efficient rHSV-based complementation system for the production of multiple rAAV vector serotypes, *Gene Ther*, **16**, 229–239.

Montagnon, B.J. (1989). Polio and rabies vaccines produced in continuous cell lines: A reality for Vero cell line, *Dev Biol Stand*, **70**, 27–47.

Singh, V. (1999). Disposable bioreactor for cell culture using wave-induced agitation, *Cytotechnology*, **30**, 149–158.

Smith, R.H., Levy, J.R., Kotin, R.M. (2009). A simplified baculovirus-AAV expression vector system coupled with one-step affinity purification yields high-titer rAAV stocks from insect cells, *Mol Ther*, **17**, 1888–1896.

Thomas, D.L., Wang, L., Niamke, J., *et al.* (2009). Scalable recombinant adeno-associated virus production using recombinant herpes simplex virus type 1 coinfection of suspension-adapted mammalian cells, *Hum Gene Ther*, **20**, 861–870.

Thorne, B.A., Takeya, R.K., Peluso, R.W. (2009). Manufacturing recombinant adeno-associated viral vectors from producer cell clones, *Hum Gene Ther*, **20**, 707–714.

Reviews

Altaras, N.E., Aunins, J.E., Evans, R.K., *et al.* (2005). Production and formulation of adenovirus vectors, *Adv Biochem Eng Biotechnol*, **99**, 193–260.

Ansorge, S., Henry, O., Kamen, A. (2009). Recent progress in lentiviral vector mass production, *Biochem Eng J*, **48**, 362–377.

Butler, M. (1988). 'A comparative review of microcarriers available for the growth of anchorage-dependent animal cells', in Spier, R.E., Griffiths, J.B. (eds.), *Animal Cell Biotechnology*, Vol. 3, London: Academic Press, pp. 283–303.

Galibert, L., Merten, O.-W. (2011). Latest developments in the large scale production of adeno-associated virus vectors in insect cells (in view of the treatment of neuro-muscular diseases), *J Invertebr Pathol*, **107**, S80–S93.

Griffiths, J.B. (1985). 'Cell biology: experimental aspects', in Spier, R.E., Griffiths, J.B. (eds), *Animal Cell Biotechnology*, Vol. 1, London: Academic Press, pp. 49–83.

Merten, O.-W. (1993). 'Cultures cellulaires', in Julien, R., Cenatiempo, Y. (eds.) *Biotechnologies d'Aujourd'hui, Dix domaines stratégiques à l'aube du troisième millénaire*, Limoges: Presse de l'Université de Limoges, pp. 251–295.

Merten, O.-W. (2006). Introduction to animal cell culture technology — past, present and future, *Cytotechnology*, **50**, 1–7.

Merten, O.-W. (2010). 'Cell detachment', in Flickinger, M.C. (ed.) *Encyclopedia of Industrial Biotechnology, Bioprocess, Bioseparation, and Cell Technology*, New York: John Wiley & Sons, pp. 1–22.

Rodrigues, T., Carrondo, M.J.T., Alves, P.M., *et al.* (2007). Purification of retroviral vectors for clinical application: Biological implications and technological challenges, *J Biotechnol*, **127**, 520–541.

Schweizer, M., Merten, O.-W. (2010). Large-scale production means for the manufacturing of lentiviral vectors, *Curr Gene Therapy*, **10**, 474–486.

Segura, M.D., Kamen, A., Garnier, A. (2006). Downstream processing of oncoretroviral and lentiviral gene therapy vectors, *Biotechnol Adv*, **24**, 321–337.

Varley, J., Birch, J. (1999). Reactor design for large scale suspension animal cell culture, *Cytotechnology*, **29**, 177–205.

Werner, R.G., Walz, F., Noé, W., *et al.* (1992). Safety and economic aspects of continuous mammalian cell culture, *J Biotechnol*, **22**, 51–68.

Woodside, S.M., Bowen, B.D., Piret, J.M. (1998). Mammalian cell retention devices for stirred perfusion bioreactors, *Cytotechnology*, **28**, 163–175.

31

PRODUCTION AND PURIFICATION OF PLASMID VECTORS

Martin Schleef [a]

31.1 The Plasmid DNA Production Strain *Escherichia coli*

The manufacturing of plasmid DNA starts with the transformation of the appropriate and characterized host cell with a fully characterized vector plasmid. These cells are typically *Escherichia coli* cells. The resulting "genetically modified organism" (GMO) is to be checked carefully for the expected characteristics. The cultivation of *E. coli* cells is a fast process in comparison with other strains. Typically, growth overnight is sufficient to obtain large amounts of cell paste with a high amount of plasmid. However, this depends on the proper cultivation parameters (Schleef and Schmidt, 2004). Small-scale plasmid preparation for research applications or clone screening is discussed below.

For long-term use of a specific clone or large-scale production, a tested and verified *E. coli* clone will be transferred into the good manufacturing practice (GMP) environment for GMP-conformal processing. This includes the generation of a master cell bank (MCB) and working cell bank (WCB), which are required for reproducible large-scale cultivation of the bacterial biomass. The cell banks have to be fully characterized to be of sufficient quality for further manufacturing. Table 31.1 summarizes the quality control assays to be performed with this cell bank material for lot release.

E. coli host cells have a genome of 4.6 mega base pairs (Mbp) (Blattner *et al.*, 1997). Derivatives of the strain *E. coli* K-12 (Tatum and Lederberg, 1947) are

[a] PlasmidFactory GmbH & Co. KG, Meisenstrasse 96, D-33739 Bielefeld, Germany
Email: martin.schleef@plasmidfactory.com

TABLE 31.1 Quality Controls for the Quality Assurance and Product Release of Plasmid DNA

Test	Analytical Method
DNA concentration	UV-absorption (260 nm)
Appearance	Visual inspection
DNA homogeneity (covalently closed circle (ccc) content)	Densitometry after agarose gel electrophoresis (AGE)
DNA homogeneity (ccc content)	Capillary gel electrophoresis (CGE)
DNA purity	UV-Scan (220–320 nm)
RNA	Visual inspection after AGE
Endotoxin (lipopolysaccharide (LPS))	Limulus amebocyte lysate (LAL) test
Bacterial chromosomal DNA	Quantitative polymerase chain reaction (PCR)
Total protein	Bicinchoninic acid (BCA) test
Host cell-derived protein	Enzyme-linked immunosorbent assay (ELISA)
Bioburden	Presence of bacterial and fungal contaminants (bioburden: Total aerobic microbial counts (TAMC), total yeast and moulds count (TYMC))
Plasmid identity	Restriction digestion and agarose gel
Plasmid identity	Sequencing (double strand)

usually chosen to amplify plasmids. Within the K-12 bacteria, various sub-strains have been characterized so far and an overview has been presented (Hayes, 2003).

The system biological approach to bacteria — based on advanced sequencing technology — since bioinformatics enabled us to do so, is able to predict and evaluate the activities of genes (genomics), proteins (proteomics), metabolites (metabolomics) and cell phenotypes (phenomics) (Bochner, 2009). Recently the idea of modifying the genomes of bacterial producing cells came up and the first data was published (Knabben *et al.*, 2010, Qian *et al.*, 2009).

31.2 Manufacturing Processes ("Kit or Service?")

A plasmid manufacturing process must start at the laboratory or pilot scale. In cases where the plasmid DNA is subject to any application leading sooner or later to any (pre)clinical application, the type of plasmid manufacture is of relevance. This is the case since these data may be necessary to apply for later clinical applications. Hence, any component used for this is required to be already of a quality that is also available later on for GMP and/or large-scale processes.

Initial laboratory use requires plasmid purification within hours — or at maximum days — to perform initial research work. This is typically the case after cloning steps, e.g., in modifying a certain expression cassette and testing by restriction analysis and subsequent expression studies if the intended gene modification was successful or if the experimental approach needs to be repeated or modified and repeated. For these cases the so-called "plasmid kits" are a sufficient and helpful tool. Purity is acceptable and the whole processing is quite fast. A typical purification process is based on an alkaline lysis procedure, shown in Fig. 31.1. However, this type of DNA frequently contains significant amounts of bacterial chromosomal DNA (up to >20% of total DNA) as can be seen within the slot of an agarose gel when staining the gel (for example, see Fig. 13.3, lane 4). Although the appearance of DNA within an analytical agarose gel does not show any negative influence below the gel slot (except a slight "smear" in cases of silica-based anion exchange chromatography tips), the total DNA being applied to eukaryotic cells in transfection (including the aforesaid contamination) is wrongly quantified if this is performed by spectrophotometry. This value may be necessary for the calculation of the ratio between DNA and a complexing agent in, e.g., lipofection. A wrong ratio in assembling three components of a triple-transfection in lentivirus (LV) production or cotransfection in AAV production may result.

Further contaminants were already mentioned within this chapter and Chapter 13 and need to be evaluated. In any quantification of transfection rates the contamination profile should — if a "perfect DNA" is not available — at least be identical to earlier lots in order to be able to use the data obtained.

Since 2000, the production at even small scales of plasmid DNA is offered as a service (Schleef *et al.*, 2010) like outsourcing sequencing, gene synthesis and cloning up to production of cell lines with specific features.

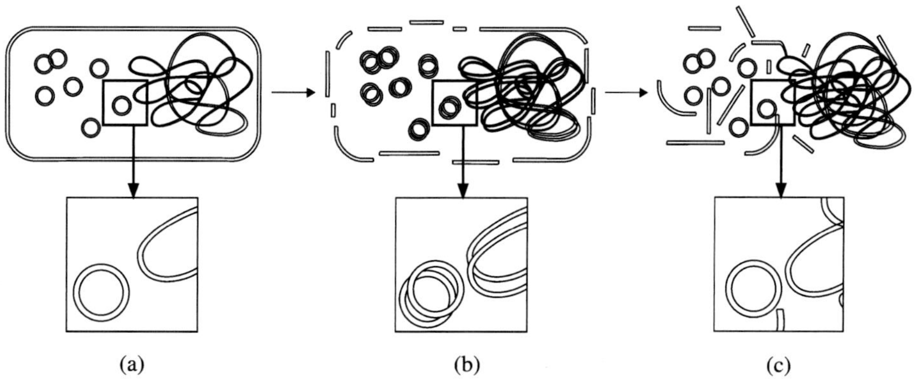

FIGURE 31.1 ■ Bacterial alkaline lysis. The intact bacterial cell (a) carrying the bacterial chromosome and some plasmids is destroyed and all DNA dissociated by sodium dodecyl sulfate (SDS)/alkaline conditions (b). A neutralization step (c) leads to aggregates of the bacterial chromosomal DNA, proteins, cell wall residues and SDS but also re-associated intact plasmids that can subsequently be separated from the white flaky SDS material.

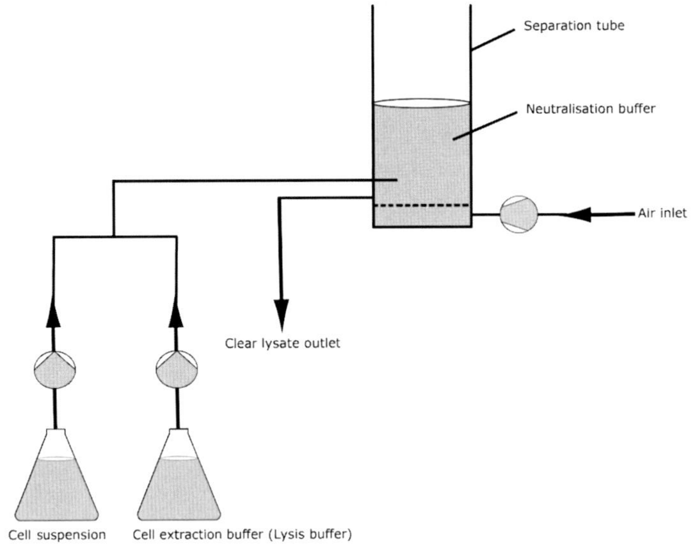

FIGURE 31.2 ■ In-line alkaline lysis for large-scale DNA production.

The up-scaling process requires investigation and it is not just a multiplication of relevant factors but a time-consuming evaluation of steps to be performed. For example, modification of temperature within a large fermenter requires a fast temperature transfer to all parts of the vessel and alkaline lysis is not shaking a very large bottle instead of a small one. Here, for example, in-line lysis and mixing devices may help to overcome such bottlenecks. A productive in-line lysis was presented by Voss and Flaschel (2010). The process is equivalent to a laboratory-scale alkaline lysis with respect to its chemical components (Fig. 31.1), but completely different with respect to the scheme and the mixing forces applied (Fig. 31.2).

All processes start with cultivation, followed by harvesting bacterial cells and performing a step to separate the cellular components from the plasmid DNA contained. This alkaline lysis has been discussed previously at a laboratory or an industrial scale, but after this lysis the plasmid DNA needs to be further purified. A significant amount of bacterial chromosomal DNA, lipopolysaccarides or RNA is present within such primary lysates. Ribonucleases (RNases) are able to hydrolyse phosphodiester bonds within RNA molecules. These enzymes are needed in cells to modify and process RNA molecules. In general, RNA has a short half-life, but is still a substantial contaminant in plasmid preparations. This RNA may block the binding capacity of, for example, anion exchange chromatography. Therefore, an enzymatic digestion of RNA prior to the chromatographic step is usually applied. If this is carried out in pharmaceutical manufacturing processes, the question must be raised as to whether the quality of such processing aids, which are normally derived from animal sources, is sufficient and can be accepted.

Since RNAse A is typically prepared from bovine pancreas, the origin of these glands has to be assessed and certified. However, recent technology allows the purification of plasmid DNA at any scale free of any enzyme, including RNAse (Schleef and Schmidt, 2004).

From the "cleared lysate" obtained by a filtration step of the alkaline lysate, the plasmid molecules should be separated from soluble biomolecules (e.g., host chromosomal DNA, RNA, nucleotides, lipids, residual proteins, amino acids and saccharides), salts and buffer components. A typical approach is chromatography including gravity-flow-tip-based anion chromatographic techniques known from plasmid kits. For clinical applications no other way — except the approach with kit-based systems — is possible at present. The design of the chromatographic process depends on the required purity of the plasmid product. So far, only a few techniques are used in pharmaceutical-grade processing of plasmid DNA. Anion exchange chromatographic techniques were initially described for certain processes (Schleef, 1999; Colpan *et al.*, 1995; Schorr *et al.*, 1999; Voss *et al.*, 2005; Ferreira *et al.*, 2001), but independent from the type of process and its robustness, a certain set of specifications need to be fulfilled with each such lot of product. Size-exclusion chromatography for small scales, reverse-phase chromatography or hydrophobic interaction chromatography are additional approaches applied. Other approaches involve plasmid DNA purification systems for the selective removal of contaminants or improved binding capacity for the intended plasmid product (e.g., monolithic stationary phase) (Horn *et al.*, 1998; Lemmens *et al.*, 2003; Strancar *et al.*, 2002). A very efficient chromatographic way to purify plasmids from contaminant is affinity chromatography. A target sequence has to be present on the plasmid in order to allow its retention on the affinity chromatography column. The first system to be introduced used a column linked to an oligonucleotide able to form a triple helix with a specific double-stranded sequence on the plasmids (Wils *et al.*, 1997). Other systems were introduced, in which a recombinant protein linked to the column was able to reversibly attach to a plasmid or minicircle sequence, as this is performed in our laboratory to obtain highly purified DNA molecules free of non-binding DNA molecules and other contaminants.

31.3 Large-Scale Manufacturing of Plasmid DNA

Diverse aspects of producing plasmid DNA still need to be addressed in order to develop processes for the safe and economic production of large amounts of plasmid DNA. The term "large-scale" was applied for production of between 10 and 1000 mg until the end of the 1990s, but recent large-scale technology for plasmid DNA requires 100 g up to kg scales (Bussey *et al.*, 1998; Schleef, 1999; Hoare *et al.*,

2005; Urthaler *et al.*, 2005). The lysis technology shown in Fig. 31.2 is a prerequisite for these developments. In all cases the question remains to be solved as to whether contamination, at present undetectable using today's analytic technology, can be excluded (see Firozi *et al.* (2010) and our work in Wooddell *et al.* (2011)).

31.4 "Difficult Sequences" in Plasmid DNA

In general, how a bacterial cell is cultivated and for what purpose is of relevance. If the DNA to be isolated is only needed for the analysis of the plasmid DNA contained, to evaluate the correct size or to use it for sequencing, PCR or even cloning, a regular kit preparation is sufficient and the cultivation is performed within shaker flasks (<200 ml). If larger amounts or higher and reproducible quality are essential, the scale-up is not a larger bottle or hundreds of them, but a fermentation device. The protocol for this fermentation influences productivity significantly (for example, see Schmidt *et al.*, 2004).

In cases where a plasmid is not stable, the type of cultivation is the screw to be turned to ensure completeness of the plasmid within the cells and their daughter cells throughout the cultivation process. Examples are those where the inverted terminal repeat (ITR)-containing transfer- or vector-plasmids for adeno-associated virus (AAV) production are (at least partially) deleted through cultivation, leading to a mixture of intact and deleted plasmids and to the phenomenon that this "creeping" effect is not identified immediately but results in a greater and greater decrease in virus titers in AAV production. The same is true for LV production, where the lentiviral plasmid contains long terminal repeats (LTRs). In the case of AAV, the helper/packaging plasmids are quite large (in the case of pDG, pDP, pDF, etc., over 20 kb) but even a 10 kb plasmid is already "low-copy". Furthermore, the specific viral sequences contained in frequent cases (mainly in LV transfer plasmids) result in a low productivity with standard techniques, followed by a higher burden of contaminants if simply isolated with laboratory equipment — e.g., the content of bacterial chromosomal DNA and lipopolysaccharide (LPS)-endotoxin is in such cases extremely high.

A cultivation process we initially developed for repetitive fragile DNA within plasmids, excludes the loss of such repetitive sequence stretches. Starting from there, the isolation of pure supercoiled covalently closed circular (ccc) DNA is the basis for a highly productive system for messenger RNA (mRNA) production, where stretches of A or T tend to get shorter with progressive cultivation time. This is equally true for sequences containing "difficult" motifs to be cultivated within bacteria. A typical example is the presence of hairpin structures as these are present within short hairpin RNA (shRNA) sequences. Starting with a reduced productivity these plasmids are additionally not easily sequenced.

31.5 Quality and Regulatory Requirements

The quality of a DNA is extremely variable depending on the type of purification. This is not in all cases a problem. For example the quality is not relevant if a special cloning approach in molecular biology is applied to search for the "right clone" and perform a quick preparation of DNA from potentially transformed *E. coli* cells. If this DNA is needed for subsequent transfection the issue becomes evident and at least contaminants such as bacterial LPS-endotoxin need to be reduced to avoid toxic effects influencing the expression level negatively or causing unintended side effects.

Any non-reproducibility is a further problem for comparing data from different transfection experiments. Due to the fact that a shaker flask — other than a fermentor — is only able to grow in a batch mode until one essential growth component runs out, it is hard, if not impossible, to repeat such a cultivation exactly, ending up with two different biomasses with different yields of DNA, LPS and other contamination profiles.

To avoid all this it is necessary and economic for DNA of relevance to produce this by a technology as explained above (cell bank, fermentor, chromatography) and to analyze it after each cycle of production.

DNA for clinical use (GMP) or even earlier for preclinical use requires at least the same quality and in addition certain documentation. Guidelines for (pre) clinical DNA-vaccine, cell or gene therapy trials were already available early and the fields of application have been reviewed (Schleef, 2001, 2005; Meager and Robertson, 1999; Kneuer, 2005). Recommended informative sources are summarized by the European Medicines Agency (EMA) and the World Health Organization (WHO). Ethical aspects of gene and cell therapies were recently reviewed (King and Cohen-Haguenauer, 2008).

In Chapter 13, further aspects influencing the safety and future regulatory compliance of non-viral vectors are discussed at the level of their technical principles (e.g., antibiotic-resistance-marker-free miniplasmid or minicircle DNA).

Acknowledgment

The author thanks Marco Schmeer for critical discussion, Janine Conde-Lopez for support with figures, and Emilie Couprianoff for editorial support. The author acknowledges the support of the German Federal Ministry of Education and Research (BMBF) for grants BioChancePLUS (0313749) and Nano-4-Life (13N9063), MOLEDA STREP, the research team of PlasmidFactory, Bielefeld, Germany for contributing to the work and the whole manufacturing team of PlasmidFactory for their discussion. Part of this work has also been supported by the CliniGene Network of Excellence funded by the European Commission FP6 Research Programme under contract LSHB-CT-2006-018933.

References

Blattner, F.R., Plunkett, G., Bloch, C.A., *et al.* (1997). The complete genome sequence of *Escherichia coli* K-12, *Science*, **277**, 1453–1474.

Bochner, B.R. (2009). Global phenotypic characterization of bacteria, *FEMS Microbiol Rev*, **33**, 191–205.

Bussey, L., Adamson, R., Atchley A. (1998). Methods for purifying nucleic acids, World Intellectual Property Organization Patent WO/1998/05673.

Colpan, M., Schorr, J., Moritz, P. (1995). Process for producing endotoxin-free or endotoxin-poor nucleic acids and/or oligonucleotides for gene therapy, World Intellectual Property Organization Patent WO/1995/21177.

Ferreira, G.N.M., Prazeres, D.M.F., Cabral, J.M.S., *et al.* (2001). 'Plasmid manufacturing — an overview', in Schleef, M. (ed.) *Plasmids for Therapy and Vaccination*, Weinheim: Wiley-VCH, pp. 193–236.

Firozi, P., Zhang, W., Chen, L., *et al.* (2010). Identification and removal of colanic acid from plasmid DNA preparations, implications for gene therapy, *Gene Ther*, **17**, 1484–1499.

Hayes, F. (2003). 'The function and organization of plasmids', in Casali, N., Preston, A. (eds.) *E. coli Plasmid Vectors*, Totowa NJ: Humana Press, pp. 1–17.

Hoare, M., Levy, M.S., Bracewell, D.G., *et al.* (2005). Bioprocess engineering issues that would be faced in producing a DNA vaccine at up to 100 m^3 fermentation scale for an influenza pandemic, *Biotechnol Prog*, 21, 1577–1592.

Horn, N., Budahazi, G., Marquet, M. (1998). Purification of plasmid DNA during column chromatography, US Patent 5707812.

King, N.M.P., Cohen-Haguenauer, O. (2008). En route to ethical recommendations for gene transfer clinical trials, *Mol Ther*, **16**, 432–438.

Knabben, I., Regestein, L., Marquering, F., *et al.* (2010). High cell-density processes in batch mode of a genetically engineered *Escherichia coli* strain with minimized overflow metabolism using a pressurized bioreactor, *J Biotechnol*, **150**, 73–79.

Kneuer, C. (2005). 'DNA as a pharmaceutical — regulatory aspects', in Schleef, M. (ed.) *DNA Pharmaceuticals — Formulation and Delivery in Gene Therapy, DNA vaccination and immunotherapy*, Weinheim: Wiley-VCH, pp. 7–22.

Lemmens, R., Olsson, U., Nyhammar, T., *et al.* (2003). Supercoiled plasmid DNA: selective purification by thiophilic/aromatic adsorption, *J Chromatogr B*, **784**, 291–300.

Meager, A., Robertson, J.S. (1999). Regulatory and standardization issues for DNA and vectored vaccines, *Curr Res Mol Ther*, **1**, 262–265.

Qian, Z.-G., Xia, X.-X., Lee, S.Y. (2009). Metabolic engineering of *Escherichia coli* for the production of putrescine: A four carbon diamine, *Biotechnol Bioeng* **104**, 651–662.

Schleef, M. (1999). 'Issues of large-scale plasmid manufacturing', in Rehm, H.J., Reed, G., Pühler, A., Stadler, P. (eds.) *Biotechnology Vol. 5a: Recombinant Proteins, Monoclonal Antibodies and Therapeutic Genes*, Weinheim: Wiley-VCH, pp. 443–470.

Schleef, M. (ed.) (2001). *Plasmids for Therapy and Vaccination*, Weinheim: Wiley-VCH.

Schleef, M. (ed.) (2005). *DNA Pharmaceuticals*, Weinheim: Wiley-VCH.

Schleef, M., Blaesen, M., Schmeer, M., *et al.* (2010). Production of non viral vectors, *Curr Gene Ther*, **10**, 487–507.

Schleef, M., Schmidt, T. (2004). Animal-free production of ccc-supercoiled plasmids for research and clinical applications, *J Gene Med*, **6**, 45–53.

Schmidt, T., Friehs, K., Flaschel, E., *et al.* (2004). Method for the isolation of ccc plasmid DNA, European patent 1144656.

Schorr, J., Moritz, P., Schleef, M. (1999). 'Production of plasmid DNA in industrial quantities according to cGMP guidelines', in Lowrie, D.B., Whalen, R.G. (eds.) *DNA Vaccines: Methods and Protocols*, Totowa NJ: Humana Press, pp. 11–21.

Strancar, A., Podgornik, A., Barut, M., *et al.* (2002). Short monlithic columns as stationary phases for biochromatography, *Adv Biochem Eng Biotechnol*, **76**, 49–85.

Tatum, E.L., Lederberg, J. (1947). Gene recombination in the bacterium *Escherichia coli*, *J Bacteriol*, **53**, 673–684.

Urthaler, J., Buchinger, W., Necina, R. (2005). Improved downstream process for the production of plasmid DNA for gene therapy, *Act Biochim Polonica*, **52**, 703–711.

Voss, C., Flaschel, E. (2010). Method for producing extra-chromosomal nucleic acid molecules, *US Patent* 7842481.

Voss, C., Schmidt, T., Schleef, M. (2005). 'From bulk to delivery: plasmid manufacturing and storage', in Schleef, M. (ed.) *DNA Pharmaceuticals: Formulation and Delivery in Gene Therapy, DNA Vaccination and Immunotherapy*, Weinheim: Wiley-VCH, pp. 23–42.

Walther, W., Stein, U., Fichtner, I., *et al.* (2001). Non-viral *in vivo* gene delivery into tumors using a novel low volume jet-injection technology, *Gene Ther*, **8**, 173–180.

Wils, P., Escriou, V., Warnery, A., *et al.* (1997). Efficient purification of plasmid DNA for gene transfer using triple-helix affinity chromatography, *Gene Ther*, **4**, 323–330.

Wooddell, C.I., Subbotin, V.M., Sebestyén, M.G., *et al.* (2011). Muscle damage after delivery of naked plasmid DNA into skeletal muscles is batch dependent, *Hum Gene Ther*, **22**, 225–235.

INDEX

Manufactured by Amazon.ca
Bolton, ON

29695561R00378